OPERATIONS RESEARCH

PRINCIPLES AND PRACTICE

Second Edition

OPERATIONS RESEARCH

PRINCIPLES AND PRACTICE

A. Ravindran

School of Industrial Engineering
University of Oklahoma
Norman, Oklahoma

Don T. Phillips

Department of Industrial Engineering
Texas A & M University
College Station, Texas

James J. Solberg

School of Industrial Engineering
Purdue University
West Lafayette, Indiana

John Wiley & Sons

New York • Chichester
Brisbane • Toronto • Singapore

Library of Congress Cataloging-in-Publication Data

Phillips, Don T.
Operations research.

Includes bibliograpies and index.
1. Operations research. I. Ravindran, A., 1944–
II. Solberg, James J., 1942– III. Title.
T57.6.P48 1986 001.4'24 86-5561
ISBN 0-471-08608-8

Printed in the United States of America

10 9 8 7 6 5 4 3 2 1

To Bhuvana, Candy, and Elizabeth

ABOUT THE AUTHORS

A. ("RAVI") RAVINDRAN, is Professor and Director of the School of Industrial Engineering at the University of Oklahoma, Norman. He received a B.S. degree in electrical engineering from Birla Institute of Technology and Science, Pilani (India), and M.S. and Ph.D. in Industrial Engineering and Operations Research from the University of California at Berkeley. He was a member of the Industrial Engineering faculty at Purdue University for several years before joining the University of Oklahoma. Dr. Ravindran won the Institute of Industrial Engineers' Operations Research Division Award for Research in 1983. He has also received several teaching awards at Purdue University and the University of Oklahoma. His research interests include mathematical programming, multicriteria optimization, goal programming, metal cutting, health planning, quality control, energy models, and transportation analysis. He is the author and coauthor of more than 50 publications in operations research. He has also coauthored a book on *Engineering Optimization—Methods and Applications* published by Wiley. He serves on the editorial boards of *Computers and Industrial Engineering* and *IEEE Transactions on Engineering Management* journals.

DON T. PHILLIPS is Professor of Industrial Engineering at Texas A & M University, College Station, Texas. He received a Bachelor of Science degree in Industrial Engineering from Lamar University in 1965 and a Ph.D. in Industrial Engineering from the University of Arkansas in 1968. Dr. Phillips has taught courses in Industrial Engineering at the University of Arkansas, the University of Texas at Austin, and Purdue University. He has worked for both Monsanto Chemical Company and Jefferson Chemical Company. At Texas A & M, Dr. Phillips has been given a TEES Research Fellow Award three times, and has served as the H. D. Brockett Professor of Engineering. He has also served on the Board of Trustees of the Institute of Industrial Engineers (IIE), the IIE Research Advisory Council, and as a member of the IIE Transactions Editorial Board. A nationally known lecturer and consultant, Dr. Phillips has authored or coauthored six college text books and more than 100 publications. An active Industrial consultant in Manufacturing Systems Engineering and Systems Simulation Methodologies, Dr. Phillips is a registered Professional Engineer.

JAMES J. SOLBERG is Professor of Industrial Engineering at Purdue University, West Lafayette, Indiana. He received a B.A. degree from Harvard University in mathematics in 1964, a M.A. in mathematics and a M.S. in industrial engineering from the University of Michigan in 1967, and a Ph.D. in industrial engineering from the University of Michigan in 1969. He was a member of the faculty of the University of Toledo before joining Purdue in 1971. In addition to several teaching awards, Dr. Solberg won the David F. Baker Distinguished Research Award from the Institute of Industrial Engineers in 1982. The author or coauthor of more than 50 publications, his research interests include stochastic processes, mathematical modeling, graphs, and networks, and manufacturing systems. He was responsible for a queueing network model of manufacturing flow, called CAN-Q, which is now widely used throughout the world. He is currently the director of a major cross-disciplinary research program, the Engineering Research Center for Intelligent Manufacturing Systems.

PREFACE

We are deeply gratified by the enthusiastic response given to the first edition of our book by our colleagues and students. We took great care in preparing the second edition. We began by sending a detailed questionnaire describing the planned revisions by chapter to all the schools that used our text. An excellent response was received with many constructive comments. These responses were extremely valuable to us during the preparation of the second edition. We added some new material, included more applications in all the chapters, revised the explanation and presentation of some topics, increased the number of exercises, and achieved a greater independence of chapters as far as is possible.

Examples of new materials include a complete new chapter on *decision analysis* (Chapter 5), the computer solution of linear programming problems, the use of sensitivity analysis output in Chapter 2, the addition of *minimal spanning tree* in Chapter 3, a new section on *goal programming—theory, algorithm and applications* in Chapter 4, the addition of *network of queues* in Chapter 7, a discussion of microcomputer based simulation languages and their applications in Chapter 9, and the inclusion of second order gradient based optimization techniques in Chapter 11.

An example of effective independence of chapters is the treatment of *queueing models* (Chapter 7). In the first edition, this discussion required the knowledge of *Markov Processes* discussed in Chapter 6. In the second edition, queueing models, their steady-state equations and relevant results are derived independently of Markov Processes. This change is in direct response to the comments from the instructors who have used the text.

To make room for the new chapter on decision analysis (Chapter 5), the probability review has been moved to Appendix B. In response to the comments received on the questionnaire, more applications have been added throughout the book, and more "drill-type" exercises have been included. In addition, the exercises at the end of each chapter have been rearranged so that they begin with short-answer review questions on principles, are followed by simple formulation and drill-type numerical problems, and end with more difficult "mind expanding" exercises.

A new feature of the second edition is a companion microcomputer software for several of the OR techniques discussed in the text and a diskette that will be provided free to all the adopters of our book. The software has been written by

Professor Jorge Haddock of Clemson University. Instructors will be given permission to copy the diskette and the manual for student distribution. A detailed solutions manual is also available to the instructor.

Our principal objective is to present the material in a way that would immediately make sense to a beginning student. Often this required a juxtaposition of what might otherwise be regarded as the natural ordering of general theory followed by specific examples. We have observed that beginners rely heavily on examples and will understand the theory far more easily if it is presented as a generalization of one or more specific examples. It seems that, for the most part, students taking operations research courses have had adequate preparation in the mechanics of calculus, linear algebra, and probability. What they have greatest difficulty with is formulation and interpretation. That is, they can "do" the mathematics, but frequently do not understand the meaning of what they are doing. Hence, we have found it helpful to include quite a bit of verbal explanation of mathematical material. Purists may object to the liberties we have taken in providing loose and imprecise statements of perfectly well-defined and precise mathematics, but our embarrassment in so doing is overcome by our conviction that the student needs that kind of help. So although our presentation is far less concise and elegant than we could have made it, we feel strongly that the extra verbiage pays dividends.

Notwithstanding our desire to improve the readability of the material, we felt that it was important to keep the size of the book down. The sheer bulk of some textbooks, particularly when you consider their density, is overwhelming to beginning students. When it takes a student an hour to digest what is written on one page and the book is a thousand pages long, it is not surprising that he or she would feel disheartened. It is a lot easier for students to stay motivated, especially in the early stages, if they can sense that they are making substantial progress. Of course, there *is* a great deal to be learned and, as we assembled this book, we began to understand how our predecessors could have ended up with so much material. We too experienced an almost compulsive urge to include more. Even with merciless editing (which involved removing several major topics that happened to be personal favorites of the authors) the book turned out to be longer than we had originally intended. Nevertheless, we feel that it is possible to cover most of this book in one academic year.

To avoid creating the impression that our treatment of these subjects is conclusive, and to open the door to sources of additional information, we have provided selected references at the end of each chapter. We have also provided a few bibliographic comments to guide readers in their research. Obviously, these brief bibliographies are only starting points, which in turn will lead to further sources.

Throughout the text, we have kept the use of higher mathematics at an intermediate level. For example, Chapters 1 through 4 on linear programming and networks require only linear algebra. Chapters 5 through 10 require a basic knowledge of probability. We have included in Appendix B the basic concepts on probability that are necessary. Although the treatment in Appendix B on probability is reasonably complete, it is probably too concise to serve as an introduction to the topic by itself.

Individual instructors will, of course, exercise their judgment to select and to rearrange certain material. Because we know this will happen, we decided at the outset that a highly integrated text would be undesirable. Thus we have deliberately sought to maintain independence among chapters to permit maximum flexibility in their use. We have used Chapters 2, 3, and 4 for a graduate level, introductory course on linear programming. For a similar undergraduate course, we omit Chapter 4. For a graduate level course on probabilistic models of OR, we use Chapters 6 to 8 and parts of 9 and 10. For the corresponding undergraduate course, we attempt to cover the same topics, but must delete some of the more specialized material in each chapter. Some of our colleagues have told us that they use Chapters 2, 3, 7, 8, 9, and 10 for a survey course in operations research. Some have also used Chapters 2, 3, 4, 10, and 11 for an overview course on optimization techniques. Teachers being an individualistic lot, we are confident that instructors will find unique ways to adapt this book to suit their own needs as they perceive them.

It is perhaps obvious that the authors are indebted to the many researchers who have developed the underlying concepts that permeate this text. Although far too numerous to mention, we have tried to recognize their contributions through bibliographic references at the end of each chapter. In addition to the above, several individuals have directly contributed to the composition of the second edition. We are specifically indebted to Dr. Herbert Moskowitz of Purdue University for writing the chapter on decision analysis and to Dr. Jorge Haddock of Clemson University for developing the companion microcomputer software for the book. Special thanks are owed to Dr. J. W. Schmidt, C.B.M., Inc., for the third example in Chapter 9; Dr. R. S. Schecter of the University of Texas; and to Dr. Charles Beightler, Dr. R. M. Crisp, and Dr. W. L. Meier for the material on geometric programming in Chapter 11. The pleasant personality and excellent typing skills of Patti Coffill at the University of Oklahoma made it easier to revise several parts of the book. Finally, we are grateful to the instructors who have adopted our first edition and for their encouragement and helpful suggestions that made the second edition a reality.

Norman, Oklahoma — A. (Ravi) Ravindran
College Station, Texas — Don T. Phillips
West Lafayette, Indiana — James J. Solberg

CONTENTS

CHAPTER 1
THE NATURE OF OPERATIONS RESEARCH

1.1 THE HISTORY OF OPERATIONS RESEARCH

In order to understand what operations research (OR) is today, one must know something of its history and evolution. Although particular models and techniques of OR can be traced back to much earlier origins, it is generally agreed that the discipline began during World War II. Many strategic and tactical problems associated with the Allied military effort were simply too complicated to expect adequate solutions from any one individual, or even a single discipline. In response to these complex problems, groups of scientists with diverse educational backgrounds were assembled as special units within the armed forces. Because of the diversity of its membership, one of the earliest groups in Britain came to be known as ''Blackett's circus.''

Partly because the scientists involved were talented men, partly because of the pressures of wartime necessity, and partly because of the synergism generated from the interactions of different disciplines, these teams of scientists were remarkably successful in improving the effectiveness of complex military operations. Examples of typical projects were radar deployment policies, antiaircraft fire control, fleet convoy sizing, and detection of enemy submarines. By 1941, each of the three wings of the British Armed Forces were utilizing such scientific teams. As the dramatic success of the idea became amply demonstrated, other allied nations adopted the same approach and organized their own teams. Because the problems assigned to these groups were in the nature of military operations, their work became known as operational research in the United Kingdom, and as operations research elsewhere. The American effort, although it began at a later date, produced many fundamental advances in the mathematical techniques for analyzing military problems. For further details of early activities in operations research, an excellent summary is given in Trefethen (5).

After the war, many of the scientists who had been active in the military OR groups turned their attention to the possibilities of applying a similar approach to civilian problems. Some returned to universities and concentrated their efforts on providing a sound foundation for many of the techniques that had been hastily developed earlier, while others devoted renewed efforts to developing new techniques. Many individuals moved into various sectors of the private economy, where they adapted methods developed by others to the unique problems of particular industries.

In terms of applications, the first civilian organizations to seize upon the OR methodology were, generally, large profit-making corporations. For example, petroleum companies were among the first to make regular use of linear programming on a large scale for production planning. It was logical that "big business" would take the lead in adopting OR. To any profit-oriented organization, OR offered a way to obtain a competitive advantage; but in the early years when all OR work was in the nature of basic research, only the large companies could afford it. Later, as researchers began to recognize common categories of problems (inventory, allocation, replacement, scheduling, etc.) and the techniques for dealing with such problems became standardized, smaller companies were able to benefit from the pool of accumulated knowledge without investing heavily in research. With a few notable exceptions—such as the work on traffic control conducted at the New York Port Authority by Leslie Edie and others in the early 1950s—applications of OR in service-oriented industries and in the public sector did not begin to flourish until the mid 1960s. Today, however, service organizations such as banks, hospitals, libraries, and judicial systems recognize that OR can aid in improving the effectiveness with which they deliver their respective services. In addition, federal, state, and local government agencies are using OR, particularly in their planning and policy-making activities. In fact, work on some of these specialized applications of OR has multiplied so rapidly in recent years that subspecialties, based upon the area of application, appear to be developing. Recent operations research conferences have included special sessions on such topics as "OR in community health planning," "OR models of the criminal justice system," "mass transit studies," "travel and tourism," "energy," "education models," and "OR applications in sports."

An important factor in the rapid spread and sustained success of the OR approach to problem solving was the concurrent development of electronic computers. The computer was from the beginning an invaluable tool, enabling the OR analyst to perform otherwise intractable calculations. Indeed, many of the problem-solving methods now regarded as standard would be unthinkably impractical to implement without modern computers. By generating practical uses for increasingly larger and faster machines. OR has both benefited from and contributed to the explosive growth of computer capability that has occurred over the past three and a half decades.

By the early 1950s, civilian OR activities had reached a level of development that began to suggest that a unique discipline was in formation. The Operations Research Society of America (ORSA) was founded in 1952 to serve the professional needs of scientists working in the OR area. A parallel movement resulted, in 1953, in the formation of The Institute of Management Sciences (TIMS). The journals of these two organizations, *Operations Research* and *Management Science*, as well as regular conferences of the members, helped to draw together the many diverse results into some semblance of a coherent body of knowledge.

Beginning about the same time and continuing into the early 1960s, more and more colleges and universities in the United States introduced first individual courses, then whole programs, into their curricula. Graduate programs leading to advanced degrees at both the M.S. and Ph.D level were approved in many major universities. For good or bad, it happened that little uniformity was observed in deciding where within the academic structure OR belonged. Depending on unique development patterns at each institution, OR programs sometimes appeared within departments of industrial engineering, sometimes in business schools, and occasionally in mathematics or economics. In keeping with the original interdisciplinary character of the work, some universities established interdisciplinary committees to administer OR programs; but because such academic "orphans" tend to be inherently unstable, they have usually been either officially or unofficially absorbed into more traditional parts of the university structure. Of course, the parent discipline tends to impart a particular unique characteristic to the OR program, and the consequent lack of uniformity in academic programs has acted against achieving "definition" in the field. Perhaps it is all for the best.

It is interesting to note that the modern perception of OR as a body of established models and techniques—that is, a discipline in itself—is quite different from the original concept of OR as an *activity*, which was performed by interdisciplinary teams. An evolution of this kind is to be expected in any emerging field of scientific inquiry. In initial formative years, there are no experts, no traditions, no literature. As problems are successfully solved, the body of specific knowledge grows to a point where it begins to require specialization even to know what has been previously accomplished. The pioneering efforts of one generation become the standard practice of the next. Still, it ought to be remembered that at least a portion of the record of success of OR can be attributed to its ecumenical nature. It is in the best traditions of the field to adopt the procedures of any discipline that can make a contribution to the solution of the problem at hand. If the initial open-mindedness of OR ever degenerates into orthodoxy, if the methods ever begin to outweigh the objectives, then the field will have lost one of its most vital precepts.

1.2 THE MEANING OF OPERATIONS RESEARCH

From the historical and philosophical summary just presented, it should be apparent that the term "operations research" has a number of quite distinct variations of meaning. To some, OR is that certain body of problems, techniques, and solutions that has been accumulated under the name of OR over the past 30 years, and we apply OR when we recognize a problem of that certain genre. To others, it is an activity or process—something we do, rather than know—which by its very nature is applied. Perhaps in time the meaning will stabilize, but at this point it would be premature to exclude any of these interpretations. It would also be counterproductive to attempt to make distinctions between "operations research" and the "systems approach." While these terms are sometimes viewed as distinct, they are often conceptualized in such a manner as to defy separation. Any attempt to draw boundaries between them would in practice be arbitrary.

How, then, can we define operations research? The Operational Research Society of Great Britain has adopted the following definition:

> Operational research is the application of the methods of science to complex problems arising in the direction and management of large systems of men, machines, materials and money in industry, business, government, and defense. The distinctive approach is to develop a scientific model of the system, incorporating measurements of factors such as chance and risk, with which to predict and compare the outcomes of alternative decisions, strategies or controls. The purpose is to help management determine its policy and actions scientifically.

The Operations Research Society of America has offered a shorter, but similar, description:

Operations research is concerned with scientifically deciding how to best design and operate man–machine systems, usually under conditions requiring the allocation of scarce resources.

Although both of these definitions leave something to be desired, they are about as specific as one would want to be in defining such a broad area. It is noteworthy that both definitions emphasize the *motivation* for the work; namely, to aid decision makers in dealing with complex real-world problems. Even when the methods seem to become so abstract as to lose real-world relevance, the student may take some comfort in the fact that the ultimate goal is always some useful application. Both definitions also mention *methodology*, describing it only very generally as "scientific." That term is perhaps a bit too general, inasmuch as the methods of science are so diverse and varied. A more precise description of the OR methodology would indicate its reliance on "models." Of course, that term would itself require further elaboration, and it is to that task that we now turn our attention

1.3 MODELS IN OPERATIONS RESEARCH

The essence of the operations research activity lies in the construction and use of models. Although modeling must be learned from individual experimentation, we will attempt here to discuss it in broad, almost philosophical terms. This overview is worth having, and setting a proper orientation in advance may help to avoid misconceptions later.

First, one should realize that some of the connotations associated with the word "model" in common English usage are not present in the OR use of the word. A model in the sense intended here is just a simplified representation of something real. This usage does carry with it the implication that a model is always, necessarily, a representation that it is less than perfect.

Why model? There are many conceivable reasons why one might prefer to deal with a substitute for the "real thing" rather than with the "thing" itself. Often, the motivation is economic—to save money, time, or some other valuable commodity. Sometimes it is to avoid risks associated with the tampering of a real object. Sometimes the real environment is so complicated that a representative model is needed just to understand it, or to communicate with others about it. Such models are quite prevalent in the life sciences, physical chemistry, and physics.

Given that one has something real, which we will call the "real system," and that there is some understandable reason for wanting to deal with it—that is, a "problem" related to the real system which calls for definite "conclusions"—the modeling process can be depicted as in Fig. 1.1. The broken line on the left represents what might be termed the "direct approach," for which we are seeking a substitute.

The first step is construction of the model itself, which is indicated by the line labeled "Formulation." This step requires a set of coordinated decisions as to what aspects of the real system should be incorporated in the model, what aspects can be ignored, what assumptions can and should be made, into what form the model should be cast, and so on. In some instances, formulation may require no

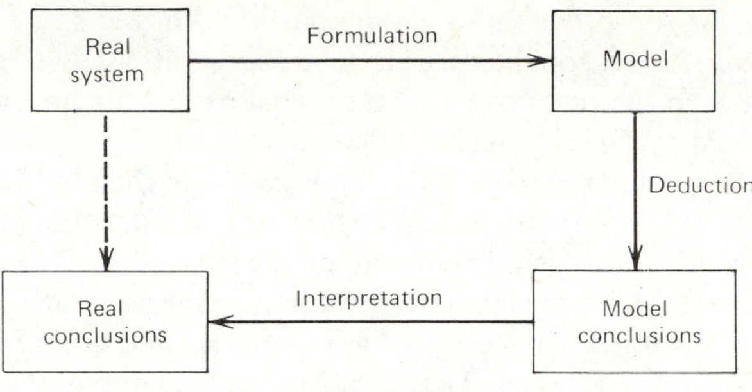

Figure 1.1
The Modeling Process.

particular creative skill, but in most cases—certainly the interesting ones—it is decidedly an art. The selection of the essential attributes of the real system and the omission of the irrelevant ones require a kind of selective perception that cannot be defined by any precise algorithm.

It is apparent, then, that the formulation step is characterized by a certain amount of arbitrariness, in the sense that equally competent researchers, viewing the same real system, could come up with completely different models. At the same time, the freedom to select one's assumptions cannot be taken to imply that one model is any better than another. A discussion of the boundary between reasonable and unreasonable assumptions would take us into philosophical issues which we could not hope to resolve. In fact, as one ponders the problem of model formulation, one discloses all sorts of metaphysical questions, such as how to define the precise boundary between the model and its referent, how to distinguish between what is real and what is only our perception, the implications of the resultant cause and effects of modeling procedure, and so on. These issues are mentioned only to reemphasize that it is quite meaningless to speak of the "right" way to formulate a model. The formative stages of the modeling process might be repeated and analyzed many times before the proper course of action becomes readily apparent. Once the problem formulation and definition is agreed upon, a more scientific step in the modeling process is begun.

Returning to Fig. 1.1, the step labeled "Deduction" involves techniques that depend on the nature of the model. It may involve solving equations, running a computer program, expressing a sequence of logical statements—whatever is necessary to solve the problem of interest relative to the model. Provided that the assumptions are clearly stated and well defined, this phase of modeling should *not* be subject to differences of opinion. The logic should be valid and the mathematics should be rigorously accurate. All reasonable people should agree that the model conclusions follow from the assumptions, even if they do not all agree with the necessary assumptions. It is simply a matter of abiding by whatever formal rules of manipulation are prescribed by the methods in use. It is a vital part of the modeling process to rationalize, analyze, and conceptualize all components of the deductive process.

The final step, labeled "Interpretation," again involves human judgment. The model conclusions must be translated to real-world conclusions cautiously, in full cognizance of possible discrepancies between the model and its real-world referent. Aspects of the real system which were either deliberately or unintentionally overlooked when the model was formulated may turn out to be important.

Since there is no way to prove that a model has *not* omitted some important factor, there is room for reasonable people to disagree about the relevance of the model conclusions to the real system, and to what extent the interpretation phase should be tempered by direct intuitive judgments.

The most important point revealed by Fig. 1.1 is that the ties between the model and the system it represents are at best ties of plausible association, and that no one, no matter how competent, can create perfection for that situation. It is part of the nature of models as simplified representations of real systems that there can be no absolute criteria by which to determine their acceptability. In short, you cannot *prove* a model. That is not to say, of course, that there exist no criteria by which to distinguish good models from poor ones, nor to say that model validation is not an integral part of the total modeling effort. There can be more specific criteria for particular kinds of models, but generally it can be said that a model is good insofar as it is *useful*, relative to the purpose for which it was intended.

The process of acquiring the conviction that a model actually "works" is commonly called *validation*. When the people involved (the users) are persuaded that a model is useful within some basic context, they will speak of it as a valid model. Its validity is of course restricted to the understood context. Even within that context, some people may refuse to accept the model's validity because they have not yet been persuaded. Thus, "validation" is a considerably weaker term than "proof" or "verification."

To further clarify the modeling approach to problem solving, contrast it to the experimentally based "scientific method" of the natural sciences. Figure 1.2 depicts the latter approach. Here, the first step is the development of a hypothesis which is arrived at, generally by induction, following a period of informal observation. At that point an experiment is devised to test the hypothesis. If the experimental results contradict the hypothesis, the hypothesis is revised and retested. The cycle continues until a verified hypothesis, or *theory*, is obtained. The result of the process is something that purports to be "truth," "knowledge," or "a law of nature." In contrast to model conclusions, theories are independently verifiable statements about factual matters. Models are invented; theories are discovered.

It might also be noted that there exist other formalized procedures for reaching conclusions about the real world. The system of trial by jury, as a means to decide between guilt or innocence of accused criminals, is one example. The determination of policy within a group by vote according to personal preference is

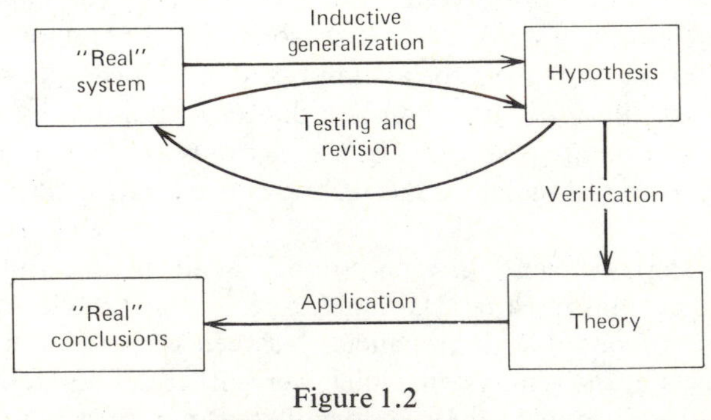

Figure 1.2
The Scientific Method.

another. Thus, modeling is a very important but certainly not unique method to assist us in dealing with complicated real-world problems.

1.4 PRINCIPLES OF MODELING

Having established a general framework for what is actually meant by modeling, let us proceed to a set of general principles useful in providing guidance to the formulation of models within the context of OR. Both developers and users of OR models should be consciously aware of the inherent limitations of the approach, and should exercise constant vigilance against falling into the common traps indicated by the following ten principles:

1. *Do not build a complicated model when a simple one will suffice.* Considering that this principle is just a restatement of Occam's razor, which goes back to the early 14th century, it is surprising how often it is ignored. The reasons are understandable—people like to illustrate their capabilities, they want to give the client his money's worth, and so on. Even with the best of motives, it is easy to get carried away by the sheer challenge of a difficult problem, and thereby spend far more time and money on refining a model than the problem is worth. Even aside from issues of cost, it is apparent that given two models, each of which adequately performs the intended function, the simpler is the more useful and is therefore to be preferred.

   In particular, the principle implies a condemnation of brute force methods of modeling. For example, one might list every variable of possible significance to a particular system, then run an enormous regression analysis to derive a predicting equation; or a systems analyst might build enormous simulations by including every conceivable parameter. In model building, "bigger and more complicated" does not necessarily mean "better."

   The principle also seems to contradict an understood axiom of mathematical analysis which says that one should first simplify a problem by introducing as many assumptions as are necessary until the mathematics becomes tractable, then "enrich" the model by weakening the assumptions in ingenious ways until the mathematics is no longer tractable. Such a procedure will always produce the most powerful and general model, but the power and generality of a model has little to do with its usefulness in dealing with a *particular* problem. In some cases, the strongest model one can construct may fall short of what would be required to have a *useful* model; in other cases, it may give rise to far more detail than is worth having. Building the strongest model possible is a common guiding principle for mathematicians who are attempting to extend the theory or to develop techniques that have broad applicability. However, in the actual practice of building models for specific purposes, the best advice is to "keep it simple."
2. *Beware of molding the problem to fit the technique.* OR professionals are often criticized (and sometimes rightly so) of distorting reality to suit the tools they prefer to use. For example, experts on linear programming methods may tend to view every problem they encounter as requiring a linear programming solution. In reality, not all optimization problems involve only linear functions. Furthermore, not all operations research problems involve optimization. As a matter of fact, not all real-world problems call for operations research! Of course, everyone sees reality in his or her own terms, so the field of operations research is not unique in this respect. Lawyers, psychologists, and political scientists are just as often guilty of "viewing the world through tinted glasses." Being human, we will tend to rely on the methods with which we are most comfortable and have been most successful with in the past. As best we can, however, we should strive not to shape the problem to preselected techniques, but rather to select the kind of model and techniques that are most appropriate to the problem. Our freedom to operate in this fashion is restricted, of course, by the breadth of our knowledge

of techniques. We certainly cannot use techniques in which we have no competence, and we cannot hope to be competent in all techniques.

We might wish to divide operations researchers into three main categories: *technique developers, teachers,* and *problems solvers*. Recognizing that one may play different roles at different times, as well as simultaneously, we might say that the problem solvers have an ongoing responsibility to broaden their working knowledge of available techniques in order to avoid the biases necessitated by a limited repertoire. At the same time, one should recognize that the technique developers who are seeking to extend the base of available techniques, and the teachers who bear the responsibility of conveying new developments to the practitioners and interpreting technical matters to people outside the profession, must operate according to different principles because the objectives are different. In particular, one should be prepared to tolerate the "I've-found-a-cure-but-I'm-trying-to-find-a-disease-to-fit-it" behavior in technique developers and teachers. The activity is legitimate in these circumstances because the purpose is not to produce a valid model of a real system, but to illustrate the technique or to establish its validity. Although this procedure is actually a reverse application of the scientific method, it is often a necessary step in translating theory into practice.

3. *The deduction phase of modeling must be conducted rigorously.* The reason for requiring rigorous deduction is that one wants to be sure that if model conclusions are inconsistent with reality, then the defect lies in the assumptions. In other words, if the deduction has not been carried out rigorously, the model will be unable to distinguish between external errors in formulation and internal errors in logic. One application of this principle is that one must be extremely careful when programming computers. Hidden "bugs" are particularly dangerous when they do not prevent the program from running, but simply produce results which are inconsistent with the model's intent.
4. *Models should be validated prior to implementation.* As mentioned earlier, it would be futile to attempt to establish with certainty that a model is appropriate, but this fact does not absolve one from the responsibility of checking it against reasonable standards of appropriateness. There are a number of commonly employed techniques for doing this, depending on the nature of the model.

One method for validating predictive models is "retrospective testing," in which the model is compared against some historical standard to see if it would have accurately predicted what has since been observed to occur. For example, if a model is constructed to forecast the monthly sales of a commodity, it could be tested using historical data to compare the forecasts it would have produced to the actual sales. A similar idea, useful in cases where the model is intended to represent a class of real things, is to test it against members of the class that were not used in formulating the model. For example, if a regression model is to be fit to data, some of the data might be held back for later testing. Another technique, which is sometimes useful in validating certain kinds of descriptive models, is to systematically vary parameters of the real system and observe whether the model successfully "tracks" the changes. Or, the model might be subjected to artificially constructed test situations which are deliberately designed to expose weaknesses. If it performs adequately in extreme situations, one can feel some confidence that it will work well under more normal circumstances.

If the model cannot be validated prior to implementation, then perhaps it can be implemented in phases for validation. For example, a new model for inventory control may be implemented for a certain selected group of items while the older system is retained for the majority of items. As the model proves itself, more items can be placed within its jurisdiction.

In a word of caution, it should be pointed out that validation can be carried too far. One might reach the point where an enormous amount of effort is required to increase model confidence only a small amount. Depending upon the importance of a model, it may be preferable to tolerate a lower confidence level. In some cases, it may be sufficient to know that someone else did something similar and it worked for them.

Finally, it is worth remembering that real things change in time. A highly satisfactory model may very well degrade with age. Depending on how such factors affect model performance and validity, an implemented model may require anything from constant surveillance to periodic reevaluation.

5. *A model should never be taken too literally.* This principle is obvious when the model is relatively unsophisticated, but is easily forgotten when it is elaborate. For example, suppose that one were to construct an elaborate computer model of the American economy, with many competent researchers spending a great deal of time and money installing all sorts of complicated interactions and relationships. Under these circumstances, is is easy to believe that the model duplicates the real system. Those who develop the model begin to believe it because their attention has been so directed to the model, that the model has *become* the real thing to them. In other words, they may become incapable of seeing the real problem in any terms other than those of the model. Those who are not so closely involved may be so overwhelmed by the technology and effort brought to bear that they just assume that the model is, or ought to be, "correct" because of its very complexity. As a consequence, the model may be both attacked and defended as if it were supposed to mirror reality exactly, when actually it should be evaluated in terms of its practical usefulness.

   This danger continues to increase as models become larger and more sophisticated, as they must to deal with increasingly complicated problems. The deduction phase is enlarged and is perhaps spread out over a longer period of time. Aside from the fact that the likelihood of logical errors is necessarily increased, the sheer gap between the model's assumptions and its conclusions increases the likelihood that in the interpretation phase the limitations of the model will be obscured. The only preventive measures are to be meticulous in expressing the original assumptions at the time they are formulated, and again when the conclusions are reached.

6. *A model should neither be pressed to do, nor criticized for failing to do, that for which it was never intended.* Now that a vast literature of operations research has accumulated, it is both natural and sensible to seek to adapt existing models to our problems, rather than view each new problem as a totally new situation. There is nothing wrong with this approach, provided that we fully understand the model in its original context. It has already been noted that a model is shaped not only by the system it represents and the tools it employs, but also by the motivations of the model builders. Because it is so easy to attribute one's own motives to someone else, there is a definite danger of making a false presumption that some model which is reported to serve its purpose well in a different but related context will also perform well for us.

   One example of this error would be the use of forecasting models to predict so far into the future that the data on which the forecasts are based have no relevance. A model might have been shown to provide excellent short-term forecasts, but that limited kind of validity provides no assurance that is capable of producing good long-term forecasts. Another example is in the use of certain network methods to describe the activities involved in a complex project. These may provide excellent descriptive and control models for projects for which there is a good base of experience, such as construction projects. However, in describing, say, a research project in which future activities may depend on factors which are not even imagined in the present, these network models give somewhat poorer representations of reality.

   Just as one must avoid stretching a model beyond its capabilities, one must also refrain from dismissing models as categorically useless when we find they do not fulfill our expectations.

7. *Beware of overselling a model.* This point is particularly significant for the OR professional because most nontechnical benefactors of an operations researcher's work are not likely to understand his methods. When a model is sold as "factual" rather than as an "integrated body of plausible assumptions that lead to useful conclusions," and it later becomes known that the implied real-world actions were somehow in error, the ensuing backlash may be out of proportion to the error. Those who deal with models

professionally can shrug off such failures as cases in which some important factor was neglected or overlooked. At any rate, they have no cause to lose faith in, or to feel betrayed by, the methods they employ. Others, however, may take the simplistic view that operations research is "no good."

The increased technicality of one's methods also increases the burden of responsibility on the OR professional to distinguish clearly between his or her role as model manipulator and model interpreter. Despite a firm conviction that the position one advocates is right, the OR professional has an obligation to concede that models, by their very nature, cannot offer conclusive evidence of anything. Simple integrity requires that any claim of certainty be limited to the deductive phase of modeling. That is, although it may be possible to prove that the conclusions necessarily follow from the assumptions, the most one can say about the assumptions is that they are *believed* to be adequate. In those cases where the assumptions seem undeniable, the argument will not be weakened by this admission; in those cases where the assumptions can be challenged, it would be dishonest to use the model to conceal them. In either case, OR analysts have a long-range vested interest in preserving the faith of others in the objectivity of their methods, and should therefore strive to present their conclusions with honest candor.

8. *Some of the primary benefits of modeling are associated with the process of developing the model.* Generally speaking, a model is never as useful to anyone else as it is to those who are involved in developing it. The model itself never contains the full knowledge and understanding of the real system that the builder must acquire in order to successfully model it, and there is no practical way to communicate this knowledge and understanding adequately. In extreme cases, the *sole* benefits may occur while the model is being developed; that is, the model may have no further value once it is completed. An example of an extreme case of this kind might occur when a small group of people attempts to develop a formal plan for some project. The plan is the final model, but the real problem may be to achieve a consensus on what the objectives ought to be. Once the consensus is achieved, the formal plan may be dispensable.

   The obvious corollary to this principle is that it is almost always desirable for the ultimate user to be involved throughout the model construction and validation period. In addition to the extra subtleties of understanding that users may gain from the activity, their presence will help to keep the model in tune with their needs. This procedure may also help to avoid the "stillborn model syndrome," in which fine technical work is allowed to languish, unimplemented, for lack of commitment on the part of the intended user.

9. *A model cannot be any better than the information that goes into it.* A well-known maxim of computer programming stated in abbreviated form as GIGO, or "Garbage In, Garbage Out," is equally applicable to modeling. It means that a computer, or a model, can only manipulate the data provided to it; it cannot recognize and correct for deficiencies in input.

   Another thing models cannot do is to *generate* information. Sometimes people get the idea that computer simulation models, for example, can produce more information than is put into them. Of course, they can produce almost unlimited quantities of data, but these data are just the direct consequences of the assumptions built into the program. Models may *condense* data, or *convert* it to more useful forms; they do not have the capacity to create it. It can certainly make one uncomfortable to be forced to make a decision without adequate information. Under these circumstances, it may be tempting to resort to modeling as an aid. It is unrealistic, however, to expect the model to supply the missing information. It is also unrealistic to expect it to somehow compensate for the absence of information and point the way to the solution that would be reached if the information were available. Although many OR models incorporate representations of uncertainty in the form of probabilities, accounting for uncertainty is not at all equivalent to eliminating it, or even reducing it. In some situations, instead of exerting one's efforts through model construction, one would be better off just gathering more information about the real system.

10. *Models cannot replace decision makers.* One of the most common misconceptions about the purpose of OR models is the idea that they are supposed to provide "optimal solutions," free of human subjectivity and error. Implicit in this notion is the concept that decision making can be automated once all the appropriate considerations have been properly defined. It is then simply a matter of finding the right formula and implementing the results. In the light of previous discussion, the falsity of this notion should be self-evident. No competent operations researcher would ever project such a view.

    There are virtually always some neglected aspects to be considered, along with the output produced by the model, before committing to a course of action. In the model formulation itself, there are many decisions to be made with respect to what aspects of the problem are important, what assumptions are reasonable, and so on. All of these decisions are subjective in nature. Often a problem is influenced by nonquantifiable factors which can only be listed for consideration. Sometimes it is necessary to compromise among multiple objectives, or to trade off one human value against another when there is not even a common scale by which to measure them. All of these real-world complications call for decision-making capabilities of a uniquely human nature. Only the most routine decisions are amenable to "automation," and even these would require human overseers to monitor performance and override the system when things go wrong.

    OR models can aid decision makers and thereby permit better decisions to be made. It would be going too far, however, to say that they make the job of decision making easier. If anything, the challenges are greater because of the expanded technical capabilities required to make good use of the modeling approach. Certainly, the role of experience, intuition, and judgment in decision making is undiminished.

REFERENCES

1. Ackoff, R. L., *Scientific Method: Optimizing Applied Research Decisions*, Wiley, N.Y., 1962.
2. Ackoff, R. L., and P. Rivett, *A Manager's Guide to Operations Research*, Wiley, N.Y., 1962.
3. Beer, Stafford, *Management Science: The Business Use of Operations Research*, Doubleday, N.Y., 1967.
4. Rivett, Patrick, *Principles of Model Building*, Wiley, N.Y., 1972.
5. Trefethen, Florence N., "A History of Operations Research," in *Operations Research for Management*, Joseph F. McCloskey and F. N. Trefethen, Eds., Johns Hopkins Press, Baltimore, Md., 1954.

CHAPTER 2
LINEAR PROGRAMMING

2.1
INTRODUCTION

Programming problems in general are concerned with the use or allocation of scarce resources—labor, materials, machines, and capital—in the "best" possible manner so that costs are minimized or profits are maximized. In using the term "best" it is implied that some choice or a set of alternative courses of actions is available for making the decision. In general, the best decision is found by solving a mathematical problem. The term linear programming merely defines a particular class of programming problems that meet the following conditions:

1. The decision variables involved in the problem are nonnegative (i.e., positive or zero).
2. The criterion for selecting the "best" values of the decision variables can be described by a linear function of these variables, that is, a mathematical function involving only the first powers of the variables with no cross products. The criterion function is normally referred to as the *objective function*.
3. The operating rules governing the process (e.g., scarcity of resources) can be expressed as a set of linear equations or linear inequalities. This set is referred to as the *constraint set*.

The last two conditions are the reasons for the use of the term linear programming.

Linear programming techniques are widely used to solve a number of military, economic, industrial, and social problems. Three primary reasons for its wide use are:

1. A large variety of problems in diverse fields can be represented or at least approximated as linear programming models.
2. Efficient techniques for solving linear programming problems are available, and
3. Ease through which data variation (Sensitivity Analysis) can be handled through linear programming models.

At this point, we should point out that the solution procedures are iterative in nature, and hence even for moderate size problems, one has to resort to a digital computer for solution. This could be a serious disadvantage if the answer is not worth more than the cost to obtain it. In other words, the cost of analysis using linear programming may offset the saving that may result. But with the advancement in computer technology, the solution of large linear programming problems by digital computer has not only become feasible but inexpensive as well.

2.2 FORMULATION OF LINEAR PROGRAMMING MODELS

The three basic steps in constructing a linear programming model are as follows:

Step I. Identify the unknown variables to be determined (decision variables) and represent them in terms of algebraic symbols.

Step II. Identify all the restrictions or constraints in the problem and express them as linear equations or inequalities which are linear functions of the unknown variables.

Step III. Identify the objective or criterion and represent it as a linear function of the decision variables, which is to be maximized or minimized.

We shall illustrate these basic steps by formulating a number of linear programming problems. Model building is not a science but primarily an art and comes mainly by practice. Hence the reader is advised to work out many of the exercises given at the end of this chapter on problem formulation.

EXAMPLE 2.2-1 PRODUCT-MIX PROBLEM

The Handy-Dandy Company wishes to schedule the production of a kitchen appliance that requires two resources—labor and material. The company is considering three different models and its production engineering department has furnished the following data:

	Model		
	A	*B*	*C*
Labor (hours per unit)	7	3	6
Material (pounds per unit)	4	4	5
Profit ($ per unit)	4	2	3

The supply of raw material is restricted to 200 pounds per day. The daily availability of labor is 150 hours. Formulate a linear programming model to determine the daily production rate of the various models in order to maximize the total profit.

Formulation

Step I. Identify the Decision Variables. The unknown activities to be determined are the daily rate of production for the three models.

Representing them by algebraic symbols,

x_A—Daily production of model A
x_B—Daily production of model B
x_C—Daily production of model C

Step II. Identify the Constraints. In this problem the constraints are the limited availability of the two resources—labor and material. Model A requires 7 hours of labor for each unit, and its production quantity is x_A. Hence, the labor requirement for model A alone will be $7x_A$ hours (assuming a linear relationship). Similarly, models B and C will require $3x_B$ and $6x_C$ hours, respectively. Thus, the total requirement of labor will be $7x_A+3x_B+6x_C$, which should not exceed the available 150 hours. So, the labor constraint becomes,

$$7x_A+3x_B+6x_C \leqq 150$$

Similarly, the raw material requirements will be $4x_A$ pounds for model A, $4x_B$ pounds for model B, and $5x_C$ pounds for model C. Thus, the raw material constraint is given by

$$4x_A+4x_B+5x_C \leqq 200$$

In addition, we restrict the variables x_A, x_B, x_C to have only nonnegative values. This is called the *nonnegativity constraint*, which the variables must satisfy. Most practical linear programming problems will have this nonnegative restriction on the decision variables. However, the general framework of linear programming is not restricted to nonnegative values, and methods for handling variables without sign restrictions will be discussed later.

Step III. Identify the Objective. The objective is to maximize the total profit from sales. Assuming that a perfect market exists for the product such that all that is produced can be sold, the total profit from sales becomes

$$Z=4x_A+2x_B+3x_C$$

Thus, the linear programming model for our product mix problem becomes:

Find numbers x_A, x_B, x_C which will maximize

$$Z=4x_A+2x_B+3x_C$$

subject to the constraints

$$7x_A+3x_B+6x_C \leqq 150$$
$$4x_A+4x_B+5x_C \leqq 200$$
$$x_A \geqq 0, \qquad x_B \geqq 0, \qquad x_C \geqq 0.$$

EXAMPLE 2.2-2 A JOB-TRAINING PROBLEM

A machine tool company conducts a job-training program for machinists. Trained machinists are used as teachers in the program at a ratio of one for every ten trainees. The training program lasts for one month. From past experience it has been found that out of ten trainees hired, only seven complete the program successfully (the unsuccessful trainees are released).

Trained machinists are also needed for machining and the company's requirements for the next three months are as follows:

January	100
February	150
March	200

In addition, the company requires 250 trained machinists by April. There are 130 trained machinists available at the beginning of the year. Payroll costs per month are:

Each trainee	\$400
Each trained machinist (machining or teaching)	700
Each trained machinist idle (Union contract forbids firing trained machinists)	500

Set up the linear programming problem that will produce the minimum cost hiring and training schedule and meet the company's requirements.

Formulation

First, we note that every month a trained machinist can do one of the following: (1) work a machine, (2) teach, or (3) stay idle.

Since the number of trained machinists machining is fixed, the only (unknown) decision variables are the number teaching and the number idle for each month. Thus, the variables to be determined are:

x_1—trained machinists teaching in January
x_2—trained machinists idle in January
x_3—trained machinists teaching in February
x_4—trained machinists idle in February
x_5—trained machinists teaching in March
x_6—trained machinists idle in March

The constraints require that a sufficient number of trained machinists be available each month for machining. This can be met by writing the following equation for each month:

$$\begin{aligned}&\text{Number machining} + \text{Number teaching} + \text{Number idle}\\ &= \text{Total trained machinists available at the beginning of the month}\end{aligned}$$

For example, for the month of January the constraint becomes

$$100 + x_1 + x_2 = 130$$

For February, the total number of trained machinists available will be the sum of trained machinists in January and those coming from the training program. In January, there are $10x_1$ trainees in the program, and out of those only $7x_1$ successfully complete and become trained machinists. Thus, the constraint for February becomes:

$$150 + x_3 + x_4 = 130 + 7x_1$$

Similarly, for March,

$$200 + x_5 + x_6 = 130 + 7x_1 + 7x_3$$

Since the company requires 250 trained machinists by April, we need the constraint

$$130 + 7x_1 + 7x_3 + 7x_5 = 250$$

Of course, all the variables are restricted to be nonnegative.

While writing the objective function, the cost of machinists doing machine work need not be included as it is a constant term. The relevant costs are only the cost of the training program (cost of trainees and teachers) and the cost of idle machinists. Thus, the objective function

$$\text{Minimize:} \quad Z = 400(10x_1 + 10x_3 + 10x_5) + 700(x_1 + x_3 + x_5) + 500(x_2 + x_4 + x_6)$$

Thus, the linear programming problem becomes

Minimize:

$$Z = 4700x_1 + 500x_2 + 4700x_3 + 500x_4 + 4700x_5 + 500x_6$$

$$\begin{aligned} \text{Subject to:} \quad & x_1 + x_2 = 30 \\ & 7x_1 - x_3 - x_4 = 20 \\ & 7x_1 + 7x_3 - x_5 - x_6 = 70 \\ & 7x_1 + 7x_3 + 7x_5 = 120 \end{aligned}$$

$$x_1 \geqq 0, \quad x_2 \geqq 0, \quad x_3 \geqq 0, \quad x_4 \geqq 0, \quad x_5 \geqq 0, \quad x_6 \geqq 0$$

EXAMPLE 2.2-3 ADVERTISING MEDIA SELECTION

An advertising company wishes to plan an advertising campaign in three different media—television, radio, and magazines. The purpose of the advertising program is to reach as many potential customers as possible. Results of a market study are given below:

	Television			
	Daytime	Prime Time	Radio	Magazines
Cost of an advertising unit	\$40,000	\$75,000	\$30,000	\$15,000
Number of potential customers reached per unit	400,000	900,000	500,000	200,000
Number of women customers reached per unit	300,000	400,000	200,000	100,000

The company does not want to spend more than \$800,000 on advertising. It further requires that (1) at least 2 million exposures take place among women; (2) advertising on television be limited to \$500,000; (3) at least 3 advertising units be bought on daytime television, and two units during prime time; and (4) the number of advertising units on radio and magazine should each be between 5 and 10.

Formulation

Let x_1, x_2, x_3, and x_4 be the number of advertising units bought in daytime television, prime-time television, radio, and magazines, respectively.

The total number of potential customers reached (in thousands) $= 400x_1 + 900x_2 + 500x_3 + 200x_4$. The restriction on the advertising budget is represented by

$$40{,}000x_1 + 75{,}000x_2 + 30{,}000x_3 + 15{,}000x_4 \leqq 800{,}000$$

The constraint on the number of women customers reached by the advertising campaign becomes

$$300{,}000x_1 + 400{,}000x_2 + 200{,}000x_3 + 100{,}000x_4 \geqq 2{,}000{,}000$$

The constraints on television advertising are

$$\begin{aligned} 40{,}000x_1 + 75{,}000x_2 &\leqq 500{,}000 \\ x_1 &\geqq 3 \\ x_2 &\geqq 2 \end{aligned}$$

Since advertising units on radio and magazines should each be between 5 and 10, we get the following constraints:

$$\begin{aligned} 5 \leqq x_3 \leqq 10 \\ 5 \leqq x_4 \leqq 10 \end{aligned}$$

The complete linear programming problem with some minor simplification is given below:

$$\begin{aligned} \text{Maximize:} \quad & Z = 400x_1 + 900x_2 + 500x_3 + 200x_4 \\ \text{Subject to:} \quad & 40x_1 + 75x_2 + 30x_3 + 15x_4 \leqq 800 \\ & 30x_1 + 40x_2 + 20x_3 + 10x_4 \geqq 200 \\ & 40x_1 + 75x_2 \leqq 500 \\ & x_1 \geqq 3 \\ & x_2 \geqq 2 \\ & x_3 \geqq 5 \\ & x_3 \leqq 10 \\ & x_4 \geqq 5 \\ & x_4 \leqq 10 \end{aligned}$$

Note: The first three constraints were simplified so that all the constraint coefficients do not have a large variation in magnitudes. This reduces the round-off errors in computer solutions.

EXAMPLE 2.2-4 AN INSPECTION PROBLEM

A company has two grades of inspectors, 1 and 2, who are to be assigned for a quality control inspection. It is required that at least 1800 pieces be inspected per 8-hour day. Grade 1 inspectors can check pieces at the rate of 25 per hour, with an accuracy of 98%. Grade 2 inspectors check at the rate of 15 pieces per hour, with an accuracy of 95%.

The wage rate of a Grade 1 inspector is $4.00 per hour, while that of a Grade 2 inspector is $3.00 per hour. Each time an error is made by an inspector, the cost to the company is $2.00. The company has available for the inspection job eight Grade 1 inspectors, and ten Grade 2 inspectors. The company wants to determine the optimal assignment of inspectors, which will minimize the total cost of the inspection.

Formulation

Let x_1 and x_2 denote the number of Grade 1 and Grade 2 inspectors assigned for inspection. Since the number of available inspectors in each grade is limited, we have the following constraints:

$$\begin{aligned} x_1 &\leqq 8 \qquad \text{(Grade 1)} \\ x_2 &\leqq 10 \qquad \text{(Grade 2)} \end{aligned}$$

The company requires at least 1800 pieces to be inspected daily. Thus, we get

$$8(25)x_1 + 8(15)x_2 \geqq 1800$$

or

$$200x_1 + 120x_2 \geqq 1800$$

To develop the objective function, we note that the company incurs two types of costs during inspection: wages paid to the inspector, and the cost of his inspection errors. The hourly cost of each Grade 1 inspector is

$$\$4 + 2(25)(0.02) = \$5 \text{ per hour}$$

Similarly, for each Grade 2 inspector

$$\$3 + 2(15)(0.05) = \$4.50 \text{ per hour}$$

Thus the objective function is to minimize the daily cost of inspection given by

$$Z = 8(5x_1 + 4.50x_2) = 40x_1 + 36x_2$$

The complete formulation of the linear programming problem thus becomes

$$\begin{aligned}
&\text{Minimize:} && Z = 40x_1 + 36x_2 \\
&\text{Subject to:} && x_1 \leqq 8 \\
& && x_2 \leqq 10 \\
& && 5x_1 + 3x_2 \geqq 45 \\
& && x_1 \geqq 0, \quad x_2 \geqq 0
\end{aligned}$$

EXAMPLE 2.2-5 HYDROELECTRIC POWER SYSTEMS PLANNING

An agency controls the operation of a system consisting of two water reservoirs with one hydroelectric power generation plant attached to each as shown in Fig. 2.1. The planning horizon for the system is broken into two periods. When the reservoir is at full capacity, additional inflowing water is spilled over a spillway. In addition, water can also be released through a spillway as desired for flood protection purposes. Spilled water does not produce any electricity.

Assume that on an average 1 kilo-acre-foot (KAF) of water is converted to 400 megawatt hours (MWh) of electricity by power plant A and 200 MWh by power plant B. The capacities of power plants A and B are 60,000 and 35,000 MWh per period. During each period, up to 50,000 MWh of electricity can be sold at \$20.00/MWh, and excess power above 50,000 MWh can only be sold for \$14.00/MWh. The following table gives additional data on the reservoir operation and inflow in kilo-acre-feet:

	Reservoir A	Reservoir B
Capacity	2000	1500
Predicted inflow		
Period 1	200	40
Period 2	130	15
Minimum allowable level	1200	800
Level at the beginning of period 1	1900	850

Develop a linear programming model for determining the optimal operating policy that will maximize the total revenue from electricity sales.

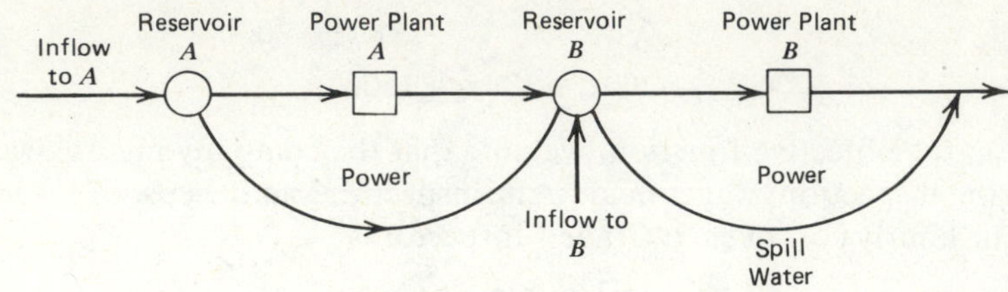

Figure 2.1
Hydroelectric system.

Formulation

This example illustrates how certain nonlinear objective functions can be handled by LP methods. The nonlinearity in the objective function is due to the differences in the unit megawatt-hour revenue from electricity sales depending on the amount sold. The nonlinearity is apparent if a graph is plotted between total revenue and electricity sales. This is illustrated by the graph of Fig. 2.2, which is a piecewise linear function, since it is linear in the regions (0, 50000) and (50000, ∞). Hence, by partitioning the quantity of electricity sold into two parts, the part that is sold for \$20.00/MWh and the part that is sold for \$14.00/MWh, the objective function could be represented as a linear function.

Let us first formulate the problem of reservoir operation for period 1 and then extend it to period 2. The variables are

PH1	Power sold at \$20/MWh, MWh
PL1	Power sold at \$14/MWh, MWh
XA1	Water supplied to power plant A, KAF
XB1	Water supplied to power plant B, KAF
SA1	Spill water drained from reservoir A, KAF
SB1	Spill water drained from reservoir B, KAF
EA1	Reservoir A level at the end of period 1, KAF
EB1	Reservoir B level at the end of period 1, KAF

The total power produced in period 1 is 400XA1 + 200XB1, while the total power sold is PH1 + PL1. Thus, the power generation constraint becomes

$$400\text{XA1} + 200\text{XB1} = \text{PH1} + \text{PL1}$$

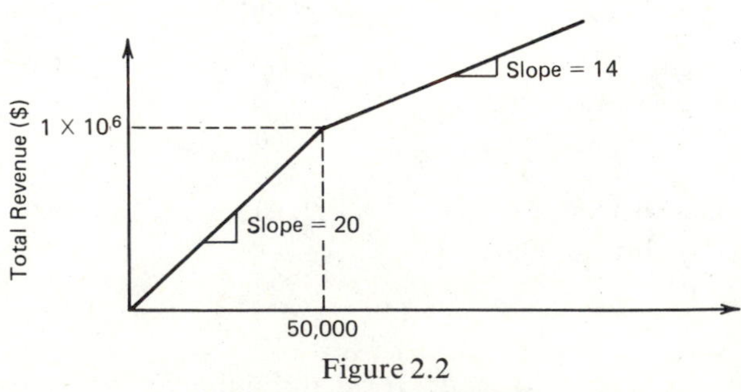

Figure 2.2
Electricity sales MWH.

Since the maximum power that can be sold at the higher rate is 50,000 MWh, we get $\mathrm{PH1} \leqslant 50{,}000$. The capacity of power plant A, expressed in water units (KAF) is

$$\frac{60{,}000}{400} = 150$$

Hence,

$$\mathrm{XA1} \leqslant 150$$

Similarly,

$$\mathrm{XB1} \leqslant 87.5$$

The conservation of water flow for reservoir A is given by

Water supplied to power plant A + spill water + ending reservoir level
= beginning reservoir level + predicted inflow

$$\mathrm{XA1} + \mathrm{SA1} + \mathrm{EA1} = 1900 + 200 = 2100$$

Since the capacity of reservoir A is 2000 KAF and the minimum allowable level is 1200, we get the following additional constraints:

$$\mathrm{EA1} \leqslant 2000$$
$$\mathrm{EA1} \geqslant 1200$$

Similarly, for reservoir B, we get

$$\mathrm{XB1} + \mathrm{SB1} + \mathrm{EB1} = 850 + 40 + \mathrm{XA1} + \mathrm{SA1}$$
$$800 \leqslant \mathrm{EB1} \leqslant 1500$$

We can now develop the model for period 2, by defining the appropriate decision variables PH2, PL2, XA2, XB2, SA2, SB2, EA2, and EB2.

The power generation constraints for period 2 are given by

$$400\mathrm{XA2} + 200\mathrm{XB2} = \mathrm{PH2} + \mathrm{LH2}$$
$$\mathrm{PH2} \leqslant 50{,}000$$
$$\mathrm{XA2} \leqslant 150$$
$$\mathrm{XB2} \leqslant 87.5$$

The water flow constraints are given by

$$\mathrm{XA2} + \mathrm{SA2} + \mathrm{EA2} = \mathrm{EA1} + 130$$
$$1200 \leqslant \mathrm{EA2} \leqslant 2000$$
$$\mathrm{XB2} + \mathrm{SB2} + \mathrm{EB2} = \mathrm{EB1} + 15 + \mathrm{XA2} + \mathrm{SA2}$$
$$800 \leqslant \mathrm{EB2} \leqslant 1500$$

In addition to the above constraints, we have the nonnegativity restrictions on all the decision variables.

The objective function is to maximize the total revenue from sales, given by

$$\text{Maximize:} \quad Z = 20(\mathrm{PH1} + \mathrm{PH2}) + 14(\mathrm{PL1} + \mathrm{PL2})$$

Thus, the final model has 16 decision variables and 20 constraints, excluding the nonnegativity constraints.

EXAMPLE 2.2-6

A machine shop has one drill press and five milling machines, which are to be used to produce an assembly consisting of two parts, 1 and 2. The productivity of each machine for the two parts is given below:

	Production time in minutes per piece	
Part	Drill	Mill
1	3	20
2	5	15

It is desired to maintain a balanced loading on all machines such that no machine runs more than 30 minutes per day longer than any other machine (assume that the milling load is split evenly among all five milling machines).

Divide the work time of each machine to obtain the maximum number of completed assemblies assuming an 8-hour working day.

Formulation

Let x_1 = number of part 1 produced per day and,
x_2 = number of part 2 produced per day.
The load on each milling machine (in minutes) $= (20x_1 + 15x_2)/5 = 4x_1 + 3x_2$, whereas the load on the drill press (in minutes) $= 3x_1 + 5x_2$. Thus the time restriction on each milling machine is

$$4x_1 + 3x_2 \leqq (8)(60) = 480$$

Similarly, for the drill press

$$3x_1 + 5x_2 \leqq 480$$

The machine balance constraint can be represented by

$$|(4x_1 + 3x_2) - (3x_1 + 5x_2)| \leqq 30$$

or

$$|x_1 - 2x_2| \leqq 30$$

This is a nonlinear constraint which can be replaced by the following two linear constraints:

$$x_1 - 2x_2 \leqq 30$$
$$-x_1 + 2x_2 \leqq 30$$

The number of completed assemblies cannot exceed the smaller value of part I and part 2 produced. Thus, the objective function is to maximize Z = minimum (x_1, x_2). This is again a nonlinear function. However, another trick can be used to represent it as a linear function. Let y = minimum of (x_1, x_2) where y represents the number of completed assemblies.

This means that

$$x_1 \geqq y$$
$$x_2 \geqq y$$

and the objective is to maximize $Z=y$. Thus, the complete linear programming formulation becomes

$$\begin{array}{ll} \text{Maximize:} & Z=y \\ \text{Subject to:} & 4x_1+3x_2 \leqq 480 \\ & 3x_1+5x_2 \leqq 480 \\ & x_1-2x_2 \leqq 30 \\ & -x_1+2x_2 \leqq 30 \\ & x_1 - y \geqq 0 \\ & x_2 - y \geqq 0 \\ & x_1 \geqq 0, \quad x_2 \geqq 0, \quad y \geqq 0 \end{array}$$

2.3 GRAPHICAL SOLUTION OF LINEAR PROGRAMS IN TWO VARIABLES

In the last section some examples were presented to illustrate how practical problems can be formulated mathematically as linear programming problems. The next step after formulation is to solve the problem mathematically to obtain the best possible solution. In this section, a graphical procedure to solve linear programming problems involving only two variables is discussed. Though in practice such small problems are usually not encountered, the graphical procedure is presented to illustrate some of the basic concepts used in solving large linear programming problems.

EXAMPLE 2.3-1

Recall the inspection problem given by Example 2.2-4:

$$\begin{array}{ll} \text{Minimize:} & Z=40x_1+36x_2 \\ \text{Subject to:} & x_1 \leqq 8 \\ & x_2 \leqq 10 \\ & 5x_1+3x_2 \geqq 45 \\ & x_1 \geqq 0, \quad x_2 \geqq 0 \end{array}$$

In this problem, we are interested in determining the values of the variables x_1 and x_2 that will satisfy all the restrictions and give the least value for the objective function. As a first step in solving this problem, we want to identify all possible values of x_1 and x_2 that are nonnegative and satisfy the constraints. For example, a solution $x_1=8$, $x_2=10$ is positive and satisfies all the constraints. Such a solution is called a *feasible solution*. The set of all feasible solutions is called the *feasible region*. Solution of a linear program is merely finding the best feasible solution in the feasible region. The best feasible solution is called an *optimal solution* to the linear programming problem. In our example, an optimal solution is a feasible solution which minimizes the objective function $40x_1+36x_2$. The value of the objective function corresponding to an optimal solution is called the *optimal value* of the linear program.

To represent the feasible region in a graph, every constraint is plotted, and all values of x_1, x_2 that will satisfy these constraints are identified. The nonnegativity constraints imply that all feasible values of the two variables will lie in the first quadrant. The constraint $5x_1+3x_2 \geqq 45$ requires that any feasible solution (x_1, x_2) to the problem should be on one side of the straight line $5x_1+3x_2=45$. The proper side is found by testing whether the origin satisfies the constraint or not. The line $5x_1+3x_2=45$ is first plotted by taking two convenient points (e.g., $x_1=0$, $x_2=15$, and $x_1=9$, $x_2=0$).

The proper side is indicated by an arrow directed above the line since the origin does not satisfy the constraint. Similarly, the constraints $x_1 \leqq 8$, and $x_2 \leqq 10$ are plotted. The feasible region is given by the shaded region ABC as shown in Fig. 2.3. Obviously there is an infinite number of feasible points in this region. Our objective is to identify the feasible point with the lowest value of Z.

Observe that the objective function, given by $Z=40x_1+36x_2$, represents a straight line if the value of Z is fixed *a priori*. Changing the value of Z essentially translates the entire line to another straight line parallel to itself. In order to determine an optimal solution, the objective function line is drawn for a convenient

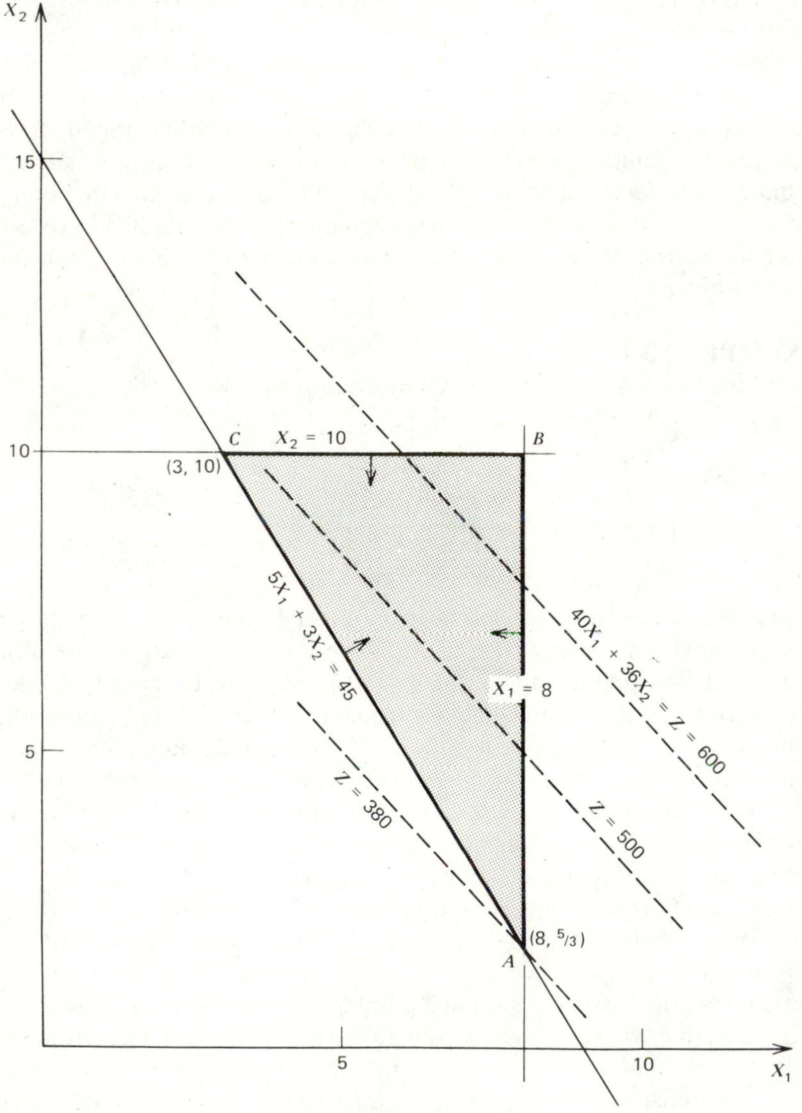

Figure 2.3
Feasible region of Example 2.3-1.

value of Z such that it passes through one or more points in the feasible region. Initially Z is chosen as 600. By moving this line closer to the origin the value of Z is further decreased (see Fig. 2.3). The only limitation on this decrease is that the straight line $40x_1 + 36x_2 = Z$ contains at least one point in the feasible region ABC. This clearly occurs at the corner point A given by $x_1 = 8$, $x_2 = \frac{5}{3}$. This is the best feasible point giving the lowest value of Z as 380. Hence, $x_1 = 8$, $x_2 = \frac{5}{3}$ is an optimal solution, and $Z = 380$ is the *optimal value* for the linear program.

Thus for the inspection problem the optimal utilization is achieved by using eight Grade 1 inspectors and 1.67 Grade 2 inspectors. The fractional value $x_2 = \frac{5}{3}$ suggests that one of the Grade 2 inspectors is only utilized for 67% of the time. If this is not possible, the normal practice is to round off the fractional values to get an optimal integer solution as $x_1 = 8$, $x_2 = 2$. (In general, rounding off the fractional values will not produce an optimal integer solution.)

Unique Optimal Solution

In Example 2.3-1, the solution $x_1 = 8$, $x_2 = \frac{5}{3}$ is the only feasible point with the lowest value of Z. In other words, the values of Z corresponding to the other feasible solutions in Fig. 2.3 exceed the optimal value of 380. Hence for this problem, the solution $x_1 = 8$, $x_2 = \frac{5}{3}$ is the *unique optimal solution*.

Alternative Optimal Solutions

In some linear programming problems, there may exist more than one feasible solution such that their objective function values are equal to the optimal value of the linear program. In such cases, all of these feasible solutions are optimal solutions, and the linear program is said to have *alternative or multiple optimal solutions*. To illustrate this, consider the following linear programming problem:

EXAMPLE 2.3-2

Maximize: $\quad Z = x_1 + 2x_2$

Subject to:

$$x_1 + 2x_2 \leqq 10$$
$$x_1 + x_2 \geqq 1$$
$$x_2 \leqq 4$$
$$x_1 \geqq 0, \qquad x_2 \geqq 0$$

The feasible region is shown in Fig. 2.4. The objective function lines are drawn for $Z = 2$, 6, and 10. The optimal value for the linear program is 10, and the corresponding objective function line $x_1 + 2x_2 = 10$ coincides with side BC of the feasible region. Thus, the corner point feasible solutions $x_1 = 10$, $x_2 = 0$ (B), and $x_1 = 2$, $x_2 = 4$ (C), and all other feasible points on the line BC are optimal solutions.

Unbounded Solution

Some linear programming problems may not have an optimal solution. In other words, it is possible to find better feasible solutions continuously improving the objective function values. This would have been the case if the constraint $x_1 + 2x_2 \leqq 10$ were not given in Example 2.3-2. In this case, moving farther away from the origin increases the objective function $x_1 + 2x_2$ and maximum Z would be $+\infty$. When there exists no finite optimum, the linear program is said to have an *unbounded solution*.

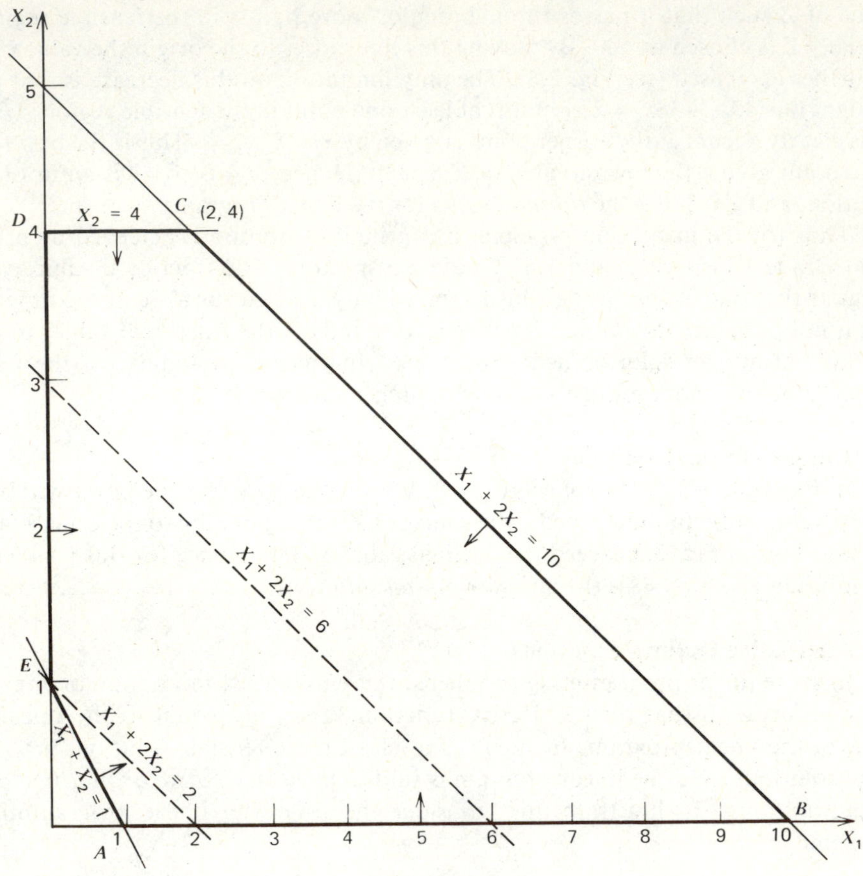

Figure 2.4
Feasible region of Example 2.3-2.

It is inconceivable for a practical problem to have an unbounded solution since this implies that one can make infinite profit from a finite amount of resources! If such a solution is obtained in a practical problem it generally means that one or more constraints have been omitted inadvertently during the initial formulation of the problem. These constraints would have prevented the objective function from assuming infinite values.

Conclusion

In Example 2.3-1, the optimal solution was unique and occurred at the corner point A in the feasible region. In Example 2.3-2, we had multiple optimal solutions to the problem that included two corner points B and C. In either case, one of the corner points of the feasible region was always an optimal solution. As a matter of fact, the following property is true for any linear programming problem:

Property 1. If there exists an optimal solution to a linear programming problem, then at least one of the corner points of the feasible region will always qualify to be an optimal solution.

This is the fundamental property on which an iterative procedure called the *simplex method* for solving linear programming problems is based. Even though the feasible region of a linear programming problem contains an infinite number of points, an optimal solution can be determined by merely examining the finite number of corner points in the feasible region. For instance, in Example 2.3-2, the objective function was to maximize

$$Z = x_1 + 2x_2$$

The corner points of the feasible region were $A(1,0)$, $B(10,0)$, $C(2,4)$, $D(0,4)$, and $E(0,1)$. Evaluating their Z values, we get $Z(A)=1$, $Z(B)=10$, $Z(C)=10$, $Z(D)=8$, and $Z(E)=2$. Since the maximum value of Z occurs at corner points B and C, both are optimal solutions according to Property 1.

2.4 LINEAR PROGRAM IN STANDARD FORM

The standard form of a linear programming problem with m constraints and n variables can be represented as follows:

Maximize (Minimize):

$$Z = c_1x_1 + c_2x_2 + \cdots + c_nx_n$$

Subject to:

$$\begin{aligned}
a_{11}x_1 + a_{12}x_2 + \cdots + a_{1n}x_n &= b_1 \\
a_{21}x_1 + a_{22}x_2 + \cdots + a_{2n}x_n &= b_2 \\
\vdots \qquad\qquad\qquad & \quad \vdots \\
a_{m1}x_1 + a_{m2}x_2 + \cdots + a_{mn}x_n &= b_m \\
x_1 \geqq 0, x_2 \geqq 0, \ldots, x_n &\geqq 0 \\
b_1 \geqq 0, b_2 \geqq 0, \ldots, b_m &\geqq 0
\end{aligned}$$

The main features of the standard form are:

1. The objective function is of the maximization or minimization type.
2. All constraints are expressed as equations.
3. All variables are restricted to be nonnegative.
4. The right-hand side constant of each constraint is nonnegative.

In matrix-vector notation, the standard linear programming problem can be expressed in a compact form as

Maximize (Minimize): $Z = \mathbf{cx}$

Subject to:

$$\begin{aligned}
\mathbf{Ax} &= \mathbf{b} \\
\mathbf{x} &\geqq 0 \\
\mathbf{b} &\geqq 0
\end{aligned}$$

where $\mathbf{A}$ is an $(m \times n)$ matrix, $\mathbf{x}$ is an $(n \times 1)$ column vector, $\mathbf{b}$ is an $(m \times 1)$ column vector, and $\mathbf{c}$ is a $(1 \times n)$ row vector.

In other words,

$$\underset{(m\times n)}{\mathbf{A}} = \begin{bmatrix} a_{11} & a_{12} & \cdots & a_{1n} \\ a_{21} & a_{22} & \cdots & a_{2n} \\ \cdot & & & \\ \cdot & & & \\ \cdot & & & \\ a_{m1} & a_{m2} & \cdots & a_{mn} \end{bmatrix}, \qquad \underset{(n\times 1)}{\mathbf{x}} = \begin{bmatrix} x_1 \\ x_2 \\ \cdot \\ \cdot \\ \cdot \\ x_n \end{bmatrix}$$

$$\underset{(m\times 1)}{\mathbf{b}} = \begin{bmatrix} b_1 \\ b_2 \\ \cdot \\ \cdot \\ \cdot \\ b_m \end{bmatrix}, \qquad \text{and} \qquad \underset{(1\times n)}{\mathbf{c}} = (c_1, c_2, \ldots, c_n)$$

In practice, $\mathbf{A}$ is called the coefficient matrix, $\mathbf{x}$ is the decision vector, $\mathbf{b}$ is the requirement vector, and $\mathbf{c}$ is the profit (cost) vector of the linear program.

The standard form and its associated general notations are illustrated by the following example:

Maximize: $Z = 5x_1 + 2x_2 + 3x_3 - x_4 + x_5$

Subject to:

$$\begin{aligned} x_1 + 2x_2 + 2x_3 + x_4 &= 8 \\ 3x_1 + 4x_2 + x_3 + x_5 &= 7 \\ x_1 \geqq 0, \ldots, x_5 &\geqq 0 \end{aligned}$$

In this problem,

$$\underset{(2\times 5)}{\mathbf{A}} = \begin{bmatrix} 1 & 2 & 2 & 1 & 0 \\ 3 & 4 & 1 & 0 & 1 \end{bmatrix}, \qquad \underset{(5\times 1)}{\mathbf{x}} = \begin{bmatrix} x_1 \\ x_2 \\ x_3 \\ x_4 \\ x_5 \end{bmatrix}$$

$$\underset{(2\times 1)}{\mathbf{b}} = \begin{bmatrix} 8 \\ 7 \end{bmatrix}, \qquad \text{and} \qquad \underset{(1\times 5)}{\mathbf{c}} = (5 \quad 2 \quad 3 \quad -1 \quad 1)$$

Reduction of a General Linear Program to a Standard Form

The simplex method for solving linear programming problems requires the problem to be expressed in standard form. But not all linear programming problems come in standard form. Very often the constraints are expressed as inequalities rather than equations. In some problems all the decision variables may not be nonnegative. Hence the first step in solving a linear program is to convert it to a problem in standard form.

Handling Inequality Constraints

Since the standard form requires all the constraints to be equations, inequality constraints have to be converted to equations. This can be done by the introduction of new variables to represent the slack between the left-hand side and right-hand side of each inequality. The new variables are called *slack or surplus variables*.

To illustrate the use of slack variables, consider Example 2.2-4. Here the constraint on the availability of Grade 1 inspectors was

$$x_1 \leqq 8$$

This can be converted to an equation by introducing a slack variable, say x_3, as follows:

$$x_1 + x_3 = 8$$

Note that x_3 is also nonnegative, and it represents the number of Grade 1 inspectors not utilized. Similarly the second constraint becomes

$$x_2 + x_4 = 10, \qquad \text{where} \qquad x_4 \geqq 0$$

To illustrate the use of surplus variables consider the third constraint on the minimum number of pieces to be inspected:

$$200x_1 + 120x_2 \geqq 1800$$

By introducing a surplus variable, say x_5, the above inequality could be expressed as an equation as follows:

$$200x_1 + 120x_2 - x_5 = 1800$$

where x_5 is nonnegative and represents the number of extra pieces inspected over the daily minimum.

It should be emphasized here that slack and surplus variables are as much a part of the original problem as the variables used in the formulation of the linear program. These variables can remain positive throughout, and their values in the optimal solution give useful information about the problem.

Handling Variables Unrestricted in Sign

In some situations, it may become necessary to introduce a variable that can assume both positive and negative values. For example, consider an investment problem where an investor has $100 cash on hand and can borrow money if necessary for investment. If x_1 denotes the amount invested, then the investment constraint can be written as:

$$x_1 + s_1 = 100$$

Here, s_1 can assume positive or negative values depending on whether s_1 represents money saved or money borrowed.

Since the standard form requires all the variables to be nonnegative, an unrestricted variable is generally replaced by the difference of two nonnegative variables. In other words, the following transformation of variables is used for s:

$$s_1 = x_5 - x_6$$
$$x_5 \geqq 0 \qquad x_6 \geqq 0$$

The value of s_1 is positive or negative depending on whether x_5 is larger or smaller than x_6.

As an illustration, consider the following nonstandard linear program:

EXAMPLE 2.4-1

Maximize: $\quad Z = x_1 - 2x_2 + 3x_3$

Subject:

$$x_1 + x_2 + x_3 \leqq 7$$
$$x_1 - x_2 + x_3 \geqq 2$$
$$3x_1 - x_2 - 2x_3 = -5$$
$$x_1, x_2 \geqq 0$$
$$x_3 \text{ unrestricted in sign}$$

To convert the above problem to standard form,

1. Replace x_3 by $x_4 - x_5$ where $x_4, x_5 \geqq 0$,
2. Multiply both sides of the last constraint by -1,
3. Introduce slack and surplus variables x_6 and x_7 to constraints 1 and 2 respectively, and
4. Assign zero profit for x_6 and x_7 so that the objective function is not altered.

Thus, the above problem reduces to the following standard form:

$$\begin{array}{ll} \text{Maximize:} & Z = x_1 - 2x_2 + 3x_4 - 3x_5 \\ \text{Subject to:} & x_1 + x_2 + x_4 - x_5 + x_6 = 7 \\ & x_1 - x_2 + x_4 - x_5 - x_7 = 2 \\ & -3x_1 + x_2 + 2x_4 - 2x_5 = 5 \\ & x_1, x_2, x_4, x_5, x_6, x_7 \geqq 0 \end{array}$$

Let us review the basic definitions using the standard form of the linear programming problem given by

$$\begin{array}{ll} \text{Maximize:} & Z = \mathbf{cx} \\ \text{Subject to:} & \mathbf{Ax} = \mathbf{b}, \\ & \mathbf{x} \geqq 0 \end{array}$$

1. A *feasible solution* is a nonnegative vector $\mathbf{x}$ satisfying the constraints $\mathbf{Ax} = \mathbf{b}$.
2. *Feasible region*, denoted by S, is the set of all feasible solutions. Mathematically,

   $$S = \{\mathbf{x} \mid \mathbf{Ax} = \mathbf{b}, \mathbf{x} \geqq 0\}$$

   If the feasible set S is empty, then the linear program is said to be *infeasible*.
3. *An optimal solution* is a vector $\mathbf{x}^o$ such that it is feasible and its value of the objective function ($\mathbf{cx}^o$) is larger than that of any other feasible solution. Mathematically, $\mathbf{x}^o$ is optimal if and only if $\mathbf{x}^o \in S$, and $\mathbf{cx}^o \geqq \mathbf{cx}$ for all $\mathbf{x} \in S$.
4. *Optimal value* of a linear program is the value of the objective function corresponding to the optimal solution. If Z^o is the optimal value, then $Z^o = \mathbf{cx}^o$.
5. *Alternative optimum.* When a linear program has more than one optimal solution, it is said to have alternate optimal solutions. In this case, there exists more than one feasible solution having the same optimal value (Z^o) for their objective functions.
6. *Unique optimum.* The optimal solution of a linear program is said to be unique when there exists no other optimal solution.
7. *Unbounded solution.* When a linear program does not possess a finite optimum (i.e., max $Z \rightarrow +\infty$), it is said to have an unbounded solution.

2.5 SOLVING SYSTEMS OF LINEAR EQUATIONS

The central mathematical problem of linear programming is to find the solution of a system of linear equations which maximizes *or* minimizes a given linear objective function. Systems of linear equations can be solved by the classical Gauss-Jordon elimination procedure. In this section, a brief review of this procedure is given through an example.

EXAMPLE 2.5-1

Consider the following system of two equations in five unknowns denoted by

$$(S_1) \qquad \begin{array}{ll} x_1 - 2x_2 + x_3 - 4x_4 + 2x_5 = 2 & (2.1) \\ x_1 - x_2 - x_3 - 3x_4 - x_5 = 4 & (2.2) \end{array}$$

Since there are more unknowns than equations, this system will have more than one solution. (As a matter of fact, the existence of multiple solutions to the constraint equations makes the solution of linear programs nontrivial.) The collection of all possible solutions to the system is called the *solution set*.

Definition

Two systems of equations are said to be *equivalent* if both systems have the same solution set. In other words, a solution to one system is automatically a solution to the other system and vice versa.

The method of solving a system of equations is to get an equivalent system which is easy to solve. By solving the simple system, we simultaneously get the solutions to the original system.

There are two types of elementary row operations that one can use to obtain equivalent systems.

1. Multiply any equation in the system by a positive or negative number.
2. Add to any equation a constant multiple (positive, negative, or zero) of any other equation in the system.

An equivalent system to system S_1 can be obtained by multiplying Eq. 2.1 by -1 and adding to Eq. 2.2 as follows:

$$(S_2) \qquad \begin{aligned} x_1 - 2x_2 + x_3 - 4x_4 + 2x_5 &= 2 \qquad (2.3) \\ x_2 - 2x_3 + x_4 - 3x_5 &= 2 \qquad (2.4) \end{aligned}$$

From system S_2, another equivalent system can be obtained by multiplying Eq. 2.4 by 2, and adding to Eq. 2.3. This gives system S_3:

$$(S_3) \qquad \begin{aligned} x_1 - 3x_3 - 2x_4 - 4x_5 &= 6 \qquad (2.5) \\ x_2 - 2x_3 + x_4 - 3x_5 &= 2 \qquad (2.6) \end{aligned}$$

Since systems S_1, S_2, and S_3 are equivalent, a solution to one automatically gives a solution to the other two. In this case, it is easy to write out all the possible solutions to system S_3. For example, setting $x_3 = x_4 = x_5 = 0$ gives $x_1 = 6$, $x_2 = 2$, which is a solution to all the three systems. Other solutions to system S_3 may be obtained by choosing arbitrary values for x_3, x_4, x_5 and finding the corresponding values of x_1 and x_2 from Eqs. 2.5 and 2.6. All of these solutions are solutions to the original system. Systems like S_3 are called *canonical systems*. In Example 2.5-1, the equivalent canonical system was obtained by eliminating the coefficients of x_1 and x_2 from Eqs. 2.2 and 2.1, respectively. The variables x_1 and x_2 are called the *basic variables* of the canonical system, but any other two variables could have been made the basic variables if their coefficients were eliminated by suitable elementary operations. At times, the introduction of slack variables for obtaining the standard form automatically produces a canonical system.

Definition

A variable x_1 is said to be a *basic* variable in a given equation if it appears with a unit coefficient in that equation and zeros in all other equations.

Those variables which are not basic are called *nonbasic variables*.

By applying the elementary row operations, a given variable can be made a basic variable. This is called a *pivot operation*.

Pivot Operation

A *pivot operation* is a sequence of elementary operations that reduces a given system to an equivalent system in which a specified variable has a unit coefficient in one equation and zeros elsewhere.

In order to get a canonical system, a sequence of pivot operations must be performed on the original system so that there exists at least one basic variable in each equation. The number of basic variables is decided by the number of equations in the system.

Basic Solution

The solution obtained from a canonical system by setting the nonbasic variables to zero and solving for the basic variables is called a *basic solution*.

In Example 2.5-1, a basic solution is given by $x_1 = 6$, $x_2 = 2$, $x_3 = 0$, $x_4 = 0$, and $x_5 = 0$. This is easily obtained by inspection from the canonical system S_3. Since the values of all the variables are nonnegative, this basic solution is also a feasible solution. We call such basic solutions *basic feasible solutions.*

Basic Feasible Solution

A *basic feasible solution* is a basic solution in which the values of the basic variables are nonnegative.

In Example 2.5-1, a canonical system was obtained using x_1 and x_2 as basic variables. The choice of x_1 and x_2 as basic variables was purely arbitrary. We could have obtained a different canonical system (with a different basic solution) by choosing x_3 and x_4 as basic variables. As a matter of fact, any two variables could have been selected as basic variables out of the possible five variables to obtain a canonical system and a basic solution. This means that the number of basic solutions possible is

$$\binom{5}{2} = \frac{5!}{2!\,3!} = 10$$

This can be extended to the constraint equations of a standard linear program developed in Section 2.4. With m constraints and n variables, the maximum number of basic solutions to the standard linear program is finite and is given by

$$\binom{n}{m} = \frac{n!}{m!\,(n-m)!}$$

By definition, every basic feasible solution is also a basic solution. Hence, the maximum number of basic feasible solutions is also limited by the above expression.

At the end of Section 2.3 it was pointed out that whenever there is an optimal solution to a linear program, one of the corner points of the feasible region is always optimal. It can be shown that every corner point of the feasible region corresponds to a basic feasible solution of the constraint equations. This means that an optimal solution to a linear program can be obtained by merely examining its basic feasible solutions. This will be a finite process since the number of basic feasible solutions cannot exceed $\dfrac{n!}{m!(n-m)!}$.

A naive approach to solve a linear program (which has an optimal solution) would be to generate all possible basic feasible solutions through canonical reduction, and determine which basic feasible solution gives the best objective function value. But, the *simplex method* for solving linear programs does this in a more efficient manner by examining only a fraction of the total number of basic feasible solutions!

The details of the simplex method will be developed in the following sections.

2.6 PRINCIPLES OF THE SIMPLEX METHOD

The *simplex method* as developed by G.B. Dantzig is an iterative procedure for solving linear programming problems expressed in standard form. In addition to the standard form, the simplex method requires that the constraint equations be expressed as a canonical system from which a basic feasible solution can be readily obtained. The general steps of the simplex method are as follows:

1. Start with an initial basic feasible solution in canonical form.
2. Improve the initial solution if possible by finding another basic feasible solution with a better objective function value. At this step the simplex method implicitly eliminates from consideration all those basic feasible solutions whose objective function values are worse than the present one. This makes the procedure more efficient than the naive approach mentioned earlier.
3. Continue to find better basic feasible solutions improving the objective function values. When a particular basic feasible solution cannot be improved further, it becomes an optimal solution and the simplex method terminates.

We shall begin by illustrating the basic principles of the simplex method with the help of an example.

EXAMPLE 2.6-1

Maximize: $Z = 5x_1 + 2x_2 + 3x_3 - x_4 + x_5$

Subject to:

$$x_1 + 2x_2 + 2x_3 + x_4 \qquad = 8 \tag{2.7}$$

$$3x_1 + 4x_2 + x_3 \qquad + x_5 = 7 \tag{2.8}$$

$$x_1 \geqq 0, \quad x_2 \geqq 0, \quad x_3 \geqq 0, \quad x_4 \geqq 0, \quad x_5 \geqq 0$$

The above linear programming problem is in standard form since (1) all the variables are nonnegative, (2) all the constraints are equations, and (3) the right-hand side constants are positive. Moreover, the variable x_4 appears only in Eq. 2.7 with a unit coefficient. Hence x_4 is a basic variable in that equation. Similarly, x_5 is a basic variable in Eq. 2.8. Thus, we have a canonical system with x_4 and x_5 as basic variables. The corresponding basic solution is given by $x_1 = x_2 = x_3 = 0$, $x_4 = 8$, $x_5 = 7$. Since all the variables have nonnegative values, the above solution is also a basic feasible solution, and its objective function value is given by

$$Z = 5(0) + 2(0) + 3(0) - 1(8) + 1(7) = -1$$

It should be pointed out here that some linear-programming problems may not have a readily available canonical system and a basic feasible solution. In such problems one has to find a basic feasible solution in canonical form before initiating the simplex method. Such complications will be discussed later in Section 2.9.

Improving a Basic Feasible Solution

Given the initial basic feasible solution as $x_1 = x_2 = x_3 = 0$, $x_4 = 8$, $x_5 = 7$ with $Z = -1$, the simplex method checks whether it is possible to find a better basic feasible solution with a larger value of Z. This is done by first examining whether the present solution is optimal. In the cases where it is not optimal, the simplex method obtains an *adjacent basic feasible solution* with a larger value of Z (or at least as large).

Definition

An *adjacent basic feasible solution* differs from the present basic feasible solution in exactly one basic variable.

In order to obtain an adjacent basic feasible solution, the simplex method makes one of the basic variables a nonbasic variable, and in its place a nonbasic variable is brought in as a basic variable. The problem is to select the appropriate basic and nonbasic variables such that an exchange between them will give the maximum improvement to the objective function.

In any basic feasible solution, the basic variables can assume positive values while the nonbasic variables are always held at zero. Hence, making a nonbasic variable a basic variable is equivalent to increasing its value from zero to a positive quantity. Of course, the choice is made based on which nonbasic variable can improve the value of Z. This is determined by increasing the nonbasic variable by one unit and examining the resulting change in the value of the objective function.

To illustrate, consider the nonbasic variable x_1. Let us increase its value from 0 to 1 and study its effect on the objective function. Since we are interested in examining adjacent basic feasible solutions only, the values of the other two nonbasic variables, x_2 and x_3, will not change. After setting their values to zero, Eqs. 2.7 and 2.8 can be rewritten as,

$$x_1 + x_4 \qquad = 8 \tag{2.9}$$
$$3x_1 \qquad + x_5 = 7 \tag{2.10}$$

As x_1 increases to 1, the value of the basic variable x_4 will decrease to 7 to satisfy Eq. 2.9. Similarly, the value of x_5 will decrease to 4 as x_1 increases to 1. Hence the new feasible solution is given by

$$x_1 = 1 \qquad x_2 = 0, \qquad x_3 = 0, \qquad x_4 = 7, \qquad \text{and} \qquad x_5 = 4$$

The new value of the objective function is

$$Z = 5(1) + 2(0) + 3(0) - 1(7) + 1(4) = 2$$

Hence, the net change in the value of Z per unit increase in x_1

$$= \text{New value of } Z - \text{old value of } Z$$
$$= 2 - (-1) = 3$$

This value is called the *relative profit* of the nonbasic variable x_1 as opposed to its actual profit of 5 units in the objective function.

Since the relative profit on x_1 is positive, the objective function can be increased further by increasing x_1. Hence the initial basic feasible solution is not optimal. The relative profit coefficient implies that Z will be increased by 3 units for every unit increase on the nonbasic variable x_1. Naturally one would like to increase x_1 as much as possible so as to get the largest increase in the objective function.

Looking back at the constraints, it is clear that x_1 cannot be increased indefinitely. As x_1 increases, both the basic variables x_4 and x_5 will decrease, and their values must remain nonnegative for the solution to remain feasible. From Eq. 2.9, we see that x_4 will become negative if x_1 is increased beyond 8. Similarly from Eq. 2.10, x_5 will turn negative if x_1 were to increase beyond 7/3. Thus x_4 limits the increase on x_1 to 8, while x_5 limits it to 7/3. In order to maintain all the variables nonnegative, the maximum increase on x_1 is given by the lower of the two limits. Thus, the maximum increase on $x_1 = \text{minimum of } [8/1, 7/3] = 7/3$.

In some problems it is possible for a nonbasic variable to have a negative or zero coefficient in some constraints. In these constraints the corresponding basic variables will not decrease or become negative as we increase the nonbasic variable. Hence, these basic variables will not limit the increase of the nonbasic variable at all.

A unit increase in x_1 increases Z by 3 units. Since x_1 can be increased to a maximum of 7/3, the net increase in the objective function $=3(7/3)=7$. Also, when x_1 is increased to 7/3, the basic variable x_5 turns zero and is made a nonbasic variable. The nonbasic variable x_1 is made a basic variable in the second constraint. Thus the new basic feasible solution will be

$$x_1=\frac{7}{3}, \qquad x_2=0, \qquad x_3=0, \qquad x_4=\frac{17}{3}, \qquad x_5=0, \qquad \text{and} \qquad Z=6$$

The new canonical system corresponding to the improved basic feasible solution is obtained by performing a pivot operation on the variable x_1 as follows:

1. Divide Eq. 2.8 by 3 to reduce the coefficient of x_1 to 1.
2. Multiply Eq. 2.8 by $-1/3$ and add it to Eq. 2.7 to eliminate x_1.

$$\frac{2}{3}x_2+\frac{5}{3}x_3+x_4-\frac{1}{3}x_5=\frac{17}{3}$$

$$x_1+\frac{4}{3}x_2+\frac{1}{3}x_3 \qquad +\frac{1}{3}x_5=\frac{7}{3}$$

Once again the simplex method checks whether the above basic feasible solution is optimal by calculating the relative profits for all the nonbasic variables. If any of them turns out to be positive, then a new basic feasible solution with an improved value of Z is obtained as before. This process is repeated until the relative profits of all the nonbasic variables are negative or zero. This implies that the current basic feasible solution cannot be improved further and hence becomes an optimal solution to the linear programming problem.

Condition of Optimality

In a *maximization* problem, a basic feasible solution is optimal if the relative profits of its nonbasic variables are all *negative or zero*.

Summary of the Simplex Method

In summary, the basic steps of the simplex method for a *maximization* problem are as follows:

Step I. Start with an initial basic feasible solution in canonical form.

Step II. Check whether the current basic feasible solution is optimal. At this solution, the relative profits of all the nonbasic variables are computed. These represent the net change in the objective function value per unit increase in each nonbasic variable. If these coefficients are negative or zero, the current solution is optimal. Otherwise, go to Step III.

Step III. Select a nonbasic variable to be the new basic variable in the solution. A general rule is to select the nonbasic variable with the largest relative profit so that it may give a larger increase in the value of Z.

Step IV. Determine the basic variable which will be replaced by the nonbasic variable. For this we examine each of the constraints to determine how far the nonbasic variable can be increased. For those constraints in which the nonbasic variable has a positive coefficient, the limit is given by the ratio of the right-hand side constant to that positive coefficient. For the other constraints the limit is set to ∞. The constraint with the lowest limit is determined, and the basic variable in that constraint will be replaced by the nonbasic variable. Since the determination of the variable to leave the basic set involves the calculation of ratios and selection of the minimum ratio, this rule is generally called the *minimum ratio rule*.

Step V. Find the new canonical system and the basic feasible solution by pivot operation. Return to Step II.

2.7 SIMPLEX METHOD IN TABLEAU FORM

In the previous section, we studied the basic principles of the simplex method for solving linear programming problems. These involved successively finding improved basic feasible solutions until the optimum is reached. The various steps of the simplex method can be carried out in a more compact manner by using a tableau form to represent the constraints and the objective function. In addition, by developing some simple formulas, the various calculations can be made mechanical. The use of the tableau form has made the simplex method more efficient and convenient for computer implementation.

The tableau representation is nothing more than writing the problem in a detached coefficient form. To illustrate this, the initial basic feasible solution of Example 2.6-1 can be expressed by the following tableau:

	C_j	5	2	3	−1	1	
C_B	Basis	x_1	x_2	x_3	x_4	x_5	Constants
−1	x_4	1	2	2	1	0	8
1	x_5	3	4	1	0	1	7

Here *basis* refers to the basic variables in the current basic feasible solution. The values of the basic variables are given under the column *constants*. The symbol C_j denotes the coefficient of the variable x_j in the objective function, while C_B denotes the coefficients of the basic variables only.

From the above table, the basic feasible solution is immediately written as $x_4=8$, $x_5=7$, and $x_1=x_2=x_3=0$. The value of the objective function is given by the inner product of the vectors C_B and *constants* as follows:

$$Z=(-1, 1)\begin{pmatrix}8\\7\end{pmatrix}=-8+7=-1$$

In order to check if the above basic feasible solution is optimal, we need the relative profits of all the nonbasic variables. These can be calculated easily by a simple formula known as the *inner product rule*. The relative profit coefficient of the variable x_j, denoted by $\bar{C}_j$, is given by

$$\bar{C}_j=C_j-\left[\begin{array}{c}\text{inner product of } C_B\text{, and the column}\\ \text{corresponding to } x_j \text{ in the canonical system}\end{array}\right]$$

For example,

$$\bar{C}_1 = 5 - (-1, 1)\begin{pmatrix}1\\3\end{pmatrix} = 5 - (-1+3) = 3$$

$$\bar{C}_2 = 2 - (-1, 1)\begin{pmatrix}2\\4\end{pmatrix} = 2 - (-2+4) = 0$$

$$\bar{C}_3 = 3 - (-1, 1)\begin{pmatrix}2\\1\end{pmatrix} = 3 - (-2+1) = 4$$

If the above calculations were carried out for x_4 and x_5, their $\bar{C}_j$ values would be zero since they are the basic variables. It is easy to verify that the inner product rule is a compact way of doing the relative profit calculations given earlier in Section 2.6

Note that the coefficient of x_2 in the objective function is 2, which is larger than those of the present basic variables x_4 and x_5. But any increase in x_2 will not contribute to an increase in the total profit Z because its relative profit is zero. We also note that the unit profit of x_3 is less than that of x_1; but x_3 contributes to a larger (per unit) increase in the Z value than x_1. This shows that the mere use of the objective function coefficient (C_j) to measure the worth of a nonbasic variable is incorrect. The relative profit coefficient (denoted by $\bar{C}_j$) reflects the true change in Z from its present value since it also takes into account the cost of the resources used by the nonbasic activity.

Now, the initial simplex tableau can be constructed by appending the *relative profit row* or the $\bar{C}$ row to the previous table.

Tableau 1
(INITIAL SOLUTION)

	C_j	5	2	3	−1	1	
C_B	Basis	x_1	x_2	x_3	x_4	x_5	Constants
−1	x_4	1	2	2	1	0	8
1	x_5	3	4	1	0	1	7
	$\bar{C}$ Row	3	0	4	0	0	$Z = -1$

Since there are some positive values in the $\bar{C}$ row, the current basic feasible solution is not optimal. The nonbasic variable x_3 gives the greatest per unit increase in Z and hence it will be chosen as the new variable to enter the basis.

In order to decide which basic variable is going to be replaced, we apply the minimum ratio rule as discussed in Section 2.6 by calculating the limits for each constraint as follows:

Row Number	Basic Variable	Upper Limit on x_3
1	x_4	$8/2 = 4$ (minimum)
2	x_5	$7/1 = 7$

The minimum ratio is obtained in the first row, which is generally called the *pivot row*. Thus when the nonbasic variable x_3 is increased to its maximum of 4 units, the basic variable in the pivot row (i.e., x_4) reduces to zero and becomes a nonbasic variable. The new basis contains x_3 and x_5 as basic variables. The new canonical system is obtained by performing a pivot operation as follows:

1. Divide the pivot row by 2 to make the coefficient of x_3 unity.
2. Multiply the pivot row by $-1/2$, and add it to the second row to eliminate x_3.

From Tableau 2 the new basic feasible solution is given by $x_1=0$, $x_2=0$, $x_3=4$, $x_4=0$, $x_5=3$, and the value of Z is 15. In order to check whether this solution is optimal, the new relative profit coefficients have to be calculated. This can be done by applying the inner product rule as before. On the other hand, it is also possible to calculate the $\bar{C}$-row coefficients through the pivot operation. Since x_3 will be the new basic variable, its relative profit coefficient in Tableau 2 should become zero. This can be done by multiplying the first row of Tableau 1 (the pivot row) by -2, and adding it to the $\bar{C}$ row. This will automatically give the new $\bar{C}$ row for Tableau 2!

Tableau 2

	C_j	5	2	3	-1	1	
C_B	Basis	x_1	x_2	x_3	x_4	x_5	Constants
3	x_3	1/2	1	1	1/2	0	4
1	x_5	5/2	3	0	$-1/2$	1	3
$\bar{C}$ Row		1	-4	0	-2	0	$Z=15$

Since $\bar{C}_1$ is positive, Tableau 2 is not optimal. An improvement in the objective function may be obtained by making x_1 a basic variable. Once again the minimum ratio rule is used to determine which basic variable will leave the basis.

Row Number	Basic Variable	Upper Limit on x_1
1	x_3	$4/(1/2)=8$
2	x_5	$3/(5/2)=6/5$ (minimum)

The minimum ratio occurs in the second row. The basic variable x_5 will be replaced by x_1 to form the new basis. A pivot operation gives the following basic feasible solution in Tableau 3:

Tableau 3

	C_j	5	2	3	-1	1	
C_B	Basis	x_1	x_2	x_3	x_4	x_5	Constants
3	x_3	0	2/5	1	3/5	$-1/5$	17/5
5	x_1	1	6/5	0	$-1/5$	2/5	6/5
$\bar{C}$ Row		0	$-26/5$	0	$-9/5$	$-2/5$	$Z=81/5$

All the coefficients of the $\bar{C}$ row are nonpositive. This implies that no further improvement in the objective function is possible. Hence, the current basic feasible solution $x_1=6/5$, $x_2=0$, $x_3=17/5$, $x_4=0$, $x_5=0$ is an optimal solution, and $Z=81/5$ is the optimal value for the linear program.

Summary of the Computational Steps

In summary, the computational steps of the simplex method in tableau form for a *maximization problem* are as follows:

Step I. Express the problem in standard form.

Step II. Start with an initial basic feasible solution in canonical form and set up the initial tableau. (In Section 2.9, we discuss how to find an initial basic feasible solution in canonical form if none exists by inspection.)

Step III. Use the inner product rule to find the relative profit coefficients ($\bar{C}$ row).

Step IV. If all the $\bar{C}_j$ coefficients are nonpositive, the current basic feasible solution is optimal. Otherwise, select the nonbasic variable with the most positive $\bar{C}_j$ value to enter the basis.

Step V. Apply the minimum ratio rule to determine the basic variable to leave the basis.

Step VI. Perform the pivot operation to get the new tableau and the basic feasible solution.

Step VII. Compute the relative profit coefficients by using the pivot operation or the inner product rule. Return to Step IV.

Each sequence of Steps IV to VII is called an *iteration* of the simplex method. Thus each iteration gives a new tableau and an improved basic feasible solution. The efficiency of the simplex method depends on the number of basic feasible solutions it examines before reaching the optimal solution. Hence the number of iterations is an important factor in simplex calculations.

EXAMPLE 2.7-1

Let us solve by the simplex method the following problem:

$$\begin{aligned} \text{Maximize:} \quad & Z=3x_1+2x_2 \\ \text{Subject to:} \quad & -x_1+2x_2 \leqq 4 \\ & 3x_1+2x_2 \leqq 14 \\ & x_1-x_2 \leqq 3 \\ & x_1 \geqq 0, \quad x_2 \geqq 0 \end{aligned}$$

Converting the problem to standard form by the addition of slack variables, we obtain

$$\begin{aligned} \text{Maximize:} \quad & Z=3x_1+2x_2 \\ \text{Subject to:} \quad & -x_1+2x_2+x_3 = 4 \\ & 3x_1+2x_2 \qquad +x_4 = 14 \\ & x_1-x_2 \qquad\qquad +x_5=3 \\ & x_1 \geqq 0, \quad x_2 \geqq 0, \quad x_3 \geqq 0, \\ & x_4 \geqq 0, \quad x_5 \geqq 0 \end{aligned}$$

We have a basic feasible solution in canonical form with x_3, x_4, and x_5 as basic variables.

The initial tableau is given below:

Tableau 1

C_B	Basis \ C_j	3	2	0	0	0	Constants
		x_1	x_2	x_3	x_4	x_5	
0	x_3	−1	2	1	0	0	4
0	x_4	3	2	0	1	0	14
0	x_5	①	−1	0	0	1	3
	$\bar{C}$ Row	3	2	0	0	0	$Z=0$

The nonbasic variable x_1 has the largest relative profit in the $\bar{C}$ row. Hence x_1 enters the basis. Applying the minimum ratio rule, the ratios are $(\infty, 14/3, 3/1)$. The minimum ratio is 3 and x_1 replaces x_5 in the basis. This is indicated by circling the coefficient of x_1 in the pivot row. The circled element is called the *pivot element*. Performing the pivot operation, we get

Tableau 2

C_B	Basis \ C_j	3	2	0	0	0	Constants
		x_1	x_2	x_3	x_4	x_5	
0	x_3	0	1	1	0	1	7
0	x_4	0	⑤	0	1	−3	5
3	x_1	1	−1	0	0	1	3
	$\bar{C}$ Row	0	5	0	0	−3	$Z=9$

The movement from Tableau 1 to Tableau 2 is illustrated in Fig. 2.5. The feasible region of Example 2.7-1 is denoted *ABCDE*. The dotted lines correspond to the objective function lines at $Z=2$ and $Z=9$. Note that the basic feasible solution represented by Tableau 1 corresponds to the corner point *A*. The $\bar{C}$ row in Tableau 1 indicates that either x_1 or x_2 may be a basic variable to increase *Z*. From Fig. 2.5 it is clear that either x_1 or x_2 can be increased to increase *Z*. Having decided to increase x_1, it is clear that we cannot increase x_1 beyond three units (point *B*) in order to remain in the feasible region. This was essentially the minimum ratio value that was obtained by the simplex method. Thus, Tableau 2 is simply corner point *B* in Fig. 2.5.

Still the $\bar{C}$ row has a positive element indicating that the nonbasic variable x_2 can improve the objective function further. To apply the minimum ratio rule, we

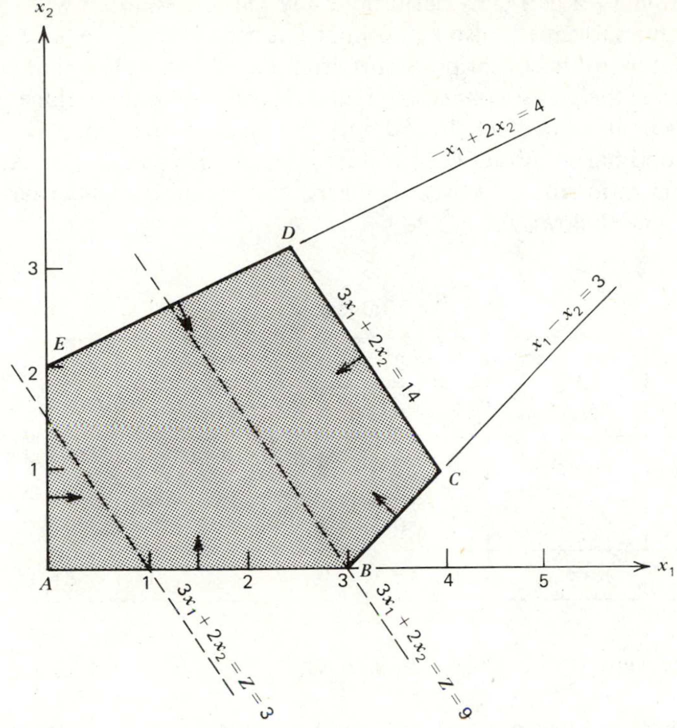

Figure 2.5
Feasible region for Example 2.7-1.

find the minimum of (7/1, 5/5, ∞). This implies that x_2 replaces the basic variable x_4. The next basic feasible solution after the pivot operation is given below:

Tableau 3

C_B	Basis \ C_j	3	2	0	0	0	Constants
		x_1	x_2	x_3	x_4	x_5	
0	x_3	0	0	1	−1/5	8/5	6
2	x_2	0	1	0	1/5	−3/5	1
3	x_1	1	0	0	1/5	2/5	4
	$\bar{C}$ Row	0	0	0	−1	0	$Z = 14$

Tableau 3 corresponds to corner point C in Fig. 2.5. Since none of the coefficients in the $\bar{C}$ row are positive, this tableau is optimal. An optimal solution is given by $x_1 = 4$, $x_2 = 1$, $x_3 = 6$, $x_4 = 0$, $x_5 = 0$, and the optimal value of Z is 14.

Alternate Optima

In Tableau 3, the nonbasic variable x_5 has a relative profit of zero. This means that any increase in x_5 will produce no change in the objective function value. In other words x_5 can be made a basic variable and the resulting basic feasible solution will also have Z as 14. By definition, any feasible solution whose value of Z equals the optimal value is also an optimal solution. Hence, we have an alternate optimal solution to this linear program. From Fig. 2.5 it is clear that corner point D, as well as all the points connecting C and D, are also optimal since they all give the same optimal value of 14. In order to obtain the corner point D, which is an alternative optimal solution, we introduce x_5 as a basic variable in Tableau 3. By the minimum ratio rule, x_3 leaves the basis, and we have an alternative optimal tableau as shown below:

Tableau 4

	C_j	3	2	0	0	0	
C_B	Basis	x_1	x_2	x_3	x_4	x_5	Constants
0	x_5	0	0	5/8	−1/8	1	15/4
2	x_2	0	1	3/8	1/8	0	13/4
3	x_1	1	0	−1/4	1/4	0	5/2
	$\bar{C}$ Row	0	0	0	−1	0	$Z=14$

Thus the alternate optimal solution is given by $x_1=5/2$, $x_2=13/4$, $x_3=0$, $x_4=0$, and $x_5=15/4$.

In general, an alternate optimal solution is indicated whenever there exists a nonbasic variable whose relative profit ($\bar{C}_j$ coefficient) is zero in the optimal tableau.

Unique Optimum

In Example 2.6-1, the optimal solution given by Tableau 3 is unique because all the nonbasic variables have a negative value for their relative profits. This means that any increase in x_2, x_4, or x_5 will result in an immediate decrease in the objective function value. Hence, it is not possible to find another feasible solution whose value of Z equals 81/5.

Minimization Problems

Let us now consider how to solve minimization problems. Recall that the coefficients in the $\bar{C}$ row give the net change in the value of Z per unit increase in the nonbasic variable. A negative coefficient in the $\bar{C}$ row indicates that the corresponding nonbasic variable (when increased) will reduce the value of the objective function. Hence, in minimization problems, only those nonbasic variables with negative $\bar{C}_j$ values are eligible to enter the basis and improve the objective function. The optimal solution is obtained when all the coefficients in the $\bar{C}$ row are nonnegative. Thus all the seven steps of the simplex method outlined earlier can be used for solving minimization problems with a minor modification in Step IV as follows:

Modified Step IV. If all the coefficients in the $\bar{C}$ row are positive or zero, the current basic feasible solution is optimal. Otherwise, select the nonbasic variable with the lowest (most negative) value in the $\bar{C}$ row to enter the basis.

Note: The *Minimum Ratio Rule* (Step V) does not change.

An Alternative Approach to Minimization Problems

Another method of solving a minimization problem is to convert it to an equivalent maximization problem, and then use the simplex method as outlined for a maximization problem. The conversion to a maximization problem is easily done by multiplying the objective function of the minimum problem by minus one. For example, consider a minimization problem:

$$\begin{array}{ll} \text{Minimize:} & Z = 40x_1 + 36x_2 \\ \text{Subject to:} & x_1 \leqq 8 \\ & x_2 \leqq 10 \\ & 5x_1 + 3x_2 \geqq 45 \\ & x_1 \geqq 0, \quad x_2 \geqq 0 \end{array}$$

This is equivalent to the following maximization problem:

$$\begin{array}{ll} \text{Maximize:} & Z' = -40x_1 - 36x_2 \\ \text{Subject to:} & x_1 \leqq 8 \\ & x_2 \leqq 10 \\ & 5x_1 + 3x_2 \geqq 45 \\ & x_1 \geqq 0, \quad x_2 \geqq 0 \end{array}$$

The optimal solutions to both the problems will be the same, while their optimal values will differ by a minus sign. In other words,

$$\text{the minimum value of } Z = -(\text{the maximum value of } Z')$$

2.8 COMPUTATIONAL PROBLEMS

There are a number of computational problems that may arise during the actual application of the simplex method for solving a linear programming problem. Some of these problems have already been discussed. In this section, we discuss other complications that may occur and how to resolve them.

Ties in the Selection of the Nonbasic Variable

The selection of a nonbasic variable to enter the basis is done by determining which nonbasic variable gives the largest *per unit* improvement in the objective function. In other words, in a maximization problem, the variable with the largest positive value in the $\bar{C}$ row is chosen. In case there exists more than one variable with the same largest value in the $\bar{C}$ row, then we have a tie for selecting the nonbasic variable. The general rule is to select any one of them arbitrarily, since selecting the nonbasic variable that gives the largest *per unit* improvement in Z need not necessarily give the largest *total* improvement in Z and minimize the number of simplex iterations.

Ties in the Minimum Ratio Rule and Degeneracy

While applying the minimum ratio rule it is possible for two or more constraints to give the same least ratio value. This results in a tie for selecting which basic variable should leave the basis. This may introduce further complications leading to a reduction in the efficiency of the simplex method. To illustrate this, consider the following simplex tableau obtained for a maximization problem:

Tableau 1

C_B	Basis \ C_j	0	0	0	2	0	3/2	Constants
		x_1	x_2	x_3	x_4	x_5	x_6	
0	x_1	1	0	0	①	−1	0	2
0	x_2	0	1	0	2	0	1	4
0	x_3	0	0	1	1	1	1	3
	$\bar{C}$ Row	0	0	0	2	0	3/2	$Z=0$

Selecting the nonbasic variable x_4 to enter the basis, we observe that the first two constraints give the same minimum ratio. This means that when x_4 is increased to 2, both the basic variables x_1 and x_2 will reduce to zero even though only one of them could be made a nonbasic variable. Deciding arbitrarily to remove x_1 from the basis, we get the new basic feasible solution as shown below:

Tableau 2

C_B	Basis \ C_j	0	0	0	2	0	3/2	Constants
		x_1	x_2	x_3	x_4	x_5	x_6	
2	x_4	1	0	0	1	−1	0	2
0	x_2	−2	1	0	0	②	1	0
0	x_3	−1	0	1	0	2	1	1
	$\bar{C}$ Row	−2	0	0	0	2	3/2	$Z=4$

The new basic feasible solution is $x_1=0$, $x_2=0$, $x_3=1$, $x_4=2$, $x_5=0$, $x_6=0$, and $Z=4$. An interesting observation here is that a basic variable (x_2) has assumed a zero value (like a nonbasic variable). Such a basic feasible solution in which one *or* more basic variables are zero is called a *degenerate* basic feasible solution. In contrast, a basic feasible solution in which all the basic variables are positive is said to be *nondegenerate*.

A tie in the minimum ratio rule is the main cause of degeneracy in the solutions, and it introduces further complications in the simplex calculations. Continuing the simplex method, we find that variable x_5 will enter the basis. Application of the minimum ratio rule indicates that the basic variable x_2 will have to be replaced by x_5. But the minimum ratio is zero, implying that x_5 cannot be increased to a positive value. This means that the objective function is not going to be improved in the next basis as shown in Tableau 3.

We find that the value of Z is still 4 even though we went through a change in the basis and obtained a new basic feasible solution. Moreover Tableau 3 is nonoptimal and x_6 can improve the value of Z. Once again, we find that the minimum ratio is zero, and we will be going through another change in the basis without an increase in the value of Z!

Tableau 3

C_B	Basis \ C_j	0	0	0	2	0	3/2	Constants
		x_1	x_2	x_3	x_4	x_5	x_6	
2	x_4	0	1/2	0	1	0	1/2	2
0	x_5	−1	1/2	0	0	1	1/2	0
0	x_3	1	−1	1	0	0	0	1
$\bar{C}$ Row		0	−1	0	0	0	1/2	$Z=4$

Naturally the question arises whether it is safe to assume that Z cannot be increased any further and call the present solution optimal. Such an assumption will be wrong. As long as there is a positive value in the $\bar{C}$ row, an increase in the objective function is possible, and the simplex calculations have to be continued. As a matter of fact, the maximum value of Z for this problem is 5 and an optimum tableau with $Z=5$ will eventually be obtained with all the $\bar{C}$ row coefficients nonpositive!

This example illustrated the possibility that under degeneracy new tableaus may be obtained with no real improvement in the objective function. This means that costly simplex calculations will have to be performed in the tableaus without getting any real return. This naturally reduces the computational efficiency of the simplex method. But a more important question is whether it is possible for the simplex method to go on indefinitely without improving the objective function. In fact, several examples have been constructed to show that such a thing is theoretically possible. In such situations the simplex method may get into an infinite loop and will fail to reach the optimal solution. This phenomenon is called *classical cycling* or simply *cycling* in the simplex algorithm.

Fortunately such cycling does not happen in practice in spite of the fact that many practical problems have degenerate solutions. In other words, it is possible that the simplex method may not improve the objective function for a few iterations; but eventually it will get out of that loop and reach the optimal solution. However, in a small number of cases, practical problems have been observed to fail to converge on a computer. This is mainly due to certain characteristics of the computer hardware and the way in which the simplex method is programmed. Gass (11) has called this *computer cycling*. Such problems invariably converge when solved on a different computer system. Thus whenever ties occur in the minimum ratio rule, an arbitrary decision is made regarding which basic variable should leave, ignoring the theoretical consequences of degeneracy and cycling. But modifications to the simple minimum ratio rule have been developed to prevent the simplex method from getting into an infinite loop. Because of the extra computational efforts involved in applying these modified rules, they are not used in practice.

Unbounded Solutions

Another complication in the minimum ratio rule may occur when it is not able to determine the basic variable to leave. This happens when none of the constraint coefficients of the nonbasic variable (selected to enter the basis) is positive. This means that no finite ratios can be formed, and in effect the minimum ratio rule fails. To see the significance of such a condition, consider the following linear programming problem:

EXAMPLE 2.8-1

Maximize: $Z = 2x_1 + 3x_2$

Subject to:

$$x_1 - x_2 + x_3 \qquad = 2$$
$$-3x_1 + x_2 \qquad + x_4 = 4$$
$$x_1 \geqq 0, \quad x_2 \geqq 0, \quad x_3 \geqq 0, \quad x_4 \geqq 0$$

The initial tableau is given below:

	c_j	2	3	0	0	
C_B	Basis	x_1	x_2	x_3	x_4	Constants
0	x_3	1	−1	1	0	2
0	x_4	−3	(1)	0	1	4
	$\bar{C}$ Row	2	3	0	0	$Z=0$

The nonbasic variable x_2 enters the basis to replace the basis variable x_4. The new tableau becomes

	c_j	2	3	0	0	
C_B	Basis	x_1	x_2	x_3	x_4	Constants
0	x_3	−2	0	1	1	6
3	x_2	−3	1	0	1	4
	$\bar{C}$ Row	11	0	0	−3	$Z=12$

The new tableau is not optimal and the nonbasic variable x_1 can enter the basis to increase Z. But the minimum ratio rule fails as there is no positive entry in the x_1 column. In other words, as x_1 increases, both the basic variables x_3 and x_2 will also increase, and hence can never become zero to limit the increase of x_1. This means that x_1 can be increased indefinitely. Since each unit increase on x_1 increases Z by 11 units, the objective function can be increased indefinitely, and we have an unbounded solution to the linear programming problem. Thus the failure of the minimum ratio rule at any simplex tableau indicates that the problem has an unbounded solution. (Recall the practical significance of an unbounded solution discussed in Section 2.3.)

2.9 FINDING A FEASIBLE BASIS

A major requirement of the simplex method is the availability of an initial basic feasible solution in canonical form. Without it the initial simplex tableau cannot be formed. In all the examples we discussed so far, a canonical system with a basic feasible solution was readily available. This may not be so in every problem. As a matter of fact, in many practical problems one may not even know whether there exists a feasible solution to the constraints. Such problems will be discussed in this section.

There are two basic approaches to finding an initial basic feasible solution:

1. *By Trial and Error.* Here a basic variable is chosen arbitrarily for each constraint, and the system is reduced to canonical form with respect to those basic variables. If the resulting canonical system gives a basic feasible solution (i.e., the final right-hand-side constants are nonnegative), then the initial tableau can be set up to start the simplex method. It is also possible that during the canonical reduction some of the right-hand-side constants may become negative. In that case the basic solution obtained will be infeasible, and the simplex method cannot be started. Of course, one can repeat the process by trying a different set of basic variables for the canonical reduction and hope for a basic feasible solution. Now it is clearly obvious that the trial and error method is very inefficient and expensive. In addition, if a problem does not possess a feasible solution, it will take a long time to realize this.
2. *Use of Artificial Variables.* This is a systematic way of getting a canonical system with a basic feasible solution when none is available by inspection. First the linear programming problem is converted to standard form such that all the variables are nonnegative, the constraints are equations, and all the right-hand-side constants are nonnegative. Then each constraint is examined for the existence of a basic variable. If none is available, a new variable is added to act as the basic variable in that constraint. In the end, all the constraints will have a basic variable, and by definition we have a canonical system. Since the right-hand-side elements are nonnegative, an initial simplex tableau can be formed readily. Of course the additional variables have no relevance or meaning to the original problem. They are merely added so that we will have a ready canonical system to start the simplex method. Hence these variables are termed as *artificial variables* as opposed to the real decision variables in the problem. Eventually they will be forced to zero lest they unbalance the equations. To illustrate the use of artificial variables, consider the following linear programming problem:

EXAMPLE 2.9-1

Minimize: $Z = -3x_1 + x_2 + x_3$

Subject to:

$$\begin{aligned} x_1 - 2x_2 + x_3 &\leq 11 \\ -4x_1 + x_2 + 2x_3 &\geqq 3 \\ 2x_1 - x_3 &= -1 \end{aligned}$$

$$x_1 \geqq 0, \quad x_2 \geqq 0, \quad x_3 \geqq 0$$

First the problem is converted to the standard form as follows:

Minimize: $Z = -3x_1 + x_2 + x_3$

Subject to:

$$x_1 - 2x_2 + x_3 + x_4 = 11 \tag{2.11}$$

$$-4x_1 + x_2 + 2x_3 - x_5 = 3 \tag{2.12}$$

$$-2x_1 + x_3 = 1 \tag{2.13}$$

$$x_1 \geqq 0, \quad x_2 \geqq 0, \quad x_3 \geqq 0, \quad x_4 \geqq 0, \quad x_5 \geqq 0$$

In Eq. 2.11 the slack variable x_4 is a basic variable. Since there are no basic variables in the other equations we add artificial variables x_6 and x_7 to Eqs. 2.12 and 2.13, respectively. To retain the standard form x_6 and x_7 will be restricted to be nonnegative. Thus we now have an "artificial system" given by:

$$\begin{aligned} x_1 - 2x_2 + x_3 + x_4 &= 11 \\ -4x_1 + x_2 + 2x_3 - x_5 + x_6 &= 3 \\ -2x_1 + x_3 + x_7 &= 1 \end{aligned}$$

$$x_1 \geqq 0, \ldots, x_7 \geqq 0$$

The artificial system has a basic feasible solution in canonical form given by $x_1 = x_2 = x_3 = 0$, $x_4 = 11$, $x_5 = 0$, $x_6 = 3$, $x_7 = 1$. But this is not a feasible solution to the original problem due to the presence of the artificial variables x_6 and x_7 at positive values. On the other hand, it is easy to see that any basic feasible solution to the artificial system in which the artificial variables x_6 and x_7 are zero is automatically a basic feasible solution to the original problem. Hence the objective is to reduce the artificial variables to zero as soon as possible. This can be accomplished in two ways, and each one gives rise to a variant of the simplex method.

The Big *M* Simplex Method

In this approach, the artificial variables are assigned a very large cost in the objective function. The simplex method, while trying to improve the objective function, will find the artificial variables uneconomical to maintain as basic variables with positive values. Hence they will be quickly replaced in the basis by the real variables with smaller costs. For hand calculations it is not necessary to assign a specific cost value to the artificial variables. The general practice is to assign the letter M as the cost in a minimization problem, and $-M$ as the profit in a maximization problem with the assumption that M is a very large positive number.

To illustrate the *big M simplex method* consider the linear program given by Example 2.9-1. In order to drive the artificial variables to zero, a large cost will be assigned to x_6 and x_7 so that the objective function becomes

Minimize: $$Z = -3x_1 + x_2 + x_3 + Mx_6 + Mx_7$$
where M is a very large positive number

The initial simplex tableau can now be constructed using x_4, x_6, and x_7 as the basic variables.

Tableau 1

	C_j	-3	1	1	0	0	M	M	
C_B	Basis	x_1	x_2	x_3	x_4	x_5	x_6	x_7	Constants
0	x_4	1	-2	1	1	0	0	0	11
M	x_6	-4	1	2	0	-1	1	0	3
M	x_7	-2	0	①	0	0	0	1	1
$\bar{C}$ Row		$-3+6M$	$1-M$	$1-3M$	0	M	0	0	$Z=4M$

The $\bar{C}$-row coefficients are calculated using the inner product rule as follows:

$\bar{C}_j = C_j -$ (inner product of C_B and the column corresponding to x_j in Tableau 1)

For example,

$$\bar{C}_1 = -3 - (0, M, M)\begin{pmatrix} 1 \\ -4 \\ -2 \end{pmatrix} = -3 + 6M$$

The value of Z is very high in Tableau 1 since M is very large. The $\bar{C}$ row indicates that the nonbasic variable x_3 can further reduce Z. By the application of the minimum ratio rule, x_7 leaves the basis and the new canonical system is given by Tableau 2.

Tableau 2

	C_j	-3	1	1	0	0	M	M	
C_B	Basis	x_1	x_2	x_3	x_4	x_5	x_6	x_7	Constants
0	x_4	3	-2	0	1	0	0	-1	10
M	x_6	0	①	0	0	-1	1	-2	1
1	x_3	-2	0	1	0	0	0	1	1
	$\bar{C}$ Row	-1	$1-M$	0	0	M	0	$3M-1$	$Z=M+1$

Tableau 2 does not give a basic feasible solution to the original problem due to the presence of the artificial variable x_6 at positive value. Continuing the simplex calculations, we find that x_2 enters the basis to replace x_6.

Tableau 3

	C_j	-3	1	1	0	0	M	M	
C_B	Basis	x_1	x_2	x_3	x_4	x_5	x_6	x_7	Constants
0	x_4	③	0	0	1	-2	2	-5	12
1	x_2	0	1	0	0	-1	1	-2	1
1	x_3	-2	0	1	0	0	0	1	1
	$\bar{C}$ Row	-1	0	0	0	1	$M-1$	$M+1$	$Z=2$

Now both the artificial variables x_6 and x_7 have been reduced to zero. Thus Tableau 3 represents a basic feasible solution to the original problem. Of course, this is not an optimal solution since x_1 can reduce the objective function further by replacing x_4 in the basis.

Tableau 4

	C_j	-3	1	1	0	0	M	M	
C_B	Basis	x_1	x_2	x_3	x_4	x_5	x_6	x_7	Constants
-3	x_1	1	0	0	1/3	$-2/3$	2/3	$-5/3$	4
1	x_2	0	1	0	0	-1	1	-2	1
1	x_3	0	0	1	2/3	$-4/3$	4/3	$-7/3$	9
	$\bar{C}$ Row	0	0	0	1/3	1/3	$M-1/3$	$M-2/3$	$Z=-2$

Tableau 4 is optimal, and the unique optimal solution is given by $x_1=4$, $x_2=1$, $x_3=9$, $x_4=0$, $x_5=0$, and minimum $Z=-2$.

REMARKS

1. An artificial variable is added merely to act as a basic variable in a particular equation. Once it is replaced by a real (decision) variable, there is no need to retain the artificial variable in the simplex tableaus. In other words, we could have omitted the column corresponding to the artificial variable x_7 in Tableaus 2, 3, and 4. Similarly, the column corresponding to x_6 could have been dropped from Tableaus 3 and 4.

2. When the big M simplex method terminates with an optimal tableau, it is sometimes possible for one or more artificial variables to remain as basic variables at positive values. This implies that the original problem is *infeasible* since no basic feasible solution is possible to the original system if it includes even one artificial variable at a positive value. In other words, the original problem without the artificial variables does not have a feasible solution. Infeasibility is due to the presence of inconsistent constraints in the formulation of the problem. In economic terms, this means that the resources of the system are not sufficient to meet the expected demands.

3. For computer solutions, M has to be assigned a specific value. Usually the largest value that can be represented in the computer is assumed.

The Two-Phase Simplex Method

This is another approach to handle the artificial variables whenever they are added. Here the linear programming problem is solved in two phases.

Phase 1. This phase consists of finding an initial basic feasible solution to the original problem. In other words, the removal of the artificial variables is taken up first. For this an artificial objective function is created which is the sum of all the artificial variables. The artificial objective is then minimized using the simplex method. If the minimum value of the artificial problem is zero, then all the artificial variables have been reduced to zero, and we have a basic feasible solution to the original problem. (Note: if the sum of nonnegative variables is zero, then each variable must be identically equal to zero). We then go to Phase 2.

In case the minimum value of the artificial problem is positive, then at least one of the artificial variables is positive. This means that the original problem without the artificial variables is infeasible, and we terminate.

Phase 2. In this phase, the basic feasible solution found at the end of Phase 1 is optimized with respect to the original objective function. In other words, the final tableau of Phase 1 becomes the initial tableau for Phase 2 after changing the objective function. The simplex method is once again applied to determine the optimal solution.

We shall illustrate the two-phase simplex method using Example 2.9-1. First the Phase 1 linear program is created with symbol W to represent the artificial objective function.

Phase 1 Problem

Minimize: $W = x_6 + x_7$

Subject to:

$$
\begin{aligned}
x_1 - 2x_2 + x_3 + x_4 &= 11 \\
-4x_1 + x_2 + 2x_3 - x_5 + x_6 &= 3 \\
-2x_1 + x_3 + x_7 &= 1 \\
x_1 \geqq 0, \ldots, x_7 &\geqq 0
\end{aligned}
$$

The original objective function, $Z = -3x_1 + x_2 + x_3$, is temporarily set aside during the Phase 1 solution. The initial basic feasible solution for the Phase 1 problem is given below:

Tableau 1
(PHASE 1)

	C_j	0	0	0	0	0	1	1	
C_B	Basis	x_1	x_2	x_3	x_4	x_5	x_6	x_7	Constants
0	x_4	1	−2	1	1	0	0	0	11
1	x_6	−4	1	2	0	−1	1	0	3
1	x_7	−2	0	(1)	0	0	0	1	1
	$\bar{C}$ Row	6	−1	−3	0	1	0	0	$W=4$

The objective function can be reduced further by replacing x_7 by x_3 as follows:

Tableau 2
(PHASE 1)

	C_j	0	0	0	0	0	1	1	
C_B	Basis	x_1	x_2	x_3	x_4	x_5	x_6	x_7	Constants
0	x_4	3	−2	0	1	0	0	−1	10
1	x_6	0	(1)	0	0	−1	1	−2	1
0	x_3	−2	0	1	0	0	0	1	1
	$\bar{C}$ Row	0	−1	0	0	1	0	3	$W=1$

The above tableau is not optimal. The variable x_2 enters the basis to replace the artificial variable x_6.

Tableau 3
(PHASE 1)

	C_j	0	0	0	0	0	1	1	
C_B	Basis	x_1	x_2	x_3	x_4	x_5	x_6	x_7	Constants
0	x_4	3	0	0	1	−2	2	−5	12
0	x_2	0	1	0	0	−1	1	−2	1
0	x_3	−2	0	1	0	0	0	1	1
	$\bar{C}$ Row	0	0	0	0	0	1	1	$W=0$

We now have an optimal solution to the Phase 1 linear program given by $x_1=0$, $x_2=1$, $x_3=1$, $x_4=12$, $x_5=0$, $x_6=0$, $x_7=0$ and minimum $W=0$. Since the artificial variables x_6 and $x_7=0$, this tableau represents a basic feasible solution to the original problem. We now begin Phase 2 of the simplex method to find the optimal solution to the original problem. The initial tableau for Phase 2 is constructed by deleting the column corresponding to the artificial variables, and computing the new $\bar{C}$-row coefficients with respect to the original objective function, $Z = -3x_1 + x_2 + x_3$, as shown below:

Tableau 1
(PHASE 2)

	C_j	-3	1	1	0	0	
C_B	Basis	x_1	x_2	x_3	x_4	x_5	Constants
0	x_4	③	0	0	1	-2	12
1	x_2	0	1	0	0	-1	1
1	x_3	-2	0	1	0	0	1
	$\bar{C}$ Row	-1	0	0	0	1	$Z = 2$

In the above tableau,

$$\bar{C}_1 = -3 - (0,1,1)(3,0,-2)^{\mathrm{T}} = -1$$
$$\bar{C}_5 = 0 - (0,1,1)(-2,-1,0)^{\mathrm{T}} = 1$$

and

$$\bar{C}_2 = \bar{C}_3 = \bar{C}_4 = 0$$

Since the objective is to minimize Z, Tableau 1 is not optimal. The nonbasic variable x_1 replaces the basic variable x_4 to reduce the value of Z further.

Tableau 2
(PHASE 2)

	C_j	-3	1	1	0	0	
C_B	Basis	x_1	x_2	x_3	x_4	x_5	Constants
-3	x_1	1	0	0	1/3	$-2/3$	4
1	x_2	0	1	0	0	-1	1
1	x_3	0	0	1	2/3	$-4/3$	9
	$\bar{C}$ Row	0	0	0	1/3	1/3	$Z = -2$

An optimal solution has been reached, and it is given by $x_1 = 4$, $x_2 = 1$, $x_3 = 9$, $x_4 = 0$, $x_5 = 0$, and minimum $Z = -2$.

Comparing the Big M simplex method and the two-phase simplex method, we observe the following:

1. The basic approach to both methods is the same. Both add the artificial variables to get the initial canonical system and then drive them to zero as soon as possible.
2. The sequence of tableaus and the basis changes are identical.
3. The number of iterations are the same.
4. The Big M method solves the linear program in one pass while the two-phase method solves it in two stages as two linear programs.

A drawback of the big M method is the presence of unusually large numbers (M) which sometimes create computational problems in a digital computer.

2.10 COMPUTER SOLUTION OF LINEAR PROGRAMS

Many practical problems formulated as linear programs run into hundreds of constraints and thousands of decision variables. These invariably have to be solved using a digital computer. In the tableau form, the simplex method is an iterative procedure and can be applied mechanically to any problem. Hence, it is ideally suited for computer implementation.

Computer Codes

Commercial LP computer codes are available from many computer manufacturers and private companies who specialize in marketing LP software for major computer systems. Depending on their capabilities, these codes vary in their complexity, ease of use, and cost. The need to solve large LP problems has led to the development of very complex and sophisticated LP computer codes, called mathematical programming systems (e.g., IBM MPSX-370, CDC APEX III, Management Science Systems MPS-III). These have sophisticated data-handling and analytical tools and can solve problems of the order of 8,000-16,000 constraints and an unlimited number of variables. An LP problem with 50,000 constraints has been successfully solved by MPS-III.

Linear programming problems with 5000 or more constraints are considered large, and their successful solution depends on problem structure and any special properties a problem may have. Typical LP problems in practice would have 500–1500 constraints and several thousand variables. These can be solved without much difficulty by any of the advanced LP computing systems, which are available on a fee basis in addition to the normal computer rental charges. If the user is planning to solve the LP problem frequently, then it may be cost effective to acquire the LP software. For infrequent use, it is better to employ a local qualified computer service bureau or rent a remote terminal connected to a major computer center that has the LP software available for its clients. An excellent survey of modern LP code characteristics, including data management, is given by Orchard-Hays (19). White (24) presents a status report on computing algorithms for mathematical programming systems. All commercial computer codes use the simplex algorithm and many of its variants as the basic method for solving LP problems. For a more detailed discussion on LP software, including matrix generators and report writers, the reader is referred to Ravindran (20).

With the ever increasing power of the personal computers, several LP software packages for microcomputers are now available [e.g., Jensen (15), Schrage (22)]. Interested readers should also refer to the past issues of the *Industrial Engineering* magazine for listings of several microcomputer codes for solving LP problems.

Computer Implementation of the Simplex Method

The early computer codes for the simplex method performed calculations on the entire tableau as discussed in the earlier sections. This required fast access to all elements of the tableau to reduce computational effort and time. This was not possible unless the entire tableau was kept in the fast memory of the computer.

But the limited core memory precluded this, and the solution of large linear programs became inefficient and expensive. Hence, further refinements were made in the simplex calculations so that the simplex method could be implemented more efficiently on a computer. This led to the development of the *revised simplex method*, which is currently implemented in all commercial computer codes.

The revised simplex method uses the same basic principles as the regular simplex method, but the entire simplex tableau need not be calculated at each iteration. The relevant information required to move from one basic feasible solution to another is generated directly from the original system of equations, so that less computational effort and time is needed in the simplex calculations. Currently all commercial computer codes use the revised simplex method for solving LP problems. It is possible to solve larger problems because the revised method uses less of central memory. Chapter 4 discusses in detail the revised simplex method and its advantages.

There are also a few minor modifications in computer implementation of the simplex method. One involves the rule for selecting the nonbasic variable to enter the basis. As we pointed out earlier, selecting the nonbasic variable that gives the largest *per unit* improvement in Z need not necessarily give the largest *total* improvement in Z. Hence, codes for large computers do not compute the entire $\bar{C}$ row and then select the nonbasic variable with the most positive (in maximization problems) or the most negative (in minimization problems) $\bar{C}_j$ coefficient. In a large problem with thousands of nonbasic variables, this could take a considerable amount of computer time. Instead, the elements of the $\bar{C}$ row are calculated one at a time, and the first nonbasic variable that shows a possible improvement in Z is selected as the variable to enter. This eliminates further calculations on the $\bar{C}$ row and an expensive search. However, this may result in an increase in the total number of iterations to solve the problem, but the reduction in the time needed for each iteration generally offsets this deficiency.

Computational Efficiency of the Simplex method

It has been pointed out by many researchers in mathematical programming that the simplex method, viewed purely from a theoretical standpoint, is not an efficient method, since it examines adjacent basic feasible solutions only (i.e., changing one basic variable at a time). It is felt that the method can move faster to an optimal solution if it examines nonadjacent solutions as well (i.e., changing more than one basic variable at a time). But many of the suggested variants to the simplex method did not produce any appreciable change in the total computational time. Hence the basic simplex method is still considered to be the best procedure for solving LP problems. A survey of many of the variants of the simplex method is given by Barnes and Crisp (1).

The computational efficiency of the simplex method depends on the number of iterations (basic feasible solutions) required to reach the optimal solution, and the total computer time needed to solve the problem. Much effort has been spent on studying computational efficiency with regard to the number of constraints and the decision variables in the problem.

Empirical experience with thousands of practical problems shows that the number of iterations of a standard linear program with m constraints and n variables varies between m and $3m$, the average being $2m$. A practical upper bound for the number of iterations is $2(m+n)$, although occasionally problems have violated this bound.

Computational time is found to vary approximately in relation to the cube of the number of constraints in the problem (m^3). For example, if Problem A has twice as many constraints as Problem B, then the computer time for Problem A will be about eight times that of Problem B.

The computational efficiency of the simplex method is more sensitive to the number of constraints than to the number of variables. Hence the general recommendation is to keep the number of constraints as small as possible by avoiding unnecessary or redundant constraints in the formulation of the LP problem.

Unlike the simplex method which moved along the boundary of the feasible region, an interior LP algorithm was developed in 1979 by the Russian mathematician Khachian (17). Even though Khachian's algorithm was theoretically an important breakthrough, it did not turn out to be promising for computer implementation [see Dantzig (5)]. Recently, a lot of interest has been generated in a new interior LP method developed by an AT&T researcher, N. Karmarkar (16). Tests of Karmarkar's algorithm are currently underway and the early results look promising.

2.11 SENSITIVITY ANALYSIS IN LINEAR PROGRAMMING

In all LP models the coefficients of the objective function and the constraints are supplied as input data or as parameters to the model. The optimal solution obtained by the simplex method is based on the values of these coefficients. In practice the values of these coefficients are seldom known with absolute certainty, because many of them are functions of some uncontrollable parameters. For instance, future demands, the cost of raw materials, or the cost of energy resources cannot be predicted with complete accuracy before the problem is solved. Hence the solution of a practical problem is not complete with the mere determination of the optimal solution.

Each variation in the values of the data coefficients changes the LP problem, which may in turn affect the optimal solution found earlier. In order to develop an overall strategy to meet the various contingencies, one has to study how the optimal solution will change with changes in the input (data) coefficients. This is known as *sensitivity analysis* or *post-optimality analysis*.

Other reasons for performing a sensitivity analysis are:

1. Some data coefficients or parameters of the linear program may be controllable; for example, availability of capital, raw material, or machine capacities. Sensitivity analysis enables us to study the effect of changes in these parameters on the optimal solution. If it turns out that the optimal value (profit/cost) changes (in our favor) by a considerable amount for a small change in the given parameters, then it may be worthwhile to implement some of these changes. For example, if increasing the availability of labor by allowing overtime contributes to a greater increase in the maximum return, as compared to the increased cost of overtime labor, then we might want to allow overtime production.
2. In many cases, the values of the data coefficients are obtained by carrying out statistical estimation procedures on past figures, as in the case of sales forecasts, price estimates, and cost data. These estimates, in general, may not be very accurate. If we can identify which of the parameters affect the objective value most, then we can obtain better estimates of these parameters. This will increase the reliability of our model and the solution.

We shall now illustrate the practical uses of sensitivity analysis with the help of an example. Methods for performing a sensitivity analysis that will provide the ranges on the input data coefficients will be discussed in Chapter 4.

EXAMPLE 2.11-1

A factory manufactures three products, which require three resources—labor, materials, and administration. The unit profits on these products are \$10, \$6, and \$4, respectively. There are 100 hr of labor, 600 lb of material, and 300 hr of administration available per day. In order to determine the optimal product mix, the following LP model is formulated and solved:

Maximize: $Z = 10x_1 + 6x_2 + 4x_3$

Subject to:

$$\begin{aligned} x_1 + x_2 + x_3 &\leqslant 100 && \text{(labor)} \\ 10x_1 + 4x_2 + 5x_3 &\leqslant 600 && \text{(material)} \\ 2x_1 + 2x_2 + 6x_3 &\leqslant 300 && \text{(administration)} \\ x_1, x_2, x_3 &\geqslant 0 \end{aligned}$$

where x_1, x_2, and x_3 are the daily production levels of products 1, 2, and 3, respectively.

A computer output of the solution of Example 2.11-1 is given in Table 2.1. We note from the optimal solution that the optimal product mix is to produce products 1 and 2 only at levels 33.33 and 66.67 units, respectively.

The *shadow prices* give the net impact in the maximum profit if additional units of certain resources can be obtained. Labor has the maximum impact, providing \$3.33 increase in profit for each additional hour of labor. Of course, the shadow prices on the resources apply as long as their variations stay within the prescribed ranges on RHS constants given in Table 2.1. In other words, a \$3.33/hr increase in profit is achievable as long as the labor hours are not increased beyond 150 hr. Suppose it is possible to increase the labor hours by 25% by scheduling overtime that incurs an additional labor cost of \$50. To see whether it is profitable to schedule overtime, we first determine the net increase in maximum profit due to 25 hr of overtime as (25)(\$3.33) = \$83.25. Since this is more than the total cost of overtime, it is economical to schedule overtime. It is important to note that when any of the RHS constants is changed, the optimal solution will change. However, the optimal product mix will be unaffected as long as the RHS constant varies within the specified range. In other words, we will still be making products 1 and 2 only, but their quantities may change.

The ranges on the objective function coefficients given in Table 2.1 exhibit the sensitivity of the *optimal solution* with respect to changes in the unit profits of the three products. It shows that the optimal solution will not be affected as long as the unit profit of product 1 stays between \$6 and \$15. Of course, the maximum profit will be affected by the change. For example, if the unit profit on product 1 increases from \$10 to \$12, the optimal solution will be the same but the maximum profit will increase to \$733.33 + (12 − 10)(\$33.33) = \$799.99.

Note that product 3 is not economical to produce. Its *opportunity cost* measures the negative impact of producing product 3 to the maximum profit. Hence, a further decrease in its profit contribution will not have any impact on the optimal solution or maximum profit. Also, the unit profit on product 3 must increase to \$6.67 (present value + opportunity cost) before it becomes economical to produce.

Table 2.1
SENSITIVITY ANALYSIS OUTPUT FOR EXAMPLE 2.11-1

Optimal Solution:	$x_1 = 33.33$, $x_2 = 66.67$, $x_3 = 0$
Optimal Value:	maximum profit = \$733.33
Shadow Prices:	For row 1 = \$3.33, for row 2 = 0.67, for row 3 = 0
Opportunity Costs:	For $x_1 = 0$, for $x_2 = 0$, for $x_3 = 2.67$

RANGES ON OBJECTIVE FUNCTION COEFFICIENTS

Variable	Lower Limit	Present Value	Upper Limit
x_1	6	10	15
x_2	4	6	10
x_3	$-\infty$	4	6.67

RANGES ON RIGHT-HAND-SIDE (RHS) CONSTANTS

Row	Lower Limit	Present Value	Upper Limit
1	60	100	150
2	400	600	1000
3	200	300	∞

Simultaneous Variations in Parameters (2)

The sensitivity analysis output on profit and RHS ranges is obtained by varying only one of the parameters and holding all others fixed at their current values. However, it is possible to use the sensitivity analysis output when several parameters are changed simultaneously. This is done with the help of the "*100% rule.*"

100% Rule (For Objective Function Coefficients)

$$\sum_j \frac{\delta c_j}{\Delta c_j} \leqslant 1 \tag{2.14}$$

where δc_j is the actual increase (decrease) in the objective function coefficient of variable x_j, and Δc_j is the maximum increase (decrease) allowed by sensitivity analysis. As long as inequality 2.14 is satisfied, the optimal solution to the LP problem will not change.

For example, suppose the unit profit on product 1 decreases by \$1 but increases by \$1 for both products 2 and 3. This simultaneous variation satisfies the *100% rule*. Since $\delta c_1 = -1$, $\Delta c_1 = -4$, $\delta c_2 = 1$, $\Delta c_2 = 4$, $\delta c_3 = 1$, $\Delta c_3 = 2.67$, and

$$\frac{-1}{-4} + \frac{1}{4} + \frac{1}{2.67} = 0.875 < 1$$

Hence, the optimal solution will not change, but the maximum profit will change by $(-1)(\$33.33) + 1(\$66.67) + 1(\$0) = \$33.34$.

100% Rule (For RHS Constants)

$$\sum_i \frac{\delta b_i}{\Delta b_i} \leqslant 1 \tag{2.15}$$

where δb_i is the actual increase (decrease) in the RHS constant of the ith constraint, and Δb_i is the maximum increase (decrease) allowed by sensitivity analysis. If inequality 2.15 is satisfied, then the optimal product mix remains the same and the shadow prices apply, but the optimal solution and maximum profit will change. Of course, the net change in the maximum profit can be obtained using the shadow prices.

For example, consider the simultaneous variation of 10 hr decrease in labor availability, 100 lb increase in material, and 50 hr decrease in administration. This implies that $\delta b_1 = -10$, $\delta b_2 = 100$, $\delta b_3 = -50$, while $\Delta b_1 = -40$, $\Delta b_2 = 400$, and $\Delta b_3 = -100$. The 100% rule is satisfied, since

$$\frac{-10}{-40} + \frac{100}{400} + \frac{-50}{-100} = 1$$

Hence, the optimal basis and the product mix will not be affected. Of course, the optimal solution will change. But the net change in the maximum profit can be obtained using the shadow prices:

$$\begin{aligned}\text{Change in optimal profit} &= (\$3.33)(-10) + (\$0.67)(100) - (0)(50) \\ &= \$33.70\end{aligned}$$

Warning. Failure of the 100% rule does not necessarily imply that the LP solution will be affected. In other words, inequality 2.14 may be violated while the optimal solution remains unchanged.

2.12 APPLICATIONS

Linear programming models are used widely to solve a variety of military, economic, industrial, and social problems. In a 1976 survey of American companies (7), linear programming came out as the most often used technique (74%) among all the optimization methods. It has been reported in *Science News* (23) that about one-fourth of computer time spent on scientific computations in recent years has been devoted to solving LP problems and their many variations. The oil companies are among the foremost users of very large LP models, using them in petroleum refining, distribution, and transportation. The number of LP applications has grown so much in the last 20 years that it will be impossible to survey all the different applications. Instead, the reader is referred to two excellent textbooks, Gass (9) and Salkin and Saha (21), which are devoted solely to LP applications in diverse areas like defense, industry, retail business, agriculture, education, and environment. Many of the applications also contain a discussion of the experiences in using the LP model in practice.

Discussion of various computer solutions to large LP problems in industry and computational features of LP software are discussed in Driebeek (6), Daellenbach and Bell (3), and Bradley, Hax, and Magnanti (2). An excellent bibliography on LP applications is available in Gass (9), which contains a list of references arranged by area (e.g., agriculture, industry, military, production, transportation). In the area of industrial application the references have been further categorized by industry (e.g., chemical, coal, airline, iron and steel, paper, petroleum, railroad). For another bibliography on LP applications, the readers may refer to a survey by Gray and Cullinan-James (12). For more recent applications of LP in practice, the reader should check the recent issues of *Interfaces, AIIE Transactions, Decision Sciences, European Journal of Operational Research, Manage-*

ment Science, Operations Research, Operational Research (U.K.), Naval Research Logistics Quarterly, and *OpSearch* (India).

2.13 ADDITIONAL TOPICS IN LINEAR PROGRAMMING

In this section, we briefly discuss some of the advanced topics in linear programming. These include *duality theory, the dual simplex method,* and *integer programming*. For a full discussion, refer to Chapter 4.

Duality Theory

From both the theoretical and practical points of view, the theory of duality is one of the most important concepts in linear programming. The basic idea behind the duality theory is that every linear program has an associated linear program called its *dual* such that a solution to one gives a solution to the other. There are a number of important relationships between the solution to the original problem and its dual. These are useful in investigating the general properties of the optimal solution to a linear program and in testing whether a feasible solution is optimal. In addition, the optimal dual solution can be interpreted as the price one pays for the constraint resources, which is known as the *shadow price*. This plays an important role in postoptimality analysis (Section 2.11). Moreover, the concept of duality and the various duality theorems are extremely useful in the study of advanced mathematical programming topics. For a complete discussion of duality theory and its application, see Section 4.2.

The Dual Simplex Method

The dual simplex method is a modified version of the simplex method using duality theory. There are instances when the dual simplex method has an obvious advantage over the regular simplex method. It plays an important part in sensitivity analysis, parametric programming, solution of integer programming problems, many of the variants of the simplex method, and solution of some nonlinear programming problems. Details of the dual simplex method are discussed in Section 4.3.

Integer Programming

In practice, many linear programming problems do require integer solutions for some of the variables. For instance, it is not possible to employ a fractional number of workers or produce a fractional number of cars. The term (linear) integer programming refers to the class of linear programming problems wherein some or all of the decision variables are restricted to integers. But solutions of integer programming problems are generally difficult, too time consuming, and expensive. Hence a practical approach is to treat all the integer variables as continuous and to solve the associated linear program by the simplex method. We may be fortunate to get some of the values of the variables as integers automatically, but when the simplex method produces fractional solutions for some integer variables, they are generally rounded off to the nearest integer such that the constraints are not violated. This is very often used in practice, and it generally produces a good integer solution close to the optimal integer solution, especially when the values of the integer variables are large.

There are situations when a model formulation may require the use of binary integer variables which only can take values 0 or 1. For these integer variables, rounding off produces poor integer solutions, and one does need other techniques to determine the optimal integer solution directly. Some of these techniques and many practical applications of integer programming are discussed in Section 4.6.

Recommended Readings

For a complete history of linear programming and its developments refer to Dantzig (4). This book also contains an activity analysis approach using "black-box" techniques for formulating linear programming problems. An excellent bibliography on the applications of linear programming in practice is available in Gass (10). Gale (8) discusses many linear programming models of production, exchange, and matrix games.

An excellent treatment of matrices, vectors, and matrix-vector operations is available in Hadley (13). For a mathematical treatment of the simplex method from a vector space theory, refer to Gale (8) and Hadley (14). Topics of degeneracy and its resolution are discussed at length in Dantzig.

Discussion of various computer solutions to large linear-programming problems in industries, and computational features of many LP computer codes are discussed in Driebeek (6), and Daellenbach and Bell (3).

The text by Murty (18) is the most complete book available on linear programming and may be referred to for any advanced material on this subject.

REFERENCES

1. Barnes, J. W., and R. M. Crisp, Jr., "Linear Programming: A Survey of General Purpose Algorithms," *Amer. Inst. Ind. Eng. Trans.*, *7*(3), 212–221, (1975).
2. Bradley, S. P., A. C. Hax, and T. L. Magnanti, *Applied Mathematical Programming,* Addison-Wesley, Reading, Mass., 1977.
3. Daellenbach, H. G., and E. J. Bell, *User's Guide to Linear Programming,* Prentice-Hall, Englewood Cliffs, N.J., 1970.
4. Dantzig, G. B., *Linear Programming and Extensions,* Princeton University Press, Princeton, N.J., 1963.
5. Dantzig, G. B., "Comments on Khachian's Algorithm for Linear Programming," Tech. Rep. SOL 79-22, Department of Operations Research, Stanford University, Stanford, Calif., 1979.
6. Driebeek, N. J., *Applied Linear Programming,* Addison-Wesley, Reading, Mass., 1969.
7. Fabozzi, E. J., and J. Valente, "Mathematical Programming in American Companies: A Sample Survey," *Interfaces, 7*(1), 93–98 (Nov. 1976).
8. Gale, D., *The Theory of Linear Economic Models,* McGraw-Hill, N.Y., 1960.
9. Gass, S. I., *An Illustrated Guide to Linear Programming,* McGraw-Hill, N.Y., 1970.
10. Gass, S. I., *Linear Programming,* 4th ed., McGraw-Hill, N.Y., 1975.
11. Gass, S. I., "Comments on the Possibility of Cycling with the Simplex Method," *Oper. Res., 27*(4), 848–852 (1979).
12. Gray, P., and C. Cullinan-James, "Applied Optimization—A Survey," *Interfaces, 6*(3), 24–41 (May 1976).
13. Hadley, G., *Linear Algebra,* Addison-Wesley, Reading, Mass., 1961.
14. Hadley, G., *Linear Programming,* Addison-Wesley, Reading, Mass., 1962.
15. Jensen, P., *Micro Solve/Operations Research Package,* Revised ed., Holden Day, Inc., Oakland, Calif., 1985.

16. Karmarkar, N., "A New Polynomial-Time Algorithm for Linear Programming," Internal Report, Mathematical Sciences Division, AT&T Bell Labs, Murray Hill, N.J., 1984.

17. Khachian, L. G., "A Polynomial Algorithm in Linear Programming," *Soviet Math. Dokl., 20*(1), 191–194 (1979).

18. Murty, K. G., *Linear Programming,* Wiley, N.Y., 1983.

19. Orchard-Hays, W., "On the Proper Use of a Powerful MPS," in *Optimization Methods for Resource Allocation* (R. W. Cottle and J. Krarup, Eds.), English Universities Press, London, 1974, pp. 121–149.

20. Ravindran, A., "Linear Programming," in *Handbook of Industrial Engineering,* (G. Salvendy, Ed.), Wiley, N.Y., 1982, Chap. 14.2, pp. 14.2.1–14.2.11.

21. Salkin, H. M., and J. Saha, *Studies in Linear Programming,* North Holland/American Elsevier, N.Y., 1975.

22. Schrage, L., *Linear, Integer, and Quadratic Programming with LINDO*, The Scientific Press, Palo Alto, Calif., 1984.

23. Steen, L. A., "Linear Programming: Solid New Algorithm," *Science News, 116,* 234–236 (Oct. 6, 1979).

24. White, W. W., "A Status Report on Computing Algorithms for Mathematical Programming," *Comput. Surv. 5*(3), 135–166 (1973).

EXERCISES

1. What is the difference between the simplex method and exhaustive enumeration of feasible corner points of the constrained region?

2. What is the function of the minimum ratio rule in the simplex method?

3. How do you recognize that an LP problem is unbounded while using the simplex method?

4. What is the function of the inner product rule in the simplex method?

5. What are artificial variables and why do we need them? How do they differ from slack/surplus variables?

6. Describe a systematic procedure for finding a feasible solution to a system of linear inequalities.

7. What is sensitivity analysis and why do we perform it?

8. What are shadow prices? What are their practical uses?

9. When do we use the "100% rule" and why?

10. How does an addition of constant K to the objective function of an LP affect (i) its optimal solution, (ii) its optimal value?

11. The Mighty Silver Ball Company manufactures three kinds of pinball machines, each requiring a different manufacturing technique. The Super Deluxe Machine requires 17 hours of labor, 8 hours of testing, and yields a profit of \$300. The Silver Ball Special requires 10 hours of labor, 4 hours of testing, and yields a profit of \$200. The Bumper King requires 2 hours of labor, 2 hours of testing, and yields a profit of \$100. There are 1000 hours of labor and 500 hours of testing available.

    In addition, a marketing forecast has shown that the demand for the Super Deluxe is no more than 50 machines, demand for the Silver Ball Special no more than 80, and demand for Bumper King no more than 150.

    The manufacturer wants to determine the optimal production schedule that will maximize his total profit. Formulate this as a linear programming problem.

12. A company has manufacturing facilities at three locations (A, B, and C) and a single product is to be shipped to 5 customers. The capacities of plants A, B, C are 100, 70, and 50 units, respectively. The firm must ship at least 60 units to customer 1 and 40 units to customer 2. Customers 3, 4, and 5 would require at least 10 units each but want to buy as many of the remaining units as possible. The net profit associated with shipping a unit from each plant to every customer is given below:

		Customer				
		1	2	3	4	5
Plants	A	6	7	8	6	9
	B	10	8	9	5	3
	C	2	9	5	10	6

The company wants to know how many units to ship to each customer from plants A, B, and C in order to maximize their profits. Formulate this as an LP problem.

13. Consider the following road network connecting six cities:

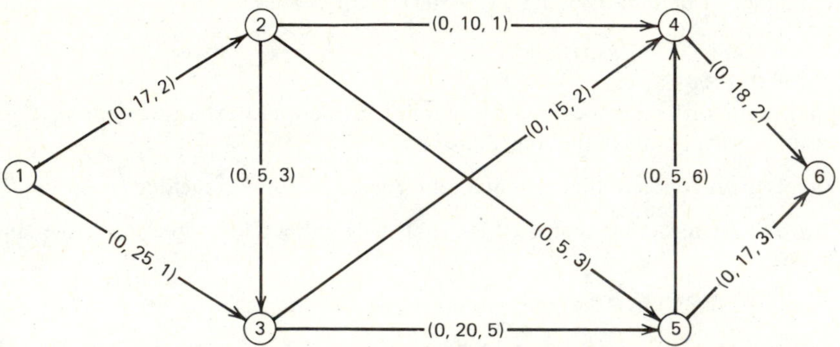

Each circle in the above picture represents a city. You can ship material from city i to city j only if there is a directed arc (road) from city i to city j as

$$(i) \longrightarrow (l_{ij}, K_{ij}, C_{ij}) \longrightarrow (j)$$

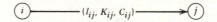

The three numbers are the least amount you can ship along this arc, the maximum tonnage you can ship, and the cost in dollars per ton shipped along this arc, respectively. There are 30 tons of material at city 1 and all of it should be shipped to city 6 at minimal total cost. All materials originate from city 1 and end up in city 6. At any of the intermediate cities the material reaching should be equal to material leaving. Formulate the above problem as a linear program.

14. Consider the problem of scheduling the weekly production of a certain item for the next 4 weeks. The production cost of the item is \$10 for the first 2 weeks, and \$15 for the last 2 weeks. The weekly demands are 300, 700, 900, and 800 units, which must be met. The plant can produce a maximum of 700 units each week. In addition the company can employ overtime during the second and third weeks. This increases the weekly production by an additional 200 units, but the cost of production increases by \$5 per item. Excess production can be stored at a cost of \$3 an item per week. How should the production be scheduled so as to minimize the total costs? Formulate this as a linear programming problem.

15. Solve the following LP problems graphically:

(a) Maximize: $Z = 6x_1 + 5x_2$

Subject to:

$$\begin{aligned} 3x_1 + x_2 &\leqq 160 \\ x_1 \quad &\leqq 40 \\ x_2 &\leqq 130 \\ x_1 \quad &\geqq 80 \\ x_1, x_2 &\geqq 0 \end{aligned}$$

(b) Maximize: $Z = 7x_1 + 10x_2$

Subject to:

$$\begin{aligned} x_1 \quad &\leqq 36 \\ x_2 &\leqq 12 \\ x_1 + 4x_2 &\leqq 60 \\ 2x_1 + x_2 &\geqq 30 \\ x_1 - x_2 &\geqq 0 \\ x_1, x_2 &\geqq 0 \end{aligned}$$

16. Transform the following linear program to the standard form:

Minimize: $Z = -3x_1 + 4x_2 - 2x_3 + 5x_4$

Subject to:

$$\begin{aligned} 4x_1 - x_2 + 2x_3 - x_4 &= -2 \\ x_1 + x_2 + 3x_3 - x_4 &\leqq 14 \\ -2x_1 + 3x_2 - x_3 + 2x_4 &\geqq 2 \end{aligned}$$

$x_1 \geqq 0, \qquad x_2 \geqq 0, \qquad x_3 \leqq 0, \qquad x_4$ unrestricted in sign

17. Consider a system of two equations in five unknowns as follows:

$$\begin{aligned} x_1 + 2x_2 + 10x_3 + 4x_4 - 2x_5 &= 5 \\ x_1 + x_2 + 4x_3 + 3x_4 + x_5 &= 8 \\ x_1, \ldots, x_5 &\geqq 0 \end{aligned}$$

(a) Reduce the system to canonical form with respect to (x_1, x_2) as basic variables. Write down the basic solution. Is it feasible? Why or why not?
(b) What is the maximum number of basic solutions possible?
(c) Find a canonical system which will give a basic feasible solution to the above system by trial and error.

18. Consider the following linear program:

Maximize: $Z = 2x_1 - x_2 + x_3 + x_4$

Subject to:

$$\begin{aligned} -x_1 + x_2 + x_3 \qquad + x_5 \qquad &= 1 \\ x_1 + x_2 \qquad + x_4 \qquad\qquad &= 2 \\ 2x_1 + x_2 + x_3 \qquad\qquad + x_6 &= 6 \\ x_1, \ldots, x_6 &\geqq 0 \end{aligned}$$

(a) Write down the initial basic feasible solution by inspection.
(b) Find a feasible solution by increasing the nonbasic variable x_1 by one unit, while holding x_2 and x_3 as zero. What will be the net change in the objective function?
(c) What is the maximum increase in x_1 possible, subject to the constraints?
(d) Find the new basic feasible solution when x_1 is increased to its maximum value found in (c).
(e) Is the new basic feasible solution obtained in (d) optimal? Why or why not?

19. Use the simplex method to solve:

Maximize: $Z = x_1 + 3x_2$

Subject to:

$$\begin{aligned} x_1 \quad &\leqq 5 \\ x_1 + 2x_2 &\leqq 10 \\ x_2 &\leqq 4 \\ x_1, x_2 &\geqq 0 \end{aligned}$$

Plot the feasible region using x_1 and x_2 as coordinates. Follow the solution steps of the simplex method graphically by interpreting the shift from one basic feasible solution to the next in the feasible region.

20. Use the simplex method to solve:

Minimize: $Z = 3x_1 + x_2 + x_3 + x_4$

Subject to:
$$-2x_1 + 2x_2 + x_3 = 4$$
$$3x_1 + x_2 + x_4 = 6$$
$$x_1, x_2, x_3, x_4 \geqq 0$$

Find an alternative optimal solution if one exists.

21. Use the simplex method to solve:

Maximize: $Z = x_1 + 2x_2 + 3x_3 + 4x_4$

Subject to:
$$x_1 + 2x_2 + 2x_3 + 3x_4 \leqq 20$$
$$2x_1 + x_2 + 3x_3 + 2x_4 \leqq 20$$
$$x_1, \ldots, x_4 \geqq 0$$

Is the optimal solution unique? Why or why not?

22. Use the big M simplex method to solve:

Minimize: $Z = 6x_1 + 3x_2 + 4x_3$

Subject to:
$$x_1 \geqq 30$$
$$x_2 \leqq 50$$
$$x_3 \geqq 20$$
$$x_1 + x_2 + x_3 = 120$$
$$x_1, x_2, x_3 \geqq 0$$

23. Use the two-phase simplex method to solve:

Maximize: $Z = 3x_1 + 4x_2 + 2x_3$

Subject to:
$$x_1 + x_2 + x_3 + x_4 \leqq 30$$
$$3x_1 + 6x_2 + x_3 - 2x_4 \leqq 0$$
$$x_2 \geqq 4$$
$$x_1, \ldots, x_4 \geqq 0$$

24. Given the following linear program (P-1):

Minimize: $Z = 2x_1 - x_2 + 2x_3$

Subject to:
$$-x_1 + x_2 + x_3 = 4$$
$$-x_1 + x_2 - x_3 \leqq 6$$
$$x_1 \leqq 0, \quad x_2 \geqq 0,$$
$$x_3 \text{ (unrestricted in sign)}$$

(a) Convert the given problem (P-1) to a standard linear program in nonnegative variables.
(b) Solve the standard linear program obtained in (a) by the big M simplex method.
(c) From (b), write down the optimal solution to the given problem (P-1) carefully. (You must give the values of x_1, x_2, and x_3). What is the minimum value of Z?

25. Use the simplex method to verify that the following problem has no optimal solution:

Maximize: $Z = x_1 + 2x_2$

Subject to:
$$-2x_1 + x_2 + x_3 \leqq 2$$
$$-x_1 + x_2 - x_3 \leqq 1$$
$$x_1, x_2, x_3 \geqq 0$$

From the final simplex tableau, construct a feasible solution whose value of the objective function is greater than 2000.

26. Use the two-phase simplex method to find a basic feasible solution to the following linear inequalities:

$$\begin{aligned} -6x_1 + x_2 - x_3 &\leqq 5 \\ -2x_1 + 2x_2 - 3x_3 &\geqq 3 \\ 2x_2 - 4x_3 &= 1 \\ x_1, x_2, x_3 &\geqq 0 \end{aligned}$$

27. Use the big M simplex method to show that the following linear program is infeasible:

Minimize: $Z = 2y_1 + 4y_2$

Subject to:

$$\begin{aligned} 2y_1 - 3y_2 &\geqq 2 \\ -y_1 + y_2 &\geqq 3 \\ y_1 \geqq 0, \quad y_2 &\geqq 0 \end{aligned}$$

28. A company that manufactures three products, A, B, and C, using three machines, M1, M2, and M3, wants to determine the optimal production schedule that maximizes the total profit. Product A has to be processed by machines M1, M2, and M3, product B requires M1 and M3, while product C requires M1 and M2. The unit profits on the three products are \$4, \$2, and \$5, respectively. The following linear program is formulated to determine the optimal product-mix:

Maximize: $Z = 4x_1 + 2x_2 + 5x_3$

Subject to:

$$\begin{aligned} x_1 + 2x_2 + x_3 &\leqq 430 \quad \text{(Machine 1)} \\ 3x_1 + 2x_3 &\leqq 460 \quad \text{(Machine 2)} \\ x_1 + 4x_2 &\leqq 450 \quad \text{(Machine 3)} \\ x_1, x_2, x_3 &\geqq 0 \end{aligned}$$

where x_1, x_2, and x_3 are the amounts of products A, B, and C and the constraints reflect the available capacities of M1, M2, and M3. The computer prints out the following solution:

optimal solution: $x_1 = 0, x_2 = 100, x_3 = 230$
optimal value: $\max Z = 1350$
shadow prices: 1.0, 2.0 and 0.0 for constraints 1, 2 and 3, respectively
opportunity costs: 3.0, 0 and 0 for variables x_1, x_2, x_3, respectively

RANGES ON OBJECTIVE FUNCTION COEFFICIENTS

Variable	Lower Limit	Present Value	Upper Limit
x_1	$-\infty$	4.9	7.0
x_2	0	2.0	10.0
x_3	3.0	5.0	∞

RANGES ON RHS CONSTANTS

Row	Lower Limit	Present Value	Upper Limit
1	230	430	455
2	410	460	860
3	400	450	∞

Using the above information, answer the following questions:

(a) Because of an increase in the cost of raw material used for product C, its unit profit drops to \$4. Determine the new optimal solution and the maximum profit.

(b) Suppose it is possible to increase the capacity of one of the machines. Which one would you recommend for expansion and why?

(c) Due to an improvement in product design, the unit profit on product A can be increased to \$6. Is it worthwhile producing product A now? Explain.
(d) Suppose the capacity of machine 2 can be increased by another 200 minutes at a cost of \$250. Is it economical to do so? Explain.
(e) Because of an increase in the cost of energy for operating the machines, the units profits of A, B, and C decrease by \$2.0, \$0.50, and \$1.0, respectively. How will it affect the optimal solution and maximum profit?

29. Consider a diet problem in which a college student is interested in finding a minimum cost diet that provides at least 21 units of Vitamin A and 12 units of Vitamin B from five foods of the following properties:

Food	1	2	3	4	5
Vitamin A content	1	0	1	1	2
Vitamin B content	0	1	2	1	1
Cost per unit (cents)	20	20	31	11	12

The corresponding linear program is:

$$\text{Minimize:} \quad Z = 20x_1 + 20x_2 + 31x_3 + 11x_4 + 12x_5$$

$$\begin{aligned} \text{Subject to:} \quad & x_1 + x_3 + x_4 + 2x_5 \geqq 21 \quad \text{(Vitamin A constraint)} \\ & x_2 + 2x_3 + x_4 + x_5 \geqq 12 \quad \text{(Vitamin B constraint)} \\ & x_1, x_2, x_3, x_4, x_5 \geqq 0 \end{aligned}$$

where x_j = amount of food "j" bought for $j = 1, 2, \ldots, 5$. Solution by the LP computer program is given below:

optimal solution: $x_1 = x_2 = x_3 = 0$, $x_4 = 3$, $x_5 = 9$
optimal value: Min $Z = 141$
shadow prices: 1.0 and 10.0 for constraints 1 and 2, respectively
opportunity costs: 19.0, 10.0, 10.0, 0, 0 for variables x_1, x_2, x_3, x_4 and x_5, respectively

RANGES ON OBJECTIVE FUNCTION COEFFICIENTS

Variable	Lower Limit	Present Value	Upper Limit
x_1	1	20	∞
x_2	10	20	∞
x_3	21	31	∞
x_4	6	11	12
x_5	11	12	22

RANGES ON RHS CONSTANTS

Row	Lower Limit	Present Value	Upper Limit
1	12	21	24
2	10.5	12	21

Using the above information, answer the following:
(a) For what range of values of C_3 (cost of food 3) is the current solution still optimal.

(b) Determine the optimal solution and the minimum cost of the diet when the cost of food 1 is reduced to 5 cents per unit.

(c) What is the optimal solution and the minimum diet cost if the unit cost of food 4 is reduced to 7 cents.

(d) A student's favorite is food 2, and she would like to include 2 units of it in the diet. Explain how this will affect the minimum cost diet.

(e) Suppose the vitamin B requirement is increased from 12 to 15 units. A local pharmacist offers the additional 3 units of vitamin B at a cost of 12 cents per unit. Would you accept the offer? Why or why not?

(f) Suppose the vitamin A requirement is reduced by 3 units and the vitamin B requirement is increased by 4 units. How does this affect the minimum cost diet? Explain.

30. A refinery has four different crudes which are to be processed to yield four products: gasoline, heating oil, jet fuel, and lube oil. There are maximum limits both on product demand (what can be sold) and crude availability. A schematic of the processing operation is given below.

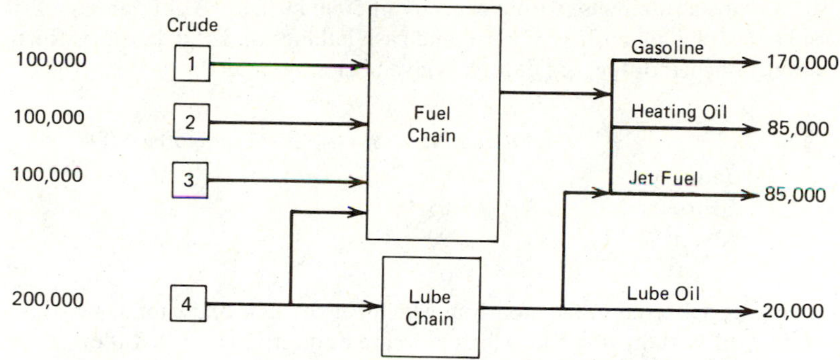

Given the tabulated profits, costs, and yields (see Table 2.2),

(a) Set up the model appropriate for scheduling the refinery for maximum profit.

(b) Suppose the given max demands are changed to be minimum requirements. Set up the model appropriate for scheduling the refinery for minimum cost operation.

Table 2.2
PROFITS, COSTS, AND YIELDS

		1	2	3	4 Fuel Process	4 Lube Process	Product Value $/bbl
Yields (bbl Product per bbl Crude)	Gasoline	0.6	0.5	0.3	0.4	0.4	45.00
	Heating Oil	0.2	0.2	0.3	0.3	0.1	30.00
	Jet Fuel	0.1	0.2	0.3	0.2	0.2	15.00
	Lube Oil	0	0	0	0	0.2	60.00
	Other[a]	0.1	0.1	0.1	0.1	0.1	
Crude cost $/bbl		15.00	15.00	15.00	25.00	25.00	
Operating cost $/bbl		5.00	8.50	7.50	3.00	2.50	

[a]Refers to losses in processing.

31. A hospital administrator has the following minimal daily requirements for nursing personnel:

Period	*Clock Time (24-Hour Day)*	*Minimal Number of Nurses Required*
1	6 AM–10 AM	60
2	10 AM– 2 PM	70
3	2 PM– 6 PM	60
4	6 PM–10 PM	50
5	10 PM– 2 AM	20
6	2 AM– 6 AM	30

Nurses report to the hospital wards at the beginning of each period and work for 8 consecutive hours. The hospital wants to determine the minimal number of nurses to employ so that there will be a sufficient number of nursing personnel available for each period. Formulate this as a linear programming problem.

32. The *ABC* Company is in the commodity trading business. It buys and sells corn for cash. It owns a warehouse with a capacity of 5000 bushels. As of January 1, it has an initial stock of 1000 bushels of corn, and cash balance of $20,000. The estimated corn prices per bushel for the next quarter is given below we added:

	Buying Price ($)	*Selling Price ($)*
January	2.85	3.10
February	3.05	3.25
March	2.90	2.95

The corn is delivered in the month in which it is bought and cannot be sold until the next month. Both buying and selling corn are done strictly on "cash on delivery" basis. The company would like to have a final inventory of 2000 bushels of corn at the end of the quarter. What buying and selling policy would maximize the total net return for the three-month period? Formulate this as a linear programming problem.

33. A scientist has observed a certain quantity Q as a function of variable t. He is interested in determining a mathematical relationship relating t and Q, which takes the form

$$Q(t) = at^3 + bt^2 + ct + d$$

from the results of his n experiments $(t_1, Q_1), (t_2, Q_2), \ldots, (t_n, Q_n)$. He discovers that the values of the unknown coefficients a, b, c, and d must be nonnegative, and should add up to 1. To account for errors in his experiments, he defines an error term,

$$e_i = Q_i - Q(t_i)$$

He wants to determine the best values for the coefficients a, b, c, and d using the following criterion functions:

Criterion 1 Minimize: $Z = \sum_{i=1}^{n} |e_i|$

Criterion 2 Minimize: $(\underset{i}{\text{maximum}}\ |e_i|)$

where $|e_i|$ is the absolute value of the error associated with the ith experiment. Show that the scientist's problem reduces to a linear programming problem under both Criterion 1 and Criterion 2.

34. A company makes two levels of purity of a specialty solvent that is sold in gallon containers. Product A is of higher purity than product B with profits of \$0.40 per gal made on A and \$0.30 per gal made on B.

Product A requires *twice* the processing time of B; if the company produced only B, it could make 1000 gal per day. However, process through-put limitations permit only a combined 800 gal per day of both A and B to be produced. Contract sales require that at least 200 gal per day of B be produced.

Assuming all of the product can be sold, what volumes of A and B should be produced? Formulate this as an LP problem and solve by graphical means.

35. Two products, A and B, are made involving two chemical operations for each. Each unit of product A requires 2 hours on Operation 1 and 3 hours on Operation 2. Each unit of product B requires 3 hours on Operation 1 and 4 hours on Operation 2. Available time for Operation 1 is 16 hours, and for Operation 2, 24 hours.

The production of product B also results in a by-product C at no extra cost. Though some of this by-product can be sold at a profit, the remainder has to be destroyed.

Product A sells for \$4 profit per unit, while product B sells for \$10 profit per unit. By-product C can be sold at a unit profit of \$3, but if it cannot be sold, the destruction cost is \$2 per unit. Forecasts show that up to 5 units of C can be sold. The company gets 2 units of C for each unit of B produced.

The problem is to determine the production quantity of A and B, keeping C in mind, so as to make the largest profit. Formulate as an LP problem and solve by the simplex algorithm.

36. In the tableau for the maximization problem below, the values of the six constants α_1, α_2, α_3, β_1, ρ_1, ρ_2, are unknown (assume there are no artificial variables):

Basis	x_1	x_2	x_3	x_4	x_5	x_6	Constants
x_3	4	α_1	1	0	α_2	0	β
x_4	-1	-5	0	1	-1	0	2
x_6	α_3	-3	0	0	-4	1	3
$\bar{C}$ Row	ρ_1	ρ_2	0	0	-3	0	

$$x_1, \ldots, x_6 \geqq 0$$

State restrictions on the six unknowns (α_1, α_2, α_3, β_1, ρ_1, ρ_2) which would make the following statements true about the given tableau:

(a) The current solution is optimal but alternate optimum exists.
(b) The current solution is infeasible. (State which variable).
(c) One of the constraints is inconsistent.
(d) The current solution is a degenerate basic feasible solution. (Which variable causes degeneracy?)
(e) The current solution is feasible but the problem has no finite optimum.
(f) The current solution is the unique optimum solution.
(g) The current solution is feasible but the objective can be improved by replacing x_6 by x_1. What will be the total change in the objective function value after the pivot?

37. Each of the following tableaus represents the end of an iteration to a *maximization* problem. Select *one or more* of the following conditions that best describe the results indicated by each tableau, and then answer any questions in parentheses.

(a) Improvement in the value of the objective function is still possible. (Which variable should be brought into solution? Which variable should be removed? What is the total improvement in the objective function?)

(b) The original problem is infeasible. (Why?)
(c) The solutuion is degenerate. (Which variable causes degeneracy?)
(d) The solution represented by the tableau is not a basic feasible solution.
(e) The unique optimal solution has been obtained.
(f) One of the optimal solutions has been obtained, but an alternative optimum exists. (Find the alternative optimum solution.)
(g) The optimal solution to the original problem is unbounded. (Which variable causes this condition?) (Note: M is a very large positive number.)

Tableau 1

C_j Basis	3 x_1	1 x_2	1 x_3	7 x_4	b
x_1	1	0	1	2	2
x_2	0	1	0	1	1
$\bar{C}$ Row	0	0	−2	0	

Tableau 2

C_j Basis	5 x_1	3 x_2	$-M$ x_3	10 x_4	b
x_1	1	3	0	2	10
x_3	0	−1	1	−3	0
$\bar{C}$ Row	0	$-12-M$	0	$-3M$	

Tableau 3

C_j Basis	−10 x_1	−5 x_2	−6 x_3	$-M$ x_4	b
x_4	−3	0	0	1	10
x_2	1	1	−2	0	20
$\bar{C}$ Row	$-3M-5$	0	−16	0	

Tableau 4

C_j Basis	0 x_1	0 x_2	3 x_3	−1 x_4	b
x_1	1	0	1	2	0
x_2	0	1	2	1	5
$\bar{C}$ Row	0	0	3	−1	

Tableau 5

C_j Basis	−2 x_1	−3 x_2	8 x_3	$-M$ x_4	b
x_1	1	0	−3	4	5
x_2	0	1	0	2	10
$\bar{C}$ Row	0	0	2	$-M+14$	

38. Consider the standard linear programming problem:

Minimize: $Z = \mathbf{cx}$

Subject to: $\mathbf{Ax} = \mathbf{b}, \quad \mathbf{x} \geqq 0$

Let the vectors $\mathbf{x}^{(1)}$ and $\mathbf{x}^{(2)}$ be two optimal solutions to the above problem. Show that the vector $\mathbf{x}(\lambda) = \lambda\mathbf{x}^{(1)} + (1 - \lambda)\mathbf{x}^{(2)}$ is also an optimal solution for any value of λ between 0 and 1. (Note: the above result is very useful in linear programming. Once we have two optimal solutions to a linear program, then we can generate an infinite number of optimal solutions by varying λ between 0 and 1.)

CHAPTER 3
NETWORK ANALYSIS

In this chapter we discuss a variety of flow problems in a network. A net work is a system of lines or channels connecting different points. Some examples of networks are communication lines, railroad networks, pipeline systems, road networks, shipping lines, and aviation networks. In all these networks we will be interested in sending some specified commodity from certain supply points to some demand points. For example, in a pipeline system we may like to send water, oil, or gas from supply stations to demanding customers. Many of the network flow problems can be formulated as linear programs and their solutions may be obtained by using the simplex method. But a number of special network flow techniques have been developed that are generally more efficient than the simplex method. This chapter will be devoted to the discussion of some of the special network flow problems and their solution techniques.

3.1 SOME EXAMPLES OF NETWORK FLOW PROBLEMS

1. A nationwide chain store is interested in supplying a special product to its retail outlets from its various warehouses. What shipping plan would minimize the total cost of transportation?
2. In a machine shop a batch of jobs is to be assigned to a group of machines. What assignment of jobs to machines will maximize the total efficiency of the shop?
3. A pipe network distributes water from several pumping stations to various customers. If the capacities of the pipes are known, what is the maximum flow that is possible from the stations to the customers?
4. If we were to assign one-way traffic signs to a road network, what assignment would maximize the flow of traffic or the number of highway users?
5. A trucking firm has a table of distances between cities. It wants to find the shortest route and the shortest distance between all pairs of cities so as to design efficient service routes for its trucks.
6. A project consists of a large number of activities that must be done in a specified sequence. The project manager wants to determine how these activities should be scheduled and coordinated so as to minimize the project duration.

3.2 TRANSPORTATION PROBLEMS

Transportation problems are generally concerned with the distribution of a certain product from several sources to numerous localities at minimum cost.

Suppose there are m warehouses where a commodity is stocked, and n markets where it is needed. Let the supply available in the warehouses be $a_1, a_2, \ldots, a_m$, and the demands at the markets be $b_1, b_2, \ldots, b_n$. The unit cost of shipping from warehouse i to market j is $\$c_{ij}$. (If a particular warehouse cannot supply a certain market, we set the appropriate c_{ij} at $+\infty$.) We want to find an optimal shipping schedule that minimizes the total cost of transportation from all the warehouses to all the markets.

Linear Programming Formulation

To formulate the transportation problem as a linear program, we define x_{ij} as the quantity shipped from warehouse i to market j. Since i can assume values from $1, 2, \ldots, m$, and j from $1, 2, \ldots, n$, the number of decision variables is given by the product of m and n. The complete formulation is given below:

Minimize:

$$Z = \sum_{i=1}^{m} \sum_{j=1}^{n} c_{ij} x_{ij} \qquad \text{(total cost of transportation)}$$

Subject to:

$$\sum_{j=1}^{n} x_{ij} \leqq a_i \qquad \text{(supply restriction at warehouse } i\text{) for } i = 1, 2, \ldots, m$$

$$\sum_{i=1}^{m} x_{ij} \geqq b_j \qquad \text{(demand requirement at market } j\text{) for } j = 1, 2, \ldots, n$$

$$x_{ij} \geqq 0 \qquad \text{(nonnegative restrictions) for all pairs } (i, j)$$

The supply constraints guarantee that the total amount shipped from any warehouse does not exceed its capacity. The demand constraints guarantee that the total amount shipped to a market meets the minimum demand at the market. Excluding the nonnegativity constraints, the total number of constraints is $(m + n)$.

It is obvious that the market demands can be met if and only if the total supply at the warehouses is at least equal to the total demand at the markets. In other words, $\sum_{i=1}^{m} a_i \geqq \sum_{j=1}^{n} b_j$. When the total supply equals the total demand $\left(\text{i.e., } \sum_{i=1}^{m} a_i = \sum_{j=1}^{n} b_j\right)$, every available supply at the warehouses will be shipped to meet the minimum demands at the markets. In this case, all the supply and demand constraints would become strict equations, and we have a *standard transportation problem* given by

Minimize:

$$Z = \sum_{i=1}^{m} \sum_{j=1}^{n} c_{ij} x_{ij}$$

Subject to:

$$\sum_{j=1}^{n} x_{ij} = a_i \qquad \text{for } i = 1, 2, \ldots, m$$

$$\sum_{i=1}^{m} x_{ij} = b_j \qquad \text{for } j = 1, 2, \ldots, n$$

$$x_{ij} \geqq 0 \qquad \text{for all } (i, j)$$

We shall develop a technique for solving a standard transportation problem only. This means that any nonstandard problem, where the supplies and demands do not balance, must be converted to a standard transportation problem before it can be solved. This conversion can be achieved by the use of a dummy warehouse or a dummy market as shown below:

1. Consider an unbalanced transportation problem where the total supply exceeds the total demand. To convert this to a standard problem, a dummy market is created to absorb the excess supply available at the warehouses. The unit cost of shipping from any warehouse to the dummy market is assumed to be zero since in reality the dummy market does not exist and no physical transfer of goods takes place. Thus the unbalanced transportation problem is equivalent to the following standard problem:

Minimize: $$Z=\sum_{i=1}^{m}\sum_{j=1}^{n+1} c_{ij}x_{ij}$$

Subject to:
$$\sum_{j=1}^{n+1} x_{ij}=a_i \qquad \text{for } i=1,2,\ldots,m$$
$$\sum_{i=1}^{m} x_{ij}=b_j \qquad \text{for } j=1,2,\ldots,n,n+1$$
$$x_{ij}\geqq 0 \qquad \text{for all } (i,j)$$

Where $j=n+1$ is the dummy market with demand $b_{n+1}=\sum_{i=1}^{m} a_i-\sum_{j=1}^{n} b_j$, and $c_{i,n+1}=0$ for all $i=1, 2, \ldots, m$. Note that the value of $x_{i,n+1}$ denotes the unused supply at warehouse i.

2. Consider now the situation where the total demand exceeds the total supply. Even though all the demands cannot be met, one may like to find a least-cost shipping schedule which will supply as much as possible to the markets. Here a dummy warehouse is created to supply the shortage. The equivalent standard problem becomes

Minimize: $$Z=\sum_{i=1}^{m+1}\sum_{j=1}^{n} c_{ij}x_{ij}$$

Subject to:
$$\sum_{j=1}^{n} x_{ij}=a_i \qquad \text{for } i=1,2,\ldots,m,m+1$$
$$\sum_{i=1}^{m+1} x_{ij}=b_j \qquad \text{for } j=1,2,\ldots,n$$
$$x_{ij}\geqq 0 \qquad \text{for all } (i,j)$$

where $i=m+1$ denotes the dummy warehouse with supply $a_{m+1}=\sum_{j=1}^{n} b_j-\sum_{i=1}^{m} a_i$, and $c_{m+1,j}=0$ for all $j=1, 2, \ldots, n$. Here $x_{m+1,j}$ denotes the amount of shortage at market j.

An important feature of the standard transportation problem is that it can be expressed in the form of a table, which displays the values of all the data coefficients (a_i, b_j, c_{ij}) associated with the problem. In fact, the constraints and the objective function of the transportation model can be read off directly from the table. A *transportation table* for three warehouses and four markets is shown on p. 76.

The supply constraints can be obtained by merely equating the sum of all the variables in each row to the warehouse capacities. Similarly the demand constraints are obtained by equating the sum of all the variables in each column to the market demands.

In principle, we shall solve the transportation problem by using the simplex method of linear programming. But the special structure of the transportation matrix (all the coefficients of the variables are 0 or 1) gives simple selection rules while choosing a nonbasic variable or dropping a basic variable.

		Markets				
		M_1	M_2	M_3	M_4	Supplies
	W_1	x_{11} / c_{11}	x_{12} / c_{12}	x_{13} / c_{13}	x_{14} / c_{14}	a_1
Warehouses	W_2	x_{21} / c_{21}	x_{22} / c_{22}	x_{23} / c_{23}	x_{24} / c_{24}	a_2
	W_3	x_{31} / c_{31}	x_{32} / c_{32}	x_{33} / c_{33}	x_{34} / c_{34}	a_3
Demands		b_1	b_2	b_3	b_4	

Finding an Initial Basic Feasible Solution

The standard transportation problem has $(m+n)$ constraints and (mn) variables. In general the number of basic variables in a basic feasible solution is given by the number of constraints. But in the transportation problem, the number of variables that can take positive values is limited to $(m+n-1)$ since one of the constraints is redundant. To see this, add up all the supply constraints. This gives $\sum_{i=1}^{m}\sum_{j=1}^{n} x_{ij}=\sum_{i=1}^{m} a_i$. summing up the demand constraints, we get $\sum_{i=1}^{m}\sum_{j=1}^{n} x_{ij}=\sum_{j=1}^{n} b_j$. Since $\sum_{i=1}^{m} a_i=\sum_{j=1}^{n} b_j$, these two equations are identical, and we have only $(m+n-1)$ independent constraints.

Because of the special structure of the transportation matrix, it is very easy to find an initial basic feasible solution to start the simplex method without the use of artificial variables. We shall illustrate this with an example.

EXAMPLE 3.2-1

Consider a transportation problem with three warehouses and four markets. The warehouse capacities are $a_1=3$, $a_2=7$, and $a_3=5$. The market demands are $b_1=4$, $b_2=3$, $b_3=4$, and $b_4=4$. The unit cost of shipping is given by the following table:

	M_1	M_2	M_3	M_4
W_1	2	2	2	1
W_2	10	8	5	4
W_3	7	6	6	8

Since $\sum_{i=1}^{3} a_i=\sum_{j=1}^{4} b_j=15$, we have a standard transportation problem. The transportation table is given by

	M_1	M_2	M_3	M_4	Supplies
W_1	x_{11} [2]	x_{12} [2]	x_{13} [2]	x_{14} [1]	3
W_2	x_{21} [10]	x_{22} [8]	x_{23} [5]	x_{24} [4]	7
W_3	x_{31} [7]	x_{32} [6]	x_{33} [6]	x_{34} [8]	5
Demands	4	3	4	4	

Hereafter we shall refrain from writing the variable names (x_{ij}) explicitly in the table.

A basic feasible solution to this example will have at most six (i.e., $3+4-1$) positive variables. There are a number of ways to find an initial basic feasible solution. We shall describe three of the most important ones here.

Northwest Corner Rule

This rule generates a feasible solution with no more than $(m+n-1)$ positive values. The variables that occupy the northwest corner positions in the transportation table are chosen as the basic variables. Thus x_{11} is selected as the first basic variable, and is assigned a value as large as possible consistent with the supply and demand restrictions. In other words, set $x_{11} = \min\ (a_1, b_1) = \min\ (3,4) = 3$. This means that supply in warehouse 1 is exhausted and no further supply to the other markets is possible. Hence we set the variables x_{12}, x_{13}, x_{14} as nonbasic at zero. In addition, 3 units of market 1 demand has been satisfied. Hence, the remaining demand at market 1 is only 1 unit. This is illustrated by the following table.

	M_1	M_2	M_3	M_4	
W_1	③	0	0	0	0
W_2					7
W_3					5
	1	3	4	4	

The value of the basic variable is circled for easy identification.

We then select the next northwest corner variable as a basic variable. This corresponds to x_{21}; value of $x_{21} = \min\ (7, 1) = 1$. Market 1 demand is now fully satisfied and supply in warehouse 2 reduces to 6 units.

	M_1	M_2	M_3	M_4	
W_1	③	0	0	0	0
W_2	①				6
W_3	0				5
	0	3	4	4	

The next northwest corner variable is x_{22} and we let $x_{22} = \min(6, 3) = 3$. Continuing in the same manner we will finally get

	M_1	M_2	M_3	M_4
W_1	③	0	0	0
W_2	①	③	③	0
W_3	0	0	①	④

We have six positive variables in the solution. The row sums are equal to the warehouse capacities, and the column sums to market demands. Thus we have a basic feasible solution to the transportation problem as shown in the following:

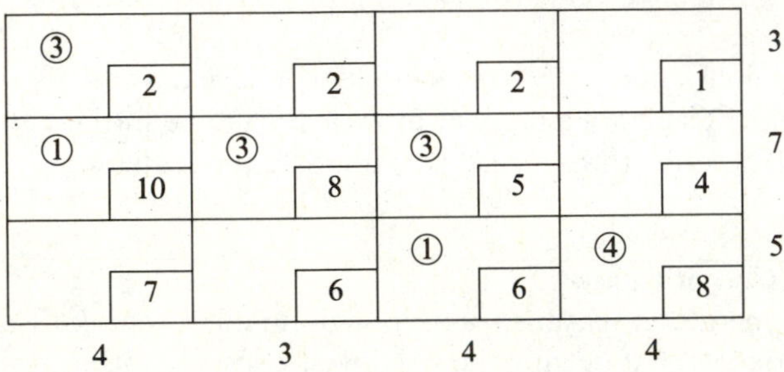

				Supply
③ / 2	2	2	1	3
① / 10	③ / 8	③ / 5	4	7
7	6	① / 6	④ / 8	5
4	3	4	4	

For the above basic feasible solution, the total cost of transportation is given by

$$Z = (3 \times 2) + (1 \times 10) + (3 \times 8) + (3 \times 5) + (1 \times 6) + (4 \times 8) = 93$$

The basic feasible solution obtained by the northwest corner rule may be far from optimal since the transportation costs are completely ignored. Now we shall discuss another method of determining an initial basic feasible solution which takes into account the shipping costs as well.

The Least Cost Rule

The only difference between the two rules is the criterion used for selecting the successive basic variables. In the least cost rule, the variable with the lowest shipping cost will be chosen as the basic variable. We shall illustrate this rule by finding a basic feasible solution to Example 3.2-1.

The transportation table is scanned for the smallest c_{ij}, and this corresponds to the variable x_{14}. This is chosen as the first basic variable and we let $x_{14} = \min(3, 4) = 3$. Row 1 is deleted from further consideration since supply in warehouse 1 is exhausted. The resulting supply-demand configuration is given by the following table:

				Supply
0 / 2	0 / 2	0 / 2	③ / 1	0
10	8	5	4	7
7	6	6	8	5
4	3	4	1	

Out of the remaining unassigned cells, the variable x_{24} has the lowest cost, and we set $x_{24} = \min(7, 1) = 1$. Proceeding in a similar manner, the initial basic feasible solution becomes

0 [2]	0 [2]	0 [2]	③ [1]	3
② [10]	0 [8]	④ [5]	① [4]	7
② [7]	③ [6]	0 [6]	0 [8]	5
4	3	4	4	

The total cost of transportation equals 79. We now have a better starting solution as compared to the one obtained by the northwest corner rule. The additional effort involved in scanning the transportation matrix for the smallest c_{ij} is generally offset by the reduction in the number of iterations required to reach the optimal solution.

In general the least cost rule provides a better starting solution as compared to the northwest corner rule. But this is not guaranteed in all problems. In fact, examples have been constructed wherein the opposite is true.

Vogel's Approximation Method (VAM)

A proven superior method for finding an initial solution is Vogel's Approximation Method. VAM computes a *penalty* for each row and column if the smallest cost cell is not selected from that row or column. Vogel defines the penalty as the absolute difference between the smallest and next smallest cost in a given row or column. (Note: If two or more cells tie for the minimum cost, the penalty is set at zero). Next, the row or column with the *largest* penalty is identified and the variable that has the smallest cost in that row or column is selected as the next basic variable. (Note: Ties may be broken arbitrarily).

EXAMPLE 3.2-2

Using Example 3.2-1, the penalties for each row and column at the first step are shown below:

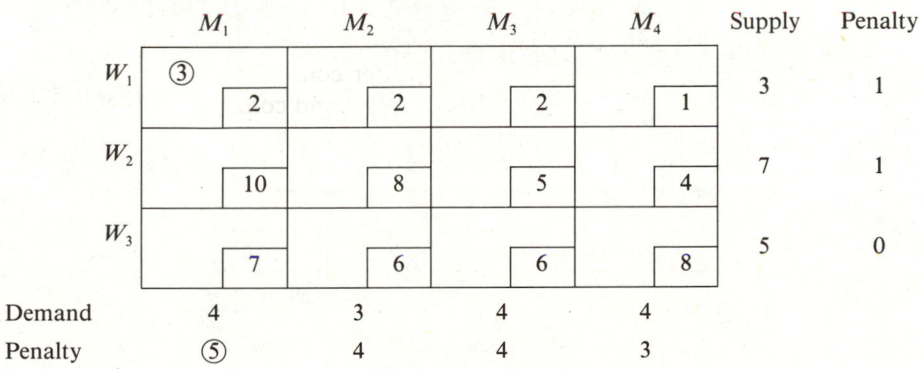

	M_1	M_2	M_3	M_4	Supply	Penalty
W_1	③ [2]	[2]	[2]	[1]	3	1
W_2	[10]	[8]	[5]	[4]	7	1
W_3	[7]	[6]	[6]	[8]	5	0
Demand	4	3	4	4		
Penalty	⑤	4	4	3		

Note that there is a tie for the minimum cost in row 3 and hence the penalty is set at zero. Since column 1 has the largest penalty, variable x_{11} is selected as the first basic variable. Set $x_{11} = \min(3,4) = 3$. Row 1 will be deleted from further consideration, since the supply in warehouse 1 is now exhausted. The next set of penalties at Step 2 are shown below:

	M_1	M_2	M_3	M_4	Supply	Penalty
W_2	10	8	5	④ 4	7	1
W_3	7	6	6	8	5	0
Demand	1	3	4	4		
Penalty	3	2	1	④		

Since column 4 has the largest penalty, we set $x_{24} = 4$. Column 4 will now be deleted and the next set of penalties are given by

	M_1	M_2	M_3	Supply	Penalty
W_2	10	8	③ 5	3	③
W_3	7	6	6	5	0
Demand	1	3	4		
Penalty	3	2	1		

We now have a tie for the largest penalty (row 1 and column 1). Selecting arbitrarily row 1, we set $x_{23} = 3$ and delete row 1. That leaves the transportation table with just one row with supply and demands as shown below:

	M_1	M_2	M_3	
W_3	① 7	③ 6	① 6	5
	1	3	1	

We simply set $x_{31} = 1$, $x_{32} = 3$ and $x_{31} = 1$. Thus the initial basic feasible solution by. VAM is given by

③ 2	2	2	1	3
10	8	③ 5	④ 4	7
① 7	③ 6	① 6	8	5
4	3	4	4	

The total cost of shipping equals 68. We now have a better starting solution compared with the ones found by northwest corner and least cost rules!

Note. For unbalanced transportation problems wherein dummy rows or columns are added, Goyal (13) has suggested that in place of zero costs for dummy cells, use the largest unit cost for that problem. This provides a better starting solution with VAM.

Degeneracy

A basic feasible solution of a transportation problem is said to be *degenerate* if one or more basic variables assume a zero value. An initial solution could become degenerate whenever the remaining supply and demand are equal corresponding to a variable selected as the next basic variable. To illustrate this, consider the following transportation problem:

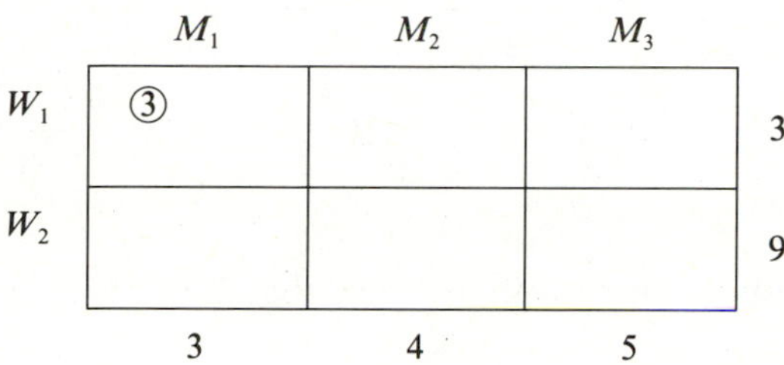

	M_1	M_2	M_3	
W_1	③			3
W_2				9
	3	4	5	

Applying the northwest corner rule, x_{11} is selected as the first basic variable. We set $x_{11} = \min(3,3) = 3$. Both the supply in warehouse 1 and demand in market 1 are now exhausted. However, we can only delete either row 1 or column 1 from consideration and *not both*. Otherwise, we are not guaranteed to get the required number of basic variables, namely, $(m+n-1)$ or 4 in this example. Arbitrarily deleting column 1, we get the following reduced table:

	M_2	M_3	
W_1	⓪		0
W_2			9
	4	5	

By the northwest corner rule x_{12} is selected as the next basic variable; however, its value is $x_{12} = \min(0,4) = 0$. Hence, we delete row 1 from consideration and set $x_{22} = 4$ and $x_{23} = 5$ in the next step. Thus the initial degenerate basic feasible solution with four basic variables is given by

③	⓪		3
	④	⑤	9
3	4	5	

Improving the Initial Basic Feasible Solution

In Example 3.2-1, an initial basic feasible solution using the least cost rule is $x_{14}=3$, $x_{21}=2$, $x_{23}=4$, $x_{24}=1$, $x_{31}=2$, $x_{32}=3$, and all other $x_{ij}=0$. The value of the objective function is 79. The corresponding transportation table is given in Table 3.1.

Table 3.1

	M_1	M_2	M_3	M_4	
W_1	2	2	2	③ 1	3
W_2	② 10	8	④ 5	① 4	7
W_3	② 7	③ 6	6	8	5
	4	3	4	4	

In Table 3.1, the empty cells correspond to nonbasic variables.

The Stepping-Stone Method

This determines whether the initial solution found by the least cost rule is optimum. We know from the simplex method that a given solution minimizes the objective function only if the relative cost coefficients of the nonbasic variables (net change in Z per unit increase in the nonbasic variables) are greater than or equal to zero.

The relative cost coefficients are computed by increasing a nonbasic variable by one unit, and computing the resulting change in the total transportation cost. To illustrate this let us increase the nonbasic variable x_{11} from 0 to 1. To satisfy the constraints (that is, row sums should add up to the various supplies, and column sums to the various demands), x_{14} will decrease by 1, x_{24} will increase by 1, and x_{21} will decrease by 1. Schematically these changes may be represented as follows:

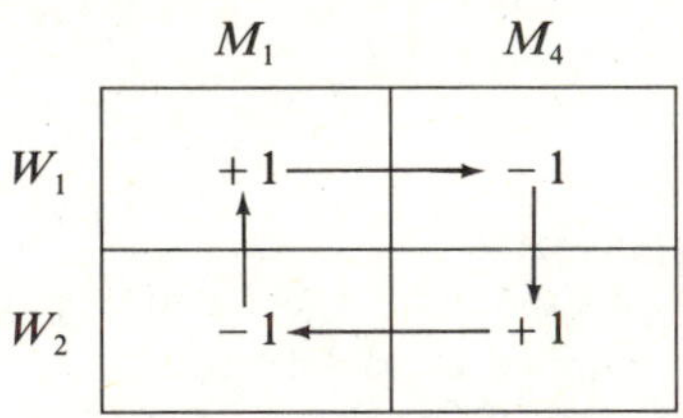

Note that the adjustment is done only on the values of the basic variables. This results in a change in the total shipping cost. Thus, the net change in Z per unit increase in x_{11}, denoted by $\bar{c}_{11}$, is given by

$$\begin{aligned}\bar{c}_{11} &= c_{11}(\text{change in } x_{11}) + c_{14}(\text{change in } x_{14}) \\ &\quad + c_{24}(\text{change in } x_{24}) + c_{21}(\text{change in } x_{21}) \\ &= 2(+1) + 1(-1) + 4(+1) + 10(-1) = -5\end{aligned}$$

This implies that the objective function will decrease by 5 units for every unit increase in the nonbasic variable x_{11}. Hence to minimize Z, x_{11} may be increased by making it a basic variable.

As a rule we calculate all the other $\bar{c}_{ij}$ coefficients, and the basic variable with the lowest relative cost ($\bar{c}_{ij}$) becomes the next basic variable. Similar to the inner product rule used in the simplex method for calculating the relative costs, there is a simple way to calculate all the $\bar{c}_{ij}$ coefficients directly. This procedure is called the *u-v Method* or the *MODI* (*Modified Distribution*) *Method*.*

u-v Method

For any basic feasible solution, find numbers u_i for warehouse i, and v_j for market j such that

$$u_i + v_j = c_{ij} \qquad \text{for every basic } x_{ij} \tag{3.1}$$

These numbers can be positive, negative, or zero. Then,

$$\bar{c}_{ij} = c_{ij} - (u_i + v_j) \qquad \text{for all nonbasic } x_{ij} \tag{3.2}$$

If all the $\bar{c}_{ij}$ are nonnegative, then the current basic feasible solution is optimal. If not, there exists a nonbasic variable x_{pq} such that $\bar{c}_{pq} = \min \bar{c}_{ij} < 0$, and x_{pq} is made a basic variable to improve the value of the objective function.

To apply the u-v method to Example 3.2-1, we have to compute seven numbers u_1, u_2, u_3, v_1, v_2, v_3, and v_4. In the current basic feasible solution (Table 3.1), the basic variables are x_{14}, x_{21}, x_{23}, x_{24}, x_{31}, and x_{32}. Using Eq. 3.1 we get the following six equations:

$$\begin{aligned} u_1 + v_4 &= 1 \\ u_2 + v_1 &= 10 \\ u_2 + v_3 &= 5 \\ u_2 + v_4 &= 4 \\ u_3 + v_1 &= 7 \\ u_3 + v_2 &= 6 \end{aligned}$$

Since the system has six equations in seven unknowns, there exists an infinite number of possible solutions. To get a particular solution we can set any of the variables zero and solve for the rest.

Setting $u_1 = 0$, we get $v_4 = 1$, $u_2 = 3$, $v_1 = 7$, $v_3 = 2$, $u_3 = 0$, and $v_2 = 6$. This computation can easily be accomplished using Table 3.1 directly rather than writing the six equations separately.

Optimality Test

For every nonbasic variable x_{ij}, the unit cost c_{ij} is compared with the sum of u_i and v_j. If all the $u_i + v_j \leqq c_{ij}$, then we have an optimal solution; otherwise, the nonbasic variable with the least relative cost value is chosen.

In our example,

$$\bar{c}_{pq} = \min (c_{ij} - u_i - v_j) = c_{11} - (u_1 + v_1) = -5$$

Hence the nonbasic variable x_{11} is introduced into the basis as the new basic variable. To determine the maximum increase in x_{11}, we assign x_{11} an unknown nonneg-

*The u-v method is based on the complementary slackness properties between the solutions to the transportation problem and its dual. For details, refer to Chapter 4 (Section 4.2 and Exercise 38).

ative value θ. To satisfy the constraints, θ has to be added or subtracted from the basic variables so that the row sums and column sums are equal to the corresponding supplies and demands. Referring to Table 3.1, we see that both the basic variables x_{14} and x_{21} are decreased by θ while x_{24} is increased by θ as shown below:

	M_1	M_2	M_3	M_4	
W_1	$+\theta$			$3-\theta$	3
W_2	$2-\theta$		4	$1+\theta$	7
W_3	2	3			5
	4	3	4	4	

Now θ is increased as long as the solution remains nonnegative. The maximum value of θ is limited by those basic variables which start decreasing with θ. The basic variable which becomes zero first is removed from the basis.* In this case the maximum θ is equal to 2, and the basic variable x_{21} is replaced by x_{11}. Table 3.2 gives the new basic feasible solution, with a new set of values for u_i and v_j.

Table 3.2

	$v_1=2$	$v_2=1$	$v_3=2$	$v_4=1$	
$u_1=0$	② 2	2	2	① 1	3
$u_2=3$	10	8	④ 5	③ 4	7
$u_3=5$	② 7	③ 6	6	8	5
	4	3	4	4	

The new value of the objective function is 69. The solution given by Table 3.2 is not optimal, since

$$\min (c_{ij} - u_i - v_j) = c_{33} - (u_3 + v_3) = -1$$

Now x_{33} is introduced as a basic variable at a nonnegative value θ. This produces the following change in the values of the basic variables:

	M_1	M_2	M_3	M_4	
W_1	$2+\theta$			$1-\theta$	3
W_2			$4-\theta$	$3+\theta$	7
W_3	$2-\theta$	3	$+\theta$		5
	4	3	4	4	

The maximum value of θ is 1, and x_{33}, replaces x_{14} in the basis. The new basic feasible solution is given by Table 3.3.

*In some problems, it is possible for more than one basic variable to become zero, simultaneously. Under such cases, the rule is to select any one of them (but only one) to leave the basis.

Table 3.3

	$v_1=2$	$v_2=1$	$v_3=1$	$v_4=0$	
$u_1=0$	③ 2	2	2	1	3
$u_2=4$	10	8	③ 5	④ 4	7
$u_3=5$	① 7	③ 6	① 6	8	5
	4	3	4	4	

The new value of the objective function is 68. Table 3.3 represents a unique optimal solution, since $\bar{c}_{ij}>0$ for all nonbasic variables. The optimal shipping schedule is to ship

3 units from W_1 to M_1;
3 units from W_2 to M_3;
4 units from W_2 to M_4;
1 unit from W_3 to M_1;
3 units from W_3 to M_2;
1 unit from W_3 to M_3.

The least cost of shipping is 68.

Applications

A number of practical problems can be formulated as transportation problems for solution. We shall present two specific applications here.

EXAMPLE 3.2-3

A canning company operates two canning plants. Three growers are willing to supply fresh fruits in the following amounts:

Smith	200 tons at \$10 per ton
Jones	300 tons at \$9 per ton
Richard	400 tons at \$8 per ton

Shipping costs in dollars per ton are

From	To	
	Plant *A*	Plant *B*
Smith	2	2.5
Jones	1	1.5
Richard	5	3

Plant capacities and labor costs are

	Plant *A*	Plant *B*
Capacity	450 tons	550 tons
Labor cost	\$25/ton	\$20/ton

The canned fruits are sold at \$50 per ton to the distributors. The company can sell at this price all they can produce. How should the company plan its operations at the two plants so as to maximize its profits?

To formulate this as a linear program, define:

$$
\begin{aligned}
x_{SA} &= \text{quantity shipped from Smith to Plant } A \\
x_{SB} &= \text{quantity shipped from Smith to Plant } B \\
x_{JA} &= \text{quantity shipped from Jones to Plant } A \\
x_{JB} &= \text{quantity shipped from Jones to Plant } B \\
x_{RA} &= \text{quantity shipped from Richard to Plant } A \\
x_{RB} &= \text{quantity shipped from Richard to Plant } B
\end{aligned}
$$

The supply constraints are given by

$$
\begin{aligned}
x_{SA} + x_{SB} &\leqq 200 \\
x_{JA} + x_{JB} &\leqq 300 \\
x_{RA} + x_{RB} &\leqq 400
\end{aligned}
$$

The constraints on plant capacities are

$$
\begin{aligned}
x_{SA} + x_{JA} + x_{RA} &\leqq 450 \\
x_{SB} + x_{JB} + x_{RB} &\leqq 550
\end{aligned}
$$

All the variables are restricted to be nonnegative. To compute the net profit for each of the variables, we have to subtract from the selling price the cost of fresh fruits, shipping, and labor costs. For example, the profit on variable x_{SA} (fruits bought from Smith and processed at Plant A) is given by

$$P_{SA} = 50 - (10 + 2 + 25) = \$13 \text{ per ton}$$

Similarly the other profits can be calculated, and the objective function is to maximize

$$Z = 13x_{SA} + 17.5x_{SB} + 15x_{JA} + 19.5x_{JB} + 12x_{RA} + 19x_{RB}$$

The above linear program is an unbalanced transportation problem where the total supply is less than the total demand. Table 3.4 gives the equivalent (standard) transportation formulation with the addition of a dummy grower.

Table 3.4

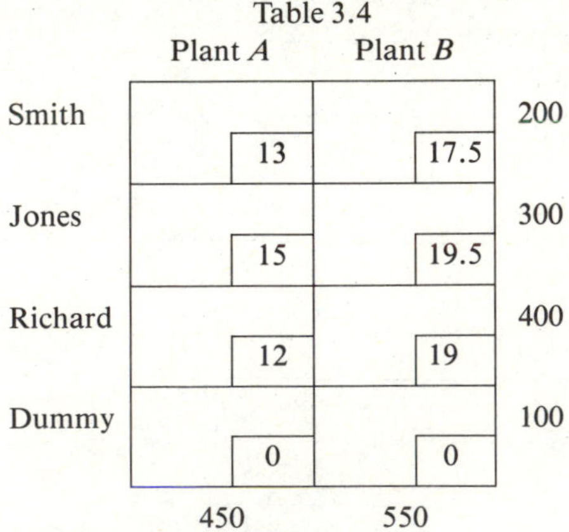

	Plant A	Plant B	
Smith	13	17.5	200
Jones	15	19.5	300
Richard	12	19	400
Dummy	0	0	100
	450	550	

Since the objective is to maximize the total profit, the variable selection rule of the transportation algorithm needs some modification. Those nonbasic variables whose $c_{ij} > u_i + v_j$ (i.e., $\bar{c}_{ij} > 0$) are eligible to enter the basis to improve the objective function. When all the $\bar{c}_{ij} \leq 0$, the solution becomes optimal. The final solution is left as an exercise.

EXAMPLE 3.2-4

Consider the problem of scheduling the weekly production of a certain item for the next 4 weeks. The production cost of the item is $10 for the first 2 weeks, and $15 for the last 2 weeks. The weekly demands are 300, 700, 900, and 800, which must be met. The plant can produce a maximum of 700 units each week. In addition the company can employ overtime during the second and third week. This increases the weekly production by an additional 200 units, but the production cost increases by $5 per item. Excess production can be stored at a unit cost of $3 per week. How should the production be scheduled so as to minimize the total costs?

To formulate this as a transportation problem, we consider the production periods as warehouses, and weekly demands as markets. Since overtime production is possible during the second and third weeks, there are in all six supply points. The decision variables are

x_{1j} = Normal production in week 1 for use in week j for $j = 1, 2, 3, 4$
x_{2j} = Normal production in week 2 for use in week j for $j = 2, 3, 4$
x_{3j} = Overtime production in week 2 for use in week j for $j = 2, 3, 4$
x_{4j} = Normal production in week 3 for use in week j for $j = 3, 4$
x_{5j} = Overtime production in week 3 for use in week j for $j = 3, 4$
x_{64} = Normal production in week 4 for use in week 4

Since the total production (normal and overtime) exceeds the total demand, we create a dummy market to absorb the excess supply. Table 3.5 gives the corresponding transportation problem.

Table 3.5

		Demands					
		Week 1	Week 2	Week 3	Week 4	Dummy	
	Week 1	10	13	16	19	0	700
	Week 2 (normal)	M	10	13	16	0	700
	Week 2 (overtime)	M	15	18	21	0	200
Supplies	Week 3 (normal)	M	M	15	18	0	700
	Week 3 (overtime)	M	M	20	23	0	200
	Week 4 (normal)	M	M	M	15	0	700
		300	700	900	800	500	

REMARKS

1. Some of the cost elements in the above table are set at M (an infinitely large value) to denote shipments that are not possible. For example, production during weeks 2, 3, and 4 cannot possibly supply the first week's demand.

2. Weekly storage cost of \$3 per item is added to the production cost whenever an item is stored to meet future demands.

This problem can now be solved by the transportation algorithm to determine the optimal production schedule. This is left as an exercise for the reader.

3.3 ASSIGNMENT PROBLEMS

Standard Assignment Problems

A certain machine shop has n machines denoted by $M_1, M_2, \ldots, M_n$. A group of n different jobs $(J_1, J_2, \ldots, J_n)$ is to be assigned to these machines. For each job the machining cost depends on the machine to which it is assigned. Let c_{ij} represent the cost of doing job j on machine i. (If a particular job cannot be done on a machine we set the appropriate c_{ij} to a very large number). Each machine can work only on one job. The problem is to assign the jobs to the machines which will minimize the total cost of machining.

A naive approach to solve this problem is to enumerate all possible assignments of jobs to machines. For each assignment the total cost may be computed, and the one with the least cost is picked as the best assignment. This will be an inefficient and expensive approach since the number of possible assignments is $n!$ Even for $n = 10$, there are 3,628,800 possible assignments!

To formulate this as a linear programming problem, define

$$x_{ij} = \begin{cases} 1 & \text{if job } j \text{ is assigned to machine } i \\ 0 & \text{otherwise} \end{cases}$$

Since each machine is assigned exactly to one job, we have

$$\sum_{j=1}^{n} x_{ij} = 1 \qquad \text{for } i = 1, 2, \ldots, n$$

Similarly, each job is assigned exactly to one machine.

$$\sum_{i=1}^{n} x_{ij} = 1 \qquad \text{for } j = 1, 2, \ldots, n$$

The objective is to minimize

$$Z = \sum_{i=1}^{n} \sum_{j=1}^{n} c_{ij} x_{ij}$$

The above is actually the formulation of a standard transportation problem with n warehouses and n markets where the supply $a_i = 1$ for $i = 1, \ldots, n$, and the demand $b_j = 1$ for $j = 1, \ldots, n$. The corresponding transportation matrix is given below:

Jobs

		J_1	J_2	$\cdots$	J_n	
	M_1	c_{11}	c_{12}		c_{1n}	1
	M_2	c_{21}	c_{22}		c_{2n}	1
	$\cdot$					$\cdot$
Machines	$\cdot$					$\cdot$
	$\cdot$					$\cdot$
	M_n	c_{n1}	c_{n2}		c_{nn}	1
		1	1	$\cdots$	1	

Nonstandard Assignment Problems

Consider a machine shop with M machines and N jobs where $M \neq N$. To convert this to an (equivalent) standard assignment problem with equal number of jobs and machines, we create dummy jobs or dummy machines.

Suppose we have more machines than jobs ($M > N$). We then create $(M - N)$ dummy jobs so that there will be M machines and M jobs. We set the machining cost of the dummy jobs as zero so that the objective function will be unaltered. When a dummy job gets assigned to a machine, that machine stays idle. Similarly, when we have more jobs than machines ($N > M$) then some jobs cannot be assigned. In this case, $(N - M)$ dummy machines are created whose machining cost will be zero for all jobs.

Solution by the Transportation Algorithm

Since any assignment problem can be formulated as a standard transportation problem, the transportation technique discussed in Section 3.2 can be used to find the optimal assignment. However, the transportation algorithm is generally not recommended for solving the assignment problem due to the presence of degeneracy in every basic feasible solution. Since a standard assignment problem with M machines and M jobs can only have M of the x_{ij}'s equal to one, every basic feasible solution (which has $M + M - 1$ basic variables) will contain $M - 1$ basic variables at zero values. As discussed earlier in Section 2.8, degeneracy can lead to nonproductive basis changes and possibly cycling in the transportation algorithm.

Hungarian Method

A more efficient way of solving the assignment problem has been developed based on a mathematical property due to the Hungarian mathematician König (hence, the name of the method). We will assume in the Hungarian method that all the cost elements (c_{ij}) are nonnegative. The basic principle of the method is that the optimal assignment is not affected if a constant is added or subtracted from any row or column of the standard assignment cost matrix. For example, if the cost of doing any job on machine 1 is reduced by \$$k$, then the objective function of the assignment problem becomes

$$\text{Minimize:} \qquad Z = \sum_{j=1}^{n} (c_{1j} - k)x_{1j} + \sum_{i=2}^{n} \sum_{j=1}^{n} c_{ij}x_{ij}$$

$$= \sum_{i=1}^{n} \sum_{j=1}^{n} c_{ij}x_{ij} - k \sum_{j=1}^{n} x_{ij}$$

However, $\sum_{j=1}^{n} x_{ij}$ equals 1, since machine 1 must be assigned exactly to one of the jobs. Hence, the new objective function becomes

$$Z = (\text{original objective}) - k$$

We know from Chapter 2 that the addition of a constant to the objective function of a linear program does not affect its optimal solution, namely, the optimal values of the x_{ij}'s.

In essence the solution procedure is to subtract a sufficiently large cost from the various rows or columns in such a way that an optimal assignment is found by inspection. We initiate the algorithm by examining each row (column) of the cost matrix to identify the smallest element. This quantity is then subtracted from all the elements in that row (column). This produces a cost matrix containing at least one zero element in each row (column). Now try to make a feasible assignment using the cells with zero costs. If it is possible then we have an optimal assignment. This is because the cost elements (c_{ij}) are nonnegative and the minimum value of the objective function $\sum_i \sum_j c_{ij}x_{ij}$ cannot be less than zero. Hence, an assignment with zero cost has to be optimal.

EXAMPLE 3.3-1

Find the optimal assignment of four jobs and four machines when the cost of assignment is given by the following table:

	J_1	J_2	J_3	J_4
M_1	10	9	8	7
M_2	3	4	5	6
M_3	2	1	1	2
M_4	4	3	5	6

Solution

Since the number of jobs and machines are equal, we have a standard assignment problem.

Examining the rows first, a reduced cost matrix can be obtained by subtracting,

1. 7 from the first row,
2. 3 from the second row,
3. 1 from the third row, and
4. 3 from the fourth row.

This produces the following cost matrix:

	J_1	J_2	J_3	J_4
M_1	3	2	1	0
M_2	0	1	2	3
M_3	1	0	0	1
M_4	1	0	2	3

From the above table a feasible assignment using only the cells with zero costs is $M_1 \to J_4$, $M_2 \to J_1$, $M_3 \to J_3$, and $M_4 \to J_2$. Hence, this is an optimal assignment.

In general, it may not always be possible to find a feasible assignment using cells with zero costs. To illustrate this, consider the following assignment problem:

EXAMPLE 3.3-2

Find an optimal solution to an assignment problem with the following cost matrix:

	J_1	J_2	J_3	J_4
M_1	10	9	7	8
M_2	5	8	7	7
M_3	5	4	6	5
M_4	2	3	4	5

Solution

First the minimum element in each row is subtracted from all the elements in that row. This gives the following reduced-cost matrix.

	J_1	J_2	J_3	J_4
M_1	3	2	0	1
M_2	0	3	2	2
M_3	1	0	2	1
M_4	0	1	2	3

Since both the machines M_2 and M_4 have a zero cost corresponding to job J_1 only, a feasible assignment using only cells with zero costs is not possible. To get additional zeros subtract the minimum element in the fourth column from all the elements in that column.

	J_1	J_2	J_3	J_4
M_1	3	2	0	0
M_2	0	3	2	1
M_3	1	0	2	0
M_4	0	1	2	2

Only three jobs can be assigned using the zero cells, and so a feasible assignment is still not possible. In such cases, the procedure draws a minimum number of lines through some selected rows and columns in such a way that all the cells with zero costs are covered by these lines. The minimum number of lines needed is equal to the maximum number of jobs that can be assigned using the zero cells.*

*This property was in fact the theorem proved by König.

In our example, this can be done with three lines as follows.

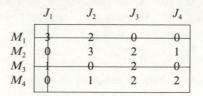

	J_1	J_2	J_3	J_4
M_1	3	2	0	0
M_2	0	3	2	1
M_3	1	0	2	0
M_4	0	1	2	2

Now select the smallest element that is not covered by the lines. In our case it is 1. Subtract this number from all the elements that are *not covered*. Then add this number to all those *covered* elements that are at the *intersection of two lines*. This gives the following reduced cost matrix.

	J_1	J_2	J_3	J_4
M_1	4	2	0	0
M_2	0	2	1	0
M_3	2	0	2	0
M_4	0	0	1	1

Note that the above procedure is equivalent to subtracting the element 1 from the second and fourth rows, and adding it to the first column of the cost matrix. Hence, the optimal assignment is once again not altered.

A feasible assignment is now possible and an optimal solution is to assign $M_1 \rightarrow J_3$, $M_2 \rightarrow J_1$, $M_3 \rightarrow J_4$, and $M_4 \rightarrow J_2$. The total cost is given by $7+5+5+3=20$. An alternate optimal solution is $M_1 \rightarrow J_3$, $M_2 \rightarrow J_4$, $M_3 \rightarrow J_2$, and $M_4 \rightarrow J_1$. In case a feasible set could not obtained at this step, one has to repeat the step of drawing lines to cover the zeros and continue until a feasible assignment is obtained.

How to Solve Maximization Problems

To solve an assignment problem whose objective is to maximize, the problem should first be converted to a minimization problem before the application of the Hungarian Method.

Step I. Convert the problem to a minimization problem by multiplying all the elements (c_{ij}) of the assignment matrix by -1.

Step II. If some of the elements of the cost matrix are negative, add a sufficiently large positive number to the corresponding rows and columns so that all the cost elements would become nonnegative.

Step III. We now have an assignment problem with a minimization objective and all cost elements nonnegative. The Hungarian Method can now be applied directly.

3.4 MAXIMAL-FLOW PROBLEMS

In this section we consider the problem of shipping a certain homogeneous commodity from a specified point, called the *source*, to a particular destination, called the *sink*. Unlike the transportation problems, there may not be a direct *link* or *arc* connecting the source and the sink. The flow network will generally consist of some intermediate nodes, known as *transshipment points*, through which the flows are rerouted.

An example of a flow network is given by Fig. 3.1. The source node is denoted by the symbol s and the sink node by n. Nodes 1 and 2 are the intermediate nodes. There are six arcs connecting the various nodes, denoted by $(s, 1)$, $(s, 2)$, $(1, 2)$, $(2, 1)$, $(1, n)$, and $(2, n)$.

Flow in arc (i, j) denoted by f_{ij}, is the quantity shipped from node i to node j. We will assume that all the arcs are *directed*. In other words, flow in arc (i, j) is possible from node i to node j only. Let f denote the total amount shipped from the source to the sink, and k_{ij} denote the *capacity* of arc (i, j), which is the maximum flow possible from i to j.

The Maximal-Flow Problem

Consider the flow network represented by Fig. 3.1. Given the capacities k_{ij} on flows on each arc (i, j), and that the flow must satisfy conservation (i.e., the total flow into a node must be equal to the total flow out of that node), we want to determine the maximum flow f that can be sent from the source node s to the sink node n.

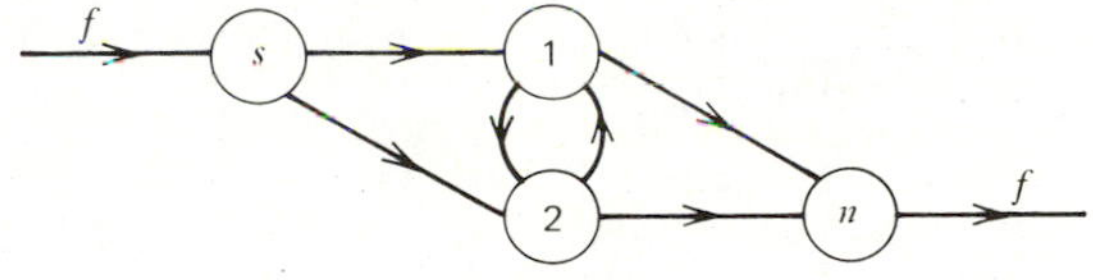

Figure 3.1

The linear programming formulation of the maximal flow problem is given:

Maximize: $Z = f$

Subject to:

$$f_{s1} + f_{s2} = f \tag{3.3}$$
$$f_{12} + f_{1n} = f_{s1} + f_{21} \tag{3.4}$$
$$f_{21} + f_{2n} = f_{s2} + f_{12} \tag{3.5}$$
$$f_{1n} + f_{2n} = f \tag{3.6}$$
$$0 \leq f_{s1} \leq k_{s1} \tag{3.7}$$
$$0 \leq f_{s2} \leq k_{s2} \tag{3.8}$$
$$0 \leq f_{12} \leq k_{12} \tag{3.9}$$
$$0 \leq f_{21} \leq k_{21} \tag{3.10}$$
$$0 \leq f_{1n} \leq k_{1n} \tag{3.11}$$
$$0 \leq f_{2n} \leq k_{2n} \tag{3.12}$$

Equation 3.3 represents the conservation of flow at the source node; Eqs. 3.4 and 3.5 represent the conservation of flow at the intermediate nodes 1 and 2, respectively; Eq. 3.6 represents the conservation at the sink node, while Eqs. 3.7 to 3.12 restrict the flows (f_{ij}) so as to be nonnegative with finite upperbounds. Although the maximal flow problem can be solved by the simplex method, there exists an efficient network method to find the maximum flow directly. We shall first introduce some basic concepts from network theory that are fundamental to the new procedure.

Definitions

Forward Arcs. At any node i, all the arcs that are leaving node i are called forward arcs with respect to node i. Similarly, all the arcs entering a node are called *backward arcs* for that node. It should be noted here that an arc may be a forward arc with respect to some node, and a backward arc with respect to some other node.

In Fig. 3.1, the arcs $(1, n)$ and $(1, 2)$ are forward arcs for node 1 while the arcs $(s, 1)$ and $(2, 1)$ are its backward arcs. But for node 2, the arc $(1, 2)$ is a backward arc while $(2, 1)$ is a forward arc.

A *path* connecting the source and the sink is a sequence of arcs starting from the source node and ending in the sink node. For example, in the same figure, a path connecting s and n is given by the sequence of arcs $(s, 1)$, $(1, 2)$, $(2, n)$. Note that there exists more than one path connecting s and n in Fig. 3.1.

A *cycle* is a path whose beginning and ending nodes are the same.

Let N denote the collection of all the nodes in the network. Then, a *cut* separating the source and the sink is a partition of the nodes in the network into two subsets S and $\bar{S}$ such that the source node is in S and the sink node is in $\bar{S}$.

For example, in Fig. 3.1, a cut $(S, \bar{S})$ is given by $S = (s, 1, 2)$, $\bar{S} = (n)$. Note that $S \cup \bar{S} = N = (s, 1, 2, n)$, $S \cap \bar{S} =$ the empty set, $s \in S$, and $n \in \bar{S}$. A pictorial representation of the cut is given below:

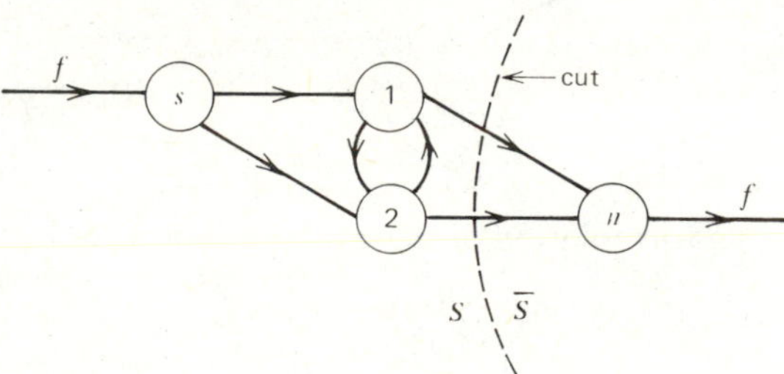

Another cut separating s and n is shown below where $S = (s, 2)$ and $\bar{S} = (1, n)$:

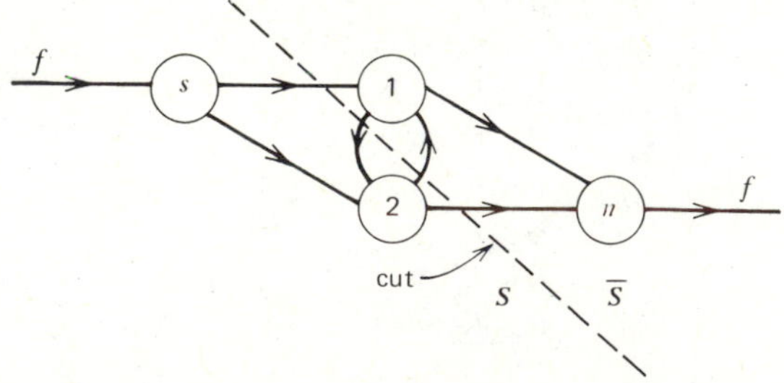

Similarly a number of other cuts may be found for the network. The *capacity* of a cut denoted by $K(S, \bar{S})$ is the sum of all the capacities of the arcs from the nodes in S to those in $\bar{S}$. For example, the capacity of the cut, $(S, \bar{S})$ where $S = (s, 1, 2)$, $\bar{S} = (n)$ is $K_{1n} + K_{2n}$. But the capacity of the cut where $S = (s, 2)$ and $\bar{S} = (1, n)$ is $K_{s1} + K_{21} + K_{2n}$. (Since arc (1, 2) is directed from a node in $\bar{S}$ to a node in S, it is not included in the summation.)

The cut with the smallest capacity is called a *minimal* cut. From the pictorial representation of the cuts, it is rather obvious that if all the arcs of a cut are removed from the network then there exists no path from the source to the sink; hence, no flow would be possible. In other words, any flow from s to n must flow through the arcs in the cut, and consequently the flow f is limited by the capacity of that cut. This relation between flows and cuts is given by the following lemma:

Lemma 1

For any directed network, if f is the flow from the source to the sink, and $(S, \bar{S})$ is a cut, then the value of f is less than or equal to the capacity of that cut $K(S, \bar{S})$.

Since Lemma 1 is true for *any* cut in the network, we can state that any feasible flow from source to sink cannot exceed the capacity of any cut. Thus the maximal flow across the network is limited by the capacity of the minimal cut. The following important theorem states that it is always possible to find a feasible flow from s to n *equal* to the capacity of the minimal cut.

Max-Flow Min-Cut Theorem [Ford and Fulkerson (3)]. For any network the value of the maximal flow from source to sink is equal to the capacity of the minimal cut.

Using the max-flow min-cut theorem one can find the maximal flow in a network by finding the capacities of all the cuts, and choosing the minimum capacity. Though this gives the maximal value of f, it does not specify how this flow is routed through various arcs. So we describe a different procedure known as the *maximal flow algorithm* whose validity is based on the max-flow min-cut theorem. The basic principle of this procedure is to find a path through which a positive flow can be sent from the source node to the sink node. Such a path is termed a *flow augmenting path*. This path is used to send as much flow as possible from s to n. This process is repeated until no such flow augmenting path can be found at which time we have found the maximal flow.

Labeling Routine. This is used to find a flow augmenting path from the source to the sink. We start from the source node s. We say that node j can be *labeled* if a positive flow can be sent from s to j. In general, from any node i, we can *label* node j, if one of the following conditions is satisfied:

1. The arc connecting the nodes i and j is a forward arc [i.e., the existing arc is (i, j)], and the flow in arc (i, j) is less than its capacity (i.e., $f_{ij} < K_{ij}$).
2. The arc connecting i and j is a backward arc [i.e., the existing arc is (j, i)], and the flow in arc (j, i) is greater than zero. (i.e., $f_{ji} > 0$)

We continue this labeling routine until the sink node is labeled. We then have a flow augmenting path.

Max-Flow Algorithm. The algorithm is initiated with a feasible flow on all arcs, satisfying capacity restrictions, and conservation of flows at all nodes. To improve this flow, we initially label node s, and then apply the labeling routine to label another node. When the sink is labeled, we have a flow augmenting path from s to n through which a positive flow can be sent. Now we retrace the flow augmenting path with the help of the labels on the nodes and compute the maximal flow δ that can be sent in the path. Then increase the flow by δ on all the forward arcs in the path, and decrease the flow by δ on all the backward arcs in the path. Repeat the procedure by finding another flow augmenting path from s to n

using the labeling routine. The algorithm terminates when no flow augmenting path can be found, at which time we have the maximal flow possible from s to n.

We shall illustrate the max-flow algorithm through an example problem.

EXAMPLE 3.4-1

Compute the maximal flow f from s to n in the following network, where the numbers of the arcs represent their capacities:

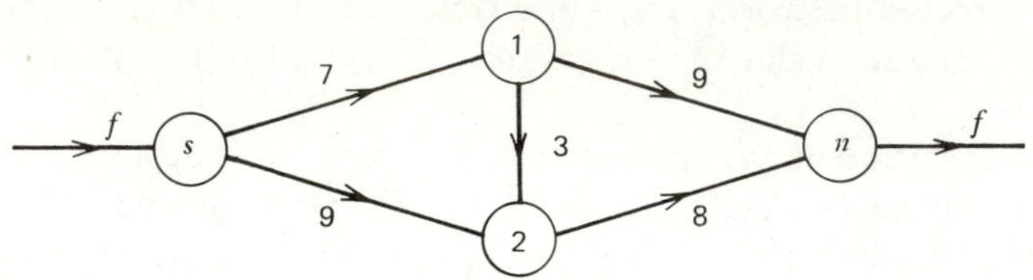

Prestep. The max-flow algorithm is initiated with zero flows on all arcs. In Fig. 3.2, the numbers on the arcs (i, j) represent (f_{ij}, k_{ij}).

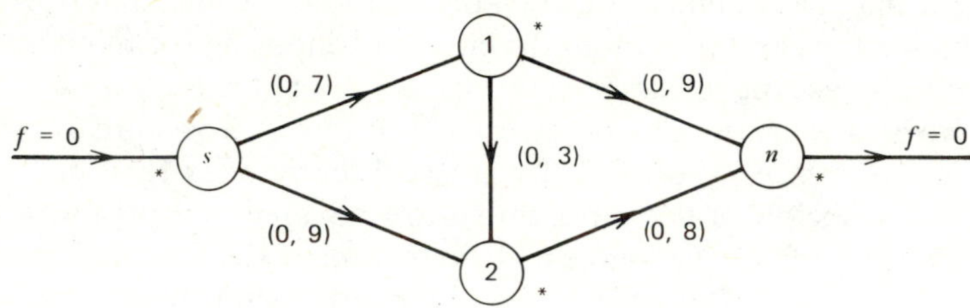

Step I (Fig. 3.2). To find a flow augmenting path from s to n, node s is initially labeled. (Labels are denoted by asterisks.) From s, we can label node 1 since $(s, 1)$ is a forward arc, carrying a flow f_{s1} which is less than its capacity. From node 1, node 2 is labeled through the forward arc (1, 2) and from node 2 the sink is labeled. We now have a flow augmenting path consisting only of forward arcs as shown below:

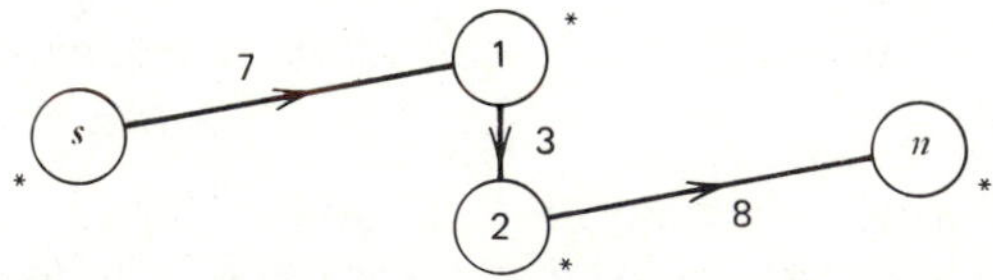

The numbers on the arcs indicate the maximum flow that is possible in each arc. Thus the maximal flow that can be sent through this flow augmenting path is 3 units. This increases f by 3 units, and the flow on all (forward) arcs in the path by 3 units. The new flow configuration is given by Fig. 3.3.

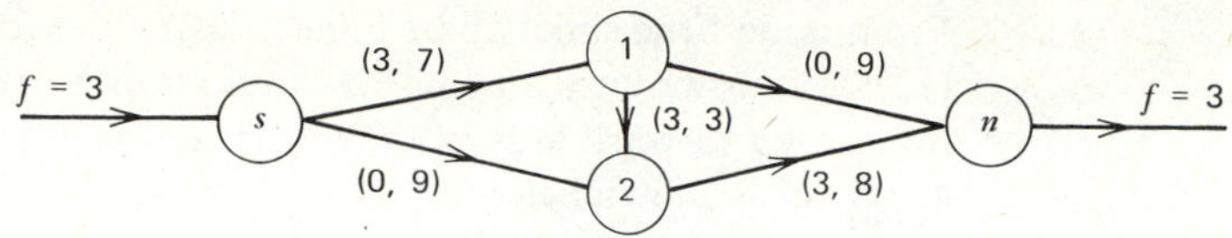

Figure 3.3

Step II (Fig. 3.3). Repeat the labeling routine and a new flow augmenting path is

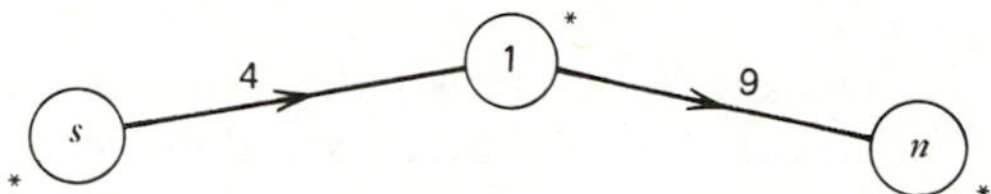

The maximum flow that can be sent through this path is 4 units. This increases the flow across the network (f) to 7 units. The new flow configuration is given by Fig. 3.4.

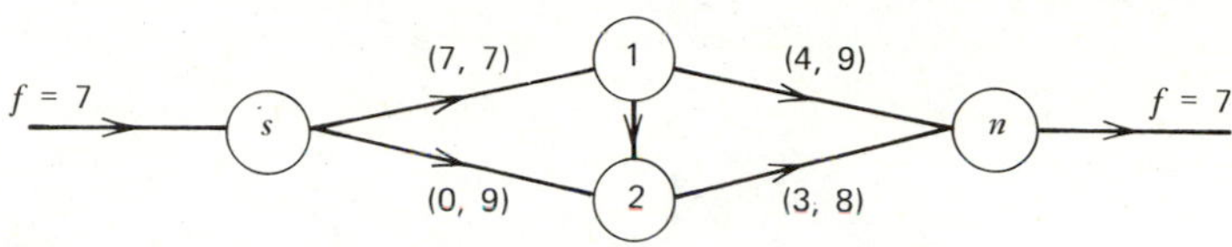

Figure 3.4

Step III (Fig. 3.4). Now node 1 cannot be labeled from node s since $(s, 1)$ is a forward arc whose flow is equal to its capacity. But a new flow augmenting path can be found as follows:

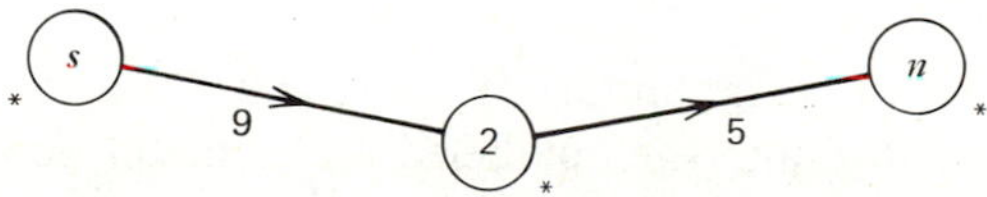

This increases the total flow from s to n by 5 units as shown in Fig. 3.5.

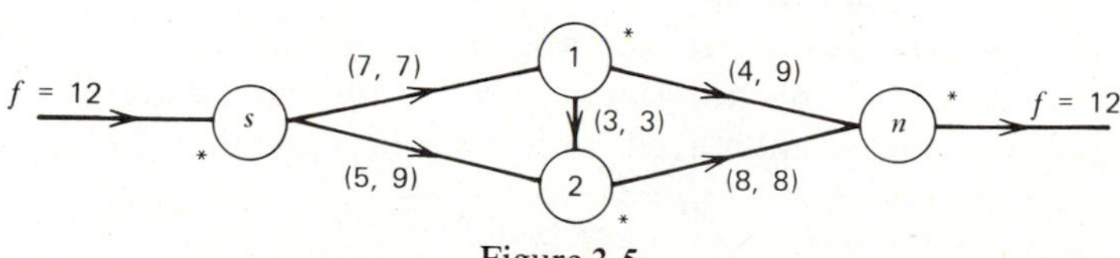

Figure 3.5

Step IV (Fig. 3.5). Starting from node s, node 2 can be labeled, but the sink cannot be labeled from 2 since the flow in arc $(2, n)$ has reached its capacity. But node 1 can be labeled from node 2 since $(1, 2)$ is a backward arc carrying a positive flow.

From node 1 the sink can be labeled using the forward arc $(1, n)$. Thus we have a new flow augmenting path consisting of two forward arcs $(s, 2)$ and $(1, n)$, and a backward arc $(1, 2)$ as shown below:

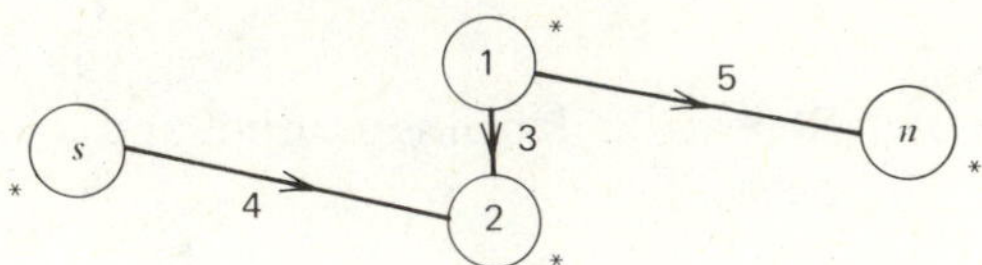

To increase the flow f through this path, we increase the flow on the forward arcs and decrease it in the backward arcs. Thus the maximum increase in f possible is 3 units, and the new assignment of flows is given in Fig. 3.6.

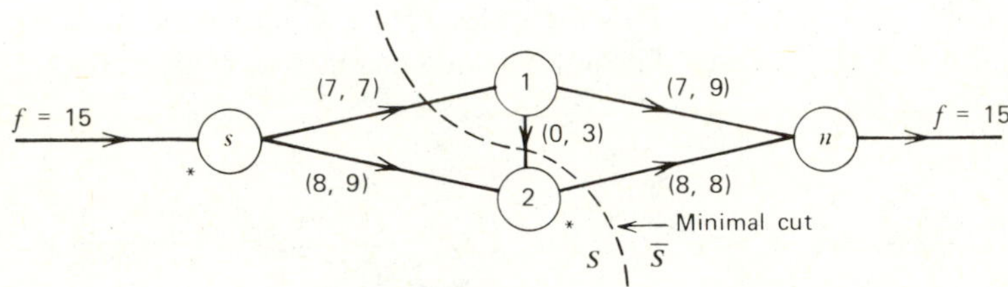

Figure 3.6

Step V (Fig. 3.6). Even though node 2 can be labeled from node s, the sink can never be labeled. Hence, no new flow augmenting paths are possible, and we have found the maximal flow possible which is 15 units.

REMARKS

To verify if $f = 15$ is the maximum flow, we can define a cut $(S, \bar{S})$ by placing all the labeled nodes in S and the unlabeled nodes in $\bar{S}$ (see Fig. 3.6). This gives $S = (s, 2)$, $\bar{S} = (1, n)$ and the capacity of cut $K(S, \bar{S}) = 15$. Since f cannot exceed the capacity of any cut by Lemma 1, $f = 15$ is the maximal flow possible, and the cut shown in Fig. 3.6 is the minimal cut.

Extensions and Applications

1. *Undirected Arcs.* Consider a network that contains arcs with no specified direction of flow. These arcs are generally called *undirected arcs.* When an arc connecting nodes i and j is undirected, with capacity K, we interpret that as

$$f_{ij} \leqq K$$
$$f_{ji} \leqq K$$
$$f_{ij} \cdot f_{ji} = 0$$

In other words, a maximum of K units of flow is possible between nodes i and j in either direction, but flow is permitted in one of the directions only. Recall that the maximal flow algorithm discussed earlier can be applied to directed networks only where the direction of flow is specified in all arcs. To find the maximal flow in an undirected network, we first convert the network to an equivalent directed network, and then apply the labeling method.

EXAMPLE 3.4-2
Consider a street network as shown below:

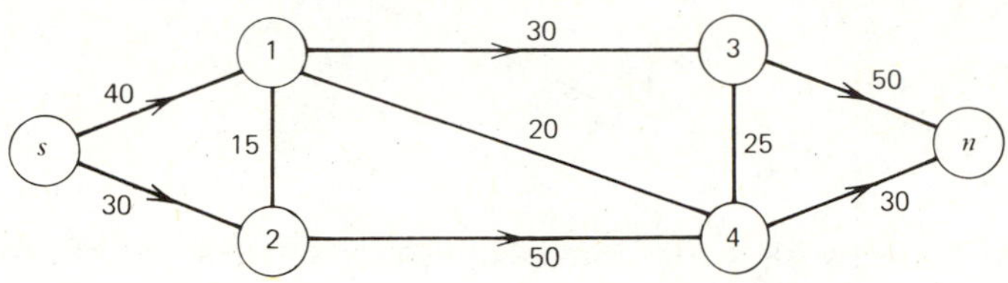

The numbers on the arcs represent the traffic flow capacities. The problems is to place one-way signs on streets not already oriented so as to maximize the traffic flow from the point s to the point n.

Solution
The trick is to replace each undirected arc by a pair of oppositely directed arcs with the same capacities. This gives a directed network as shown in Fig. 3.7:

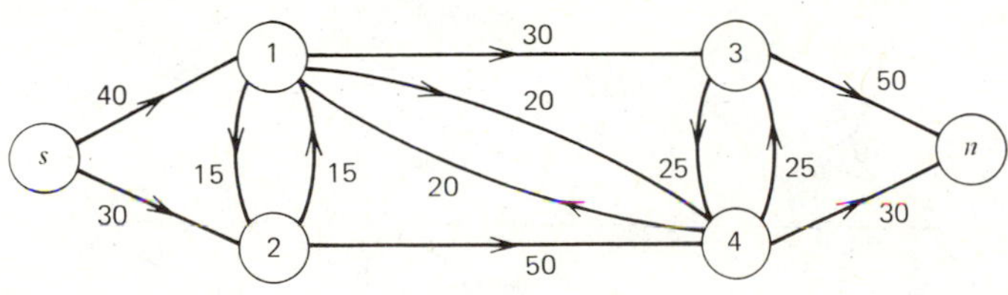

Figure 3.7

The maximal flow algorithm is applied to this network to determine the largest traffic flow from s to n. After the optimal flows have been found we cancel out the arc flows in the opposite directions so as to get the direction of flow in each of the undirected arcs. For instance, if the arc connecting i and j was undirected, and $f_{ij} > f_{ji}$, then a flow of $(f_{ij} - f_{ji})$ should be directed from i to j.

The numerical solution is left as an exercise for the reader.

2. *Multiple Sources and Sinks (Transshipment Problems).* Consider a network with several supply points and demand points. The problem is to maximize the flow from all the sources to all the destinations. The max-flow algorithm can be applied to solve this problem by converting to a single source-single sink situation. This can be done by creating an imaginary super source and an imaginary super sink. From the super source a directed arc will be created to every one of the real sources such that the super source becomes the supplier to the real sources. Similarly from each one of the real sink, a directed arc to the super sink will be created. We then apply the max-flow algorithm to maximize the flow from the super source to the super sink. This will be equivalent to maximizing the flow from all the sources to all the sinks.

To illustrate this, consider a transshipment problem wherein we have a network with multiple sources and multiple sinks with limited supplies and demands as shown below:

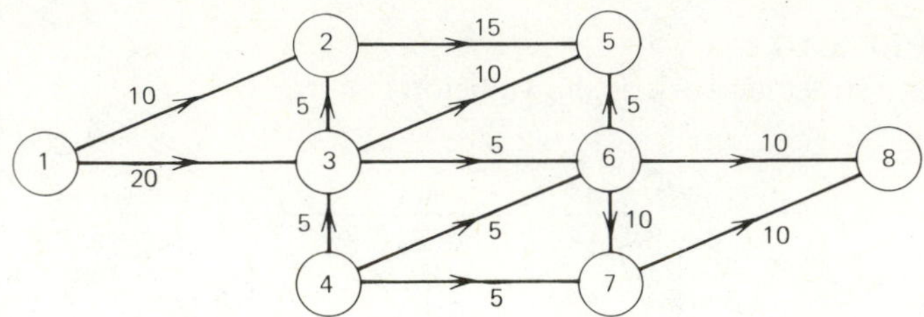

In the above network, nodes 1 and 4 are sources with supplies $s_1 = 20$, $s_4 = 20$. Nodes 5 and 8 are sinks with demands $d_5 = 15$, $d_8 = 20$. The numbers on the arcs represent the arc capacities. We wish to determine whether the transshipment problem is feasible, that is, whether it will be possible to meet the demands with the available supplies.

Solution

First the problem is converted to a problem of maximizing the flow from a single source to a single sink. An equivalent network (Fig. 3.8) is created with an imaginary source (s) and an imaginary sink (n).

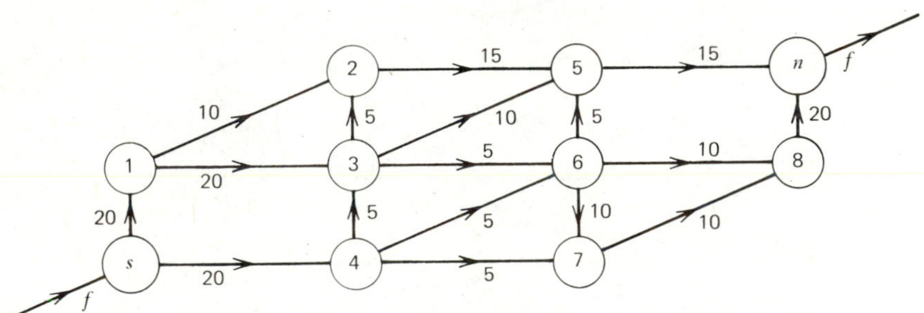

Figure 3.8
A transshipment problem as a max-flow problem.

In Fig. 3.8 the imaginary supply arcs (s, 1) and (s, 4) have capacities equal to the supplies in nodes 1 and 4, respectively. Similarly, arcs (5, n) and (8, n) have capacities equal to their respective demands. Even though the total supply exceeds the total demand, the transshipment problem may not be feasible due to the capacity restrictions on the intermediate arcs.

We now maximize the flow (f) from the imaginary source to the imaginary sink in the directed network of Fig. 3.8. Omitting the intermediate steps of the max-flow algorithm, the optimal distribution of flows is shown in Fig. 3.9. The numbers on arc (i, j) denote (f_{ij}, K_{ij}). The maximal flow possible is 30, and the minimal cut is given by the subsets $S = (s, 1, 2, 3, 4, 5)$ and $\bar{S} = (6, 7, 8, n)$. Since the maximum flow is less than the total demand of 35 units, it is not possible to satisfy all the demands at the sinks. Therefore, the transshipment problem is not feasible.

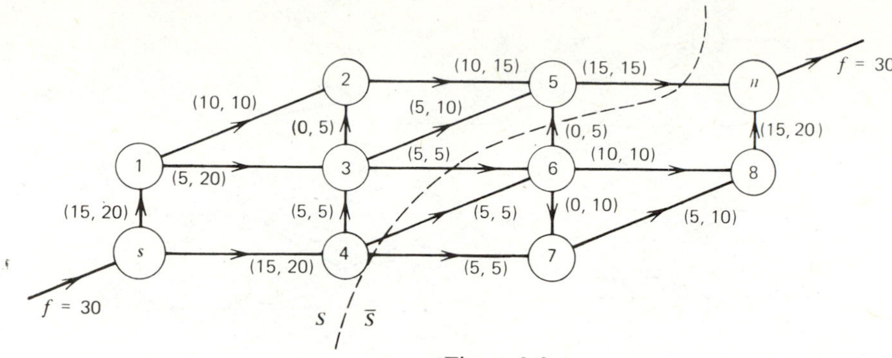

Figure 3.9
Solution of the transshipment problem.

REMARK

When the value of maximal flow f equals the sum of all the demands at the sinks, then the transshipment problem becomes feasible.

3. *Manpower Scheduling.* Another application of the maximum flow model is in determining a feasible assignment of personnel to different projects over a planning period. Assume that the XYZ company has four projects A, B, C, D to be completed during the next 6 months. The projects have varying start and completion dates and require different man-months of engineering efforts as shown below:

Job	Earliest Start	Latest Finish	Required Efforts (man-months)
A	Month 1	Month 4	6
B	Month 1	Month 6	8
C	Month 2	Month 5	3
D	Month 1	Month 6	4

The company has four engineers who can be assigned to these projects. However, no more than two engineers may be assigned to any one project on a given month. Moreover, no engineer should be assigned to more than one project at any time. The problem is to determine whether there exists a feasible assignment that will meet the project deadlines.

We shall show that this problem can be formulated as a transshipment problem that in turn can be solved by the maximal flow algorithm. Construct a transshipment network with six sources (one for each month) and four sinks (one for each project). The supply at each source is 4 man-months of engineers' effort available each month, and the demands are the required man-months of effort for each project. Every arc connecting a source and a sink represents a feasible assignment of engineers (man-months) for any given month to a certain project and has a capacity of 2 man-months. Figure 3.10 illustrates the transshipment network. Note that arcs exist from node 1 to nodes A, B, and D only, since project C cannot be started until month 2.

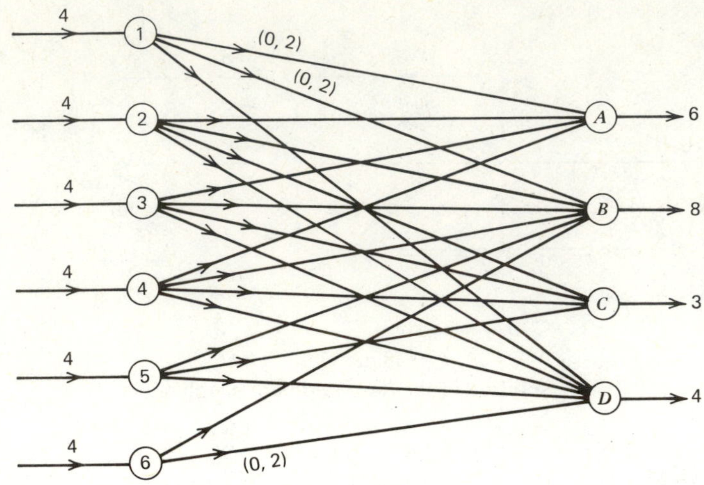

Figure 3.10
Transshipment network for project scheduling.

Similarly, no arc exists from nodes 5 and 6 to node A, since A has to be completed by the end of month 4. The transshipment problem can now be solved by creating a super source s connected to nodes 1, 2, . . . , 6, and a super sink n connected to nodes A, B, C, D. If the maximal flow equals 21 (total man-months of efforts for all projects), there exists a feasible schedule subject to the manpower constraints.

3.5 SHORTEST-ROUTE PROBLEMS

Another important network flow problem from an applied standpoint is that of determining the shortest path from the source node to the sink node. Here we are given a network of n nodes denoted by $(1, 2, \ldots, n)$. Corresponding to each arc (i, j), there is a nonnegative number d_{ij} called the distance or transit time from node i to node j. In case there is no way of getting from i to j directly, we set $d_{ij} = +\infty$. It is possible to have $d_{ij} \neq d_{ji}$. The problem is to find the length of the shortest path and the shortest route from the source node 1 to the sink node n.

One way of solving this problem is to reinterpret it as a shipping problem in which we wish to send one unit of flow from node 1 to node n, and the unit cost of shipping from i to j is d_{ij}. Then the shortest route problem can be formulated as a linear program for solution. However there are better and more elegant ways of solving the shortest-route problem. We now describe one of the most efficient algorithms, given by Dijkstra.

Dijkstra's Algorithm [Dreyfus (2)]

It is assumed that the direct distance between any two nodes (d_{ij}) in the network of n nodes is given, and all the distances are nonnegative. The algorithm proceeds by assigning to all nodes a *label* that is either *temporary* or *permanent*. A temporary label represents an upper bound on the shortest distance from node 1 to that node; while a permanent label is the actual shortest distance from node 1 to that node.

Initially the source node 1 is given a permanent label of zero. All other nodes $(2, 3, \ldots, n)$ are assigned temporary labels equal to the direct distance from node

1 to the node in question. Any node that cannot be reached directly from node 1 is assigned a temporary label of ∞, while all the other nodes receive temporary labels equal to d_{ij}. The algorithm then makes these tentative node labels, one at a time, permanent labels. As soon as the sink node receives a permanent label, the shortest distance from the source node to the sink node is immediately known.

Iterative Steps of the Algorithm.

Prestep. Initialize by assigning a permanent label of zero to the source node. All other node labels are temporary and are equal to the direct distance from the source node to that node. Select the minimum of these temporary labels and declare it permanent. In case of ties, choose any one.

Step I. Suppose node K has been assigned a permanent label most recently. Now consider the remaining nodes with temporary labels. Compare one at a time the temporary label of each node, to the *sum* of the permanent label of node K and the *direct distance* from node K to the node under consideration. Assign the minimum of these two distances as the new temporary label for that node. (If the old temporary label is still minimal, then it will remain unchanged during this step.)

Step II. Select the minimum of all the temporary labels, and declare it permanent. In case of ties, select any one of them (*but exactly one*), and declare it permanent. If this happens to be the sink node then terminate. Otherwise return to Step I.

To find the sequence of nodes in the shortest path from node 1 to node n, a label indicating the node from which each permanently labeled node was labeled should be available. Then by retracing the path backwards from the sink node to the source node, the minimal path may be constructed. An alternative method is to determine which nodes have permanent labels that differ by exactly the length of the connecting arc. Again by retracing the path backwards from n to 1, the shortest path may be found.

We illustrate Dijkstra's algorithm with an example.

EXAMPLE 3.5-1

Consider an undirected network shown in Fig. 3.11, where numbers along the arcs (i, j) represent distances between nodes i and j. Assume that the distance from i to j is the same as from j to i (i.e., all arcs are two-way streets). The problem is to determine the shortest distance and the length of the shortest path from node 1 to node 6.

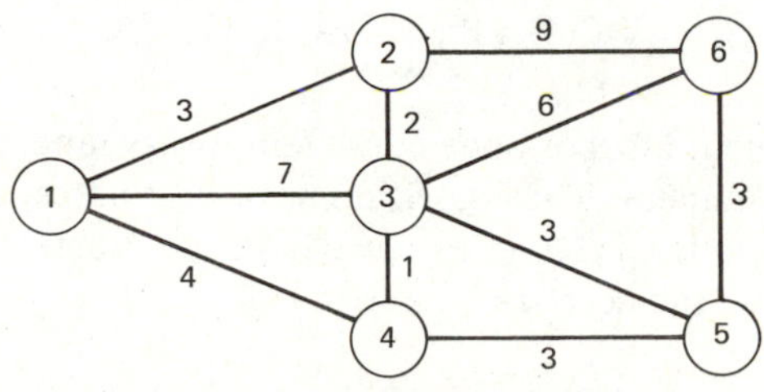

Figure 3.11

Solution

Initially node 1 is labeled permanently as zero, and all other nodes are given temporary labels equal to their direct distance from node 1. Thus the node labels at Step 0, denoted by $L(0)$, are

$$L(0) = [0, 3, 7, 4, \infty, \infty]$$

*

(An asterisk indicates a permanent label.)

At Step 1 the smallest of the temporary labels is made permanent. Thus node 2 gets a permanent label equal to 3, and it is the shortest distance from node 1 to node 2. To understand the logic behind this step, consider any other path from node 1 to node 2 through an intermediate node $j = 3, 4, 5, 6$. The shortest distance from node 1 to node j will be at least equal to 3 and d_{j2} is nonnegative, since all the distances are assumed to be nonnegative. Hence any other path from node 1 to node 2 cannot have a distance less than 3, and the shortest distance from node 1 to node 2 is 3. Thus at Step 1 the node labels are

$$L(1) = [0, 3, 7, 4, \infty, \infty]$$

* *

For each of the remaining nodes j ($j = 3, 4, 5, 6$), compute a number that is the sum of the permanent label of node 2 and the direct distance from node 2 to node j. Compare this number with the temporary label of node j, and the smaller of the two values becomes the new tentative label for node j. For example, the new temporary label for node 3 is given by

$$\text{minimum of } (3 + 2, 7) = 5$$

Similarly, for nodes 4, 5, and 6, the new temporary labels are 4, ∞, and 12, respectively. Once again the minimum of the new temporary labels is made permanent. Thus at Step 2, node 4 gets a permanent label as shown below:

$$L(2) = [0, 3, 5, 4, \infty, 12]$$

* *   *

Now using the permanent label of node 4, the new temporary labels of nodes 3, 5, and 6 are computed as 5, 7, and 12, respectively. Node 3 gets a permanent label and the node labels at Step 3 are

$$L(3) = [0, 3, 5, 4, 7, 12]$$

* * * *

It should be emphasized here that at each step, only the node that has been recently labeled permanent is used for further calculations. Thus at Step 4 the permanent label of node 3 is used to update the temporary labels of nodes 5 and 6 (if possible). Node 5 gets a permanent label and the node labels at Step 4 are

$$L(4) = [0, 3, 5, 4, 7, 11]$$

* * * * *

Using the permanent label of node 5, the temporary label of node 6 is changed to 10 and is made permanent. The algorithm now terminates, and the shortest distance from node 1 to node 6 is 10. In fact, we have the shortest distance from node 1 to every other node in the network as shown below:

$$L(5) = [0, 3, 5, 4, 7, 10]$$

* * * * * *

To determine the sequence of nodes in the shortest path from node 1 to node 6, we work backwards from node 6 . Node j ($j = 1, 2, 3, 4, 5$) precedes node 6 if the difference between the permanent labels of nodes 6 and j equals the length of the arc from j to 6. This gives node 5 as its immediate predecessor. Similarly node 4 precedes node 5, and the immediate predecessor of node 4 is node 1. Thus the shortest path from node 1 to node 6 is 1→4→5→6.

REMARK

If we are interested in finding the shortest path between every pair of nodes in the network, then we have to repeat Dijkstra's algorithm four times taking nodes 2, 3, 4, 5 as the source node.

Applications

A number of seemingly different problems can be formulated as shortest route problems for solution. We shall present two applications here.

An Equipment Replacement Problem. Most equipment requires increased maintenance and running costs with age. By replacing the equipment at frequent intervals, this cost could be reduced. But this reduction is achieved at the expense of increased capital costs incurred during every replacement. One of the most important problems faced by the management is to decide how often to replace the equipment so as to minimize the total costs, which include the capital cost, the cost of maintenance, and running costs. This can be formulated as a shortest route problem in a directed network, and Dijkstra's algorithm can be used for its solution.

As an illustration, consider a company planning its equipment replacement during the next 5 years. Let K_j represent the purchase price of the equipment in year j, and S_j the salvage value after j years of use. The maintenance and running cost of the equipment during its jth year of operation is c_j. Since costs increase with the age of the equipment, we assume $c_{j+1} > c_j$ for all $j = 1, 2, 3, 4, 5$. To formulate the problem of determining the optimal replacement policy as a shortest route problem, we construct a directed network as shown in Fig. 3.12.

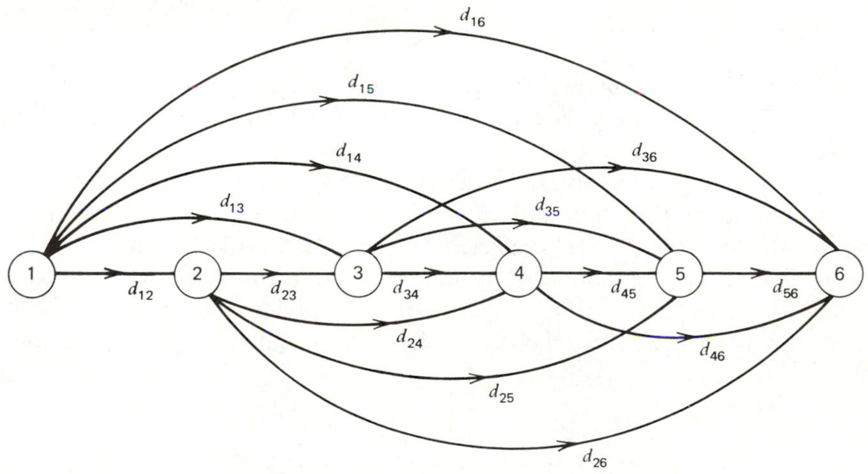

Figure 3.12
An equipment replacement problem as a shortest route problem.

Nodes 1 and 6 represent the start and the end of the planning period. Each intermediate node j ($j=2, 3, 4, 5$) represents the beginning of year j (or the end of year $j-1$) where an equipment replacement is possible. From every node i there exists a connecting (directed) arc to node j only if $j>i$. This corresponds to the situation where having replaced the equipment in year i, the next replacement is possible only in the later years.

The distance between node i and node j is of the form:

$$d_{ij}=K_i-S_{j-i}+\sum_{t=1}^{j-i} c_t \qquad \text{for } j>i$$
$$=\infty \qquad \text{for } j\leq i$$

The distance function d_{ij} represents the purchase cost minus any salvage value plus the maintenance and running costs of the equipment that is purchased at the start of year i and is to be replaced at the beginning of year j. (A value of ∞ indicates that there exists no arc from i to j.)

Every path from node 1 to node 6 in the network represents a possible replacement policy for the company. For example, the path ①→②→③→④→⑤→⑥ corresponds to replacing the equipment every year so that the total cost of this policy $=\sum_{i=1}^{5} K_i-5S_1+5c_1$. Another policy is to use the same equipment all 5 years, which corresponds to the path ①→⑥, using the arc (1, 6). The cost of this policy is given by

$$d_{16}=K_1-S_5+\sum_{t=1}^{5} c_t$$

Thus determining the shortest path from node 1 to node 6 is equivalent to finding the minimal cost policy for the equipment replacement problem.

Compact Book Storage in Libraries (10). Consider the problem of storing books by their size. Suppose the heights and thicknesses of all books in a collection are given. Let the book heights be arranged in ascending order of its n known heights $H_1, H_2, \ldots, H_n$:

$$H_1<H_2<\cdots<H_n$$

(Any book of height H_i can be shelved in a shelf of height $\geq H_i$.) Since the thickness of each book is known, the required length of shelving of each height class i can be computed and is denoted by L_i.

If the books are stored upright using only one shelf height (corresponding to the tallest book) for the whole collection, then the total shelf area needed is the product of the total length, and the height of the tallest book. Instead, if the collection is divided by height into two or more groups, it can be easily seen that the total shelf area needed will be less than that of the undivided collection.

The cost of constructing shelves of different heights and lengths is given below:

For each shelf height H_i,

$K_i=$ fixed cost independent of the shelf area
$C_i=$ variable cost per unit area

For example, let the collection be placed in two different shelves of heights H_m and H_n ($H_m < H_n$) (i.e., books of height H_m or less are placed in shelf H_m, and the rest in shelf H_n). Then the total cost of shelving the collection will be

$$\left[K_m + C_m H_m \sum_{i=1}^{m} L_i\right] + \left[K_n + C_n H_n \sum_{i=m+1}^{n} L_i\right]$$

The problem is to determine the optimal set of shelf heights, and their respective lengths, which will minimize the total shelving cost.

We will show here that the compact book storage problem can be formulated as a network flow problem. Consider a directed network of $(n+1)$ nodes $(0, 1, 2, \ldots, n)$ where the nodes correspond to the various book heights in the collection. (Here, node 0 corresponds to height zero and node n to the height of the tallest book.) A distance function on the set of paths connecting node 0 to node n will be proposed in terms of the shelving cost, where each intermediate node in the path is a possible partition of the set of all shelf heights. To make the network model compatible with the storage problem, the following assumptions are made:

1. $0 = H_0 < H_1 < H_2 < \cdots < H_n$
2. From every node i, there exists a connecting (directed) arc to node j, only if $j > i$. This corresponds to the situation in the book storage problem where having chosen a shelf of height H_i, the next shelf must be of height greater than H_i.

Because of this assumption, the number of arcs in the network is $n(n+1)/2$.

3. The distance function between node i and node j is of the form:

$$\begin{aligned} d_{ij} &= K_j + C_j H_j \sum_{k=i+1}^{j} L_k \qquad &\text{for } j > i \\ &= +\infty \qquad &\text{for } j \leq i \end{aligned}$$

The distance function between i and j represents the fixed partition cost K_j plus the cost of shelving books of height H_j or less (but greater than H_i) in the shelf of height H_j. (A value of $+\infty$ indicates that there exists no connecting arc between those two nodes.)

It may be seen clearly that as a consequence of the Assumptions 1, 2, and 3, finding a shortest path between the source node 0 and the sink node n in the above network is equivalent to determining the number of different shelves and their respective heights which minimizes the shelving cost for a given collection. For example, a minimum path solution of the form $(0, 5, 10, n)$ means, to go from node 0 to node n, the shortest route is to use the intermediate nodes 5 and 10. This says then, store all the books of height H_5 or less on the shelf of height H_5, books of height H_{10} or less (but over H_5) on the shelf of height H_{10} and the rest on the shelf of height H_n. The shortest distance between 0 and n in the network is the sum of $d_{0,5}$, $d_{5,10}$, and $d_{10,n}$. From Assumption 3, it is equivalent to the total cost of shelving the collection using three shelves of height H_5, H_{10}, and H_n.

3.6 MINIMAL SPANNING TREE PROBLEMS

Consider a network problem where our primary interest is to select a set of arcs such that (i) a *path* exists from any node to any other node and (ii) there is no *cycle* in the network. Such a collection of nodes is called a *spanning tree*. For example, for the network shown in Fig. 3.11, the arcs (1,2), (1,3), (2,6), (1,4), (4,5) form a spanning tree. Note that the spanning tree has the property that the addition of one more arc to the tree will result in a cycle.

We define the *length* of a spanning tree as the sum of the distances of all arcs in the tree. For example, for the tree identified for Fig. 3.11, the length of that tree is computed as 26. Since there is more than one spanning tree for a given network, the one with the least distance is called the *minimal spanning tree*.

Algorithm

The minimal spanning tree problem can be solved by a simple and straightforward procedure like the following:

Step I. Select an arbitrary node initially. Identify a node that is "closest" to the selected node and include the arc connecting these two nodes in the spanning tree.

Step II. Out of the remaining unconnected nodes, determine the one that is closest to a node already selected in the spanning tree. Include the arc connecting these two nodes in the spanning tree. Whenever there is a tie for the closest node, it is broken arbitrarily.

Step II is repeated until all the nodes in the network have been selected in the spanning tree.

EXAMPLE 3.6-1

Consider a network of 6 nodes and 10 arcs given in Fig. 3.13.

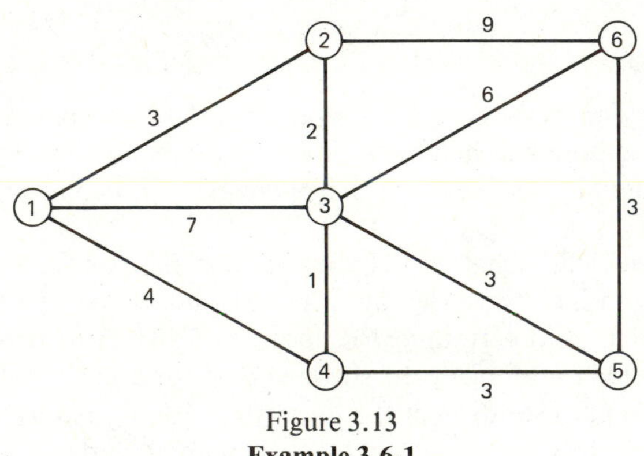

Figure 3.13
Example 3.6-1

To determine the minimal spanning tree, we proceed as follows:

Step 0. Let the set G denote the arcs in the minimal spanning tree.

Step I. Choose node 1 as the initial node. Since node 2 is the closest node, arc (1, 2) becomes part of the spanning tree. Hence, $G = \{(1, 2)\}$.

Step II. Consider the arcs connected to nodes 1 and 2. Choose the node that is closest to either node 1 or node 2. At this step, node 3 will be selected, arc (2, 3) will be added to the spanning tree, and $G = \{(1, 2), (2, 3)\}$.

Step III. Considering the arcs connected to nodes 1, 2, and 3, we find that arc (3, 4) is closest to node 3. Thus, $G = \{(1, 2), (2, 3), (3, 4)\}$.

Step IV. Both arcs (3, 5) and (4, 5) are closest to the arcs in the set G. Choosing arc (3, 5) arbitrarily, we get $G = \{(1, 2), (2, 3), (3, 4), (3, 5)\}$.

Step V. Arc (5, 6) is selected at this step and the minimal spanning tree is given by $G = \{(1, 2), (2, 3), (3, 4), (3, 5), (5, 6)\}$.

Figure 3.14 illustrates the minimal spanning tree graphically.

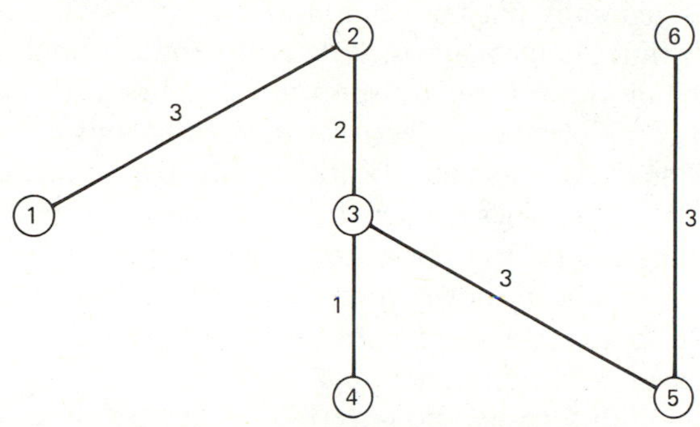

Figure 3.14
Minimal spanning tree for Example 3.6-1.

Applications

Consider a communication network where we want to install phone lines between cities such that every city is connected to every other city either directly or indirectly through other cities. Given the cost of installing the phone lines, the problem of selecting the pairs of cities for installing the direct lines at the lowest cost reduces to a minimal spanning tree problem. The same concept can be applied to the planning of a road network, railroad tracks, and the like.

3.7 PROJECT MANAGEMENT

Management of big projects that consist of a large number of activities pose complex problems in *planning, scheduling*, and *control*, especially when the project activities have to be performed in a specified technological sequence. With the help of *PERT* (Program Evaluation and Review Technique), and *CPM* (Critical Path Method), the project manager can

1. Plan the project ahead of time and foresee possible sources of troubles and delays in completion.
2. Schedule the project activities at the appropriate times to conform with proper job sequence so that the project is completed as soon as possible.
3. Coordinate and control the project activities so as to stay on schedule in completing the project.

Thus both PERT and CPM are aids to efficient project management. They differ in their approach to the problem and the solution technique. The nature of the project generally dictates the proper technique to be used.

Origin and Use of PERT

PERT was developed in the U.S. Navy during the late 1950s to accelerate the development of the Polaris Fleet Ballistic Missile. The development of this weapon involved the coordination of the work of thousands of private contractors and other government agencies. The coordination by PERT was so successful that the entire project was completed 2 years ahead of schedule. This has resulted in further applications of PERT in other weapons development programs in the Navy, Air Force, and Army. Nowadays it is extensively used in industries and other service organizations as well.

The time required to complete the various activities in a research and development project is generally not known *a priori*. Thus PERT incorporates uncertainties in activity times in its analysis. It determines the probabilities of completing various stages of the project by specified deadlines. It also calculates the expected time to complete the project. An important and extremely useful by-product of PERT analysis is its identification of various "bottlenecks" in a project. In other words, it identifies the activities that have high potential for causing delays in completing the project on schedule. Thus, even before the project has started, the project manager knows where he or she can expect delays. The manager can then take the necessary preventive measures to reduce possible delays so that the project schedule is maintained.

Because of its ability to handle uncertainties in job times, PERT is mostly used in research and development projects.

Origin and Use of CPM

Critical Path Method closely resembles PERT in many aspects but was developed independently by E. I. du Pont de Nemours Company. Actually, both techniques, PERT and CPM, were developed almost simultaneously. The major difference between the two techniques is that CPM does not incorporate uncertainties in job times. Instead it assumes that activity times are proportional to the amount of resources allocated to them, and by changing the level of resources the activity times and the project completion time can be varied. Thus CPM assumes prior experience with similar projects from which the relationships between resources and job times are available. CPM then evaluates the trade-off between project costs and project completion time.

CPM is mostly used in construction projects where there is prior experience in handling similar projects.

Applications of PERT and CPM

A partial list of applications of PERT and CPM techniques in project management is as follows:

1. Construction projects (e.g., buildings, highways, houses, and bridges).
2. Preparation of bids and proposals for large projects.
3. Maintenance planning of oil refineries, ship repairs, and other large operations.
4. Development of new weapons systems and new manufactured products.
5. Manufacture and assembly of large items such as airplanes, ships, and computers.
6. Simple projects such as home remodeling, moving to a new house, and home cleaning and painting.

Project Network

Analysis by PERT/CPM techniques uses the network formulation to represent the project activities and their ordering relations. Construction of a project network is done as follows:

1. Arcs in the network represent individual jobs in the project.

2. Nodes represent specific points in time which mark the completion of one or more jobs in the project.
3. Direction on the arc is used to represent job sequence. It is assumed that any job directed toward a node must be completed before any job directed away from that node can begin.

We illustrate the construction of project networks with a few examples.

EXAMPLE 3.7-1

Consider seven jobs *A, B, C, D, E, F,* and *G* with the following job sequence:

Job *A* precedes *B* and *C*
Jobs *C* and *D* precede *E*
Job *B* precedes *D*
Jobs *E* and *F* precede *G*

The project network is shown below:

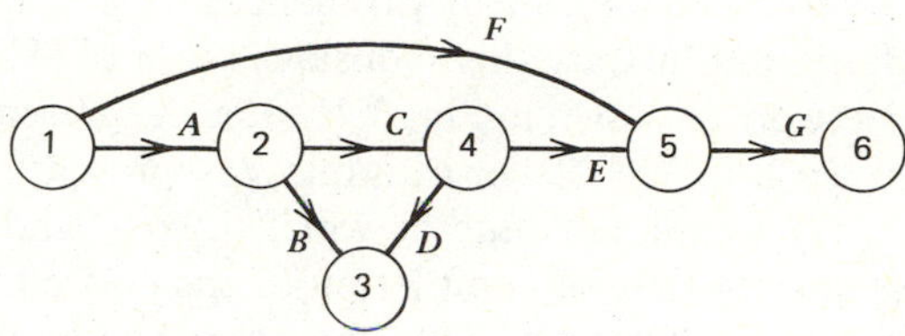

In the above network every arc (i, j) represents a specific job in the project. Node 1 represents the start of the project, whereas node 6 denotes the project's completion time. The intermediate nodes represent the completion of various stages of the project. The nodes of the project network are generally called *events*.

Definition. An *event* is a specific point in time that marks the completion of one or more activities, well recognizable in the project.

EXAMPLE 3.7-2

Consider a project with five jobs *A, B, C, D,* and *E* with the following job sequence:

Job *A* precedes *C* and *D*
Job *B* precedes *D*
Jobs *C* and *D* precede *E*

The completion times for *A, B, C, D,* and *E* are 3, 1, 4, 2, 5 days, respectively. The project network is shown in Fig. 3.15.

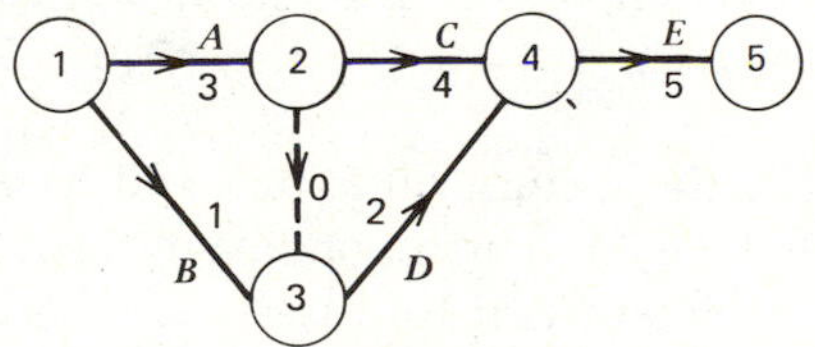

Figure 3.15
Project network for Example 3.7-2.

Arc (2, 3) (dotted line in Fig. 3.15) represents a *dummy job* that does not exist in reality in the project. The dummy job is necessary so as to avoid ambiguity in the job sequence. The completion time of the dummy job is always zero, and it is added in the project network whenever we want to avoid an arc (i, j) representing more than one job in the project. In Fig. 3.15, event 3 represents the completion of job B and the dummy job. Since the dummy job is completed as soon as A is completed, event 3 in essence marks the completion of jobs A and B.

Simplified Project Management Problem

The analysis of a simplified project management problem is useful to both PERT and CPM. In the simplified problem the completion times of all the project activities and their technological sequence are known. The management wants to determine the minimum time in which the project can be completed, and to identify the crucial jobs whose delay can delay the entire project.

Solution by Linear Programming

The simplified project management problem can be solved by formulating it as a linear program. To illustrate this, consider the problem given in Example 3.7-2. The project network is shown in Fig. 3.15. Let t_i represent the time at which event i occurs where $i = 1, 2, 3, 4, 5$. For example, t_5 represents when the project is completed, while t_4 represents the time at which jobs C and D are completed. Thus $(t_5 - t_1)$ represents the time of completion of the entire project, and the objective is to minimize this duration. The linear programming formulation becomes:

$$\begin{array}{lrl} \text{Minimize:} & Z = t_5 - t_1 & \\ \text{Subject to:} & t_2 - t_1 \geqq 3 & \\ & t_3 - t_1 \geqq 1 & \\ & t_3 - t_2 \geqq 0 & \\ & t_4 - t_2 \geqq 4 & \\ & t_4 - t_3 \geqq 2 & \\ & t_5 - t_4 \geqq 5 & \\ & t_i \geqq 0 & \text{for all } i = 1, 2, \ldots, 5 \end{array}$$

We have one constraint for every arc in the project network. The constraint for arc (i, j),

$$t_j - t_i \geqq t_{ij}$$

ensures that the time *available* for completing job (i, j) should be greater than or equal to the time *required* to complete job (i, j).

The above linear program may be solved by the simplex method, and the optimal value of Z gives the minimum completion time for the project. Actually, the linear programming problem can be solved by inspection by setting $t_1 = 0$, and choosing the values of t_i as small as possible to satisfy the constraints. Thus, an optimal solution by inspection is $t_1 = 0$, $t_2 = 3$, $t_3 = 3$, $t_4 = 7$, $t_5 = 12$, and minimum $Z = t_5 - t_1 = 12$ days.

Now we can identify those constraints that will be satisfied as equations in the optimal solution, and the jobs corresponding to those constraints are the *critical jobs*. For example, the constraint for arc (1, 2) is satisfied as an equation by $t_2 = 3$ and $t_1 = 0$. Hence, A is a critical job. For arc (1, 3), the corresponding constraint is a strict inequality since $t_3 - t_1 = 3 - 0 > 1$. Hence, B is not a critical job, and hence in Example 3.7-2 the critical jobs are A, C, and E, while jobs B and D are not critical.

Definition. A path in the project network connecting the starting event (node) and the ending event such that it passes through the critical jobs is called a *Critical Path*.

In Fig. 3.15, the critical path is given by

It can be shown that finding a critical path in a project network is equivalent to finding the *longest path* in the network.

For a large project, the number of constraints will be too many for an easy solution by linear programming. A direct approach using the project network is available to solve the simplified project management problem.

Solution by Network Analysis

Definition. The *earliest time* of node j, denoted by U_j, is the earliest time at which event j can occur. We know that event j can occur as soon as all the jobs (arcs) directed towards node j are completed. For example, in the following diagram, event j occurs as soon as jobs A, B, and C are completed.

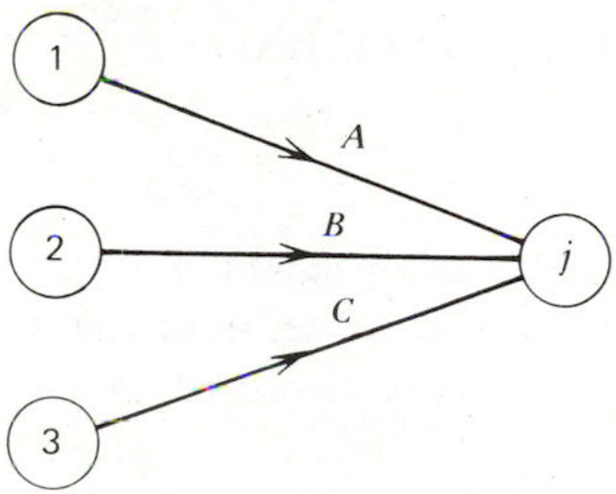

The earliest time of node j is then given by, $U_j = \max\,(U_1 + t_{1j},\ U_2 + t_{2j},\ U_3 + t_{3j})$ where t_{1j}, t_{2j}, t_{3j} are the completion times of jobs A, B, and C. Thus the general formula for calculating U_j is

$$U_j = \max_i\,(U_i + t_{ij})$$

where the index i ranges over all nodes for which arc (i, j) exists, and t_{ij} is the completion time of the job represented by arc (i, j). Note that the earliest time of the last event in the project network gives the earliest time of completing the project.

For the project network shown in Fig. 3.15, the U_j are calculated as follows:

Set

$$U_1 = 0$$

Then,

$$\begin{aligned}
U_2 &= U_1 + t_{12} = 3 \\
U_3 &= \max\,[(U_2 + t_{23}),\ (U_1 + t_{13})] \\
&= \max\,(3,\ 1) = 3 \\
U_4 &= \max\,[(U_2 + t_{24}),\ (U_3 + t_{34})] \\
&= \max\,(7,\ 5) = 7 \\
U_5 &= U_4 + t_{45} = 12
\end{aligned}$$

Hence, the minimum duration of the project is 12 days from the start. Of course, this does not identify the critical jobs nor the critical path. For that we need to calculate the *latest time* of an event.

Definition. The *latest time* of node i, denoted by V_i, is the latest time at which event i can occur without delaying the completion of the project beyond its earliest time.

To illustrate the concept of latest time, consider the following diagram:

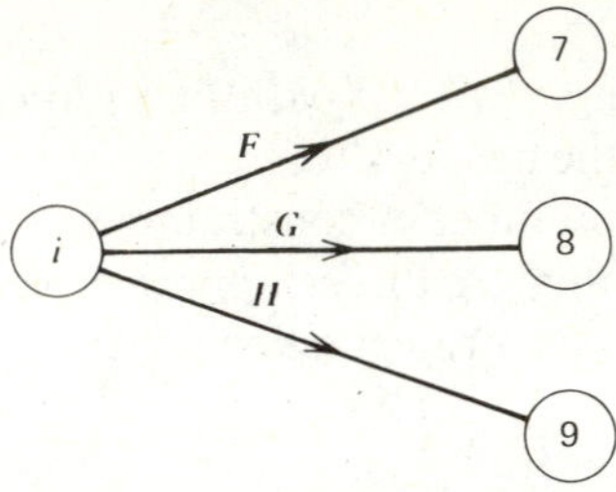

The project will not be delayed if the three jobs F, G, and H are completed by V_7, V_8, V_9, respectively. This is possible if we let

$$V_i = \min\,[V_7 - t_{i7},\ V_8 - t_{i8},\ V_9 - t_{i9}]$$

Hence the general formula to calculate V_i becomes

$$V_i = \min_j\,(V_j - t_{ij})$$

where the index j ranges over all nodes for which arc (i, j) exists.

To calculate the latest times of the events in the project network (Fig. 3.15), we set the latest time of the last event equal to its earliest time and work backwards.

Thus,

$$\begin{aligned}
V_5 &= U_5 = 12 \\
V_4 &= V_5 - t_{45} = 7 \\
V_3 &= V_4 - t_{34} = 5 \\
V_2 &= \min\,[(V_4 - t_{24}), (V_3 - t_{23})] \\
&= \min\,(3, 5) = 3 \\
V_1 &= \min\,[(V_2 - t_{12}), (V_3 - t_{13})] \\
&= \min\,(0, 4) = 0
\end{aligned}$$

The difference between the latest time and the earliest time of an event is called the *slack time* of that event. The slack time denotes how much delay can be tolerated in reaching that event without delaying the project completion date.

For the project network shown in Fig. 3.15, the slack times of events 1, 2, 3, 4, and 5 are given by 0, 0, 2, 0 and 0, respectively. Those events that have zero slack times are the *critical events*, where every care must be taken to stay on schedule if the project is to be completed on time.

The *critical path* for a project is a path through the project network such that the jobs on this path have zero slack times. The *critical jobs* are the arcs (jobs) in the critical path.

In Example 3.7-2, the critical path is 1→2→4→5, and the critical jobs are A, C, and E.

Table 3.6 summarizes the results of network analysis for Example 3.7-2.

Table 3.6
RESULTS OF NETWORK ANALYSIS OF EXAMPLE 3.7-2

Event	Earliest Time	Latest Time	Slack Time	Remark
1	0	0	0	Critical
2	3	3	0	Critical
3	3	5	2	Noncritical
4	7	7	0	Critical
5	12	12	0	Critical

To prepare a *project schedule* in terms of the activities, it is essential to have the starting time and the ending time of all jobs. From the event times it is possible to get the following information on each one of the activities in the project:

1. The earliest starting time.
2. The latest starting time.
3. The earliest finishing time.
4. The latest finishing time.
5. The slack time.

To illustrate this, consider an arc (i,j) representing job J in the project as shown below:

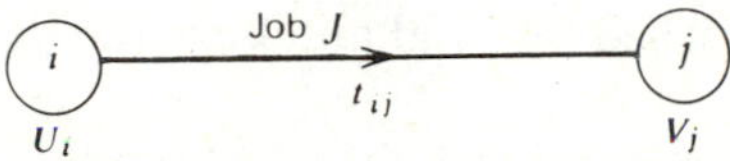

Let U_i denote the earliest occurrence time of event i, V_j denote the latest occurrence time of event j, and t_{ij} be the completion time of job J. It is clear that job J may be started as early as U_i, but must be completed no later than V_j. Thus $(V_j - t_{ij})$ gives the latest starting time of job J, and $(U_i + t_{ij})$ is the earliest completion time of job J. In other words, $(V_j - U_i)$ is the maximum time available for job J. Thus the *slack time* of job J (maximum delay in completion) is given by $(V_j - U_i - t_{ij})$. If job J is a critical job then $V_j - U_i = t_{ij}$.

Table 3.7 gives the project schedule for the network shown in Fig. 3.15.

Table 3.7
PROJECT SCHEDULE FOR EXAMPLE 3.7-2

Job	Expected Duration (days)	Earliest Start	Latest Start	Earliest Finish	Latest Finish	Slack Time (Maximum Delay)	Remark
A	3	0	0	3	3	0	Critical
B	1	0	4	1	5	4	Noncritical
C	4	3	3	7	7	0	Critical
D	2	3	5	5	7	2	Noncritical
E	5	7	7	12	12	0	Critical

Critical Path Method (CPM)

The basic assumption in CPM is that the activity times are proportional to the level of resources allocated to them. By assigning additional resources (capital, people, materials, and machines) to an activity, its duration can be reduced to a certain extent. Shortening the duration of an activity is known as *crashing* in the CPM terminology. The additional cost incurred in reducing the activity time is called the *crashing cost*.

We learned earlier that the duration of the critical activities determines the project completion time. Thus by crashing the critical jobs, the project duration can be reduced. Of course, crashing the critical activities increases the total direct cost of the project, but the reduction in project duration may result in other advantages or returns that may offset this increased cost. These may include indirect costs such as equipment rental, supervisory personnel, supplies, and other costs that are directly proportional to the project duration. There may be other economic benefits in completing the project ahead of schedule. For instance, a new product may capture a larger share of the market if it is introduced before its competitor's. In addition some project contracts may specify bonuses or penalties for completing the project sooner or later than stipulated. Thus the total cost of the project is the sum of the direct costs (proportional to the activity times) and the indirect costs (proportional to the project duration). The critical path Method essentially studies the trade-off between the total cost of the project and its completion time.

For the critical path analysis it is assumed that every job has a *normal completion time* (maximum time) if no additional resources were assigned, and a *crash completion time* (minimum time) with the maximum amount of resources. In addition, a cost versus time relationship is available for every job in the project. The project management problem is to determine the amount by which the various jobs are to be crashed that will minimize the total cost of the project (direct and indirect).

There are two basic approaches to the CPM problem:

1. An Enumerative Method (for small projects only).
2. Mathematical Programming Methods (for large projects).

The Enumerative Method

For small projects one could determine the optimal project schedule by this method. The basic idea behind this approach is that the project length can be reduced by reducing the duration of the critical jobs. Hence, the critical jobs are crashed as long as the cost of crashing is less than the reduction in overhead costs. The main difficulty with this approach is that the critical path of the project changes once one starts crashing the critical jobs. There are other practical problems in applying this method to a large project, and these are discussed at the end of this section. Let us illustrate the enumerative method with an example.

EXAMPLE 3.7-3

Consider a project consisting of eight jobs (A, B, C, D, E, F, G, H). About each job we know the following:

Job	Predecessors	Normal Time (days)	Crash Time (days)	Cost of Crashing per Day ($)
A	—	10	7	4
B	—	5	4	2
C	*B*	3	2	2
D	*A, C*	4	3	3
E	*A, C*	5	3	3
F	*D*	6	3	5
G	*E*	5	2	1
H	*F, G*	5	4	4

Given the overhead costs as $5 per day, we want to determine the optimal duration of the project in terms of both the crashing and overhead costs and to develop an optimal project schedule.

Solution

The project network is shown in Fig. 3.16.

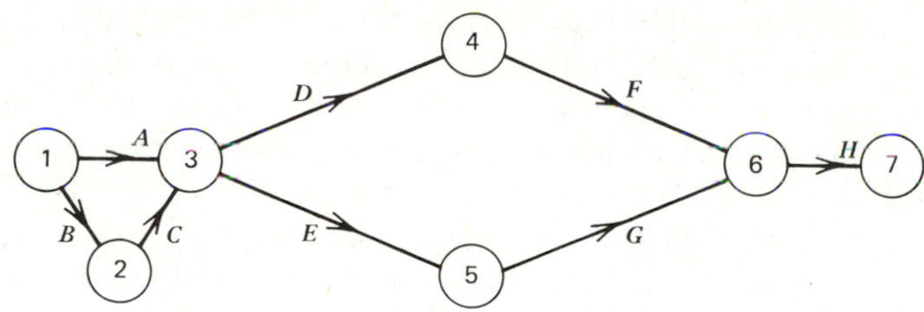

Figure 3.16
Project network for Example 3.7-3.

If all the jobs are done at their normal times, the project duration (length of the longest path) is 25 days. Hence, under a "no crashing" schedule,

$$\text{Total cost} = \text{overhead costs} + \text{crashing costs} = \$5(25) + 0 = \$125.$$

If all the jobs were crashed to their minimum time, then the project duration is 17 days. Under this schedule, the total cost $= \$5(17) + 47 = \$132$. The project management problem is to determine the optimal duration of jobs that will minimize the total cost.

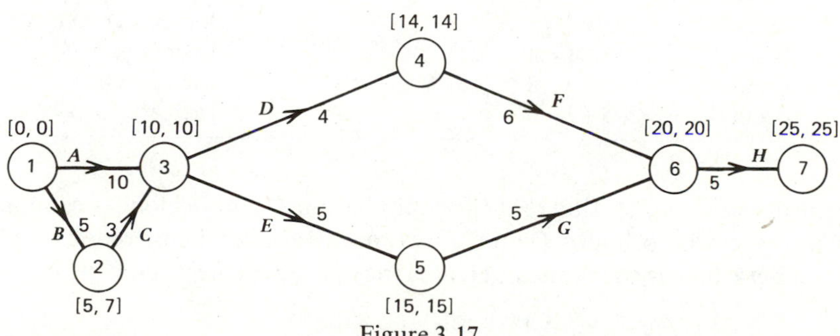

Figure 3.17

Using the normal times for the jobs, the earliest and latest occurrence times of the events (U_i, V_i) are computed as shown in Fig. 3.17. The project completion time is 25 days, and the project network has two critical paths as shown below:

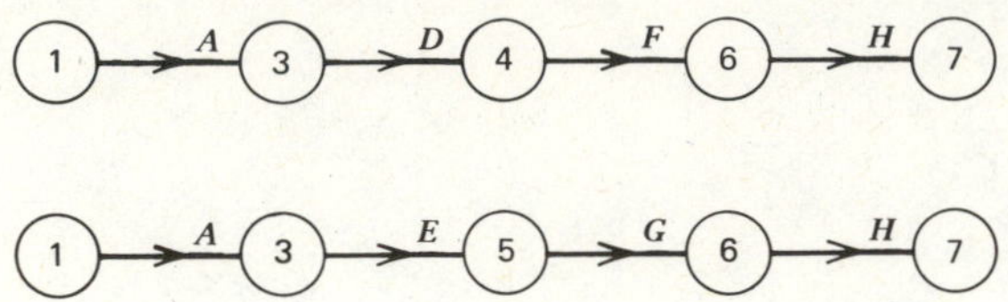

Between the nodes 3 and 6 we have *parallel critical paths*. The critical activities are jobs *A, D, E, F, G,* and *H*. The total cost of the project under normal time is $125.

To reduce the project duration, it is necessary to reduce the duration of the critical jobs. Consider the critical job *H*, which can be crashed by 1 day at a cost of $4. This reduces the project duration by 1 day at a savings of $5 in overhead costs. Hence, job *H* is crashed to its lowest limit of 4 days and the total cost is reduced to $124.

Consider the crashing of job *A* now. This also results in a net savings of $1 in total cost for each day of crashing. But job *A* cannot be crashed to its minimum value of 7 days because when *A* is crashed to 8 days, jobs *B* and *C* also become critical. This results in parallel critical paths between nodes 1 and 3, and any reduction in *A* alone will not reduce the project duration. Hence, job *A* is crashed only to 8 days, and the total cost is reduced to $122.

To reduce the project duration by one more day, we must crash job *A* by 1 day and either jobs *B* or *C* by 1 day. The total cost of crashing *A* and *B* is $6, which is more than the savings in overhead costs. Similarly crashing *A* and *C* is also not economical.

Now consider the critical jobs *D, E, F,* and *G*. Since we have parallel critical paths between the nodes 3 and 6, we have to crash one job in the path

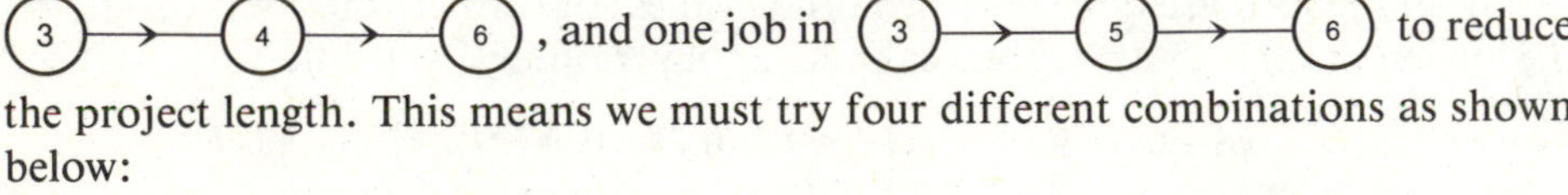

3 → 4 → 6, and one job in 3 → 5 → 6 to reduce the project length. This means we must try four different combinations as shown below:

Jobs	Increase in Crashing Costs ($)	Decrease in Overhead Costs ($)	Net Change in Total Cost ($)
D and *E*	$3+3=6$	5	Increases by $1
D and *G*	$3+1=4$	5	Decreases by $1
F and *E*	$5+3=8$	5	Increases by $3
F and *G*	$5+1=6$	5	Increases by $1

From the above table, we find that the combination *D* and *G* alone is economical, and when we crash both jobs *D* and *G* by one day, the total cost reduces to $121. No further crashing is economical. Hence, the optimal project schedule is

Crash job *A* to 8 days
Crash job *D* to 3 days

Crash job G to 4 days
Crash job H to 4 days

Jobs B, C, E, and F are completed at normal times 5, 3, 5, and 6 days, respectively. The optimal length of the project is 21 days, and the minimum project cost is $121.

Very large projects will generally contain many parallel critical paths, and each critical path may have a large number of jobs. Examining all possible combinations of jobs in the parallel paths by the enumerative method will not only be inefficient but also expensive. For example, if we have two parallel paths with 10 critical jobs in each, we will have to examine 100 combinations of jobs for possible crashing!

Mathematical Programming Methods

For large projects, the mathematical programming methods are more efficient in determining the optimal project schedule. In this section we shall discuss some linear programming models for the critical path analysis. Once again, it is assumed that a cost-versus-time relationship is available for every job in the project as shown in Fig. 3.18. We denote by k_{ij} the normal completion time of job (i, j) if no additional resources were assigned, while 1_{ij} denotes the crash completion time with the maximum amount of resources. C_{ij} represents the unit cost of shortening the duration of job (i, j). If t_{ij} is the completion time of job (i, j), then t_{ij} is an unknown variable between 1_{ij} and k_{ij}, and the cost of crashing is given by

$$C_{ij}(k_{ij} - t_{ij})$$

Let t_i be the unknown event times ($i = 1, 2, \ldots, n$) for a project consisting of n events where events 1 and n denote the start and the end of the project.

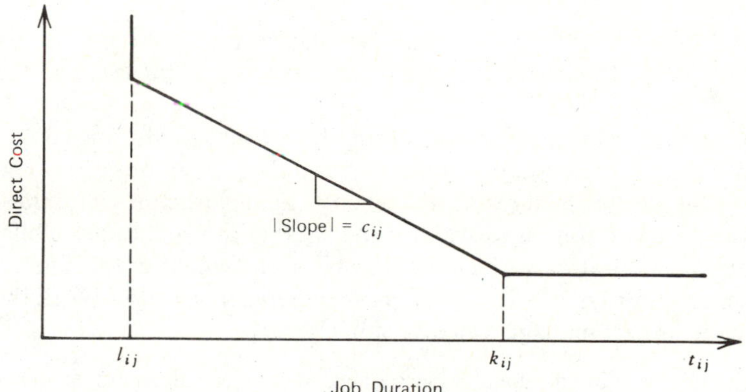

Figure 3.18
Job duration versus direct cost.

We shall now develop three important models in the critical path analysis that are useful for project management. In all these models, we will assume that the normal time, crash time, and the crashing cost are available for all the activities in the project.

Model I. Given that the project must be completed by time T, we want to determine how the project activities are to be expedited such that the total cost of crashing is minimized. This problem can be formulated as a linear programming problem as follows:

$$\text{Minimize:} \qquad Z = \sum_{(i,j)} C_{ij}(k_{ij} - t_{ij})$$

$$\begin{aligned} \text{Subject to:} \qquad t_j - t_i &\geqq t_{ij} && \text{for all jobs } (i, j) \\ 1_{ij} \leqq t_{ij} &\leqq k_{ij} && \text{for all jobs } (i, j) \\ t_n - t_1 &\leqq T \\ t_i &\geqq 0 && \text{for all } i = 1, 2, \ldots, n \end{aligned}$$

The above problem may be solved by the simplex method. The optimal value of Z gives the minimum crashing cost. From the optimal values of t_{ij}, we can determine which jobs are expedited, and by how much. It should be pointed out here that for the above linear program to be feasible, the value of T must be greater than or equal to the length of the critical path with all the jobs at their crash (minimum) times.

Model II. Suppose an additional budget of B dollars is available for crashing the project activities. We want to determine how these additional resources may be allocated in the best possible manner so as to minimize the project completion time.

The linear programming model of this problem is given as follows:

$$\text{Minimize:} \qquad Z = t_n - t_1$$

$$\begin{aligned} \text{Subject to:} \qquad t_j - t_i &\geqq t_{ij} && \text{for all jobs } (i, j) \\ 1_{ij} \leqq t_{ij} &\leqq k_{ij} && \text{for all jobs } (i, j) \\ \sum_{(i,j)} C_{ij}(k_{ij} - t_{ij}) &\leqq B \\ t_i &\geqq 0 && \text{for all } i = 1, 2, \ldots, n \end{aligned}$$

The solution to this linear program gives the least project duration that can be achieved by the additional budget B, the activities to be crashed, and their durations.

By using the linear programming Model I or II repeatedly, one could obtain a relationship between the total crashing cost and the project duration. Figure 3.19 gives a typical plot of the direct (activity) costs against the project duration. $T_{\max}$ denotes the project duration with all the jobs at their normal times, while $T_{\min}$ denotes the project duration with all the jobs reduced to their crash times. The cost function shown in Fig. 3.19 is called a *piecewise linear function*. With the help of this curve the project manager can determine

1. The minimum cost of additional resources needed to meet a given project deadline.
2. The optimal allocation of scarce resources to achieve the maximum reduction in project duration.

As seen in Fig. 3.19, the direct cost of completing the project activities increases when the project duration is reduced. But the indirect costs discussed earlier reduce with a reduction in project duration. Hence, it will be of interest to study how the total cost (direct + indirect costs) varies with the project duration. For various project lengths the indirect cost is added to the direct cost, and a plot of points is obtained to get a relationship between the project length and the total project cost. Such a plot is shown in Fig. 3.20. This U-shaped curve is called a *project cost curve*. With the help of this curve, a project manager can select the

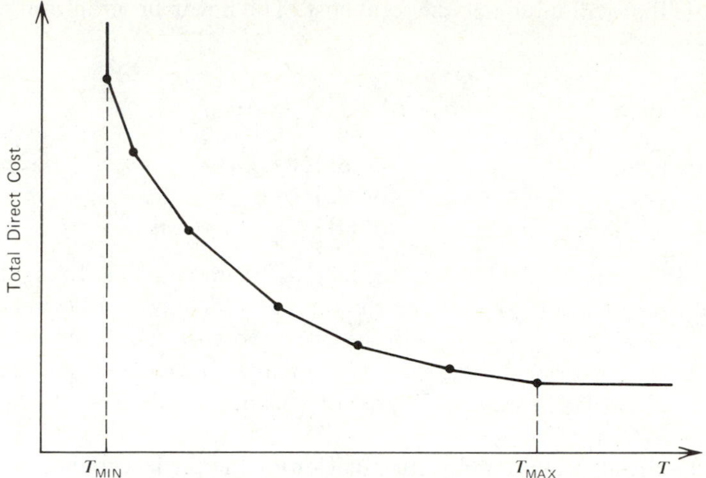

Figure 3.19
Direct cost versus project duration.

optimal project duration (T^*) that will minimize the total costs. Corresponding to the optimal value of T, he or she can then determine the optimal durations of all the jobs, the cost of crashing, and the critical path. From this information the *optimal project schedule* can be prepared.

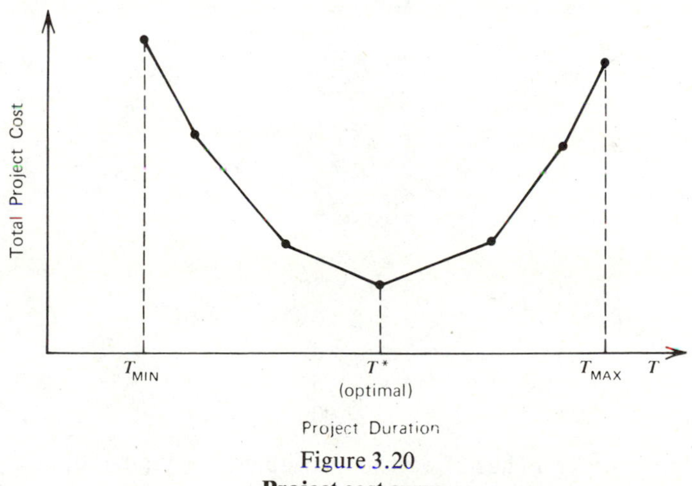

Figure 3.20
Project cost curve.

Model III. If the indirect cost of the project varies linearly with the project duration, then one can determine the optimal length of the project (T^*) and the optimal project schedule by solving a linear programming problem.

Let the indirect (overhead) costs, proportional to the project duration, be denoted by F per unit time. Then the indirect cost is given by $F(t_n - t_1)$, where $(t_n - t_1)$ is the unknown length of the project. The direct cost is given by $\sum_{(i,j)} C_{ij}(k_{ij} - t_{ij})$, where t_{ij} is the unknown length of job (i, j). The problem is to determine the optimal schedule that will minimize the total cost. The linear programming formulation becomes

$$\text{Minimize:} \quad Z = F(t_n - t_1) + \sum_{(i,j)} C_{ij}(k_{ij} - t_{ij})$$

$$\begin{aligned} \text{Subject to:} \quad t_j - t_i &\geqq t_{ij} && \text{for all jobs } (i, j) \\ 1_{ij} \leqq t_{ij} &\leqq k_{ij} && \text{for all jobs } (i, j) \\ t_i &\geqq 0 && \text{for all } i = 1, 2, \ldots, n \end{aligned}$$

We can illustrate Model III with the help of Example 3.7-3. Let t_{ij} denote the completion time of job (i, j) and t_i be the time at which event i occurs. Then the duration of the project is $(t_7 - t_1)$, and the overhead cost equals $5(t_7 - t_1)$. The direct costs are the costs of crashing each job, which are proportional to how far the jobs are expedited. For instance, the cost of crashing job A is $4(10 - t_{13})$ whereas for job B it is $2(5 - t_{12})$.

Thus the linear programming formulation of the project management problem becomes

$$\begin{aligned} \text{Minimize:} \quad Z = {} & 5(t_7 - t_1) + 4(10 - t_{13}) \\ & + 2(5 - t_{12}) + 2(3 - t_{23}) + 3(4 - t_{34}) \\ & + 3(5 - t_{35}) + 5(6 - t_{46}) + 1(5 - t_{56}) \\ & + 4(5 - t_{67}) \end{aligned}$$

$$\begin{aligned} \text{Subject to:} \quad & t_3 - t_1 \geqq t_{13} \\ & t_2 - t_1 \geqq t_{12} \\ & t_3 - t_2 \geqq t_{23} \\ & t_4 - t_3 \geqq t_{34} \\ & t_5 - t_3 \geqq t_{35} \\ & t_6 - t_4 \geqq t_{46} \\ & t_6 - t_5 \geqq t_{56} \\ & t_7 - t_6 \geqq t_{67} \\ & 7 \leqq t_{13} \leqq 10 \\ & 4 \leqq t_{12} \leqq 5 \\ & 2 \leqq t_{23} \leqq 3 \\ & 3 \leqq t_{34} \leqq 4 \\ & 3 \leqq t_{35} \leqq 5 \\ & 3 \leqq t_{46} \leqq 6 \\ & 2 \leqq t_{56} \leqq 5 \\ & 4 \leqq t_{67} \leqq 5 \\ & t_1, \ldots, t_7 \geqq 0 \end{aligned}$$

The above linear program has 15 decision variables. Setting $t_1 = 0$, an optimal solution is found by the simplex method as $t_2 = 5$, $t_3 = 8$, $t_4 = 11$, $t_5 = 13$, $t_6 = 17$, $t_7 = 21$, $t_{13} = 8$, $t_{12} = 5$, $t_{23} = 3$, $t_{34} = 3$, $t_{35} = 5$, $t_{46} = 6$, $t_{56} = 4$, and $t_{67} = 4$. The optimal project length is 21 days, and the minimum cost of the project is \$121. This means that job A is crashed by 2 days, while jobs D, G, and H are each crashed by 1 day. All the jobs in the project are critical.

Program Evaluation and Review Technique (PERT)*

Thus far in our analysis the probability considerations in the management of a project were not included. CPM assumed that the job times are known but can

*This section assumes some knowledge of probability theory. Appendix B contains a review of probability theory.

be varied by changing the level of resources. However, in all the research and development projects, many activities are performed only once. Hence, no prior experience with similar activities is available. The management of such projects is done by PERT, which takes into account uncertainties in the completion times of the various activities.

For each activity in the project network, PERT assumes three time estimates on its completion time. They include (1) *a most probable time* denoted by m, (2) *an optimistic time* denoted by a, and (3) *a pessimistic time* denoted by b.

The most probable time is the time required to complete the activity under normal conditions. To include uncertainties, a range of variation in job time is provided by the optimistic and pessimistic times. The optimistic estimate is a good guess on the minimum time required when everything goes according to plan whereas the pessimistic estimate is a guess on the maximum time required under adverse conditions such as mechanical breakdowns, minor labor troubles, or shortage of or delays in delivery of material. It should be remarked here that the pessimistic estimate does not take into consideration unusual and prolonged delays or other catastrophes. Because both these estimates are only qualified guesses, the actual time for an activity could lie outside this range. (From a probabilistic view point we can only say that the probability of a job time falling outside this range is very small.)

Most PERT analysis assumes a Beta distribution for the job times as shown in Fig. 3. 21, where μ represents the average length of the job duration. The value of μ depends on how close the values of a and b are relative to m.

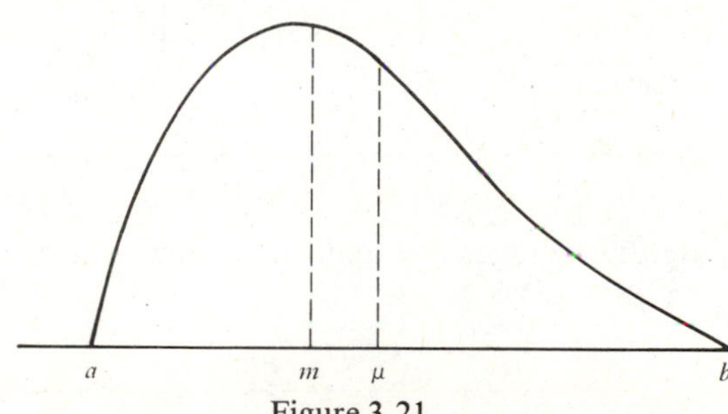

Figure 3.21
Beta distribution for job time.

The expected time to complete an activity is approximated as

$$\mu = \frac{a + 4m + b}{6} \tag{3.13}$$

Since the actual time may vary from its mean value, we need the variance of the job time. For most unimodal distributions (with single peak values), the end values lie within three standard deviations from the mean value. Thus the spread of the distribution is equal to six times the standard deviation value (σ).

Thus $6\sigma = b - a$, or $\sigma = (b - a)/6$. The variance of the job time equals

$$\sigma^2 = \left(\frac{b - a}{6}\right)^2 \tag{3.14}$$

With the three time estimates on all the jobs, PERT calculates the average time and the variance of each job using Eqs. 3.13 and 3.14. Treating the average times as the actual job times, the critical path is found. The duration of the project (T) is given by the sum of all the job times in the critical path. But the job times are random variables.* Hence, the project duration T is also a random variable, and we can talk of the average length of the project and its variance.

The expected length of the project is the sum of all the average times of the jobs in the critical path. Similarly, the variance of the project duration is the sum of all the variances of the jobs in the critical path assuming that all the job times are independent.

EXAMPLE 3.7-4

Consider a project consisting of nine jobs ($A, B, \ldots, I$) with the following precedence relations and time estimates:

Job	Predecessors	Optimistic Time (a)	Most Probable Time (m)	Pessimistic Time (b)
A	—	2	5	8
B	A	6	9	12
C	A	6	7	8
D	B, C	1	4	7
E	A	8	8	8
F	D, E	5	14	17
G	C	3	12	21
H	F, G	3	6	9
I	H	5	8	11

First we compute the average time and the variance for each one of the jobs. They are tabulated as follows:

Job	Average Time	Standard Deviation	Variance
A	5	1	1
B	9	1	1
C	7	1/3	1/9
D	4	1	1
E	8	0	0
F	13	2	4
G	12	3	9
H	6	1	1
I	8	1	1

Figure 3.22 gives the project network, where the numbers on the arcs indicate the average job times. Using the average job times, the earliest and latest times of each event are calculated. The critical path is found as $1 \rightarrow 2 \rightarrow 4 \rightarrow 5 \rightarrow 6 \rightarrow 7 \rightarrow 8$. The critical jobs are $A, B, D, F, H,$ and I.

*A variable whose value depends on chance.

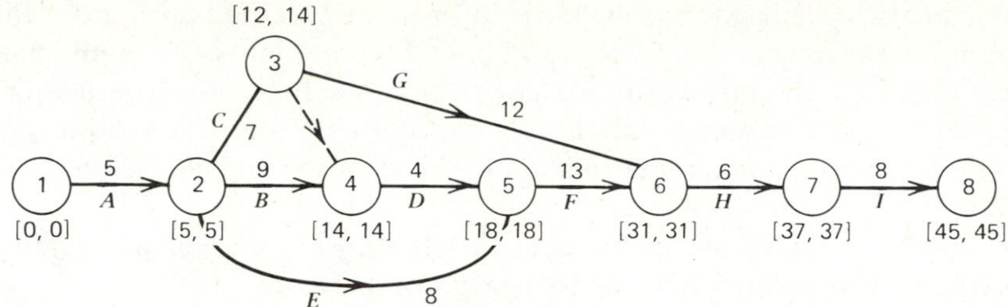

Figure 3.22
Project network of Example 3.7-4.

Let T denote the project duration. Then the expected length of the project is

$$\begin{aligned} E(T) &= \text{Sum of the expected times of jobs } A, B, D, F, H, \text{ and } I \\ &= 5+9+4+13+6+8=45 \text{ days} \end{aligned}$$

The variance of the project duration is

$$\begin{aligned} V(T) &= \text{Sum of the variances of jobs } A, B, D, F, H, \text{ and } I \\ &= 1+1+1+4+1+1=9 \end{aligned}$$

The standard deviation of the project duration is

$$\sigma(T)=\sqrt{V(T)}=3$$

Probabilities of Completing the Project

The project length T is the sum of all the job times in the critical path. PERT assumes that all the job times are independent, and are identically distributed. Hence, by the *Central Limit Theorem*,* T has a normal distribution with mean $E(T)$, and variance $V(T)$. Figure 3.23 exhibits a normal distribution with mean μ and variance σ^2.

In our example T is distributed normal with mean 45 and standard deviation 3.† For any normal distribution, the probability that the random variable lies within one standard deviation from the mean is 0.68. Hence, there is a 68% chance that the project duration will be between 42 and 48 days. Similarly there is a 99.7% chance that T will lie within three standard deviations (i.e., between 36 and 54).

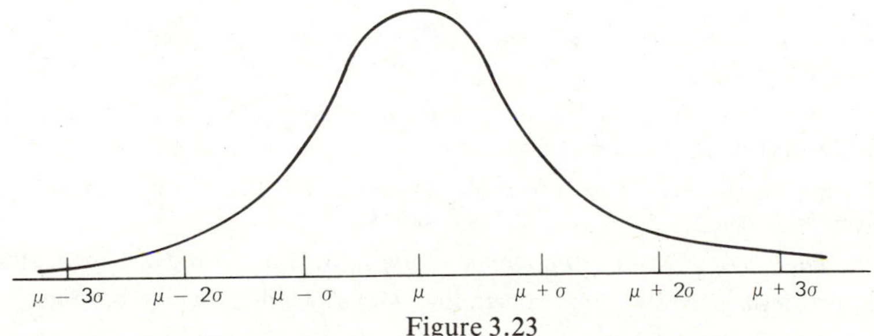

Figure 3.23
A normal distribution with mean μ and standard deviation σ.

**Central Limit Theorem:* Let $X_1, X_2, \ldots, X_N$ be independent and identically distributed random variables. Then for large N, the sum of the random variables, $S_N = X_1 + X_2 + \cdots + X_N$, is normally distributed with mean $\Sigma_{i=1}^N E(X_i)$ and variance $\Sigma_{i=1}^N V(X_i)$.

†Theoretically this is not correct, since we have a very small project and the central limit theorem does not apply. But for illustration we will assume that T is distributed normal.

We can also calculate the probabilities of meeting specified project deadlines. For example, the management wants to know the probability of completing the project by 50 days. In other words, we have to compute Prob $(T \leqq 50)$ where $T \sim N(45, 3^2)$. This can be obtained from the tables of normal distribution; however, the tables are given for a standard normal only whose mean is 0 and standard deviation is 1.

From probability theory the random variable $Z = [T - E(T)]/\sigma(T)$ is distributed normally with mean 0 and standard deviation 1. Hence,

$$\text{Prob}\,(T \leqq 50) = \text{Prob}\left(Z \leqq \frac{50-45}{3}\right) = \text{Prob}\,(Z \leqq 1.67) = 0.95$$

Thus there is a 95% chance that the project will be completed within 50 days.

Suppose we want to know the probability of completing the project 4 days sooner than expected. This means we have to compute

$$\text{Prob}\,(T \leqq 41) = \text{Prob}\left(Z \leqq \frac{41-45}{3}\right) = \text{Prob}\,(Z \leqq -1.33) = 0.09$$

Hence, there is only a small 9% chance that the project will be completed in 41 days.

Note: When multiple critical paths exist, the variance of each critical path may be different even though the expected values are the same. In such a circumstance, it is recommended that the largest variance of T *be used for probability estimates.*

Recommended Readings

Capacitated transportation problems, and transshipment problems are discussed fully in Dantzig (1). A primal-dual network method to solve the transportation problems is available in Dantzig (1), and Ford and Fulkerson (3). Dreyfus (2) presents an excellent appraisal of the various shortest-route algorithms available in the literature. Application of network analysis to solve the compact book storage problems in libraries is fully discussed in Ravindran (10).

Project management problems with finite resources are discussed at a practical level in Wiest and Levy (13). For advanced topics in network analysis like multicommodity flows, the reader may refer to Ford and Fulkerson (3), Hu (9), and Geoffrion (6).

REFERENCES

1. Dantzig, G. B., *Linear Programming and Extensions*, Princeton University Press, Princeton, N.J., 1963.
2. Dreyfus, S. E., "An Appraisal of Some Shortest-Path Algorithms," *Operations Research*, *17* (3), (1969).
3. Ford, L. R., and D. R. Fulkerson, *Flows in Networks*, Princeton University Press, Princeton, N.J., 1962.
4. Gale, D., *The Theory of Linear Economic Models*, McGraw-Hill, New York, 1960.
5. Gass, S. I., *Linear Programming*, Fourth Ed., McGraw-Hill, New York, 1975.
6. Geoffrion, A. M., ed., *Perspectives on Optimization: A Collection of Expository Articles*, Addison-Wesley, Reading, Mass., 1972.
7. Goyal, S. K., "Improving VAM for Unbalanced Transportation Problems," *J. Operational Res. Society, 35* (12):1113–1114 (1984).
8. Hillier, F. S., and Gerald J. Lieberman, *Introduction to Operations Research*, Third Edition, Holden-Day, San Francisco, Calif., 1980.

9. Hu, T. C., *Integer Programming and Network Flows*, Addison-Wesley, Reading, Mass., 1969.
10. Ravindran, A., "On Compact Book Storage in Libraries," *Opsearch, 8* (4), (1971).
11. Taha, H. A., *Operations Research An Introduction,* 2nd Ed. Macmillan, New York, 1976.
12. Wagner, H. M., *Principles of Operations Research,* 2nd Ed., Prentice-Hall, Englewood Cliffs, N.J., 1975.
13. Wiest, J. D., and F. K. Levy, *A Management Guide to PERT/CPM*, 2nd Ed., Prentice-Hall, Englewood Cliffs, N.J., 1977.

EXERCISES

1. What are dummy warehouses and dummy markets? When are they used and why?
2. Explain why the transportation algorithm is not appropriate for solving the assignment problem?
3. What are the practical uses of max-flow-min-cut theorem?
4. What are the differences between a *transportation* problem and a *transshipment* problem?
5. Explain the difference between the shortest route problem and the minimal spanning tree problem.
6. What are the essential differences between PERT and CPM?
7. What is the meaning of *critical path* in project management?
8. Show that finding a critical path in a project network is equivalent to determining the longest path in the network.
9. We have three reservoirs with daily supplies of 15, 20, and 25 million liters of fresh water, respectively. On each day we must supply four cities *A, B, C, D* whose demands are 8, 10, 12, and 15, respectively. The cost of pumping per million liters is given below:

		Cities			
		A	*B*	*C*	*D*
	1	2	3	4	5
Reservoirs	2	3	2	5	2
	3	4	1	2	3

Use the transportation algorithm to determine the cheapest pumping schedule if excess water can be disposed of at no cost.

10. Solve Example 3.2-4 by the transportation algorithm.
11. Faced with a court order to desegregate its schools, a county school board decides to redistribute its minority students through bussing. The plan calls for bussing 50 students from each of the three cities White, Black, and Brown to the four schools East, West, North, and South. For a perfect desegregation, the schools need 20, 40, 30, and 60 minority students respectively. The dollar cost of bussing each student is given as follows:

		School			
		East	West	North	South
	White	7	6	5	4
City	Black	9	7	3	6
	Brown	8	8	7	3

The school board wishes to meet the court order with the least cost.

(a) Set up the transportation table for the above problem.
(b) Find an initial basic feasible solution by Northwest corner rule.
(c) Determine the optimal bussing plan using the u-v method.
(d) Because of a "detour" near the East school for road construction, the bussing cost from every city to that school increases by $1. Explain how this will affect your optimal solution found in (c).

12. Solve Example 3.2-3 using the transportation algorithm.

13. To stimulate interest and provide an atmosphere for intellectual discussion, an engineering faculty decides to hold special seminars on four contemporary topics—ecology, energy, transportation, and bioengineering. Such seminars should be held once per week in the afternoons. However, scheduling these seminars (one for each topic, and not more than one seminar per afternoon) has to be done carefully so that the number of students unable to attend is kept to a minimum. A careful study indicates that the number of students who cannot attend a particular seminar on a specific day is as follows:

	Ecology	Energy	Transportation	Bioengineering
Monday	50	40	60	20
Tuesday	40	30	40	30
Wednesday	60	20	30	20
Thursday	30	30	20	30
Friday	10	20	10	30

(a) Show that the problem of determining the optimal schedule of seminars is equivalent to an assignment problem.
(b) Using (a), find an optimal schedule of seminars.

14. A group of four boys and four girls are planning on a one day picnic. The extent of mutual happiness between boy i and girl j when they are together is given by the following matrix (data obtained from their previous dating experiences):

		Girl			
		1	2	3	4
Boy	1	11	1	5	8
	2	9	9	8	1
	3	10	3	5	10
	4	1	13	12	11

The problem is to decide the proper matching between the boys and the girls during the picnic that will maximize the sum of all the mutual happiness of all the couples. Formulate this as an assignment problem and solve.

15. A batch of four jobs can be assigned to five different machines. The setup time for each job on various machines is given by the following table:

		Machine				
		1	2	3	4	5
Job	1	10	11	4	2	8
	2	7	11	10	14	12
	3	5	6	9	12	14
	4	13	15	11	10	7

Find an optimal assignment of jobs to machines which will minimize the total setup time.

16. Consider the following directed network where s is the source node, n is the sink node, and the numbers along the arcs denote the capacities of flows:

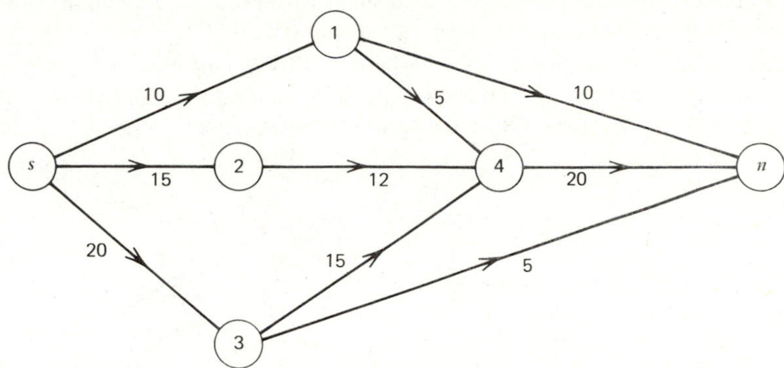

(a) Illustrate the following notions with an example for the above network:
 (i) A *path* connecting the source and the sink.
 (ii) A *cut* separating the source and the sink.
 (iii) The *capacity* of a cut.
(b) Find the maximum flow from the source to the sink using the labeling method.
(c) Find the minimum cut, and verify the max-flow min-cut theorem.

17. Consider a street network as shown below:

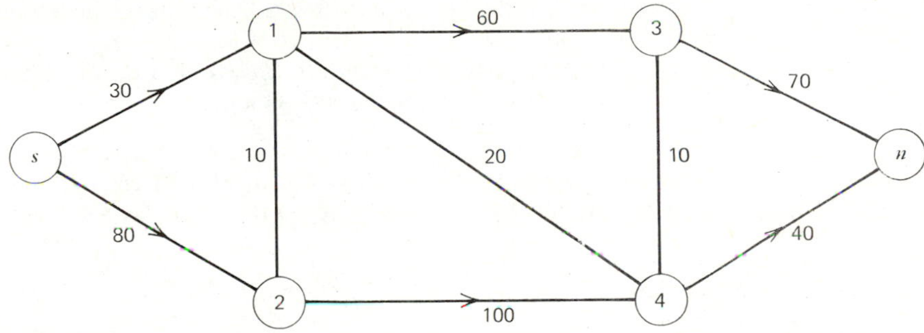

The numbers on the arcs represent the traffic flow capacities. The problem is to place one-way signs on streets not already oriented so as to maximize the traffic flow from the point s to the point n. Solve by the labeling method.

18. A certain commodity is to be shipped from three warehouses to four markets. The warehouse supplies are 20, 20, and 100 units. The market demands are 20, 20, 60, and 20 units. It is not possible to ship from all the warehouses to all the markets. The following table gives the capacities of various routes (a zero capacity implies that there exists no direct route between those two points):

		Market				
		1	2	3	4	Supplies
Warehouse	1	30	10	0	40	20
	2	0	0	10	50	20
	3	20	10	40	5	100
Demands		20	20	60	20	

The problem is to determine whether it is possible to meet the market demands with the available supplies.

(a) Show that the above problem is equivalent to finding the maximum flow from a single source to a single sink in an equivalent network.

(b) Solve (a) by the labeling method, and interpret your solution.

19. Use Dijkstra's algorithm to determine the shortest distance, and the shortest path from node 0 to every other node in the network given below:

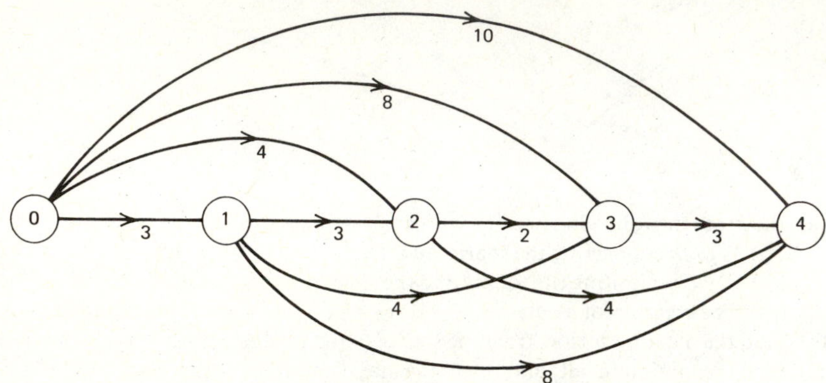

20. Consider an undirected network with 8 nodes and 13 arcs. The distance in kilometers between nodes i and j, denoted by d_{ij}, is as follows: $d_{12}=5$, $d_{13}=1$, $d_{14}=7$, $d_{23}=4$, $d_{34}=2$, $d_{35}=10$, $d_{36}=9$, $d_{46}=3$, $d_{56}=3$, $d_{57}=4$, $d_{67}=9$, $d_{68}=8$, and $d_{78}=6$. Find the shortest distance and the shortest path between the nodes 1 and 8.

21. Use Dijkstra's algorithm to solve the compact book storage problem discussed in Section 3.5 with the following data: $H_1=15$ cm, $H_2=20$ cm, $H_3=25$ cm, $H_4=30$ cm, $L_1=5$ cm, $L_2=20$ cm, $L_3=45$ cm, $L_4=30$ cm, $K_1=K_2=\$100$, $K_3=K_4=\$200$, $C_1=\$6$, $C_2=\$8$, $C_3=\$9$, and $C_4=\$12$.

22. Consider the network given in Exercise 20. Use the algorithm described in Section 3.6 to determine the minimal spanning tree.

23. Determine the minimal spanning tree for the following network:

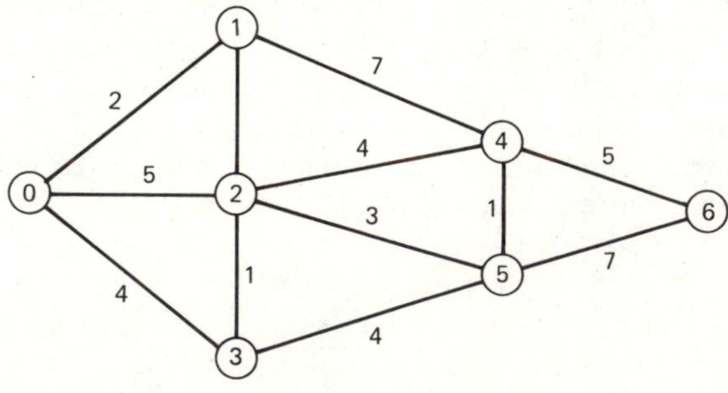

24. Consider a project with eight jobs A, B, C, D, E, F, G, and H having the following job sequence ($X \to Y$ implies job X precedes job Y): $A \to C$, $B \to D$, $C \to H$, $A \to E$, $D \to F$, $B \to E$, $F \to G$, $E \to G$, $G \to H$. Draw the project network.

25. Draw the network for a project consisting of 16 jobs *A, B, C, D, . . . , M, N, O, P* with the following job sequence.

$$A, B, C, D, \to E, F, G$$
$$E, F, G \to H$$
$$H \to I$$
$$I \to J, K, L, M, N$$
$$J, K, L, M, N \to O$$
$$G, O \to P$$

26. Consider a project consisting of nine jobs (*A, B, . . . , I*) with the following precedence relations and time estimates:

Job	Predecessor	Time (days)
A	—	15
B	—	10
C	*A, B*	10
D	*A, B*	10
E	*B*	5
F	*D, E*	5
G	*C, F*	20
H	*D, E*	10
I	*G, H*	15

(a) Draw the project network for this problem designating the jobs by arcs and events by nodes.
(b) Determine the earliest completion time of the project, and identify the critical path.
(c) Determine a project schedule listing the earliest and latest starting times of each job. Also identify the critical jobs.

27. Consider the project given in Exercise 26. Discuss how the project schedule will be affected by the following events:
(a) Job *E* is delayed, and it takes 15 days for completion.
(b) Job *H* is delayed by 10 more days.
(c) Jobs *F* and *G* are completed 1 day ahead of schedule.

28. Consider the project network given below:

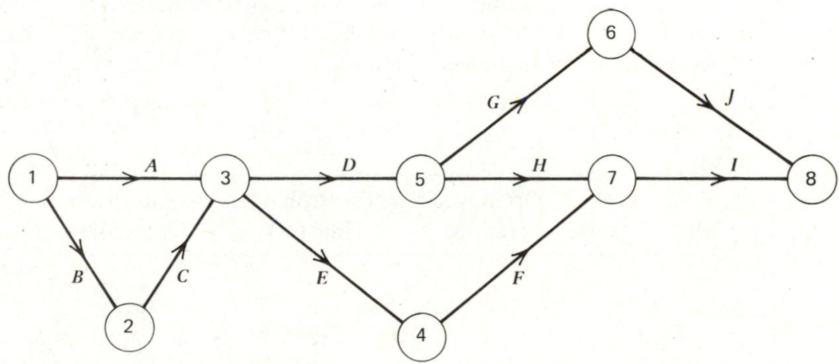

The data for normal times, crash times, and crashing costs are given as follows:

Job	Normal Time (days)	Crash Time (days)	Cost of Crashing Per Day ($)
A	10	7	4
B	5	4	2
C	3	2	2
D	4	3	3
E	5	3	3
F	6	3	5
G	5	2	1
H	6	4	4
I	6	4	3
J	4	3	3

Let T represent the earliest completion time of the project.

(a) Determine the maximum and the minimum value of T.

(b) Set up the linear program to solve the CPM problem if the project is to be completed in 21 days at minimum cost.

(c) Given the overhead costs as $5 per day, determine the optimal duration of the project in terms of both the crashing and the overhead costs by direct enumeration method.

29. Consider the project network given below:

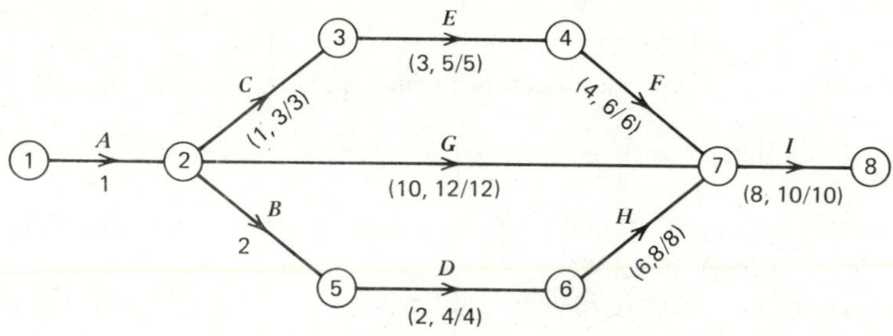

Numbers along the arcs represent: crash time (days), normal time (days) per unit cost of crashing ($). *Note:* jobs *A* and *B* cannot be crashed.

(a) Use the direct enumeration method to determine the least cost crashing schedule that will meet the project deadline of 20 days. What is the total crashing cost?

(b) Given that we have $50 available for crashing, which jobs would you crash and by how much? What is the least completion time of the project?

(c) Construct an LP model to answer part (b). You must define your variables clearly, write out the constraints and the objective function.

30. Consider a project consisting of 7 jobs ($A, \ldots, G$) with the following precedence relations, and time estimates

Job	Predecessors	Optimistic Time (a)	Most Probable Time (m)	Pessimistic Time (b)
A	—	2	5	8
B	*A*	6	9	12
C	*A*	5	14	17
D	*B*	5	8	11
E	*C, D*	3	6	9
F	—	3	12	21
G	*E, F*	1	4	7

(a) Draw the project network for the above problem.
(b) Determine the expected duration, and variance of each job.
(c) What is the expected length of the project, and its variance?
(d) Compute the probabilities of completing the project
  (i) 3 days earlier than expected.
  (ii) No more than 5 days later than expected.

31. Consider the production planning problem discussed in Example 3.2-4. Assume that "backorders" are allowed, that is, if we do not have sufficient stock, we can backorder so that they can be produced and supplied in a later week. However, this incurs a backorder penalty cost of \$4 per unit per week. Reformulate the problem as a transportation problem assuming that all backorders must be filled by the end of the fourth week. Solve by the transportation algorithm and interpret your solution.

32. A company has signed the following contract to supply customers A and B for the next 3 months:

	June	July	August
Customer A	30	20	15
Customer B	20	20	10

They can produce 40 units per month at a cost of \$100 per unit on regular time and an additional 10 units per month on overtime at a cost of \$120 per unit. They can store units at a cost of \$10 per unit per month. The contract allows the company to fall short on its supply commitment to customer A during the months of June and July, but this incurs a penalty cost of \$5 per unit; however, all shipments are to be completed by August. No shortages are allowed for customer B and his demands must be satisfied on the months specified. The company wishes to determine the optimal production schedule that will minimize the total cost of production, storage, and shortage. Formulate this as a transportation problem and determine an initial basic feasible solution.

33. The following matrix (S) gives the shortest distance between all pairs of nodes in a network of six nodes.

$$\underset{(6\times 6)}{S} = \begin{bmatrix} - & 10 & 20 & 5 & 12 & 6 \\ 10 & - & 25 & 12 & 2 & 15 \\ 15 & 10 & - & 15 & 20 & 10 \\ 20 & 15 & 13 & - & 11 & 9 \\ 10 & 25 & 15 & 11 & - & 15 \\ 10 & 10 & 20 & 9 & 15 & - \end{bmatrix}$$

(*Note. The* (i, j)*th element in the matrix* S *give the shortest distance from node* i *to node* j).

If C_i denotes the length of the shortest cycle starting and ending at node i ($C_i > 0$), compute the values of C_1, C_2, C_3, C_4, C_5, and C_6.

(*Hint. First prove that in order to compute* C_i *it is sufficient to consider one intermediate node*).

CHAPTER 4
ADVANCED TOPICS IN LINEAR PROGRAMMING

The simplex method discussed in Chapter 2 performs calculations on the entire tableau during each iteration. However, updating all the elements in the tableau during a basis change is not really necessary for using the simplex method. The only information needed in moving from one tableau (basic feasible solution) to the next tableau is the following:

1. The relative profit coefficients ($\bar{c}$ row).
2. The column corresponding to the nonbasic variable entering the basis (the pivot column).
3. The current basic variables and their values (right-hand-side constants).

The information contained in the other columns of the tableau plays no role in the simplex process. Hence the solution of large linear programming problems on a digital computer will become very inefficient and costly if the simplex method were to be used in its full tableau form. Hence many refinements in the simplex calculations were carried out so that the implementation of the simplex method on a computer could be done more efficiently. This led to the development of the *revised simplex method* or *simplex method with multipliers*, which is implemented in all the commercial computer codes.

4.1 THE REVISED SIMPLEX METHOD

The revised simplex method uses the same basic principles of the regular simplex method. But at each iteration the entire tableau is never calculated. The relevant information it needs to move from one basic feasible solution to another is directly generated from the original equations.

In order to illustrate the basic principles of the revised simplex method, recall the linear program discussed in Section 2.7.

EXAMPLE 4.1-1

Maximize: $Z = 5x_1 + 2x_2 + 3x_3 - x_4 + x_5$

Subject to:

$$x_1 + 2x_2 + 2x_3 + x_4 = 8$$
$$3x_1 + 4x_2 + x_3 + x_5 = 7$$
$$x_1, \ldots, x_5 \geqq 0$$

For easy reference the three tableaus of the (regular) simplex method are reproduced in Table 4.1. To go from one basic feasible solution to the next, the simplex method needs the following:

1. The $\bar{c}$ row; to determine the nonbasic variable to enter the basis.
2. The pivot column, and the right-hand-side constants; to perform the minimum ratio test, and determine the basic variable to leave the basis.

Table 4.1
SIMPLEX TABLEAUS FOR EXAMPLE 4.1-1

		c_j	5	2	3	-1	1	
Tableau	c_B	Basis	x_1	x_2	x_3	x_4	x_5	Constants
1	-1	x_4	1	2	$\textcircled{2}$	1	0	8
	1	x_5	3	4	1	0	1	7
		$\bar{c}$ row	3	0	4	0	0	$Z=-1$
2	3	x_3	$\frac{1}{2}$	1	1	$\frac{1}{2}$	0	4
	1	x_5	$\textcircled{\frac{5}{2}}$	3	0	$-\frac{1}{2}$	1	3
		$\bar{c}$ row	1	-4	0	-2	0	$Z=15$
3	3	x_3	0	$\frac{2}{5}$	1	$\frac{3}{5}$	$-\frac{1}{5}$	$\frac{17}{5}$
	5	x_1	1	$\frac{6}{5}$	0	$-\frac{1}{5}$	$\frac{2}{5}$	$\frac{6}{5}$
		$\bar{c}$ row	0	$-\frac{26}{5}$	0	$-\frac{9}{5}$	$-\frac{2}{5}$	$Z=\frac{81}{5}$

For example, to go from Tableau 2 to Tableau 3 in Table 4.1, we need the value of $\bar{c}_1$, the column corresponding to x_1, and the right-hand-side constants. None of the other information contained in Tableau 2 is used. Hence there is no need to carry the entire tableau.

The revised simplex method works on the principle that any tableau corresponding to a basic feasible solution can be generated directly from the original equations by matrix-vector operations. To illustrate this, let the original columns corresponding to x_1, x_2, x_3, x_4, x_5 be denoted by column vectors $\mathbf{P}_1, \mathbf{P}_2, \mathbf{P}_3, \mathbf{P}_4, \mathbf{P}_5$, and the right-hand-side constants by the column vector $\mathbf{b}$. Thus,

$$\mathbf{P}_1=\begin{pmatrix}1\\3\end{pmatrix}, \quad \mathbf{P}_2=\begin{pmatrix}2\\4\end{pmatrix}, \quad \mathbf{P}_3=\begin{pmatrix}2\\1\end{pmatrix},$$

$$\mathbf{P}_4=\begin{pmatrix}1\\0\end{pmatrix}, \quad \mathbf{P}_5=\begin{pmatrix}0\\1\end{pmatrix} \quad \text{and} \quad \mathbf{b}=\begin{pmatrix}8\\7\end{pmatrix}$$

Tableau 2 of Table 4.1 in which x_3 and x_5 are basic variables, may be generated directly by matrix theory as follows. Define a *basis matrix* $\mathbf{B}$ whose elements are the original columns corresponding to the basic variables x_3 and x_5. Thus

$$\mathbf{B} = [\mathbf{P}_3, \mathbf{P}_5] = \begin{bmatrix} 2 & 0 \\ 1 & 1 \end{bmatrix}$$

The inverse of the basis matrix denoted by $\mathbf{B}^{-1}$ (see Appendix A for a review of matrices) is given by

$$\mathbf{B}^{-1} = \begin{bmatrix} \frac{1}{2} & 0 \\ -\frac{1}{2} & 1 \end{bmatrix}$$

Then from matrix theory any column in Tableau 2 can be obtained by premultiplying the original columns by the inverse of the basis matrix. Suppose $\bar{\mathbf{P}}_j$ denotes the updated column corresponding to the variable x_j and $\bar{\mathbf{b}}$ denotes the new constants in Tableau 2. Then, $\bar{\mathbf{P}}_j = \mathbf{B}^{-1}\mathbf{P}_j$ for all $j = 1, 2, 3, 4, 5$, and $\bar{\mathbf{b}} = \mathbf{B}^{-1}\mathbf{b}$. For example,

$$\bar{\mathbf{P}}_1 = \mathbf{B}^{-1}\mathbf{P}_1 = \begin{bmatrix} \frac{1}{2} & 0 \\ -\frac{1}{2} & 1 \end{bmatrix} \begin{bmatrix} 1 \\ 3 \end{bmatrix} = \begin{bmatrix} \frac{1}{2} \\ \frac{5}{2} \end{bmatrix},$$

$$\bar{\mathbf{P}}_2 = \mathbf{B}^{-1}\mathbf{P}_2 = \begin{bmatrix} \frac{1}{2} & 0 \\ -\frac{1}{2} & 1 \end{bmatrix} \begin{bmatrix} 2 \\ 4 \end{bmatrix} = \begin{bmatrix} 1 \\ 3 \end{bmatrix},$$

$$\bar{\mathbf{b}} = \mathbf{B}^{-1}\mathbf{b} = \begin{bmatrix} \frac{1}{2} & 0 \\ -\frac{1}{2} & 1 \end{bmatrix} \begin{bmatrix} 8 \\ 7 \end{bmatrix} = \begin{bmatrix} 4 \\ 3 \end{bmatrix}$$

Note that the vector $\bar{\mathbf{b}}$ gives the values of the basic variables x_3 and x_5.

Recall that there are two key steps in the simplex method, namely, the selection of a nonbasic variable to enter the basis and the basic variable to leave. Let us see how these key steps are carried out in the revised simplex method when it does not carry the entire tableau at each iteration.

The selection of a nonbasic variable is based on the values of the relative profit coefficients ($\bar{c}_j$). For example, in the regular simplex method, the $\bar{c}_j$ values in Tableau 2 will be calculated as follows:

$$\bar{c}_1 = c_1 - \mathbf{c}_B\bar{\mathbf{P}}_1 = 5 - (3, 1) \begin{bmatrix} \frac{1}{2} \\ \frac{5}{2} \end{bmatrix} = 1$$

$$\bar{c}_2 = c_2 - \mathbf{c}_B\bar{\mathbf{P}}_2 = 2 - (3, 1) \begin{bmatrix} 1 \\ 3 \end{bmatrix} = -4$$

$$\bar{c}_4 = c_4 - \mathbf{c}_B\bar{\mathbf{P}}_4 = -1 - (3, 1) \begin{bmatrix} \frac{1}{2} \\ -\frac{1}{2} \end{bmatrix} = -2$$

In general, for the nonbasic variable x_j,

$$\bar{c}_j = c_j - \mathbf{c}_B \bar{\mathbf{P}}_j \tag{4.1}$$

From the matrix theory it was shown earlier that

$$\bar{\mathbf{P}}_j = \mathbf{B}^{-1}\mathbf{P}_j$$

Hence, $\bar{c}_j = c_j - \mathbf{c}_B\mathbf{B}^{-1}\mathbf{P}_j$. Let the vector $\boldsymbol{\pi}$ denote $\mathbf{c}_B\mathbf{B}^{-1}$. The elements of vector $\boldsymbol{\pi}$ are called the *simplex multipliers*. Thus Eq. 4.1 reduces to

$$\bar{c}_j = c_j - \boldsymbol{\pi}\mathbf{P}_j \qquad \text{for all } j \tag{4.2}$$

where c_j is the original profit coefficient, and $\mathbf{P}_j$ is the original column coefficients of the variable x_j. For example, in Tableau 2,

$$\boldsymbol{\pi} = (\pi_1, \pi_2) = \mathbf{c}_B\mathbf{B}^{-1} = (3, 1)\begin{bmatrix} \frac{1}{2} & 0 \\ -\frac{1}{2} & 1 \end{bmatrix} = (1, 1)$$

and

$$\bar{c}_1 = c_1 - \boldsymbol{\pi}\mathbf{P}_1 = 5 - (1, 1)\begin{bmatrix} 1 \\ 3 \end{bmatrix} = 1$$

$$\bar{c}_2 = c_2 - \boldsymbol{\pi}\mathbf{P}_2 = 2 - (1, 1)\begin{bmatrix} 2 \\ 4 \end{bmatrix} = -4$$

$$\bar{c}_4 = c_4 - \boldsymbol{\pi}\mathbf{P}_4 = -1 - (1, 1)\begin{bmatrix} 1 \\ 0 \end{bmatrix} = -2$$

Since $\bar{c}_1 > 0$, we select x_1 to be the new basic variable. Thus using Eq. 4.2, the $\bar{c}$ row for any tableau can be calculated directly from the original equations and the profit coefficients. The only other information needed is the inverse of the basis matrix corresponding to that tableau.

Having chosen x_1 as the variable to enter the basis, we have to find the basic variable to leave the basis by the minimum ratio rule. In order to apply this rule, we need the new right-hand-side constants, and the column corresponding to x_1 in Tableau 2. As shown earlier, these can be obtained directly from the original columns by premultiplying them by the inverse of the basis matrix. Thus, the pivot column,

$$\bar{\mathbf{P}}_1 = \mathbf{B}^{-1}\mathbf{P}_1 = \begin{bmatrix} \frac{1}{2} & 0 \\ -\frac{1}{2} & 1 \end{bmatrix}\begin{bmatrix} 1 \\ 3 \end{bmatrix} = \begin{bmatrix} \frac{1}{2} \\ \frac{5}{2} \end{bmatrix}$$

The new constants,

$$\bar{\mathbf{b}} = \mathbf{B}^{-1}\mathbf{b} = \begin{bmatrix} \frac{1}{2} & 0 \\ -\frac{1}{2} & 1 \end{bmatrix}\begin{bmatrix} 8 \\ 7 \end{bmatrix} = \begin{bmatrix} 4 \\ 3 \end{bmatrix}$$

Applying the minimum ratio rule, the minimum ratio is found in the second row, and the basic variable x_5 will be replaced by x_1. Hence the new set of basic variables are x_3 and x_1.

Once again we can form the new basis matrix

$$\mathbf{B}=(\mathbf{P}_3, \mathbf{P}_1)=\begin{bmatrix} 2 & 1 \\ 1 & 3 \end{bmatrix}$$

and compute

$$\mathbf{B}^{-1}=\begin{bmatrix} \frac{3}{5} & -\frac{1}{5} \\ -\frac{1}{5} & \frac{2}{5} \end{bmatrix}$$

The new right-hand-side constants are given by

$$\mathbf{B}^{-1}\mathbf{b}=\begin{bmatrix} \frac{3}{5} & -\frac{1}{5} \\ -\frac{1}{5} & \frac{2}{5} \end{bmatrix}\begin{bmatrix} 8 \\ 7 \end{bmatrix}=\begin{bmatrix} \frac{17}{5} \\ \frac{6}{5} \end{bmatrix}$$

In other words, the new basic feasible solution is $x_3=17/5$, $x_1=6/5$, $x_2=x_4=x_5=0$. To check the optimality of this solution, we need the $\bar{c}_j$ coefficients for x_2, x_4, and x_5. Using Eq. 4.2,

$$\bar{c}_j=c_j-\boldsymbol{\pi}\mathbf{P}_j \qquad \text{where} \qquad \boldsymbol{\pi}=\mathbf{c}_B\mathbf{B}^{-1}$$

The new simplex multipliers are given by

$$\boldsymbol{\pi}=(\pi_1, \pi_2)=\mathbf{c}_B\mathbf{B}^{-1}=(3, 5)\begin{bmatrix} \frac{3}{5} & -\frac{1}{5} \\ -\frac{1}{5} & \frac{2}{5} \end{bmatrix}=\left(\frac{4}{5}, \frac{7}{5}\right)$$

Then,

$$\bar{c}_2=c_2-\boldsymbol{\pi}\mathbf{P}_2=2-\left(\frac{4}{5}, \frac{7}{5}\right)\begin{bmatrix} 2 \\ 4 \end{bmatrix}=-\frac{26}{5}$$

$$\bar{c}_4=c_4-\boldsymbol{\pi}\mathbf{P}_4=-1-\left(\frac{4}{5}, \frac{7}{5}\right)\begin{bmatrix} 1 \\ 0 \end{bmatrix}=-\frac{9}{5}$$

$$\bar{c}_5=c_5-\boldsymbol{\pi}\mathbf{P}_5=1-\left(\frac{4}{5}, \frac{7}{5}\right)\begin{bmatrix} 0 \\ 1 \end{bmatrix}=-\frac{2}{5}$$

Since all the $\bar{c}_j$ coefficients are negative, the current solution is optimal.

To summarize, any information contained in a simplex tableau may be obtained directly from the original equations by knowing the inverse of the basis matrix corresponding to that tableau. The inverse of the basis matrix can be computed from the original equations by merely knowing the current basic variables in that tableau. Thus, the revised simplex method has the capability to generate any information that is available in the regular simplex method. But it generates only the relevant information that is needed to perform the simplex steps.

In the actual implementation of the revised simplex method, the inverse of the basis is not computed explicitly by inverting the matrix of basic columns. Inverting a matrix is generally time consuming and costly on a digital computer. Hence the basis inverse at each step is obtained by a simple pivot operation on the previous inverse. To illustrate this refer to the three simplex tableaus given in Table

4.1. In the initial basic feasible solution (Tableau 1), x_4 and x_5 are the basic variables. The initial basis matrix (that corresponds to the columns $\mathbf{P}_4$ and $\mathbf{P}_5$) is the identity matrix since

$$[\mathbf{P}_4, \mathbf{P}_5] = \begin{bmatrix} 1 & 0 \\ 0 & 1 \end{bmatrix} = \mathbf{I} \qquad \text{(identity matrix).}$$

In any subsequent tableau, the new column coefficients corresponding to x_4 and x_5 are obtained by premultiplying $\mathbf{P}_4$ and $\mathbf{P}_5$ by the inverse of the current basis matrix. That is, $\bar{\mathbf{P}}_4 = \mathbf{B}^{-1}\mathbf{P}_4$ and $\bar{\mathbf{P}}_5 = \mathbf{B}^{-1}\mathbf{P}_5$. Hence,

$$[\bar{\mathbf{P}}_4, \bar{\mathbf{P}}_5] = \mathbf{B}^{-1}[\mathbf{P}_4, \mathbf{P}_5] = \mathbf{B}^{-1}\mathbf{I} = \mathbf{B}^{-1}$$

In other words, the columns corresponding to x_4 and x_5 in any tableau contains the inverse of the basis for that tableau! For example, in Tableau 2 (Table 4.1) the columns corresponding to x_4 and x_5 is given by $\begin{bmatrix} 1/2 & 0 \\ -1/2 & 1 \end{bmatrix}$, which was the computed basis inverse for that tableau. This is true for Tableau 3 as well. This implies that the new basis inverse may be easily obtained by carrying the columns corresponding to the initial basic variables, and updating them by the pivot operation.

Similarly the right-hand-side constants of every tableau are always needed for determining the basic feasible solution, and later for the minimum ratio rule. Instead of computing it using the basis inverse, we can carry the right-hand-side constants, and update their values for each iteration by the pivot operation. Thus the revised simplex method uses a reduced simplex tableau that contains the columns corresponding to the initial basic variables, the right-hand-side constants, and the current basic variables. For example, Table 4.2 shows how the second tableau of the revised simplex method for Example 4.1-1 may look like.

Table 4.2

Basis	$\mathbf{B}^{-1}$		$\bar{\mathbf{b}}$
x_3	$\frac{1}{2}$	0	4
x_5	$-\frac{1}{2}$	1	3

From $\mathbf{B}^{-1}$, the simplex multipliers and the $\bar{c}_j$ elements are calculated. This results in selecting x_1 to enter the basis. To perform the minimum ratio test, the column corresponding to x_1 in Tableau 2 (pivot column) is required. This is computed as follows:

$$\bar{\mathbf{P}}_1 = \mathbf{B}^{-1}\mathbf{P}_1 = \begin{bmatrix} \frac{1}{2} & 0 \\ -\frac{1}{2} & 1 \end{bmatrix} \begin{bmatrix} 1 \\ 3 \end{bmatrix} = \begin{bmatrix} \frac{1}{2} \\ \frac{5}{2} \end{bmatrix}$$

Using $\bar{\mathbf{P}}_1$ and $\bar{\mathbf{b}}$, the ratio test is performed. This identifies x_5 as the basic variable to leave. Thus x_1 becomes the new basic variable in the second constraint. This means that the pivot column $\begin{bmatrix} 1/2 \\ 5/2 \end{bmatrix}$ should be reduced to $\begin{bmatrix} 0 \\ 1 \end{bmatrix}$ by pivot operation. This is done as follows:

1. Multiply the second row by $-1/5$ and add it to the first row.
2. Divide the second row by $5/2$.

The pivot operation is carried out to the reduced tableau shown in Table 4.2. The new tableau is shown in Table 4.3.

Table 4.3

Basis	$\mathbf{B}^{-1}$		$\bar{\mathbf{b}}$
x_3	$\frac{3}{5}$	$-\frac{1}{5}$	$\frac{17}{5}$
x_1	$-\frac{1}{5}$	$\frac{2}{5}$	$\frac{6}{5}$

The new basic feasible solution is $x_3=17/5$, $x_1=6/5$, $x_2=x_4=x_5=0$. The new inverse of the basis is $\begin{bmatrix} 3/5 & -1/5 \\ -1/5 & 2/5 \end{bmatrix}$. Using the new basis inverse, the simplex multipliers and $\bar{c}_j$ coefficients are computed to check the optimality of Table 4.3.

General Steps of the Revised Simplex Method

Recall the matrix form of the standard linear program given in Section 2.4.

Minimize: $Z = \mathbf{cx}$

Subject to: $\mathbf{Ax} = \mathbf{b}$

$\mathbf{x} \geqq 0$

where

$$\underset{(m\times n)}{\mathbf{A}} = \begin{bmatrix} a_{11} & a_{12} & \cdots & a_{1n} \\ a_{21} & a_{22} & \cdots & a_{2n} \\ \cdot & \cdot & & \cdot \\ \cdot & \cdot & & \cdot \\ \cdot & \cdot & & \cdot \\ a_{m1} & a_{m2} & \cdots & a_{mn} \end{bmatrix}$$

$$\underset{(m\times 1)}{\mathbf{b}} = \begin{bmatrix} b_1 \\ b_2 \\ \cdot \\ \cdot \\ \cdot \\ b_m \end{bmatrix} \qquad \underset{(n\times 1)}{\mathbf{x}} = \begin{bmatrix} x_1 \\ x_2 \\ \cdot \\ \cdot \\ \cdot \\ x_n \end{bmatrix} \qquad \underset{(1\times n)}{\mathbf{c}} = (c_1, c_2, \ldots, c_n)$$

Let the columns corresponding to the matrix $\mathbf{A}$ be denoted by $\mathbf{P}_1, \mathbf{P}_2, \ldots, \mathbf{P}_n$ where

$$\underset{(m\times 1)}{\mathbf{P}_1} = \begin{bmatrix} a_{11} \\ a_{21} \\ \cdot \\ \cdot \\ \cdot \\ a_{m1} \end{bmatrix} \qquad \underset{(m\times 1)}{\mathbf{P}_{m+1}} = \begin{bmatrix} a_{1,m+1} \\ a_{2,m+1} \\ \cdot \\ \cdot \\ \cdot \\ a_{m,m+1} \end{bmatrix}, \ldots, \quad \text{and} \quad \underset{(m\times 1)}{\mathbf{P}_n} = \begin{bmatrix} a_{1n} \\ a_{2n} \\ \cdot \\ \cdot \\ \cdot \\ a_{mn} \end{bmatrix}$$

Suppose we have a basic feasible solution to the linear program using $x_1, x_2, \ldots, x_m$ as the basic variables. Then the basis matrix is given by,

$$\underset{(m\times m)}{\mathbf{B}} = [\mathbf{P}_1, \mathbf{P}_2, \ldots, \mathbf{P}_m] = \begin{bmatrix} a_{11} & a_{12} & \cdots & a_{1m} \\ a_{21} & a_{22} & \cdots & a_{2m} \\ \cdot & \cdot & & \cdot \\ \cdot & \cdot & & \cdot \\ \cdot & \cdot & & \cdot \\ a_{m1} & a_{m2} & & a_{mm} \end{bmatrix}$$

Let

$$\underset{(m\times m)}{\mathbf{B}^{-1}} = \begin{bmatrix} \beta_{11} & \beta_{12} & \cdots & \beta_{1m} \\ \beta_{21} & \beta_{22} & \cdots & \beta_{2m} \\ \cdot & \cdot & & \cdot \\ \cdot & \cdot & & \cdot \\ \cdot & \cdot & & \cdot \\ \beta_{m1} & \beta_{m2} & & \beta_{mm} \end{bmatrix}$$

Let the vector $\mathbf{x}$ be partitioned as $\underset{(n\times 1)}{\mathbf{x}} = \begin{bmatrix} \mathbf{x}_B \\ \mathbf{x}_N \end{bmatrix}$, where $\mathbf{x}_B$ corresponds to the basic variables, and $\mathbf{x}_N$ to the nonbasic variables. Thus,

$$\underset{(m\times 1)}{\mathbf{x}_B} = \begin{bmatrix} x_1 \\ x_2 \\ \cdot \\ \cdot \\ \cdot \\ x_m \end{bmatrix} \quad \text{and} \quad \underset{(n-m\times 1)}{\mathbf{x}_N} = \begin{bmatrix} x_{m+1} \\ x_{m+2} \\ \cdot \\ \cdot \\ \cdot \\ x_n \end{bmatrix}$$

The current basic feasible solution is given by

$$\mathbf{x}_B = \mathbf{B}^{-1}\mathbf{b} = \begin{bmatrix} \beta_{11}b_1 + \beta_{12}b_2 + \cdots + \beta_{1m}b_m \\ \beta_{21}b_1 + \beta_{22}b_2 + \cdots + \beta_{2m}b_m \\ \cdot \\ \cdot \\ \cdot \\ \beta_{m1}b_1 + \beta_{m2}b_2 + \cdots + \beta_{mm}b_m \end{bmatrix}$$

$$= \begin{bmatrix} \bar{b}_1 \\ \bar{b}_2 \\ \cdot \\ \cdot \\ \cdot \\ \bar{b}_m \end{bmatrix}, \quad \text{and} \quad \mathbf{x}_N = \begin{bmatrix} 0 \\ 0 \\ \cdot \\ \cdot \\ \cdot \\ 0 \end{bmatrix}$$

Let $\mathbf{c}_B$ denote the cost coefficients of the basic variables. Then the value of the objective with respect to the basis $\mathbf{B}$ becomes

$$Z = \mathbf{cx} = \mathbf{c}_B\mathbf{x}_B = c_1\bar{b}_1 + c_2\bar{b}_2 + \cdots + c_m\bar{b}_m$$

To verify whether the present solution is optimal, the revised simplex method computes the simplex multipliers,

$$\underset{(1\times m)}{\boldsymbol{\pi}} = (\pi_1, \pi_2, \ldots, \pi_m) = \mathbf{c}_B\mathbf{B}^{-1}$$

Thus

$$\begin{aligned}
\pi_1 &= c_1\beta_{11} + c_2\beta_{21} + \cdots + c_m\beta_{m1} \\
\pi_2 &= c_1\beta_{12} + c_2\beta_{22} + \cdots + c_m\beta_{m2} \\
&\vdots \\
\pi_m &= c_1\beta_{1m} + c_2\beta_{2m} + \cdots + c_m\beta_{mm}
\end{aligned}$$

The relative cost coefficients are given by,

$$\bar{c}_j = c_j - \boldsymbol{\pi}\mathbf{P}_j \qquad \text{for } j = m+1, m+2, \ldots, n$$

Thus

$$\begin{aligned}
\bar{c}_{m+1} &= c_{m+1} - \boldsymbol{\pi}\mathbf{P}_{m+1} = c_{m+1} - (\pi_1 a_{1,m+1} + \pi_2 a_{2,m+1} + \cdots + \pi_m a_{m,m+1}) \\
&\vdots \\
\bar{c}_n &= c_n - \boldsymbol{\pi}\mathbf{P}_n = c_n - (\pi_1 a_{1n} + \pi_2 a_{2n} + \cdots + \pi_m a_{mn})
\end{aligned}$$

Since we have a minimization problem, the solution is optimal when all the $\bar{c}_j \geqq 0$. Otherwise, select the nonbasic variable with the most negative value for $\bar{c}_j$ to enter the basis.

This step is sometimes called the *pricing out routine* in the computer code. During this step, the simplex multipliers are generally stored in the computer memory. The data matrix **A** and cost coefficients c_j are kept externally to the computer on a magnetic tape or disk and are read in one column at a time to compute the $\bar{c}_j$ values. (As mentioned earlier, computer codes for large problems do not compute all the $\bar{c}_j$ elements. As soon as a negative $\bar{c}_j$ is found, the corresponding nonbasic variable x_j becomes the new basic variable.)

From the pricing out routine let us say that the nonbasic variable x_n is selected to enter the basis. Once the entering column $\mathbf{P}_n$ has been selected, it has to be transformed in terms of the current basis (tableau). This gives the pivot column which is needed in the ratio test. Thus the pivot column becomes:

$$\underset{(m\times 1)}{\bar{\mathbf{P}}_n} = \mathbf{B}^{-1}\mathbf{P}_n = \begin{bmatrix} \beta_{11}a_{1n} + \beta_{12}a_{2n} + \cdots + \beta_{1m}a_{mn} \\ \beta_{21}a_{1n} + \beta_{22}a_{2n} + \cdots + \beta_{2m}a_{mn} \\ \vdots \\ \beta_{m1}a_{1n} + \beta_{m2}a_{2n} + \cdots + \beta_{mm}a_{mn} \end{bmatrix} = \begin{bmatrix} \bar{a}_{1n} \\ \bar{a}_{2n} \\ \vdots \\ \bar{a}_{mn} \end{bmatrix}$$

The minimum ratio test is performed to identify the basic variable to leave the basis. Suppose that

$$\frac{\bar{b}_2}{\bar{a}_{2n}} = \underset{\bar{a}_{in}>0}{\text{Minimum}} \left(\frac{\bar{b}_i}{\bar{a}_{in}} \right) \qquad \text{for } i = 1, 2, \ldots, m$$

Since the minimum ratio occurs in the second row (corresponding to the basic variable x_2), x_n replaces x_2 from the basis. A pivot operation is carried out on the basis inverse ($\mathbf{B}^{-1}$) and the right-hand-side constants ($\bar{\mathbf{b}}$) using $\bar{a}_{2n}$ as the pivot element. This gives the new basis inverse, and the constants with respect to the new basic variables $(x_1, x_n, x_3, x_4, \ldots, x_m)$.

Let the new right-hand-side constants be denoted by

$$\mathbf{b}^* = \begin{bmatrix} b_1^* \\ b_2^* \\ \cdot \\ \cdot \\ \cdot \\ b_m^* \end{bmatrix}$$

and the new inverse of the basis matrix be

$$\underset{(m \times m)}{(\mathbf{B}^*)^{-1}} = \begin{bmatrix} \beta_{11}^* & \cdots & \beta_{1m}^* \\ \cdot & & \cdot \\ \cdot & & \cdot \\ \cdot & & \cdot \\ \beta_{m1}^* & \cdots & \beta_{mm}^* \end{bmatrix}$$

The pivot operation to get $\bar{\mathbf{b}}^*$ and $(\mathbf{B}^*)^{-1}$ is described below:

$$\begin{aligned} &b_2^* = \bar{b}_2/\bar{a}_{2n} \\ &b_i^* = \bar{b}_i - (\bar{a}_{in}\bar{b}_2/\bar{a}_{2n}) && \text{for all } i = 1, 3, \ldots, m \\ &\beta_{2j}^* = \beta_{2j}/\bar{a}_{2n} && \text{for } j = 1, 2, \ldots, m \\ &\beta_{ij}^* = \beta_{ij} - (\bar{a}_{in}\beta_{2j}/\bar{a}_{2n}) && \text{for } j = 1, 2, \ldots, m \text{ and } i = 1, 3, \ldots, m \end{aligned}$$

The new basic feasible solution is $x_1 = b_1^*$, $x_n = b_2^*$, $x_3 = b_3^*$, $\ldots$, $x_m = b_m^*$, $x_2 = x_{m+1} = \cdots = x_{n-1} = 0$. With the help of the new basis inverse the simplex multipliers and $\bar{c}_j$ elements are once again computed to check the optimality of the new basic feasible solution. This process is continued until the optimal solution is reached.

We shall now illustrate the revised simplex method using the example of Section 2.9.

EXAMPLE 4.1-2

Minimize: $Z = -3x_1 + x_2 + x_3$

Subject to:

$$\begin{aligned} x_1 - 2x_2 + x_3 &\leqq 11 \\ -4x_1 + x_2 + 2x_3 &\geqq 3 \\ 2x_1 - x_3 &= -1 \\ x_1, x_2, x_3 &\geqq 0 \end{aligned}$$

In standard form, the problem reduces to

Minimize: $Z = -3x_1 + x_2 + x_3$

Subject to:

$$\begin{aligned} x_1 - 2x_2 + x_3 + x_4 &= 11 \\ -4x_1 + x_2 + 2x_3 - x_5 &= 3 \\ -2x_1 + x_3 &= 1 \\ x_1, \ldots, x_5 &\geqq 0 \end{aligned}$$

Since there are no basic variables in the second and third equations, artificial variables x_6 and x_7 are added as shown below:

$$\begin{aligned} x_1 - 2x_2 + x_3 + x_4 \qquad\qquad &= 11 \\ -4x_1 + x_2 + 2x_3 \quad - x_5 + x_6 \quad &= 3 \\ -2x_1 \quad + x_3 \qquad\qquad + x_7 &= 1 \\ x_1, \ldots, x_7 &\geqq 0 \end{aligned}$$

Using the Big M method, the objective function becomes

$$\text{Minimize } Z = -3x_1 + x_2 + x_3 + Mx_6 + Mx_7$$

Let $\mathbf{P}_1, \ldots, \mathbf{P}_7$ and $\mathbf{b}$ denote the columns corresponding to $x_1, \ldots, x_7$ and the right-hand side. Thus,

$$\mathbf{P}_1 = \begin{bmatrix} 1 \\ -4 \\ -2 \end{bmatrix}, \quad \mathbf{P}_2 = \begin{bmatrix} -2 \\ 1 \\ 0 \end{bmatrix}, \quad \mathbf{P}_3 = \begin{bmatrix} 1 \\ 2 \\ 1 \end{bmatrix}, \quad \mathbf{P}_4 = \begin{bmatrix} 1 \\ 0 \\ 0 \end{bmatrix}, \quad \mathbf{P}_5 = \begin{bmatrix} 0 \\ -1 \\ 0 \end{bmatrix}$$

$$\mathbf{P}_6 = \begin{bmatrix} 0 \\ 1 \\ 0 \end{bmatrix}, \quad \mathbf{P}_7 = \begin{bmatrix} 0 \\ 0 \\ 1 \end{bmatrix}, \quad \text{and} \quad \mathbf{b} = \begin{bmatrix} 11 \\ 3 \\ 1 \end{bmatrix}$$

Since (x_4, x_6, x_7) form the initial basis,

$$\underset{(3\times 3)}{\mathbf{B}} = [\mathbf{P}_4, \mathbf{P}_6, \mathbf{P}_7] = \begin{bmatrix} 1 & 0 & 0 \\ 0 & 1 & 0 \\ 0 & 0 & 1 \end{bmatrix} = \mathbf{I}$$

Hence,

$$\mathbf{B}^{-1} = \mathbf{I} \qquad \text{and} \qquad \bar{\mathbf{b}} = \mathbf{B}^{-1}\mathbf{b} = \mathbf{b}$$

The initial tableau of the revised simplex method is given below (the last two columns are added later):

Tableau 1

Basis	$\mathbf{B}^{-1}$			Constants	Variable to Enter	Pivot Column
x_4	1	0	0	11		1
x_6	0	1	0	3	x_3	2
x_7	0	0	1	1		①

The simplex multipliers are

$$\boldsymbol{\pi} = (\pi_1, \pi_2, \pi_3) = \mathbf{c}_B\mathbf{B}^{-1} = (0, M, M) \begin{bmatrix} 1 & 0 & 0 \\ 0 & 1 & 0 \\ 0 & 0 & 1 \end{bmatrix} = (0, M, M)$$

Since $\bar{c}_j = c_j - \boldsymbol{\pi}\mathbf{P}_j$ for $j = 1, 2, 3, 5$ we get,

$$\bar{c}_1 = -3 - (0, M, M) \begin{bmatrix} 1 \\ -4 \\ -2 \end{bmatrix} = 6M - 3$$

$$\bar{c}_2 = \quad 1 - (0, M, M) \begin{bmatrix} -2 \\ 1 \\ 0 \end{bmatrix} = 1 - M$$

$$\bar{c}_3 = \quad 1 - (0, M, M) \begin{bmatrix} 1 \\ 2 \\ 1 \end{bmatrix} = 1 - 3M$$

$$\bar{c}_5 = \quad 0 - (0, M, M) \begin{bmatrix} 0 \\ -1 \\ 0 \end{bmatrix} = M$$

Since $\bar{c}_3$ is most negative, x_3 enters the basis. The pivot column is

$$\bar{\mathbf{P}}_3 = \mathbf{B}^{-1}\mathbf{P}_3 = \begin{bmatrix} 1 & 0 & 0 \\ 0 & 1 & 0 \\ 0 & 0 & 1 \end{bmatrix} \begin{bmatrix} 1 \\ 2 \\ 1 \end{bmatrix} = \begin{bmatrix} 1 \\ 2 \\ 1 \end{bmatrix}$$

(The entering variable x_3 and the pivot column elements are now entered in Tableau 1.) Applying the minimum ratio rule, the ratios are (11/1, 3/2, 1/1). Hence x_3 replaces the artificial variable x_7 in the basis. This is shown by circling the pivot element in Tableau 1. Using the pivot column, the pivot operation is done on Tableau 1 as follows:

1. Multiply row 3 by -1, and add it to row 1.
2. Multiply row 3 by -2, and add it to row 2.

The new $\mathbf{B}^{-1}$, and the constants are given below:

Tableau 2

Basis	$\mathbf{B}^{-1}$			Constants	Variable to Enter	Pivot Column
x_4	1	0	-1	10		-2
x_6	0	1	-2	1	x_2	①
x_3	0	0	1	1		0

The simplex multipliers corresponding to Tableau 2 are

$$\boldsymbol{\pi} = (0, M, 1) \begin{bmatrix} 1 & 0 & -1 \\ 0 & 1 & -2 \\ 0 & 0 & 1 \end{bmatrix} = (0, M, -2M + 1)$$

The $\bar{c}_j$ elements are given by $\bar{c}_1 = -1$, $\bar{c}_2 = 1 - M$ and $\bar{c}_5 = M$. ($\bar{c}_7$ is not calculated because x_7 is an artificial variable). Since $\bar{c}_2$ is the most negative, x_2 enters the basis and the pivot column becomes

$$\bar{\mathbf{P}}_2 = \mathbf{B}^{-1}\mathbf{P}_2 = \begin{bmatrix} 1 & 0 & -1 \\ 0 & 1 & -2 \\ 0 & 0 & 1 \end{bmatrix} \begin{bmatrix} -2 \\ 1 \\ 0 \end{bmatrix} = \begin{bmatrix} -2 \\ 1 \\ 0 \end{bmatrix}$$

The ratios are $(\infty, 1/1, \infty)$. Hence x_2 replaces the artificial variable x_6 in the basis. (We shall discard x_6 from further consideration.) Performing the pivot operation we obtain the new basic feasible solution shown in Tableau 3.

Tableau 3

Basis	$\mathbf{B}^{-1}$			Constants	Variable to Enter	Pivot Column
x_4	1	2	−5	12		③
x_2	0	1	−2	1	x_1	0
x_3	0	0	1	1		−2

The simplex multipliers of Tableau 3 are

$$\pi = (0, 1, 1) \begin{bmatrix} 1 & 2 & -5 \\ 0 & 1 & -2 \\ 0 & 0 & 1 \end{bmatrix} = (0, 1, -1)$$

The $\bar{c}_j$ elements of the nonbasic variables are given by,

$$\bar{c}_1 = -1 \qquad \text{and} \qquad \bar{c}_5 = 1$$

Thus x_1 enters the basis. The pivot column becomes

$$\bar{\mathbf{P}}_1 = \begin{bmatrix} 1 & 2 & -5 \\ 0 & 1 & -2 \\ 0 & 0 & 1 \end{bmatrix} \begin{bmatrix} 1 \\ -4 \\ -2 \end{bmatrix} = \begin{bmatrix} 3 \\ 0 \\ -2 \end{bmatrix}$$

By the minimum ratio rule, x_1 replaces x_4 in the basis, and the new tableau after the pivot operation is given below:

Tableau 4

Basis	$\mathbf{B}^{-1}$			Constants
x_1	$\frac{1}{3}$	$\frac{2}{3}$	$-\frac{5}{3}$	4
x_2	0	1	−2	1
x_3	$\frac{2}{3}$	$\frac{4}{3}$	$-\frac{7}{3}$	9

The simplex multipliers of Tableau 4 are

$$\pi = [-3, 1, 1] \begin{bmatrix} \frac{1}{3} & \frac{2}{3} & -\frac{5}{3} \\ 0 & 1 & -2 \\ \frac{2}{3} & \frac{4}{3} & -\frac{7}{3} \end{bmatrix} = \left[-\frac{1}{3}, \frac{1}{3}, \frac{2}{3} \right]$$

and

$$\bar{c}_4 = 0 - \left[-\frac{1}{3}, \frac{1}{3}, \frac{2}{3}\right] \begin{bmatrix} 1 \\ 0 \\ 0 \end{bmatrix} = \frac{1}{3}$$

$$\bar{c}_5 = 0 - \left[-\frac{1}{3}, \frac{1}{3}, \frac{2}{3}\right] \begin{bmatrix} 0 \\ -1 \\ 0 \end{bmatrix} = \frac{1}{3}$$

Hence Tableau 4 is optimal, and the unique optimal solution is given by

$$x_1 = 4, \qquad x_2 = 1, \qquad x_3 = 9, \qquad x_4 = 0, \qquad \text{and} \qquad x_5 = 0$$

The optimal value of the objective function is

$$z = \mathbf{c}_B \bar{\mathbf{b}} = (-3, 1, 1) \begin{pmatrix} 4 \\ 1 \\ 9 \end{pmatrix} = -2$$

Since Example 4.1-2 has been solved by the standard simplex method in Section 2.9, it is suggested that the reader compare the two to see the major differences between them.

Advantages of the Revised Simplex Method Over the Standard Simplex Method

1. There is a reduced amount of computations when the number of variables in the linear program is much larger than the number of constraints. The revised simplex method works with a reduced tableau whose size is determined by the number of constraints. Hence, the amount of computations is considerably reduced.
2. Less new information is stored in the computer memory from one iteration to the next. Since the revised simplex method works with a reduced tableau, it stores only the basic variables, the basis inverse, and the constants. The data matrix A and the objective function coefficients are kept externally on a file (tape or disk) so that larger problems can be solved. Many commercial codes store even the basis inverse on tape using what is known as the *product form of the inverse*.
3. Less space is needed for recording data. The original data is usually given in fixed decimals of three or four digits. Since the revised simplex method works only with the original data, the data can be stored compactly and more accurately.
4. Less accumulation of round-off errors occurs because no calculations are done on a column until it is ready to enter the basis. Many commercial codes periodically get the basis inverse directly by inverting the original columns instead of using the pivot operation. This further reduces the accumulation of round-off errors.
5. For problems with a large percentage of zero coefficients (practical problems often have 90% or more zero elements), the revised simplex method performs fewer multiplications involving nonzero elements as compared to the standard simplex method, in which the zero elements disappear during the initial pivot operations.
6. The theory of the revised simplex method, especially the importance of the basis inverse and the simplex multipliers, is helpful in understanding the other advanced topics in linear programming like duality theory and sensitivity analysis.

We conclude this section with the remark that the above advantages of the revised simplex method are mainly for computation by a digital computer. For hand calculations, the above may not be significant because of the amount of orderly bookkeeping, and scratch work needed in the revised simplex method.

4.2 DUALITY THEORY AND ITS APPLICATIONS

From both the theoretical and practical points of view, the theory of duality is one of the most important and interesting concepts in linear programming. The basic idea behind the duality theory is that every linear programming problem has an associated linear program called its *dual* such that a solution to the original linear program also gives a solution to its dual. Thus, whenever a linear program is solved by the simplex method, we are actually getting solutions for two linear programming problems!

We shall introduce the concept of a dual with the following linear program.

EXAMPLE 4.2-1

Maximize: $Z = x_1 + 2x_2 - 3x_3 + 4x_4$

Subject to:

$$x_1 + 2x_2 + 2x_3 - 3x_4 \leqq 25$$
$$2x_1 + x_2 - 3x_3 + 2x_4 \leqq 15$$
$$x_1, \ldots, x_4 \geqq 0$$

The above linear program has two constraints and four variables. The dual of this problem is written as

Minimize: $W = 25y_1 + 15y_2$

Subject to:

$$y_1 + 2y_2 \geqq 1$$
$$2y_1 + y_2 \geqq 2$$
$$2y_1 - 3y_2 \geqq -3$$
$$-3y_1 + 2y_2 \geqq 4$$
$$y_1 \geqq 0, y_2 \geqq 0$$

y_1 and y_2 are called the dual variables. The original problem is called the *primal* problem. Comparing the primal and the dual problems, we observe the following relationships:

1. The objective function coefficients of the primal problem have become the right-hand-side constants of the dual. Similarly, the right-hand-side constants of the primal have become the cost coefficients of the dual.
2. The inequalities have been reversed in the constraints.
3. The objective is changed from maximization in primal to minimization in dual.
4. Each column in the primal corresponds to a constraint (row) in the dual. Thus the number of dual constraints is equal to the number of primal variables.
5. Each constraint (row) in the primal corresponds to a column in the dual. Hence there is one dual variable for every primal constraint.
6. The dual of the dual is the primal problem.

In both of the primal and the dual problems, the variables are nonnegative and the constraints are inequalities. Such problems are called *symmetric dual linear programs.*

Definition. A linear program is said to be in *symmetric form*, if all the variables are restricted to be nonnegative, and all the constraints are inequalities (in a maximization problem the inequalities must be in ‘‘less than or equal to’’ form; while in a minimization problem they must be ‘‘greater than or equal to’’).

Symmetric Dual Linear Programs

We shall now give the general representation of the primal-dual problems in symmetric form.

Primal

Maximize: $Z = c_1x_1 + c_2x_2 + \cdots + c_nx_n$

Subject to:

$$
\begin{aligned}
a_{11}x_1 + a_{12}x_2 + \cdots + a_{1n}x_n &\leqq b_1 \\
a_{21}x_1 + a_{22}x_2 + \cdots + a_{2n}x_n &\leqq b_2 \\
\vdots \qquad & \qquad \vdots \\
a_{m1}x_1 + a_{m2}x_2 + \cdots + a_{mn}x_n &\leqq b_m \\
x_1, x_2, \ldots, x_n &\geqq 0
\end{aligned}
$$

Dual

Minimize: $W = b_1y_1 + b_2y_2 + \cdots + b_my_m$

Subject to:

$$
\begin{aligned}
a_{11}y_1 + a_{21}y_2 + \cdots + a_{m1}y_m &\geqq c_1 \\
a_{12}y_1 + a_{22}y_2 + \cdots + a_{m2}y_m &\geqq c_2 \\
\vdots \qquad & \qquad \vdots \\
a_{1n}y_1 + a_{2n}y_2 + \cdots + a_{mn}y_m &\geqq c_n \\
y_1, y_2, \ldots, y_m &\geqq 0
\end{aligned}
$$

In matrix notation the symmetric dual linear programs are:

Primal

Maximize: $Z = \mathbf{cx}$

Subject to:

$$
\begin{aligned}
\mathbf{Ax} &\leqq \mathbf{b} \\
\mathbf{x} &\geqq \mathbf{0}
\end{aligned}
$$

Dual

Minimize: $W = \mathbf{yb}$

Subject to:

$$
\begin{aligned}
\mathbf{yA} &\geqq \mathbf{c} \\
\mathbf{y} &\geqq \mathbf{0}
\end{aligned}
$$

where $\mathbf{A}$ is an $(m \times n)$ matrix, $\mathbf{b}$ is an $(m \times 1)$ column vector, $\mathbf{c}$ is a $(1 \times n)$ row vector, $\mathbf{x}$ is an $(n \times 1)$ column vector, and $\mathbf{y}$ is an $(1 \times m)$ row vector.

The general rules for writing the dual of a linear program in symmetric form are summarized below:

1. Define one (nonnegative) dual variable for each primal constraint.
2. Make the cost vector of the primal the right-hand-side constants of the dual.
3. Make the right-hand-side vector of the primal the cost vector of the dual.
4. The transpose of the coefficient matrix of the primal becomes the constraint matrix of the dual.
5. Reverse the direction of the constraint inequalities.
6. Reverse the optimization direction, that is, change minimizing to maximizing and vice versa.

Rules for writing the dual of the asymmetric linear programs will be developed later in this section.

Economic Interpretation of the Dual Problem

A midwestern manufacturer is faced with the problem of transporting his goods from two warehouses to three retail outlets at minimum cost. The supplies at the warehouses are 300 and 600 units. The demands at the retail outlets are 200, 300, and 400 units, respectively. The unit cost of transportation (in dollars) from a warehouse (W) to a retail outlet (R) is given below:

	R_1	R_2	R_3
W_1	2	4	3
W_2	5	3	4

The manufacturer's problem is to determine the least-cost shipping schedule which will meet the demands with the available supplies. The manufacturer's problem is actually a transportation problem which was discussed in detail in Section 3.2. If x_{ij} denotes the quantity shipped from warehouse i to the retail outlet j ($i=1, 2$ and $j=1, 2, 3$), then the linear programming formulation becomes:

$$
\begin{aligned}
\text{Minimize:}\quad & Z=2x_{11}+4x_{12}+3x_{13}+5x_{21}+3x_{22}+4x_{23}\\
\text{Subject to:}\quad & x_{11}+x_{12}+x_{13} \geqq \leqq 300
\end{aligned}
$$

For the above linear program to be in symmetric form, all the constraints must be in "greater than or equal to" form. Hence we multiply the first two (supply) constraints by -1. The manufacturer's problem in symmetric form, which we will call the *primal* problem is given below:

Primal

$$
\begin{aligned}
\text{Minimize:}\quad & Z=2x_{11}+4x_{12}+3x_{13}+5x_{21}+3x_{22}+4x_{23}\\
\text{Subject to:}\quad & -x_{11}-x_{12}-x_{13} \geqq -300\\
& -x_{21}-x_{22}-x_{23} \geqq -600\\
& x_{11}+x_{21} \geqq 200\\
& x_{12}+x_{22} \geqq 300\\
& x_{13}+x_{23} \geqq 400\\
& x_{11},\ldots,x_{23} \geqq 0
\end{aligned}
$$

The dual of the above problem becomes:

Dual

$$
\begin{aligned}
\text{Maximize:}\quad & \\
& W=-300y_1-600y_2+200y_3+300y_4+400y_5\\
\text{Subject to:}\quad & -y_1+y_3 \leqq 2\\
& -y_1+y_4 \leqq 4\\
& -y_1+y_5 \leqq 3\\
& -y_2+y_3 \leqq 5\\
& -y_2+y_4 \leqq 3\\
& -y_2+y_5 \leqq 4\\
& y_1, y_2, y_3, y_4, y_5 \geqq 0
\end{aligned}
$$

Let us now give an economic interpretation of the dual problem. Imagine that a nationwide moving company approaches the manufacturer with the proposition that it will buy all the 300 units at warehouse 1 paying \$$y_1$ per unit, and all the 600 units at warehouse 2 paying \$$y_2$ per unit. It will then deliver 200, 300, and 400 units at the retail outlets 1, 2, and 3 selling them back to the manufacturer at a unit price of \$$y_3$, \$$y_4$, and \$$y_5$, respectively. The moving company then uses the dual constraints to convince the manufacturer that employing him is cheaper than transporting the goods on his own. For example, consider the transportation of goods from warehouse 1 to the retail outlet 1. The manufacturer's cost of transporting 1 unit is \$2. Instead, if he employs the moving company his net cost is only \$$(y_3 - y_1)$. Since $y_3 - y_1 \leqq 2$ (first dual constraint), employing the moving company seems attractive for transporting from warehouse 1 to retail outlet 1. Similarly each dual constraint implies that the transportation cost in any route is as expensive or more than the net cost of selling and buying. Hence, the manufacturer will accept the nationwide moving company's proposition with the hope that he will save some money in transportation. But, the moving company will fix the values of $y_1, \ldots, y_5$ in such a way that the dual constraints are satisfied and its net profit is maximized. The moving company's return is given by

$$-300y_1 - 600y_2 + 200y_3 + 300y_4 + 400y_5$$

which is the dual objective function. Thus the dual problem is the moving company's problem which is trying to maximize its return.

The duality theory says that the optimal values of both the primal and the dual problems are always equal. Hence, the manufacturer does not really save any money since he will be paying the minimum transportation cost to the moving company. But, he is saved from the trouble of solving the (primal) linear program to determine the least transportation cost. At the same time, the nationwide moving company has bagged the business of moving the goods at maximum profit.

Now we shall turn our attention to some of the duality theorems that give important relationships between the primal and the dual solutions.

THEOREM 1 WEAK DUALITY THEOREM

Consider the symmetric primal-dual linear programs, max $Z = \mathbf{cx}$, $\mathbf{Ax} \leqq \mathbf{b}$, $\mathbf{x} \geqq \mathbf{0}$, and min $W = \mathbf{yb}$, $\mathbf{yA} \geqq \mathbf{c}$, $\mathbf{y} \geqq \mathbf{0}$. The value of the objective function of the minimum problem (dual) for any feasible solution is always greater than or equal to that of the maximum problem (primal).

Proof

Let $\mathbf{x}^0$ and $\mathbf{y}^0$ be feasible solution vectors to the primal and the dual problems, respectively. We have to prove

$$\mathbf{y}^0\mathbf{b} \geqq \mathbf{cx}^0$$

Since $\mathbf{x}^0$ is feasible for the primal,

$$\begin{aligned} \mathbf{Ax}^0 &\leqq \mathbf{b} \\ \mathbf{x}^0 &\geqq \mathbf{0} \end{aligned} \tag{4.3}$$

Similarly, since $\mathbf{y}^0$ is feasible for the dual,

$$\begin{aligned} \mathbf{y}^0\mathbf{A} &\geqq \mathbf{c} \\ \mathbf{y}^0 &\geqq \mathbf{0} \end{aligned} \tag{4.4}$$

Multiplying both sides of Inequality 4.3 by $\mathbf{y}^0$ (actually, we take the inner product with respect to $\mathbf{y}^0$), we get

$$\mathbf{y}^0\mathbf{A}\mathbf{x}^0 \leqq \mathbf{y}^0\mathbf{b} \tag{4.5}$$

Similarly, multiplying both sides of Inequality 4.4 by $\mathbf{x}^0$,

$$\mathbf{y}^0\mathbf{A}\mathbf{x}^0 \geqq \mathbf{c}\mathbf{x}^0 \tag{4.6}$$

Inequalities 4.5 and 4.6 imply that

$$\mathbf{y}^0\mathbf{b} \geqq \mathbf{y}^0\mathbf{A}\mathbf{x}^0 \geqq \mathbf{c}\mathbf{x}^0$$

From the weak duality theorem we can infer the following important results:

Corollary 1. The value of the objective function of the maximum (primal) problem for any (primal) feasible solution is a lower bound to the minimum value of the dual objective.

Corollary 2. Similarly the objective function value of the minimum problem (dual) for any (dual) feasible solution is an upper bound to the maximum value of the primal objective.

Corollary 3. If the primal problem is feasible and its objective is unbounded (ie., max $Z \to +\infty$), then the dual problem cannot have a feasible solution.

Corollary 4. Similarly, if the dual problem is feasible, and is unbounded (i.e., min $W \to -\infty$), then the primal problem is infeasible.

To illustrate the weak duality theorem, consider the following primal dual problems:

EXAMPLE 4.2-2

Primal

Maximize: $Z = x_1 + 2x_2 + 3x_3 + 4x_4$

Subject to:

$$\begin{aligned} x_1 + 2x_2 + 2x_3 + 3x_4 &\leqq 20 \\ 2x_1 + x_2 + 3x_3 + 2x_4 &\leqq 20 \\ x_1, \ldots, x_4 &\geqq 0 \end{aligned}$$

Dual

Minimize: $W = 20y_1 + 20y_2$

Subject to:

$$\begin{aligned} y_1 + 2y_2 &\geqq 1 \\ 2y_1 + y_2 &\geqq 2 \\ 2y_1 + 3y_2 &\geqq 3 \\ 3y_1 + 2y_2 &\geqq 4 \\ y_1, y_2 &\geqq 0 \end{aligned}$$

$x_1^0 = x_2^0 = x_3^0 = x_4^0 = 1$ is feasible for the primal and $y_1^0 = y_2^0 = 1$ is feasible for the dual. The value of the primal objective is

$$Z = \mathbf{c}\mathbf{x}^0 = 10$$

The value of the dual objective is

$$W = \mathbf{y}^0\mathbf{b} = 40$$

Note that $\mathbf{c}\mathbf{x}^0 < \mathbf{y}^0\mathbf{b}$ which satisfies the weak duality theorem.

Using Corollary 1, the minimum value of W for the dual objective function cannot go below 10. Similarly, from Corollary 2, the maximum value of Z for the primal problem cannot exceed 40.

The converse of the results of Corollaries 3 and 4 are also true.

Corollary 5. If the primal problem is feasible, and the dual is infeasible, then the primal is unbounded.

Corollary 6. If the dual problem is feasible and the primal is infeasible, then the dual is unbounded.

EXAMPLE 4.2-3

Consider the following primal-dual problems:

Primal

Maximize: $Z = x_1 + x_2$

Subject to:

$$-x_1 + x_2 + x_3 \leqq 2$$
$$-2x_1 + x_2 - x_3 \leqq 1$$
$$x_1, x_2, x_3 \geqq 0$$

Dual

Minimize: $W = 2y_1 + y_2$

Subject to:

$$-y_1 - 2y_2 \geqq 1$$
$$y_1 + y_2 \geqq 1$$
$$y_1 - y_2 \geqq 0$$
$$y_1, y_2 \geqq 0$$

$x_1 = x_2 = x_3 = 0$ is a feasible solution to the primal problem. But the dual problem is infeasible since the constraint $-y_1 - 2y_2 \geqq 1$ is inconsistent. (For all nonnegative values of y_1 and y_2, the left-hand side is nonpositive while the right-hand side is strictly positive.) Hence, by Corollary 5, the primal problem has an unbounded solution such that maximum Z tends to infinity. This can be easily verified by solving the primal problem by the simplex method.

THEOREM 2 OPTIMALITY CRITERION THEOREM

If there exist feasible solutions $\mathbf{x}^0$ and $\mathbf{y}^0$ for the symmetric dual linear programs such that the corresponding values of their objective functions are equal, then these feasible solutions are in fact optimal solutions to their respective problems.

Proof

Let $\mathbf{x}$ be any other feasible solution to the primal problem. Then by Theorem 1,

$$\mathbf{cx} \leqq \mathbf{y}^0\mathbf{b}$$

But it is given that $\mathbf{cx}^0 = \mathbf{y}^0\mathbf{b}$. Hence $\mathbf{cx} \leqq \mathbf{cx}^0$ for all feasible solutions to the primal problem. Then by definition, $\mathbf{x}^0$ is optimal for the primal. A symmetrical argument proves the optimality of $\mathbf{y}^0$ for the dual problem.

Illustration

Consider the primal-dual pair given in Example 4.2-2, $x_1^0 = 0$, $x_2^0 = 0$, $x_3^0 = 4$, and $x_4^0 = 4$ is a feasible solution to the primal, while $y_1^0 = 1.2$, and $y_2^0 = 0.2$ is feasible for the dual. The value of Z for the primal = 28 = the value of the W for the dual. Hence by Theorem 2, the above feasible solutions are optimal solutions to the primal and the dual problems, respectively.

THEOREM 3 MAIN DUALITY THEOREM

If both the primal and the dual problems are feasible, then they both have optimal solutions such that their optimal values of the objective functions are equal.

Proof

When both the primal and the dual problems are feasible, then by Corollaries 1 and 2 of Theorem 1, we have a lower bound on the minimum value of W, and an upper bound on the maximum value of Z. In other words, neither the primal nor the dual can have an unbounded solution. Therefore they both must have optimal solutions, but the difficult part of Theorem 3 is to show that at the optimum both the problems will have the same value for the objective function. The proof of this is rather involved and hence is omitted here. Interested readers may refer to Gale (13).

THEOREM 4 COMPLEMENTARY SLACKNESS THEOREM

Consider the symmetric dual problems in matrix form:

Primal

Maximize: $Z = \mathbf{cx}$

Subject to: $\mathbf{Ax} \leq \mathbf{b}$

$\mathbf{x} \geqq \mathbf{0}$

Dual

Minimize: $W = \mathbf{yb}$

Subject to: $\mathbf{yA} \geqq \mathbf{c}$

$\mathbf{y} \geqq \mathbf{0}$

where $\mathbf{A}$ is an $(m \times n)$ matrix, $\underset{(m\times 1)}{\mathbf{b}}$ and $\underset{(n\times 1)}{\mathbf{x}}$ are column vectors, and $\underset{(1\times n)}{\mathbf{c}}$ and $\underset{(1\times m)}{\mathbf{y}}$ are row vectors. Let $\mathbf{x}^0$ and $\mathbf{y}^0$ be feasible for the primal and the dual problems respectively. Then $\mathbf{x}^0$ and $\mathbf{y}^0$ are optimal to their respective problems if and only if

$$(\mathbf{y}^0\mathbf{A} - \mathbf{c})\mathbf{x}^0 + \mathbf{y}^0(\mathbf{b} - \mathbf{A}\mathbf{x}^0) = 0$$

Proof

Let the column vector $\underset{(m\times 1)}{\mathbf{u}} = \begin{pmatrix} u_1 \\ u_2 \\ \cdot \\ \cdot \\ u_m \end{pmatrix}$ represent the slack vector for the primal, and the row vector $\underset{(1\times n)}{\mathbf{v}} = (v_1, v_2, \ldots, v_n)$ be the slack vector of dual. Since $\mathbf{x}^0$ and $\mathbf{y}^0$ are feasible solutions we have

$$\mathbf{A}\mathbf{x}^0 + \mathbf{u}^0 = \mathbf{b}; \qquad \mathbf{x}^0, \mathbf{u}^0 \geqq \mathbf{0} \tag{4.7}$$

$$\mathbf{y}^0\mathbf{A} - \mathbf{v}^0 = \mathbf{c}; \qquad \mathbf{y}^0, \mathbf{v}^0 \geqq \mathbf{0} \tag{4.8}$$

($\mathbf{u}^0$ and $\mathbf{v}^0$ represent the values of the slack variables $\mathbf{u}$ and $\mathbf{v}$ corresponding to the feasible solutions $\mathbf{x}^0$ and $\mathbf{y}^0$.)

Multiplying Eq. 4.7 by $\mathbf{y}^0$ (i.e., taking the inner product with $\mathbf{y}^0$) we get

$$\mathbf{y}^0\mathbf{A}\mathbf{x}^0 + \mathbf{y}^0\mathbf{u}^0 = \mathbf{y}^0\mathbf{b} \tag{4.9}$$

Similarly, multiplying Eq. 4.8 by $\mathbf{x}^0$ we get

$$\mathbf{y}^0\mathbf{A}\mathbf{x}^0 - \mathbf{v}^0\mathbf{x}^0 = \mathbf{c}\mathbf{x}^0 \tag{4.10}$$

Subtracting Eq. 4.10 from Eq. 4.9, we obtain

$$\mathbf{y}^0\mathbf{u}^0 + \mathbf{v}^0\mathbf{x}^0 = \mathbf{y}^0\mathbf{b} - \mathbf{c}\mathbf{x}^0 \tag{4.11}$$

To prove Theorem 4 we have to show that $\mathbf{x}^0$ and $\mathbf{y}^0$ are optimal to the primal and the dual problems if and only if

$$\mathbf{v}^0\mathbf{x}^0 + \mathbf{y}^0\mathbf{u}^0 = 0 \tag{4.12}$$

Part 1. We will assume that $\mathbf{x}^0$ and $\mathbf{y}^0$ are optimal solutions, and prove that Eq. 4.12 is true. Since $\mathbf{x}^0$ and $\mathbf{y}^0$ are optimal, $\mathbf{c}\mathbf{x}^0 = \mathbf{y}^0\mathbf{b}$ by Theorem 3. Hence, Eq. 4.11 reduces to Eq. 4.12.

Part 2. We will assume that Eq. 4.12 is true and prove that $\mathbf{x}^0$ and $\mathbf{y}^0$ are optimal to the primal and the dual, respectively. Since Eq. 4.12 is true, Eq. 4.11 reduces to

$$\mathbf{y}^0\mathbf{b} = \mathbf{c}\mathbf{x}^0$$

By Theorem 2 it follows that $\mathbf{x}^0$ and $\mathbf{y}^0$ are optimal solutions.

Complementary Slackness Conditions

Equation 4.12 of the complementary slackness theorem can be further simplified to

$$v_j^0 x_j^0 = 0 \qquad \text{for all } j = 1, 2, \ldots, n \tag{4.13}$$
$$y_i^0 u_i^0 = 0 \qquad \text{for all } i = 1, 2, \ldots, m \tag{4.14}$$

by observing the following:

1. $\mathbf{x}^0, \mathbf{u}^0, \mathbf{v}^0, \mathbf{y}^0 \geqq 0$ and hence $\mathbf{v}^0\mathbf{x}^0 \geqq 0$ and $\mathbf{y}^0\mathbf{u}^0 \geqq 0$.
2. If the sum of nonnegative terms equals zero, then each term is zero.

Equations 4.13 and 4.14 are generally called the *complementary slackness conditions*. In words the complementary slackness condtions can be stated as follows:

1. If a primal variable (x_j^0) is positive, then the corresponding dual constraint will be satisfied as an equation at the optimum (i.e., $v_j^0 = 0$).
2. If a primal constraint is a strict inequality at the optimum (i.e., $u_i^0 > 0$), then the corresponding dual variable (y_i^0) must be zero at the optimum.
3. If a dual variable (y_i^0) is positive, then the corresponding primal constraint will be satisfied as an equation at the optimum (i.e., $u_i^0 = 0$).
4. If a dual constraint is a strict inequality (i.e., $v_j^0 > 0$), then the corresponding primal variable (x_j^0) must be zero at the optimum.

Illustration

Consider Example 4.2-2. With the addition of slack variables, the primal-dual problems may be stated as follows:

Primal

Maximize: $\qquad Z = x_1 + 2x_2 + 3x_3 + 4x_4$

Subject to:

$$x_1 + 2x_2 + 2x_3 + 3x_4 + u_1 = 20$$
$$2x_1 + x_2 + 3x_3 + 2x_4 + u_2 = 20$$
$$x_1, x_2, x_3, x_4, u_1, u_2 \geqq 0$$

Dual

Minimize: $$W = 20y_1 + 20y_2$$

Subject to:

$$y_1 + 2y_2 - v_1 = 1$$
$$2y_1 + y_2 - v_2 = 2$$
$$2y_1 + 3y_2 - v_3 = 3$$
$$3y_1 + 2y_2 - v_4 = 4$$
$$y_1, y_2, v_1, v_2, v_3, v_4 \geqq 0$$

The complementary slackness conditions imply that at the optimum $u_1^0 y_1^0 = 0$, $u_2^0 y_2^0 = 0$, $x_1^0 v_1^0 = 0$, $x_2^0 v_2^0 = 0$, $x_3^0 v_3^0 = 0$, and $x_4^0 v_4^0 = 0$. From these conditions it is possible to determine the primal optimal solution from the dual optimal solution, and vice versa.

Without the slack variables the dual problem has just two variables, and hence can be solved by the graphical methods discussed in Section 2.3. The optimal solution is found as $y_1^0 = 1.2$, $y_2^0 = 0.2$, and min $W = 28$. By applying the complementary slackness conditions, the primal optimal solution is determined as follows:

1. $y_1^0 = 1.2 > 0$ implies that $u_1^0 = 0$
2. $y_2^0 = 0.2 > 0$ implies that $u_2^0 = 0$
3. $y_1^0 + 2y_2^0 = 1.6 > 1$ implies that $v_1^0 > 0$ and $x_1^0 = 0$
4. $2y_1^0 + y_2^0 = 2.6 > 2$ implies that $v_2^0 > 0$ and $x_2^0 = 0$
5. $2y_1^0 + 3y_2^0 = 3$ implies that $v_3^0 = 0$. Since $v_3^0 x_3^0 = 0$, x_3^0 could be positive or zero.
6. Similarly $3y_1^0 + 2y_2^0 = 4$ and $v_4^0 = 0$. Hence x_4^0 is positive or zero.

In other words Conditions 5 and 6 do not give any new information in determining the primal optimal solution. But at the optimum, Conditions 1 through 4 imply that,

$$2x_3^0 + 3x_4^0 = 20$$
$$3x_3^0 + 2x_4^0 = 20$$

Solving the two equations in two unknowns, we get the optimal primal solution as $x_1^0 = 0$, $x_2^0 = 0$, $x_3^0 = 4$, and $x_4^0 = 4$. The maximum value of Z equals 28 which corresponds to the minimum value of W, verifying Theorem 3.

The complementary slackness theorem is also very useful in testing some hypotheses on the nature of optimal solutions to the linear programs. For instance, we can check the hypothesis whether both the primal constraints are strict inequalities at the optimum; in other words, all the available resources are not fully utilized.

Mathematically this implies that $u_1^0 > 0$ and $u_2^0 > 0$. The complementary slackness conditions imply $y_1^0 u_1^0 = 0$ and $y_2^0 u_2^0 = 0$. Hence if the hypothesis is true, $y_1^0 = y_2^0 = 0$ must be optimal for the dual. But we find that this is not true since $y_1^0 = y_2^0 = 0$ is infeasible; hence the hypothesis is false.

Applications of the Complementary Slackness Conditions

A partial list of the general applications of the complementary slackness conditions is given below:

1. Used in finding an optimal primal solution from the given optimal dual solution and vice versa.

2. Used in verifying whether a feasible solution is optimal for the primal problem. (Here we assume the given feasible solution as optimal, and try to construct an optimal dual solution using the complementary slackness conditions. If we are successful, then the given feasible solution is in fact optimal for the primal.)
3. Used in investigating the general properties of the optimal solutions to primal and dual by testing different hypotheses.
4. The Kuhn-Tucker optimality conditions of nonlinear programming are direct extensions of the complementary slackness conditions, and are extremely useful in advanced mathematical programming.

Thus far our discussion has been limited to symmetric primal-dual problems. Before considering the asymmetric problems, let us summarize the essential characteristics of the symmetric primal-dual pair.

	Primal	Dual
$\mathbf{A}$	Constraint matrix	Transpose of the constraint matrix
$\mathbf{b}$	Right-hand-side constants	Cost (price) vector
$\mathbf{c}$	Price (cost) vector	Right-hand-side constants
Objective function	Maximize $Z=\mathbf{cx}$	Minimize $W=\mathbf{yb}$
Constraint inequalities	$\mathbf{Ax}\leqq\mathbf{b}$	$\mathbf{yA}\geqq\mathbf{c}$
Decision variables	$\mathbf{x}\geqq\mathbf{0}$	$\mathbf{y}\geqq\mathbf{0}$

Notes

1. If the primal problem is a minimization problem, then we have the following primal-dual relationships:

Objective function	Minimize $Z=\mathbf{cx}$	Maximize $W=\mathbf{yb}$
Constants	$\mathbf{Ax}\geqq\mathbf{b}$	$\mathbf{yA}\leqq\mathbf{c}$
	$\mathbf{x}\geqq\mathbf{0}$	$\mathbf{y}\geqq\mathbf{0}$

2. The dual of the dual is the primal problem.

Economic Interpretation of the Dual Solution

In an economic sense, the optimal dual solution can be interpreted as the price one pays for the constraint resources. By Theorem 3 (main duality) the optimal values of the objective functions of the primal and dual are equal. If $\mathbf{x}^0$ and $\mathbf{y}^0$ are the respective optimal solutions then $Z_0=\mathbf{cx}^0=\mathbf{y}^0\mathbf{b}=W_0$. In other words, the optimal value of the linear program (primal or dual), is given by

$$Z_o=y_1^0b_1+y_2^0b_2+\cdots+y_m^0b_m$$

where $b_1, b_2, \ldots, b_m$ represent the limited quantities of the resources $1, 2, \ldots, m$, and $y_1^0, y_2^0, \ldots, y_m^0$ are the optimal values of the dual variables. Suppose we assume that the level of resource 1 (i.e., b_1) can be altered. Then, for small variations in the value of b_1, say Δb_1, the net change in the optimal value of the linear program Z_0 is given by $y_1^0(\Delta b_1)$.

In other words, the optimal value of the dual variable for each primal constraint gives the net change in the optimal value of the objective function for unit increase in the right-hand-side constants. Hence, these are called *shadow prices* on the constraint resources. These shadow prices could be used to determine whether it is economical to get additional resources at premium prices. The application of shadow price in the post-optimality analysis is discussed in detail in Chapter 2, Section 2.11.

To illustrate the application of the optimal dual solution, consider Example 4.2-2. Assume that the constraints correspond to raw materials A and B used in making four products, x_1, x_2, x_3, and x_4. Let the objective function represent the total profit. The solution of the primal problem by the simplex method gives the maximum profit as $Z=28$. Since the dual problem involves only two variables, it can be solved graphically and the optimal dual solution becomes $y_1=1.2$, $y_2=0.2$ and minimum $W=28$. Thus, $y_1=1.2$ represents the shadow price of raw material A, that is, the net change in the maximum value of Z per unit increase in raw material A. Similarly, $y_2=0.2$ represents the shadow price of raw material B. In other words, the maximum price one can pay for additional units of raw materials A and B is limited to 1.2 and 0.2, respectively.

Asymmetric Primal-Dual Problems

Since not all the linear programs come in symmetric form, we shall discuss the primal-dual relationships for the asymmetric problems in this section.

EXAMPLE 4.2-4

Consider a primal problem in asymmetric form as follows:

$$\begin{aligned} \text{Maximize:} \quad & Z=4x_1+5x_2 \\ \text{Subject to:} \quad & 3x_1+2x_2 \leqq 20 && (4.15) \\ & 4x_1-3x_2 \geqq 10 && (4.16) \\ & x_1+\ \ x_2=5 && (4.17) \\ & x_1 \geqq 0,\ x_2 \text{ unrestricted in sign} \end{aligned}$$

Since we know how to write the dual of a symmetric problem, let us convert the above problem to symmetric form. This means that all the constraints must be "less than or equal to" type inequalities (since the primal is a maximization problem), and all the variables are nonnegative. This can be accomplished as follows:

1. Inequality 4.16 is multiplied by -1.
2. Equation 4.17 is replaced by a pair of inequalities, $x_1+x_2 \leqq 5$ and $x_1+x_2 \geqq 5$.
3. The unrestricted variable x_2 is replaced by the difference of two nonnegative variables, x_3 and x_4.

Thus the symmetric form of the primal problem becomes

$$\begin{aligned} \text{Maximize:} \quad & Z=4x_1+5x_3-5x_4 \\ \text{Subject to:} \quad & 3x_1+2x_3-2x_4 \leqq 20 \\ & -4x_1+3x_3-3x_4 \leqq -10 \\ & x_1+\ x_3-\ x_4 \leqq 5 \\ & -x_1-\ x_3+\ x_4 \leqq -5 \\ & x_1,\ x_3, x_4 \geqq 0 \end{aligned}$$

Symmetric Dual

$$\begin{aligned} \text{Minimize:} \quad & W=20w_1-10w_2+5w_3-5w_4 \\ \text{Subject to:} \quad & 3w_1-4w_2+w_3-w_4 \geqq 4 \\ & 2w_1+3w_2+w_3-w_4 \geqq 5 \\ & -2w_1-3w_2-w_3+w_4 \geqq -5 \\ & w_1,\ w_2,\ w_3,\ w_4 \geqq 0 \end{aligned}$$

Comparing the above dual problem with the original primal given by the Eqs. 4.15, 4.16, and 4.17, we find that none of the characteristics of the primal-dual pair listed earlier is satisfied. We do not have the transpose of the coefficient matrix for the dual constraints, the original right-hand-side vector is not the cost vector of the dual, and so on.

To patch things up, suppose we let $y_1 = w_1$, $y_2 = -w_2$, $y_3 = w_3 - w_4$, and replace the last two inequalities of the dual by an equation. This gives the following modified dual problem:

Minimize: $W = 20y_1 + 10y_2 + 5y_3$

Subject to:

$$3y_1 + 4y_2 + y_3 \geqq 4$$

$$2y_1 - 3y_2 + y_3 = 5$$

$$y_1 \geqq 0,\ y_2 \leqq 0,\ y_3 \text{ unrestricted in sign}$$

Comparing the above dual with the original primal we find that all the essential characteristics of the primal-dual pair are satisfied except for the direction of inequalities on the constraints and the sign restrictions on the variables. Thus for any linear program (symmetric or asymmetric), the dual always satisfies the following characteristics:

1. The coefficient matrix of the dual is the transpose of the coefficient matrix of the primal.
2. The cost vector of the dual is the right-hand-side vector of the primal.
3. The right-hand-side vector of the dual is the cost vector of the primal.
4. If the primal is a maximization problem, then the dual becomes a minimization problem and vice versa.

Table 4.4 summarizes the primal-dual correspondence for all linear programming problems where the primal is a *maximization* problem. (If the primal is a minimization problem, the primal-dual table should be altered accordingly.)

Table 4.4
PRIMAL-DUAL TABLE

Primal (Maximize)	Dual (Minimize)
A Coefficient matrix	Transpose of the coefficient matrix
b Right-hand-side vector	Cost vector
c Price vector	Right-hand-side vector
ith constraint is an equation	The dual variable y_i is unrestricted in sign
ith constraint is $\leqq$ type	The dual variable $y_i \geqq 0$
ith constraint is $\geqq$ type	The dual variable $y_i \leqq 0$
x_j is unrestricted	jth dual constraint is an equation
$x_j \geqq 0$	jth dual constraint is $\geqq$ type
$x_j \leqq 0$	jth dual constraint is $\leqq$ type

We shall illustrate the rules given in the primal-dual table with a few examples.

EXAMPLE 4.2-5
Write the dual of

Maximize: $Z = x_1 + 4x_2 + 3x_3$

Subject to:

$$2x_1+3x_2-5x_3 \geqq 2$$
$$3x_1- x_2+6x_3 \geqq 1$$
$$x_1+ x_2+ x_3=4$$
$$x_1 \geqq 0,\ x_2 \leqq 0,\ x_3 \text{ unrestricted in sign}$$

Dual

Minimize:
$$W=2y_1+y_2+4y_3$$

Subject to:

$$2y_1+3y_2+y_3 \geqq 1$$
$$3y_1- y_2+y_3 \leqq 4$$
$$-5y_1+6y_2+y_3=3$$
$$y_1 \geqq 0,\ y_2 \leqq 0,\ y_3 \text{ unrestricted in sign}$$

EXAMPLE 4.2-6

Minimize:
$$Z=2x_1+x_2-x_3$$

Subject to:

$$x_1+x_2-x_3=1$$
$$x_1-x_2+x_3 \geqq 2$$
$$x_2+x_3 \leqq 3$$
$$x_1 \geqq 0,\ x_2 \leqq 0,\ x_3 \text{ unrestricted}$$

Dual

Maximize:
$$W=y_1+2y_2+3y_3$$

Subject to:

$$y_1+y_2 \leqq 2$$
$$y_1-y_2+y_3 \geqq 1$$
$$-y_1+y_2+y_3=-1$$
$$y_1 \text{ unrestricted in sign},\ y_2 \geqq 0,\ y_3 \leqq 0$$

Theorems 1, 2, 3, and 4 of the duality theory apply to the asymmetric primal-dual pair as well. For instance, in Example 4.2-6, $x_1^0=2$, $x_2^0=0$, $x_3^0=1$ is a feasible solution to the primal; while $y_1^0=1$, $y_2^0=0$, $y_3^0=0$ is feasible for the dual. By Theorem 1 (Weak Duality) the objective function value for the minimum problem should be greater than or equal to that of the maximum problem, and we find, $\mathbf{cx}^0=3>\mathbf{y}^0\mathbf{b}=1$. Since both the primal and dual are feasible, both must have optimal solutions by Theorem 3.

Similarly, in Example 4.2-5, $x_1^0=0$, $x_2^0=0$, $x_3^0=4$ is feasible for the primal with $Z=12$, and $y_1^0=0$, $y_2^0=0$, $y_3^0=3$ is feasible for the dual with $W=12$. The values of the objective function of the primal and dual are equal, and by Theorem 2, $\mathbf{x}^0=(0, 0, 4)$ is optimal for the primal, and $\mathbf{y}^0=(0, 0, 3)$ is optimal for the dual. Note also that complementary slackness conditions are satisfied between the primal and the dual optimal solutions. Substituting $\mathbf{x}^0=(0, 0, 4)$ in the first two inequalities of the primal, we get

1. $2x_1^0+3x_2^0-5x_3^0=-20<2$, which implies $y_1^0=0$
2. $3x_1^0-x_2^0+6x_3^0=24>1$, which implies $y_2^0=0$

Hence, $y_3^0=3$ is the optimal solution for the dual.

Consider the linear programming problem in standard form:

Maximize:
$$Z=\mathbf{cx}$$

Subject to:

$$\mathbf{Ax}=\mathbf{b}$$
$$\mathbf{x} \geqq \mathbf{0}$$

The dual is given by

Minimize: $W = \mathbf{yb}$

Subject to: $\mathbf{yA} \geqq \mathbf{c}$

$\mathbf{y}$ unrestricted in sign

The complementary slackness conditions to be satisfied at optimality are

$$(\mathbf{yA} - \mathbf{c})\mathbf{x} = 0$$

Computing the Optimal Dual Solution

The optimal dual solution may be computed using the complementary slackness theorem as discussed earlier. It is also possible to obtain the dual solution directly from the optimal simplex tableau of the primal problem.

Consider the standard linear program.

Minimize: $Z = \mathbf{cx}$

Subject to: $\mathbf{Ax} = \mathbf{b}, \ \mathbf{x} \geqq \mathbf{0}$

Let $\mathbf{P}_j$ denote the jth column of the $\mathbf{A}$ matrix, and $\mathbf{B}$ denote the optimal basis matrix. Applying the principles of the revised simplex method, the optimal primal solution is given by

$$\mathbf{x}^0 = \begin{pmatrix} \mathbf{x}_B \\ \mathbf{x}_N \end{pmatrix} = \begin{pmatrix} \mathbf{B}^{-1}\mathbf{b} \\ \mathbf{0} \end{pmatrix}$$

where $\mathbf{x}_B$ and $\mathbf{x}_N$ are the basic and the nonbasic variables in the optimal solution. The minimum value of $Z = \mathbf{cx}^0 = \mathbf{c}_B\mathbf{x}_B = \mathbf{c}_B\mathbf{B}^{-1}\mathbf{b}$. Since $\mathbf{B}$ represents an optimal basis, the relative cost coefficients ($\bar{c}_j$) corresponding to this basis must be nonnegative. In other words,

$$\bar{c}_j = c_j - \boldsymbol{\pi}\mathbf{P}_j \geqq 0 \qquad \text{for all } j \tag{4.18}$$

where $\boldsymbol{\pi} = \mathbf{c}_B\mathbf{B}^{-1}$ is the vector of simplex multipliers. In matrix notation, the set of inequalities above can be written as

$$\mathbf{c} - \boldsymbol{\pi}\mathbf{A} \geqq \mathbf{0} \qquad \text{or} \qquad \boldsymbol{\pi}\mathbf{A} \leqq \mathbf{c} \tag{4.19}$$

Inequality 4.19 is nothing but the constraints of the dual linear program. Thus, the optimal simplex multipliers satisfy the dual constraints. The value of the dual objective function corresponding to this feasible solution is

$$W = \mathbf{yb} = \boldsymbol{\pi}\mathbf{b} = \mathbf{c}_B\mathbf{B}^{-1}\mathbf{b}$$

which is equal to the minimum value of Z. Hence by the optimality criterion theorem, the optimal simplex multipliers of the primal problem are in fact the optimal values of the dual variables!

Illustration

Recall Example 4.1-2 from Section 4.1 Expressing it in standard form, we get

Primal

Minimize: $Z = -3x_1 + x_2 + x_3$

Subject to:

$$\begin{aligned} x_1 - 2x_2 + x_3 + x_4 \quad &= 11 \\ -4x_1 + x_2 + 2x_3 \quad - x_5 &= 3 \\ -2x_1 \quad + x_3 \quad &= 1 \\ x_1, x_2, x_3, x_4, x_5 &\geqq 0 \end{aligned}$$

Dual

$$\begin{array}{ll} \text{Maximize:} & W = 11y_1 + 3y_2 + y_3 \\ \text{Subject to:} & y_1 - 4y_2 - 2y_3 \leqq -3 \\ & -2y_1 + y_2 \leqq 1 \\ & y_1 + 2y_2 + y_3 \leqq 1 \\ & y_1 \leqq 0 \\ & -y_2 \leqq 0 \\ & y_1, y_2, y_3 \text{ unrestricted in sign} \end{array}$$

Using the revised simplex method, the optimal solution to the primal problem was obtained in Section 4.1 as $x_1 = 4$, $x_2 = 1$, $x_3 = 9$, and minimum $Z = -2$. Since x_1, x_2, x_3 are the optimal basic variables, the optimal basis matrix is given by

$$\mathbf{B} = [\mathbf{P}_1, \mathbf{P}_2, \mathbf{P}_3] = \begin{bmatrix} 1 & -2 & 1 \\ -4 & 1 & 2 \\ -2 & 0 & 1 \end{bmatrix}$$

The optimal simplex multipliers are

$$\boldsymbol{\pi} = \mathbf{c}_B\mathbf{B}^{-1} = (-3, 1, 1)\begin{bmatrix} \frac{1}{3} & \frac{2}{3} & -\frac{5}{3} \\ 0 & 1 & -2 \\ \frac{2}{3} & \frac{4}{3} & -\frac{7}{3} \end{bmatrix} = \left(-\frac{1}{3}, \frac{1}{3}, \frac{2}{3}\right)$$

It can be easily verified that $\pi_1 = -1/3$, $\pi_2 = 1/3$, $\pi_3 = 2/3$, satisfy the dual constraints, and the value of the dual objective corresponding to this feasible solution is

$$W = 11\left(-\frac{1}{3}\right) + 3\left(\frac{1}{3}\right) + 1\left(\frac{2}{3}\right) = -2$$

which corresponds to the optimal value of the primal problem. Hence the solution $y_1 = -1/3$, $y_2 = 1/3$, and $y_3 = 2/3$ is optimal for the dual. In other words, the simplex multipliers corresponding to the optimal (primal) tableau give the optimal solution to the dual problem.

4.3 THE DUAL SIMPLEX METHOD

Consider a linear program in standard form which we call the primal problem.

$$\begin{array}{ll} \text{Minimize:} & Z = \mathbf{cx} \\ \text{Subject to:} & \mathbf{Ax} = \mathbf{b} \\ & \mathbf{x} \geqq \mathbf{0} \end{array}$$

where **A** is an $(m \times n)$ matrix. Let the columns of the **A** matrix be denoted by the vectors $\mathbf{P}_1, \mathbf{P}_2, \ldots, \mathbf{P}_n$.

Recall that a basis **B** for the primal problem is an $(m \times m)$ matrix consisting of any m independent columns of the **A** matrix. Let $\mathbf{x}_B$ denote the basic variables corresponding to the basis **B**.

Primal Feasible Basis

A basis **B** is called a primal feasible basis if and only if $\mathbf{B}^{-1}\mathbf{b} \geqq \mathbf{0}$. If a basis **B** is primal feasible, then the values of the basic variables are given by the vector $\mathbf{B}^{-1}\mathbf{b}$

and the basic feasible solution becomes $\mathbf{x}_B = \mathbf{B}^{-1}\mathbf{b}$ and $\mathbf{x}_N = \mathbf{0}$ where $\mathbf{x}_N$ denotes the nonbasic variables. The value of the objective function corresponding to this feasible basis is given by

$$Z = \mathbf{c}_B \mathbf{B}^{-1}\mathbf{b}$$

where $\mathbf{c}_B$ corresponds to the cost coefficients of the basic variables.

Optimality Conditions

In order to check whether the feasible basis **B** is an optimal basis, we have to compute the relative cost coefficients ($\bar{c}_j$). Using the revised simplex method,

$$\bar{c}_j = c_j - \pi \mathbf{P}_j \qquad \text{for } j = 1, 2, \ldots, n \tag{4.20}$$

where $\pi = \mathbf{c}_B \mathbf{B}^{-1}$ are the simplex multipliers. The primal feasible basis is optimal when,

$$\bar{c}_j \geqq 0 \qquad \text{for all } j = 1, 2, \ldots, n \tag{4.21}$$

Now consider the dual of the standard linear program:

Dual

$$\begin{aligned} &\text{Maximize:} && W = \mathbf{y}\mathbf{b} \\ &\text{Subject to:} && \mathbf{y}\mathbf{A} \leqq \mathbf{c} \\ & && \mathbf{y} \text{ unrestricted in sign} \end{aligned}$$

The dual constraints $\mathbf{yA} \leqq \mathbf{c}$ can be rewritten as

$$\mathbf{y}(\mathbf{P}_1, \mathbf{P}_2, \ldots, \mathbf{P}_n) \leqq (c_1, c_2, \ldots, c_n) \quad \text{or} \quad \mathbf{y}\mathbf{P}_j \leqq c_j$$

or (4.22)

$$c_j - \mathbf{y}\mathbf{P}_j \geqq 0 \qquad \textit{for } j = 1, 2, \ldots, n$$

Comparing Inequality 4.22 with the Eqs. 4.20 and 4.21, we note that checking the optimality conditions in the (revised) simplex method is nothing but verifying whether the simplex multipliers satisfy the dual constraints! Thus, if the primal feasible basis **B** is also an optimal basis to the primal problem, then the simplex multipliers $\boldsymbol{\pi} = \mathbf{c}_B \mathbf{B}^{-1}$ satisfy,

$$c_j - \boldsymbol{\pi}\mathbf{P}_j \geqq 0 \qquad \text{for all } j = 1, 2, \ldots, n$$

This implies that $\boldsymbol{\pi}$ is feasible to the dual problem. The value of the dual objective function $W = \boldsymbol{\pi}\mathbf{b} = \mathbf{c}_B \mathbf{B}^{-1}\mathbf{b}$, which is equal to the value of the primal objective function. Hence, by the optimality criterion theorem (Theorem 2) $\boldsymbol{\pi}$ is optimal for the dual problem.

Dual Feasible Basis

A basis **B** to the primal problem,

$$\begin{aligned} &\text{Minimize:} && Z = \mathbf{c}\mathbf{x} \\ &\text{Subject to:} && \mathbf{A}\mathbf{x} = \mathbf{b},\ \mathbf{x} \geqq \mathbf{0} \end{aligned}$$

is *dual feasible* if and only if

$$\mathbf{c} - \mathbf{c}_B \mathbf{B}^{-1}\mathbf{A} \geqq \mathbf{0}$$

Note that the definition of a dual feasible basis is the same as the Eqs. 4.20 and 4.21 for verifying whether a feasible basis **B** is optimal. In other words, when a basis **B** to the primal problem is both primal feasible and dual feasible, it becomes an optimal basis. The optimal solution to the primal becomes $\mathbf{x}_B = \mathbf{B}^{-1}\mathbf{b}$, and $\mathbf{x}_N = \mathbf{0}$, while the optimal solution to the dual problem becomes $\mathbf{y} = \mathbf{c}_B\mathbf{B}^{-1}$. The optimal values of both the problems are equal since $Z^0 = W^0 = \mathbf{c}_B\mathbf{B}^{-1}\mathbf{b}$, verifying the main duality theorem.

To sum up, the main crux in solving a linear program is to find a basis **B** (or canonical tableau) which is both primal feasible and dual feasible. The simplex method (which we have studied up to now) does this by going from one primal feasible basis to another until the basis also becomes dual feasible. Hence this approach may be called the *primal simplex method*. Instead of searching for an optimal solution by moving from one primal feasible point to another, one can start with a dual feasible basis (a canonical tableau where the optimality conditions are satisfied), and search for a primal feasible basis by moving from one dual feasible tableau to another dual feasible tableau. This approach is called the *dual simplex method*. Generally, by simplex method, we always refer to the primal simplex method.

Details of the Dual Simplex Method

Recall the standard linear program,

$$\begin{aligned} \text{Minimize:} \quad & Z = \mathbf{cx} \\ \text{Subject to:} \quad & \mathbf{Ax} = \mathbf{b} \\ & \mathbf{x} \geqq \mathbf{0} \end{aligned}$$

In essence, the dual simplex method uses the same tableau as the primal simplex method. However, in all the tableaus the relative cost row ($\bar{c}_j$ coefficients) are maintained nonnegative. (In a maximization problem, the $\bar{c}_j$ coefficients will be maintained nonpositive). But the right-hand-side constants need not be nonnegative. The algorithm then proceeds to make the right-hand-side elements nonnegative, while at the same time preserving the $\bar{c}$-row coefficients as nonnegative. In other words, we always have a basic solution which is dual feasible but is not primal feasible from one iteration to the next. The algorithm terminates when all the right-hand-side constants are made nonnegative. We then have a tableau which is both primal feasible and dual feasible, and hence optimal.

Let us assume that there exists a dual feasible basis consisting of the first m columns of the **A** matrix. In other words we have a basic solution using $x_1, x_2, \ldots, x_m$ as basic variables. Table 4.5 gives the equivalent canonical system.

Table 4.5

Basis	$x_1 \cdots\cdots x_r \cdots\cdots x_m$			x_{m+1}	$\cdots\cdots x_s$	$\cdots\cdots x_n$	Constants
x_1	1			$y_{1,m+1}$	$\cdots\cdots y_{1s}$	y_{1n}	$\bar{b}_1$
x_r		1		$y_{r,m+1}$	$\cdots\cdots y_{rs}$	$\cdots\cdots y_{rn}$	$\bar{b}_r$
x_m			1	$y_{m,m+1}$	y_{ms}	y_{mn}	$\bar{b}_m$
$\bar{c}$ Row	0	0	0	$\bar{c}_{m+1}$	$\cdots\cdots \bar{c}_s$	$\cdots\cdots \bar{c}_n$	

In the above tableau, the y_{ij}'s represent the modified coefficients of the **A** matrix after the canonical reduction. Since the tableau is dual feasible, the relative cost coefficients $\bar{c}_j$ are nonnegative. If the constants $\bar{b}_1, \ldots, \bar{b}_m$ are nonnegative, then the tableau is also primal feasible, and hence optimal. Otherwise the basic solution given by $x_1=\bar{b}_1, \ldots, x_r=\bar{b}_r, \ldots, x_m=\bar{b}_m, x_{m+1}=\cdots=x_n=0$, is infeasible to the given problem, and hence not optimal even though the optimality conditions are satisfied.

Now the dual simplex method moves to an adjacent basic solution (a canonical tableau) by replacing a basic variable by a nonbasic variable.

Selection of a Basic Variable to Leave the Basis

This is done by choosing a basic variable which is making the present solution infeasible; in other words, a basic variable whose solution value is negative. In general, the basic variable with the most negative value for $\bar{b}_i$ will be chosen to leave the basis.

Let $\bar{b}_r=\min_i (\bar{b}_i)<0$. Hence the basis variable x_r is to be replaced, and row r becomes the pivot row.

Selection of a Nonbasic Variable to Enter the Basis

The pivot column is chosen such that it satisfies the following two conditions:

1. The primal infeasibility should be reduced (at least should not get worse). In other words, we want a positive right-hand-side constant at least in row r in the next tableau. This means that only those nonbasic variables (x_j) with negative coefficients in row $r(y_{rj}<0)$ are eligible to enter the basis.
2. The next tableau after the pivot operation must still be dual feasible. This can be guaranteed if the nonbasic variable to enter the basis is selected by the following ratio rule:

$$\underset{y_{rj}<0}{\text{Maximum}}\left[\frac{\bar{c}_j}{y_{rj}}\right] \qquad \text{for } j=m+1, \ldots, n \tag{4.23}$$

Let the maximum ratio correspond to the nonbasic variable x_s. In other words,

$$\frac{\bar{c}_s}{y_{rs}}=\max_{y_{rj}<0}\frac{\bar{c}_j}{y_{rj}}.$$

This implies that x_s replaces x_r in the basis. The new tableau is obtained by a pivot operation using y_{rs} as the pivot element. Note that the new basic variable in row r is x_s whose value is $\bar{b}_r/y_{rs}$. Since both $\bar{b}_r$ and y_{rs} are negative, row r will have a positive constant as required by Condition 1.

To prove that the ratio rule satisfies Condition 2, observe that the new $\bar{c}_j$ coefficients after the pivot operation are given by

$$\text{new } \bar{c}_j=(\text{old } \bar{c}_j)-\left[\frac{y_{rj}}{y_{rs}}\right](\text{old } \bar{c}_s)$$

This can be rewritten as

$$\text{new } \bar{c}_j=y_{rj}\left[\frac{\bar{c}_j}{y_{rj}}-\frac{\bar{c}_s}{y_{rs}}\right] \tag{4.24}$$

We have to show that if the Ratio Rule 4.23 is used, then the new $\bar{c}_j$ given by Eq. 4.24 will always be nonnegative. Consider the following two cases:

Case 1. For those $y_{rj}\geqq 0$, $\bar{c}_j/y_{rj}\geqq 0$, since $\bar{c}_j\geqq 0$. Since y_{rs} is the pivot element, $y_{rs}<0$, and $\bar{c}_s/y_{rs}\leqq 0$. Hence, the term within the brackets in Eq. 4.24 is nonnegative. This implies that the new $\bar{c}_j$ will be nonnegative.

Case 2. Now consider those y_{rj} elements which are negative. By the ratio rule,

$$\frac{\bar{c}_s}{y_{rs}}=\max_{y_{rj}<0}\left[\frac{\bar{c}_j}{y_{rj}}\right].$$

Hence, the term within the brackets in Eq. 4.24 will be nonpositive. Since y_{rj} is negative, the new $\bar{c}_j$ will be nonnegative, and the ratio rule guarantees that the new tableau after the pivot operation will be dual feasible.

Once again the dual simplex method checks whether the right-hand-side constants are nonnegative. If the condition is not met, then the algorithm is continued until a tableau which is both dual feasible and primal feasible is obtained.

Let us illustrate the steps of the dual simplex method with an example.

EXAMPLE 4.3-1

Minimize: $Z = x_1 + 4x_2 + 3x_4$

Subject to:

$$x_1 + 2x_2 - x_3 + x_4 \geqq 3$$
$$-2x_1 - x_2 + 4x_3 + x_4 \geqq 2$$
$$x_1, x_2, x_3, x_4 \geqq 0$$

Introducing x_5 and x_6 as slack variables the standard form becomes

Minimize: $Z = x_1 + 4x_2 + 3x_4$

Subject to:

$$x_1 + 2x_2 - x_3 + x_4 - x_5 \qquad = 3$$
$$-2x_1 - x_2 + 4x_3 + x_4 \qquad - x_6 = 2$$
$$x_1, \ldots, x_6 \geqq 0$$

Note that we can get a canonical system using x_5 and x_6 as basic variables by simply multiplying both equations by -1. Though this basic solution is infeasible to the primal, it is feasible for the dual since the $\bar{c}_j$ coefficients will be nonnegative. Thus we have a dual feasible tableau using x_5 and x_6 as basic variables as shown below:

Tableau 1

	c_j	1	4	0	3	0	0	
c_B	Basis	x_1	x_2	x_3	x_4	x_5	x_6	Constants
0	x_5	(−1)	−2	1	−1	1	0	−3
0	x_6	2	1	−4	−1	0	1	−2
	$\bar{c}$ Row	1	4	0	3	0	0	

The basic solution given by $x_5 = -3$, $x_6 = -2$, $x_1 = x_2 = x_3 = x_4 = 0$ is infeasible, though it satisfies the optimality conditions. Since the basic variable x_5 has the most negative value, it will be chosen to leave the basis. To determine the nonbasic variable to enter, we note that only x_1, x_2, and x_4 are eligible since they have negative coefficients in row 1. Forming the ratios for these nonbasic variables we get the following:

Nonbasic Variable	y_{ij}	$\bar{c}_j$	Ratios
x_1	−1	1	−1
x_2	−2	4	−2
x_4	−1	3	−3

The maximum ratio occurs corresponding to the nonbasic variable x_1. Hence, x_1 replaces x_5 in the basis. This is indicated by circling the pivot element (-1). The pivot operation is performed in the usual manner as follows:

1. Divide the pivot row (row 1) by -1.
2. Multiply the pivot row by 2, and add it to the second row.
3. Multiply the pivot row by 1 and add it to the $\bar{c}$ row.

The new tableau is given below:

Tableau 2

c_B	Basis \ c_j	1	4	0	3	0	0	Constants
		x_1	x_2	x_3	x_4	x_5	x_6	
1	x_1	1	2	-1	1	-1	0	3
0	x_6	0	-3	(-2)	-3	2	1	-8
	$\bar{c}$ Row	0	2	1	2	1	0	

The tableau is still not feasible to the primal since the basic variable x_6 has a negative value. Hence x_6 is chosen to leave the basis. To determine the nonbasic variable to enter, the ratios are computed for the nonbasic variables x_2, x_3, and x_4 as $-2/3$, $-1/2$, and $-2/3$, respectively. The maximum ratio corresponds to x_3, and x_3 replaces x_6 in the basis. The new tableau after the pivot operation is given below:

Tableau 3

c_B	Basis \ c_j	1	4	0	3	0	0	Constants
		x_1	x_2	x_3	x_4	x_5	x_6	
1	x_1	1	7/2	0	5/2	-2	$-1/2$	7
0	x_3	0	3/2	1	3/2	-1	$-1/2$	4
	$\bar{c}$ Row	0	1/2	0	1/2	2	1/2	$Z=7$

Tableau 3 is both primal feasible and dual feasible. Hence we have an optimal solution to the primal given by $x_1=7$, $x_2=0$, $x_3=4$, $x_4=0$, $x_5=0$, $x_6=0$ and the optimum value of Z is 7.

Identifying Primal Infeasibility in the Dual Simplex Method

In the dual simplex method there always exists a feasible solution to the dual. Hence, when the dual simplex method is applied to a linear program the primal problem either has an optimum solution or is infeasible. The dual simplex method recognizes the primal infeasibility when the ratio rule fails to identify the nonbasic variable to enter. In other words, when all the elements in the pivot row are nonnegative, the dual simplex method terminates with the conclusion that the primal problem has no feasible solution. To justify this, consider the dual simplex tableau in general form as shown in Table 4.5. Let the constant in row $r(\bar{b}_r)$ be negative, and all elements of row $r(\bar{y}_{rj})$ be nonnegative. Writing the constraint r in expanded form, we get

$$x_r + y_{r,m+1}x_{m+1} + \cdots + y_{rn}x_n = \bar{b}_r \tag{4.25}$$

For all nonnegative values of x_j, the left-hand side of Eq. 4.25 is nonnegative, whereas the right-hand side is negative. Hence, there exists no nonnegative solution satisfying Eq. 4.25. In other words, constraint r is inconsistent, and hence the primal problem is infeasible.

Solving a Maximization Problem by the Dual Simplex Method

In a maximization problem, the relative cost coefficients $(\bar{c}_j)$ must be nonpositive for optimality. Assume that Table 4.5 represents a dual feasible tableau of a maximization problem. Hence, all the elements of $\bar{c}$ row will be nonpositive ($\leqq 0$). Assume that $\bar{b}_r < 0$ and x_r is chosen to leave the basis. The nonbasic variable to enter the basis is chosen in such a way that the $\bar{c}$ row elements remain nonpositive at subsequent iterations. This can be guaranteed by using the following ratio rule:

$$\min_{y_{rj}<0} \left[\frac{\bar{c}_j}{y_{rj}} \right]$$

The minimum ratio identifies the nonbasic variable to enter. The validity of this ratio rule may be proved in a manner similar to the one used for the minimization problem.

Applications of the Dual Simplex Method

In general it is not always easy to find a dual feasible basis. Many practical problems do not have a canonical tableau which is either primal feasible or dual feasible. Hence, as a rule the primal simplex method is preferred over the dual simplex method for solving the general linear programing problem. But there are instances when the dual simplex method has an obvious advantage over the primal simplex method. These are problems in which a dual feasible tableau is readily available to start the dual simplex method. A list of such applications of the dual simplex method is given below:

1. Sensitivity analysis and parametric programming (Sections 4.4 and 4.5).
2. Most of the integer programming algorithms.
3. Some nonlinear programming algorithms.
4. Some variants of the simplex method such as the primal-dual algorithm, and the self-dual parametric algorithm.

4.4 SENSITIVITY ANALYSIS IN LINEAR PROGRAMMING

Sensitivity analysis refers to the study of the changes in the optimal solution and optimal value of Z due to changes in the input data coefficients. In Section 2.11, we assumed that the sensitivity coefficients have already been provided by the computer code and we studied how to *use* the information given under sensitivity analysis. In this section, we will study how to *get* the sensitivity coefficients for performing a sensitivity analysis.

For a discussion of the sensitivity analysis, we shall confine ourselves to the following changes in the data, and how to handle these changes:

1. Changes in the cost coefficients (c_j).
2. Changes in the right-hand-side constants (b_i).
3. Changes in the constraint or coefficient matrix (A).
   a. Adding new activities or variables.
   b. Changing existing columns.
   c. Adding new constraints.

In this section, we shall see how to minimize the additional computations necessary to study the above changes. In many cases, it will not be necessary to solve the problem all over again. We shall present the discussion with the help of an illustrative example. The basic principles involved in performing a sensitivity analysis will be sufficiently developed so that the reader will have no difficulty in extending them to other problems or to the general case.

EXAMPLE 4.4-1 A PRODUCT-MIX PROBLEM

The Dependable Company plans production on three of their products—*A, B,* and *C*. The unit profits on these products are \$2, \$3, and \$1, respectively, and they require two resources—labor and material. The company's operations research department formulates the following linear programming model for determining the optimal product mix:

Maximize: $Z = 2x_1 + 3x_2 + x_3$

Subject to:

$$\frac{1}{3}x_1 + \frac{1}{3}x_2 + \frac{1}{3}x_3 \leqq 1 \quad \text{(labor)}$$

$$\frac{1}{3}x_1 + \frac{4}{3}x_2 + \frac{7}{3}x_3 \leqq 3 \quad \text{(material)}$$

$$x_1, x_2, x_3 \geqq 0$$

where x_1, x_2, x_3 are the number of products *A, B,* and *C* produced.

The initial simplex tableau with the addition of slack variables x_4 and x_5 is given in Table 4.6.

Table 4.6

	c_j	2	3	1	0	0	
c_B	Basis	x_1	x_2	x_3	x_4	x_5	Constants
0	x_4	$\frac{1}{3}$	$\frac{1}{3}$	$\frac{1}{3}$	1	0	1
0	x_5	$\frac{1}{3}$	$\frac{4}{3}$	$\frac{7}{3}$	0	1	3
	$\bar{c}$ Row	2	3	1	0	0	$Z=0$

Since this is already in canonical form, no artificial variables are needed. The simplex method comes out, after some iterations, with the optimal tableau shown in Table 4.7.

Table 4.7

	c_j	2	3	1	0	0	
c_B	Basis	x_1	x_2	x_3	x_4	x_5	Constants
2	x_1	1	0	-1	4	-1	1
3	x_2	0	1	2	-1	1	2
	$\bar{c}$ Row	0	0	-3	-5	-1	$Z=8$

From the optimal tableau, we see that the optimal product mix is to produce 1 unit of product A and 2 units of product B for a total profit of \$8. By performing a sensitivity analysis, it is possible to obtain valuable information regarding alternative production schedules in the neighborhood of the optimal solution. Quite often, this information will be more significant and useful than the determination of the optimal product mix itself. As a matter of fact, one of the reasons for the extensive use of linear programming in practice is its ability to provide a sensitivity analysis along with the optimal solution!

Variations in the Objective Function Coefficients (c_j)

Variations in the coefficients of the objective function may occur due to a change in profit or cost of either a basic activity or a nonbasic activity. We shall treat these cases separately.

Case 1. Changing the Objective Function Coefficient of a Nonbasic Variable. In the optimal product mix shown in Table 4.7, product C is not produced because of its low profit of \$1 per unit (c_3). One may be interested in finding the range on the values of c_3 such that the current optimal solution remains optimal. It is clear that when c_3 decreases it has no effect on the present optimal solution. However, when the profit is increased beyond a certain value, product C may become profitable to produce.

As a rule, the sensitivity of the current optimal solution can be best obtained by studying how the optimal tableau given in Table 4.7 changes due to variations in the input data. When the value of c_3 changes, the value of the relative profit coefficient of the nonbasic variable x_3 ($\bar{c}_3$) changes in the optimal tableau. Table 4.7 is optimal as long as $\bar{c}_3$ is nonpositive. In the present optimal tableau $\mathbf{c}_B=(c_1, c_2)=(2, 3)$. Hence,

$$\bar{c}_3=c_3-(2, 3)\begin{pmatrix}-1\\2\end{pmatrix}=c_3-4$$

For Table 4.7 to be optimal, $\bar{c}_3=c_3-4\leqq 0$, or $c_3\leqq 4$. In other words, as long as the unit profit on product C is less than \$4, it is not economical to produce product C.

Suppose if the unit profit on product C is increased to \$6, then $\bar{c}_3=+2$, and the current product mix is not optimal. The maximum profit can be increased further by producing product C. In other words, Table 4.7 is nonoptimal since x_3 can enter the basis to increase Z. By the minimum ratio rule, x_2 leaves the basis. The new optimal solution can be determined by applying the simplex method as shown below:

Basis	x_1	x_2	x_3	x_4	x_5	Constants
x_1	1	0	−1	4	−1	1
x_2	0	1	②	−1	1	2
			↑			
$\bar{c}$ Row	0	0	2	−5	−1	$Z=8$
x_1	1	$\frac{1}{2}$	0	$\frac{7}{2}$	$-\frac{1}{2}$	2
x_3	0	$\frac{1}{2}$	1	$-\frac{1}{2}$	$\frac{1}{2}$	1
$\bar{c}$ Row	0	−1	0	−4	−2	$Z=10$

Hence, the new (optimal) product mix is to produce 2 units of product A, and 1 unit of product C with a maximum profit of \$10.

In general, the new optimal solution is found in just one iteration, but it should not be taken for granted at all times.

Case 2. Changing the Objective Function Coefficient of a Basic Variable. Suppose we want to determine the effect of changes on the unit profit of product $A(c_1)$. It is intuitively clear that when c_1 decreases below a certain level, it may not be profitable to include product A in the optimal product mix. Even when c_1 increases, it is possible that it may change the optimal product mix at some level. This happens because product A may become so profitable that the optimal mix may include only product A. Hence, there is an upper and a lower limit on the variation of c_1 within which the optimal solution given by Table 4.7 is not affected.

To determine the range on c_1, observe that a change in c_1 changes the profit vector of the basic variables ($\mathbf{c}_B$) since $\mathbf{c}_B=(c_1, c_2)$. It can be verified that the relative profit coefficients of the basic variables, namely $\bar{c}_1$ and $\bar{c}_2$, will not be affected, and they will still remain at zero value. However, the relative profits of the nonbasic variables, namely $\bar{c}_3$, $\bar{c}_4$, $\bar{c}_5$, will change. But as long as these $\bar{c}_j$ remain nonpositive, Table 4.7 is still optimal. We can express the values of $\bar{c}_3$, $\bar{c}_4$, $\bar{c}_5$, as a function of c_1 as follows:

$$\bar{c}_3=1-(c_1, 3)\begin{pmatrix}-1\\2\end{pmatrix}=c_1-5$$

$$\bar{c}_4=0-(c_1, 3)\begin{pmatrix}4\\-1\end{pmatrix}=-4c_1+3$$

$$\bar{c}_5=0-(c_1, 3)\begin{pmatrix}-1\\1\end{pmatrix}=c_1-3$$

From the above calculations, $\bar{c}_3\leqq 0$ as long as $c_1\leqq 5$. Similarly, each nonbasic variable puts a limit (lower or upper) on the value of c_1. Thus,

$$\bar{c}_4\leqq 0 \quad \text{implies} \quad c_1\geqq \frac{3}{4}, \text{ and}$$

$$\bar{c}_5\leqq 0 \quad \text{implies} \quad c_1\leqq 3$$

Table 4.7 will remain optimal as long as the variation on c_1 is within the limits imposed by all the nonbasic variables. Hence, if the range on c_1 is chosen as [3/4, 3], then all the $\bar{c}_j$ will remain nonpositive, and the present solution $x_1=1$, $x_2=2$, $x_3=0$ is still optimal. Of course as c_1 changes, the optimal value of the objective function will change. For example, when $c_1=1$, the optimal solution is given by $x_1=1$, $x_2=2$, $x_3=0$, but the maximum profit = \$7. When the value of c_1 goes beyond the range provided by the sensitivity analysis, Table 4.7 will no longer be optimal as one of the nonbasic $\bar{c}_j$ will become positive. Once again, we can apply the simplex method to determine the new optimal solution as discussed in Case 1.

Case 3. Changing the Price of Both the Basic and the Nonbasic Variables. A simple case of this can be easily solved. For example, the profits on all three products are changed such that the objective function becomes $Z=x_1+4x_2+2x_3$. The effect on the optimal product mix can be determined by checking whether the $\bar{c}$ row in Table 4.7 remains nonpositive. $\bar{c}_1=\bar{c}_2=0$, while

$$\bar{c}_3=2-(1,4)\begin{pmatrix}-1\\2\end{pmatrix}=-5<0$$

$$\bar{c}_4=0-(1,4)\begin{pmatrix}4\\-1\end{pmatrix}=0$$

$$\bar{c}_5=0-(1,4)\begin{pmatrix}-1\\1\end{pmatrix}=-3<0$$

Hence the optimal solution does not change, and the optimal product mix is $x_1=2$, $x_2=2$, $x_3=0$, and the maximum $Z=\$9$. Of course, we now have an indication for an alternate optimal solution since $\bar{c}_4=0$.

Changing the Right-Hand-Side Constants (b_i)

Suppose that an additional one unit of labor is made available, and the company is interested in determining how this affects the optimal product mix.

The addition of one more unit of labor changes the vector of right-hand-side constants in the initial simplex tableau. In other words, the vector of constants in Table 4.6 change from $\begin{pmatrix}1\\3\end{pmatrix}$ to a vector $\begin{pmatrix}2\\3\end{pmatrix}$. It is clear that this change has no effect in the optimal tableau given by Table 4.7 except for changes in the values of the constants. Even after the change, if the new right-hand-side constants remain nonnegative, then the solution given by Table 4.7 is still a basic feasible solution. Because the $\bar{c}$-row coefficients are the same (namely, nonpositive) this tableau also becomes an optimal solution to the problem. Therefore in order to study the effect of variation in the right-hand-side constants, it is sufficient to verify whether the new vector of constants in the final tableau stays nonnegative. To do this, there is no need to solve the problem again. In Section 4.1 we have seen that any column in the final tableau (including the right-hand-side vector) can be obtained by multiplying the corresponding column in the initial tableau by the inverse of the basic columns. In this case, the basic columns are the columns corresponding to the variables (x_1, x_2) in the initial tableau. Hence, the basis matrix correspondng to Table 4.7 is given by

$$\mathbf{B}=\begin{bmatrix}\frac{1}{3} & \frac{1}{3}\\ \frac{1}{3} & \frac{4}{3}\end{bmatrix}$$

One can compute the inverse of the basis matrix directly by the pivot method or the adjoint method. But in the discussion of the revised simplex method (Section 4.1), we have observed that the columns corresponding to the initial basic variables in any simplex tableau give the inverse of the basis matrix corresponding to that tableau. Since x_4 and x_5 are the initial basic variables (Table 4.6), the columns corresponding to x_4 and x_5 in Table 4.7 give the inverse of the basis matrix **B**. This implies

$$\mathbf{B}^{-1}=\begin{bmatrix} 4 & -1 \\ -1 & 1 \end{bmatrix}$$

The values of the new right-hand-side constants in Table 4.7 due to the increased labor is given by

$$\bar{\mathbf{b}}=\begin{bmatrix} 4 & -1 \\ -1 & 1 \end{bmatrix}\begin{bmatrix} 2 \\ 3 \end{bmatrix}=\begin{bmatrix} 5 \\ 1 \end{bmatrix}$$

Thus, when the value of the constants in Table 4.6 change to $\begin{bmatrix} 2 \\ 3 \end{bmatrix}$, the new values of the constants in Table 4.7 become $\begin{bmatrix} 5 \\ 1 \end{bmatrix}$ which is a positive vector. Hence, Table 4.7 still remains optimal, and the new optimal product mix is $x_1 = 5$, $x_2 = 1$, $x_3 = 0$, and maximum value of $Z = \$13$. Note here that both the optimal solution, and the optimal value of the objective function have changed due to a variation in the availability of labor. But, the optimal basis has not changed; in other words, it is still optimal to produce only the two products A and B. The only difference lies in the quantity of A and B produced.

Suppose the extra unit of labor can be obtained by allowing overtime which costs an additional \$4 to the company. The company may want to find out whether it is profitable to use overtime labor. This may be found by comparing the increased profit by employing overtime to its added cost. In our example, the increased profit is \$13 – \$8 = \$5, which is more than the cost of overtime (\$4); it is therefore profitable to get the additional 1 unit of labor.

The increased profit of \$5-per-unit increase in labor availability is called the *shadow price* for the labor constraint. Knowing the shadow prices of various constraints helps in determining how much one can afford to pay for increases in the constrained resources. In Section 4.2, we showed that the optimal dual solution corresponds to the shadow prices on the various constraints. In Example 4.4-1, the optimal dual solution (y_1^0, y_2^0) is given by

$$\begin{aligned}(y_1^0, y_2^0) &= \text{optimal simplex multipliers corresponding to Table 4.7} \\ &= \mathbf{c}_B\mathbf{B}^{-1} \\ &= (2, 3)\begin{bmatrix} 4 & -1 \\ -1 & 1 \end{bmatrix} = (5, 1)\end{aligned}$$

Thus the optimal dual solution is $y_1^0 = 5$ and $y_2^0 = 1$. In other words, the shadow price on the labor constraint is \$5, which corresponds to the one computed by the sensitivity analysis. Similarly, the shadow price on the materials constraint is \$1. It is important to note here that the shadow prices reflect the net change in the optimal value of Z per unit increase on the constraint resources, as long as the variation in the constraint resources does not change the optimal basis. Hence, to use the shadow prices meaningfully, one has to compute the range on the variation of a constraint resource such that the optimal basis (product mix) remains the same.

To illustrate, let us compute how far the availability of labor can be varied (increased or decreased) such that the present optimal basis (product mix) still remains optimal. Let b_1 denote the amount of labor available, and $\mathbf{b}^*$ denote the new vector of constants in the initial tableau. Hence,

$$\mathbf{b}^* = \begin{bmatrix} b_1 \\ 3 \end{bmatrix}$$

For the simplex tableau given by Table 4.7 to be optimal, we should have $\mathbf{B}^{-1}\mathbf{b}^* \geqq 0$. Since $\mathbf{B}^{-1} = \begin{bmatrix} 4 & -1 \\ -1 & 1 \end{bmatrix}$, we get

$$\mathbf{B}^{-1}\mathbf{b}^* = \begin{bmatrix} 4 & -1 \\ -1 & 1 \end{bmatrix} \begin{bmatrix} b_1 \\ 3 \end{bmatrix} = \begin{bmatrix} 4b_1 & -3 \\ -b_1 & +3 \end{bmatrix}$$

$\mathbf{B}^{-1}\mathbf{b}^*$ is nonnegative as long as

$$4b_1 - 3 \geqq 0 \quad \text{or} \quad b_1 \geqq \frac{3}{4}$$

$$-b_1 + 3 \geqq 0 \quad \text{or} \quad b_1 \leqq 3$$

This means that x_1 and x_2 will remain in the optimal product mix as long as the labor availability varies between 3/4 of a unit and 3 units. But the optimal solution, and the maximum profit will change. Thus we have a range of optimal solutions as given below:

For all $3/4 \leqq b_1 \leqq 3$, the optimal solution is

$$x_1 = 4b_1 - 3$$
$$x_2 = -b_1 + 3$$
$$x_3 = 0$$

The maximum profit $Z = 2(4b_1 - 3) + 3(-b_1 + 3)$

$$= \$5b_1 + 3$$

Let us now consider the case when the labor availability is increased to 4 units. This means that the initial values of the right-hand-side constants in Table 4.6 will change to $\begin{bmatrix} 4 \\ 3 \end{bmatrix}$. The new values of the constants in the final tableau (Table 4.7) will be given by

$$\bar{\mathbf{b}} = \mathbf{B}^{-1}\begin{pmatrix} 4 \\ 3 \end{pmatrix} = \begin{bmatrix} 4 & -1 \\ -1 & 1 \end{bmatrix} \begin{bmatrix} 4 \\ 3 \end{bmatrix} = \begin{bmatrix} 13 \\ -1 \end{bmatrix}$$

This implies that Table 4.7 is no longer optimal since the basic solution $x_1 = 13$, $x_2 = -1$, $x_3 = x_4 = x_5 = 0$ is infeasible. To find the new optimal product mix, the simplex tableau of Table 4.7 is reproduced with the new values of the constants in Table 4.8.

Table 4.8

	c_j	2	3	1	0	0	
c_B	Basis	x_1	x_2	x_3	x_4	x_5	Constants
2	x_1	1	0	−1	4	−1	13
3	x_2	0	1	2	(−1)	1	−1
	$\bar{c}$ Row	0	0	−3	−5	−1	

Even though the simplex tableau corresponding to Table 4.8 is infeasible for the primal problem, it is feasible for the dual since all the relative cost coefficients are nonpositive. Hence, the new optimal solution can be obtained by the *dual simplex method*. Since the basic variable x_2 is negative, it will leave the basis, and since x_4 is the only nonbasic variable with a negative coefficient, it will replace x_2 in the basis. Table 4.9 gives the new tableau after the pivot operation.

Table 4.9

	c_j	2	3	1	0	0	
c_B	Basis	x_1	x_2	x_3	x_4	x_5	Constants
2	x_1	1	4	7	0	3	9
0	x_4	0	−1	−2	1	−1	1
	$\bar{c}$ Row	0	−5	−13	0	−6	$Z=18$

Table 4.9 is optimal since the right-hand-side constants are positive. The new optimal product mix when the labor availability is increased to 4 units is given by $x_1=9, x_2=0, x_3=0$, and the maximum profit is \$18.

Variations in the Constraint Matrix (A)

As mentioned earlier, the constraint matrix or the coefficient matrix (**A**) may be changed by

1. adding new variables or activities
2. changing the resources requirements of the existing activities
3. adding new constraints

We shall discuss each of these cases separately.

Case (1): Adding a New Activity. Suppose the company's R & D department has come out with a new product D which requires 1 unit of labor and 1 unit of material. The new product has sufficient market and can be sold at a unit profit of \$3. The company wants to know whether it is economical to manufacture product D.

Inclusion of a new product in our possible product mix is mathematically equivalent to adding a variable (say x_6), and a column $\begin{pmatrix}1\\1\end{pmatrix}$ in the initial tableau (Table 4.6). The present optimal product mix given by Table 4.7 is optimal as long as the relative profit coefficient of the new product, namely $\bar{c}_6$, is nonpositive. From the revised simplex method, we know that

$$\bar{c}_6=c_6-\boldsymbol{\pi}\mathbf{P}_6$$

where $c_6=\$3$, $\mathbf{P}_6=\begin{pmatrix}1\\1\end{pmatrix}$, and $\boldsymbol{\pi}$ is the simplex multiplier corresponding to Table 4.7. Note that

$$\begin{aligned}\boldsymbol{\pi}&=\mathbf{c}_B\mathbf{B}^{-1}\\&=(2,3)\begin{bmatrix}4&-1\\-1&1\end{bmatrix}=(5,1)\end{aligned}$$

Hence,

$$\bar{c}_6 = 3 - (5, 1)\binom{1}{1} = -3$$

This indicates that producing product D will not improve the present value of the maximum profit.

In case it turns out that the new activity can contribute to an increased profit (because of its $\bar{c}_j$ value being positive), then the simplex method will be applied to determine the new optimal solution.

Case (2): Variation in the Resources Requirements of the Existing Activities. When the labor or the material requirements of a nonbasic activity (e.g., product C) change, its effect on the optimal solution can be studied by following the same steps as given in Case (1). On the other hand, if the constraint coefficients of a basic activity (e.g., product A or product B) change, then the basis matrix itself is affected, which in turn may affect all the quantities given in Table 4.7. It is then possible for Table 4.7 to be neither primal feasible nor dual feasible. Under such circumstances, it may be better to solve the linear program over again.

Case (3): Adding New Constraints. Consider the addition of an administrative services constraint to the problem wherein the products A, B, and C require 1, 2, and 1 hour of administrative services, while the available administrative hours are 10. This amounts to adding a new constraint of the form,

$$x_1 + 2x_2 + x_3 \leqq 10$$

to the original formulation of the problem. To study its effect on the present optimal solution, it is sufficient to verify whether the present optimal product mix satisfies the new constraint. It can be shown mathematically that the present optimal solution stays optimal as long as it satisfies the new constraint. In our case it does satisfy the administrative services constraint, and hence the optimal product mix is not altered.

Suppose the available administrative hours are only 4, then the new constraint becomes,

$$x_1 + 2x_2 + x_3 \leqq 4.$$

The present optimal solution $x_1 = 1$, $x_2 = 2$, $x_3 = 0$ violates this constraint. Hence Table 4.7 is no longer optimal. In order to find the new optimal solution, let us add the new constraint as the third row in Table 4.7. Using x_6 as the slack variable in the new constraint, the modified tableau is shown below:

	c_j	2	3	1	0	0	0	
c_B	Basis	x_1	x_2	x_3	x_4	x_5	x_6	Constants
2	x_1	1	0	−1	4	−1	0	1
3	x_2	0	1	2	−1	1	0	2
0	x_6	1	2	1	0	0	1	4
	$\bar{c}$ Row	0	0	−3	−5	−1	0	

The modified tableau is not in canonical form since the basic variables x_1 and x_2 have positive coefficients in the third row. To eliminate the coefficients of x_1 and x_2, we can multiply the first row by -1, the second row by -2, and add them to the third row. Table 4.10 gives the new canonical tableau after the row operations. Note that the $\bar{c}$ row is not affected in this process since the new basic variable x_6 is a slack variable.

Table 4.10

	c_j	2	3	1	0	0	0	
c_B	Basis	x_1	x_2	x_3	x_4	x_5	x_6	Constants
2	x_1	1	0	-1	4	-1	0	1
3	x_2	0	1	2	-1	1	0	2
0	x_6	0	0	-2	-2	(-1)	1	-1
	$\bar{c}$ Row	0	0	-3	-5	-1	0	

Since Table 4.10 is dual feasible, the dual simplex method is applied to find the new optimal solution. The basic variable x_6 leaves the basis. The ratios are formed for the nonbasic variables x_3, x_4, and x_5 as 3/2, 5/2, and 1, respectively. Since the minimum ratio corresponds to x_5, the basic variable x_6 is replaced by x_5. The new tableau is given in Table 4.11.

Table 4.11

	c_j	2	3	1	0	0	0	
c_B	Basis	x_1	x_2	x_3	x_4	x_5	x_6	Constants
2	x_1	1	0	1	6	0	-1	2
3	x_2	0	1	0	-3	0	1	1
0	x_5	0	0	2	2	1	-1	1
	$\bar{c}$ Row	0	0	-1	-3	0	-1	$Z=7$

Table 4.11 is optimal and the new optimal product mix is to produce 2 units of product A and one unit of product B. The maximum profit has been reduced from \$8 to \$7 due to the addition of the new constraint. This is true of any linear program. In other words, whenever a new constraint is added to a linear program the old optimal value will always be better or equal to the new optimum value. Thus the addition of a new constraint cannot improve the optimal value of any linear programming problem.

The ability to add more constraints to a linear programming problem has other practical applications. We have noted earlier that the computational effort in solving a linear program varies in proportion to the cube of the number of constraints. Hence to reduce the computation time, one may be able to identify and set aside from past experience some constraints which may not be very critical in determining the optimal solution. These constraints are generally called *inactive* or *secondary constraints*. These may include resources which are under the control of the firm or those which can be obtained easily. A significant reduction in the computational time (and hence the cost of solution) can be achieved by solving the linear program without the secondary constraints. After the optimal solution has been determined, the secondary constraints are added to verify whether the optimal solution satisfies these constraints. If one or more constraints are violated, then the dual simplex method is applied to determine the new optimal solution. Of course, the overall savings in computational effort will depend on how good the initial judgments are made while identifying the secondary constraints.

4.5 PARAMETRIC PROGRAMMING

The discussion of sensitivity analysis in the previous section considered the effects of variations in the input coefficients when these coefficients are changed one at a time. When simultaneous variations occur in the input data, one can use the 100% rule described in Section 2.11. In this section, we consider the effects of simultaneous changes in data, where the coefficients change as a function of one parameter. Hence, this is called *parametric programming*, and is simply an extension of sensitivity analysis. We shall consider the following parametric problems: (1) the cost problem, and (2) the right-hand-side problem.

The Parametric Cost Problem

Consider a linear program of the form:

$$\begin{aligned} &\text{Minimize:} && Z = (\mathbf{c} + \lambda \mathbf{c}^*)\mathbf{x} \\ &\text{Subject to:} && \mathbf{A}\mathbf{x} = \mathbf{b} \\ &&& \mathbf{x} \geqq \mathbf{0} \end{aligned}$$

where $\mathbf{c}$ is the given cost vector, $\mathbf{c}^*$ is the given variation vector, and λ is an unknown positive or negative parameter. Varying the value of λ changes the cost coefficients of all the variables. We are interested in finding the family of optimal solutions for all values of λ in the range $-\infty$ to $+\infty$.

To give a practical application of the parametric cost problem, consider a manufacturer of different products. Each of these products require a basic raw material at varying amounts. Suppose the manufacturer knows in advance that the cost of this basic raw material is going to fluctuate widely during the coming year, and he is interested in finding the effect of this variation on the optimal product mix. In this case, the variation vector $\mathbf{c}^*$ represents the quantity of this raw material used by the different products, and λ denotes the variation in the raw material cost. As λ varies, the cost of production of all the products changes by different amounts. A solution of the parametric cost problem will provide the manufacturer different optimal policies to follow depending on the cost of this raw material.

The parametric cost problem is solved by using the simplex method and sensitivity analysis. The parametric linear program is first solved by the simplex method for a fixed value of λ which is usually taken as zero. Let $\mathbf{B}$ represent the optimal basis matrix for $\lambda = 0$. This implies that the relative cost coefficients with respect to the basis $\mathbf{B}$ are nonnegative, and are given by $\bar{c}_j = c_j - \mathbf{c}_B \bar{\mathbf{P}}_j$, where $\bar{\mathbf{P}}_j$ is the jth column (corresponding to the variable x_j) in the optimal tableau, and $\mathbf{c}_B$ is the cost vector of the basic variables. When λ varies from zero to a positive or negative value, the relative cost coefficient of the variable x_j becomes

$$\begin{aligned} \bar{c}_j(\lambda) &= (c_j + \lambda c_j^*) - (\mathbf{c}_B + \lambda \mathbf{c}_B^*)\bar{\mathbf{P}}_j \\ &= (c_j - \mathbf{c}_B \bar{\mathbf{P}}_j) + \lambda (c_j^* - \mathbf{c}_B^* \bar{\mathbf{P}}_j) \\ &= \bar{c}_j + \lambda \bar{c}_j^* \end{aligned} \qquad (4.26)$$

Since the vectors **c** and **c*** are known, $\bar{c}_j$ and $\bar{c}_j^*$ can be computed. Then for any value of λ, the relative cost coefficients are given by Eq. 4.26. For a simplex tableau to be optimal, $\bar{c}_j(\lambda)$ must be nonnegative. In other words, for a given tableau, one can determine the range of values of λ for which that tableau is optimal.

To illustrate the parametric cost technique, consider Example 4.4-1, with the variation cost vector as $\mathbf{c}^* = (1, -1, 0, 0)$.

EXAMPLE 4.5-1

Maximize: $Z = (2+\lambda)x_1 + (3-\lambda)x_2 + (1+\lambda)x_3$

Subject to:
$$\frac{1}{3}x_1 + \frac{1}{3}x_2 - \frac{1}{3}x_3 + x_4 = 1$$
$$\frac{1}{3}x_1 + \frac{4}{3}x_2 + \frac{7}{3}x_3 + x_5 = 3$$
$$x_1, \ldots, x_5 \geqq 0$$

For $\lambda = 0$, the above linear program is the same as the one given in Example 4.4-1, and its optimal solution is given in Table 4.7 (Section 4.4). Since we have a maximization problem, a basic feasible solution is optimal when the relative profit coefficients are nonpositive. For nonzero values of λ, the relative profits become linear functions of λ. To study the effect of this variation, we add a new relative profit row called the $\bar{c}^*$ row. The expanded simplex tableau is shown in Table 4.12.

Table 4.12

		c_j^*	1	−1	1	0	0	
		c_j	2	3	1	0	0	
c_B^*	c_B	Basis	x_1	x_2	x_3	x_4	x_5	Constants
1	2	x_1	1	0	−1	4	−1	1
−1	3	x_2	0	1	2	−1	1	2
$\bar{c}$ Row			0	0	−3	−5	−1	$Z = 8$
$\bar{c}^*$ Row			0	0	4	−5	2	$Z^* = -1$

The $\bar{c}^*$ row is calculated just like the $\bar{c}$ row except that the vector **c** is replaced by **c***. For example,

$$\bar{c}_3 = c_3 - \mathbf{c}_B\bar{\mathbf{P}}_3 = 1 - (2, 3)\begin{pmatrix} -1 \\ 2 \end{pmatrix} = -3$$

$$\bar{c}_3^* = c_3^* - \mathbf{c}_B^*\bar{\mathbf{P}}_3 = 1 - (1, -1)\begin{pmatrix} -1 \\ 2 \end{pmatrix} = 4$$

Table 4.12 represents a basic feasible solution given by $x_1 = 1$, $x_2 = 2$, $x_3 = x_4 = x_5 = 0$, whose value of the objective function is $Z(\lambda) = Z + \lambda Z^* = 8 - \lambda$. Its relative profit coefficients are given by

$$\bar{c}_j(\lambda) = \bar{c}_j + \lambda\bar{c}_j^* \qquad \text{for } j = 1, 2, 3, 4, 5$$

For $\lambda = 0$, Table 4.12 represents an optimal tableau, and it remains optimal for other values of λ as long as the $\bar{c}_j(\lambda) \leqq 0$ for $j = 3, 4, 5$. Thus, one can determine the range of λ for which Table 4.12 is optimal as follows:

$$\bar{c}_3(\lambda) = -3 + 4\lambda \leqq 0 \quad \text{or} \quad \lambda \leqq 3/4$$
$$\bar{c}_4(\lambda) = -5 - 5\lambda \leqq 0 \quad \text{or} \quad \lambda \geqq -1$$
$$\bar{c}_5(\lambda) = -1 + 2\lambda \leqq 0 \quad \text{or} \quad \lambda \leqq 1/2$$

In other words, $x_1 = 1$, $x_2 = 2$, $x_3 = x_4 = x_5 = 0$ is an optimal solution for all values of λ between -1 and $1/2$. The optimal value of Z is given by $8 - \lambda$.

As λ exceeds $1/2$, the relative profit coefficient of the nonbasic variable x_5, namely $\bar{c}_5(\lambda)$, turns positive making Table 4.12 nonoptimal. Applying the simplex method, x_5 enters the basis to replace x_2. The new tableau after the pivot operation is given in Table 4.13.

Table 4.13

		c_j^*	1	−1	1	0	0	
		c_j	2	3	1	0	0	
c_B^*	c_B	Basis	x_1	x_2	x_3	x_4	x_5	Constants
1	2	x_1	1	1	1	3	0	3
0	0	x_5	0	1	2	−1	1	2
		$\bar{c}$ Row	0	1	−1	−6	0	$Z=6$
		$\bar{c}^*$ Row	0	−2	0	−3	0	$Z^*=3$

Table 4.13 represents an optimal tableau as long as $\bar{c}_2(\lambda)$, $\bar{c}_3(\lambda)$, and $\bar{c}_4(\lambda)$ remain nonpositive. They will be nonpositive if λ is not less than $1/2$. Thus for all $\lambda \geqq 1/2$, the optimal solution is given by $x_1 = 3$, $x_2 = x_3 = x_4 = 0$, $x_5 = 2$, and the optimal value of $Z = 6 + 3\lambda$.

Similarly we can obtain the family of optimal solutions for $\lambda < -1$ from Table 4.12.

Figure 4.1 gives a plot of the optimal value of Z for different values of λ between -1 and ∞ for the parametric cost problem of Example 4.5-1.

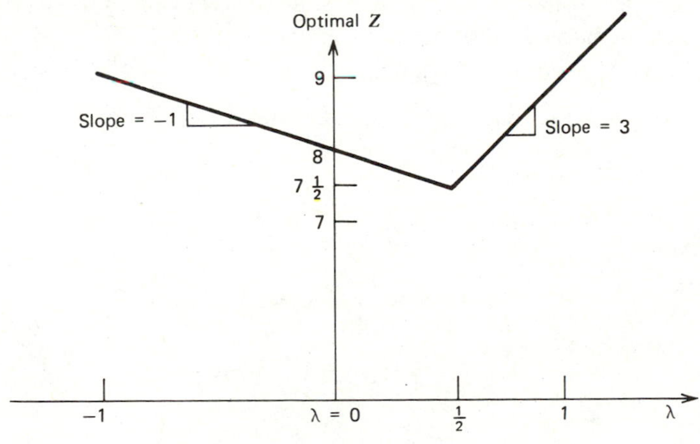

Figure 4.1
Parametric cost curve of Example 4.5-1

Parametric Right-Hand-Side Problem

The right-hand-side constants in a linear programming problem represent the limits on the resources and the outputs. It is not necessary that all the resources be independent of one another. It is quite possible in a practical problem that a shortage of one resource may be accompanied by shortages in other resources at varying levels. This is true of the outputs as well. For example, consider a manufacturer of electrical appliances. A shortage of electricity (or increased utility cost) may affect the demands for all his products at varying degrees depending on their energy consumptions. In all these problems we are considering simultaneous changes in the right-hand-side constants when they are functions of one parameter, and studying their effects on the optimal solution.

Consider a parametric right-hand-side problem of the form:

$$\begin{aligned} \text{Maximize:} \quad & Z = \mathbf{cx} \\ \text{Subject to:} \quad & \mathbf{Ax} = \mathbf{b} + \alpha\mathbf{b}^* \\ & \mathbf{x} \geqq \mathbf{0} \end{aligned}$$

where

$\mathbf{b}$ is a known right-hand-side vector,
$\mathbf{b}^*$ is the variation vector, and
α is an unknown parameter

As the value of α changes, the values of the right-hand-side constants change. We are interested in determining the family of optimal solutions for all values of α from $-\infty$ to $+\infty$.

For $\alpha = 0$, let $\mathbf{B}$ be the optimal basis matrix. The optimal solution is then given by $\mathbf{x}_B = \mathbf{B}^{-1}\mathbf{b}$, and $\mathbf{x}_N = \mathbf{0}$, where $\mathbf{x}_B$ and $\mathbf{x}_N$ are the set of basic and nonbasic variables. As the parameter α is varied, the values of the basic variables change, and their new values are given by

$$\mathbf{x}_B = \mathbf{B}^{-1}(\mathbf{b} + \alpha\mathbf{b}^*) = \mathbf{B}^{-1}\mathbf{b} + \alpha\mathbf{B}^{-1}\mathbf{b}^* = \bar{\mathbf{b}} + \alpha\bar{\mathbf{b}}^*$$

A change in α has no effect on the values of the relative profit coefficients $\bar{c}_j$. In other words, the $\bar{c}_j$ values remain nonpositive. Hence as long as $\bar{\mathbf{b}} + \alpha\bar{\mathbf{b}}^*$ is a nonnegative vector, the solution $\mathbf{x}_B = \bar{\mathbf{b}} + \alpha\bar{\mathbf{b}}^*$, and $\mathbf{x}_N = \mathbf{0}$ is feasible and optimal. For a given basis $\mathbf{B}$, the values of $\mathbf{b}$ and $\mathbf{b}^*$ can be calculated. The basis $\mathbf{B}$ is optimal as long as $\bar{\mathbf{b}} + \alpha\bar{\mathbf{b}}^* \geqq \mathbf{0}$. From this, we can determine the range of values of the parameter α for which the basis $\mathbf{B}$ is optimal.

To illustrate the parametric analysis on the right-hand-side constants, consider Example 4.4-1 with the variation right-hand-side vector as $\mathbf{b}^* = \begin{pmatrix} 1 \\ -1 \end{pmatrix}$.

EXAMPLE 4.5-2

$$\begin{aligned} \text{Maximize:} \quad & Z = 2x_1 + 3x_2 + x_3 \\ \text{Subject to:} \quad & \frac{1}{3}x_1 + \frac{1}{3}x_2 + \frac{1}{3}x_3 + x_4 = 1 + \alpha \\ & \frac{1}{3}x_1 + \frac{4}{3}x_2 + \frac{7}{3}x_3 + x_5 = 3 - \alpha \\ & x_1, \ldots, x_5 \geqq 0 \end{aligned}$$

For $\alpha = 0$, Example 4.5-2 is the same as Example 4.4-1, for which the optimal tableau is given in Table 4.7 (Section 4.4). As α changes, the values of the right-hand-side constants will change because of the variation vector $\mathbf{b}^*$. This is shown in the expanded Tableau given by Table 4.14.

Table 4.14

c_B	Basis \ c_j	2	3	1	0	0		
		x_1	x_2	x_3	x_4	x_5	$\bar{\mathbf{b}}$	$\bar{\mathbf{b}}^*$
2	x_1	1	0	−1	4	−1	1	5
3	x_2	0	1	2	−1	1	2	−2
	$\bar{c}$ Row	0	0	−3	−5	−1	$Z=8$	$Z^*=4$

The vectors $\bar{\mathbf{b}}$ and $\bar{\mathbf{b}}^*$ are computed using the inverse of the basis corresponding to Table 4.14. In other words,

$$\bar{\mathbf{b}} = \mathbf{B}^{-1}\mathbf{b} = \begin{pmatrix} 4 & -1 \\ -1 & 1 \end{pmatrix} \begin{pmatrix} 1 \\ 3 \end{pmatrix} = \begin{pmatrix} 1 \\ 2 \end{pmatrix}$$

$$\bar{\mathbf{b}}^* = \mathbf{B}^{-1}\mathbf{b}^* = \begin{pmatrix} 4 & -1 \\ -1 & 1 \end{pmatrix} \begin{pmatrix} 1 \\ -1 \end{pmatrix} = \begin{pmatrix} 5 \\ -2 \end{pmatrix}$$

For a fixed α, the values of the basic variables in Table 4.14 are given by

$$x_1 = \bar{b}_1 + \alpha \bar{b}_1^* = 1 + 5\alpha$$
$$x_2 = \bar{b}_2 + \alpha \bar{b}_2^* = 2 - 2\alpha$$

But the $\bar{c}_j$ values are not affected as long as the basis consists of the variables x_1 and x_2. As α varies, the values of the basic variables x_1 and x_2 will change, and Table 4.14 stays optimal as long as the basis (x_1, x_2) remains feasible. This implies that

$$x_1 = 1 + 5\alpha \geqq 0 \quad \text{or} \quad \alpha \geqq -1/5$$
$$x_2 = 2 - 2\alpha \geqq 0 \quad \text{or} \quad \alpha \leqq 1$$

In other words, Table 4.14 is optimal as long as α varies between $-1/5$ and 1. Thus for all $-1/5 \leqq \alpha \leqq 1$, the optimal solution is given by

$$x_1 = 1 + 5\alpha, \qquad x_2 = 2 - 2\alpha, \qquad x_3 = x_4 = x_5 = 0,$$

and the optimal value of $Z = 8 + 4\alpha$.

As α exceeds 1, the basic variable x_2 becomes negative. Although this makes Table 4.14 infeasible for the primal, it is feasible for the dual since all the $\bar{c}$-row coefficients are nonpositive. We can thus apply the dual simplex method to determine the new optimal solution for $\alpha > 1$. The new tableau is obtained by replacing x_2 with x_4, and is shown in Table 4.15.

Table 4.15

c_B	Basis \ c_j	2	3	1	0	0		
		x_1	x_2	x_3	x_4	x_5	$\bar{\mathbf{b}}$	$\bar{\mathbf{b}}^*$
2	x_1	1	4	7	0	3	9	−3
0	x_4	0	−1	−2	1	−1	−2	2
	$\bar{c}$ Row	0	−5	−13	0	−6	$Z=18$	$Z^*=-6$

The basic solution corresponding to Table 4.15 is given by

$$x_1 = 9 - 3\alpha, \qquad x_2 = 0, \qquad x_3 = 0, \qquad x_4 = -2 - 2\alpha, \qquad x_5 = 0,$$

and the value of $Z = 18 - 6\alpha$. The above solution is optimal as long as the basic variables x_1 and x_4 are nonnegative. This means that Table 4.15 is optimal for all values of the parameter α between 1 and 3.

For α greater than 3, the basic variable x_1 turns negative. Since there is no negative coefficient in the first row, we conclude that the primal problem is infeasible. Hence there exists no optimal solution for $\alpha > 3$ since the first constraint of Table 4.15 becomes inconsistent.*

To determine the optimal solution for α less than $-1/5$, we can apply the dual simplex method once more to Table 4.14.

4.6 INTEGER PROGRAMMING

This section will be devoted to the study of integer linear programming problems. An *integer linear programming problem*, henceforth called an *integer program*, is a linear programming problem wherein some or all the decision variables are restricted to be integer valued. A *pure integer program* is one where all the variables are restricted to be integers. A *mixed integer program* restricts some of the variables to be integers whereas others can assume continuous (fractional) values.

The reason for considering integer programs is that many practical problems require integer solutions. In resolving such problems, one could simply solve the linear program while ignoring the integer restrictions, and then either round off or truncate the fractional values of the LP optimal solution to get an integer solution. Of course, while doing this, one has to be careful that the resulting solution stays feasible. Such an approach is frequently used in practice, especially when the values of the variables are very large so that rounding or truncating produces negligible change. But in dealing with problems where the integer variables assume small values, rounding and truncating may produce a solution far from the true optimal integer solution. In addition, for large problems such a procedure can become computationally expensive. For instance, if the optimal LP solution is $x_1 = 2.4$ and $x_2 = 3.5$, then one has to try four combinations of integer values to x_1 and x_2 that are closest to their continuous values, namely, (2, 3), (2, 4), (3, 3), and (3, 4). The one that is feasible and is closest to the LP optimal value of the objective function will be an approximate integer solution. With just 10 integer variables, we have to try $2^{10} = 1024$ combinations of integer solutions! Even after examining all such combinations, we cannot guarantee an optimal integer solution to the problem.

Applications of Integer Programming

We shall illustrate the importance of developing techniques to solve integer programs by showing how a number of real-world problems can be formulated as integer programming problems.

EXAMPLE 4.6-1 A CAPITAL BUDGETING PROBLEM

A company is planning its capital spending for the next T periods. There are N projects that compete for the limited capital B_i, available for investment in pe-

*$\alpha > 3$ corresponds to the situation where the Dependable Company's material availability becomes less than zero, and no feasible production is possible.

riod i. Each project requires a certain investment in each period once it is selected. Let a_{ij} be the required investment in project j for period i. The value of the project is measured in terms of the associated cash flows in each period discounted for inflation. This is called the net present value (NPV). Let v_j denote the NPV for project j. The problem is to select the proper projects for investment that will maximize the total value (NPV) of all the projects selected.

Formulation

To formulate this as an integer program, we introduce a binary variable for each project to denote whether it is selected or not.

Let

$$x_j = 1 \quad \text{if project } j \text{ is selected}$$
$$x_j = 0 \quad \text{if project } j \text{ is not selected}$$

It is then clear that the following pure integer program will represent the capital budgeting problem:

$$\text{Maximize:} \qquad Z = \sum_{j=1}^{N} v_j x_j$$

$$\text{Subject to:} \qquad \sum_{j=1}^{N} a_{ij} x_j \leqq B_i \quad \text{for } i = 1, \ldots, T$$

$$0 \leqq x_j \leqq 1,\ x_j \text{ a binary variable for all } j = 1, \ldots, N$$

EXAMPLE 4.6-2 FIXED CHARGE PROBLEM

Consider a production planning problem with N products such that the jth product requires a fixed production or set up cost K_j, independent of the amount produced, and a variable cost C_j per unit, proportional to the quantity produced. Assume that every unit of product j requires a_{ij} units of resource i and there are M resources. Given that product j, whose sales potential is d_j, sells for $\$p_j$ per unit and no more than b_i units of resource i are available ($i = 1, 2, \ldots, M$), the problem is to determine the optimal product mix that maximizes the net profit.

Formulation

The total cost of production (fixed plus variable) is a nonlinear function of the quantity produced. But, with the help of binary (0-1) integer variables, the problem can be formulated as an integer linear program.

Let the binary integer variable δ_j denote the decision to produce or not to produce product j. In other words,

$$\delta_j = \begin{cases} 1 & \text{if product } j \text{ is produced} \\ 0 & \text{otherwise} \end{cases}$$

Let x_j ($\geqslant 0$) denote the quantity of product j produced. Then the cost of producing x_j units of product j is $K_j\delta_j + C_j x_j$, where $\delta_j = 1$ if $x_j > 0$ and $= 0$ if $x_j = 0$. Hence, the objective function is

$$\text{Maximize:} \quad Z = \sum_{j=1}^{N} p_j x_j - \sum_{j=1}^{N} (K_j\delta_j + C_j x_j)$$

The supply constraint for the ith resource is given by

$$\sum_{j=1}^{N} a_{ij} x_j \leqslant b_i \qquad \text{for } i = 1, 2, \ldots, M$$

The demand constraint for the jth product is given by

$$x_j \leq d_j \delta_j \qquad \text{for } i = 1, 2, \ldots, N$$
$$x_j \geq 0 \qquad \text{and} \qquad \delta_j = 0 \text{ or } 1 \qquad \text{for all } j$$

Note that x_j can be positive only when $\delta_j = 1$, in which case its production is limited by d_j and the fixed production cost K_j is included in the objective function.

EXAMPLE 4.6-3 A WAREHOUSE LOCATION PROBLEM

A retail firm is planning to expand its activities in an area by opening two new warehouses. Three possible sites are under consideration as shown in Fig. 4.2. Four customers have to be supplied whose demands are D_1, D_2, D_3, and D_4.

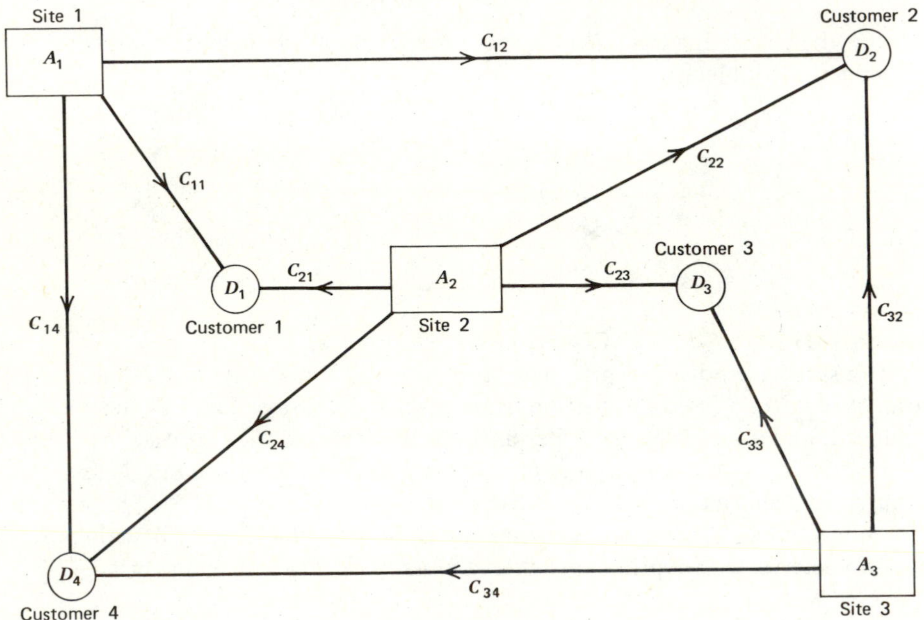

Figure 4.2
A warehouse location problem.

Assume that any two sites can supply all the demands but site 1 can supply customers 1, 2, and 4 only; site 3 can supply customers 2, 3, and 4; while site 2 can supply all the customers. The unit transportation cost from site i to customer j is C_{ij}. For each warehouse site we have the following data:

Site	Capacity	Initial Capital Investment ($)	Unit Operating Cost ($)
1	A_1	K_1	P_1
2	A_2	K_2	P_2
3	A_3	K_3	P_3

The optimization problem is to select the proper sites for the two warehouses which will minimize the total costs of investment, operation, and transportation.

Formulation

Each warehouse site has a fixed capital cost independent of the quantity stored, and a variable cost proportional to the quantity shipped. Thus the total cost of opening and operating a warehouse is a nonlinear function of the quantity stored. Through the use of binary integer variables the warehouse location problem can be formulated as an integer program.

Let the binary integer variable δ_i denote the decision to select or not to select site i. In other words,

$$\delta_i = \begin{cases} 1 & \text{if site } i \text{ is selected} \\ 0 & \text{if otherwise} \end{cases}$$

Let x_{ij} denote the quantity shipped from site i to customer j.

The supply constraint for site 1 is given by

$$x_{11} + x_{12} + x_{14} \leqq A_1\delta_1 \text{(site 1)}$$

When $\delta_1 = 1$, site 1 is selected with capacity A_1 and the quantity shipped from site 1 cannot exceed A_1. When $\delta_1 = 0$, the nonnegative variables x_{11}, x_{12}, and x_{14} will automatically become zero, implying no possible shipment from site 1.

Similarly for sites 2 and 3, we obtain

$$x_{21} + x_{22} + x_{23} + x_{24} \leqq A_2\delta_2 \text{ (site 2)}$$
$$x_{32} + x_{33} + x_{34} \leqq A_3\delta_3 \text{ (site 3)}$$

To select exactly two sites we need the following constraint:

$$\delta_1 + \delta_2 + \delta_3 = 2$$

Since the δ_1's can assume values 0 or 1 only, the new constraint will force two of the δ_i's to be one.

The demand constraints can be written as

$$x_{11} + x_{21} = D_1 \quad \text{(customer 1)}$$
$$x_{12} + x_{22} + x_{32} = D_2 \quad \text{(customer 2)}$$
$$x_{23} + x_{33} = D_3 \quad \text{(customer 3)}$$
$$x_{14} + x_{24} + x_{34} = D_4 \quad \text{(customer 4)}$$

To write the objective functions, we note that the total cost of investment, operation, and transportation for site 1 is

$$K_1\delta_1 + P_1(x_{11} + x_{12} + x_{14}) + C_{11}x_{11} + C_{12}x_{12} + C_{14}x_{14}$$

When site 1 is not selected, δ_1 will be zero. This will force x_{11}, x_{12}, and x_{14} to become zero. Similarly, the cost functions for sites 2 and 3 can be written. Thus the complete formulation of the warehouse location problem reduces to the following mixed integer program:

$$\begin{aligned} \text{Minimize:} \quad Z = {} & K_1\delta_1 + P_1(x_{11} + x_{12} + x_{14}) \\ & + C_{11}x_{11} + C_{12}x_{12} + C_{14}x_{14} + K_2\delta_2 \\ & + P_2(x_{21} + x_{22} + x_{23} + x_{24}) \\ & + C_{21}x_{21} + C_{22}x_{22} + C_{23}x_{23} + C_{24}x_{24} \\ & + K_3\delta_3 + P_3(x_{32} + x_{33} + x_{34}) \\ & + C_{32}x_{32} + C_{33}x_{33} + C_{34}x_{34} \end{aligned}$$

Subject to:

$$x_{11}+x_{12}+x_{14} \leqq A_1\delta_1$$
$$x_{21}+x_{22}+x_{23}+x_{24} \leqq A_2\delta_2$$
$$x_{32}+x_{33}+x_{34} \leqq A_3\delta_3$$
$$\delta_1+\delta_2+\delta_3=2$$
$$x_{11}+x_{21}=D_1$$
$$x_{12}+x_{22}+x_{32}=D_2$$
$$x_{23}+x_{33}=D_3$$
$$x_{14}+x_{24}+x_{34}=D_4$$
$$0 \leqq \delta_i \leqq 1 \quad \text{and} \quad \delta_i \text{ integer for } i=1, 2, 3$$
$$x_{ij} \geqq 0 \quad \text{for all } (i, j)$$

EXAMPLE 4.6-4 A JOB SEQUENCING PROBLEM

Three products A, B, and C are to be produced using four machines. The technological sequence and the processing time on the machines for the three products are shown below:

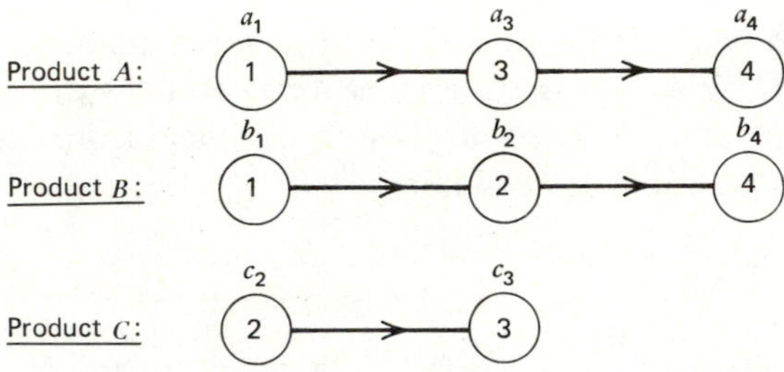

For instance, product A is processed on machine 1 first for a_1 hours, then on machine 3 for a_3 hours, and finally on machine 4 for a_4 hours. Each machine can work on only one product at a time. Moreover, each product requires a different set of tools for machining, which requires each machine to complete the processing of one product before taking up the next one.

In addition, it is required to complete product B in no more than d hours from the starting time. The problem is to determine the sequence in which the various products are processed on the machines so as to complete all the products in the least possible time.

Formulation

Let x_{Aj} denote the time (measured in hours from zero datum) when the processing of product A is started on machine j for $j=1, 3, 4$. Similarly x_{Bj} for $j=1, 2, 4$ and x_{Cj} for $j=2, 3$ are defined.

The first set of constraints enforces the technological sequence in which the machining is to be done for the three products. For product A, the processing on machine 1 is done first, followed by machine 3, and then by machine 4. This means

$$x_{A1}+a_1 \leqq x_{A3} \tag{4.27}$$

and

$$x_{A3}+a_3 \leqq x_{A4} \tag{4.28}$$

Similarly, for products B and C, we need

$$x_{B1}+b_1 \leqq x_{B2} \tag{4.29}$$
$$x_{B2}+b_2 \leqq x_{B4} \tag{4.30}$$
$$x_{C2}+c_2 \leqq x_{C3} \tag{4.31}$$

The next set of pertinent constraints is the noninterference constraints which guarantee that no machine work on more than one product at a time. For instance, machine 1 can work on either product A or product B at any given time. This is equivalent to the statement that either product A precedes product B on machine 1 or vice versa. Thus we have an "either-or" type constraint for noninterference on machine 1 given by

$$x_{A1}+a_1 \leqq x_{B1}$$

or

$$x_{B1}+b_1 \leqq x_{A1}$$

With the help of a binary integer variable, the either-or constraint can be reduced to the following two constraints:

$$x_{A1}+a_1-x_{B1} \leqq M\delta_1 \tag{4.32}$$
$$x_{B1}+b_1-x_{A1} \leqq M(1-\delta_1) \tag{4.33}$$

where $0 \leqq \delta_1 \leqq 1$, δ_1 is an integer, and M is a large positive number. Note that when $\delta_1=1$, the first constraint becomes $x_{A1}+a_1-x_{B1} \leqq M$ and is inactive; while the second constraint reduces to $x_{B1}+b_1-x_{A1} \leqq 0$ implying product B precedes product A on machine 1. On the other hand, when $\delta_1=0$, the first constraint is active implying that product A precedes product B. Thus with the help of the binary integer variable both possibilities are simultaneously included in the problem.

In like fashion, for machines 2, 3, and 4 we obtain:

$$x_{B2}+b_2-x_{C2} \leqq M\delta_2 \tag{4.34}$$
$$x_{C2}+c_2-x_{B2} \leqq M(1-\delta_2) \tag{4.35}$$
$$x_{A3}+a_3-x_{C3} \leqq M\delta_3 \tag{4.36}$$
$$x_{C3}+c_3-x_{A3} \leqq M(1-\delta_3) \tag{4.37}$$
$$x_{A4}+a_4-x_{B4} \leqq M\delta_4 \tag{4.38}$$
$$x_{B4}+b_4-x_{A4} \leqq M(1-\delta_4) \tag{4.39}$$
$$0 \leqq \delta_2 \leqq 1,\ 0 \leqq \delta_3 \leqq 1,\ 0 \leqq \delta_4 \leqq 1,\ \delta_2,\ \delta_3, \text{ and } \delta_4 \text{ are integers.}$$

The time constraint for product B becomes

$$x_{B4}+b_4 \leqq d \tag{4.40}$$

To write the objective function, observe that product A will be completed at time $x_{A4}+a_4$, product B at $x_{B4}+b_4$, and product C at $x_{C3}+c_3$. If y represents the time when all the three products are completed, then the objective is to minimize y where

$$y=\max(x_{A4}+a_4,\ x_{B4}+b_4,\ x_{C3}+c_3).$$

This nonlinear function is equivalent to the following constraints:

$$y \geqq x_{A4}+a_4 \tag{4.41}$$
$$y \geqq x_{B4}+b_4 \tag{4.42}$$
$$y \geqq x_{C3}+c_3 \tag{4.43}$$

Thus the complete formulation of the mixed integer program is to minimize y, subject to the Eqs. (4.27) through (4.43), $0 \leqq \delta_i \leqq 1$, δ_i integer for $i=1, 2, 3, 4$, and all other variables just nonnegative.

EXAMPLE 4.6-5 HANDLING NONLINEAR 0–1 INTEGER PROBLEM

Consider a nonlinear (binary) integer programming problem:

Maximize: $\quad Z = x_1^2 + x_2 x_3 - x_3^3$

Subject to: $\quad -2x_1 + 3x_2 + x_3 \leqq 3$

$$x_1, x_2, x_3 \in (0, 1)$$

The above nonlinear integer problem can be converted to a linear integer programming problem for solution. Observe the fact that for any positive k and a binary variable x_j, $x_j^k = x_j$. Hence, the objective function immediately reduces to $Z = x_1 + x_2 x_3 - x_3$. Now consider the product term $x_2 x_3$. For binary values of x_2 and x_3, the product $x_2 x_3$ is always 0 or 1. Now introduce a binary variable y_1 such that $y_1 = x_2 x_3$. When $x_2 = x_3 = 1$, we want the value of y_1 to be 1, while for all other combinations y_1 should be zero. This can be achieved by introducing the following two constraints:

$$x_2 + x_3 - y_1 \leqq 1$$
$$-x_2 - x_3 + 2y_1 \leqq 0$$

Note that when $x_2 = x_3 = 1$, the above constraints reduce to $y_1 \geqq 1$ and $y_1 \leqq 1$ implying $y_1 = 1$. When $x_2 = 0$ or $x_3 = 0$ or both are zero, the second constraint, $y_1 \leqq (x_2 + x_3)/2$ forces y_1 to be zero.

Thus the equivalent linear (binary) integer program becomes

Maximize: $\quad Z = x_1 + y_1 - x_3$

Subject to: $\quad -2x_1 + 3x_2 + x_3 \leqq 3$

$$x_2 + x_3 - y_1 \leqq 1$$
$$-x_2 - x_3 + 2y_1 \leqq 0$$

x_1, x_2, x_3, y_1 are (0, 1) variables.

A drawback of the above procedure for handling product terms is that an integer variable is introduced for each product term. It has been observed in practice that the solution time for integer programming problems increases with the number of integer variables. An alternate procedure has been suggested by Glover and Woolsey (14) that introduces a continuous variable rather than an integer variable. This procedure replaces $x_2 x_3$ by a continuous variable x_{23} and introduces three new constraints as follows:

$$x_2 + x_3 - x_{23} \leqq 1$$
$$x_{23} \leqq x_2$$
$$x_{23} \leqq x_3$$
$$x_{23} \geqq 0$$

where x_{23} replaces the product term $x_2 x_3$.

Whenever x_2, x_3, or both are zero, the last two constraints force x_{23} to be zero. When $x_2 = x_3 = 1$, all the three constraints together force x_{23} to be 1. The primary disadvantage of this procedure is that it adds more constraints than the previous method.

REMARKS

The procedure for handling the product of two binary variables can be easily extended to the product of any number of variables. For example, consider the product terms $x_1x_2 \ldots x_k$. Under the first procedure, a binary variable y_1 will replace $x_1x_2 \ldots x_k$ and the following two constraints will be added:

$$\sum_{j=1}^{k} x_j - y_1 \leqq k - 1$$

$$-\sum_{j=1}^{k} x_j + ky_1 \leqq 0$$

$$y_1 \in (0, 1)$$

$$x_j \in (0, 1)$$

Under the second procedure, the product terms $x_1x_2 \ldots x_k$ will be replaced by a nonnegative variable x_o and the following $(k+1)$ constraints will be added:

$$\sum_{j=1}^{k} x_j - x_o \leqq k - 1$$

$$x_o \leqq x_j \qquad \text{for all } j = 1, \ldots, k$$

SOLUTION OF INTEGER PROGRAMMING PROBLEMS

The Branch and Bound Algorithm

The branch and bound algorithm is the most widely used method for solving both pure and mixed integer programming problems in practice. Most commercial computer codes for solving integer programs are based on this approach. Basically the branch and bound algorithm is just an efficient enumeration procedure for examining all possible integer feasible solutions.

We discussed earlier that a practical approach to solving an integer program is to ignore the integer restrictions initially and solve the problem as a linear program. If the LP optimal solution contains fractional values for some integer variables, then by the use of truncation and rounding-off procedures, one can attempt to get an approximate optimal integer solution. For instance, if there are two integer variables x_1 and x_2 with fractional values 3.5 and 4.4, then one could examine the four possible integer solutions (3, 4), (4, 4), (4, 5), (3, 5) obtained by truncation and rounding methods. We also observe that the true optimal integer solution may not correspond to any of these integer solutions, since it is possible for x_1 to have an optimal (integer) value less than 3 or greater than 5. Hence, to obtain the true optimal integer solution one has to consider all possible integer values of x_1 which are smaller and larger than 3.5. In other words, the optimal integer solution must satisfy

either $\qquad x_1 \leqq 3$

or $\qquad x_1 \geqq 4.$

When a problem contains a large number of integer variables, it is essential to have a systematic method that will look into all possible combinations of integer solutions obtained from the LP optimal solution. The branch and bound algorithm essentially does this in the most efficient manner.

Basic Principles

To illustrate the basic principles of the branch and bound method, consider the following mixed integer program (MIP):

EXAMPLE 4.6-6

Maximize: $Z = 3x_1 + 2x_2$

Subject to:

$$x_1 \leqq 2$$
$$x_2 \leqq 2$$
$$x_1 + x_2 \leqq 3.5$$
$$x_1, x_2 \geqq 0 \text{ and integer}$$

The initial step is to solve the MIP as a linear program by ignoring the integer restrictions on x_1 and x_2. Call this linear program LP-1. Since we only have two variables, a graphical solution of LP-1 is presented in Fig. 4.3. The LP optimal solution is $x_1 = 2$, $x_2 = 1.5$, and the maximum value of the objective function $Z_o = 9$. Since x_2 takes a fractional value, we do not have an optimal solution for the MIP problem. But observe that the optimal integer solution cannot have an objective function value larger than 9, since the imposition of integer restrictions on x_2 can only make the LP solution worse. (Recall the fact from Section 4.4 that addition of new constraints to a linear program cannot improve the original optimal value of Z.) Thus we have an *upper bound* on the maximum value of Z for the integer program given by the optimal value of LP–1.

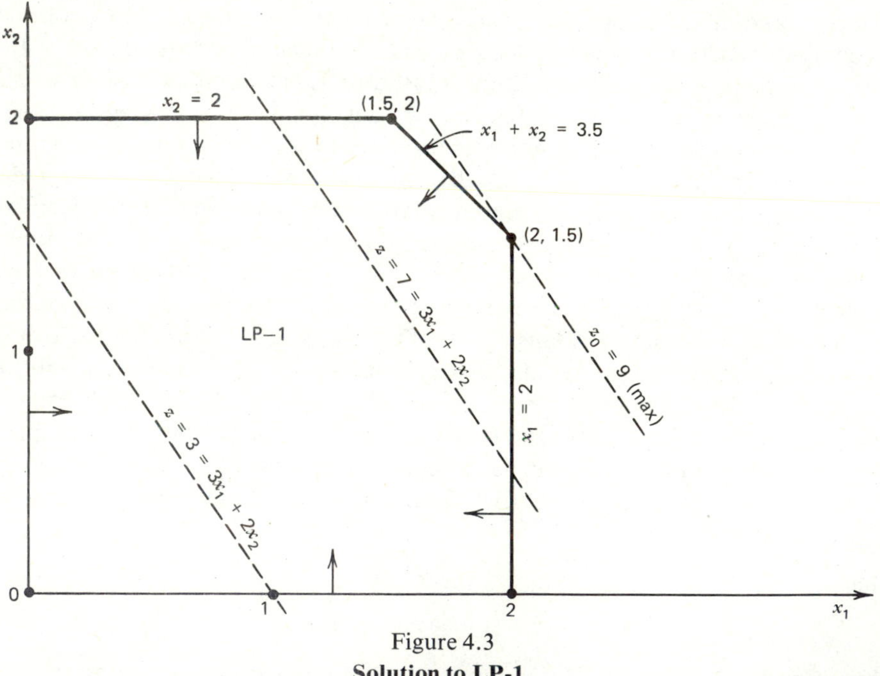

Figure 4.3
Solution to LP-1.

The next step of the branch and bound method is to examine other integer values of x_2, which are larger or smaller than 1.5. This is done by adding a new constraint either $x_2 \leqq 1$ or $x_2 \geqq 2$ to the original linear program (LP–1). This creates two new linear programs (LP–2 and LP–3) as follows:

LP–2	*LP–3*
Maximize: $Z=3x_1+2x_2$	Maximize: $Z=3x_1+2x_2$
Subject to: $x_1 \leqq 2$	Subject to: $x_1 \leqq 2$
$x_2 \leqq 2$	$x_2 \leqq 2$
$x_1+x_2 \leqq 3.5$	$x_1+x_2 \leqq 3.5$
(new constraint) $x_2 \leqq 1$	(new constraint) $x_2 \geqq 2$
$x_1, x_2 \geqq 0$	$x_1, x_2 \geqq 0$

The feasible regions corresponding to LP–2 and LP–3 are shown graphically in Figs. 4.4 and 4.5, respectively (note that the feasible region for LP–3 is just the straight line *AB*). Observe also that the feasible regions of LP–2 and LP–3 satisfy the following:

1. The optimal solution to LP–1 ($x_1=2$, $x_2=1.5$) is infeasible to both LP–2 and LP–3. Thus the old fractional optimal solution will not be repeated.
2. Every integer (feasible) solution to the original problem (MIP) is contained in either LP–2 or LP–3. Thus none of the feasible (integer) solutions to MIP is lost due to the creation of two new linear programs.

The optimal solution to LP–2 (Fig. 4.4) is $x_1=2$, $x_2=1$, and $Z_o=8$. Thus we have a feasible (integer) solution to the MIP problem. Even though LP–2 may contain other integer solutions, their values of the objective function cannot be larger than 8. Hence, $Z_0=8$ is a lower bound on the maximum value of Z for the mixed integer program. In other words, the optimal value of Z for the mixed integer problem cannot be lower than 8. Since we had computed earlier the upper bound as 9, we cannot call the LP–2 solution as the optimal integer solution without examining LP–3.

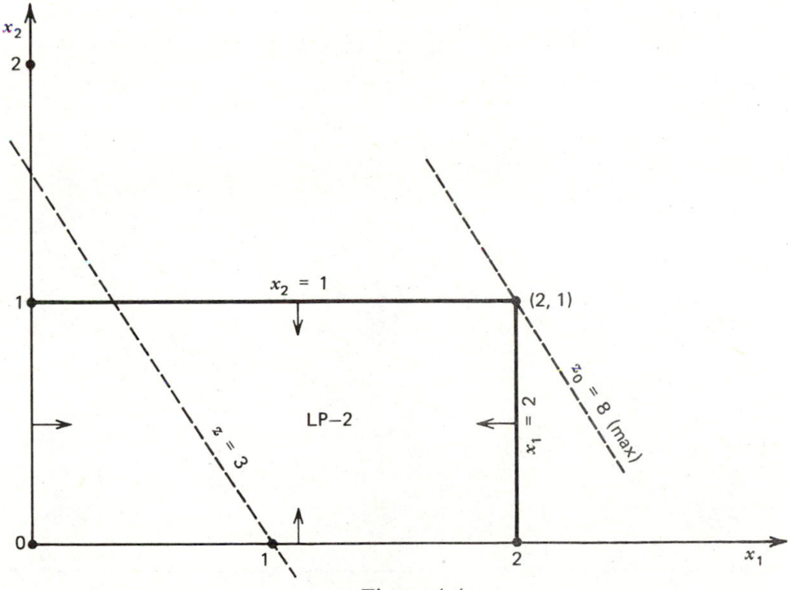

Figure 4.4
Solution to LP-2.

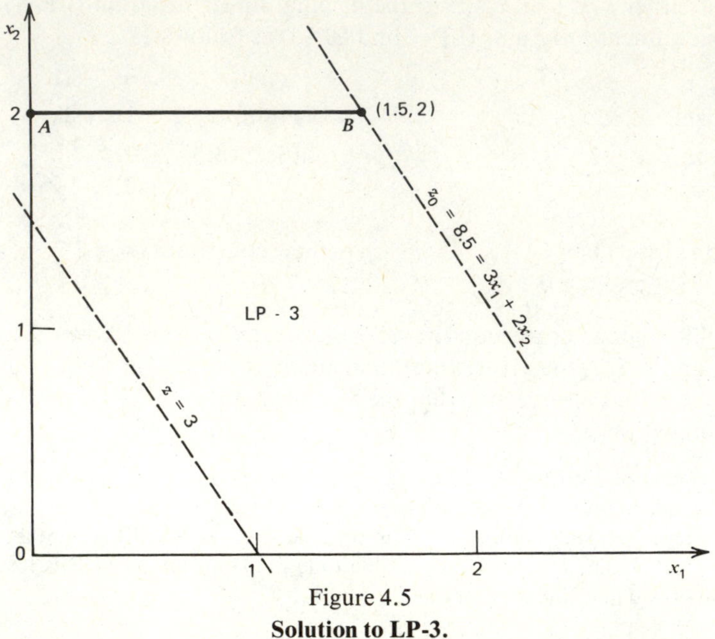

Figure 4.5
Solution to LP-3.

The optimal solution to LP–3 (Fig. 4.5) is $x_1 = 1.5$, $x_2 = 2$ and $Z_o = 8.5$. This is not feasible for the mixed integer program, since x_1 is taking a fractional value. But the maximum Z value (8.5) is larger than the lower bound (8). Hence, it is necessary to examine whether there exists an integer solution in the feasible region of LP–3 whose value of Z is larger than 8. To determine this we add the constraint either $x_1 \leqq 1$ or $x_1 \geqq 2$ to LP–3. This gives two new linear programs LP–4 and LP–5. The feasible region for LP–4 is the straight line DE shown in Figure 4.6, while LP–5 becomes infeasible.

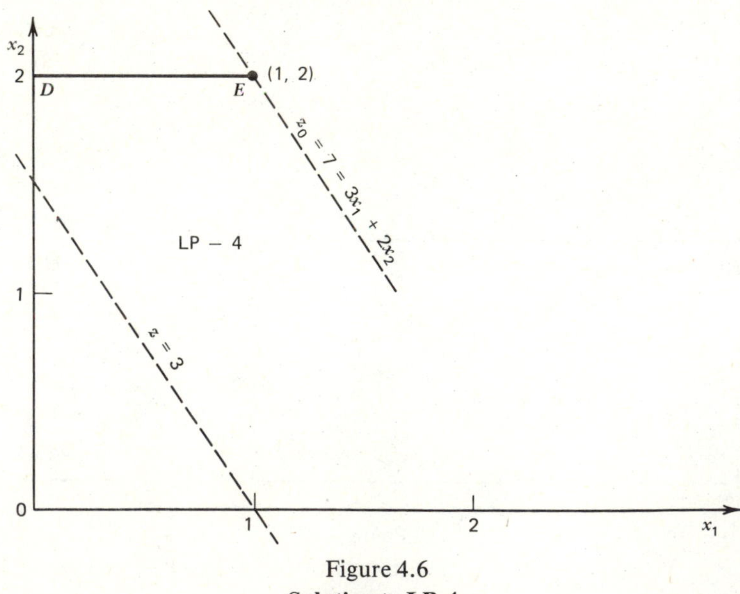

Figure 4.6
Solution to LP-4.

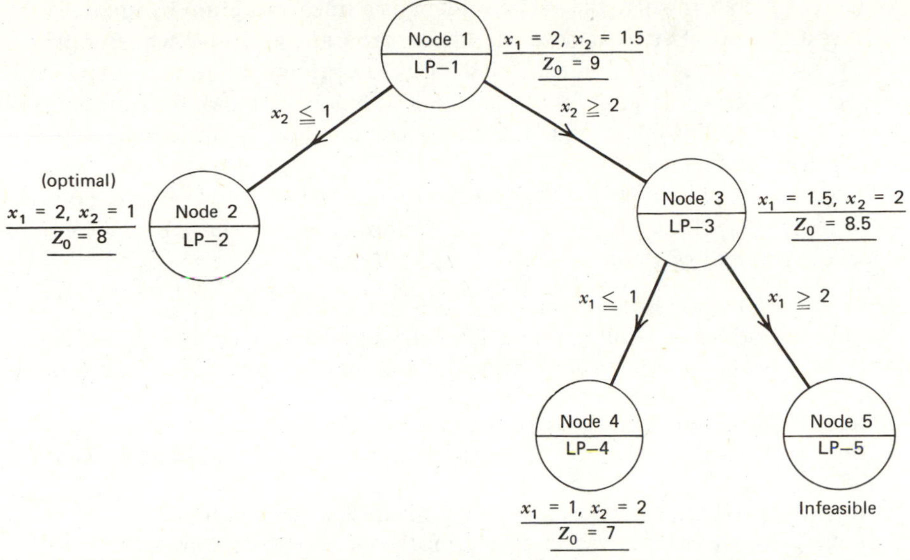

Figure 4.7
A network representation of the branch and bound method for Example 4.6-6

The optimal solution to LP–4 (Figure 4.6) is given by $x_1=1$, $x_2=2$, and $Z_o=7$. This implies that every integer solution in the feasible region of LP–3 cannot have an objective function value better than 7. Hence, the integer solution obtained while solving LP–2, namely $x_1=2$, $x_2=1$ and $Z_o=8$, is the optimal integer solution to the mixed integer problem.

The sequence of linear programming problems solved under the branch and bound procedure for Example 4.6-6 may be represented in the form of a *network* or *tree* diagram as shown in Figure 4.7. Node 1 represents the equivalent linear programming problem (LP–1) of the mixed integer program ignoring the integer restrictions. From node 1 we *branch* to node 2 (LP–2) with the help of the integer variable x_2 by adding the constraint $x_2 \leqq 1$ to LP–1. Since we have an integer optimal solution for node 2 no further branching from node 2 is necessary. Once this type of decision can be made, we say that node 2 has been *fathomed*. Branching on $x_2 \geqq 2$ from node 1 results in LP–3 (node 3). Since the optimal solution to LP–3 is fractional, we branch further from node 3 using the integer variable x_1. This results in the creation of nodes 4 and 5. Both have been fathomed since LP–4 has an integer solution while LP-5 is infeasible. The best integer solution obtained at a fathomed node (in this case, node 2) becomes the optimal solution to the mixed integer program.

Details of the Algorithm

Consider a mixed integer programming problem (MIP) of the following form:

$$\begin{aligned} &\text{Maximize:} && Z = \mathbf{cx} \\ &\text{Subject to:} && \mathbf{Ax} = \mathbf{b} \\ &&& \mathbf{x} \geqq \mathbf{0} \\ &&& x_j \text{ is an integer for } j \in \mathbf{I} \end{aligned}$$

where **I** is the set of all integer variables.

The first step is to solve the MIP problem as a linear program by ignoring the integer restrictions. Let us denote the linear program as LP–1 whose optimal value of the objective function is Z_1. Assume the optimal solution to LP–1 contains some integer variables at fractional values. Hence we do not have an optimal solution to the MIP problem. But Z_1 is an upper bound on the maximum value of Z for the MIP problem.

The next step is to partition the feasible region of LP–1 by branching on one of the integer variables at fractional value. A number of rules have been proposed to select the proper branching variable. They include:

1. Selecting the integer variable with the largest fractional value in the LP solution.
2. Assigning priorities to the integer variables such that we branch on the most important variable first. The importance of an integer variable may be based on one or more of the following criteria:
   a. It represents an important decision in the model.
   b. Its cost or profit coefficient in the objective function is very large compared to the others.
   c. Its value is very critical to the model based on the experience of the user.
3. Arbitrary selection rules; for instance, selecting the variable with the lowest index first.

Suppose that the integer variable x_j is selected for further branching and its fractional value is β_j in the LP – 1 solution. Now we create two new linear programming problems LP – 2 and LP – 3 by introducing the constraints $x_j \leqq \lfloor \beta_j \rfloor$ and $x_j \geqq \lceil \beta_j \rceil$ respectively, where $\lfloor \beta_j \rfloor$ is the largest integer less than β_j, while $\lceil \beta_j \rceil$ is the smallest integer greater than β_j (see Fig. 4.8). In other words,

LP – 2		*LP – 3*	
Maximize	$Z = \mathbf{cx}$	Maximize	$Z = \mathbf{cx}$
Subject to	$\mathbf{Ax} = \mathbf{b}$	Subject to	$\mathbf{Ax} = \mathbf{b}$
	$x_j \leqq \lfloor \beta_j \rfloor$		$x_j \geqq \lceil \beta_j \rceil$
	$\mathbf{x} \geqq \mathbf{0}$		$\mathbf{x} \geqq \mathbf{0}$

Since only one new constraint has been added, we can use the dual simplex method (discussed in Section 4.3) to find the new optimal solution to LP–2 and LP–3. Assume that the optimal solutions to LP–2 and LP–3 are still fractional, and hence infeasible to the MIP problem with integer restrictions.

The next step is to select either LP–2 or LP–3 and branch from that by adding a new constraint. Here again a number of rules have been proposed for selecting the proper node (LP problem) from which to branch. They include:

1. *Using the optimal value of the objective function.* Considering each of the nodes which can be selected for further branching, we choose the one whose LP optimal value is the largest (for a maximization problem). The rationale for this rule is that the LP feasible region with the largest Z value may contain better integer solutions. For instance any integer solution obtained by branching from LP–2 cannot have a Z value better than the optimal value of Z for LP–2.
2. *Last-in-First-Out Rule.* The LP problem which was solved most recently is selected (arbitrarily) for further branching.

Once the proper node (LP region) is selected for further branching, we branch out by choosing an integer variable with a fractional value. This process of branching and solving a sequence of linear programs is continued until an integer solution is obtained for one of the linear programs. The value of Z for this integer solution becomes a *lower bound* on the maximum value of Z for the MIP problem. At this point we can eliminate from consideration all those nodes (LP regions) whose values of Z are not better than the lower bound. We say that these nodes have been *fathomed* because it is not possible to find a better integer solution from these LP regions than what we have now.

As an illustration, consider the tree diagram given in Fig. 4.8. With the solution of LP–4 we have a lower bound on the maximum value of Z for the MIP problem given by Z_4. In other words, the optimal solution to the MIP problem cannot have a Z value smaller than Z_4. Further branching from node 4 is not necessary since any subsequent LP solution can only have a Z value less than Z_4. In other words, node 4 has ben *fathomed*. Node 5 has also been fathomed since the additional constraints render the LP problem infeasible. This only leaves nodes 6 and 7 for possible branching. Suppose $Z_6 < Z_4$ and $Z_7 > Z_4$. This means node 6 has also been fathomed (implicitly), since none of the integer solutions under node 6 can produce a better value than Z_4. However, it is possible for the LP region of node 7 to contain an integer solution better than that of node 4 since $Z_7 > Z_4$. Hence, we select node 7 for further branching and continue. In this manner, an intermediate node (LP problem) is explicitly or implicitly fathomed whenever it satisfies one of the following conditions:

1. The LP optimal solution of that node is integer valued; that is, it is feasible for the MIP problem.
2. The LP problem is infeasible.
3. The optimal value of Z for the LP problem is not better than the current lower bound.

The branch and bound algorithm continues to select a node for further branching until all the nodes have been fathomed. The fathomed node with the

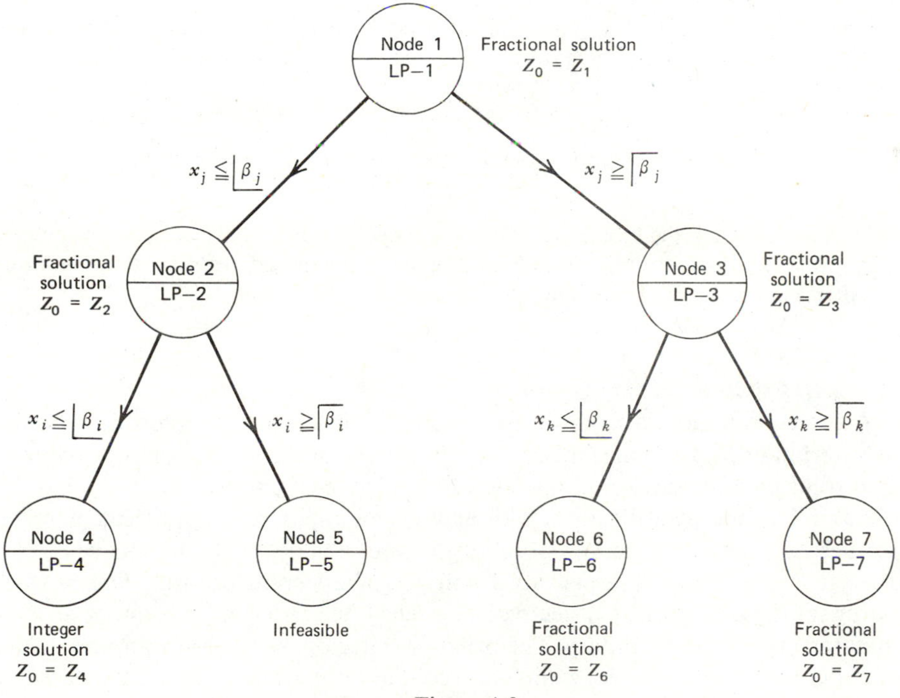

Figure 4.8
Branch and bound algorithm.

largest value of Z gives the optimal solution to the mixed integer program. Hence, the efficiency of the branch and bound algorithm depends on how soon the successive nodes are fathomed. The Fathoming Conditions 1 and 2 generally take considerable time to reach. Condition 3 cannot be used until a lower bound for the MIP problem is found. However, a lower bound is not available until a feasible (integer) solution to the MIP problem is obtained (Condition 1)! Thus it is always helpful if a feasible integer solution to MIP problem can be found before the start of the branch and bound procedure. This will provide the initial lower bound to the MIP problem until a better lower bound is found by the branch and bound algorithm. In many practical problems, the present way of operating the system may provide an initial solution.

Guidelines on Problem Formulation

The solution time for solving the integer programming problem is very sensitive to the way the problem is formulated initially. From practical experience in solving a number of integer programs by the branch and bound method, the IBM research staff have come up with some suggestions on model formulations. We describe them briefly in the following points. For more information, see the IBM Reference Manual (17). The following guidelines should not be considered as restrictions but rather as suggestions that have often reduced computational time in practice.

1. Keep the number of integer variables as small as possible. One way to do this is to treat all integer variables whose values will be at least 20 as continuous variables.
2. Provide a good (tight) lower and upper bound on the integer variables when possible.
3. Unlike the general linear programming problem, the addition of new constraints to an MIP problem will generally reduce the computational time, especially when the new constraints contain integer variables.
4. If there is no critical need for obtaining an exact optimal integer solution, then considerable savings in computational time may be obtained by accepting the first integer solution which is within 3% of the continuous optimum. In other words, for a maximization problem we can terminate the branch and bound procedure whenever

$$\frac{\text{upper bound} - \text{lower bound}}{\text{upper bound}} < 3\%$$

5. The order in which the integer variables are chosen for branching affects the solution time. It is recommended that the integer variables be processed in a priority order based on their economic significance and user experience.

4.7 GOAL PROGRAMMING

All our discussion of optimization methods until now have considered only *one* criterion or performance measure to define the optimum. It is not possible to find a solution that, say, simultaneously minimizes cost and maximizes reliability and minimizes energy utilization. This again is an important simplification of reality, because in many practical situations it would be desirable to achieve a solution that is "best" with respect to a number of different criteria. One way of treating multiple competing objectives is to select one criterion as primary and the remaining criteria as secondary. The primary criterion is then used an an optimization performance measure, while the secondary criteria are assigned acceptable minimum or maximum values and are treated as problem constraints. However, if careful considerations were not given while selecting the acceptable levels, a feasible design that satisfies all the constraints may not exist. This problem is overcome by a technique called *goal programming*, which is fast becoming a practical method for handling multiple criteria.

In goal programming, all the objectives are assigned target levels for achievement and a relative priority on achieving these levels. Goal programming treats these targets as *goals to aspire for* and not as absolute constraints. It then attempts to find an optimal solution that comes as "close as possible" to the targets in the order of specified priorities. In this section, we shall discuss how to formulate goal programming models, their solution methods, and their applications.

Before we discuss the formulation of goal programming problems, we should discuss the difference between the terms "real constraints" and "goal constraints" (or simply "goals") as used in goal programming models. The real constraints are absolute restrictions on the decision variables, while the goals are conditions one would like to achieve but are not mandatory. For instance, a real constraint given by,

$$x_1 + x_2 = 3$$

requires all possible values of $x_1 + x_2$ to always equal 3. As opposed to this, a goal requiring $x_1 + x_2 = 3$ is not mandatory, and we can choose values of $x_1 + x_2 \geq 3$ as well as $x_1 + x_2 \leq 3$. In a goal constraint, positive and negative deviational variables are introduced as follows:

$$x_1 + x_2 + d_1^- - d_1^+ = 3$$
$$d_1^+,\ d_1^- \geq 0$$

Note that if $d_1^- > 0$, then $x_1 + x_2 < 3$ and if $d_1^+ > 0$, then $x_1 + x_2 > 3$. By assigning suitable weights w_1^- and w_1^+ on d_1^- and d_1^+ in the objective function the model will try to achieve the sum $(x_1 + x_2)$ as close as possible to 3. If the goal were to satisfy $x_1 + x_2 \geq 3$, then only d_1^- is assigned a positive weight in the objective, while the weight on d_1^+ is set to zero.

The general *goal programming* model can be expressed as follows.

Minimize: $$Z = \sum_{i=1}^{m} (w_i^+ d_i^+ + w_i^- d_i^-) \tag{4.44}$$

Subject to: $$\sum_{j=1}^{n} a_{ij} x_j + d_i^- - d_i^+ = b_i \quad \text{for all } i \tag{4.45}$$
$$x_j,\ d_i^-,\ d_i^+ \geqq 0 \quad \text{for all i and j} \tag{4.46}$$

Eq. 4.44 represents the objective function, which minimizes the weighted sum of the deviational variables. The system of equations (4.45) represents the goal constraints, relating the decision variables (x_j) to the targets (b_i), and the set of inequalities (4.46) represents the standard nonnegativity restrictions on all variables.

If the relative weights (w_i^+ and w_i^-) can be specified by the management, then the goal programming problem reduces to a simple linear program. Unfortunately, it is difficult or almost impossible in many cases to secure a numerical approximation to the weights. In reality, goals are usually *incompatible* (i.e., *incommensurable*) and some goals can be achieved only at the expense of some other goals. Hence, goal programming uses *ordinal ranking* or *preemptive priorities* to the goals by assigning incommensurable goals to different priority levels and weights to goals at the same priority level. The objective function (4.44) takes the form:

$$\text{Minimize:} \qquad Z = \sum_k P_k \sum_i (w_{ik}^+ d_i^+ + w_{ik}^- d_i^-)$$

where P_k represents priority k with the assumption that P_k is much larger than P_{k+1} and w_{ik}^+ and w_{ik}^- are the weights assigned to the ith deviational variables at priority k. In this manner, lower priority goals are considered only after attaining the higher priority goals. Thus, *preemptive* goal programming or simply goal programming is essentially a sequential optimization process, in which successive optimizations are carried out on the alternate optimal solutions of the previously optimized goals at higher priority.

Goal Programming Formulation

EXAMPLE 4.7-1

Suppose a company has two machines for manufacturing a product. Machine 1 makes two units per hour, while machine 2 makes three per hour. The company has an order for 80 units. Energy restrictions dictate that only one machine can operate at one time. The company has 40 hours of regular machining time, but overtime is available. It costs \$4.00 to run machine 1 for one hour, while machine 2 costs \$5.00 per hour. The company's goals, in order of importance, are to:

1. Meet the demand of 80 units exactly
2. Limit machine overtime to 10 hours
3. Use the 40 hours of normal machining time
4. Minimize costs.

Formulation

Letting x_j represent the number of hours Machine j is operating, the goal programming model is:

$$\text{Minimize:} \qquad Z = P_1(d_1^- + d_1^+) + P_2 d_3^+ + P_3(d_2^- + d_2^+) + P_4 d_4^+ \tag{4.47}$$

$$\text{Subject to:} \qquad 2x_1 + 3x_2 + d_1^- - d_1^+ = 80 \tag{4.48}$$

$$x_1 + x_2 + d_2^- - d_2^+ = 40 \tag{4.49}$$

$$d_2^+ + \quad + d_3^- - d_3^+ = 10 \tag{4.50}$$

$$4x_1 + 5x_2 + d_4^- - d_4^+ = 0 \tag{4.51}$$

$$x_i, d_i^-, d_i^+ \geqq 0 \qquad \text{for all } i$$

where P_1, P_2, P_3, and P_4 represent the preemptive priority factors such that $P_1 >> P_2 >> P_3 >> P_4$. Note that the target for cost is set at an unrealistic level of zero. Since the goal is to minimize d_4^+, this is equivalent to minimizing the total cost of production.

In the above formulation, Eq. 4.50 does not conform to the general model given by Eqs. 4.44 to 4.46, where no goal constraint involves a deviational variable defined earlier. However, if $d_2^+ > 0$, then $d_2^- = 0$ and from Eq. 4.49, we get

$$d_2^+ = x_1 + x_2 - 40$$

which when substituted into Eq. 4.50 yields,

$$x_1 + x_2 + d_3^- - d_3^+ = 50 \tag{4.52}$$

Thus, Eq. 4.52 can replace Eq. 4.50 and the problem fits the general model, where each deviational variable appears in only one goal constraint, and has at most one positive weight in the objective function.

EXAMPLE 4.7-2 AN ECONOMIC MACHINING PROBLEM WITH TWO COMPETING OBJECTIVES (33)

Consider a single-point, single-pass turning operation in metal cutting wherein an optimum set of cutting speed and feed rate is to be chosen which balances the conflict between metal removal rate and tool life as well as being within the restrictions of horsepower, surface finish, and other cutting conditions. In developing the mathematical model of this problem, the following constraints will be considered for the machining parameters:

Constraint 1: Maximum Permissible Feed

$$f \leqq f_{\max}$$

where f is the feed in inches per revolution. A cutting force restriction or surface finish requirements usually determine $f_{\max}$. [See Armarego and Brown (1).]

Constraint 2: Maximum Cutting Speed Possible

If v is the cutting speed in surface feet per minute, then

$$v \leqq v_{\max}$$

where

$$v_{\max} = \frac{\pi D N_{\max}}{12}$$

D = mean work piece diameter, in.
$N_{\max}$ = maximum spindle speed available on the machine, rpm.

Constraint 3: Maximum Horsepower Available

If $P_{\max}$ is the maximum horsepower available at the spindle, then

$$v f^{\alpha} \leq \frac{P_{\max}(33000)}{c_t d_c^{\beta}}$$

where α, β, and c_t are constants (1). The depth of cut in inches, d_c, is fixed at a given value. For a given $P_{\max}$, c_t, β, and d_c, the right-hand side of the above constraint will be a constant. Hence, the horsepower constraint can be simply written as

$$v f^{\alpha} \leq \text{constant}$$

Constraint 4: Non-negativity Restrictions on Feed Rate and Speed

$$v, f \geqq 0$$

In optimizing metal cutting there are a number of optimality criteria which can be used. Suppose we consider the following objectives in our optimization: (i) Maximize Metal Removal Rate (MRR); (ii) Maximize Tool Life (TL). The expression for metal removal rate (MRR) is

$$\text{MRR} = 12 v f d_c \text{ cu in/min}$$

The tool life (TL) for a given depth of cut is given by:

$$\text{TL} = \frac{A}{v^{1/n} f^{1/n_1}}$$

where A, n, and n_1 are constants. We note that the MRR objective is directly proportional to feed and speed, while the TL objective is inversely proportional to feed and speed. In general, there is no single solution to a problem formulated in this way, since MRR and TL are competing objectives and their respective maxima must include some compromise between the maximum of MRR and the maximum of TL.

A GOAL PROGRAMMING MODEL

The management considers that a given single-point, single-pass turning operation will be operating at an acceptable efficiency level if the following goals are met as closely as possible:

1. The metal removal rate must be greater than or equal to a given rate M_1 (cu in/min).
2. The tool life must equal T_1 (min).

In addition, management requires that a higher priority be given to achieving the first goal than the second.

The goal programming approach may be illustrated by expressing each of the goals as goal constraints as shown below. Taking the MRR goal first,

$$12vfd_c + d_1^- - d_1^+ = M_1$$

where d_1^- represents the amount by which the MRR goal is underachieved, and d_1^+ represents any overachievement of the MRR goal. Similarly, the TL goal can be expressed as

$$\frac{A}{v^{1/n} f^{1/n_1}} + d_2^- - d_2^+ = T_1$$

Since the objective is to have a metal removal rate of at least M_1, the objective function must be set up so that a high penalty will be assigned to the underachievement variable d_1^-. No penalty will be assigned to d_1^+. To achieve a tool life of T_1, penalties must be associated with both d_2^- and d_2^+ so that both of these variables are minimized to their fullest extent. The relative magnitudes of these penalties must reflect the fact that the first goal is considered to be more important than the second. Accordingly, the goal programming objective function for this problem is:

Minimize: $$Z = P_1 d_1^- + P_2(d_2^- + d_2^+)$$

where

P_1 and P_2 are nonnumerical preemptive priority factors such that

$$P_1 >> P_2 \text{ (i.e., } P_1 \text{ is infinitely larger than } P_2)$$

With this objective function every effort will be made to completely satisfy the first goal before any attempt is made to satisfy the second.

To express the problem as a linear goal programming problem, M_1 is replaced by M_2, where

$$M_2 = \frac{M_1}{12d_c}$$

The goal T_1 is replaced by T_2, where

$$T_2 = \frac{A}{T_1}$$

and logarithms are taken of the goals and constraints. The problem can then be stated as follows:

Minimize: $Z = P_1 d_1^- + P_2(d_2^- + d_2^+)$

Subject to:

(MRR goal)	$\log v + \log f + d_1^- - d_1^+ = \log M_2$
(TL goal)	$1/n \log v + 1/n_1 \log f + d_2^- - d_2^+ = \log T_2$
(f_{max} constraint)	$\log f \leqq \log f_{max}$
(V_{max} constraint)	$\log v \leqq \log v_{max}$
(H.P. constraint)	$\log v + \alpha \log f \leqq \log \text{constant}$

$$\log v, \log f, d_1^-, d_1^+, d_2^-, d_2^+, \geqq 0$$

We would like to emphasize again that the last three inequalities are real constraints on feed, speed, and H.P. that must be satisfied at all times, whereas the equations for MRR and TL are simply goal constraints. For a further discussion of this problem and its solution, see reference 33.

Partitioning Algorithm

Goal programming problems can be solved efficiently by the *partitioning algorithm* developed by Arthur and Ravindran (2), (3). It is based on the fact that the definition of preemptive priorities implies that higher order goals must be optimized before lower order goals are even considered. Their procedure consists of solving a series of linear programming sub-problems by using the solution of the higher priority problem as the starting solution for the lower priority problem.

The partitioning algorithm begins by solving the smallest subproblem S_1, which is composed of those goal constraints assigned to the highest priority P_1 and the corresponding terms in the objective function. The optimal tableau for this subproblem is then examined for alternate optimal solutions. If none exist, then the present solution is optimal for the original problem with respect to all of the priorities. The algorithm then substitutes the values of the decision variables into the goal constraints of the lower priorities to calculate their attainment levels, and the problem is solved. However, if alternate optimal solutions do exist, the next set of goal constraints (those assigned to the second highest priority) and their objective function terms are added to the problem. This brings the algorithm to the next largest subproblem in the series, and the optimization resumes. The algorithm continues in this manner until no alternate optimum exists for one of the subproblems, or until all priorities have been included in the optimization. The linear dependence between each pair of deviational variables simplifies the operation of adding the new goal constraints to the optimal tableau of the previous subproblem without the need for a dual-simplex iteration.

At the time when the optimal solution to the subproblem S_{k-1} is obtained, a variable elimination step is performed prior to the addition of goal constraints of priority k. The elimination step involves deleting all nonbasic columns which have a positive relative cost ($\bar{C}_j > 0$) in the optimal tableau of S_{k-1} from further consideration. This is based on the well-known LP result that a nonbasic variable with a positive relative cost in an optimal tableau cannot enter the basis to form an alternate optimal solution. Figure 4.9 gives a flow chart of the partitioning algorithm.

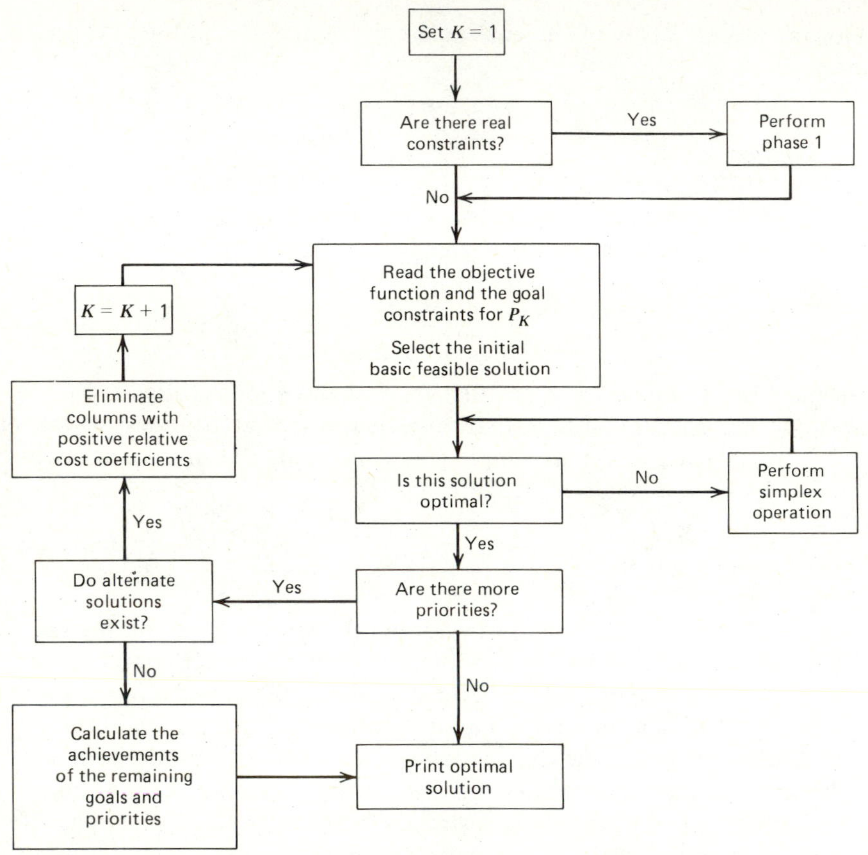

Figure 4.9
Flowchart of the partitioning algorithm.

We now illustrate the partitioning algorithm using Example 4.7-1. The subproblem S_1 for priority P_1 to be solved initially is given below:

$$\begin{aligned} S_1: \quad &\text{Minimize:} && Z_1 = d_1^- + d_1^+ \\ &\text{Subject to:} && 2x_1 + 3x_2 + d_1^- - d_1^+ = 80 \\ &&& x_1, x_2, d_1^-, d_1^+ \geqq 0 \end{aligned}$$

The solution to subproblem S_1 by the simplex method is given in Table 4.16. However, alternate optima exist to subproblem S_1 (the nonbasic variable x_1 has a relative cost of zero). Since the relative costs for d_1^+ and d_1^- are positive, they cannot enter the basis later; else they destroy the optimality achieved for priority 1. Hence, they are eliminated from the tableau from further consideration.

Table 4.16
SOLUTION TO SUBPROBLEM S_1

P_1: c_j		0	0	1	1	
c_B	Basis	x_1	x_2	d_1^-	d_1^+	b
1	d_1^+	2	③	3	−1	80
P_1:$\bar{c}$ Row		−2	−3 ↑	0	2	$Z_1 = 80$
0	x_2	2/3	1	1/3	−1/3	80/3
P_1: $\bar{c}$ Row		0	0	1	1	$Z_1 = 0$

We now add the goal constraint assigned to the second priority [Eq.(4.52)]:

$$x_1 + x_2 + d_3^- - d_3^+ = 50$$

Since x_2 is a basic variable in the present optimal tableau (Table 4.16), we perform a row operation on the above equation to eliminate x_2, and we get

$$1/3x_1 + d_3^- - d_3^+ = 70/3 \qquad (4.53)$$

Eq. 4.53 is now added to the present optimal tableau after deleting the columns corresponding to d_1^- and d_1^+. This is shown in Table 4.17. The objective function of subproblem S_2 is given by

Minimize: $Z_2 = d_3^+$

Since the right-hand side of the new goal constraint (Eq. 4.52) remained nonnegative after the row reduction (Eq. 4.53), d_3^- was entered as the basic variable in the new tableau (Table 4.17). If, on the other hand, the right-hand side had become negative, the row would be multiplied by −1 and d_3^+ would become the new basic variable. Table 4.17 indicates that we have found an optimal solution to S_2. Since alternate optimal solutions exist, we add the goal constraint and objective corresponding to priority 3. We also eliminate the column corresponding to d_3^+. The goal constraint assigned to P_3, given by

$$x_1 + x_2 + d_2^- - d_2^+ = 40$$

is added after eliminating x_2. This is shown in Table 4.18. Now, x_1 can enter the basis to improve priority 3 goal, while maintaining the levels achieved for priorities 1 and 2. d_2^- is replaced by x_1 and the next solution becomes optimal for sub-

Table 4.17
SOLUTION TO SUBPROBLEM S_2

P_2: c_j		0	0	0	1	
c_B	Basis	x_1	x_2	d_3^-	d_3^+	b
0	x_2	2/3	1	0	0	80/3
0	d_3^-	1/3	0	1	−1	70/3
P_2:$\bar{c}$ Row		0	0	0	1	$Z_2 = 0$

Table 4.18
SOLUTION TO SUBPROBLEM S_3

	$P_3: c_j$	0	0	0	1	1	
c_B	Basis	x_1	x_2	d_3^-	d_2^-	d_2^+	b
0	x_2	2/3	1	0	0	0	80/3
0	d_3^-	1/3	0	1	0	0	70/3
1	d_2^-	(1/3)	0	0	1	−1	40/3
P_3: $\bar{c}$ Row		−1/3 ↑	0	0	0	2	$Z_3=40/3$
0	x_2	0	1	0	−2	2	0
0	d_3^-	0	0	1	−1	1	10
0	x_1	1	0	0	3	3	40
P_3: $\bar{c}$ Row		0	0	0	1	1	$Z_3=0$

problem S_3 (see Table 4.18). Moreover, the solution obtained is *unique*. Hence, it is not possible to improve the goal corresponding to priority 4, and we terminate the partitioning algorithm. It is only necessary to substitute the values of the decision variables ($x_1=40$ and $x_2=0$) into the goal constraint for P_4 (Eq. 4.51) to get $d_4^+=160$. Thus, the cost goal is not achieved and the minimum cost of production is \$160.

Applications

Goal programming can be applied to a wide range of planning problems, in which the benefits are multi-dimensional and not directly comparable. A partial list of goal programming applications is given below:

Manpower planning (8), (43)
Forest management (38)
Land use planning (10)
Water resources (16)
Marine environmental protection (7)
Metal cutting (33), (34).
Accounting (19)
Academic planning (6), (23)
Portfolio selection (24)
Transportation (9), (27), (30), (40)
Marketing (26)
Production planning and scheduling (5), (12), (21), (25), (28)
Nurse scheduling (4), (32), (41)
Quality Control (36), (39)

For additional list of applications of goal programming, the reader is referred to the texts by Lee (22) and Schniederjan (37). Chapter 1 of the latter text has an extensive bibliography (199 citations) on goal programming applications categorized by areas— accounting, finance, management, marketing, health, education, military, and miscellaneous. For a categorized bibliographic survey of goal programming algorithms and applications the reader should consult the recent survey article by Zanakis and Gupta (42).

Recommended Readings

For an excellent discussion of duality theory and its applications to bi-matrix games, refer to Gale (13) and Murty (31). Dantzig (11) discusses a modification of the simplex method to solve linear programs with lower and upper bounds on variables. A complete discussion on the decomposition principle to solve very large linear programming problems is available in Dantzig (11) and Murty (31). Many variants of the simplex method, including the primal-dual algorithm and the self-dual parametric algorithm, are discussed in Hadley (15). Solution of nonlinear programming problems is discussed in Chapter 11.

For an excellent discussion of integer programming problems at an applied level, refer to McMillan (29) and Plane and McMillan (35). These also contain many practical applications of integer programming to real world problems and some case studies. Lee (22) presents an excellent discussion of goal programming and its applications. Ignizio (18) discusses advanced topics in integer goal programming algorithms. Zeleny (44) and Keeney and Raiffa (20) deal with other techniques of multiple criteria optimization.

REFERENCES

1. Armarego, E. J. A., and R. H. Brown, *The Machining of Metals*, Prentice-Hall, Englewood Cliffs, N.J., 1969.
2. Arthur, J. L., and A. Ravindran, "An Efficient Goal Programming Algorithm Using Constraint Partitioning and Variable Elimination," *Management Science*, *24*(8), 867–868, April 1978.
3. Arthur, J. L., and A. Ravindran, "PAGP—Partitioning Algorithm for (Linear) Goal Programming Problems," *ACM Trans. on Mathematical Software*, *6*, 378–386, 1980.
4. Arthur, J. L., and A. Ravindran, "A Multiple Objective Nurse Scheduling Model," *Inst. of Indus. Engrs. Trans.*, *13*(1), 55–60, 1981.
5. Arthur, J. L., and K. D. Lawrence, "Multiple Goal Production and Logistics Planning in a Chemical and Pharmaceutical Company," *Comput. & Ops. Res.*, *9*(2), 127–137, 1982.
6. Beilby, M. H., and T. H. Mott, Jr., "Academic Library Acquisitions Allocation Based on Multiple Collection Development Goals," *Computers & OR*, *10*(4), 335–344, 1983.
7. Charnes, A., W. W. Cooper, J. Harrald, K. R. Karwan, and W. A. Wallace, "A Goal Interval Programming Model for Resource Allocation in a Marine Environmental Protection Program," *J. of Environmental Economics and Management*, *3*, 347–362, 1976.
8. Charnes, A., W. W. Cooper, and R. J. Niehaus, "Dynamic Multi-attribute Models for Mixed Manpower Systems," *Naval Res. Logistics Quarter*, *22*(2) (1975).
9. Cook, W. D., "Goal Programming and Financial Planning Models for Highway Rehabilitation," *J. of Operational Research Society*, *35*(3), 217–224, 1984.
10. Dane, C. W., N. C. Meador, and J. B. White, "Goal Programming in Land-use Planning," *J. Forestry* (June 1977).
11. Dantzig, G. B., *Linear Programming and Extensions*, Princeton University Press, Princeton, N. J., 1963.
12. Deckro, R. J., J. E. Hebert, and E. P. Winkofsky, "Multiple Criteria Job-Shop Scheduling," *Computers & OR*, *9*(4), 279–285, 1984.
13. Gale, D., *The Theory of Linear Economic Models*, McGraw-Hill, New York, 1960.
14. Glover, F., and E. Woolsey, "Converting 0–1 Polynomial Programming Problem to a 0–1 Linear Program" *Operations Research*, *22*, 180–182 (1974).

15. Hadley, G., *Linear Programming*, Addison-Wesley, Reading, Mass., 1962.

16. Haimes, Y. Y., W. A. Hall, and H. T. Freedman, *Multiobjective Optimization in Water Resources Systems*. Elsevier, Amsterdam, 1975.

17. *IBM General Information Manual,* "An Introduction to Modeling Using Mixed Integer Programming," Amsterdam, 1972.

18. Ignizio, J. P., *Goal Programming and Extensions,* Health, Lexington, Ky., 1976.

19. Ijiri, Y., *Management Goals and Accounting for Control*, Rand-McNally, Chicago, Ill., 1965.

20. Keeney, R. L., and H. Raiffa, *Decisions with Multiple Objectives,* Wiley, New York, 1976.

21. Lawrence, K. D., and J. J. Burbridge, "A Multiple Goal Linear Programming Model for Coordinated and Logistics Planning," *Int. J. Prod. Res., 14*(2), 215–222, 1976.

22. Lee, S. M., *Goal Programming for Decision Analysis*, Auerback Publishers, Philadelphia, Pa., 1972.

23. Lee, S. M., and E. R. Clayton, "A Goal Programming Model for Academic Resource Allocation," *Management Science, 17*(8), (1972).

24. Lee, S. M., and A. J. Lerro, "Optimizing the Portfolio Selection for Mutual Funds," *Journal of Finance, 28*(8), (1972).

25. Lee, S. M., and L. J. Moore, "A Practical Approach to Production Scheduling," *Production and Inventory Management, 15*(1), (1974).

26. Lee, S. M., and R. Nicely, "Goal Programming for Marketing Decisions: A Case Study," *Journal of Marketing, 38*(1), (1974).

27. Lee, S. M., and L. J. Moore, "Multi-Criteria School Busing Models," *Management Science, 23*(7), 703–715 (1977).

28. Lee, S. M., E. R. Clayton, and B. W. Taylor, "A Goal Programming Approach to Multi-Period Production Line Scheduling," *Comput. & Ops. Res., 5*, 205–211, 1978.

29. McMillan, C., *Mathematical Programming*, Wiley, New York, 1970.

30. Moore, L. J., B. W. Taylor, and S. M. Lee, "Analysis of a Transhipment Problem with Multiple Conflicting Objectives," *Comput. & Ops. Res., 5*, 39–46, 1978.

31. Murty, K. G., *Linear and Combinatorial Programming*, Wiley, New York. 1976.

32. Musa, A. A., and U. Saxena, "Scheduling Nurses Using Goal Programming Techniques," *Inst. of Indus. Engrs. Trans., 16*(3), 216–221, 1984.

33. Philipson, R. H., and A. Ravindran, "Application of Goal Programming to Machinability Data Optimization," *J. of Mechanical Design: Trans. of ASME, 100*, 286–291, 1978.

34. Philipson, R. H., and A. Ravindran, "Application of Mathematical Programming to Metal Cutting," *Mathematical Programming Study, 11*, 116–134, 1979.

35. Plane, D. R., and C. McMillan, Jr., *Discrete Optimization*, Prentice-Hall, Englewood Cliffs, N.J., 1971.

36. Ravindran, A., W. Shin, J. L. Arthur, and H. Moskowitz, "Nonlinear Integer Goal Programming Models for Acceptance Sampling," *Comput. & Ops. Res.*, Vol. 13, No. 6, 1986.

37. Schniederjan, M., *Linear Goal Programming*, Petrocelli Books, Princeton, N.J., 1984.

38. Schuler, A. T., H. H. Webster, and J. C. Meadows, "Goal Programming in Forest Management," *J. Forestry* (June 1977).

39. Sengupta, S., "A Goal Programming Approach to a Type of Quality Control Problem," *J. of Operational Research Society, 32*(2), 207–212, 1981.

40. Sinha, K. C., M. Muthusubramanyam, and A. Ravindran, "Optimization Approach for Allocation of Funds for Maintenance and Preservation of the Existing Highway System," *Transportation Research Record, 826*, pp. 5–8, 1981.

41. Trivedi, V. M., "A Mixed-Integer Goal Programming Model for Nursing Service Budgeting," *Oper. Res., 29*(5), 1019-1034, 1981.

42. Zanakis, S. H., and S. K. Gupta, "A Categorized Bibliographic Survey of Goal Programming," *Omega, 13*(3), pp. 211–222, 1985.

43. Zanakis, S. H., and M. W. Maret, "A Markovian Goal Programming Approach to Aggregate Manpower Planning," *32*(1), 55–63, 1981.

44. Zeleny, M., *Multiple Criteria Decision Making*, McGraw-Hill, New York, 1982.

EXERCISES

1. What are simplex multipliers and how are they used in the revised simplex method?

2. Explain why the computational effort in solving LP problems is more sensitive to the number of constraints and not to the number of variables.

3. What are shadow prices and how are they computed? What is their relationship to the dual problem?

4. Given that a primal problem and its dual are feasible, what can you deduce about the optimality of both problems and why?

5. Explain under what conditions rounding off continuous solutions is a good strategy for obtaining integer values. Cite some situations where such a strategy will not work and explain why.

6. What is the meaning of "fathoming" a node as used in the branch and bound algorithm? Under what conditions can a node be fathomed?

7. What is the difference between a goal and a constraint as used in goal programming?

8. What are the drawbacks of using preemptive weights in goal programming?

9. Solve the following LP problems by the revised simplex method:

(a) Maximize: $Z = 2x_1 - x_2 + x_3 + x_4$

Subject to:

$$\begin{aligned} -x_1 + x_2 + x_3 \qquad + x_5 \quad &= 1 \\ x_1 + x_2 \qquad + x_4 \qquad &= 2 \\ 2x_1 + x_2 + x_3 \qquad\qquad + x_6 &= 6 \\ x_1, \ldots, x_6 &\geqq 0 \end{aligned}$$

(b) Maximize: $Z = x_1 + 2x_2$

Subject to:

$$\begin{aligned} -2x_1 + x_2 + x_3 &\leqq 2 \\ -\ x_1 + x_2 - x_3 &\leqq 1 \\ x_1, x_2, x_3 &\geqq 0 \end{aligned}$$

10. Solve the following LP problems by the two-phase or big M revised simplex method:

(a) Minimize: $Z = 40x_1 + 36x_2$

Subject to:

$$\begin{aligned} x_1 \qquad &\leqq 8 \\ x_2 &\leqq 10 \\ 5x_1 + 3x_2 &\geqq 45 \\ x_1, x_2 &\geqq 0 \end{aligned}$$

(b) Maximize: $Z = -2x_1 + x_2 - 2x_3$

Subject to:

$$-x_1 + x_2 + x_3 = 4$$
$$-x_1 + x_2 - x_3 \leqq 6$$
$$x_1 \leqq 0,\ x_2 \geqq 0$$
$$x_3 \quad \text{unrestricted in sign}$$

11. The following linear program is fed to a computer for solution:

Maximize: $Z = 3x_1 + 2x_2$

Subject to:

$$-x_1 + 2x_2 + x_3 = 4$$
$$3x_1 + 2x_2 + x_4 = 14$$
$$x_1 - x_2 + x_5 = 3$$
$$x_1, \ldots, x_5 \geqq 0$$

Using the revised simplex method, the computer has the following information corresponding to a basis **B** at some stage:

$$\mathbf{x}_B = (x_3, x_4, x_1) \quad \text{and} \quad \mathbf{B}^{-1} = \begin{bmatrix} 1 & 0 & 1 \\ 0 & 1 & -3 \\ 0 & 0 & 1 \end{bmatrix}$$

(a) Compute the current basic feasible solution and the simplex multipliers corresponding to the basis **B**.
(b) Show that the present solution is not optimal. Which variable should now be introduced into the basis?
(c) Having chosen a variable to enter the basis, now select a variable to leave the basis and describe the selection rule. Is it possible that no variable meets the criterion of your rule? If this happens in some problem, what does that indicate about the original problem?
(d) Using (c), find the new inverse of the basis and the new simplex multipliers.
(e) Write down the new basic feasible solution. Is it optimal? Why or why not?

12. Consider the following linear programming problem:

Maximize: $Z = x_1 + 2x_2 + 3x_3 + 4x_4$

Subject to:

$$x_1 + 2x_2 + 2x_3 + 3x_4 \leqq 20$$
$$2x_1 + x_2 + 3x_3 + 2x_4 \leqq 20$$
$$x_1, x_2, x_3, x_4 \geqq 0$$

Using the principles of the revised simplex method, prove that an optimal solution exists with x_3 and x_4 as basic variables.

13. Write down the dual of the following linear programs:

(a) Minimize: $Z = x_1 + 2x_2 + 3x_3 + 4x_4$

Subject to:

$$x_1 + 2x_2 + 2x_3 + 3x_4 \geqq 30$$
$$2x_1 + x_2 + 3x_3 + 2x_4 \geqq 20$$
$$x_1, x_2, x_3, x_4 \geqq 0$$

(b) Maximize: $Z = -4x_1 - 3x_2$

Subject to:

$$x_1 + x_2 \leqq 1$$
$$-x_2 \leqq -1$$
$$-x_1 + 2x_2 \leqq 1$$
$$x_1, x_2 \geqq 0$$

(c) Minimize: $Z = 40x_1 + 36x_2$

Subject to:

$$\begin{aligned} x_1 \quad &\leqq 8 \\ x_2 &\leqq 10 \\ 5x_1 + 3x_2 &\geqq 45 \\ x_1, x_2 &\geqq 0 \end{aligned}$$

(d) Minimize: $Z = 2x_1 + x_2 + 2x_3$

Subject to:

$$\begin{aligned} -x_1 + x_2 + x_3 &= 4 \\ -x_1 + x_2 - x_3 &\leqq 6 \\ x_1 \leqq 0, x_2 &\geqq 0 \\ x_3 \text{ unrestricted in sign} \end{aligned}$$

14. Consider the following linear program which we call the primal problem:

Maximize: $Z = 3x_1 + 2x_2$

Subject to:

$$\begin{aligned} -x_1 + 2x_2 &\leqq 4 \\ 3x_1 + 2x_2 &\leqq 14 \\ x_1 - x_2 &\leqq 3 \\ x_1, x_2 &\geqq 0 \end{aligned}$$

(a) Write down the dual problem.
(b) Prove using the duality theory that there exists an optimal solution to both the primal and dual problems.
(c) Compute an upper and a lower bound on the optimal values of both the problems using the weak duality theorem.

15. Prove using the duality theory that the following linear program is feasible but has no optimal solution:

Minimize: $Z = x_1 - x_2 + x_3$

Subject to:

$$\begin{aligned} x_1 \qquad - x_3 &\geqq 4 \\ x_1 - x_2 + 2x_3 &\geqq 3 \\ x_1, x_2, x_3 &\geqq 0 \end{aligned}$$

16. Consider the following primal linear program:

Maximize: $Z = x_1 + 2x_2 + x_3$

Subject to:

$$\begin{aligned} x_1 + x_2 - x_3 &\leqq 2 \\ x_1 - x_2 + x_3 &= 1 \\ 2x_1 + x_2 + x_3 &\geqq 2 \\ x_1 \geqq 0, x_2 \leqq 0, x_3 \text{ unrestricted in sign.} \end{aligned}$$

(a) Write the dual of the above problem.
(b) Prove using the duality theory that the maximum value of Z for the primal problem cannot exceed 1.

17. Use the dual simplex method to show that the following linear program is infeasible:

Maximize: $Z = -4x_1 - 3x_2$

Subject to:

$$\begin{aligned} x_1 + x_2 &\leqq 1 \\ - x_2 &\leqq -1 \\ -x_1 + 2x_2 &\leqq 1 \\ x_1, x_2 &\geqq 0 \end{aligned}$$

18. Solve by the dual simplex method:

Minimize: $Z = x_1 + 2x_2 + 3x_3 + 4x_4$

Subject to:

$$\begin{aligned} x_1 + 2x_2 + 2x_3 + 3x_4 &\geqq 30 \\ 2x_1 + x_2 + 3x_3 + 2x_4 &\geqq 20 \\ x_1, x_2, x_3, x_4 &\geqq 0 \end{aligned}$$

19. A factory manufactures three products. Three resources—technical services, labor, and administration—are required to produce these products. The following table gives the requirements on each of the resources for the three products:

Product	Resources (hours)			Unit Profit ($)
	Technical Service	Labor	Administration	
1	1	10	2	10
2	1	4	2	6
3	1	5	6	4

There are 100 hours of technical services available, 600 hours of labor, and 300 hours of administration. In order to determine the optimal product mix which will maximize the total profit, the following linear program was solved:

$$\text{Maximize:} \quad Z = 10x_1 + 6x_2 + 4x_3$$

$$\text{Subject to:} \quad \begin{aligned} x_1 + x_2 + x_3 &\leqq 100 \quad \text{(technical)} \\ 10x_1 + 4x_2 + 5x_3 &\leqq 600 \quad \text{(labor)} \\ 2x_1 + 2x_2 + 6x_3 &\leqq 300 \quad \text{(administration)} \\ x_1, x_2, x_3 &\geqq 0 \end{aligned}$$

where x_1, x_2, and x_3 are the number of product 1, product 2, and product 3 produced. The optimal solution is given by the following tableau, where x_4, x_5, and x_6 are the slack variables:

Basis \ c_j	10 x_1	6 x_2	4 x_3	0 x_4	0 x_5	0 x_6	Constants
x_2	0	1	5/6	10/6	−1/6	0	400/6
x_1	1	0	1/6	−4/6	1/6	0	200/6
x_6	0	0	4	−2	0	1	100
$\bar{c}$ Row	0	0	−16/6	−20/6	−4/6	0	$Z = 4400/6$

Using sensitivity analysis, answer the following with respect to the above optimal tableau:

(a) What should be the profit of product 3 before it becomes worthwhile to manufacture? Find the most profitable product mix if the profit on product 3 were increased to \$50/6.

(b) What is the range on the profit of product 1 so that the current solution is still optimal?

(c) It is believed that the estimates of the available hours of technical services might be wrong. The correct estimate is $100 + 10\lambda$ where λ is some unknown parameter. Find the range of values of λ within which the given product mix is still optimal.

(d) Determine the shadow prices of all the resources.

(e) The manufacturing department comes up with a proposal to produce a new product requiring 1 hour of technical service, 4 hours of labor, and 3 hours of administration. The marketing and sales department predicts that the product can be sold at a unit profit of \$8. What should be the management's decision?

(f) Suppose the company decides to produce at least 10 units of product 3; determine the optimal product mix.

20. A company which manufactures three products A, B, and C, requiring two raw materials—labor and material—wants to determine the optimal production schedule that maximizes the total profit. The following linear program was formulated to answer this:

$$\text{Maximize:} \quad Z = 3x_1 + x_2 + 5x_3$$
$$\text{Subject to:} \quad 6x_1 + 3x_2 + 5x_3 \leqq 45 \quad \text{(labor)}$$
$$3x_1 + 4x_2 + 5x_3 \leqq 30 \quad \text{(material)}$$
$$x_1, x_2, x_3 \geqq 0$$

where x_1, x_2, x_3 are the amount of products A, B, and C. The computer prints out the following optimal tableau (x_4 and x_5 are the slacks):

c_j / Basis	\$3	1	5	0	0	
	x_1	x_2	x_3	x_4	x_5	Constants
x_1	1	−1/3	0	1/3	−1/3	5
x_3	0	1	1	−1/5	2/5	3
$\bar{c}$ Row	0	−3	0	0	−1	$Z=30$

Answer the following with respect to the above optimal tableau:

(a) Find the range on the unit profit of Product A. Find the optimal solution for $c_1 = 2$.
(b) Suppose an additional 15 units of material may be obtained at a cost of \$10. Is it profitable to do so?
(c) Find the optimal solution when the available material is increased to 60 units.
(d) Due to a "technological breakthrough" the material requirements of product B are reduced to 2 units. Does this affect your optimal solution? Why or why not?
(e) Suppose a "supervision" constraint, $x_1 + x_2 + 3x_3 \leqq 20$ is added to the original problem. How does this affect the optimal primal and dual solutions? Explain fully.

21. The following tableau gives an optimal solution to a standard linear program:

$$\text{Maximize:} \quad Z = \mathbf{cx}, \text{ Subject to: } \mathbf{Ax} = \mathbf{b}, \mathbf{x} \geqq \mathbf{0}$$

c_j / Basis	2	3	1	0	0	
	x_1	x_2	x_3	x_4	x_5	**b**
x_1	1	0	1	3	−1	1
x_2	0	1	1	−1	2	2
$\bar{c}$ Row	0	0	−4	−3	−4	$Z=8$

Assume (x_4, x_5) was the initial identity matrix.

(a) How much can c_3 be increased before the current basis is no longer optimal? Find an optimal solution when $c_3 = 6$.
(b) How much can c_1 be varied so that the given basis (x_1, x_2) is still optimal?
(c) Find the largest and smallest value of λ for which the given solution is still optimum if $\mathbf{c}$ is replaced by $\mathbf{c} + \lambda\mathbf{c}^*$ where $c^* = (0, 0, 1, -1, 2)$ and $-\infty \leqq \lambda \leqq \infty$.

(d) How much can b_2 (the original value) be varied before the given basis (x_1, x_2) is no longer feasible? (Note: It is *not* necessary to compute the original value of b_2 to answer this.)

(e) Find an optimal solution by the dual simplex method when b_2 is *increased by* 2 units.

22. The following tableau gives an *optimal* solution to a standard linear program:

$$\text{Maximize: } Z=\mathbf{cx}, \text{ Subject to: } \mathbf{Ax}=\mathbf{b}, \mathbf{x}\geqq\mathbf{0}$$

Basis \ c_j	2	3	1	0	0	
	x_1	x_2	x_3	x_4	x_5	**b**
x_1	1	0	−1	3	−1	1
x_2	0	1	2	−1	1	2
$\bar{c}$ Row	0	0	−3	−3	−1	$Z=8$

(a) How much can c_2 be varied without affecting the optimal solution? Find the optimum solution when $c_2=1$.

(b) Find the range on λ for which the given basis (x_1, x_2) is still optimal if the original **b** vector is replaced by $\mathbf{b}+\lambda\mathbf{b}^*$ where $\mathbf{b}^*=\begin{bmatrix}1\\-1\end{bmatrix}$ and $-\infty<\lambda<\infty$. Also, find the optimal solution when $\lambda=1/2$. (Assume that (x_4, x_5) formed the initial basis.)

(c) Find the optimal solution when a new constraint

$$x_1+x_2\geqq 2$$

is added to the *original* problem.

(d) Compute the shadow prices of the constraints of the original problem.

23. A firm has four possible sites for locating its warehouses. The cost of locating a warehouse at site i is $\$K_i$. There are nine retail outlets, each of which must be supplied by at least one warehouse. It is not possible for any one site to supply all the retail outlets as shown in the following figure:

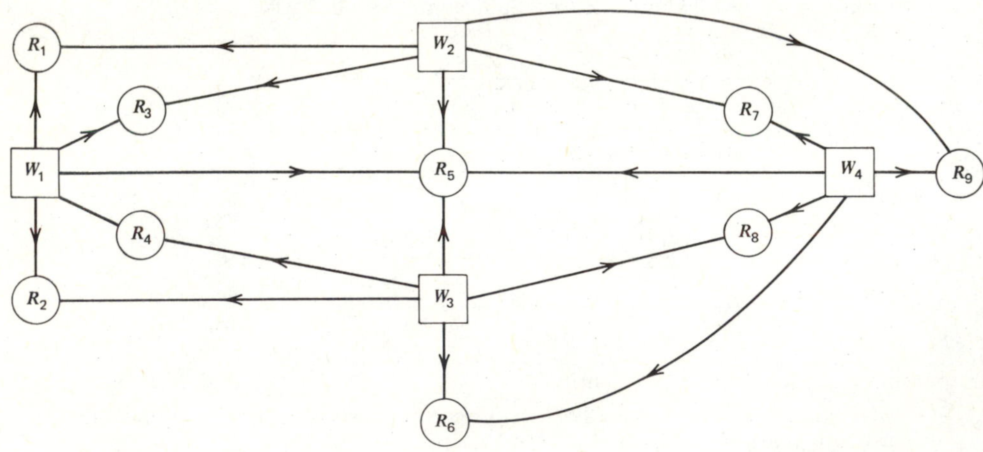

The problem is to determine the location of the warehouses such that the total cost is minimized. Formulate this as an integer program.

24. A company manufactures three products A, B, and C. Each unit of product A requires 1 hour of engineering service, 10 hours of direct labor, and 3 pounds of material. To produce one unit of product B requires 2 hours of engineering, 4 hours of direct labor, and 2 pounds of material. Each unit of product C requires 1 hour of engineering, 5 hours of direct labor, and 1 pound of material. There are 100 hours of engineering, 700 hours of labor, and 400 pounds of materials available. The cost of production is a non-linear function of the quantity produced as shown below:

Product A		*Product B*		*Product C*	
Production (units)	Unit Cost ($)	Production (units)	Unit Cost ($)	Production (units)	Unit Cost ($)
0–40	10	0–50	6	0–100	5
41–100	9	51–100	4	over 100	4
101–150	8	over 100	3		
over 150	7				

(*Note*: If 60 units of A are made, the first 40 units cost \$10 per unit and the remaining 20 units cost \$9 per unit.)

Formulate a mixed integer program to determine the minimum-cost production schedule.

25. Explain how the following conditions can be represented as linear constraints by the use of binary (0–1) integer variables:
(a) Either $x_1 + x_2 \leqq 2$ or $2x_1 + 3x_2 \geqq 8$
(b) Variable x_3 can assume values 0, 5, 9, and 12 only.
(c) If $x_4 \leqq 4$, then $x_5 \geqq 6$. Otherwise $x_5 \leqq 3$.
(d) At least two out of the following four constraints must be satisfied:

$$x_6 + x_7 \leqq 2$$
$$x_6 \leqq 1$$
$$x_7 \leqq 5$$
$$x_6 + x_7 \geqq 3$$

26. Solve the following pure integer program by the branch and bound algorithm:

Maximize: $Z = 21x_1 + 11x_2$

Subject to: $7x_1 + 4x_2 + x_3 = 13$

x_1, x_2, x_3, nonnegative integers

27. Solve the following mixed integer program by the branch and bound algorithm:

Minimize: $Z = 10x_1 + 9x_2$

Subject to:
$$x_1 \leqq 8$$
$$x_2 \leqq 10$$
$$5x_1 + 3x_2 \geqq 45$$
$$x_1 \geqq 0, x_2 \geqq 0$$
x_2 – integer

28. Consider the following Integer Program (IP):

Maximize: $Z = 5x_1 + 8x_2$

Subject to:
$$x_1 + x_2 \leqq 6$$
$$5x_1 + 9x_2 \leqq 45$$
$x_1, x_2 \geqq 0$ and integer

(a) Solve the IP as an LP problem ignoring integer restrictions.
(b) Find all the *adjacent feasible integer* solutions by rounding and/or truncating the continuous optimum found in part (a) and determine the best solution.
(c) Now solve the IP by the branch and bound algorithm. Compare the optimal integer solution with the one found in part (b). What can you conclude from the comparison?

Note: Use graphical methods for solving all the LP problems.

29. During the *maximization* of a *pure* integer programming problem by the branch and bound algorithm, we have the following branch and bound tree at a certain stage:

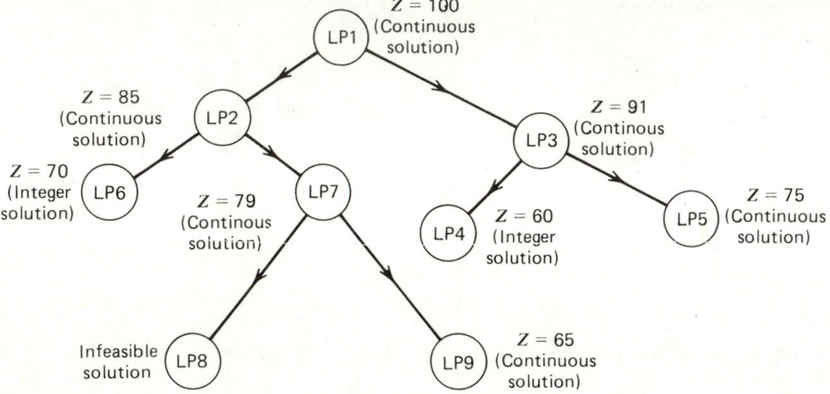

(a) What is the *best* upper bound we have on the maximum value of Z for the integer program at this stage?
(b) What is the *best* lower bound we have on the maximum Z value?
(c) Indicate the node(s) that have been *fathomed* and explain why.
(d) Identify the node(s) that have not been fathomed and explain why not.
(e) Have we reached an optimal solution to the integer program? Why or why not?
(f) What is the maximum *absolute error* on the optimal value of Z if we terminate the branch and bound algorithm at this stage? What is the *fractional error* as a percentage of the worst case optimum?

30. A company has two machines that can manufacture a product. Machine 1 makes 2 units per hour, while machine 2 makes 3 per hour. The company has an order for 160 units. Each machine can be run for 40 hours on regular time; however, overtime is available. The company's goals, in order of importance, are to:

1. Avoid falling short of meeting the demand.
2. Avoid any overtime of machine 2 beyond 10 hours.
3. Minimize the sum of overtime (assign differential weights according to the relative cost of overtime hours, assuming that the operating cost of the two machines are the same).
4. Avoid under-utilization of regular working hours (assign weights according to the productivity of the machines).

31. A company supplies a single product from two warehouses to three of its customers. The supplies at the warehouses are 120 and 70 units, respectively. The customer demands are 70, 80 and 100 units respectively. The net profit associated with shipping one unit from warehouse i to customer j is given below:

	Customer		
	1	2	3
Warehouse 1	7	9	8
Warehouse 2	3	4	6

The company has set the following goals for shipping:

1. Meet the demand of customer 3.
2. Meet at least 60% of the demand of customers 1 and 2.
3. Maximize the total profit from goods shipped.
4. Balance the percentage of demands satisfied between customers 1 and 2.

Formulate the above problem as a goal program.

32. An advertising company wishes to plan an advertising campaign in three different media—television, radio, and magazines. The purpose of the advertising program is to reach as many of the potential customers as possible. Results of a market study are given below:

	Television		Radio	Magazines
	Daytime	Prime Time		
Cost of an advertising unit	\$40,000	\$75,000	\$30,000	\$15,000
Number of potential customers reached per unit	400,000	900,000	500,000	200,000
Number of women customers reached per unit	300,000	400,000	200,000	100,000

The company has an advertising budget of \$800,000 and has the following goals to achieve in order of importance:

1. At least two million exposures should take place among women.
2. Advertising on television to be limited to \$500,000.
3. Achieve a total exposure of three million potential customers.
4. Advertising budget should not be exceeded by more than 10%.
5. The number of advertising units on radio and magazine should each be between 5 and 10.
6. At least 3 advertising units to be bought on daytime television, and 2 units during prime time.

Formulate the advertising problem as a goal program.

33. Solve the goal program formulated in Exercise 30 by the partitioning algorithm.

34. Solve the following goal program by the partitioning algorithm:

Minimize: $Z = P_1 d_1^- + P_2 d_2^+ + P_3(6d_3^+ + 5d_4^+) + P_4(6d_4^- + 5d_3^-)$

Subject to:

$$
\begin{aligned}
5x_1 + 6x_2 + d_1^- - d_1^+ &= 120 \\
x_2 + d_2^- - d_2^+ &= 11 \\
x_1 \qquad + d_3^- - d_3^+ &= 8 \\
10x_2 + d_4^- - d_4^+ &= 80 \\
x_i, d_i^-, d_i^+ \geqq 0 \text{ for all } i.
\end{aligned}
$$

35. For the standard linear program

Minimize: $Z = \mathbf{cx}$

Subject to:

$$
\begin{aligned}
\mathbf{Ax} &= \mathbf{b} \\
\mathbf{x} &\geqq \mathbf{0}
\end{aligned}
$$

prove the appropriate weak duality theorem, optimality criterion theorem, and complementary slackness theorem.

36. Consider the bounded variable linear program:

Maximize: $Z = x_1 - x_2 + x_3 - x_4$

Subject to:
$$x_1 + x_2 + x_3 + x_4 = 8$$
$$0 \leqq x_1 \leqq 8$$
$$-4 \leqq x_2 \leqq 4$$
$$-2 \leqq x_3 \leqq 4$$
$$0 \leqq x_4 \leqq 10$$

(a) Write the dual of the above problem.

(b) Show that $x_1 = 8$, $x_2 = -4$, $x_3 = 4$, and $x_4 = 0$ is an optimal solution by using the complementary slackness theorem.

37. It is given that the following standard linear program has an optimal solution:

Minimize; $Z = \mathbf{cx}$

Subject to: $\mathbf{Ax} = \mathbf{b}, \mathbf{x} \geqq \mathbf{0}$

Suppose the requirement vector $\mathbf{b}$ is changed to another vector $\mathbf{d}$; that is, we now have a modified problem:

Minimize: $Z = \mathbf{cx}$

Subject to:
$$\mathbf{Ax} = \mathbf{d}$$
$$\mathbf{x} \geqq \mathbf{0}$$

Prove that if the modified problem is feasible, then it will always have an optimal solution.

38. Recall the formulation of a standard transportation problem with m warehouses and n markets (Chapter 3):

Minimize: $Z = \sum_i \sum_j c_{ij} x_{ij}$

Subject to:
$$\sum_j x_{ij} = a_i \qquad \text{for } i = 1, 2, \ldots, m,$$
$$\sum_i x_{ij} = b_j \qquad \text{for } j = 1, 2, \ldots, n,$$

where x_{ij} is the amount shipped from i to j, c_{ij} is the unit cost of shipping, and a_i and b_j are the supply and demand at warehouse i and market j, respectively.

(a) Write the dual of the transportation problem.

(b) Write the complementary slackness conditions to be satisfied at optimality. Compare these with the u-v method for solving the transportation problem.

39. Given the cost matrix of a transportation problem with three warehouses and four markets as:

	M_1	M_2	M_3	M_4
W_1	2	2	2	1
W_2	10	8	5	4
W_3	7	6	6	8

The supplies and demands are $a_1 = 3$, $a_2 = 7$, $a_3 = 5$, $b_1 = 4$, $b_2 = 3$, $b_3 = 4$, $b_4 = 4$. Prove, using the complementary slackness theorem, that the optimal shipping schedule is to send 3 units from W_1 to M_1, 3 units from W_2 to M_3, 4 units from W_2 to M_4, 1 unit from W_3 to M_1, 3 units from W_3 to M_2, and 1 unit from W_3 to M_3.

40. An electronic system consists of three components, each of which must function in order for the system to function. The reliability of the system (probability of successful performance of the system) can be improved by installing several stand-by units for one or more of the components. The reliability of the system is given by the product of

the reliability of each component, while the reliability of each component is a function of the number of stand-by units as given below:

Number of Stand-by Units	Reliability of Component		
	1	2	3
0	0.5	0.6	0.7
1	0.6	0.75	0.9
2	0.7	0.95	1.0
3	0.8	1.0	1.0
4	0.9	1.0	1.0
5	1.0	1.0	1.0

For instance, if no stand-by unit is provided, the reliability of component 2 is 0.6; with 2 stand-by units, its reliability increases to 0.95. The cost and weight of each stand-by unit for the three components are as follows:

Component	Unit Cost ($)	Weight/Unit (kilograms)
1	20	2
2	30	4
3	40	6

We have a budget restriction of $150 and a weight restriction of 20 kg for all the stand-by units. The problem is to determine how many stand-by units should be installed for each of the three components so as to maximize the total reliability of the system. Formulate this as a *linear* integer program.

41. Convert the following nonlinear integer program to a linear integer program:

Minimize: $Z = x_1^2 - x_1x_2 + x_2$

Subject to:

$$x_1^2 + x_1x_2 \leqq 8$$
$$x_1 \leqq 2$$
$$x_2 \leqq 7$$
$$x_1, x_2 \geqq 0 \text{ and integer}$$

(*Hint: Replace* x_1 *by* $2^0\delta_1 + 2^1\delta_2$, *and* x_2 *by* $2^0\delta_3 + 2^1\delta_4 + 2^2\delta_5$, *where* $\delta_i \in (0, 1)$ *for* $i = 1, 2, \ldots, 5$).

42. Consider a production problem where it is required to produce 2000 units of a certain product on three different machines. The set-up costs, the production costs per unit, and the maximum production capacity for each machine are given below:

Machine	Set Up Cost ($)	Maximum Capacity	Production Cost
1	$100	600 units	$10 per unit for the first 300 units $7 per unit for the remaining 300 units
2	$500	800 units	$2 per unit for all 800 units
3	$300	1200 units	$6 per unit for the first 500 units $4 per unit for the remaining 700 units

The objective is to minimize the total cost of producing the required lot. Formulate the problem as an integer programming problem. You must define all your variables clearly, write out the constraints to be satisfied with a brief explanation of each and develop the objective function.

CHAPTER 5
DECISION ANALYSIS*

This chapter has been authored by Dr. Herbert Moskowitz, Professor of Management, Krannert Graduate School of Management, Purdue University, West Lafayette, Indiana 47907.

5.1 INTRODUCTION

In recent years, statisticians, engineers, economists, and students of management have placed increasing emphasis on decision making under conditions of uncertainty. This area of study has been called statistical decision theory, Bayesian decision theory, decision theory, decision analysis, and many other things. We shall henceforth use the term decision analysis as a matter of expediency and as an anchor point in discussing this topic.

This chapter will deal with the problem of making decisions under conditions of uncertainty. Much of life, of course, involves making choices under uncertainty, that is, choosing from some set of alternative courses of action in situations where we are uncertain about the actual consequences that will occur for each course of action being considered. Often, however, we must make a choice and are naturally concerned about whether it is a best or optimal choice.

In today's fast-moving technological world, the need for sound, rational decision making by business, industry, and government is vividly apparent. Consider, for example, the area of design and development of new and improved products and equipment. Typically, development from invention to commercialization is expensive and filled with uncertainty regarding both technical and commercial success. Product development problems related to research and development (R&D), production, finance, and marketing activities, of both a tactical and strategic nature, abound. In R&D, for example, decision makers might be faced with the problem of choosing whether to pursue a parallel versus a sequential strategy (i.e, pursuing two or more designs simultaneously versus developing the most promising design, and if it fails, going to the next most promising design, etc.). In production, they may have to decide on a production method or process for manufacture; or choose whether to lease, subcontract, or manufacture; or select a quality-control plan. In finance, they may have to decide whether to invest in a new plant, equipment, research programs, marketing facilities, even risky orders. In marketing, they may have to determine the pricing scheme, whether to do mar-

*Herbert Moskowitz, and Gordon P. Wright, *Operations Research Techniques for Management*, ©1979, pp. 112–137, 188–214. Adapted by permission of Prentice Hall, Inc., Englewood Cliffs, N.J.

ket research and what type and amount of it, the type of advertising campaign, and so on.

Each of these decision problems is characteristically complex and can have a significant impact on the health of a firm. It is almost impossible for any decision maker to intuitively take full account of all the factors impinging on a decision simultaneously. It thus becomes useful to find some method of separating such decision problems into parts in a way that would allow a decision maker to think through the implications of each set of factors one at a time in a rational, consistent manner.

Decision analysis provides a rich set of concepts and techniques to aid the decision maker in dealing with complex decision problems under uncertainty. The decision analysis formulation differs from classical statistical inference and decision procedures by considering explicitly both the preference structure of the individual decision maker and the uncertainties that characterize the decision situation. An exposure to the concepts of decision analysis rapidly makes one aware of the common shortcomings in more informal approaches, such as intuition. A particular benefit of the decision analysis approach is that it facilitates communication and analysis among individuals involved and affected by the decision problem.

Decision analysis is concerned with the making of rational, consistent decisions, notably under conditions of uncertainty. That is, it helps the decision maker choose the best alternative in light of the information available (which normally is incomplete and unreliable). Decision analysis enables the decision maker to analyze a complex situation with many different alternatives, states, and consequences. The major objective is to choose a course of action consistent with the basic values (tastes) and knowledge (beliefs) of the decision maker. It is a prescriptive rather than descriptive approach. That is, it presents the concepts and methods for how one should choose but does not purport to describe how people really make decisions.

We shall introduce some of the concepts of decision analysis and show, in a practical way, how to apply these principles and techniques to resolve decision problems under uncertainty. In presenting these concepts, we shall place emphasis on generating an intuitive, managerially-oriented interpretive view of the fundamental ideas and procedures, as well as their application in a variety of actual decision contexts. Decision trees, revising beliefs based on new information, and value of information are several of the key concepts to be discussed. This chapter should provide you with practical tools for formulating, analyzing, and making effective decisions under uncertainty. It should also sharpen your ability to be an aware, critical, and appropriately cautious "consumer" of such formal analysis.

We will first discuss the concepts and techniques of decision analysis when a single terminal decision must be made on the basis of a decision maker's existing information state. Such a decision is often called a *terminal decision based on prior information*; such an analysis is commonly called *prior analysis*. Sequential decisions and opportunities to acquire further information before deciding are then considered.

To facilitate our presentation, we adopt three primary policies: (1) to avoid the use of mathematics when there is no real need for it; (2) to restrict the discussion principally to discrete problems to avoid having to use calculus; and (3) to explain the various methods of decision analysis through simple, concrete illustrations.

5.2 CHARACTERISTICS OF A DECISION PROBLEM

All decision problems have certain general characteristics. These characteristics constitute the formal description of the problems and provide the structure for solutions. The decision problem under study may be represented by a model in terms of the following elements:

1. *The Decision Maker*. The decision maker is responsible for making the decision. Viewed as an entity, the decision maker may be a single individual, committee, company, nation, or the like.
2. *Alternative Courses of Action*. An important part of the decision maker's task, over which the decision maker has control, is the specification and description of the alternatives. Given that the alternatives are specified, the decision involves a choice among the alternative courses of action. When the opportunity to acquire information is available, the decision maker's problem is to choose a best information source or sources and a best overall strategy. A strategy is a set of decision rules indicating which action should be taken contingent on a specific observation received from the chosen information source(s).
3. *Events*. Events are the scenarios or states of the environment not under the control of the decision maker that may occur. Under conditions of uncertainty, the decision maker, when making the decision, does not know for certain which event will occur. The true event indeed may be a fact unknown to the decision maker.

   The events are defined to be mutually exclusive and collectively exhaustive. That is, one and only one of all possible events specified will occur. Events are also called states, states of nature, states of the world, or payoff relevant events.

   Uncertainty is measured in terms of probabilities assigned to the events. One of the distinguishing characteristics of decision analysis is that these probabilities can be subjective (reflecting the decision maker's state of knowledge or beliefs) or objective (theoretically or empirically determined) or both. The decision maker must identify and specify the events as well as assess their probabilities of occurrence.
4. *Consequences*. The consequences, which must be assessed by the decision maker, are measures of the net benefit, or payoff, received by the decision maker. The consequences that result from a decision depend not only on the decision but also on the event that occurs. Thus there is a consequence (or vector of consequences) associated with each action-event pair. Consequences are also called payoffs, outcomes, benefits, or losses. They can be conveniently summarized in a payoff matrix, or decision matrix, which displays the consequences of all action-event combinations.

   Consequences should reflect the subjective values of the decision maker, that is, the decision maker's preferences, or values, for the corresponding objective consequences. In other words, the objective consequences should be transformed into utilities, which reflect the subjective value of the consequences to the decision maker. Hence, a decision maker's utility function, a mathematical expression or curve relating the decision maker's utilities for a given set of consequences, will need to be determine through appropriate questioning. To simplify the discussion, unless otherwise stated, we will assume that the consequence values given have already been transformed into utilities. The issue of utility and utility assessment will not be taken up, but can be found in standard references on decision analysis.[1]

We now present an example that we shall use through the following section to illustrate the concepts, structure, and techniques of formally making terminal decisions under prior uncertainty.

[1]See, for example, H. Raiffa, *Decision Analysis* (Reading, Mass.: Addison-Wesley, 1968).

5.3
TERMINAL DECISIONS BASED ON PRIOR INFORMATION
EXAMPLE 5.3 STOCKING TENNIS SHIRTS

The owner of a tennis shop must decide how many tennis shirts to order for the summer season. For a particular type of shirt, he must order in batches of 100. If he orders 100 shirts, his cost is \$10 per shirt; if he orders 200, his cost is \$9 per shirt; and if he orders 300 or more shirts, his cost is \$8.50 per shirt. The selling price is \$12, but if any shirts are left unsold at the end of the summer, they will be sold for half price. For simplicity, the owner believes that demand for this shirt will be either 100, 150, or 200. Of course, he cannot sell more shirts than he stocks. If, however, he understocks, there is a goodwill loss of \$0.50 for each shirt a person wants to buy but cannot because it is out of stock. Furthermore, the owner must place the order now for the forthcoming summer season; he cannot wait to see how the demand is running for this shirt before he orders, nor can he place several orders.

Problem

For this example problem, define

1. The decision maker.
2. The alternative courses of action.
3. The events.
4. The consequences.

Solution

1. The decision maker is the tennis shop owner.
2. The alternative courses of actions are: a_1: Order 100, a_2: Order 200, and a_3: Order 300.
3. The possible events are: θ_1: Demand is 100, θ_2: Demand is 150, and θ_3: Demand is 200.
4. There is one consequence associated with each of the nine action-event pairs (totaling nine possible consequences). For example, if the tennis shop owner orders 100 shirts and demand turns out to be 100, the consequence associated with this action-event pair is that the owner will make a profit of \$200. The consequences for the other eight action-event pairs would be determined similarly.

The Payoff (Decision) Matrix

One way of representing a terminal decision problem based on prior uncertain information is through the use of a payoff or decision matrix. The matrix is used to depict the set of actions, events, and consequences associated with the decision problem. Our convention will be to structure the payoff matrix as follows. The actions are listed on the left of the matrix along the rows, the states are listed at the top of the matrix along the columns, and the possible consequences are listed in the matrix cells, one consequence associated with each action-event pair. The consequences for the tennis shop owner's problem are determined from the following consequence function. If supply is greater than demand (i.e., if $S > D$), then profit is equal to $\$D(R-C) - \$(C-6)(S-D)$, where R is the per-unit revenue (selling price), C is the relevant unit cost, and the parameter \$6 is the half-price, end-of-summer selling price for unsold goods. If demand is greater than supply (i.e., if $D > S$), then profit is equal to $\$S(R-C) - \$0.50(D-S)$, where \$0.50 is the goodwill loss per shirt.

EXAMPLE 5.3-2

Problem

Construct the decision matrix for the tennis shop owner's problem.

Solution

The relevant decision matrix is depicted in Table 5.1.

Table 5.1
PAYOFF (DECISION) MATRIX FOR THE TENNIS SHOP OWNER'S PROBLEM

	Event		
Action	θ_1 Demand is 100	θ_2 Demand is 150	θ_3 Demand is 200
a_1: Order 100	200	175	150
a_2: Order 200	0	300	600
a_3: Order 300	− 150	150	450

If the event that will occur were known with certainty beforehand by the decision maker (e.g., in the tennis shop example, θ_2: Demand will be 150), then the decision maker would merely have to look down the appropriate column in the payoff matrix (i.e., column θ_2) and choose the action that yields the highest payoff [i.e., max(175, 300, 150) = 300, which indicates that 200 shirts should be ordered.] However, in the real world, since the states of nature lie beyond the control of the decision maker, it is not possible to know with certainty which specific event will occur (or is true). The choice of an optimal course of action in the face of uncertainty is the crux of the decision maker's problem.

Although the term uncertainty refers to the decision maker's ignorance about which event will occur (or which event is the true event), the decision maker may also feel uncertain about various aspects of the so-called consequences. For instance, if the tennis shop owner was not sure of his goodwill loss, then some of the profits in the decision matrix would be uncertain. Such thoughts may lead us to expand the formulation of the problem to include greater detail and scope. Whether to do so depends on how complex and realistic we want the model to be.[2]

It should be noted that some authors differentiate between what they call decision making under risk (decision making when the state of the world is not known but probabilities for the various possible states are known) and decision making under uncertainty (decision making when the state of the world is not known and probabilities for the various possible states are not known). Under the subjective interpretation of probability, it is always possible to assess probabilities for the possible events, or states of the world. Hence, the risk versus uncertainty dichotomy is extremely artificial (in fact, it is nonexistent according to the subjective interpretation of probability), and in this text any decision-making problem in which the state of the world is not known for certain is called decision making under uncertainty. The term risk is reserved for discussions relating to utility.

Choice Criteria: Expected Value

Having examined the fundamental elements and structures of decision problems under uncertainty, we must now consider an appropriate criterion that may be used in selecting a course of action. Various nonprobabilistic and probabilistic

[2]See, for example, H. Raiffa, *Decision Analysis* (Reading, Mass.: Addison-Wesley, 1968) for a further discussion of this.

criteria have been proposed. The former have been developed, at least in part, to avoid the assessment of probabilities, thereby ignoring the probabilistic nature of decision making under uncertainty. These include: (a) dominance (inadmissibility), (b) maximin, (c) maximax, (d) Hurwicz α, and (e) minimax regret or opportunity loss. The latter include criteria such as (a) Laplace and (b) maximizing expected value (MEV), whereby the decision maker is willing to assign a probability distribution to the events. We will use the generally accepted criterion of MEV as our basis for choice.[3]

Maximizing Expected Value (MEV). A maximum expected value (MEV) criterion has us (a) assign a probability to each mutually exclusive event, with the probabilities summing to 1 (since the events should be collectively exhaustive); (b) compute the expected value of each action by multiplying each consequence value by its corresponding event probability and summing these products; (c) choose an action whose expected value is largest (smallest, if working with costs or losses).[4] In other words, the expected value of an act is the weighted average of the payoffs under that act, where the weights are the probabilities of the various events that can occur.

In many realistic decision problems, it would be reasonable to suppose that decision makers would have some idea of the likelihood of occurrence of the various events and that this knowledge would help them to choose a course of action. For example, in our illustrative problem, if the tennis shop owner felt very confident that sales (demand) would be 200 units, this would tend to move him toward ordering 200 tennis shirts (a_2). By the same reasoning, if he was highly confident that sales (demand) would be 100 units, he would order only 100 shirts. If there are many possible events and many possible courses of action, the problem becomes complex, and the decision maker clearly needs some orderly method of processing all the relevant information. Such a systematic procedure is provided by the computation of the expected value of each course of action, and the selection of the act that yields the best of these expected values.

EXAMPLE 5.3-3

Suppose that the tennis shop owner (on the basis of past data, experience, intuition, and instinct) assigns the following subjective probability distribution to the events:

Event	Probability, $P(\theta)$
θ_1: Demand is 100	0.5
θ_2: Demand is 150	0.3
θ_3: Demand is 200	0.2
Total	1.0

Note that if demand is 100, it cannot be 150 or 200 (i.e., events are mutually exclusive), and the probabilities sum to 1, as they should (i.e., events are collectively exhaustive).

Problem

Choose an optimal action based on the MEV criterion.

[3]Discussion of the other criteria mentioned above can be found in Moskowitz, H., and Wright, G.P., *Statistics for Management and Economics* (Columbus, Ohio: Merrill, 1985).

[4]As an alternative to maximizing expected value (utility), we can equivalently minimize expected opportunity loss as we shall also show.

Solution

The owner's expected profit for each act is shown in Table 5.2. Using the MEV criterion, the tennis shop owner should choose a_2: Order 200 shirts, with an expected profit of $210.

Table 5.2
CALCULATING EXPECTED PROFITS

Act a_1: Order 100 Event	Probability	Profit	Weighted Profit
θ_1: Demand is 100	0.5	$200	$100.0
θ_2: Demand is 150	0.3	175	52.5
θ_3: Demand is 200	0.2	150	30.0
Total	1.0		$182.5
Expected Profit = $182.5			

Act a_2: Order 200 Event	Probability	Profit	Weighted Profit
θ_1: Demand is 100	0.5	$ 0	$ 0
θ_2: Demand is 150	0.3	300	90.0
θ_3: Demand is 200	0.2	600	120.0
Total	1.0		$210.0
Expected Profit = $210.0			

Again, it should be emphasized that the consequences and event probabilities can be interpreted as objective or subjective. Objective consequences (values) represent "physical" quantities such as dollars, units of time, and so forth. Subjective consequences (values) represent the decision maker's relative preferences or values for the corresponding consequences. They are called subjective because they are directly related to the decision maker's preferences in the particular problem situation.

Minimizing Expected Opportunity Loss (EOL). A useful concept in the analysis of decisions under uncertainty is that of opportunity loss. As characterized by L. J. Savage (5), the noted statistician, after decisions have been made and the events have occurred, decision makers may experience regret because they now know what events have taken place and may wish that they had selected different actions. Decision makers thus might wish to minimize their expected regret, or equivalently, *expected opportunity loss* (EOL). This criterion requires the development of a regret or opportunity loss matrix. Regret or opportunity loss is defined as the difference between the actual payoff and the payoff one could have received if one knew which event was going to occur. To minimize EOL, we first convert the payoff matrix into a corresponding regret (opportunity loss) matrix. This is done by subtracting each entry in the payoff matrix from the largest entry in its column. The largest entry in a column will have zero regret (opportunity loss).

EXAMPLE 5.3-4

Problem

Develop a regret or opportunity loss matrix for the tennis shop owner's problem.

Solution

Table 5.3 shows the payoff and resulting regret matrices for our example problem. A regret (opportunity loss), computed as described previously, represents the difference in value between what we obtain for a given action and a given event and what we could obtain if we knew beforehand that the given event was in fact the true event. In the example, if the owner ordered 100 units (a_1) and demand was 150 units (0_2), then his regret would be \$125, since he could have gotten \$125 more by ordering 200 units had he known beforehand that demand would actually be for 150 units. If, however, he had ordered 200 units (a_2) and demand was 150 units, his regret would be zero, since he could not have chosen a better action if he had known in advance that demand would be 150 units.

Table 5.3
PAYOFF AND REGRET MATRICES FOR THE TENNIS SHOP OWNER'S PROBLEM

Payoff Matrix	Event		
Action	θ_1 Demand is 100	θ_2 Demand is 150	θ_3 Demand is 200
a_1: Order 100	200	175	150
a_2: Order 200	0	300	600

Regret (Opportunity Loss) Matrix	Event		
Action	θ_1 Demand is 100	θ_2 Demand is 150	θ_3 Demand is 200
a_1: Order 100	0	125	450
a_2: Order 200	200	0	0

In the event that the original decision matrix was in terms of losses (costs), then we would select the smallest loss for each event and subtract it from each row entry. Each entry in the regret table is now the opportunity loss associated with an action-event combination. The cells where the smallest loss appeared before now show zeros.

The calculation of expected opportunity loss (EOL) proceeds in a manner completely analogous to the calculation of expected value. That is, we use the probabilities of events as weights and determine the weighted average opportunity loss for each act. The goal is to select the act having the minimum EOL.

EXAMPLE 5.3-5

Problem

Using the minimum EOL criterion, determine the optimal action for the tennis shop owner's problem.

Solution

The calculation of the EOLs for the two actions in the tennis shop owner's problem is given in Table 5.4 (refer to the lower part of Table 5.3 for the opportunity losses). If the owner selects the act that minimizes EOL, he will choose a_2: Order 200. Note that this is the same act that he selected under the criterion of MEV. It can be proved that the best action according to the criterion of MEV is also best using minimization of EOL. It should also be noted that opportunity losses are not accounting losses, but represent foregone opportunities.

Table 5.4
CALCULATING EXPECTED OPPORTUNITY LOSSES

Act a_1: Order 100

Event	Probability	Opportunity Loss	Weighted Opportunity Loss
θ_1: Demand is 100	0.5	\$ 0	\$ 0
θ_2: Demand is 150	0.3	125	37.5
θ_3: Demand is 200	0.2	450	90.0
Total	1.0		\$127.5

EOL(a_1) = \$127.5

Act a_2: Order 200

Event	Probability	Opportunity Loss	Weighted Opportunity Loss
θ_1: Demand is 100	0.5	\$200	\$100.0
θ_2: Demand is 150	0.3	0	0
θ_3: Demand is 200	0.2	0	0
Total	1.0		\$100.0

EOL(a_2) = \$100.0

5.4 EXPECTED VALUE OF PERFECT INFORMATION (EVPI)

Up to this point, we have considered situations in which decision makers choose among alternative courses of action on the basis of the information they have, that is, their prior information, without attempting to acquire further information before they make their decisions. The probabilities used in computing the expected value of an action, as shown in Table 5.2, for example, are called prior probabilities to indicate that they are probabilities established prior to obtaining additional information through testing, experimentation, sampling, and so on. Choosing an optimal act based on MEV is often called prior analysis.

We shall also consider the question of deciding whether or not it is worth gathering additional information (which is rarely if ever perfectly reliable) and, if so, what information to gather and what actions or strategies to take based on the information received. Before making these decisions, it is first desirable to ask what perfect information is worth. Suppose that in our tennis shop owner's problem, perfect information about the demand for the tennis shirt in question is worth \$100 and the survey under consideration to find out what demand will be costs \$200. In this case he would not have the survey conducted, since it would be irrational to pay \$200 for information worth \$100. To make such judgments, we need to determine the expected value of perfect information (EVPI). There are three ways that EVPI can be calculated.

Method 1

We first calculate the expected value under certainty (i.e., having perfect information) and subtract from this the expected value under uncertainty (the best action chosen using MEV under the current information state).

EXAMPLE 5.4-1

Problem

Compute the EVPI for the tennis shop owner's problem using method 1.

Solution

First, we have said that under uncertainty we would take action a_2 in our tennis example, and obtain an MEV of $(0.5)(0)+(0.3)(300)+(0.2)(600)=\$210$ (Table 5.2). This is the expected value under uncertainty. To obtain the expected value under certainty, we ask what our expected value is if we can choose our action after learning the true event (demand), that is, after obtaining perfect information. Clearly, the availability of perfect information allows us to obtain a profit of \$200 if θ_1 occurs, since we would choose action a_1 over a_2 in this situation. If θ_2 occurs, we would realize a profit of \$300, since action a_2 would be preferred to a_1. And, similarly, if we knew θ_3 would occur, we would earn a profit of \$600, since we would also choose a_2 over a_1. To understand and compute the expected value under certainty, it is necessary to adopt a long-run relative frequency point of view. That is, we must weigh each of these profits by the prior probabilities of each of these events occurring. In other words, from a relative frequency viewpoint, these probabilities are now interpreted as the proportion of times a perfect predictor would forecast that each of the given events would occur if the current situation were faced repeatedly. Each time the predictor makes a forecast, the decision maker chooses the optimal payoff action. The calculation of the expected profit under certainty (i.e., with perfect information) is shown in Table 5.5 and is equal to \$310. The EVPI is thus $\$310-\$210=\$100$. This is the expected (average) profit that could be gained if the tennis shop owner had perfect information about future demand.

Table 5.5
CALCULATING EXPECTED PROFIT UNDER CERTAINTY

	Event		
Action	θ_1: Demand is 100 $P(\theta_1)=0.5$	θ_2: Demand is 150 $P(\theta_2)=0.3$	θ_3: Demand is 200 $P(\theta_3)=0.2$
a_1: Order 100	200[a]	175	150
a_2: Order 200	0	300[a]	600[a]
Expected profit under certainty $=(200)(0.5)+(300)(0.3)+(600)(0.2)=\$310$			

[a]Optimal consequence (and action) for a given event.

Method 2

Another way of computing the EVPI is by a sort of incremental analysis, which we shall illustrate by an example.

EXAMPLE 5.4-2

Problem

Compute the EVPI for the tennis shop owner's problem using an incremental analysis (method 2).

Solution

The tennis shop owner's best act under uncertainty is to choose a_2. Suppose that he knew θ_1 would occur. Then, he would wish to choose a_1 over a_2, his choice under uncertainty. The gain in profit by choosing a_1 over a_2 when θ_1 occurs is \$200, but θ_1 would occur only 50% of the time. Hence, on the average he would gain \$100 if he chose a_1 over a_2 when θ_1 was known to occur. However, if θ_2 or θ_3 occurred, the tennis shop owner would still be content with a_2, the decision he made under uncertainty. Hence, his action would change only if θ_1 were known to occur. His overall expected gain, then, under perfect information, is \$100, which is the EVPI.

Method 3

The EVPI can also be determined by calculating the minimum EOL; that is, the minimum EOL is equal to the EVPI.

EXAMPLE 5.4-3

Problem

Determine the EVPI by computing the minimum EOL for the tennis shop owner's problem.

Solution

The EOL of choosing the optimal act under conditions of uncertainty in the tennis shop owner's problem was shown earlier to be \$100 (Table 5.4). That is, this amount represents the minimum value among the expected opportunity losses associated with each action. This value is equal to the EVPI. Note that all three methods yield an EVPI of \$100.

To see intuitively why the minimum EOL equals the EVPI, consider what the EOL would be under perfect information; it would be equal to zero. That is, there would be no opportunity losses if the decision maker knew which event would occur and behaved rationally. Thus, under uncertainty the decision maker would do the best possible, that is, minimize the EOL. It can be proved mathematically that EVPI = EOL of the optimal act under uncertainty. Another term used for the EOL of the optimal act under uncertainty is cost of uncertainty. This term stresses the cost of making the best decision under uncertainty, a cost that would be eliminated if perfect information were available. Hence, the cost of uncertainty is also equal to the EVPI. In short, the EVPI, the EOL of the optimal action under uncertainty, and the cost of uncertainty are equivalent.

It is interesting to note that the expected value plus the EOL is constant for all acts and is equal to \$310 in our example, the expected value under certainty. Observe from Table 5.6 that action a_2 has a maximum expected value and a minimum EOL.

By similar reasoning, we can define what we mean by the more general term of value of information. Specifically, this is the expected value of the optimal

Table 5.6
RELATIONSHIPS AMONG EXPECTED VALUE, EXPECTED OPPORTUNITY LOSS, AND EXPECTED VALUE UNDER CERTAINTY

	Act a_1	Act a_2
EV	\$182.5	\$210
EOL	127.5	100[a]
Total[b]	\$310.0	\$310

[a]Minimum EOL, which is equal to EVPI.
[b]EV under certainty.

action under a better information state (i.e., after additional information is received) minus the expected value of the optimal action given the current (prior) information state. It is a measure of the improvement of the decision as a result of new information. We will say more about this shortly when we deal with the question of whether or not to buy more information. In summary, the EVPI is a gross measure for determining whether it is worthwhile to purchase information, or to analyze whether it is worthwhile to do so.

5.5 DECISION TREES

An alternative way to structure a decision problem pictorially is in terms of a *tree diagram*, or *decision tree*. A decision tree chronologically depicts the sequence of actions and outcomes as they unfold. For a terminal decision problem based on prior information, the first fork (square node) corresponds to the action chosen by the decision maker, and the second fork (round node) corresponds to the event. The numbers at the ends of these terminal branches are the corresponding payoffs (of losses).

EXAMPLE 5.5-1

Problem

Draw a decision tree for the tennis shop owner's problem, labeling all actions, events and their probabilities of occurrence, and consequences.

Solution

The decision tree for the tennis shop owner's problem is presented in Figure 5.1. The chronological development of the tree can be illustrated as follows. The owner can follow either branch a_1 or branch a_2; that is, he first chooses either act a_1 or a_2 (a_3 was dominated and therefore eliminated from consideration). Assuming that he follows path a_1, he comes to another juncture, which is a chance fork. Chance now determines whether the event that will occur is θ_1, θ_2, or θ_3. If chance takes him down the θ_1 path, the terminal payoff is \$200; the corresponding payoffs are indicated for the other paths. An analogous interpretation holds if he chooses to follow branch a_2. Thus, the decision diagram depicts the basic structure of the decision problem in schematic form. In Figure 5.1, additional information is superimposed on the diagram to represent the analysis and solution to the problem.

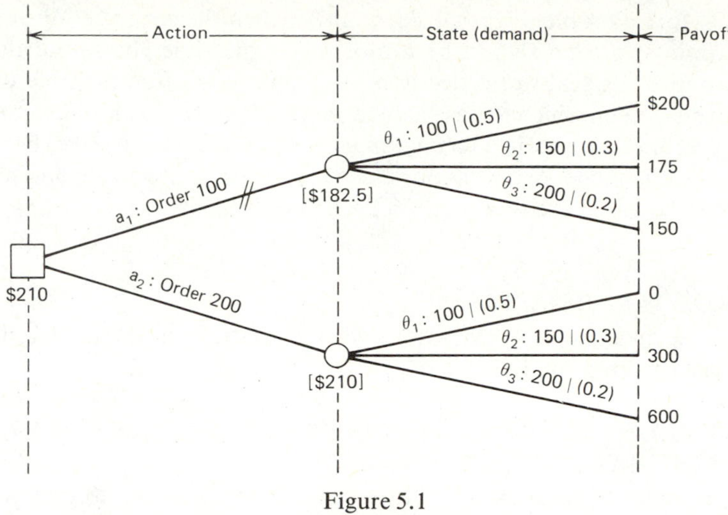

Figure 5.1
Decision tree.

The decision analysis process represented in Figure 5.1 is called backward induction. We imagine ourselves located at the right side of the tree diagram, where the payoffs are located. Let us consider first the upper three paths denoted θ_1, θ_2, and θ_3. To the right of these symbols we enter the respective probability assignments 0.5, 0.3, and 0.2 as given in the original problem. These represent the probabilities assigned by chance of following these three paths after the decision maker has selected act a_1. Moving back to the chance fork from which these three paths branch, we can calculate the expected monetary value of being located at that fork. This expected monetary value is \$182.5 and is determined in the usual way; that is,

$$\$182.5 = (0.5)(200) + (0.3)(175) + (0.2)(150)$$

This value is entered at the chance fork under discussion. It represents the value of standing at that fork after choosing act a_1, as chance is about to select one of the three paths. The analogous figure entered at the lower chance fork is \$210. Therefore, imagining ourselves as being transferred back to node 1, where the square represents a fork at which the decision maker can make a choice, we have the alternative of selecting act a_1 or selecting act a_2. Each of these acts leads us down a path at the end of which is a risky option whose expected profit has been indicated. Since following path a_2 yields a higher expected payoff than path a_1, we block off a_1 (using a double slash) as a nonoptimal course of action. Hence, a_2 is the optimal course of action, and it has the indicated expected payoff of \$210.

Thus, the decision tree diagram in Figure 5.1 reproduces in compact schematic form the analysis given in Table 5.2. An analogous diagram could be constructed in terms of opportunity losses. However, it is much more customary to use tree diagrams to portray analyses in terms of payoffs.

Tree diagrams are particularly useful in representing complex decision-making problems with sequences of actions and events over time. Such complex problems are very difficult to represent clearly in tabular (matrix) form.

Up to this point we have been concerned with what are called single-stage or static decision problems, where the decision maker selects one action at a single point in time. Often, however, decision problems involve making a sequence of decisions before a problem is resolved, that is, when the entire set of decisions at different points in time should be evaluated at the time the initial decision is made. Moreover, in real-world decision problems, it is often possible to acquire more information on which to base a decision through, for example, analysis, laboratory experimentation, field testing, market research, and the like, before a terminal decision is made. We will consider such decision problems and show how they can be structured and analyzed via decision tree analysis.

5.6 SEQUENTIAL DECISIONS

Let us use an example to show how we may analyze sequential decision problems by a decision tree analysis.

EXAMPLE 5.6-1 COMPANY ACQUISITION ALTERNATIVES: THE CASE OF HI VOLTAGE TRANSFORMER

The president of Solar Phasic Industries, Jayne Cash, is interested in buying the Hi Voltage Transformer Company. She sees the possibility of large profits occurring to Hi Voltage if business is good. If she purchases the company now, large returns can be made in 2 years, when it would be sold.

Hi Voltage's stock is currently held by two families, the Edisons and Franklins. The Edisons and Franklins have agreed to sell all their stock now for $1 million, or half now for $600,000 and the rest in 1 year based on the profit picture at that time.

Jayne Cash sees the following alternatives available to her. One is to buy all the stock now and sell at the end of 2 years. On the other hand, if she buys only half now, she can purchase the rest in 1 year and sell all at the end of 2 years. Or she could hold her initial purchase and not buy the second half, then sell at the end of the second year. A third alternative is not to purchase any stock and instead buy 2-year treasury bills.

Jayne's payoffs are influenced by business conditions. If business is good the company's first year, the price of the second 50% of the stock will be $800,000. If business is bad, the remainder of the stock will sell for $300,000. However, if business is good the first year, it may be bad during the second, and vice versa. All these events influence what the payoffs could be when Jayne plans to sell her shares of Hi Voltage.

To resolve the chances of these states of nature facing Jayne Cash, G. N. Potter, Solar's chief economist, was called in to develop estimates of future events. Potter gathered data and projected future events. He stated that in the first year, there would be a 60% chance of good business and 40% of bad. In the second year, if business were good in the first, there would be a 70% chance it would be good in the second and a 30% chance that it would be bad. If the first year were bad, there would be a 40% chance of good times and 60% of bad in the second year.

Colleen Smart, Solar's chief systems analyst, was also called on to help. She computed the net present value (NPV) of all possible alternatives available, based on the expected payoffs for each possible outcome. These are presented in Table 5.7. Jayne Cash can reap up to $800,000 or lose $700,000, depending upon what decision she makes and the events that occur in the future.

Jayne sat at her desk reviewing the information from Potter and Smart. She will have to make a decision in the next few days or lose the opportunity to buy

Table 5.7
POSSIBLE PAYOFFS FOR BUYING HI VOLTAGE STOCK

Alternative	Business Condition Year 1	Year 2	NPV (Payoff) ($1000)
Buy 100% (Single purchase)	Good	Good	800
	Good	Bad	−500
	Bad	Good	600
	Bad	Bad	−700
Buy 50%	Good	Good	300
	Good	Bad	0
	Bad	Good	100
	Bad	Bad	−100
Buy 100% (Double purchase)	Good	Good	600
	Good	Bad	−600
	Bad	Good	500
	Bad	Bad	−400
Buy treasury bills			50

into the company. The Edison and Franklin families have received other inquiries about selling.

Decision Tree Analysis

A decision tree diagram of this problem is shown in Figure 5.2. Starting from the first decision fork, at the extreme left, Cash is initially confronted with three alternative courses of action: (a) to purchase 100% of Hi Voltage's stock now (a_1);

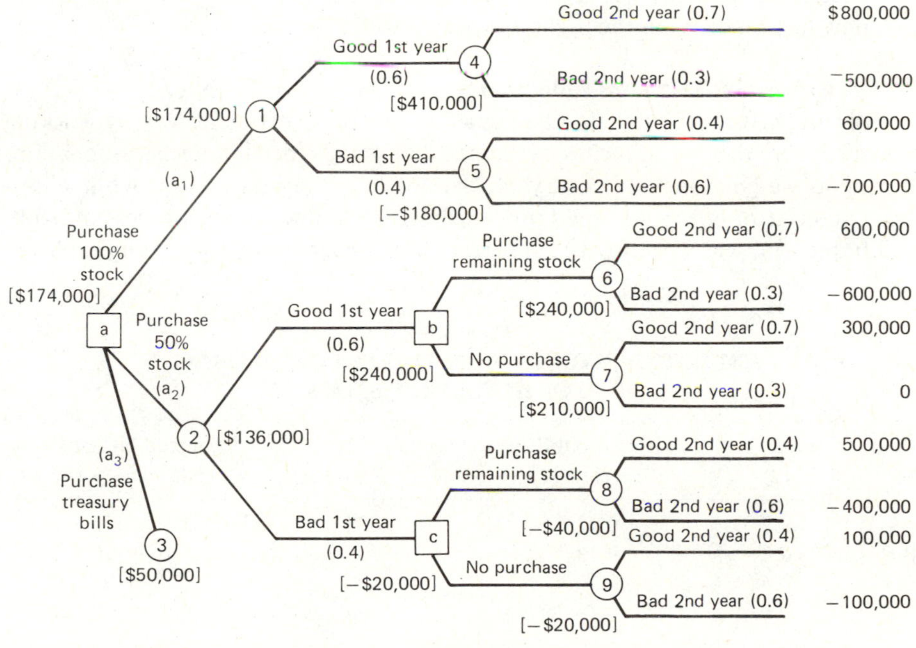

Figure 5.2
Discrete decision tree for Hi Voltage.

(b) to purchase 50% of the stock now (a_2); or (c) not purchase any Hi Voltage stock but instead to buy treasury bills (a_3). Action a_1 is a one-time (terminal) decision. If Cash chooses a_1, she has no further recourse. The payoff resulting from this decision is out of her hands because it depends on economic conditions in the first and second years, of which there are four mutually exclusive possibilities.

If Cash decides to purchase only 50% of Hi Voltage's stock now (a_2), then depending on next year's business conditions, she can then decide (next year) whether or not to purchase the remaining 50% of the stock. She has four possible strategies here: (a) to purchase the remaining stock regardless of business conditions in the first year; (b) to purchase the remaining stock only if business conditions are good in the first year; (c) to purchase the stock only if business conditions are bad in the first year; and (d) not to purchase the remaining stock regardless of business conditions in the first year.

We could argue intuitively that strategy (b) is better than the three other strategies, simply because strategies (a) and (d) make no use of the information gained in the first year, while strategy (c) misuses the information, that is, takes actions inconsistent with the information. (We shall verify this in our formal analysis.) With the information presented in the decision tree, we can proceed to find the best decision or sequence of decisions for Jayne Cash's decision problem. We shall assume that Cash's objective is to maximize expected value.

For the terminal action of purchasing 100% of Hi Voltage's stock immediately (a_1), the expected value is $174,000, as shown in Table 5.8. The expected values of the four feasible strategies associated with the initial action of purchasing 50% of Hi Voltage's stock now (a_2) are also calculated, and are shown in Table 5.9. The optimal strategy, if Cash purchased 50% of the stock initially, would be to purchase the remaining stock if business was good the first year, and not to purchase any more stock if business was bad the first year (=$136,000). Thus, of the six available sequences of actions, the optimal one is to purchase 100% of the stock now (a_1), since it has the highest expected value.

The Process of Backward Induction

We can perform the decision tree analysis in a more efficient way by working backward from the end branches of the tree diagram to the first decision fork, optimizing as we go along (i.e., carrying forward only the best action, while eliminating all inferior ones). As noted previously, this is known as the process of backward induction, where we work from right to left, each event fork being replaced

Table 5.8
EXPECTED PAYOFFS OF INITIALLY PURCHASING 100% OF HI VOLTAGE STOCK

Year 1 (1)	Year 2 (2)	Joint Probability (3) = (1) × (2)	Payoff ($1000) (4)	Expected NPV ($1000) (5) = (3) × (4)
0.6	0.7	0.42	800	336
0.6	0.3	0.18	−500	−90
0.4	0.4	0.16	600	96
0.4	0.6	0.24	−700	−168
Total		1.00		174

Table 5.9
EXPECTED PAYOFFS OF INITIALLY PURCHASING 50% OF HI VOLTAGE STOCK

Year 1 (1)	Year 2 (2)	Joint Probability (3) = (1) × (2)	Payoff ($1000) (4)	Expected NPV ($1000) (5) = (3) × (4)
Strategy 1: If good or bad 1st year, purchase remaining stock				
0.6	0.7	0.42	600	252
0.6	0.3	0.18	−600	−108
0.4	0.4	0.16	500	80
0.4	0.6	0.24	−400	− 96
Total				128
Strategy 2: If good 1st year, purchase remaining stock; if bad 1st year, no purchase				
0.6	0.7	0.42	600	252
0.6	0.3	0.18	−600	−108
0.4	0.4	0.16	100	16
0.4	0.6	0.24	−100	− 24
Total				136
Strategy 3: If good 1st year, no purchase; if bad 1st year, purchase remaining stock				
0.6	0.7	0.42	300	126
0.6	0.3	0.18	0	0
0.4	0.4	0.16	500	80
0.4	0.6	0.24	−400	− 96
Total				110
Strategy 4: If good or bad 1st year, no purchase				
0.6	0.7	0.42	300	126
0.6	0.3	0.18	0	0
0.4	0.4	0.16	100	16
0.4	0.6	0.24	−100	− 24
Total				118

by its certainty equivalent [i.e., by the "sure" sum (in this case, expected value) that the decision maker assesses as equivalent to the risky prospect represented by the event fork.] For example, from Figure 5.2 we can compute the expected value of purchasing 100% of the stock immediately as follows. First, we compute the expected values (certainty equivalents) at chance nodes 4 and 5:

$$\text{Node 4: } (800{,}000)(0.7) + (-500{,}000)(0.3) = \$410{,}000$$
$$\text{Node 5: } (600{,}000)(0.4) + (-700{,}000)(0.6) = -\$180{,}000$$

Then, we compute the expected value at chance node 1 using the expected values calculated at chance nodes 4 and 5:

$$\text{Node 1: } (410{,}000)(0.6) + (-180{,}000)(0.4) = \$174{,}000$$

Thus, the expected value of purchasing 100% of the stock immediately is $174,000. Using this backward induction, certainty equivalent approach, we have reduced the decision tree associated with action a_1 to its ultimate reduced form, shown in Figure 5.3

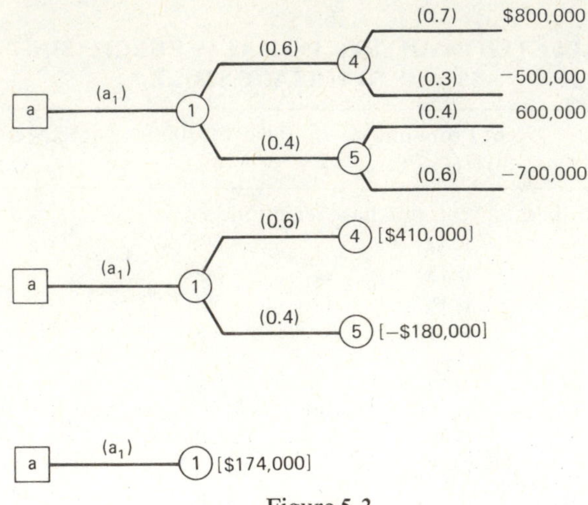

Figure 5.3
Optimal certainty equivalent of initially purchasing 100% of Hi Voltage stock.

Let us now compute the optimal strategy for purchasing 50% of the stock immediately (a_2). We first compute the expected values at chance nodes 6 and 7, and also for chance nodes 8 and 9:

$$\begin{aligned}
&\text{Node 6: } (600{,}000)(0.7) + (-600{,}000)(0.3) = \$240{,}000 \\
&\text{Node 7: } (300{,}000)(0.7) + (0)(0.3) = \$210{,}000 \\
&\text{Node 8: } (500{,}000)(0.4) + (-400{,}000)(0.6) = -\$40{,}000 \\
&\text{Node 9: } (100{,}000)(0.4) + (-100{,}000)(0.6) = -\$20{,}000
\end{aligned}$$

To determine the optimal action to take at decision node b, we compare the expected values associated with nodes 6 and 7 and choose the action whose node has the higher value, that is, we choose to purchase the remaining stock (node 6) and discard the action associated with node 7. A double slash is used to indicate that the no-purchase action should be discarded at decision node b. Thus, if business conditions are good the first year, the best action is to purchase the remaining stock in the second year.

Similarly, to determine the optimal action at decision node c, we compare the expected values at nodes 8 and 9 and choose the action whose node has the higher value, that is, we select the no-purchase action and discard the action of purchasing the remaining stock. Thus, if business conditions are bad the first year, the best action is not to purchase any more stock in the second year.

Next, we compute the expected value of chance node 2:

$$\text{Node 2: } (240{,}000)(0.6) + (-20{,}000)(0.4) = \$136{,}000$$

This means that if 50% of the stock is initially purchased, the best strategy is to purchase the remaining stock if first-year business conditions are good, and if first-year conditions are bad, not to make any further stock purchase. The expected value of this strategy is $136,000. Note again that by using this backward induction, certainty equivalent approach, we have reduced the initial action a_2 to its ultimate reduced form, as shown in Figure 5.4

To make the optimal decision at decision node a, we note the expected values (certainty equivalents) of our three initial choices: purchase 100% of the stock (node 1 = $174,000), purchase 50% of the stock (node 2 = $136,000), and pur-

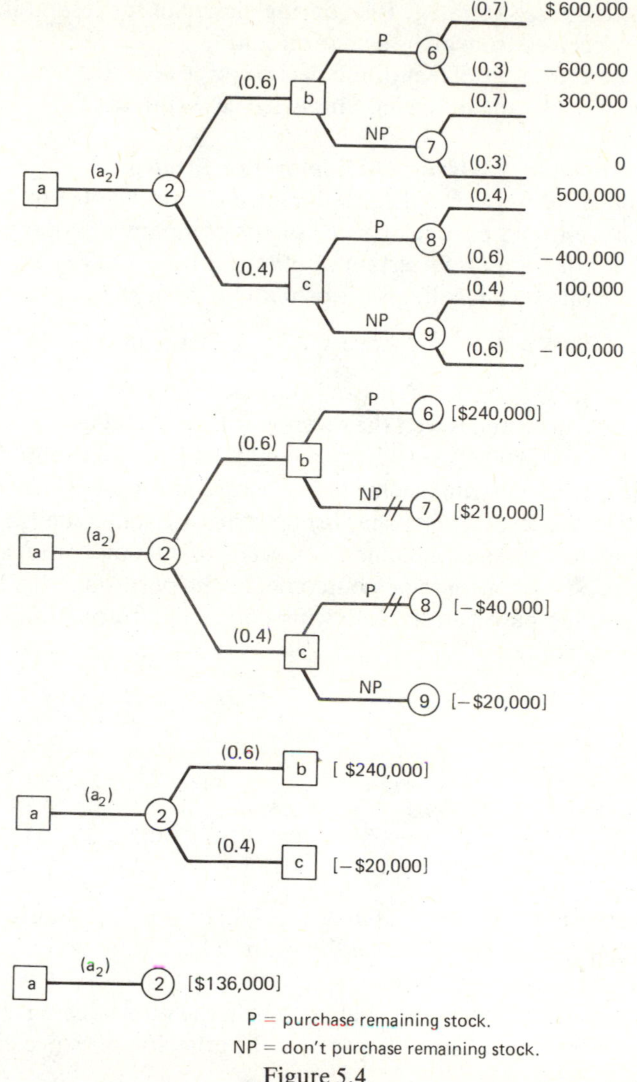

Figure 5.4

Optimal certainty equivalent of initially purchasing 50% of Hi Voltage stock.

chase treasury bills (node 3 = $50,000). The optimal decision at decision node *a* is again to purchase 100% of the stock, since it has the highest expected value.

5.7 INFORMATION ACQUISITION DECISIONS

In real-world decision problems, it is often possible to gain more information on which to base a decision through analysis, experimentation, testing, sampling, and so forth. In such situations, two basic choices must be made: (a) the choice of inquiry or information sources (including the null case, i.e., not to acquire any information), in other words, what information to acquire and when; and (b) the choice of strategy or decision rule in response to the messages, data, observations, or outcomes received from a chosen information source. Again, by strategy we mean a precise set of instructions explaining what action is to be taken under every possible message or outcome that might be received from a given information source. In essence, a strategy is a contingency plan for responding to all possible messages received from an information source.

To illustrate the development of the decision tree analysis in such decision situations, let us consider the following simplified case situation.

Cactus Petroleum: Analyzing Oil Exploration Strategies

An oil wildcatter for Cactus Petroleum must decide whether to drill (act a_1) or not drill (a_2) a well on a particular piece of the company's property in northern Oklahoma. The wildcatter is uncertain whether the hole is dry (state θ_1), wet (state θ_2), or soaking (state θ_3) but believes the probabilities for these states to be

$$P(\text{dry}) = 0.5 \qquad P(\text{wet}) = 0.3 \qquad P(\text{soaking}) = 0.2.$$

The cost of drilling is $70,000. If the well is judged to be soaking, the revenue will be $270,000. But if the well is wet, the revenue will be $120,000.

At a cost of $10,000, the wildcatter could take seismic soundings that will help determine the underlying geological structure at the site. The soundings will disclose whether the terrain below has (a) no structure (outcome NS)—a bad outcome, (b) an open structure (outcome OS)—a so-so outcome, or (c) a closed structure (outcome CS)—an encouraging outcome. In the past, oil wells that have been dry, wet, or soaking have had the following conditional probability distributions of seismic test outcomes:

Well Type	O_1: NS	O_2: OS	O_3: CS	Total
θ_1: Dry	0.6	0.3	0.1	1.0
θ_2: Wet	0.3	0.4	0.3	1.0
θ_3: Soaking	0.1	0.4	0.5	1.0

The oil wildcatter must decide what to do: to drill or not drill immediately; to take seismic soundings; and if seismic soundings are taken, what actions to choose for each possible seismic test outcome.

You are called in as a decision analyst to help the oil wildcatter with this problem. What would you recommend based on the criterion of expected value maximization?

Analysis: Optimal Action Without Experimentation and EVPI

As noted in Table 5.10, the determination of expected value for this simple two-act, two-state case merely involves the calculation of a weighted payoff for each act, that is,

$$\text{EV}(a_1) = (-70{,}000)(0.5) + (50{,}000)(0.3) + (200{,}000)(0.2) = \$20{,}000$$
$$\text{EV}(a_2) = \$0$$

In the absence of any further information, the decision maker who wished to maximize expected value would choose to drill (a_1).

More realistically, the decision maker may elect to delay the terminal decision by conducting seismic soundings in order to gather more information regarding the probabilities attached to alternative states that affect the consequences of each course of action. The concept of expected value of perfect information (EVPI) provides useful information regarding how much the decision would be improved if such a test were perfectly reliable. Thus, it provides a rough measure

Table 5.10
DECISION MATRIX FOR CACTUS PETROLEUM

	Event		
Action	θ_1: Dry $P(\theta_1)=0.5$	θ_2: Wet $P(\theta_2)=0.3$	θ_2: Soaking $P(\theta_3)=0.2$
a_1: Drill	−70,000	50,000[a]	200,000[a]
a_2: Don't drill	0[a]	0	0

[a]Optimal act for a given event.

as to whether it is worthwhile gathering more information. Using Method 1, the EVPI, which is the difference between the optimal expected value under perfect information and the optimal expected value under the current uncertainty (i.e., before testing) is

$$\begin{aligned} \text{EVPI} &= \sum_{j=1}^{3} P(\theta_j) \max_i[c(a_i, \theta_j)] - \max_i \text{EV}(a_i) \\ &= [(0)(0.5) + (50{,}000)(0.3) + (200{,}000)(0.2)] - 20{,}000 = \$35{,}000 \end{aligned}$$

where $c(a_i, \theta_j)$ is the consequence or payoff associated with a given action and state of nature (Table 5.10). Therefore, we should be willing to pay up to \$35,000 for sample information, but no more. Of course, sample information is generally not perfect, and so the amount we should expect to pay should be less than \$35,000. Since the cost of the seismic test is \$10,000, we should examine its potential value to our decision process. That is, it may be worthwhile to perform the seismic test.

Decision Tree Diagramming and Analysis: Extensive Form

Figure 5.5 represents a decision tree of this problem summarizing the pertinent payoffs and probabilities associated with the null test and the seismic test. By following the path e_0 (no test), we note that the same data are shown in Table 5.10. As before, we designate the decision forks (nodes) with squares, while the chance forks (nodes) are denoted by circles. It is clear that in the absence of experimentation, the decision maker would select a_1 (drill), leading to an expected value of \$20,000. As before, we use a double slash to eliminate the nonoptimal actions; hence, the path labeled a_2 is blocked off by a double slash. This part of the analysis is known as *prior analysis* and refers to the manner in which a terminal act is chosen on the basis of prior probabilities alone (our current information state or beliefs).

The path labeled e_1 gives the strategies associated with the opportunity to conduct a seismic test at the cost of \$10,000. Note how the tree traces out the choices and events as they would occur over time. That is, we first perform the seismic test. Three possible test outcomes can occur: NS (O_1), OS (O_2), and CS (O_3). Whichever of these test outcomes occurs, we will then have to decide whether or not to drill. After taking our action, we will then learn which state is true (θ_1, θ_2, or θ_3); this results in a certain payoff to us. Note that there are eight possible strategies associated with seismic testing. (Can you define them?)

The seismic test is not completely reliable, as indicated by the conditional probabilities stated in the problem. That is, the test outcome would not disclose without some error which event θ_1, θ_2, or θ_3 is true. (If the test were perfectly reliable, then $P(O_1/\theta_1)$, $P(O_2/\theta_2)$, and $P(O_3/\theta_3)$, where O_k and θ_j denote the test outcome and state of nature, respectively, would all be equal to 1, and all the other conditional probabilities would be equal to zero.)

To evaluate the e_1 path, we first need to determine the values of the following probabilities: (a) $P(O_k)$, the probabilities of the three possible test outcomes; and (b) $P(\theta_j/O_k)$, the posterior probabilities of the states of nature given the test outcomes. We can do this by using Bayes' theorem, that is,

$$P(\theta_j \mid O_k) = \frac{P(\theta_j \cap O_k)}{P(O_k)} = \frac{P(\theta_j)P(O_k \mid \theta_j)}{\sum_{j=1}^{3} P(\theta_j \cap O_k)}$$

The results of these computations are shown in Table 5.11.

Table 5.11
OBTAINING THE PROBABILITIES FOR CACTUS PETROLEUM: BAYES' THEOREM

$P(O_k \mid \theta_j)$

θ_j \ O_k	O_1	O_2	O_3
θ_1	0.6	0.3	0.1
θ_2	0.3	0.4	0.3
θ_3	0.1	0.4	0.5

$P(\theta_j)$

θ_j	$P(\theta_j)$
θ_1	0.5
θ_2	0.3
θ_3	0.2

$P(O_k \cap \theta_j)$

θ_j \ O_k	O_1	O_2	O_3	
θ_1	0.30	0.15	0.05	0.5
θ_2	0.09	0.12	0.09	0.3
θ_3	0.02	0.08	0.10	0.2
	0.41	0.35	0.24	1.0

$P(O_k)$

O_k	$P(O_k)$
O_1	0.41
O_2	0.35
O_3	0.24

$P(\theta_j \mid O_k)$

θ_j \ O_k	O_1	O_2	O_3
θ_1	0.73	0.43	0.21
θ_2	0.22	0.34	0.37
θ_3	0.05	0.23	0.42

Now we can place the probabilities at the appropriate points on the decision tree of Figure 5.5. Given that we decide to test, one of the three test outcomes will occur with probabilities $P(O_1)=0.41$, $P(O_2)=0.35$, and $P(O_3)=0.24$. For each of these possible test outcomes, we must choose either one of two actions: drill or not drill. Given a test outcome and a specific course of action, one of three states will occur with probabilities $P(\theta_1/O_k)$, $P(\theta_2/O_k)$, and $P(\theta_3/O_k)$. For example, if the test outcome is O_1(NS), then $P(\theta_1/O_1)=0.73$, $P(\theta_2/O_1)=0.22$, and $P(\theta_3/$

$O_1) = 0.05$. Observe and compare the values of the posterior probabilities [e.g., $P(\theta_1/O_1)$, $P(\theta_2/O_1)$, and $P(\theta_3/O_1)$] to the prior probabilities [$P(\theta_1) = 0.5$, $P(\theta_2) = 0.3$, and $P(\theta_3) = 0.2$]. Note that if the test outcome is O_1, we would expect $P(\theta_1/O_1)$ to be greater than $P(\theta_1)$, since a no-structure test indication would increase our confidence or belief that the well is really dry, as Bayes' theorem indicates.

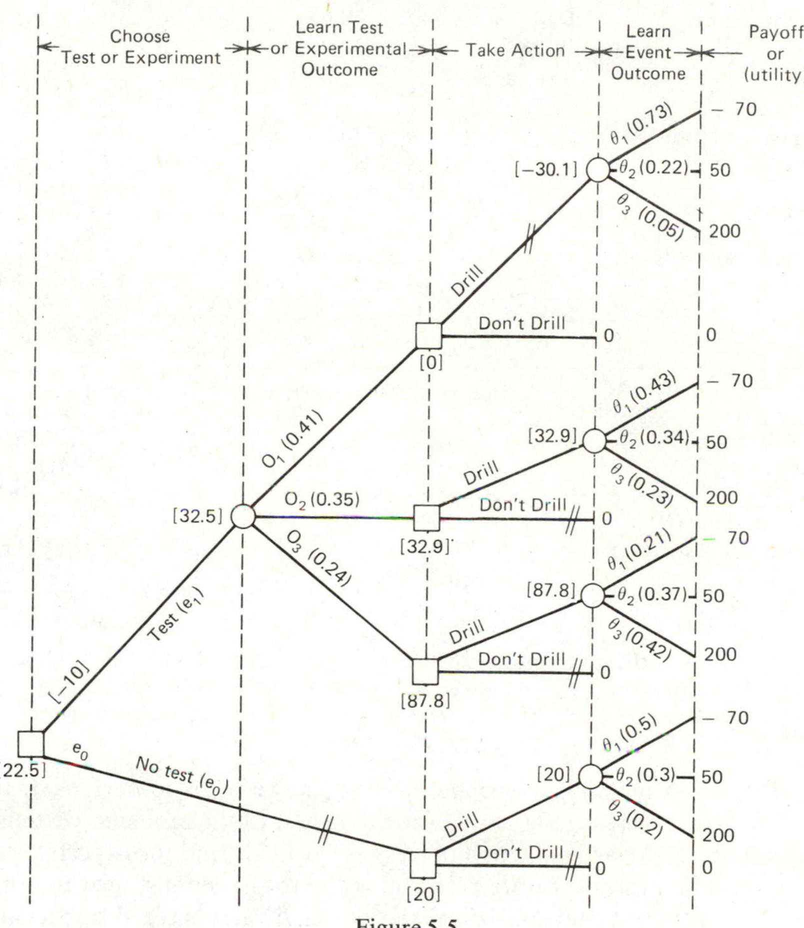

Figure 5.5

Decision tree analysis of Cactus Petroleum: Extensive form (in thousands of dollars).

We are now ready to choose our best strategy for experimentation and action. First, we compute the optimal strategy associated with performing a seismic test using the process of backward induction. That is, we start from the tips of the tree and work backward, always choosing the path with the highest expected value. The expected values of the optimal action associated with each test outcome are computed in Table 5.12 and shown in Figure 5.5. Thus, the optimal strategy for seismic testing is as follows: If O_1(NS) occurs, we should not drill; if O_2(OS) or O_3(CS) occurs, we should drill. We refer to this part of the analysis, which deals with the optimal choice and evaluation of an action subsequent to all experimentation or testing (i.e., after learning the test outcomes), as *posterior analysis*.

Table 5.12
POSTERIOR AND PREPOSTERIOR ANALYSES OF THE SEISMIC TEST

Posterior Analysis

Test Outcome, O_k	State, θ_j	Act: Drill	Act: Don't Drill	$P(\theta_j \mid O_k)$
O_1: No structure (NS)	Dry	\$ − 70,000	\$0	0.73
	Wet	50,000	0	0.22
	Soaking	200,000	0	0.05
	EV	− 30,100	0	
O_2: Open structure (OS)	Dry	− 70,000	0	0.43
	Wet	50,000	0	0.34
	Soaking	200,000	0	0.23
	EV	32,900	0	
O_3: Closed structure (CS)	Dry	− 70,000	0	0.21
	Wet	50,000	0	0.37
	Soaking	200,000	0	0.42
	EV	87,500	0	

Preposterior Analysis

Test Outcome, O_k	$P(O_k)$ (1)	EV_k (2)	$\Sigma_k P(O_k)\text{EV}_k$ (3) = (1) × (2)
O_1: No structure (NS)	0.41	\$ 0	\$ 0
O_2: Open structure (OS)	0.35	32,900	11,500
O_3: Closed structure (CS)	0.24	87,500	21,000
Total			\$32,500

Prior expectation of EV posterior to seismic tests = 32,500 − 10,000 = \$22,500
EV (test, including its cost) − EV (no test) = 22,500 − 20,000 = \$2500

The decision maker's immediate problem, however, is to decide whether or not to perform a seismic test. To analyze this aspect of the problem, we must continue with the backward induction process. That is, to find the expected value of the optimal test strategy (before performing the test), we must next multiply the probability of the test outcomes times the expected value of each respective optimal posterior act, that is,

$$(0.41)(0) + (0.35)(32{,}900) + (0.24)(87{,}500) = \$32{,}500$$

After subtracting the cost of the seismic test, we get a net expected return of \$22,500. This part of the analysis is known as *preposterior analysis* since we evaluate the value of performing the test using the optimal strategy before we have learned the particular results of the test.

Since the value of performing the test (\$22,500) is \$2500 greater than the optimal no-test alternative (\$20,000), the seismic test should be performed, and if the test outcome is O_1(NS), then we should not drill; if the test outcome is either O_2(OS), or O_3(CS), then we should drill.

A decision problem of the type just presented was analyzed as follows:

1. Structure the decision tree chronologically.
2. Assign payoffs or utilities at the tips of the branches.

3. Assign probabilities at all chance forks.
4. Compute expected values using the process of backward induction.

This four-step procedure is called the extensive form of analysis. In this form, the decision maker can be viewed as a participant in a game against chance (or nature). The decision maker makes a choice, sits back and waits for the opposition (chance) to make a move, and then makes the next choice in light of chance's countermove.

There is a second mode of analysis, called the normal form of analysis, which gives a solution identical to that obtained by the extensive form. This procedure will not be discussed, but can be found in (2), (3), or (4).

5.8 SUMMARY

Decision analysis is concerned with choosing the best act from a set of possible acts, given uncertainty as to the event that will occur. By the use of an expected value (utility) criterion, this can be accomplished in a manner consistent with the decision maker's belief and tastes (values).

We have presented methods for performing a decision analysis of terminal decision problems, sequential decision problems, and those where there is an opportunity to acquire information prior to making a final (terminal) decision. The real managerial value of a decision tree analysis is seen when the analysis is applied to such complex decision problems—where there exist many alternative actions, information sources, states of nature, future decision periods, and so on. In brief, a decision tree analysis aids and improves the decision process in at least the following ways:

1. It structures the decision process into constituent parts, forcing the decision maker to analyze the decision problem in an orderly, rational manner.
2. It helps the decision maker enumerate all feasible alternatives and outcomes (some initially not considered, and others not worth considering) into a workable set.
3. It communicates the decision-making process to organizational members, quantifying each subjective belief, each subjective preference, and each assumption in a concise manner. If a group decision is involved, this analysis facilitates the reconciliation of differences.
4. It documents the decision process for historical purposes.
5. It serves as a management by exception device, to isolate subjective inputs to which the decisions are sensitive, and that therefore need to be assessed more accurately.
6. It is often used with a computer, permitting detailed sensitivity analysis of the various subjective inputs, assumptions, and so forth, on the choices made.

Recommended Readings

Good sources of recommended readings on this topic are given below as references.

REFERENCES

1. Keeney, R. L., and Raiffa, H., *Decisions with Multiple Objectives*, Wiley, New York, 1976.
2. Moskowitz, H., and Wright, G. P., *Operations Research Techniques for Management*, Prentice-Hall, Englewood Cliffs, N.J., 1979.
3. Moskowitz, H., and Wright, G. P., *Statistics for Management and Economics,* Merrill, Columbus, Ohio, 1985.

4. Raiffa, H., *Decision Analysis*, Addison-Wesley, Reading, Mass., 1968.
5. Savage, L. J., *The Foundations of Statistics*, Wiley, New York, 1954.

EXERCISES

1. The possible price of General Motors common stock tomorrow will be either higher, the same, or lower.
   (a) Are these events mutually exclusive?
   (b) Are the events collectively exhaustive?
   (c) What is wrong with assessing these prior probabilities as 0.5, 0.4, and zero, respectively?
2. The owner of a tennis shop finds that in the first month of business, 125 wood rackets, 50 aluminum rackets, and 25 steel rackets were sold. Based solely on these data, what prior probability distribution would you formulate for the type of tennis racket sold?
3. Suppose that it is an overcast Sunday morning and you have 100 people coming for a party in the afternoon. You have a nice garden and your house is not too large; so weather permitting you would like to set up the refreshments in the garden and have the party there (it would be pleasant, and your guests would be more comfortable). On the other hand, if you set up the party for the garden and after all the guests arrive it begins to rain, the refreshments will be ruined, your guests will get wet, and you will surely wish you had decided to have the party in the house. Define the acts, events, and consequences and represent the problem in a payoff table.
4. What is a payoff matrix? A decision tree? Is there any benefit in representing a decision problem in either of these forms? Under what circumstances is a decision tree a better representation of the problem than a decision matrix?
5. What are the two basic choices that constitute the essence of decision analysis? Illustrate these choices using an example.
6. In carrying out a decision tree analysis in extensive form, prior, posterior, and preposterior analyses are performed. Define and illustrate these three types of analysis. What are the differences among them?
7. You are given the following payoff matrix:

	Event				
Action	θ_1	θ_2	θ_3	θ_4	θ_5
a_1	−100	160	40	200	0
a_2	60	80	140	40	100
a_3	20	60	−60	20	80
a_4	−20	−100	−140	−40	400

Suppose also that the following prior probability distribution has been assessed:

Event	θ_1	θ_2	θ_3	θ_4	θ_5
Probability	0.08	0.10	0.50	0.12	0.20

8. A processor of frozen vegetables has to decide what crop to plant in a particular area. Suppose there are only two strategies: to plant cabbage and to plant cauliflower. Also suppose that the states of nature can be summarized in three possibilities: perfect weather, variable weather, and bad weather. On the basis of weather records, it is de-

termined that the probability of perfect weather is 0.25, variable weather is 0.50, and bad weather is 0.25. The yields (in dollars) of the two crops under these different conditions are known, and the utility of the company can be assessed to be measured by dollar amounts as shown in the following payoff table. What action should the processor select?

	Event		
Action	θ_1 Perfect Weather	θ_2 Variable Weather	θ_3 Bad Weather
a_1: Plant cabbage	40,000	30,000	20,000
a_2: Plant cauliflower	70,000	20,000	0

9. Consider the following payoff matrix (in dollars):

	Event		
Action	θ_1	θ_2	θ_3
a_1	10	6	3
a_2	5	8	4
a_3	2	5	9

The prior probability distribution assessed by the decision maker is

Event	θ_1	θ_2	θ_3
P(event)	1/4	1/2	1/4

(a) Specify the opportunity loss matrix.
(b) Compute the EVPI three different ways.
(c) If nearly perfect information about which event is true could be bought for \$2, would you buy it?

10. Suppose that a coin is either two-headed or balanced, with heads on one side and tails on the other. We cannot inspect the coin, but at a cost of \$0.05 we can flip it once, observe whether it comes up heads or tails, and then we must decide if it is two headed or not. Furthermore, there is a penalty of \$2.00 if our decision is wrong, and no penalty (or reward) if our decision is right. Determine the optimal strategy for experimentation and action.

11. The product manager for a large firm is faced with choosing between two advertising campaigns for a product. The monetary return from sales generated by each of two campaigns (a_1 and a_2, respectively) will depend on the prevailing market conditions, which can be either "good" (θ_1) or "bad" (θ_2). The manager believes that there is a 50-50 chance for the occurrence of θ_1 versus θ_2. The monetary returns (in thousands of dollars) for each act-state combination are summarized as follows:

	State	
Act	θ_1	θ_2
a_1	800	−500
a_2	0	0

The manager may elect to delay the terminal decision in order to gather information regarding which state is more likely to occur. Suppose that the manager has the following data-gathering options prior to making a terminal choice:

e_0: Do not experiment, but make a terminal choice now.

e_1: Purchase a perfect survey service; the cost of this survey is $250,000.

e_2: Purchase a survey that is 75% reliable; the cost of this survey is $75,000.

Select an appropriate analysis and determine what the marketing manager should do.

12. The Rodney Sportswear Company has designed two new tennis short styles for next year, "Wimbledon" and "Forest Hills." The company can produce either or both or neither of the two styles. Thus, management must select one of four actions available: (a) "Wimbledon" only, (b) "Forest Hills" only, (c) both, or (d) neither. The cost of production, all of which must be borne in advance if a model design is to be produced, is $50,000 for either of the models, but it is $125,000 for both together because of the strain on capacity involved in producing two styles. The profit, including all income and costs except production cost, is $100,000 per style if the style is successful, and zero if the style is unsuccessful. Assuming that each tennis short style has a 50% chance of being commercially successful, and is independent of the success or failure of the other style, what is the best course of action?

13. The Mocheck Company has developed a new product that it is considering marketing. The cost of marketing is $500,000. If its product is preferred to that of its competitor, and if it is marketed, the company anticipates factory gross sales of $1,500,000. If it is worse than the competing product, and if it is marketed, the gross sales are estimated to be $250,000. If Mocheck does not market the product, it can and will sell the new product idea to its competitors for $500,000 if it is a superior product, and $150,000 if it is inferior. The decision whether to market or to sell the idea must be made now although the positive confirmation of the relative superiority or inferiority of the product will not be available until some weeks from now. Should Mocheck market the product?

14. Johnson's Metal (JM), a small manufacturer of metal parts, is attempting to decide whether or not to enter the competition to be a supplier of transmission housings for Protrac. To compete, the firm must design a test fixture for the production process and produce 10 housings that Protrac will test. The cost of development, including designing and building the fixture and the test housings, is $50,000. If JM gets the order, an event estimated as occurring with probability 0.4, it will be possible to sell 10,000 items to Protrac for $50 each. If JM does not get the order, the development cost is essentially lost.

    To produce the housings, JM may either use its current machines or purchase a new forge. Tooling with the current machines will cost $40,000, and the per-unit production cost is $20. However, if JM uses its current machines, it runs the risk of incurring overtime costs. The relationship between overtime costs and the status of JM's other business is as follows:

Other Business	Probability	Overtime Cost
Heavy	0.2	$200,000
Normal	0.7	100,000
Light	0.1	0

    The new forge costs $260,000, including tooling costs for the transmission housings. However, with the new forge JM would certainly not incur any overtime costs, and the production cost would be only $10 per unit. Determine the optimal set of actions for JM.

15. Suppose that your firm is faced with the decision whether to continue with regional distribution of one of its products or to expand to national distribution. As the marketing vice-president you feel that if you expand to national distribution and a large national demand develops, you can expect to make \$6 million in the time period under consideration. Should you expand distribution and encounter limited national demand, the cost of attracting new distributors would exceed the expected returns by \$1 million. On the basis of your past experience and current knowledge of the market, you estimate the likelihood of a large national demand to be 0.4. You can delay your decision as to whether or not to expand nationally pending a market survey of likely demand levels. This survey indicates whether national demand would be large or limited with 90% reliability.
    (a) Draw a decision matrix of this problem and compute the optimal (Bayes') action without the survey.
    (b) For what range of probability values for large national demand would the optimal action be to expand to national distribution?
    (c) Using the chronological decision tree (extensive form) of analysis, determine the optimal strategy for the survey.
    (d) Calculate the conditional expected values (given state θ) and overall expected value for the following strategy:

$$S_1 = \begin{cases} \text{Expand to national distribution if survey indicates} \\ \text{a large national demand.} \\ \text{Otherwise, continue regional distribution.} \end{cases}$$

    (e) What would be the maximum amount you would be willing to pay for the survey, such that you would prefer to conduct the survey and then decide.

16. A manufacturer's purchasing agent must decide to accept or reject an incoming shipment of machine parts. The agent wishes to use either of the following two acts for this situation:

a_1: Accept the shipment
a_2: Reject the shipment

The fraction of defective parts in the shipment is either 0.1 or 0.5 with a prior likelihood of each occurring being 0.5. The costs associated with the possible decisions are \$1000 if a 0.1 shipment is rejected and \$1500 if a 0.5 shipment is accepted. No costs are incurred if a 0.1 shipment is accepted or a 0.5 shipment is rejected. There is a sample cost of \$10 per part tested.
    (a) What is the optimal action without sampling?
    (b) What is the EVPI?
    (c) Determine the optimal strategy, that is, what actions to take in response to the sample outcomes. Show all your work and indicate the results on your tree.

6
RANDOM PROCESSES

6.1
INTRODUCTION

This chapter is concerned with the techniques of modeling random processes—processes that evolve through time in a manner that is not completely predictable. With a description this general, it should be obvious that real-world examples exist in abundance. Indeed, almost any conceivable context can provide an opportunity to use the tools presented here.

Whenever a mathematical technique has many potential applications, there is a danger in presenting it in terms so oriented to one application or one category of applications that others cannot be envisioned. Yet a purely abstract presentation is just as likely to leave the reader unappreciative of the potential applicability of the mathematics. Every teacher knows that students need concrete examples; the real dilemma occurs in deciding whether it is better to present the general, abstract concepts first and then illustrate them, or to give a concrete example first and then attempt to generalize.

Some very fine textbooks (particularly those intended for mathematics students) take the first approach. Presumably the authors feel that it is most important that the concepts be understood in their full abstract generality, even if that means that some students may encounter difficulty in relating the abstractions to their real-world referents. Other textbooks (often those intended for business school students or other "practitioners") take the second approach, preferring to sacrifice generality, if necessary, to ensure that whatever *is* learned can be directly applied.

Although this chapter will adopt the second approach—that of example first, theory second—the contention held here is that it is extremely important to achieve both goals. We have elected to make continuous use of one concrete example, referring to it each time a new concept is introduced. This is done only to provide a temporary pedagogical crutch in the early phases of understanding the concepts. You should earnestly try to get beyond the example as soon as possible by imagining other situations that could be handled in the same way. The problems should help considerably in this effort. Your objective should be to acquire the *skills* to model *new* situations. It is important that you understand the techniques well enough to recognize appropriate opportunities to use them while you also appreciate their limitations enough to be conservative in their use. That is you should know when and how to use them, but also when *not* to use them.

The basic methodology presented here was developed initially by the Russian mathematician, A. A. Markov, around the beginning of the 20th century. "Markov processes" form a subclass of the set of all random processes—a subclass with enough simplifying assumptions to make them easy to handle. More complicated (non-Markovian) processes are beyond the scope of this text, but certainly warrant further study by the interested student.

The first major section of the chapter deals with Markov processes in which time is measured discretely, or "counted." The concepts seem to be easier to learn when this restriction is made, perhaps because the only mathematics used is algebra. Many of the concepts established in the discrete time case hold also for continuous time Markov processes, which are the subject of the second major section. The mathematics is more difficult, however; calculus is required where algebra previously sufficed.

I. DISCRETE TIME PROCESSES

6.2 AN EXAMPLE

Suppose that you are assisting the owner of a limousine service who operates between the airport, downtown, and a cluster of suburban hotels. He operates essentially as a taxi service, except that he has only one vehicle and restricts his trips to those three areas. (We want a very simple example here.) He accepts one passenger (or a group traveling together) and takes the customer to whichever of the three allowable destinations is desired. After discharging the customer, the driver will stay at the same location to await another. What we want is a model that tracks the movements of the limousine.

Some representative examples of the kinds of questions we might want to answer are the following:

1. If the limousine is currently at the airport, what is the probability that it will be back at the airport after three fares?
2. If the limousine is downtown, how many trips on the average will occur before it next reaches the airport?
3. Over a long period of time, what fraction of the trips are to the airport?
4. At an arbitrary time sometime in the future, what is the likelihood that the limousine would be on the way to the airport?

These are questions that relate only to the random nature of the trips themselves. With additional information about, say, the average profit or distance traveled by trip type, we could ask and answer further questions. For now, we will confine our attention to just the trips.

6.3 MODELING THE PROCESS

The easiest way to measure the passing of time is to count the trips. This convention will avoid any difficulties having to do with the differences in the durations of separate trips. Of course, it also prevents us from dealing with any issues related to actual clock time. If the latter were important to us, we could count off time in, say, half-hour increments. Either way would be acceptable from the standpoint of the mathematics. In this situation, we are making the choice based on which is easier to describe. To be specific, we shall regard the nth point of time as the instant at which trip number n ends.

There are only three possible destinations in our example. It will be convenient to both name and number these for quick reference. There is no particular ordering of the possibilities in this case (although a natural ordering does appear in many applications), so let us arbitrarily make the following assignment.

Airport = 1
Downtown = 2
Hotels = 3

At each time step, a trip terminates at one of these locations. We are modeling the process *only* at these times, so all of the details of what occurs between these instants is ignored. This may seem unnatural or improper until you get used to the idea that modeling is an art of selective simplification.

Using some of the terminology of stochastic processes, the points of time at which the system is observed are called *epochs*. The values that correspond to the possible conditions observed are *states*. This, for example, if the first trip ends downtown, we would say that the state at epoch 1 is 2. A record of observed states over time is called a *realization* of the process. A realization can be portrayed in a graph as in Fig. 6.1. The epochs appear on the horizontal axis, and the states appear on the vertical. Reading the graph from left to right, we can trace the sequence of trips as they occurred in this particular realization. Each change of state is called a *transition*.

If, instead of recording observations of the past, we specify a future epoch and ask what state will occur at that time, we would have to answer in terms of probabilities. A random variable capable of assuming any of the state values can be defined for each future epoch. In other words, the entire unknown future can be described as a sequence of random variables $(X_1, X_2, X_3, \ldots)$. It is customary to include another (degenerate) random variable, X_0 in this sequence, to represent the information about the last known state. Epoch 0 is usually interpreted as "the present." The entire sequence $(X_0, X_1, X_2, X_3, \ldots)$, consisting of an ordered sequence of arbitrary random variables indexed by time and extending out to infinity is a *discrete time stochastic process*.

If you focus not on time but on the structure of allowable transitions, it is natural to construct the *transition diagram*. This is a pictorial map of the process in which states are represented by points and transitions by arrows. An arrow from state 1 to state 2, for example, would mean that if the process is in state 1 at some epoch n, then it is possible for it to be in state 2 at epoch $n+1$. In our example

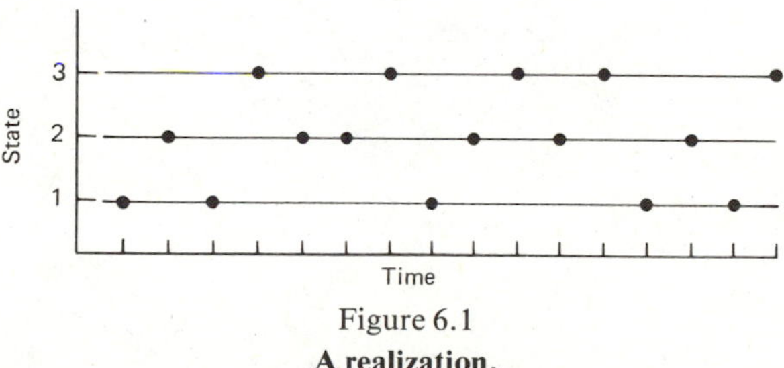

Figure 6.1
A realization.

there are three states with no restrictions on possible transitions, so the transition diagram would be as shown in Fig. 6.2

Instead of thinking of a random process as a sequence of random variables, one can visualize it as the random walk of a particle over the transition diagram. Occupying a state is equivalent to the particle being at a particular point; a transition corresponds to a "jump" or an instantaneous movement of the particle along one of the arrows.

There is a slight deliberate ambiguity in the term "transition." In some applications, we will want a transition to occur each time that the time index is incremented, even if the new state is the same as the old. Such transitions, to give them a name, are called *virtual transitions*, and are represented by loops in the transition diagram. On other occasions, we want to consider only genuine changes of state, or *real transitions*. If so, the transition diagram would have no loops.

Thus far we have no data or probability distributions. Because the process is described by a sequence of random variables, $\{X_1, X_2, X_3, \ldots\}$, we must *at least* specify the probability distribution of each X_i. However, this will not be enough.

In general, the random variables, X_i, are not independent. Whenever one has two or more dependent random variables, either a joint distribution function or a conditional distribution function is necessary to completely express the relationship. The marginal distributions of the random variables taken individually do not do the whole job. At worst, then, the complete specification of a stochastic process would require joint distribution functions, taken jointly over all of the random variables. But this would be absurdly impractical to attempt for any real-world process.

At this point we need Markov's simplifying assumption. A *Markov chain* is a discrete time stochastic process in which each random variable, X_i, depends only on the previous one, X_{i-1}, and affects only the subsequent one, X_{i+1}. The term "chain" suggests the linking of the random variables to their immediately adjacent neighbors in the sequence. In our limousine example, the Markov assumption is perfectly justified, because the destination for each successive trip is chosen by a new customer.

Compared with stochastic processes in general, Markov chains are very simple. Without completely eliminating the dependence between random variables (which would severely curtail our ability to represent real-life processes realistically), we have avoided the necessity of having to express joint distributions of ev-

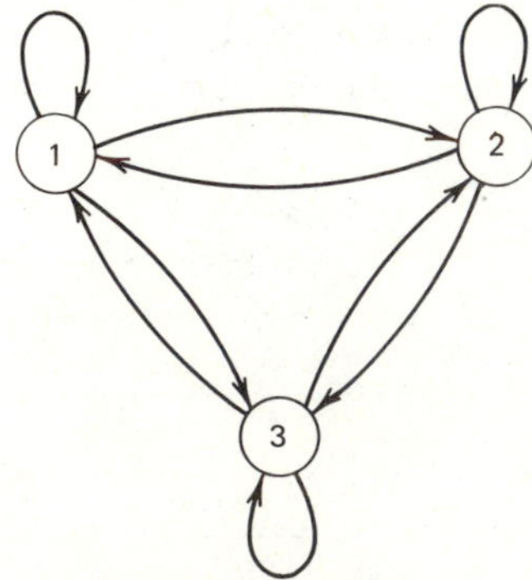

Figure 6.2
A transition diagram.

erything all at once. Instead, it will be sufficient to express joint or conditional distributions of just two neighboring random variables at a time.

Suppose that X_0 represents the known starting state, and we are interested in X_1, the state after the first trip. What we want is the probability distribution over the three possible values for X_1. But these probabilities depend on where the limousine picks up that first customer. Suppose the limousine starts at the airport, that is, $X_0 = 1$. Then the first customer might want to go to the airport ($X_1 = 1$)—an unlikely but conceivable possibility—or to the downtown ($X_1 = 2$) or to the hotels ($X_1 = 3$). There is a separate probability for each of these outcomes. Similarly, there are three more probabilities for trips starting from downtown ($X_0 = 2$) and three more for trips starting from the hotels ($X_0 = 3$). All together, there are nine probabilities corresponding to each source/destination pair.

One convenient way to record these different probabilities is in matrix form:

$$\mathbf{P} = \begin{bmatrix} p_{11} & p_{12} & p_{13} \\ p_{21} & p_{22} & p_{23} \\ p_{31} & p_{32} & p_{33} \end{bmatrix}$$

The first row of this matrix contains the three probabilities for trips starting from the airport. Generally, the first subscript of p_{ij}, which identifies the row, corresponds to the state at epoch 0. The second subscript matches the column and the state at epoch 1.

The element p_{ij} means the probability that $X_1 = j$ if you know that $X_0 = i$. Formally, it is the conditional probability, $p_{ij} = P\{X_1 = j \mid X_0 = i\}$. A handy, though somewhat inaccurate, way to think of it is as the probability of "going" from state i to j, or of making the transition from i to j. The p_{ij}'s are, in fact, called (one-step) *transition probabilities*, and the matrix $\mathbf{P}$ is called the *transition matrix*. Incidentally, the fact that each row is a probability distribution implies that the elements of each row must sum to 1.

Instead of using the matrix, we could have just labeled the arrows of the transition diagram. The arrow from i to j would receive the label p_{ij}, which can then be thought of as the probability that that arrow is used when the particle leaves point i. Row i of $\mathbf{P}$ would correspond to the set of arrows leaving point i, so the sum of the probabilities taken over this set must equal 1. For our example, the transition diagram would be as in Fig. 6.3. Of course, if any of these p_{ij} are zero (indicating

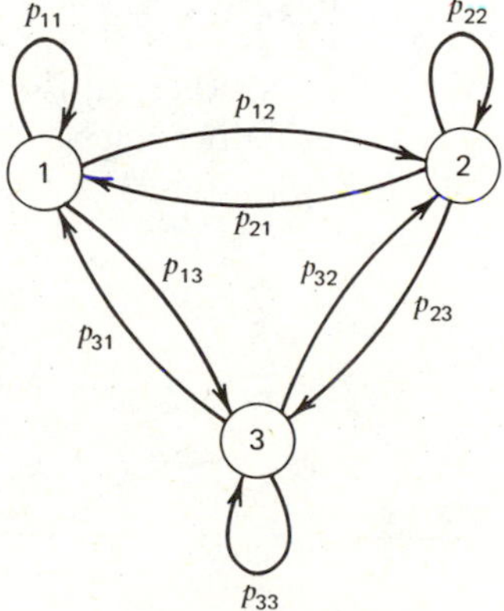

Figure 6.3

A transition diagram.

that a direct transition from i to j is not possible) the arrow from i to j would simply not appear in the diagram.

It should be apparent that the labeled transition diagram contains exactly the same information that the matrix **P** does. Given a transition matrix, we can construct a unique transition diagram; given a transition diagram we can construct a unique transition matrix. To construct either one, in other to model some real-life process, it is necessary to imagine every possible transition from one state to another and then to find the probability that is associated with each possible transition.

We have now expressed everything necessary to describe X_1, the state at time 1, taking into account the dependence on X_0, the state at time 0. By the Markov assumption, no further information about previous states is required. We now turn our attention to X_2, the state at time 2. Suppose once again that $X_0 = 1$, and we want to know the probability that $X_2 = 1$. Although X_2 is not directly dependent on X_0, it is dependent on X_1, which in turn is dependent on X_0.

There are three distinct ways that the event we are interested in could come about, corresponding to the three alternative possibilities for the state at time 1. Figure 6.4 shows these three possible sequences of transitions. There is no way other than these that the event could occur, and the three ways are mutually exclusive. Consequently, the probability that we are interested in can be broken into the sum of three probabilities, one for each possible path. Using an inefficient, but natural, notation, we might write

$$\begin{aligned} P\{X_0 = 1 \rightarrow X_2 = 1\} = & P\{X_0 = 1 \rightarrow X_1 = 1 \rightarrow X_2 = 1\} \\ & + P\{X_0 = 1 \rightarrow X_1 = 2 \rightarrow X_2 = 1\} \\ & + P\{X_0 = 1 \rightarrow X_1 = 3 \rightarrow X_2 = 1\} \end{aligned}$$

(This actually represents an abuse of conventional probability notation, but will be corrected in due course.) The problem now is to express "path" probabilities, such as $P\{X_0 = 1 \rightarrow X_1 = 3 \rightarrow X_2 = 1\}$.

We have, of course, notation to express the probability for the first leg of the path, namely p_{13}. Furthermore, the probability for the second leg, the probability of going from state 3 to 1, is (by the Markov assumption) independent of the state at time 0. Hence, we can multiply the probabilities of the two legs to get the probability of the path:

$$\begin{aligned} P\{X_0 = 1 \rightarrow X_1 = 3 \rightarrow X_2 = 1\} & = P\{X_0 = 1 \rightarrow X_1 = 3\} P\{X_1 = 3 \rightarrow X_2 = 1\} \\ & = p_{13} P\{X_1 = 3 \rightarrow X_2 = 1\} \end{aligned}$$

A similar expression can be written for the other paths.

An additional simplifying assumption becomes convenient at this point. Because $\{X_1 = 3 \rightarrow X_2 = 1\}$ is a transition from state 3 to 1, it would be nice to be able

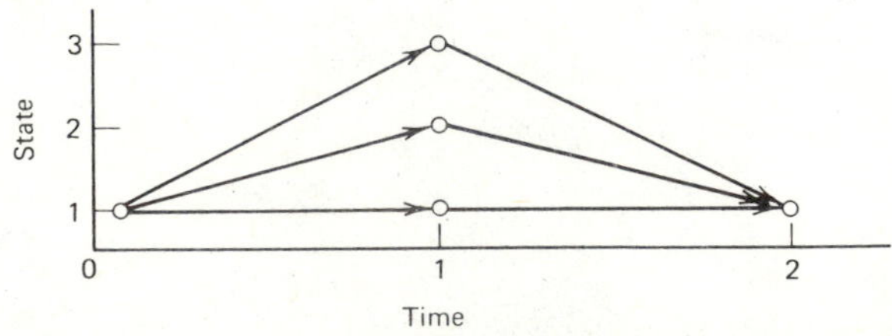

Figure 6.4
The three possible sequences.

to use the notation p_{31} for its probability, but up until now p_{31} has referred to the probability of making the transition from 3 to 1 at the *first* transition and we are now talking about the *second* transition. The obvious thing to do is to assume that transition probabilities do not change with the passage of time. The term used to describe this assumption is *stationarity*. For a stationary Markov chain, the matrix **P** will provide the one-step transition probabilities for any time, and (as we shall see) is sufficient to describe the entire process. If our process were not stationary, it would be possible to continue; but we would require a new transition matrix for each transition. The stationarity assumption is discussed later. For now, it is expedient to make the assumption and continue the development.

With the assumption of stationarity,

$$P\{X_0=1 \rightarrow X_1=3 \rightarrow X_2=1\}=p_{13}p_{31}$$

In words, the two-step path probability is given by the product of the two one-step transition probabilities comprising the path. Finding the corresponding probabilities for the other two paths, and collecting results gives

$$P\{X_0=1 \rightarrow X_2=1\}=p_{11}p_{11}+p_{12}p_{21}+p_{13}p_{31}$$

This expresses, in terms of values already known, the probability that the limousine will be at the airport after the second trip, given that it started at the airport.

By similar logic, we may obtain the probabilities for being in states 2 or 3 after the second trip.

$$P\{X_0=1 \rightarrow X_2=2\}=p_{11}p_{12}+p_{12}p_{22}+p_{13}p_{32}$$
$$P\{X_0=1 \rightarrow X_2=3\}=p_{11}p_{13}+p_{12}p_{23}+p_{13}p_{33}$$

The three expressions together give the probability distribution of the random variable X_2, given that $X_0=1$.

If the limousine is not initially at the airport but downtown, we get three different expressions for the distribution of X_2, and three more for when it begins at the hotels. There are a total of nine expressions required to specify the distribution of X_2 under the various possible initial conditions. As it was for X_1, it is convenient to use a matrix arrangement. Let $\mathbf{P}^{(2)}$ denote the matrix of these probabilities, and let $p_{ij}^{(2)}$ represent the *i, j*th entry.

$$\mathbf{P}^{(2)}=\begin{bmatrix} p_{11}^{(2)} & p_{12}^{(2)} & p_{13}^{(2)} \\ p_{21}^{(2)} & p_{22}^{(2)} & p_{23}^{(2)} \\ p_{31}^{(2)} & p_{32}^{(2)} & p_{33}^{(2)} \end{bmatrix}$$

$\mathbf{P}^{(2)}$ is called the two-step transition matrix; its elements, two-step transition probabilities. They give conditional probabilities for the states at time 2 under varying possible conditions for the state at time zero. Each row is a probability distribution. Loosely speaking, $p_{ij}^{(2)}$ is the probability of "going" from state i to state j in two steps. It is also the probability that we previously designated $P\{X_0=i= \rightarrow X_2=j\}$. The briefer notation is clearly more efficient (once you know what it means), so at this point we drop the old notation.

We know how to express the $p_{ij}^{(2)}$ in terms of the one-step transition probabilities. For example,

$$p_{11}^{(2)}=p_{11}p_{11}+p_{12}p_{21}+p_{13}p_{31}$$

We got this by considering all the ways that the event could occur. It would be preferable to have a systematic method that would yield the correct expressions without requiring a detailed consideration of the process.

One method that will work utilizes the transition diagram. Each line of the diagram represents a one-step transition. Because of stationarity, the probability associated with each line remains constant as time progresses. So a particular sequence of transitions is represented by a sequence of arrows in the transition diagram, and the probability of taking that sequence of transitions is given by the product of probabilities associated with the arrows. (Independence of events is required to multiply probabilities, but this independence is assured by the Markov assumption.) Of course, to represent a legitimate sequence of transitions, the arrows must be arranged "head-to-tail." The graph theoretic term for such a sequence is a *walk*. One can obtain a walk by placing a pencil at any point of the diagram and, without lifting the pencil, following arrows from point to point, always moving *with* the direction of the arrows. Arrows may be traversed more than once in a walk.

Thus, a short-cut way to obtain the expressions for the two-step transition probabilities is to examine the transition diagram for all two-step walks starting at the initial state and ending at the state of interest. For each walk, you multiply the probabilities associated with the arrows in that walk, then add all of the products. You should verify that this method works by using it to obtain the expressions we derived earlier.

There is another systematic way to get the two-step transition probabilities—a way so automatic that it lends itself to computer implementation. The method is based on the simple observation that the expressions previously derived are exactly what would be obtained if the matrix **P** were multiplied by itself (using, of course, matrix multiplication). In symbols,

$$\mathbf{P}^{(2)} = \mathbf{P}\mathbf{P} = \mathbf{P}^2$$

In words, the matrix of two-step transition probabilities is equal to the square of the matrix of one-step transition probabilities. This is not obvious; it is a derived relation, requiring both the Markov and stationarity assumptions. It is a most fortuitous relation because it means that a familiar, well-known matrix operation can be used to compute the probabilities for X_2.

Thus far we have described X_1 and X_2. To get the distribution of X_3 under varying possible conditions at time 0, we again have to consider the ways in which the states could occur. For example, suppose that $X_0 = 1$ and we are interested in the probability that $X_3 = 1$. Then at epoch 2, the state must have been 1, 2, or 3, and the last trip must have been to the airport. Since these are mutually exclusive, collectively exhaustive possibilities.

$$p_{11}^{(3)} = p_{11}^{(2)}p_{11} + p_{12}^{(2)}p_{21} + p_{13}^{(2)}p_{31}$$

By this logic, we can write all three-step transition probabilities in terms of the two-step and one-step transition probabilities, which are of course already known. Furthermore, we may observe that the same results would be obtained by multiplying the two-step transition matrix by the one-step transition matrix. That is,

$$\mathbf{P}^{(3)} = \mathbf{P}^{(2)}\mathbf{P}$$

But since $\mathbf{P}^{(2)} = \mathbf{P}^2$,

$$\mathbf{P}^{(3)} = \mathbf{P}^2\mathbf{P} = \mathbf{P}^3$$

So the three-step transition matrix is just the one-step matrix cubed. The elements of this matrix completely describe X_3, the random variable describing the state at time 3.

By now it should be apparent how to continue. In general, the n-step transition matrix equals the one-step transition matrix raised to the nth power,

$$\mathbf{P}^{(n)} = \mathbf{P}^n$$

This is the single most important result for Markov chains. By raising $\mathbf{P}$ to the appropriate power, we can answer any question pertaining to the probability of being in a particular state at a particular time. The complete sequence of random variables $\{X_1, X_2, X_3, \ldots\}$ which characterizes the process is now fully described.

The method of finding walks in the transition diagram also generalizes. The probability of a walk of any number of steps is given by the product of the transition probabilities associated with the arrows in the sequence. The n-step transition probability $p_{ij}^{(n)}$ is given by the sum of the probabilities of all walks of length n from i to j. If the transition diagram is large, complicated, or both; or if the number of steps is large, it is impractical to try to enumerate every such walk. Nevertheless, it is worthwhile to realize that the raising of $\mathbf{P}$ to the nth power accomplishes something that is precisely equivalent.

There is one minor extension to be considered now. We have all of the random variables $\{X_1, X_2, X_3, \ldots\}$ characterized *provided* that the state at time zero is known. Occasionally it happens that the value of X_0 is not known with certainty, but is given by a probability distribution. Suppose that such is the case, and let $p_i^{(0)}$ denote the probability that $X_0 = i$. For example, $p_1^{(0)}$ would be the probability that the state is 1 at time 0. Suppose we want to know the probability that the state will be 2 at some later time, say $n = 5$. Having the matrix $\mathbf{P}^{(5)}$, we can read off $p_{12}^{(5)}$, $p_{22}^{(5)}$, and $p_{32}^{(5)}$ from the second column. These are all probabilities of being in state 2 at time 5, but differ in the assumed state at time 0. The obvious thing to do is to take a weighted average of $p_{12}^{(5)}$, $p_{22}^{(5)}$, and $p_{32}^{(5)}$, weighting each according to the likelihood that the initial state it assumes is, in fact, the correct one. This logic would yield the expression

$$p_1^{(0)}p_{12}^{(5)} + p_2^{(0)}p_{22}^{(5)} + p_3^{(0)}p_{32}^{(5)}$$

which is, indeed, the probability that $X_5 = 2$.

Extending the notation a bit, let $p_i^{(n)} = P\{X_n = i\}$. Such a probability is called a *state probability* to distinguish it from a transition probability. It would give the probability of being in a particular state at a particular time, regardless of the state at time zero. Since matrix notation has proved so convenient for the transition probabilities, let $\mathbf{p}^{(n)}$ be the row vector of state probabilities. (Note that this is a lowercase $\mathbf{p}$; the uppercase is reserved for the transition matrix.)

Using the new notation, we may observe that the expressions we would obtain for the elements of $\mathbf{p}^{(5)}$—one of which was derived above—can also be obtained from

$$\mathbf{p}^{(5)} = \mathbf{p}^{(0)}\mathbf{P}^{(5)}$$

Furthermore, the logic will generalize to give the state probabilities at any time:

$$\mathbf{p}^{(n)} = \mathbf{p}^{(0)}\mathbf{P}^{(n)}$$

In words, the vector of state probabilities for time n is given by the vector of initial state probabilities multiplied by the n-step transition matrix.

The right-hand side of this relation involves matrix multiplication, which is ordinarily not commutative (i.e., $AB \neq BA$), so the order is important. Also it is important to remember that the $\mathbf{p}^{(n)}$ are *row* vectors. If, for reasons of convenience in writing the expressions on paper, one prefers column vectors, everything must be transposed. This would in turn reverse the order of multiplication on the right-hand side. That is,

$$\mathbf{p}^{(n)T} = \mathbf{P}^{(n)T}\mathbf{p}^{(0)T}$$

6.4 A NUMERICAL EXAMPLE

Let us now put some specific numbers into our example and perform the calculations we have prescribed. Suppose data from the field indicate that customers starting from the airport never go to the airport and are equally likely to go to the downtown or to the hotels. Those who start from the downtown area divide equally among the three destinations. The ones who start from the hotels go to the airport two thirds of the time and to downtown one third of the time.

The transition diagram resulting from these assumptions is shown in Fig. 6.5.

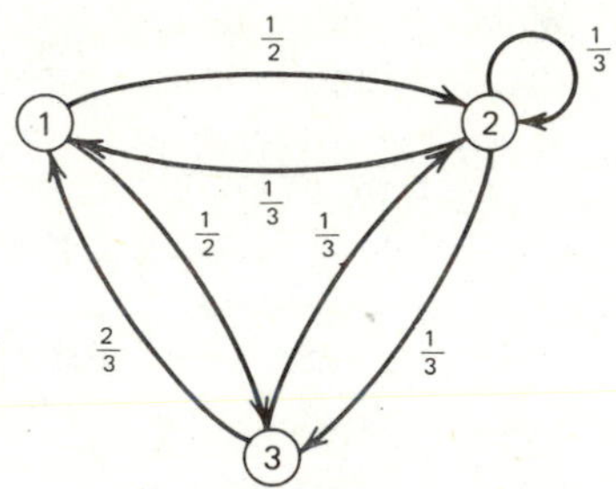

Figure 6.5
Limousine example transition diagram.

The one-step transition matrix is

$$\mathbf{P} = \begin{bmatrix} 0 & .5 & .5 \\ 0.333 & 0.333 & 0.333 \\ 0.667 & 0.333 & 0 \end{bmatrix}$$

We get the matrix of two-step transition probabilities by squaring $\mathbf{P}$.

The interpretation of these numbers is related, of course, to their positions in the matrix. For example, the value 0.5 in the first row and first column of $\mathbf{P}^{(2)}$ says that there is a probability of one half that the limousine will be back at the airport at the end of the second trip after leaving the airport.

$$\mathbf{P}^{(2)} = \mathbf{P}^2 = \begin{bmatrix} 0.500 & 0.333 & 0.167 \\ 0.333 & 0.389 & 0.278 \\ 0.111 & 0.444 & 0.444 \end{bmatrix}$$

If the initial state of the limousine is unknown, but is described by a probability distribution, we would have recourse to the formula

$$\mathbf{p}^{(2)} = \mathbf{p}^{(0)}\mathbf{P}^{(2)}$$

If, for example, the limousine were equally likely to start from any of the three locations, then

$$\mathbf{p}^{(0)} = [0.333 \quad 0.333 \quad 0.333]$$

and

$$\mathbf{p}^{(2)} = [0.315 \quad 0.389 \quad 0.296]$$

The transition matrices of higher order are

$$\mathbf{P}^{(3)} = \begin{bmatrix} 0.222 & 0.417 & 0.361 \\ 0.315 & 0.389 & 0.296 \\ 0.444 & 0.352 & 0.204 \end{bmatrix}$$

$$\mathbf{P}^{(4)} = \begin{bmatrix} 0.38 & 0.37 & 0.25 \\ 0.327 & 0.386 & 0.287 \\ 0.253 & 0.407 & 0.34 \end{bmatrix}$$

$$\mathbf{P}^{(5)} = \begin{bmatrix} 0.29 & 0.397 & 0.313 \\ 0.32 & 0.388 & 0.292 \\ 0.362 & 0.376 & 0.262 \end{bmatrix}$$

$$\mathbf{P}^{(6)} = \begin{bmatrix} 0.341 & 0.382 & 0.277 \\ 0.324 & 0.387 & 0.289 \\ 0.3 & 0.394 & 0.306 \end{bmatrix}$$

and so forth.

6.5 THE ASSUMPTIONS RECONSIDERED

Although it is easy to remember *how* to compute state and transition probabilities, it can be tempting to forget about *why* the methods work. In particular, one can easily forget that the Markov and stationarity assumptions are crucial. Although it is true that many real-life processes satisfy these two assumptions (or, at least well enough for the models to be considered useful), it is also true that many do not. There are no automatic safeguards built into the methods to prevent their misuse. Even if a process does *not* satisfy the Markov and stationarity assumptions, it is still possible to define a one-step transition matrix and use it to compute "answers." The answers would simply be meaningless. Thus, a certain amount of vigilance is required to avoid applying these methods when the circumstances are inappropriate.

Although there exist certain statistical tests to verify the reasonableness of making the Markov and stationarity assumptions [cf. Bhat(3) pp. 96–102], the best protection is provided by *understanding* the assumptions and what they imply and by *thinking* about whether they seem appropriate to the situation.

The Markov assumption is basically an independence assumption. It says that knowledge of the state at any time is sufficient to predict the future of the process from that point on; information about how that state was reached (the sequence of prior states) is superfluous. Simpler but less precise ways to say the same thing are "given the present, the future is independent of the past," and "the process is forgetful." In the case of the example we have been using, the process would *not* be Markovian if the destination of the limousine were somehow influenced by where it had been previously. Suppose, for example, that the driver had just delivered a passenger from the hotel to the airport and had learned in the process that a lot of "high tippers" would be making the same trip. In such a situation, the driver might refuse any airport customer except those wanting to go to the hotel. This situation would clearly violate the Markov assumption. Our example was selected, however, to provide a case where the assumption would most likely be satisfied. More commonly, it is satisfied (if at all) in the weaker sense that one is unable to *detect* any relationship between new assignments and past ones. One may sometimes suspect that a relationship could exist but at the same time be satisfied that it is so weak that one is content to make use of the Markov assumption.

The stationarity assumption is one of "constancy" over time. It suggests stability of the process, although of course it does not imply that the process remains in a fixed state or even that there is a sluggishness in the rate at which transitions occur. It is the *probability mechanism* that is assumed stable. Again, it is easy to imagine cases for which the assumption does not hold. If the time scale of the model is very long, that process may undergo such fundamental changes—growth, evolution, aging, policy change, and the like—that one would hesitate to call it the same process. Even over the short run, many real-life processes have "peak periods" or "slow times" during which they exhibit behavior that is different from the norm. Such changes can occur gradually or all at once, but in either case would violate the stationarity assumption. In the case of the example, one might be conscious of an increase in traffic to the airport at certain times of the day, or rush hour effects involving the downtown. If so, the transition matrix that accurately described the transition probabilities at one time would not be accurate at a later time.

The Markov and stationarity assumptions are unrelated to one another; either could hold in the absence of the other.

6.6 FORMAL DEFINITIONS AND THEORY

The development of Markov modeling techniques has thus far been heuristic. Although the formulas were made to sound plausible, no proofs were given to assure that they really are consistent with the rules of probability theory and require *only* the Markov and stationarity assumptions. Furthermore, we still lack formal definitions that are general enough to cover every conceivable application yet are specific enough to be used in proofs. This section remedies these deficiencies. At the same time, it concisely summarizes everything done so far.

Let $\{X_n, n=0, 1, 2, \ldots\}$ be a family of real valued random variables indexed by n. The family is called a *discrete parameter* (or discrete time) *stochastic process*. The value of X_n for a specific realization of the process is called the *state of the process at the* n*th step*. If the random variables X_n are discrete random variables—that is, if they take on only integer values—we call the process a *discrete state* process. Although it is possible to consider continuous state processes, this discussion will be confined to discrete state and discrete parameter processes.

In general, the random variables are dependent on one another, therefore, one must describe the process in terms of either joint or conditional probabilities.

The index n typically denotes something akin to time, so at worst X_n will be dependent on $X_{n-1}, X_{n-2}, \ldots, X_0$ (that is, the "past history" of the process), but not upon $X_{n+1}, X_{n+2}, \ldots$ (the "future" of the process).

If for all n,

$$P\{X_n=j_n \mid X_{n-1}=j_{n-1}, X_{n-2}=j_{n-2}, \ldots, X_0=j_0\}=P\{X_n=j_n \mid X_{n-1}=j_{n-1}\}$$

the process is said to be a discrete state, discrete parameter Markov process, or a *Markov chain*. In other words the conditional distribution of X_n given the whole past history of the process must equal the conditional distribution of X_n given X_{n-1}. The conditional probability $P\{X_n=j \mid X_{n-1}=i\}$ is referred to as the one-step transition probability from i to j at step n. If for all m and n, $P\{X_n=j \mid X_{n-1}=i\}=P\{X_m=j \mid X_{m-1}=i\}$, the Markov chain is said to be *stationary*. In this case, the one-step transition probabilities do not depend explicitly on the step number. For such a process, it is sufficient to specify the *one-step transition probabilities,* $p_{ij}=P\{X_1=j \mid X_0=i\}$, because the one-step transition probabilities at any step number are the same. The square matrix $\mathbf{P}$ whose elements are the p_{ij}'s is called the *one-step transition matrix*, or just the transition matrix when there is no likelihood of confusion.

The n-step transition probabilities, $p_{ij}^{(n)}$, are defined by $p_{ij}^{(n)}= P\{X_n=j \mid X_0=i\}$. In words, $p_{ij}^{(n)}$ is the probability that the process is in state j at time n, given that it was in state i at time 0. Of course, $p_{ij}^{(1)}$ is just p_{ij}. In addition, it can easily be seen from the definition that $p_{ij}^{(0)}$ must be 1 if $i=j$, and 0 otherwise.

We may derive an important relationship as follows:

$$\begin{aligned}
p_{ij}^{(n)} &= P\{X_n=j \mid X_0=i\} && \text{(definition)}\\
&= \sum_k P\{X_n=j, X_{n-1}=k \mid X_0=i\} && \text{(marginal from joint)}\\
&= \sum_k P\{X_n=j \mid X_{n-1}=k, X_0=i\}P\{X_{n-1}=k \mid X_0=i\} && \text{(joint from conditional)}\\
&= \sum_k P\{X_n=j \mid X_{n-1}=k\}P\{X_{n-1}=k \mid X_0=i\} && \text{(Markov assumption)}\\
&= \sum_k P\{X_1=j \mid X_0=k\}P\{X_{n-1}=k \mid X_0=i\} && \text{(stationarity)}\\
&= \sum_k p_{kj}p_{ik}^{(n-1)} && \text{(definition)}
\end{aligned}$$

In conclusion, then,

$$p_{ij}^{(n)} = \sum_k p_{ik}^{(n-1)} p_{kj}$$

Observing that the right-hand side looks like the definition of matrix multiplication, we note that the same equation can be represented in matrix form by

$$\mathbf{P}^{(n)} = \mathbf{P}^{(n-1)}\mathbf{P}$$

where $\mathbf{P}^{(n)}$ is the matrix whose elements are the n-step transition probabilities. This fundamental relationship specifying the n-step transition probabilities of lower order is called the *Chapman-Kolmogorov equation*. It should be apparent from the derivation that the equation would not hold if either the Markov or stationarity properties were absent. On the other hand, these are the *only* assumptions required.

In fact, what has been derived is just a special case of the general form of the Chapman-Kolmogorov equations. (It is, however, the most frequently used form.) The general form, which can be proved by an argument similar to that given for the special case, is given by

$$p_{ij}^{(n)} = \sum_k p_{ik}^{(n-m)} p_{kj}^{(m)}$$

for any m between 1 and n, or in matrix representation,

$$\mathbf{P}^{(n)} = \mathbf{P}^{(n-m)}\mathbf{P}^{(m)}$$

Applying the Chapman-Kolmogorov equations iteratively, it is easy to prove that $\mathbf{P}^{(n)} = \mathbf{P}^n$. That is, the matrix of n-step probabilities is just the matrix of one-step transition probabilities raised to the nth power (using, of course, matrix multiplication).

If the unconditional probabilities $P\{X_n = j\}$ (called the *state probabilities*) are desired, the initial conditions must be specified. Of course, if the process is known with certainty to have started in a certain state, say i, at time zero, we have only to read off the elements of the ith row of $\mathbf{P}^{(n)}$, because the condition $X_0 = i$ is known to hold. More generally, the initial conditions may be given by a distribution over the possible states at time zero. Let $p_i^{(n)} = P\{X_n = i\}$, and let $\mathbf{p}^{(n)}$ denote the row vector whose elements are the $p_i^{(n)}$. (There is a slight possibility for confusion among elements, vectors, and matrices because of the similarity of notation, but in every case the number of subscripts provides the telltale clue.) Then the initial conditions are given by the vector $\mathbf{p}^{(0)}$. To get $\mathbf{p}^{(n)}$, we may use the following argument.

$$\begin{aligned} p_j^{(n)} &= P\{X_n = j\} \\ &= \sum_i P\{X_n = j \mid X_0 = i\} P\{X_0 = i\} \\ &= \sum_i p_{ij}^{(n)} p_i^{(0)} \end{aligned}$$

Therefore,

$$p_j^{(n)} = \sum_i p_i^{(0)} p_{ij}^{(n)}$$

Recognizing the right-hand side as multiplication of vectors, we see that

$$\mathbf{p}^{(n)} = \mathbf{p}^{(0)}\mathbf{P}^{(n)}$$

In words, the vector of state probabilities for time n is given by the vector of initial state probabilities multiplied by the n-step transition matrix. Thus, we see that a Markov chain is completely specified when its transition matrix $\mathbf{P}$ and the initial conditions $\mathbf{p}^{(0)}$ are known.

6.7 FIRST-PASSAGE AND FIRST-RETURN PROBABILITIES

The probabilities treated thus far answer questions of the general form, "What is the probability of being in a certain state at a certain time?" Another type of question that is frequently of interest is, "How long will it take to reach a certain state?" The answer must involve probabilities, but the random variable is the number of transitions that occur before a specified state is reached rather than the state after a specified number of transitions.

One might be tempted to interpret $p_{ij}^{(n)}$ as the probability that n steps are required to reach state j given that the process starts in state i. If this were a correct interpretation, then the set $\{p_{ij}^{(1)}, p_{ij}^{(2)}, p_{ij}^{(3)}, \ldots\}$ would give the probability distribution of the number of steps to get from i to j. However, this is not a correct interpretation of $p_{ij}^{(n)}$.

To see what is wrong, consider $p_{ij}^{(3)}$ and suppose the event to which this probability refers does indeed happen. That is, suppose the process is in state j three steps after it was in state i. But this does not mean that you have waited three steps for state j to be reached. It might have been reached after only one or two steps, after which the process either stayed in state j or changed to another state and then returned to state j. Any of the possibilities could result in the event indicated by $X_3=j$.

When we speak of the number of steps required to reach state j, we mean the number of steps required to reach state j *for the first time*. So to get the distribution of this time, we must consider the first passage probability, $f_{ij}^{(n)}$, defined to be

$$f_{ij}^{(n)}=P\{X_n=j, X_{n-1}\neq j, X_{n-2}\neq j, \ldots, X_1\neq j \mid X_0=i\}$$

In words, $f_{ij}^{(n)}$ is the probability that the process is in state j at time n *and not before*, given that it was in state i at time 0. This may be correctly reinterpreted as the probability that n steps are required to reach state j for the first time given that the process starts in state i.

Clearly, $f_{ij}^{(1)}=p_{ij}$. By extending the logic of the explanation of $p_{ij}^{(3)}$, one can show that

$$f_{ij}^{(n)}=p_{ij}^{(n)}-\sum_{k=1}^{n-1} f_{ij}^{(k)}p_{jj}^{(n-k)}$$

Thus the $f_{ij}^{(n)}$ can be obtained iteratively if the $p_{ij}^{(n)}$ are known.

The above discussion tacitly assumed that i and j were distinct. If they are not, the formal definition would be exactly the same, but we would speak of *first return* rather than first passage. Formally,

$$f_{ii}^{(n)}=P\{X_n=i, X_{n-1}\neq i, X_{n-2}\neq i, \ldots, X_1\neq i \mid X_0=i\}$$

The equation relating $f_{ii}^{(n)}$ to the $p_{ii}^{(n)}$ would also be the same.

For purposes of displaying some numerical results, let $\mathbf{F}^{(n)}$ be the matrix whose elements are $f_{ij}^{(n)}$. Then, for the limousine example we have been using,

$$\mathbf{F}^{(2)}=\begin{bmatrix} 0.5 & 0.16667 & 0.166667 \\ 0.33333 & 0.27778 & 0.277778 \\ 0.11111 & 0.33333 & 0.444444 \end{bmatrix}$$

$$\mathbf{F}^{(3)}=\begin{bmatrix} 0.22222 & 0.16667 & 0.361111 \\ 0.14815 & 0.16667 & 0.296296 \\ 0.11111 & 0.11111 & 0.203704 \end{bmatrix}$$

$$\mathbf{F}^{(4)}=\begin{bmatrix} 0.12963 & 0.05556 & 0.074074 \\ 0.08642 & 0.09259 & 0.095679 \\ 0.04938 & 0.11111 & 0.141975 \end{bmatrix}$$

6.8 CLASSIFICATION TERMINOLOGY

Everything discussed up to this point would apply to any Markov chain, regardless of structure. However, some of the most interesting questions that one might want to answer pertain to behavior in the limit as time goes to infinity, and these are definitely affected by structural variations. For example, some Markov chains could have one or more "trapping" sets of states, which once entered cannot be escaped from. Before progressing to further calculations, we must establish some terminology to express distinctions in structure that influence the long-term behavior.

A process is said to be *irreducible* if every state can be reached from every other state, not necessarily in a single transition but through some sequence of allowable transitions. This property assures that the process cannot get "caught" within a subset of states.

A state is *recurrent* if it is certain to occur again (eventually) given that it has occurred at least once; otherwise, it is said to be *transient*. Loosely speaking, a transient state will occur only a limited number of times and then "go away" forever, whereas a recurrent state is permanent. If a process is finite (i.e., has a finite number of states) and is irreducible, then all states are recurrent. This is perhaps the most common form of Markov chain in applications. More generally, a finite process that is reducible must contain at least one recurrent state, but could contain one or more subsets of recurrent states. The other states would be transient. Figure 6.6 illustrates a reducible structure with several recurrent and transient states. It is quite unlikely that you would ever construct such a model in practice.

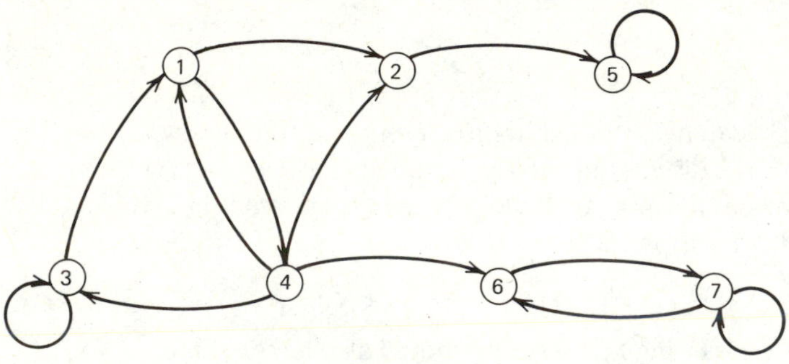

Figure 6.6
A reducible Markov chain.

In a process with an infinite number of states, it is not quite enough for the process to be irreducible to guarantee that the states are recurrent. It is possible, depending on the probability values, for the process to drift off to infinity, so that the reoccurrence of any state is not certain. An example of an infinite process where the structure alone is not enough to indicate whether the states are recurrent or transient is shown in Fig. 6.7. One could think of this Markov chain as a so-called "gambler's ruin" problem, where the probability of winning a dollar in a game of chance is p, and the probability of losing is $q=1-p$. The state of the Markov chain represents the wealth of the player.

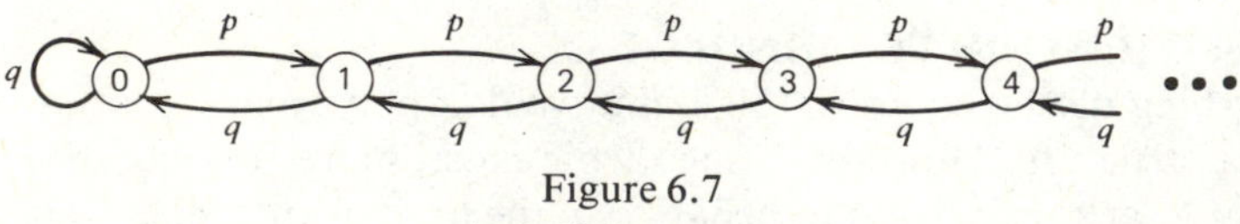

Figure 6.7
An infinite process.

If $p<q$, there is a bias against the player that tends to keep his wealth small. Even if it occasionally grows large, it is certain that he will return to the lower states. However, if $p>q$, it is possible for the player to grow increasingly wealthy and never return to the lower states. In the first incidence, all states are recurrent, and in the second all are transient.

The case $p=q$ illustrates another interesting possibility that can occur only in infinite processes. It turns out that in this case, all states are recurrent, but that the mean recurrence time (the time for a state to occur again once it has occurred) is infinite. This apparent paradox is simply one of the bizarre things that can happen when dealing with infinity. Recurrent states that have infinite mean recurrence time are called *null* states. Usually, of course, recurrent states will be *nonnull*, or *positive*.

One other anomaly can interfere with the calculations of limit results in either the finite or infinite process. Sometimes states are *periodic*, which means that they can recur only at certain epochs—epochs that are multiples of some integer larger than one. For example, if a state could recur only on even numbered epochs, it would be periodic of period two. Figure 6.8 illustrates a process in which every state is of period three. Notice that the definition does not assert that the state is certain to recur every three epochs, but only that it cannot recur except at those times.

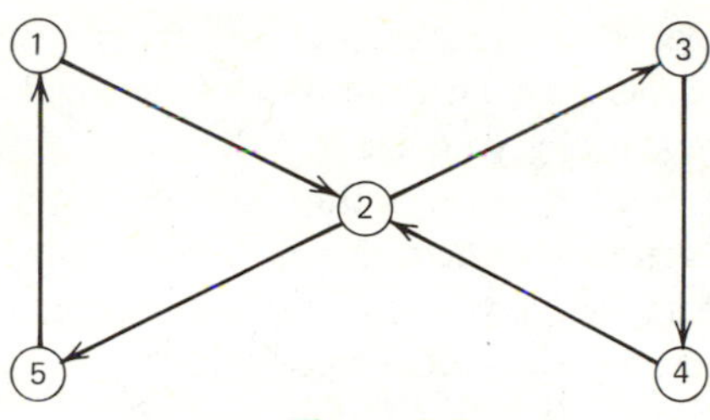

Figure 6.8
A periodic process.

All of the states of an irreducible process will share the same periodicity property. That is, if any one is periodic, they all are and must have the same period. Or if any one is aperiodic (not periodic), they all are.

The problem with periodic states is that they behave erratically even in the limit as time goes to infinity. The probability of being in the state will alternate between zero (at nonmultiples of the period) and positive values. The problem does not appear very often in applications and, even then, is more of a nuisance than a serious obstacle. If you should encounter a problem with periodic states, they are treated fully in Feller (7).

Now that we have words to describe the exceptions, we may consider the classes of Markov chains that are well behaved for purposes of limit calculations. When the process is irreducible, recurrent, positive, and aperiodic, we call the Markov chain *ergodic*. Since null states can occur only in infinite processes and finite irreducible processes can have only recurrent states, it is also sufficient to know that a finite process is irreducible and aperiodic to classify it as ergodic. Almost always, when you are modeling a real life process that does not terminate, the Markov chain will be ergodic. The exceptions arise infrequently. The limousine problem is a small, but otherwise typical, example.

If you are modeling a process that does terminate after some length of time, the model will usually take the form of a so-called *absorbing* Markov chain. In such a model, one or perhaps several states are *trapping* or absorbing states which, once entered, cannot be exited. These states represent the conditions under which the modeled process can terminate. All other states of the model are transient states.

Figure 6.9 illustrates a typical absorbing Markov chain. It might represent a production process with several stages and random branching to indicate the outcomes of inspection checks. The two absorbing states represent completion of the product as either an accepted, good product or a rejected, defective product.

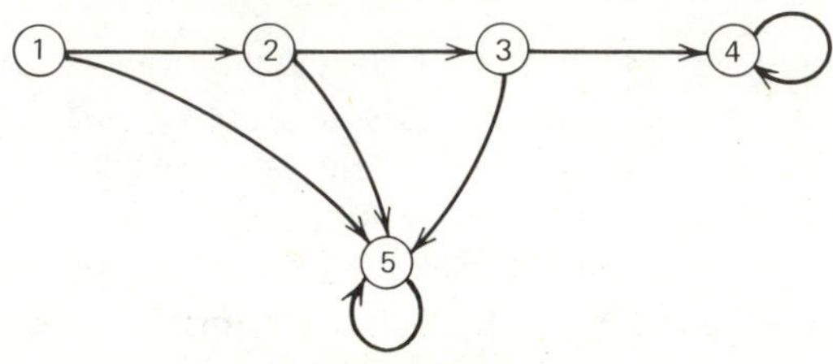

Figure 6.9
An absorbing Markov chain.

It is possible to model the same physical process as either an ergodic or absorbing Markov chain, depending on the point of view taken and the questions you want answered. For example, you could look at one of the work stations in a production process and model the randomly varying sequence of parts passing through it. Most likely, you would view this as a continuing process and end up with an ergodic chain. Alternatively, you could focus on a single workpiece, as in the model above, and end up with an absorbing chain.

6.9
ERGODIC MARKOV CHAINS

As we mentioned, the most common type of Markov chain model in practical applications is of the ergodic category. When the model is ergodic, several additional quantities (other than the transition probabilities, occupancy probabilities, and first passage probabilities) can easily be calculated. Two of the most important of these are *steady-state probabilities* and *mean first passage times*.

To introduce the concept of steady-state probabilities, let us return to the example used at the beginning of the chapter. The location of the limousine at successive epochs was tracked, probabilistically, by n-step transition probabilities, which were obtained by raising the one-step transition matrix to the appropriate power. Consider what happens to this matrix as higher and higher powers are taken. The row index indicates what state the process started in at epoch 0. As more and more time accumulates, it is logical to believe that the significance of the initial state diminishes. For example, after a hundred trips of the limousine, it would not seem to matter where the limousine started out. What this means mathematically is that, for large values of n, the rows of $\mathbf{P}^{(n)}$ should be identical. This, in fact, is exactly what happens. Secondly, we might expect that when you are predicting far into the future, it should not make much difference which particular value of n is specified. Mathematically, $\mathbf{P}^{(n)}$ and $\mathbf{P}^{(n+1)}$ are essentially the same for large n, or the transition matrix approaches stable limits as n grows large. This also happens as expected. If you review the powers of the limousine transition matrix, you can see these patterns starting to emerge after only a few transitions.

Let us define

$$\Pi_j = \lim_{n\to\infty} p_{ij}^{(n)}$$

This notation already makes use of the fact that the limits are independent of the initial state i. Then our comments above suggest that as n grows large,

$$\mathbf{P}^{(n)} \to \begin{bmatrix} \Pi_1, & \Pi_2, & \Pi_3, \ldots \\ \Pi_1, & \Pi_2, & \Pi_3, \ldots \\ \\ \Pi_1, & \Pi_2, & \Pi_3, \ldots \end{bmatrix}$$

As long as the process is ergodic, it can be proved that such limits exist and behave as we have suggested.

It would be tedious to try to find the steady-state probabilities by taking higher and higher powers of **P**, particularly when **P** is large. Fortunately we can obtain equations to determine them directly, as follows:

$$\mathbf{P}^{(n)} = \mathbf{P}^{(n-1)}\mathbf{P}$$

$$\lim_{n\to\infty} \mathbf{P}^{(n)} = \lim_{n\to\infty} \mathbf{P}^{(n-1)}\mathbf{P}$$

$$\begin{bmatrix} \Pi_1, & \Pi_2, & \Pi_3, \ldots \\ \Pi_1, & \Pi_2, & \Pi_3, \ldots \\ \Pi_1, & \Pi_2, & \Pi_3, \ldots \\ \cdot & \cdot & \cdot \\ \cdot & \cdot & \cdot \\ \cdot & \cdot & \cdot \end{bmatrix} = \begin{bmatrix} \Pi_1, & \Pi_2, & \Pi_3, \ldots \\ \Pi_1, & \Pi_2, & \Pi_3, \ldots \\ \Pi_1, & \Pi_2, & \Pi_3, \ldots \\ \cdot & \cdot & \cdot \\ \cdot & \cdot & \cdot \\ \cdot & \cdot & \cdot \end{bmatrix} [\mathbf{P}]$$

This represents many replications of the same set of equations

$$\mathbf{\Pi} = \mathbf{\Pi P}$$

or, if we wish to express this equation in terms of the more conventional column vectors

$$\mathbf{\Pi}^T = \mathbf{P}^T \mathbf{\Pi}^T$$

It turns out that this set of linear equations, though it has as many equations as unknowns, is a dependent set and therefore possesses an infinite number of solutions. (The dependency derives from the fact that every row of **P** sums to 1.) Only one of the infinite number of solutions, however, will qualify as a probability distribution. This one solution can be "forced" by requiring that the Π_i sum to 1. That is, we append the linear equation

$$\sum_{\text{all } i} \Pi_i = 1$$

to those previously expressed, and the resulting set of linear equations will possess a unique solution satisfying all the requirements of a probability distribution.

This last equation is called the *normalizing equation*. The usual practice is first to obtain an "unnormalized" solution by manipulating the $\mathbf{\Pi} = \mathbf{\Pi P}$ equations to express all of the Π_i in terms of one of them; then to use the normalizing equation to fix the value of this last one; and finally to substitute this value into the expressions for the others.

Referring once again to our numerical example, the steady-state equations would be given by

$$\begin{aligned}\Pi_1 &= \qquad\qquad 0.333\Pi_2 + 0.667\Pi_3\\ \Pi_2 &= 0.5\Pi_1 + 0.333\Pi_2 + 0.333\Pi_3\\ \Pi_3 &= 0.5\Pi_1 + 0.333\Pi_2\end{aligned}$$

Solving these by simple substitution methods will show the necessity of using the additional normalization equation, which in this case is

$$\Pi_1 + \Pi_2 + \Pi_3 = 1$$

The solution is

$$\begin{aligned}\Pi_1 &= 0.323\\ \Pi_2 &= 0.387\\ \Pi_3 &= 0.290\end{aligned}$$

Notice that the ease with which these exact steady-state probability values were obtained using the linear equations contrasts rather sharply with the effort required to raise the transition matrix to higher and higher powers, the latter yielding only approximations to the same steady-state values.

There is a sort of "physical" analog to the determination of the steady-state probabilities, which uses the transition diagram. The trick is to think of the points as small reservoirs and the arcs as connecting pipes through which liquid can flow, with valves to ensure that the flow goes only in the direction of the arrows. The probability p_{ij} associated with any arc is to be thought of as the fraction of the liquid in reservoir i that will pass to reservoir j in one transition time unit.

One unit of liquid is poured into the system according to $p^{(0)}$. If, for example, $p_1^{(0)} = 1/4$, then 1/4 of the liquid is poured in at point 1. After a while, a dynamic equilibrium is attained; the liquid continues to flow, but the amount in every reservoir remains constant. When this happens, the amount in each reservoir gives the steady-state probability for the corresponding state. They are proper probabilities because they are nonnegative and sum to 1.

The analogy can be made more exact by thinking of the liquid in terms of its molecules. The trajectory of an individual molecule describes a realization of the stochastic process. The effect of pouring in many molecules is to consider many realizations simultaneously. Hence, we are using, in effect, a statistical mechanics approach. This technique is used in chemical diffusion models, in electronics, and elsewhere.

Now when is equilibrium reached? Certainly a necessary condition is that the flow into any reservoir must equal the flow out, because if the two were not equal, the amount in the reservoir would be changing. If this condition is met for *all* reservoirs, it is sufficient. For any i,

$$\text{flow out} = \sum_j \Pi_i p_{ij} = \Pi_i \sum_j p_{ij} = \Pi_i, \quad \text{since} \sum_j p_{ij} = 1$$

$$\text{flow in} = \sum_k \Pi_k p_{ki}$$

Hence, $\Pi_i = \sum \Pi_k p_{ki}$ for all i; or, in matrix form $\mathbf{\Pi} = \mathbf{\Pi P}$.

The steady-state probabilities have several useful interpretations. If you fix a point in time in the distant future, Π_j is the probability that you will find the process in state j at that time. This is the most obvious interpretation. It can also be viewed as a time average; if you ran the process for a long time, Π_j would be the fraction of time that the process spent in state j. Or it can be viewed as an ensemble average; if you ran many identical processes simultaneously, Π_j would be the fraction of processes that you would find in state j (after a long period of time). Finally, it can be viewed as the reciprocal of the mean number of transitions between recurrences of the state.

Another commonly required quantity that is easily computed when the model is ergodic is the mean first passage time. Recall that we have already defined a first passage probability, the $f_{ij}^{(n)}$, as the probability that the process reaches state j for the first time at epoch n, given that it started in state i. This is almost, but not quite, the same as the probability that it takes n steps to reach state j, given that it started in i. The slight difference has to do with whether or not it is certain that j will *ever* be reached. If the model is constructed in such a way that eventual passage to j is certain, then we can legitimately regard the set of $f_{ij}^{(n)}$, for $n=1$, 2, 3, . . . , as the probability distribution of passage time from i to j. If, on the other hand, there is some chance that j will never be reached from i, then the same set of $f_{ij}^{(n)}$ will sum to something less than one, and therefore cannot be treated as an ordinary probability distribution. Fortunately, any model that is ergodic is guaranteed to be of the former type; that is, eventual passage from any state to any other is assured.

If N_{ij} represents the random variable for the number of epochs to reach j for the first time starting from i, then

$$P\{N_{ij}=n\}=f_{ij}^{(n)}$$

Because the $f_{ij}^{(n)}$ give the distribution of N_{ij}, the passage time from i to j, the *mean first passage time* from i to j, denoted m_{ij}, is given by

$$m_{ij}=E(N_{ij})=\sum_{n=1}^{\infty} nf_{ij}^{(n)}$$

Of course, higher moments can be expressed also. In the case where $i=j$, m_{ii} would be called the *mean recurrence time*. Recall that in the discussion of steady-state probabilities, it was pointed out that Π_i can be interpreted as the reciprocal of the mean recurrence time of state i. In symbols,

$$\Pi_i=\frac{1}{m_{ii}}$$

Although it is possible to get the first passage probabilities, and hence the distribution of first passage times, the calculation is at best tedious. If all that is desired is the *mean* first passage times, a rather simple calculation will suffice. Formally, m_{ij} is given by

$$m_{ij}=\sum_{n=1}^{\infty} nf_{ij}^{(n)}$$

but this would require the complete first passage time distribution. Instead, we may obtain a formula for m_{ij} by conditioning on the state at step 1. Given that the process is in state i at time 0, either the next state is j, in which case the passage time is exactly 1, or it is some other state k, in which case the passage time will be 1 plus the passage time from that state k to j. Weighting the possibilities by the appropriate probabilities gives

$$m_{ij}=1\,p_{ij}+\sum_{k\neq j}(1+m_{kj})p_{ik}$$

which then manipulates

$$m_{ij} = p_{ij} + \sum_{k \neq j} p_{ik} + \sum_{k \neq j} p_{ik} m_{kj}$$

$$m_{ij} = \sum_{\text{all } k} p_{ik} + \sum_{k \neq j} p_{ik} m_{kj}$$

to produce the formula

$$m_{ij} = 1 + \sum_{k \neq j} p_{ik} m_{kj}$$

This expresses m_{ij} as a linear function of the m_{kj}. By using the same relation for other m_{ij}'s, a set of linear equations may be expressed. The solution to the set gives the mean first passage times from any state into state j. Mean first recurrence times are also available in this way, but it is just as easy to get them as inverses of steady-state probabilities.

6.10 ABSORBING MARKOV CHAINS

Processes that terminate after some period of time are most conveniently modeled as absorbing chains. Sometimes it is useful to model even continuing processes as absorbing chains. For example, one might be interested in which of two states is entered first and would therefore artificially make them both absorbing states.

Because an absorbing Markov chain has a mix of absorbing states and transient states, it will make things easier to sort them into the two types and rearrange the transition matrix into the following form.

$$\mathbf{P} = \begin{array}{c} \overbrace{\hphantom{p_{11}, \; p_{12} \cdots \quad p_{kk}}}^{\text{Transient}} \; \overbrace{\hphantom{p_{1k+1} \cdots \quad 1}}^{\text{Absorbing}} \\ \left[\begin{array}{cccc|ccc} p_{11}, & p_{12} \cdots & & & p_{1k+1} \cdots & & \\ p_{21}, & p_{22} \cdots & & & & & \\ \cdot & & & & & & \\ \cdot & & & & & & \\ \cdot & & & & & & \\ p_{k1} & & & p_{kk} & p_{kk+1} \cdots & & \\ \hline 0 & 0 \ldots & & 0 & 1 \; 0 & & \\ 0 & 0 \ldots & & 0 & 0 \; 1 & & \\ \cdot & & & & & \cdot & \\ \cdot & & & & & & \cdot \\ 0 & & & 0 & 0 & & 1 \end{array}\right] \end{array} \begin{array}{l} \left.\vphantom{\begin{array}{c} a\\a\\a\\a\\a\\a \end{array}}\right\} \text{Transient} \\ \left.\vphantom{\begin{array}{c} a\\a\\a\\a\\a \end{array}}\right\} \text{Absorbing} \end{array}$$

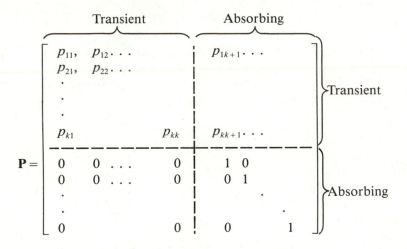

The lower right portion of the P matrix will be an identity matrix of dimension equal to the number of absorbing states. The lower left will be all zeros. There are no constraints on the elements in the first rows (the transition probabilities out of the transient states), other than that each row sum to one. However, it is useful to partition the matrix and name the portions for later reference.

$$\mathbf{P} = \left[\begin{array}{c|c} \mathbf{Q} & \mathbf{R} \\ \hline \mathbf{0} & \mathbf{I} \end{array}\right]$$

We will also use K to represent the number of transient states and speak of them as the first K states. (Actually, there could be an infinite number of transient states, so this is slightly improper.)

The first question we will address is which absorbing state will be entered. Of course, if a process has only one, the answer is trivial. If there is more than one possibility, the question is answered by a probability distribution over all of the possible absorbing states. Figure 6.10 gives a transition diagram that will illustrate the problem. States 1 and 2 are transient; 3 and 4 are both absorbing. Over a short period, states 1 and 2 may be visited a number of times; but ultimately, the process will enter either state 3 or state 4 and remain there. If chance determines that state 3 is entered, then state 4 will never occur, and vice versa. What we need is a method to determine the probability that state 3 is *ever* entered (as opposed to state 4, since one of the two must eventually occur). This is called an *absorption probability*.

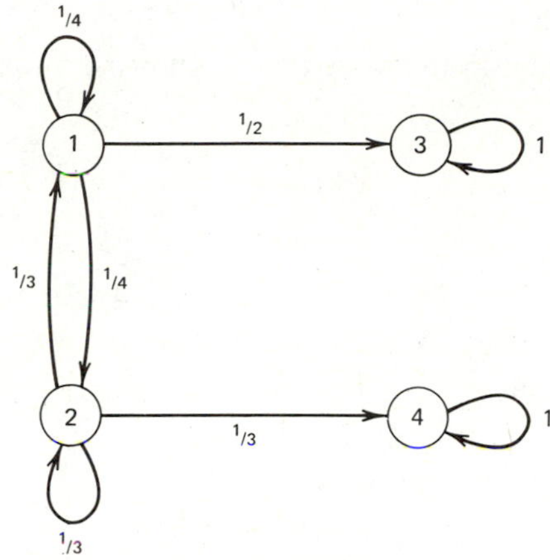

Figure 6.10
A transition diagram.

Now it should be apparent from the diagram that the probability of ever entering state 3 is greater if we start in state 1 than if we start in state 2. We are saying, then, that the absorption probabilities, unlike steady-state probabilities in an ergodic chain, *do* depend upon the initial state.

Let a_{ij} represent the probability that the process ever enters absorbing state j given that the initial state is i. It is easy to see that i must be a transient state for this probability to be greater than zero. Formally,

$$a_{ij} = \sum_{n=1}^{\infty} f_{ij}^{(n)}$$

but this is not computationally practical. An alternative method is available via the following argument.

There are two immediate possibilities leading to the ultimate occurrence of state j. Either the first transition is to state j (immediate absorption), or the first transition is to some *transient* state k and the process ultimately enters state j from that state. The probabilility of the first event is just p_{ij}, of the second is $\sum_k p_{ik}a_{kj}$, where the sum is over all transient states k. Since the two possibilities are mutually exclusive,

$$a_{ij} = p_{ij} + \sum_{k=1}^{K} p_{ik} a_{kj}.$$

This is one linear equation in several unknown absorption probabilities. A complete set of independent linear equations will be obtained if the same formula is reapplied to all possible transient states, regarding each in turn as the initial state.

By way of example, if we want a_{13} for the process illustrated in Fig. 6.10, our first equation would be

$$a_{13} = p_{13} + p_{11} a_{13} + p_{12} a_{23}.$$

This is an equation in two unknowns, a_{13} and a_{23}. If we now write the equation for a_{23},

$$a_{23} = p_{23} + p_{21} a_{13} + p_{22} a_{23}$$

we get another equation in the same two unknowns. Substituting numerical values, our equations are

$$a_{13} = \frac{1}{2} + \frac{1}{4} a_{13} + \frac{1}{4} a_{23}$$

$$a_{23} = 0 + \frac{1}{3} a_{13} + \frac{1}{3} a_{23}$$

Solving these, we obtain

$$a_{13} = \frac{4}{5}$$

$$a_{23} = \frac{2}{5}$$

Notice that even if we have no direct interest in a_{23}, it was necessary to obtain it in order to find a_{13}. Notice also that a_{13} and a_{23} do not sum to 1; there is no reason that they should. On the other hand, if the initial state is fixed, then the process must ultimately enter one of the two absorbing states, so

$$a_{13} + a_{14} = 1$$

and

$$a_{23} + a_{24} = 1$$

From this, we can deduce that

$$a_{14} = \frac{1}{5}$$

$$a_{24} = \frac{3}{5}$$

Alternatively, we could have used the formula to get two linear equations in a_{14} and a_{24} and solved for them directly.

It is convenient to express the equations for the absorption probabilities in matrix form. Let **A** be the matrix of the a_{ij}. It will not necessarily be square, but will have as many rows as there are transient states, and as many columns as there are absorbing states. Thus it has the same dimension as **R** does. In fact, examination of the equations for the a_{ij} reveals that

$$\mathbf{A} = \mathbf{R} + \mathbf{QA}$$

Manipulation of this matrix equation gives **A** in terms of **R** and **Q**, which are just portions of the transition matrix **P**.

$$\mathbf{A} - \mathbf{QA} = \mathbf{R}$$
$$(\mathbf{I} - \mathbf{Q})\mathbf{A} = \mathbf{R}$$
$$\mathbf{A} = [\mathbf{I} - \mathbf{Q}]^{-1}\mathbf{R}$$

It can be shown that $(\mathbf{I} - \mathbf{Q})$ is nonsingular, so the inverse exists, and this method can be used if all of the a_{ij} are desired. If only one or two are sought, it would be more efficient to solve the equations for them than to invert the entire $(\mathbf{I} - \mathbf{Q})$ matrix.

To illustrate the matrix method for the example previously calculated, observe that **R** and **Q** are, respectively,

$$\mathbf{R} = \begin{bmatrix} \frac{1}{2} & 0 \\ 0 & \frac{1}{3} \end{bmatrix}$$

$$\mathbf{Q} = \begin{bmatrix} \frac{1}{4} & \frac{1}{4} \\ \frac{1}{3} & \frac{1}{3} \end{bmatrix}$$

The matrix $\mathbf{I} - \mathbf{Q}$ is

$$\mathbf{I} - \mathbf{Q} = \begin{bmatrix} \frac{3}{4} & -\frac{1}{4} \\ -\frac{1}{3} & \frac{2}{3} \end{bmatrix}$$

and its inverse is

$$(\mathbf{I} - \mathbf{Q})^{-1} = \begin{bmatrix} \frac{8}{5} & \frac{3}{5} \\ \frac{4}{5} & \frac{9}{5} \end{bmatrix}$$

Finally, the matrix of absorption probabilities is

$$\mathbf{A} = (\mathbf{I} - \mathbf{Q})^{-1}\mathbf{R} = \begin{bmatrix} \frac{4}{5} & \frac{1}{5} \\ \frac{2}{5} & \frac{3}{5} \end{bmatrix}$$

which agrees with our previous calculation.

The matrix $(\mathbf{I}-\mathbf{Q})^{-1}$ is also useful in answering a question of a different sort. Suppose that you want to know how many times a transient state will be occupied before absorption occurs. This would, of course, be a random variable that depended on the initial state. Let e_{ij} be the *mean number of times that transient state j is occupied given that the initial state is i*, and let $\mathbf{E}$ be the matrix of the e_{ij}. Because i and j each range over the transient states, $\mathbf{E}$ is a square matrix.

To develop an expression for the e_{ij}, we may use the same logic already used several times—that of conditioning on the state at time 1. If $i \neq j$,

$$e_{ij} = \sum_{k=1}^{K} p_{ik} e_{kj}$$

because j will be occupied a mean of e_{kj} times if the first transition is to any transient state k, and 0 times if the first transition is to any of the absorbing states. If $i = j$,

$$e_{ii} = 1 + \sum_{k=1}^{K} p_{ik} e_{ki}$$

because i will be occupied once (it spends the first time interval in i) plus a mean of e_{ki} times if the first transition is to any transient state k. Putting these relations into matrix form,

$$\mathbf{E} = \mathbf{I} + \mathbf{QE}$$

and solving for the matrix $\mathbf{E}$,

$$\mathbf{E} - \mathbf{QE} = \mathbf{I}$$
$$(\mathbf{I} - \mathbf{Q})\mathbf{E} = \mathbf{I}$$
$$\mathbf{E} = (\mathbf{I} - \mathbf{Q})^{-1}$$

In words, then, the i, jth entry of the matrix $(\mathbf{I}-\mathbf{Q})^{-1}$ gives the mean number of times that the state j will be occupied before absorption occurs, given that the process starts in state i. For example, in the numerical example treated earlier, if the process begins in state 1, it will occupy state 2 an average of 3/5 times before absorption occurs.

When $\mathbf{E}$ is available, the mean total number of transitions until absorption is easily obtained. This will obviously depend on the initial state, so let d_i represent this mean, given that i is the initial state. In many practical applications, d_i will represent the mean duration of the process. The total expected number of transitions will equal the total number of times that any of the transient states are occupied, so

$$d_i = \sum_{j=1}^{K} e_{ij}$$

which is just the sum of elements across row i of $\mathbf{E}$. In the numerical example, if the process begins in state 1, it makes an average of 11/5 transitions before being absorbed in one of the two absorbing states.

It is quite a different question if one specifies a particular absorbing state and asks how many transitions will occur before it is entered. This question would be answered by something like a mean first passage time. In fact, if there is only one absorbing state, so that ultimate passage to that state is certain, then an ordinary mean first passage time will do. In other cases, when it is not certain that passage will ever occur, the ordinary mean first passage time would be infinite. In these cases, what is really desired is the *conditional* mean first passage time, given that passage does occur.

We have previously used the notation m_{ij} for the mean first passage time from i to j in an ergodic process. Because there is little possibility of confusion, let the same notation be used for the *conditional* mean first passage time in a reducible process. Then it can be shown that

$$a_{ij}m_{ij} = a_{ij} + \sum_{k \neq j} p_{ik}a_{kj}m_{kj}$$

Notice that if passage to j is certain to occur, as in an irreducible process, then $a_{kj} = 1$ for all k, and the equation reduces to the one derived earlier.

Assuming that the absorption probabilities a_{kj} are known, the above equation is linear in the unknowns m_{kj}. By writing a complete set, one such equation for each transient state i, a sufficient number of independent equations will be available to determine the m_{kj}.

To illustrate, suppose that we want to know the mean number of transitions that would occur in the process we used to illustrate absorption probabilities, if we know that it started in state 1 and ultimately reached state 3. We want m_{13}, so the associated equation would be,

$$a_{13}m_{13} = a_{13} + p_{11}a_{13}m_{13} + p_{12}a_{23}m_{23}$$

Since this equation has the two unknowns m_{13} and m_{23}, we need another equation. Writing the one for m_{23}, we get

$$a_{23}m_{23} = a_{23} + p_{21}a_{13}m_{13} + p_{22}a_{23}m_{23}$$

Substituting the values of the p_{ij} and the a_{ij}, which were previously calculated, we have the two equations:

$$\frac{4}{5}m_{13} = \frac{4}{5} + \left(\frac{1}{4}\right)\left(\frac{4}{5}\right)m_{13} + \left(\frac{1}{4}\right)\left(\frac{2}{5}\right)m_{23}$$
$$\frac{2}{5}m_{23} = \frac{2}{5} + \left(\frac{1}{3}\right)\left(\frac{4}{5}\right)m_{13} + \left(\frac{1}{3}\right)\left(\frac{2}{5}\right)m_{23}$$

These can be solved to yield

$$m_{13} = 1.9$$
$$m_{23} = 3.4$$

To answer our original question, then, it will take a mean of 1.9 transitions to get from state 1 to state 3, assuming that it does occur.

Table 6.1 summarizes all the Markov chain terms, notations and formulas discussed thus far.

Table 6.1
MARKOV CHAIN RESULTS

Name	Notation	Type of Model	Formula
Transition probability	$p_{ij}^{(n)}$	Any	$p_{ij}^{(n)} = \sum_k p_{ik}^{(n-1)} p_{kj}$
Occupancy probability	$p_j^{(n)}$	Any	$p_j^{(n)} = \sum_i p_i^{(0)} p_{ij}^{(n)}$
First passage (or return) probability	$f_{ij}^{(n)}$	Any	$f_{ij}^{(n)} = p_{ij}^{(n)} - \sum_{k=1}^{n-1} f_{ij}^{(k)} p_{jj}^{(n-k)}$
Steady-state probability	π_j	Ergodic	$\sum_k \pi_k p_{ki} = \pi_i$ $\sum_i \pi_i = 1$
Mean first passage time	m_{ij}	Ergodic	$m_{ij} = 1 + \sum_{k \neq j} p_{ik} m_{kj}$
Absorption probability	a_{ij}	Absorbing	$a_{ij} = p_{ij} + \sum_{\text{Trans } k} p_{ik} a_{kj}$
Mean number of entries	e_{ij}	Absorbing	$e_{ij} = \begin{cases} \sum_{\text{Trans } k} p_{ik} e_{kj}, \text{ if } i \neq j \\ 1 + \sum_{\text{Trans } k} p_{ik} e_{kj}, \text{ if } i = j \end{cases}$
Mean duration	d_i	Absorbing	$d_i = \sum_{\text{Trans } j} e_{ij}$
Conditional mean first passage time	m_{ij}	Absorbing	$a_{ij} m_{ij} = a_{ij} + \sum_{\text{Trans } k} p_{ik} a_{kj} m_{kj}$

II. CONTINUOUS TIME PROCESSES

6.11 AN EXAMPLE

Continuous time stochastic processes are similar in most respects to discrete time stochastic processes. Additional complexities occur, however, because each infinitesimal instant is available as a possible transition time. For example, it will not make sense to speak of a one-step transition matrix because time is not measured in steps.

To facilitate understanding, a simple two-state example will be solved in its entirety before the formal development is presented. Consider, then, for purposes of illustration the following situation which concerns the operation of an automatic loom used to weave cloth. Normally, the loom will operate without human intervention. Occasionally, however, a thread breaks and the shuttle may jam. There is an attendant standing by whose sole responsibility is to unjam the shuttle, tie threads, and put the loom back into operation.

Ignoring shift changes, lunch hours, and coffee breaks the system will always be in one of two states: 0—the loom is shut off, the man is working to repair it; 1—the loom is operating, the man is idle. A typical realization of the process would resemble Fig. 6.11. The upper lines, representing continuous intervals of time during which the loom is working, could be called "operating times." The lower lines could be called "repair times." One way to approach the problem of modeling this system would be to describe the operating and repair times as continuous random variables having some particular distribution. Here, we shall pursue an alternative approach.

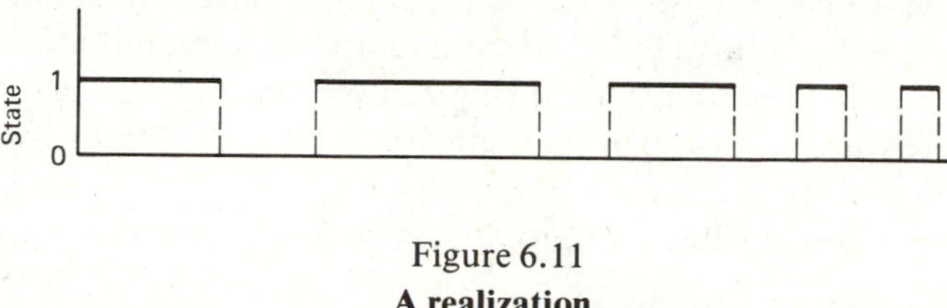

Figure 6.11
A realization.

Let $p_{ij}(t)$ represent the probability that the system is in state j at time t given that it was in state i at time 0. Since there are only two states in this example, we are speaking of four functions: $p_{00}(t)$, $p_{01}(t)$, $p_{10}(t)$, and $p_{11}(t)$. The development of the equations that determine the $p_{ij}(t)$ functions can be simplified if the following four assumptions are made:

1. The process satisfies the Markov property.
2. The process is stationary.
3. The probability of a transition from 0 to 1 (a repair) or from 1 to 0 (a breakdown) in a short interval, Δt, is proportional to Δt.
4. The probability of two or more changes of state in a short interval Δt is zero.

Before using these assumptions, let us pause to consider what they mean. When we assume that the process satisfies the Markov property, we are asserting that if at *any* time we know whether the loom is operating or shut off, we can determine all probabilities associated with the process from that time on. We do not need to know, nor does it even help us to know, the sequence of states leading up to the present one, how long the process spent in each of these states, or even how long it has been in the present state. It is common to personify this property and say that the process has no memory, or is forgetful at every point of time.

The process will be stationary if the breakdown and repair rates do not depend on the time of day. It would not be stationary if, for example, the attendant worked slower at certain times.

Assumptions 3 and 4 could not strictly hold. If Δt is small enough, however, they will be acceptable approximations. Later Δt will be reduced to zero; at this time, they will become precise. At any rate, the results will be correct, and the development is aided considerably by making these assumptions. To make assumption 3 even more specific, let the constants of proportionality be 3 for the repair transition and 2 for the breakdown transition. That is, let

$$p_{01}(\Delta t) = 3\Delta t$$
$$p_{10}(\Delta t) = 2\Delta t$$

These constants of proportionality are called the "repair rate" and "breakdown rate"; they receive further interpretation later.

Consider now the function value $p_{01}(t+\Delta t)$, or the probability that the loom is in state 1 (operating) at time $t+\Delta t$, given that it was in state 0 (being repaired) at time zero. Either the loom was being repaired at time t and was put into operation during the interval Δt, or the loom was operating at time t and continued to operate for the short interval Δt.

In symbolic terms,

$$p_{01}(t+\Delta t)=p_{00}(t)p_{01}(\Delta t)+p_{01}(t)p_{11}(\Delta t)$$

This expression, a special case of the Chapman-Kolmogorov equations for the continuous time case, requires the Markov assumption to permit multiplication of the probabilities referring to events during t and to events during Δt. It also requires stationarity to permit use of the same probability functions for the interval t and for the later interval Δt.

We now substitute our linear approximations for $p_{01}(\Delta t)$ and $p_{11}(\Delta t)$, and manipulate

$$p_{01}(t+\Delta t)=p_{00}(t)3\Delta t+p_{01}(t)[1-2\Delta t]$$

[because $p_{11}(\Delta t)=1-p_{10}(\Delta t)$]

$$p_{01}(t+\Delta t)=3p_{00}(t)\Delta t+p_{01}(t)-2p_{01}(t)\Delta t$$

$$p_{01}(t+\Delta t)-p_{01}(t)=3p_{00}(t)\Delta t-2p_{01}(t)\Delta t$$

$$\frac{p_{01}(t+\Delta t)-p_{01}(t)}{\Delta t}=3p_{00}(t)-2p_{01}(t)$$

Taking the limit of both sides as Δt goes to zero, we recognize the definition of a derivative

$$\lim_{\Delta t\to 0}\left[\frac{p_{01}(t+\Delta t)-p_{01}(t)}{\Delta t}\right]=\lim_{\Delta t\to 0}[3p_{00}(t)-2p_{01}(t)]$$

and obtain the differential equation

$$\frac{dp_{01}(t)}{dt}=3p_{00}(t)-2p_{01}(t)$$

If we were to go through the corresponding derivations for the other three transition functions, we would get

$$\frac{dp_{00}(t)}{dt}=-3p_{00}(t)+2p_{01}(t)$$

$$\frac{dp_{10}(t)}{dt}=-3p_{10}(t)+2p_{11}(t)$$

$$\frac{dp_{11}(t)}{dt}=3p_{10}(t)-2p_{11}(t)$$

Thus, we have a system of linear first-order differential equations with constant coefficients. In this particular case, they are simple enough to solve directly. Making use of whatever manipulations or solution technique we find most convenient, and using the initial conditions $p_{ij}(0)=0$ for $i\neq j$ and $p_{ij}(0)=1$ for $i=j$, we obtain the solutions:

$$p_{00}(t)=\frac{2}{5}+\frac{3}{5}e^{-5t}$$

$$p_{01}(t)=\frac{3}{5}-\frac{3}{5}e^{-5t}$$

$$p_{10}(t)=\frac{2}{5}-\frac{2}{5}e^{-5t}$$

$$p_{11}(t)=\frac{3}{5}+\frac{2}{5}e^{-5t}$$

These are conveniently expressed in the matrix form:

$$\mathbf{P}(t)=\begin{bmatrix} p_{00}(t) & p_{01}(t) \\ p_{10}(t) & p_{11}(t) \end{bmatrix}=\begin{bmatrix} \frac{2}{5}+\frac{3}{5}e^{-5t} & \frac{3}{5}-\frac{3}{5}e^{-5t} \\ \frac{2}{5}-\frac{2}{5}e^{-5t} & \frac{3}{5}+\frac{2}{5}e^{-5t} \end{bmatrix}$$

If we graph these solutions as functions of time, using the same matrix ordering, we obtain the four graphs shown in Fig. 6.12.

Notice that the functions behave as they should in order for them to represent probabilities. They lie uniformly within the interval [0, 1] for all values of $t \geqq 0$; the sum over the second subscript equals 1 for all values of $t \geqq 0$; and the initial conditions are satisfied. Observe also that the limit of each function as t goes to infinity is immediately apparent, both in the function itself and in the graph. Convergence to this value is smooth and monotonic (as opposed to discontinuous, oscillating, or both), and functions with the same second subscript but different first subscripts converge to the same value. These points, as illustrated by this particular example, are generally true in processes of this kind.

The solution is complete, but we do not yet have an interpretation of the "rates" that is sufficient to enable us to measure or estimate the rates in a real-

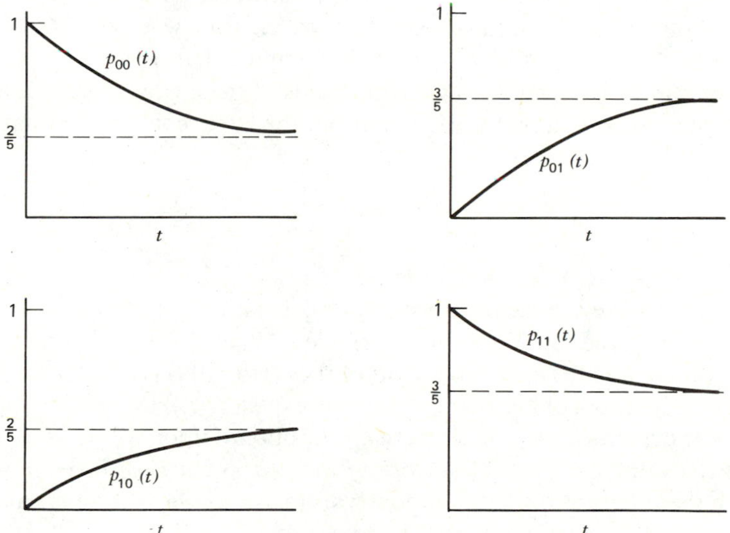

Figure 6.12
The four functions graphed.

world situation. Heretofore, we have assumed they were *given* as 2 and 3. To get this interpretation, as well as some other information about the process, consider now the repair process in isolation. We will imagine that the process begins a repair time at $t=0$, remains in state 0 until the repair is complete, then changes to and remains in state 1. There are no breakdowns in this modified version of the process. Using the same method of derivation as before,

$$p_{00}(t+\Delta t) = p_{00}(t)p_{00}(\Delta t)$$
$$= p_{00}(t)[1-3\Delta t]$$
$$p_{00}(t+\Delta t) - p_{00}(t) = -3\Delta t p_{00}(t)$$
$$\frac{p_{00}(t+\Delta t) + p_{00}(t)}{\Delta t} = -3p_{00}(t)$$
$$\frac{dp_{00}(t)}{dt} = -3p_{00}(t)$$

This is immediately solvable [again using the initial condition $p_{00}(0)=1$] as

$$p_{00}(t) = e^{-3t}$$

Now $p_{00}(t)$ is the probability that, given the loom is not operating at time zero, it is not operating at time t. Since breakdowns are not allowed, this event can also be described as "the repair time exceeds t." So the probability that the repair time is less than or equal to t, that is, the cumulative distribution function of the repair time, is $1-e^{-3t}$. This describes a *negative exponential* distribution.

In summary, then, our original four assumptions imply that the distribution of repair time is negative exponential. *No other distribution* of repair times will satisfy the assumptions. Of course, by means of a similar argument, we can deduce the same implications for the distribution of operating times. The cumulative distribution of operating times is $1-e^{-2t}$.

A negative exponentially distributed random variable having the cumulative distribution $1-e^{-\lambda t}$ has an expected value of $1/\lambda$. In this example, the repair time distribution had $\lambda=3$, so the expected repair time would be 1/3. The operating time distribution had $\lambda=2$, so the expected operating time would be 1/2. Turning the logic around, we can say that the "rates" of 3 and 2 that we used for the original derivation can be interpreted as the reciprocals of the mean repair and operating times, respectively. In a real-world situation, the latter could be estimated statistically.

6.12 FORMAL DEFINITIONS AND THEORY

A continuous time stochastic process $\{X(t)\}$ is an infinite family of random variables indexed by the continuous real variable t. That is, for any fixed t, $X(t)$ is a random variable, and the collection of all of these (for all t) is the stochastic process. We ordinarily think of t as time, so we may expect $X(t_1)$, the random variable at time t_1, to be dependent on $X(t_0)$, where $t_0<t_1$, but not upon $X(t_2)$, where $t_2>t_1$. We refer to the value of $X(t_1)$ *as the state of the process at time* t_1. Here we will assume that the $X(t)$ are discrete random variables, so we will be speaking of discrete-state, continuous-time stochastic processes.

If for all $t_n, t_{n-1}, \ldots, t_0$ satisfying $t_n>t_{n-1}> \ldots >t_0$, we have that $P[X(t_n)=j_n \mid X(t_{n-1})=j_{n-1}, \ldots, X(t_0)=j_0] = P[X(t_n)=j_n \mid X(t_{n-1})=j_{n-1}]$ we say that the process has the *Markov property*, or is a (continuous time) *Markov pro-*

cess. This property is an independence property. It says that if you have information as to the state of the process at a sequence of points in time $(t_0, t_1, \ldots, t_{n-1})$ and wish to predict the state at some future time (t_n), that only the information as to the most recent state is of value. If we think of t_{n-1} as the present, t_n as any future time, and the $t_0, \ldots, t_{n-2}$ as any times in the past, we may say that the future of a Markov process is dependent only on the present, and not at all on the past. It is sometimes said that "a Markov Process is forgetful."

A Markov process is said to be *time homogeneous* or *stationary* if

$$Pr[X(t_2)=j \mid X(t_1)=i] = Pr[X(t_2-t_1)=j \mid X(0)=i]$$

for all i, all j, all t_1 and t_2 such that $t_1 < t_2$. In words, the process is stationary if these conditional probabilities depend only on the *interval* between the events rather than on absolute time. The words "time homogeneous" and "stationary" suggest "invariance" or "sameness" in time. This connotation is correct if we apply it to the *probability* mechanism of the process. The state of the process does, of course, change in time.

A stationary Markov process is completely described by its *transition probability functions*, denoted $p_{ij}(t)$, where

$$p_{ij}(t) = Pr[X(t)=j \mid X(0)=i]$$

The sufficiency of these functions (to describe everything of interest having to do with the stochastic process) is a consequence of both the Markov and stationarity assumptions. That only one state need be known follows from the Markov assumption, and that we may arbitrarily call the time at which that state known as "zero" time follows from stationarity.

A fundamental equation for stationary Markov processes, known as the Chapman-Kolmogorov equation, will now be derived. One could argue the validity of the equation more directly, but it is given here in full detail to emphasize its reliance on the assumptions.

$$\begin{aligned}
p_{ij}(t+s) &= Pr[X(t+s)=j \mid X(0)=i] && \text{(definition)}\\
&= \sum_k Pr[X(t+s)=j,\ X(t)=k \mid X(0)=i] && \text{(marginal from joint)}\\
&= \sum_k Pr[X(t+s)=j \mid X(t)=k,\ X(0)=i]\,Pr[X(t)=k \mid X(0)=i] && \text{(joint from conditional)}\\
&= \sum_k Pr[X(t+s)=j \mid X(t)=k]\,Pr[X(t)=k \mid X(0)=i] && \text{(Markov assumption)}\\
&= \sum_k Pr[X(s)=j \mid X(0)=k]\,Pr[X(t)=k \mid X(0)=i] && \text{(Stationarity)}\\
&= \sum_k p_{kj}(s)p_{ik}(t) && \text{(definition)}\\
p_{ij}(t+s) &= \sum_k p_{ik}(t)p_{kj}(s)
\end{aligned}$$

This final line is the Chapman-Kolmogorov equation in general form. It will be used shortly in the more specialized form

$$p_{ij}(t+\Delta t) = \sum_k p_{ik}(t)p_{kj}(\Delta t)$$

which is known as the "forward" Chapman-Kolmogorov equation. There is also the somewhat less useful "backward" equation,

$$p_{ij}(\Delta t + t) = \sum_k p_{ik}(\Delta t)p_{kj}(t).$$

Now a fair amount can be known about these $p_{ij}(t)$ functions just as a consequence of the fact that they are, for all t, probabilities. For example, they are nonnegative, bounded functions, because a probability must lie between 0 and 1. The values of the functions at $t=0$ can be deduced because $p_{ij}(0) = Pr[X(0)=j \mid X(0) = i]$. Clearly, for i different from j, $p_{ij}(0)=0$, and for i equal to j, $p_{jj}(0)=1$. If we fix i and vary j over all states, the sum of the $p_{ij}(t)$ must equal 1 (for all t), because,

$$\begin{aligned}\sum_j p_{ij}(t) &= \sum_j Pr[X(t)=j \mid X(0)=i] \\ &= Pr[X(t) = \text{any of its possible states} \mid X(0)=i] \\ &= 1\end{aligned}$$

So these transition probability functions are not just arbitrary functions of time, but are quite restricted in the sense that they must satisfy all of these properties. Notice also that these properties are not *assumed*, but are a direct consequence of what the notation means.

Under the reasonable assumption that the $p_{ij}(t)$ are continuous functions (since the probabilities are not likely to suddenly jump), we may express p_{ij} for small Δt by use of MacLaurin's series,

$$p_{ij}(\Delta t) = p_{ij}(0) + p_{ij}'(0)\Delta t + o(\Delta t^2)$$

where $o(\Delta t^2)$ represents all terms of the order of $(\Delta t)^2$ or higher. If we consider this expression for $i \neq j$, and let $\lambda_{ij} = p_{ij}'(0)$, we obtain

$$p_{ij}(\Delta t) = \lambda_{ij}\Delta t + o(\Delta t^2).$$

We may think of this as a linear approximation to $p_{ij}(t)$ which is a good approximation as long as Δt is small. The λ_{ij} is called the *transition rate* from i to j. This terminology is, of course, entirely consistent with its definition as a time derivative of the transition function. Since $p_{ij}(0)=0$ and this is its minimum value, we may be certain that λ_{ij} is nonnegative.

For $i=j$, the MacLaurin's series expansion yields

$$p_{jj}(\Delta t) = 1 + p_{jj}'(0)\Delta t + o(\Delta t^2),$$

and if we again let $\lambda_{jj} = p_{jj}'(0)$, we get the linear approximation

$$p_{jj}(\Delta t) = 1 + \lambda_{jj}\Delta t + o(\Delta t^2)$$

Since we know that $p_{jj}(0)=1$ and that this is its maximum value, we may be certain that λ_{jj} is nonpositive.

If we now consider the forward Chapman-Kolmogorov equation,

$$p_{ij}(t+\Delta t) = \sum_k p_{ik}(t)p_{kj}(\Delta t)$$

for small Δt, and substitute our linear approximation, we get

$$p_{ij}(t+\Delta t) = p_{ij}(t)[1 + \lambda_{jj}\Delta t + o(\Delta t^2)] + \sum_{k\neq j} p_{ik}(t)[\lambda_{kj}\Delta t + o(\Delta t^2)]$$

Manipulating slightly,

$$\begin{aligned}\frac{p_{ij}(t+\Delta t) - p_{ij}(t)}{\Delta t} &= p_{ij}(t)\lambda_{jj} + \frac{p_{ij}(t)o(\Delta t^2)}{\Delta t} + \sum_{k\neq j}\left[p_{ik}(t)\lambda_{kj} + \frac{p_{ik}(t)o(\Delta t^2)}{\Delta t}\right] \\ &= \sum_k p_{ik}(t)\lambda_{kj} + \sum_k \frac{p_{ik}(t)o(\Delta t^2)}{\Delta t}\end{aligned}$$

Taking the limit as $\Delta t \to 0$,

$$\frac{dp_{ij}(t)}{dt} = \sum_{k} p_{ik}(t)\lambda_{kj}$$

The terms of the order of $(\Delta t)^2$ go to zero faster than Δt, so these terms drop out. The result is an *exact* (not approximate) differential equation for $p_{ij}(t)$ in terms of the $p_{ik}(t)$. It is a linear, first-order differential equation with constant coefficients—the λ_{kj}'s.

Recognizing the above sum as matrix multiplication, we may express all of the differential equations at once in the matrix form

$$\frac{d\mathbf{P}(t)}{dt} = \mathbf{P}(t)\mathbf{\Lambda},$$

where $d\mathbf{P}(t)/dt$ is the matrix whose (i, j)th element is $dp_{ij}(t)/dt$, $\mathbf{P}(t)$ is the matrix whose (i, j)th element is $p_{ij}(t)$, and $\mathbf{\Lambda}$ is the matrix whose (i, j)th element is λ_{ij}.

The elements of $\mathbf{\Lambda}$ may be further related by extending the properties of $\mathbf{p}(t)$. In particular, since for each i,

$$\sum_{j} p_{ij}(t) = 1$$

then

$$\frac{d}{dt}\left[\sum_{j} p_{ij}(t)\right]\Bigg|_{t=0} = \frac{d}{dt}[1]\Bigg|_{t=0}$$

$$\sum_{j} \frac{d}{dt} p_{ij}(t)\Big|_{t=0} = 0$$

$$\sum_{j} \lambda_{ij} = 0$$

In words, each row of $\mathbf{\Lambda}$ must sum to zero. Since every off-diagonal element is nonnegative, the diagonal element, λ_{ii}, must be equal in magnitude and opposite in sign to the sum of the others in the same row. That is,

$$\lambda_{ii} = -\sum_{j \neq i} \lambda_{ij}$$

For the two state example studied earlier, λ_{01} was 3 and λ_{10} was 2. In this case, the matrix $\mathbf{\Lambda}$ would be given by

$$\mathbf{\Lambda} = \begin{bmatrix} -3 & 3 \\ 2 & -2 \end{bmatrix}$$

and the four differential equations could be obtained from

$$\frac{d\mathbf{P}(t)}{dt} = \mathbf{P}(t)\mathbf{\Lambda}$$

The elements λ_{ij} *for* $i \neq j$, may be interpreted (and hence measured) as the parameters of negative exponential distributions. For a particular λ_{ij}, the distribution referred to is the distribution of time spent in state i when j is the next state. If T_{ij} is the random variable, $E(T_{ij}) = 1/\lambda_{ij}$, so λ_{ij} can be estimated by the inverse of a sample mean.

6.13 THE ASSUMPTIONS RECONSIDERED

As was the case in discrete time, the derivation emphasizes the assumptions, but the result is in a form that permits use without understanding. That is, the λ_{ij}'s could be measured and the equations could be expressed and solved in the complete absence of understanding—or, for that matter, awareness—of the assumptions. Furthermore, to state the assumptions is not enough. You must be sufficiently cognizant of their implications to judge whether or not they are reasonable in particular contexts. The Markov and stationarity assumptions are both strong assumptions—especially so when time is treated as continuous; there are, however, many, many situations where they are inappropriate. In fact, as you begin to appreciate how strong they are, you may begin to doubt whether any real-life process behaves as a stationary Markov process. It is an empirical fact, however, that many such real-life processes do exist. So the object now is to develop sufficient understanding to be able to distinguish between those processes which might properly be modeled as stationary Markov processes and those which should not.

We have already seen that the Markov and stationarity assumptions imply that the times between events must be negative-exponentially distributed. The parameters of these distributions, the λ_{ij} may be dependent on the state occupied, i, and the next state, j, but all of the distributions must be of the negative exponential form. No other distribution family can even be considered as a candidate for describing the times between events.

Recall that the negative exponential density function is of the form $f(t) = \lambda e^{-\lambda t}$. Figure 6.13 plots this function for three values of λ. Notice that the function intercepts the vertical axis at λ, that it diminishes monotonically to zero (asymptotically), and that the rate of convergence is proportional to λ. The total area under the curve is, of course, always equal to 1, as it must be for any density function. Varying λ, the only parameter of the distribution, cannot alter the basic form shown in Fig. 6.13. The mean is $1/\lambda$ and the variance is $(1/\lambda)^2$. Thus the mean and variance are not separately adjustable, as one may frequently desire.

In many applications, by contrast, the times between events are most naturally conceived of as having a density function of the general form shown in Fig. 6.14 (perhaps a gamma or Weibull). That is, one tends to think in terms of some nominal value, the mean, plus or minus some relatively minor variation. Or, put another way, the most likely values are considered to be clustered about the mean,

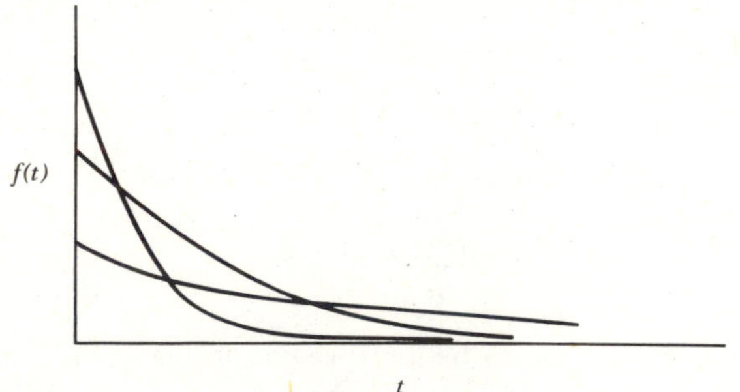

Figure 6.13
The negative exponential density function.

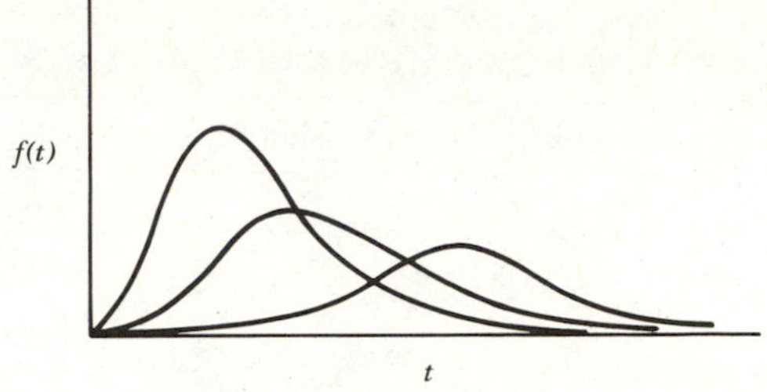

Figure 6.14
Other possible density functions.

and large deviations from the mean are viewed as increasingly unlikely. However, the form of the negative exponential density function implies that the most likely times are close to *zero*, and very *long* times are increasingly unlikely. If this characteristic of the negative exponential distribution seems incompatible with the application you have in mind, perhaps a Markov model is inappropriate.

Another characteristic of the negative exponential distribution that may help you decide whether it is appropriate is the ''forgetfulness'' property. To use a specific example, suppose that you want to represent some process in which the time between a change of state corresponds to the lifetime of an automobile tire. The state changes when the tire fails. Now there are at least two distinct kinds of tire failure. If what you have in mind is the kind of failure that is due to road hazards, that is, ''blowouts,'' then the distribution can legitimately be assumed to be negative exponential. There is no logical connection between road hazards and the age of the tire, so the fact that a blowout has not yet occurred would not carry the connotation that the remaining time to failure is any more or less than when the tire was new. In other words, the lifetime distribution is forgetful, but the only forgetful distribution is the negative exponential. On the other hand, if you are interested in ''wear-out'' failures, the process would not be forgetful. How long the tire has been in use, as well as its age, would affect our estimate of its remaining life. In this instance, a Markov model would be questionable.

6.14 STEADY-STATE PROBABILITIES

In most situations involving more than a few states, the differential equations expressing the $p_{ij}(t)$ cannot be solved without expending impractical efforts. However, their limiting values can be found indirectly. In many practical situations, these will be sufficient.

As in the discrete time case, the states must be ergodic for the procedure to work. All of the state classification terms (transient, recurrent, null, ergodic, irreducible) developed for the discrete time case apply to the continuous time case as well, with one exception: It does not make any sense to speak of periodic states when time is continuous. Quantities such as mean first passage times and absorption probabilities are defined in a manner that is analogous to the discrete time case. Formulas for these are given in Table 6.2. Only the steady-state equations will be developed here.

Table 6.2
CONTINUOUS TIME MARKOV RESULTS

Name	Notation	Type of Model	Formula
Transition probability	$p_{ij}(t)$	Any	$\frac{dp_{ij}(t)}{dt}=\sum_k p_{ik}(t)\lambda_{kj}$
Occupancy probability	$p_j(t)$	Any	$p_j(t)=\sum_i p_i(0)p_{ij}(t)$
Steady-state probability	π_j	Ergodic	$\sum_k \pi_k\lambda_{ki}=0$ $\sum_i \pi_i=1$
Mean first passage time	m_{ij}	Ergodic	$-\lambda_{ii}m_{ij}=1+\sum_{\substack{k\neq j\\ k\neq i}}\lambda_{ik}m_{kj}$
Absorption probability	a_{ij}	Absorbing	$-\lambda_{ii}a_{ij}=\lambda_{ij}+\sum_{\text{Trans } k\neq i}\lambda_{ik}a_{kj}$
Mean time in a transient state	e_{ij}	Absorbing	$-\lambda_{ii}e_{ij}=\begin{cases}\sum_{\text{Trans } k\neq i}\lambda_{ik}e_{kj}, & \text{if } i\neq j\\ 1+\sum_{\text{Trans } k\neq i}\lambda_{ik}e_{ki}, & \text{if } i=j\end{cases}$
Mean duration	d_i	Absorbing	$d_i=\sum_{\text{Trans } j}e_{ij}$
Conditional mean first passage time	m_{ij}	Absorbing	$-\lambda_{ii}a_{ij}m_{ij}=a_{ij}+\sum_{\substack{\text{Trans } k\\ k\neq i}}\lambda_{ik}a_{kj}m_{kj}$

Assuming, then, that the process at hand is irreducible and all states are ergodic, we may derive a set of linear equations determining the steady-state probabilities. These are defined, as you should expect, by

$$\Pi_j=\lim_{t\to\infty}p_{ij}(t).$$

The argument goes as follows:

$$\frac{dp_{ij}(t)}{dt}=\sum_k p_{ik}(t)\lambda_{kj}$$

$$\lim_{t\to\infty}\frac{dp_{ij}(t)}{dt}=\lim_{t\to\infty}\sum_k p_{ik}(t)\lambda_{kj}$$

$$\frac{d}{dt}\lim_{t\to\infty}p_{ij}(t)=\sum_k\lim_{t\to\infty}p_{ik}(t)\lambda_{kj}$$

$$\frac{d}{dt}\Pi_j=\sum_k\Pi_k\lambda_{kj}$$

$$0=\sum_k\Pi_k\lambda_{kj}$$

In matrix form, the final result is

$$\mathbf{0} = \mathbf{\Pi\Lambda},$$

where $\mathbf{0}$ is a row vector of zeros, $\mathbf{\Pi}$ is the row vector of steady-state probabilities, and $\mathbf{\Lambda}$ is the matrix of transition rates.

Actually the argument above contains a flaw. It is not generally true that the limit of a derivative is equal to the derivative of the limit. However, these $p_{ij}(t)$ functions are suitably well behaved to ensure that the equality will hold in this case.

There will be an equal number of equations and unknowns in the system of linear equations $\mathbf{0} = \mathbf{\Pi\Lambda}$, but, as in the discrete case, the equations will be dependent. Still assuming that the process is irreducible and the states ergodic, there will be exactly one dependency. The additional equation necessary to determine the unique solution is again as it was in the discrete time case,

$$\sum \Pi_i = 1,$$

and this is still called the normalizing equation.

To illustrate the solution procedure, consider once again the loom example. In that case, the matrix of transition rates was

$$\Lambda = \begin{bmatrix} -3 & 3 \\ 2 & -2 \end{bmatrix}$$

so the steady-state equations would be

$$(0, 0) = (\Pi_0, \Pi_1)\begin{bmatrix} -3 & 3 \\ 2 & -2 \end{bmatrix}$$

or

$$0 = -3\Pi_0 + 2\Pi_1$$
$$0 = 3\Pi_0 - 2\Pi_1$$

It is immediately apparent that these two equations are dependent. Using either one, we can conclude that

$$\Pi_1 = \frac{3}{2}\Pi_0$$

Now by the normalizing equation,

$$\Pi_0 + \Pi_1 = 1$$
$$\Pi_0 + \frac{3}{2}\Pi_0 = 1$$
$$\Pi_0\left(1 + \frac{3}{2}\right) = 1$$
$$\Pi_0 = \frac{2}{5}$$

And, finally,

$$\Pi_1 = \frac{3}{5}, \qquad \Pi_0 = \frac{2}{5}.$$

6.15 BIRTH-DEATH PROCESSES

Nearly all simple queueing models (as well as a number of nonqueueing models) are special cases of the birth-death process, which is itself a special case of the general continuous-time Markov process. A birth-death process is characterized as a Markov process in which all transitions are to the next state immediately above (a "birth") or immediately below (a "death") in the natural integer ordering of states. That is, a birth-death process does not "jump" states.

The transition diagram of a birth-death process would look like Fig. 6.15. There may be either a finite or an infinite number of states, and the transition rates associated with the lines shown are arbitrary.

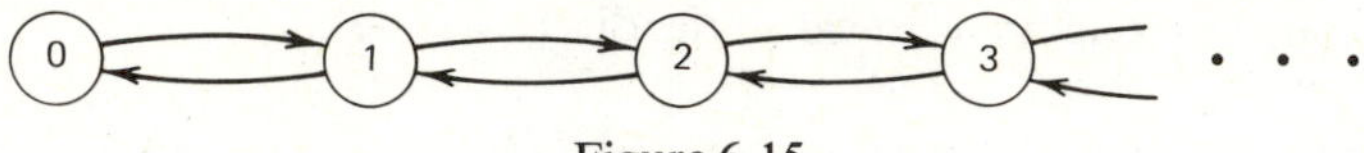

Figure 6.15
The transition diagram of a birth-death process.

The transition rate matrix is also of recognizable form. Because "jumps" are forbidden, $\lambda_{ij}=0$ except when $|i-j| \leqq 1$. This implies that Λ is tridiagonal:

$$\Lambda = \begin{bmatrix} \lambda_{00} & \lambda_{01} & & & & \\ \lambda_{10} & \lambda_{11} & \lambda_{12} & & & 0 \\ & \lambda_{21} & \lambda_{22} & \lambda_{23} & & \\ & & \lambda_{32} & \lambda_{33} & \lambda_{34} & \\ 0 & & \cdot & \cdot & \cdot & \\ & & \cdot & \cdot & \cdot & \\ & & \cdot & \cdot & \cdot & \end{bmatrix}$$

Furthermore, since the diagonal terms must equal "minus-the-sum-of-the-other-elements-in-the-same-row", we must have

$$\lambda_{ii} = -(\lambda_{ii-1} + \lambda_{ii+1})$$

That is, the matrix must be of the form:

$$\Lambda = \begin{bmatrix} -\lambda_{01} & \lambda_{01} & & \\ \lambda_{10} & -(\lambda_{10}+\lambda_{12}) & \lambda_{12} & \\ & \lambda_{21} & -(\lambda_{21}+\lambda_{23}) & \lambda_{23} \\ & \cdot & \cdot & \cdot \\ & \cdot & \cdot & \cdot \\ & \cdot & \cdot & \cdot \end{bmatrix}$$

Again, there may be either a finite or an infinite number of states, and the specific values for the transition rates are arbitrary.

It is fairly common to simplify the notation, using λ_i for the birth rate λ_{ii+1} and μ_i for the death rate λ_{ii-1}. No generality is lost, but the special notation does tend to conceal the fact that the birth-death process is just a special case of the general Markov process.

The steady-state equations for the birth-death process have the characteristic form:

$$
\begin{aligned}
0 &= -\lambda_{01}\Pi_0 + \lambda_{10}\Pi_1 \\
0 &= \lambda_{01}\Pi_0 - (\lambda_{10} + \lambda_{12})\Pi_1 + \lambda_{21}\Pi_2 \\
0 &= \lambda_{12}\Pi_1 - (\lambda_{21} + \lambda_{23})\Pi_2 + \lambda_{32}\Pi_3 \\
&\quad \vdots
\end{aligned}
$$

6.16 THE POISSON PROCESS

The Poisson process is a special case of the birth-death process. It might be more descriptive to call it a pure birth process, since the death rates, μ_i, are all zero. The birth rates, the λ_i, are all equal to a constant value λ. Thus the behavior of the Poisson process is governed by the single parameter λ, and the structure is such that the state index can only increase (see Fig. 6.16).

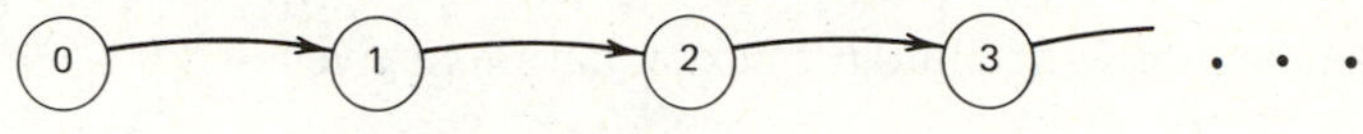

Figure 6.16
The transition diagram of a Poisson process.

The Poisson process is often used to model the kind of situation in which a count is made on the number of events occurring in a given time. For example, it is used in the next chapter, "Queueing Models," to represent the arrivals of customers to a service facility. The state at time t would correspond to the number of arrivals by time t.

The matrix of transition rates $\mathbf{\Lambda}$ would be

$$
\mathbf{\Lambda} = \begin{bmatrix}
-\lambda & \lambda & & & 0 \\
& -\lambda & \lambda & & \\
& & -\lambda & \lambda & \\
& & \cdot & \cdot & \\
0 & & \cdot & \cdot & \\
& & \cdot & \cdot &
\end{bmatrix}
$$

Since the applications usually involve counting events, and the count would ordinarily begin at zero, we will assume that the initial state is zero. The differential equations determining the $p_{0j}(t)$ would be

$$
\begin{aligned}
\frac{dp_{00}(t)}{dt} &= -\lambda p_{00}(t) \\
\frac{dp_{01}(t)}{dt} &= \lambda p_{00}(t) - \lambda p_{01}(t) \\
\frac{dp_{02}(t)}{dt} &= \lambda p_{01}(t) - \lambda p_{02}(t) \\
&\ \ \vdots
\end{aligned}
$$

In general,

$$\frac{dp_{0j}(t)}{dt} = \lambda p_{0j-1}(t) - \lambda p_{0j}(t) \qquad \text{for } j > 0$$

but note that the first equation, for $j = 0$, is a special case.

The first equation involves only $p_{00}(t)$ and is of a particularly simple form. It can be integrated directly to yield the solution

$$p_{00}(t) = e^{-\lambda t}$$

(The constant of integration is evaluated from $p_{00}(0) = 1$.) This solution can be substituted into the next equation, giving

$$\frac{dp_{01}(t)}{dt} = \lambda e^{-\lambda t} - \lambda p_{01}(t)$$

This now involves only $p_{01}(t)$. Tables of standard differential equations or integration by parts will yield the solution

$$p_{01}(t) = \lambda t e^{-\lambda t}$$

This can then be substituted into the next equation to give

$$\frac{dp_{02}(t)}{dt} = \lambda^2 t e^{-\lambda t} - \lambda p_{02}(t)$$

which can be solved to give

$$p_{02}(t) = \frac{(\lambda t)^2}{2} e^{-\lambda t}$$

Continuing in this manner, we find that the general solution is

$$p_{0j}(t) = \frac{(\lambda t)^j}{j!} e^{-\lambda t} \qquad j \geqq 0$$

(To verify that it is, indeed, the general solution, it can be substituted into the general differential equation.)

The form of the expression is that of a Poisson distribution, which accounts for the name given to the model. One way to interpret the result is to fix t, which has the effect of making λt a fixed parameter. Then the set $\{p_{00}(t), p_{01}(t), p_{02}(t), \ldots\}$ would give the probability distribution of the state at the fixed time t, which has been identified as Poisson. In terms of a count of events, we would say that the number of events occurring in a fixed time interval t is Poisson distributed with parameter λt. Furthermore, since the mean of a Poisson distribution is equal to the parameter, λt can be interpreted as the expected number of events occurring in time t.

There are other ways to characterize the same process. Suppose that T_n represents the random variable corresponding to the time between the $(n-1)$th and nth event. In particular T_1 would be the time until the first event. Then

$$\begin{aligned} p_{00}(t) &= P\{\text{no events in time } t\} \\ &= P\{T_1 > t\} \end{aligned}$$

So

$$\begin{aligned} P\{T_1 \leqq t\} &= 1 - p_{00}(t) \\ &= 1 - e^{-\lambda t} \end{aligned}$$

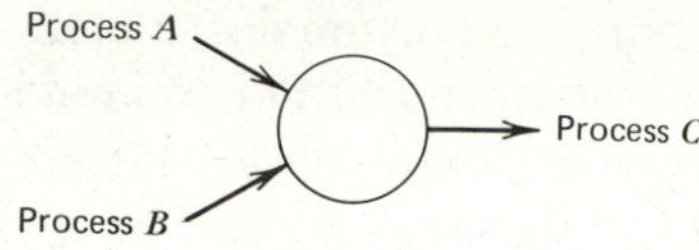

Figure 6.17
Merging two processes.

This distribution can be recognized as the negative exponential distribution with parameter λ. By a similar argument, it can be shown that

$$P\{T_n \leqq t\} = 1 - p_{n-1,n-1}(t)$$

and, by solution of the appropriate differential equation, it can further be shown that

$$p_{n-1,n-1}(t) = e^{-\lambda t}$$

Consequently,

$$P\{T_n \leqq t\} = 1 - e^{-\lambda t} \qquad \text{for all } n$$

In words, the times between events in a Poisson process are all negative-exponentially distributed with the same parameter λ. Since the mean of a negative-exponential random variable is the reciprocal of the parameter, λ can be interpreted as the reciprocal of the expected time between events.

The Poisson process has a number of mathematically convenient properties that contribute to its usefulness in modeling. First, imagine that we have two independent processes operating simultaneously, and consider the effect of "merging" these processes as shown in Fig. 6.17. More precisely, what we mean by merging the processes is technically called "superposition." The superposition of two processes, A and B, produces a third process, C, in which an event occurs whenever an event occurs in either A or B. Figure 6.18 shows how events in both A and B are "copied" in C.

The convenient fact is that the superposition of two independent Poisson processes produces a Poisson process. This fact can be quite tricky to prove if you think in terms of the times between events, but becomes almost trivial if you think in terms of counting the number of events occurring in time t. Let $N_A(t)$, $N_B(t)$, and $N_C(t)$ be the random variables counting the number of events occurring in processes A, B, and C, respectively, over the same time interval t. Then since process C counts events occurring in either A or B,

$$N_C(t) = N_A(t) + N_B(t)$$

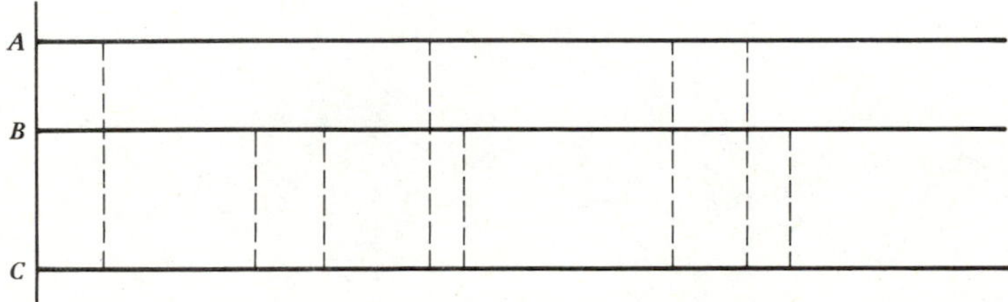

Figure 6.18
The superposition of two processes.

Of course, both $N_A(t)$ and $N_B(t)$ are Poisson distributed random variables. But we already know that the sum of independent Poisson random variables is a Poisson random variable. Since this is true for all t, C is a Poisson process.

Furthermore we know that the expectation of a sum is the sum of the expectations. Consequently, if λ_A and λ_B are the parameters or rates of occurrence of the A and B processes,

$$E[N_A(t)] = \lambda_A t$$
$$E[N_B(t)] = \lambda_B t$$

and, therefore,

$$\begin{aligned} E[N_C(t)] &= E[N_A(t) + N_B(t)] \\ &= E[N_A(t)] + E[N_B(t)] \\ &= \lambda_A t + \lambda_B t \\ &= (\lambda_A + \lambda_B)t \end{aligned}$$

In other words, the rate of occurrence in process C is equal to the sum of the rates of occurrence in processes A and B. One could not ask for a nicer result. It should also be apparent from the argument used that the result extends in the obvious way to the superposition of more than two independent Poisson processes.

This "mergibility" property of the Poisson process is useful in a variety of contexts. The processes A and B might represent, for example, the arrivals of orders for some good to each of two different stores that are supplied from the same inventory. The superposition process would represent the total demands from inventory. Or, in a queueing problem, the component processes may represent arrivals from two different sources, or perhaps of two different types (e.g., male and female). The superposition would represent all of the arrivals to the system.

Since Poisson processes merge so nicely, it is logical to ask whether they may be "separated," as in Fig. 6.19. Although we will make no attempt to prove it, the answer to this question is yes, provided that the separation is done in a particular way. Whenever an event occurs in the parent, or input, process, it will be assigned to one or the other of the output processes. The rule that determines whether a particular event is assigned to process B or to process C must be probabilistic and independent of everything else. In other words, it is as if the assignment were determined by the toss of a (possibly biased) coin. Other conceivable rules, such as alternating the assignments, will not produce Poisson processes as outputs. If λ_A is the rate for the input stream, and p_B denotes the probability that an event is assigned to stream B, then the rate for process B is $p_B\lambda_A$, just as you would probably suspect. It is also true, as you might imagine, that the result can be extended to the separation of a Poisson process into more than two processes.

There are several useful ways to interpret the "separability" property of Poisson processes. If the input process represents arrivals of customers to a bank, then the output processes might be the arrivals to individual tellers. Of course, to pre-

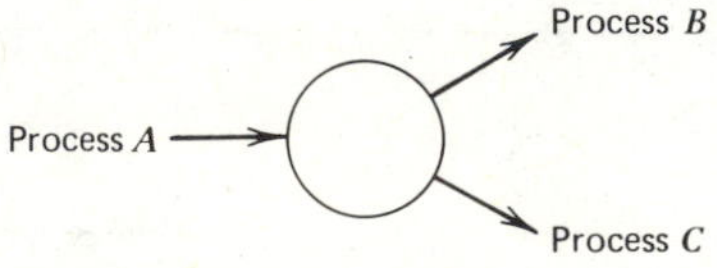

Figure 6.19
Separating two processes.

serve the Poisson process, the customers must be separated on a completely random basis. In particular we could not allow the customer to choose their tellers on the basis of how long the lines are. Another possible interpretation of the separating device is that of a filter. Perhaps the input process represents parts arriving at an inspection station which separates them into acceptable parts and rejects. Provided that the probability of acceptance is independent of everything, both the stream of accepted parts and the stream of rejects would retain the Poisson character.

Perhaps the most remarkable property of the Poisson process is indicated by a rather deep limit theorem formulated by Khintchine (11). Consider a number of independent processes, each of which generates events over time. Within each process, the times between successive events are assumed to be independent, identically distributed random variables. In Fig. 6.20, the inter-event times for the first process are indicated by T_1, T_2, and so on. It is these random variables that are assumed to be identically distributed. We do not assume that the inter-event times for different processes have the same distribution. Of course, if the inter-event times are negative-exponentially distributed for any process, then that process would be a Poisson process, but we specifically refrain from making any such assumption. The forms of the inter-event time distributions are arbitrary. The technical terminology for a process of the kind described—that is, one with independent, identically distributed inter-event times—is a *renewal process*.

Khintchine's theorem states that, provided certain reasonable conditions are satisfied, the superposition of a large number of independent renewal processes is approximately a Poisson process. Because we already know that the superposition of Poisson processes is a Poisson process, the real point of the theorem is that the component processes need not be Poisson. Regardless of the distributions of the times between events within processes, if there are enough of them, the superposition is approximately Poisson.

This theorem is somewhat analogous to the Central Limit Theorem. It will be recalled that the latter ensures that the sum of a sufficiently large number of independent random variables will be approximately normally distributed regardless of what the distributions of the constituent random variables may be. Khintchine's theorem states that the superposition (a kind of sum) of a sufficiently large number of independent renewal processes will produce a Poisson process re-

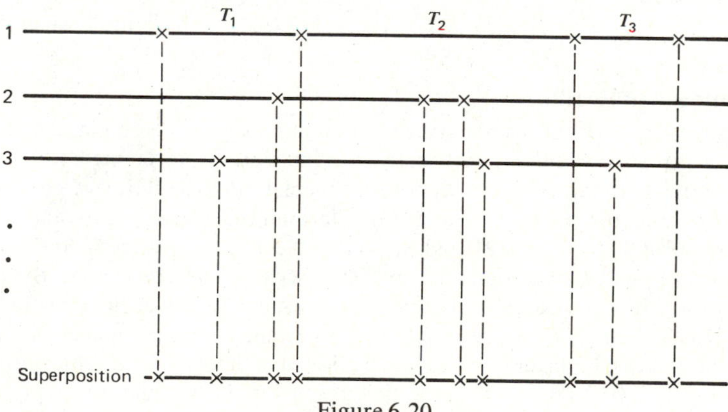

Figure 6.20
Khintchine's theorem.

gardless of what the distributions of the constituent renewal processes may be. Like the Central Limit Theorem, Khintchine's theorem "explains" why such a remarkably simple process with such convenient properties should happen to occur so often in real life. It also serves to justify the Poisson assumption in cases where the process of interest can be thought of as consisting of the superposition of a large number of independent renewal processes.

To illustrate this last point, consider the arrivals of customers to a supermarket. If you think in terms of an individual customer, the process describing his arrivals is almost certainly not Poisson. That is, it seems very unlikely that a negative-exponential distribution would adequately serve to describe the times between one customer's visits to the supermarket. On the other hand, it does not seem unreasonable to assume that successive intervals between visits are independent and identically distributed, and that the arrival patterns for different customers are independent. Since the arrivals to the store are given by the superposition of the arrival processes of all of the customers, Khintchine's theorem provides reason to expect that the arrivals do indeed form a Poisson process.

6.17 CONCLUSIONS

We have seen that the behavior over time of a stationary Markov process is completely characterized by, in the discrete case, the one step transition matrix, $\boldsymbol{P}$, and in the continuous case, the matrix of instantaneous transition rates, Λ. Given the elements of the appropriate matrix (plus, in some cases, information about the initial conditions), it is possible to calculate virtually everything of interest about the process. For the most part, the mathematics required is no worse than linear algebra. Even in the continuous case, in which differential equations may appear, they are first-order, linear differential equations with constant coefficients. In short, there are no theoretical barriers to solving practically any kind of problem that can be phrased in terms of stationary Markov processes.

Although there is some danger of overextending the techniques to model situations where the stationarity, Markov assumptions, or both, are not warranted, the methods are sufficiently robust to permit an enormous variety of real world processes to be represented and studied. The next chapter develops just one category of application.

Recommended Readings

For many years, most of the literature of random processes was so mathematically advanced that a beginner would have difficulty in gaining access to the methods. This problem, however, has been remedied in recent years; there is now a selection of good introductory textbooks available. Bhat (3), Cinlar (4), Clark and Disney (5), Cox and Miller (6), Heyman and Sobel (8), Parzen (12), and Ross (13) all offer the basic fundamentals of random processes as used in operations research. The styles of treatment, and the depths to which they probe, vary considerably; but one could begin with any of these books. Howard (9) and Kemeny and Snell (10) are also suitable for beginners; however, their treatments are so unique that there is a danger of obtaining a misleading image of the subject as a whole. Used in conjunction with one of the more standard textbooks, they have a great deal to offer. Although not introductory in nature, Bailey (1) and Bartholomew (2) are useful in suggesting applications.

REFERENCES

1. Bailey, N. T. J., *The Elements of Stochastic Processes with Applications to the Natural Sciences*, Wiley, New York, 1964.
2. Bartholomew, D. J., *Stochastic Models for Social Processes*, Wiley, New York, 1967.
3. Bhat, U. N., *Elements of Applied Stochastic Processes*, Wiley, New York, 1972.
4. Cinlar, E. *Introduction to Stochastic Processes*, Prentice-Hall, Englewood Cliffs, N.J., 1975.
5. Clarke, A. B., and R. Disney, *Probability and Random Processes for Engineers and Scientists,* 2nd ed., Wiley, New York, 1985.
6. Cox, D. R., and H. D. Miller, *Theory of Stochastic Processes*, Wiley, New York, 1965.
7. Feller, W., *An Introduction to Probability Theory and Its Applications*, Vol. I, 2nd ed., Wiley, New York, 1957.
8. Heyman, D. P., and M. J. Sobel, *Stochastic Models in Operations Research*, Vol. I, McGraw-Hill, New York, 1982.
9. Howard, R., *Dynamic Probablistic Systems,* Vols. I and II, Wiley, New York, 1971.
10. Kemeny, J. G., and L. Snell, *Finite Markov Chains*, Van Nostrand, New York, 1960.
11. Khintchine, A. Y., *Mathematical Methods in the Theory of Queueing*, Hafner, New York, 1960.
12. Parzen, E., *Stochastic Processes,* Holden-Day, New York, 1962.
13. Ross, S., *Introduction to Probability Models*, Academic, New York, 1972.

EXERCISES

1. What is a stochastic process?
2. When do we call a Markov chain stationary?
3. Explain how to get the n-step transition probabilities of a Markov chain.
4. What is the difference between *state probability* and *transition probability*?
5. What is an *ergodic* Markov chain?
6. What do you call a Markov process in which all states communicate?
7. What is the distribution of the time between transitions in a continuous time Markov process?
8. What do you call a birth-death process with zero death rate and a constant birth rate?
9. Suppose that the one-step transition matrix of a Markov chain is as follows:

$$\mathbf{P} = \begin{bmatrix} 0.5 & 0.2 & 0.3 \\ 0.1 & 0.3 & 0.6 \\ 0.8 & 0.2 & 0 \end{bmatrix}$$

   Find the matrix of two-step and three-step transition probabilities.

10. Using the same matrix as in Exercise 9, find the steady-state probabilities.
11. Again using the same matrix, find the mean first passage time from state 1 to state 2.
12. Suppose that the one-step transition matrix of a Markov chain is as follows:

$$\mathbf{P} = \begin{bmatrix} 0.5 & 0.2 & 0.2 & 0.1 \\ 0.1 & 0.3 & 0 & 0.6 \\ 0 & 0 & 1 & 0 \\ 0 & 0 & 0 & 1 \end{bmatrix}$$

   Find the absorption probability a_{14}.

13. Using the matrix of Exercise 12, find the matrix E.

14. A naturalist is observing the behavior of a frog in a small lily pond. There are four lily pads in the pond, and the frog jumps from one to another. The pads are numbered arbitrarily; the state of the system is the number of the pad that the frog is on. Transitions occur only when he jumps. Although he can jump from any pad to any other, the probability of jumping to any given pad from the one he is on is inversely proportional to the distance (that is, he is more likely to jump to a near pad than a far one). The distances are:

between 1 and 2—6/5 feet
between 1 and 3—2 feet
between 1 and 4—3/2 feet
between 2 and 3—6/7 feet
between 2 and 4—1/2 feet
between 3 and 4—3/4 feet

(a) Set up the transition matrix.
(b) If the frog starts on pad 2, what is the probability that he is on pad 3 two jumps later?
(c) Interpret the steady-state probabilities in terms of the behavior of the frog.
(d) Explain in terms of the frog, his motivation, his behavior, and so forth, what the Markov and stationarity assumptions mean.

15. A self-service elevator in a four-story building operates solely according to the buttons pushed inside the elevator. That is, a person on the outside cannot "call" the elevator to the floor he is on. Consequently, the only way to get on the elevator is for someone else to get off at your floor. Of the passengers entering the building at the ground floor and wishing to use the elevator, half go to the second floor, and the other half divide equally between the third and the fourth. Passengers above the ground floor want to go to the ground floor in 80% of the cases. Otherwise, they are equally likely to want to go to the other floors.
(a) If you walk into the building, what is the probability that you will find the elevator at the ground floor?
(b) If it is not at the ground floor, but at the second floor, how many stops will it make (on the average) before coming to the ground floor?
(c) Assuming that each stop takes 30 seconds (including travel time even if it is between floors of maximum separation) and the elevator is at the second floor, what is the probability of having to wait at the ground floor for more than 2 minutes?

16. The quality of wine obtained from a vineyard varies from year to year depending on a combination of factors, some of which are predictable and some of which are not. From historical data extending over the past hundred years, the year-to-year succession pattern was as follows:

		Graded Year			
		Poor	Fair	Good	Excellent
Previous Year	Poor	2	6	8	4
	Fair	6	12	20	2
	Good	9	20	5	0
	Excellent	3	2	1	0
		20	40	34	6

So, for example, there were only six years in which the wine was graded as excellent, and in four of those cases, the previous year's wine had been poor. Use this data to construct a Markov chain model to answer the following questions, and show how to calculate the answers.

(a) If 1980 was a "good" year, what is the expected number of years until the next "excellent" year?

(b) If 1980 was good and it turns out that 1981 is also good, what is the probability that 1983 will be excellent?

(c) What is the mean number of years between occurrences of "excellent" years?

17. A simple industrial robot has four axes of motion, controlled by four separate motors. We will refer to the motions as "Turn" (right or left), "Lift" (raise or lower), "Extend" (in or out), and "Grip" (contract or release). When the robot is programmed to perform a task, a sequence of these motions will take place. For example, to pick up an object, the arm would turn to align with the object, extend to reach it, grip it, and lift. We are interested in predicting how the actions will occur, to accelerate the controlling software. If, for example, a turn is most likely to be followed by an extension, the code that controls the extension can be readied in advance.

From a sample of robot programs, representing a set of typical tasks, the following count data was derived:

		Next motion:			
		T	L	E	G
Current motion:	T	5	40	90	15
	L	50	0	20	80
	E	30	40	20	60
	G	15	90	30	15

Construct a Markov chain and show how to address the following questions.

(a) If the arm is turning, what is the most likely action to follow the next action (without specifying the intermediate action)?

(b) If the arm is turning, how many action steps will occur on the average before the grip action is taken?

(c) If a grip occurs, what is the probability that another grip occurs within the next four actions?

(d) Over the long run, how does the usage of the lift motor compare to the usage of the extension motor?

18. Three bellboys, Matthew, Mark, and Luke, share the night-shift duty at the Paradise Hotel. On any given night, only two are on duty; the other has the night off. Rather than maintain a fixed schedule, they have adopted the following procedure: the one who is off-duty one night will be on-duty the following night, and the two who are on-duty will flip a fair coin to determine which will get the next night off. The question on everyone's mind is, of course, "When is Mark off?" To answer this question more specifically, do the following.

(a) Construct a mathematical model which describes the "duty roster" (i.e., who is on duty) as it changes over time. State your assumptions and comment on their reasonableness. You may use previously defined notation freely, but be sure to relate your model to the particular situation described (for example, you will have to define your states and your time parameter).

(b) Assuming that today is Thursday and Mark had last night (Wednesday night) off, find the probability that he will have Saturday night off.

(c) Still assuming that Mark had last night off, we can be sure that he will be working tonight, but he may have tomorrow night off. If so, we would say that he worked only one night in succession. It is possible that he will have to work two, three, or even more nights in succession. Explain how you would find the probability distribution which describes the number of nights in succession that Mark will work.

19. Paper currency is initially issued by a bank. It will then be circulated throughout the economy, changing hands perhaps many times before it is ultimately judged by a bank teller to be worn out and is removed from circulation. There are three questions we want to answer:
    (a) How many times, on the average, will a particular dollar bill be deposited in banks during its lifetime?
    (b) How many times, on the average, will it be in the hands of private citizens (as opposed to banks or other institutions such as businesses and corporations)?
    (c) Counting each transfer of the bill as a usage, what is the probability distribution of the number of usages obtained from the bill over its entire lifetime?

    Set up a Markov model that is capable of answering these questions. Indicate its structure by drawing a transition diagram. Define your states and time scale. State your assumptions and comment on them. Indicate what data is needed. Explain how to use your model to answer the questions.

20. A market survey has been conducted to determine the movements of people between types of residences. Two hundred apartment dwellers were asked if their previous residence was an apartment, a townhouse, their own home, or a rented home. Similarly, 200 townhouse dwellers were asked about their previous residences, and so on. The results of the survey are tabulated below.

Present	Previous Residence				
Residence	Apt.	Townhouse	Own House	Rented House	Total
Apartment	100	20	40	40	200
Townhouse	150	40	0	10	200
Own House	50	20	120	10	200
Rented House	100	20	20	60	200
Total	400	100	180	120	800

    The data are believed to be representative of the behavior of the population at large. Develop a model and answer the following:
    (a) What is the probability that someone who is now in an apartment will own his own home after two moves?
    (b) What is the probability that the same person will own his own home after two moves but not sooner?
    (c) How many moves will occur, on the average, before a homeowner will live in a townhouse?
    (d) What, if anything, can you say about the long-term demand for townhouses, relative to the other residence types?
    (e) Consider the validity of your assumptions.

21. The accounting department of a large firm is interested in modeling the dynamics of its accounts receivable (i.e., the money that is owed to it by its customers). When a "charge" sale occurs, a bill is sent out at the end of the month. Payment is due within 30 days, but may not occur in that time. If it is late, a penalty charge of 4% of the amount due is added. If no payment is received within 6 months of the billing date, the amount is classified as a bad debt. Thus, an individual account is described by its age in months, or by "paid," or by "bad debt."

    Realizing that data would have to be collected to determine, for example, the percentage of bills that are paid on time, make whatever assumptions seem reasonable or necessary, and model the process. You may use symbols to represent the probabilities, or you may feel free to insert some invented numerical values. Using the model,

(a) Calculate the probability that a new sale will ultimately be paid for.
(b) Calculate the probability that a bill which is not paid within the 30-day period will never be paid.
(c) Calculate the average time that a bill is outstanding (for those that are paid).
(d) If a bill for $100 is sent out, what is the expected amount of the return?

22. The Department of Agriculture wants a model to predict crop yields in Indiana. Of course, weather and other factors will have to be taken into account, but the main factor affecting how much corn (for example) will be produced is how many acres are planted with corn. Other things being equal, individual farmers try to plant those crops for which they think the market price at harvest time will be best, but other considerations are also important. They have to take into account soil conditions, crop rotation, opportunity for irrigation, and so forth. Because of the variability in the decisions made by individual farmers, the Department of Agriculture feels that a random process model would be appropriate.

    There are two major crops, corn and soybeans. All other marketable crops can be lumped into a single class called "other"; land not used for marketable crops is called "fallow." Information is available on the acreage devoted to each of the four categories during 1975 and what was planted on the same acreage during 1974. For example, of the 8 million acres planted with corn during 1975, 6 million acres were planted with corn in 1974, 1 million were planted with soybeans, one-half million were planted with some other crop, and one-half million were left fallow.

1975 Agricultural Land Usage (in millions of acres)	1974 Usage (in millions of acres)			
	Corn	Soybeans	Other	Fallow
Corn 8	6	1	0.5	0.5
Soybeans 3.2	1	1.2	0.6	0.4
Other 4	1	1	1.3	0.7
Fallow 2	1	0.8	0.2	nil

(a) Develop a model for the Department of Agriculture describing the sequence of uses of a randomly selected acre of farmland. Define states, time, transitions, and show how to use the available data. State your assumptions and comment on their reasonableness.
(b) Explain how the model could be used to:
    1. Calculate the amount of land devoted to each category at some distant future time.
    2. Predict 1977 acreages.
    3. Estimate the average number of years between fallow years.

23. This problem asks you to develop a Markov chain model to represent aging automobiles. Let there be a set of states representing the age of a car in years:

0 = car is new
1 = car is 1 year old
.
.
.
10 = car is 10 years old

(No doubt 10 years is enough to cover the lifetimes of nearly all American-made cars.) Let state 11 represent "car is junked." Because of accidents, even a new car has some chance of being junked during its first year. Ordinary wear and tear, however, has the effect of increasing the likelihood that a car will be junked during the next year as the car gets older.

(a) Show what the transition matrix would look like.
(b) If one were contemplating buying a two-year-old car to keep for three years, how would you calculate the probability that the car would last that long?
(c) How would you calculate the average lifetime of a car, if you had the numerical data to fill in the transition matrix?
(d) Critique the model.

24. The Auditor's Manual for the Internal Revenue Service dictates the following review procedures for tax returns from private individuals.
   (a) All submitted returns are subjected to a computer check for arithmetic errors and "unusual" deviations from typical returns. Seventy percent pass the test; 10% must be returned to the taxpayer for correction of errors; the remainder are sent to trained auditors for review. In addition, a spot check is made on those returns that passed the test by selecting 5% at random to be audited.
   (b) Of those returns selected for audit, 40% are judged satisfactory without involving the taxpayer. The others require that the taxpayer be contacted to provide additional information or substantiation. After hearing the taxpayer's case, the department may assess additional taxes, which happens in seven out of ten cases, or it may accept the return as submitted.
   (c) If a taxpayer is assessed additional taxes and wants to appeal, he may do so by applying for review by the IRS District Appeal Board. The chance that the Appeal Board will overturn the auditor's decision is only 1 in 50. Nevertheless a third of such taxpayers do appeal.
   (d) As a last resort, a taxpayer may go to court to obtain a favorable ruling, but his chances are only 1 in 100 of emerging without any additional tax assessment. Of those who could go to court, half do.
      1. Of a million submitted returns, how many (on the average) go to court?
      2. If a return is audited, what is the probability that the taxpayer will have to pay additional taxes?
      3. Including the computer check, what is the expected number of times that a return will be reviewed?
      4. What is the expected number of times that an *audited* return will be reviewed?

25. Consider the game of "craps" as played in Las Vegas casinos. The basic rules of the game are as follows: one player (the "shooter") rolls a pair of dice. If the outcome of that roll is a 2, 3, or 12, he immediately loses; if it is a 7 or 11, he wins. In all other cases, the number he rolls on the first toss becomes his "point," which he must try to duplicate on subsequent rolls. If he manages to roll his point before he rolls a 7, he wins; otherwise he loses. It may take several rolls to determine whether he wins or loses. After the first roll, only a 7 or his point have any significance until the win or loss is decided. Of course, there are many ways to place bets on various possibilities, but we are concerned here with just the basic rules of the game, played only once.
   1. Model the process as a Markov chain with absorbing states for "win" and "lose." Show either a transition diagram or the one-step transition matrix. Assume the dice are fair in evaluating the probabilities.
   2. Before the first roll of the dice, what is the probability the shooter will ultimately win?
   3. If on the first roll he gets a 6, what is the probability he will ultimately win?
   4. What is the expected number of rolls required to decide the win or loss?

26. If the following matrix is supposed to be the instantaneous transition rate matrix of some continuous time Markov process, determine the values of the missing entries (shown as ?):

$$\Lambda = \begin{bmatrix} ? & 1 & 0 & 0 \\ 0.8 & ? & 2.6 & 0.1 \\ 1.5 & 2 & ? & 2 \\ 1 & 0 & ? & -4 \end{bmatrix}$$

27. Suppose that the following matrix is the instantaneous transition rate matrix of a continuous time Markov process:

$$\Lambda = \begin{bmatrix} -2.5 & 1 & 1.5 \\ 4 & -4 & 0 \\ 0 & 2 & -2 \end{bmatrix}$$

Determine the steady-state probabilities.

28. Find the steady-state distribution for the birth–death process having the following transition rate matrix:

$$\Lambda = \begin{bmatrix} -1 & 1 & 0 & 0 \\ 2 & -4 & 2 & 0 \\ 0 & 2 & -4 & 2 \\ 0 & 0 & 2 & -2 \end{bmatrix}$$

29. A very simplified description of the activity of a single nerve cell (neuron) would be as follows: it "collects" input signals until a certain threshold is reached, at which time it "fires," producing an output signal that is passed along to other cells. After firing, there is a brief "recovery" period, during which it does not recognize input signals. To simplify even further, assume that the input signals occur according to a Poisson process with rate λ, that exactly N input signals must be received and recognized before the neuron fires, and that the recovery time is negative exponentially distributed with mean $1/\mu$.

(a) Develop a continuous time Markov process model, showing either a transition diagram or an instantaneous transition matrix. Be sure to define your states.

(b) Show how to compute the fraction of input signals that are recognized over the long run.

(c) Show how to compute the (steady-state mean) "firing rate" of the neuron, that is, the number of times it will fire per unit time.

30. During the course of a day, the vehicle traffic at a certain location varies randomly, changing among "light," "medium," and "heavy," where these terms have definite, measurable meanings. Imagine a three-state continuous time Markov process where changes in traffic density are represented by transitions. Assume that the only way to get from "light" to "heavy" or vice versa is by way of the "medium" state. Of course, to be Markovian, the transition times must be negative exponentially distributed.

(a) Explain how to estimate the needed parameter values for the model, referring specifically to how the traffic data would be collected and manipulated.

Assuming the values are $\lambda_{LM}=4$, $\lambda_{ML}=3$, $\lambda_{MH}=2$, $\lambda_{HM}=1$ in units of transitions per hour show how to answer these questions:

(b) If the traffic is light now, what is the expected length of time it will stay light?

(c) If the traffic is medium now, what is the expected length of time it will stay medium?

(d) If the traffic has been medium for the last 10 minutes, what is the expected length of time until it changes?

(e) If the traffic has been medium for the last 10 minutes, what is the expected length of time until it becomes light?

(d) What percentage of the time is the traffic heavy?

31. To show that the parameter identified as "time" need not be taken literally, consider the following model. Suppose that government expenditures for cancer research produce results which are only probabilistically related to the amount of expenditure. To be more specific, suppose that we have defined an index for the state of medical technology in the area of cancer. Say that the present value of this index is 10, and higher values are better. To achieve an increase of one point in this index will require an amount of research whose *expected* cost is \$5 million, but because of the uncertain nature of research, the actual cost is a random variable. Suppose that this random variable is negative exponentially distributed.
    (a) Show how to obtain the expected cost and the distribution of cost necessary to achieve an increase of three points in the index.
    (b) Show how to obtain the expected value and distribution of the index that would be achieved from an expenditure of \$18 million.

32. A truck trailer has four wheels on one axle, two on each side. Each tire is independently subject to failure (blowouts) at the rate of one failure per 20,000 miles, on the average. Assume that the failure process is Markovian, and let $R(m)$ denote the probability that none of the tires has blown after m miles.
    (a) Explain how to obtain $R(m)$.
    (b) What is the expected number of miles until the first blowout? At least one tire on each side must *not* have failed in order to keep going. That is, one blowout on either or both sides is acceptable, but two on the same side forces a stop. Assume that flat tires are not repaired, and let $Q(m)$ denote the probability that the trailer is still operable after m miles.
    (c) Explain how to obtain $Q(m)$.
    (d) What is the expected number of miles until the first forced stop?
    (e) How reasonable is the Markov assumption?

CHAPTER 7
QUEUEING MODELS

7.1 INTRODUCTION

In Britain ''queue'' is a common word that means either a waiting line or the act of joining a line. In the United States, queue is rarely used in ordinary conversation. Consequently, the subject of queueing theory may seem rather esoteric to some readers. Actually it deals with phenomena that we all encounter as part of our everyday lives. In fact, one is almost certain to encounter some form of queue, or waiting line, in every waking hour. Aside from obvious examples, such as standing in line at a ticket window or grocery checkout counter, there are many more subtle cases of queueing behavior that we experience. For example, whenever we make a telephone call, there is a brief delay as a free communication channel is located. As you drive a car, the traffic congestion you encounter at intersections or on the open highway is a form of queueing behavior. The flow of material through a manufacturing operation or of paper orders through a sequence of processing stages represent other forms of queues.

In this chapter, you will learn how to construct and solve equations that describe queueing behavior for a wide variety of situations. We cannot hope to exhaust even the elementary models, because the number of possible variations is enormous, as you will soon discover. But we will accomplish enough to permit you to apply a significant number of standard formulas, and to derive your own in many other cases. After establishing some basic results, we will examine a case study with several illuminating aspects. You will be impressed with the ability of the theory to help in a situation where common sense fails. If you want to peek ahead to Section 7.10, just for motivation, feel free to do so.

7.2 AN EXAMPLE

Let us begin with a straightforward example that possesses a full range of possibilities for modeling, but offers no difficulties in interpretation. Assume that we are studying a small retail bakery shop that sells such things as bread, rolls, cookies, and cakes to individual customers. The customers arrive at random moments throughout the day and are given service by a single clerk. The service consists of taking the order, selecting and bagging the goods, accepting payment, making change, and all of the other normal things you would expect. What is important from the current modeling perspective is not the actual activities of providing service, but rather the length of time consumed by the transaction. Just so there is no ambiguity, we will say that the service begins at the moment the customer arrives if the clerk is not already occupied at that time, or as soon as the previous customer completes service if there is any delay. A service is completed when the clerk is free to attend to the next customer.

If a customer happens to arrive at a moment when the clerk is already occupied with serving a prior arrival, then the new arrival must wait. Of course, it is possible that several customers may back up just by chance. That is, a series of random arrivals may occur very close together, or a particular service time might be unusually long, or both. To keep things simple, we will assume that customers are served in the order of their arrival. To ensure this, the bakery might employ a "take-a-number" system. Even if the shop is occasionally crowded with customers awaiting service, we will assume that there is unlimited waiting room; no arriving customer is denied access. We will later relax these assumptions, but they are both convenient and reasonable for a start.

The whole idea of developing a mathematical model to represent this situation is to permit the calculation of quantities such as the average delay time, the fraction of time the clerk is busy, or the average length of the line. These performance measures are the result of rather complex interactions among the elements of the system, and are therefore hard to estimate without the aid of a model. Of course, to be able to use any of the formulas that are developed, it will be necessary either to collect statistical data from the system or to assume values for certain input variables. It is worth mentioning in advance (since many students lose this perspective as they learn about the modeling possibilities), that there is no need to model what we can measure directly. So, in the case of an existing bakery shop, if one wanted to know the average customer delay time, the easiest and most reliable method would be to go to the shop and time the delays. The real value of a model would become apparent, however, if one wanted to evaluate the effect of a proposed change in the system. It is generally easier, cheaper, faster, and safer to use a mathematical model to predict the consequences then to make the changes and observe the results. Or, if the shop did not yet exist and it was our task to design it, we could use a model to avoid trial and error.

There is a commonly used pictorial representation of a queueing system, which for the bakery shop would take the form of Fig. 7.1.

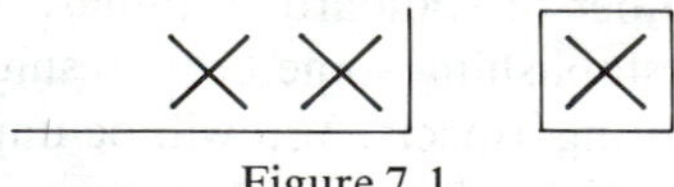

Figure 7.1
A single-server queueing system.

Here, the box represents the server, the X's are the customers, and the angled trough is the queue.

If the baker shop had, say, three clerks all serving the same line, then the picture of the system would look like Fig. 7.2.

The take-a-number system that is common in bakery shops and certain other retail businesses maintains, in effect, a single waiting line. In many other circumstances, such as grocery stores, toll booths, and fast food restaurants, each server has a separate line. Although it may seem at first like a subtle distinction, hardly worth mentioning, such differences can have a profound effect on system behavior. Perhaps you have noticed the trend in many banks and at airline ticket

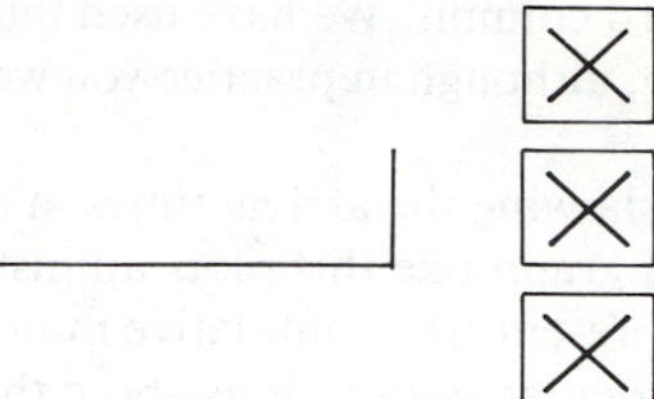

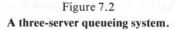

Figure 7.2
A three-server queueing system.

counters to replace separate waiting lines by a single one. At any rate, the multiple queue system would be pictured as Fig. 7.3.

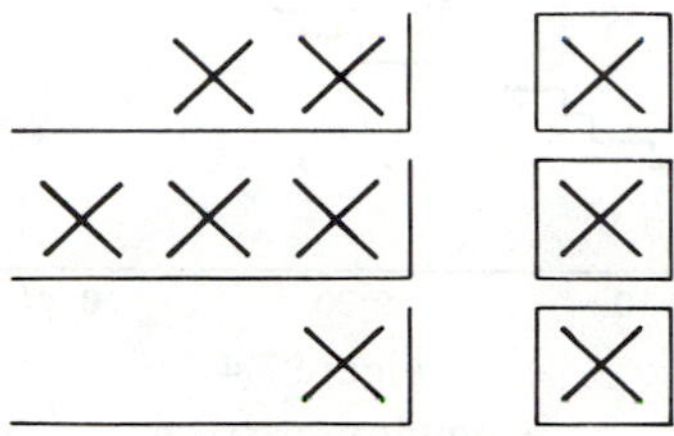

Figure 7.3
A three-queue, three-server system.

Returning to the original bakery shop example, imagine that we observed and recorded the times at which the successive customers arrive. These times might be recorded in a tabular format, as shown in Table 7.1 (the first and second columns). Imagine also that we recorded the service time intervals for each cus-

Table 7.1
BAKERY SHOP EXAMPLE

Customer	Arrival Time	Service Time Interval (in minutes)
1	9:03	4
2	9:05	1
3	9:07	2
4	9:09	3
5	9:10	1
6	9:12	2
7	9:13	1
8	9:15	2
9	9:20	2
10	9:22	2
11	9:23	3
12	9:26	1
13	9:27	1
14	9:30	2
15	9:33	4
16	9:36	4
17	9:40	2

tomer, as shown in the third column. We have used integer (minute) values in this table, to simplify graphing, although in practice you would probably measure decimal values.

We can plot the data showing the arrival times in the format of Fig. 7.4. Formally, we could say that a graph like this plots an instance of the stochastic process $\{A(t)\}$, where $A(t)$ represents the cumulative number of arrivals in time t. We have thus far made no particular assumptions about the nature of this process, except for the one that asserts that arrivals occur one at a time. That is, the steps rise just one unit each time there is an arrival event.

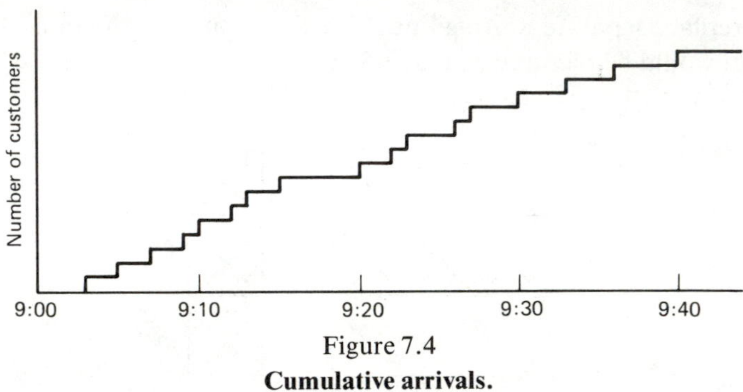

Figure 7.4
Cumulative arrivals.

It is convenient to think of the service time intervals as short blocks of time that can then be added to the graph of Fig. 7.4 to show when each customer's service began and ended. In constructing this new graph, it is important to treat the customers in order and to remember that a particular service interval cannot begin until the prior customer's is finished. The resulting graph would look like Fig. 7.5.

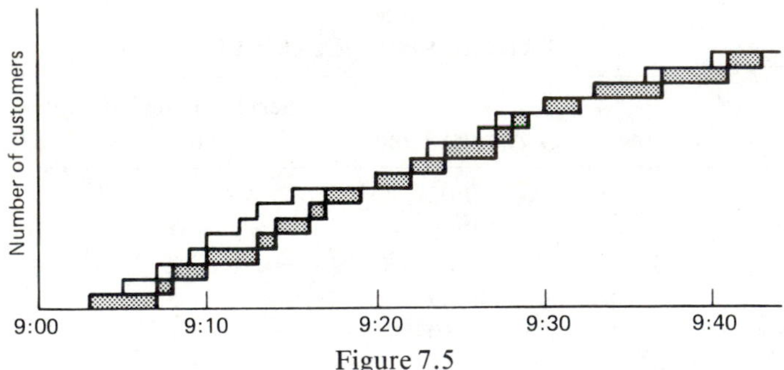

Figure 7.5
Combined arrivals and service times.

Now the picture looks like a series of stacked blocks. If you look horizontally across one level, you can see what an individual customer experienced. You can see when he or she arrived, how long a delay was encountered, when service began, how long it lasted, and when it was completed. If you look vertically at a line cutting through the blocks at a particular point in time, you can see the conditions that existed at that time. You can identify which customer, if any, was in service, and which ones, if any, were waiting. The vertical distance between the top and bottom stair—stepped lines can be interpreted as the number of customers present in the system at any time. Of course, the completions line can never rise above the arrivals line, for that would mean either (depending on how you look at it) that there was a negative number of customers in the system, or that some customer completed service before arriving.

Remember, this charting convention is for recording past performance of an observed system and what we have shown is only one possible pattern out of an infinite number of possibilities. Our real task is to develop tools to predict the performance of systems that may not even exist. Using just a little information about the arrival pattern and service requirements, and taking into account the random nature of both, we want to be able to calculate various performance measures that depend upon the overall working of the system. But before getting too involved in the details, it is worth your time to improve your insight into queueing behavior by exploring some of the possibilities using this graphical tool. To get you started, there are some brief exercises suggested at the end of the chapter.

7.3 GENERAL CHARACTERISTICS

Before delving into the mathematics of queueing models, we should clarify some of the basic concepts, assumptions, and notation. The broadest concepts will be discussed first, with gradually increasing restrictions added as we work toward specific results.

A customer is anything that arrives, occupies a server for some period of time, and departs. It need not be human, or for that matter a physical object. It could be a job, an order, a transaction, a message, or some similar abstraction. One must be careful not to place too narrow an interpretation on the word customer, even in cases where the meaning may seem obvious. For example, a family of four buying tickets to a movie would count as only one customer, because they arrive, are served, and depart as a single unit.

Similarly, a server is an abstract concept that should not automatically be assumed to correspond to anything physical. It could be a person, or a machine, or something you can see, but it need not be. A server is just whatever is occupied by a customer. The number of servers (sometimes called "channel capacity") is the maximum number of customers that can be serviced simultaneously. One server takes care of one customer. An example of where you might be misled would occur when a crew of workers take care of a single customer. In such a case, the entire crew would count as only one server. Occasionally, you may be hard-pressed to decide which of several possibilities determines the number of servers. For example, a loading dock for truck trailers may have room for 10 trailers, but only three crews to do the loading. In such cases, you will usually want to use the smaller number (three, in this instance), but you should always refer to the original definition (the number of customers that can be serviced simultaneously) to make a precise determination.

In some of the more obscure applications of queueing theory, you may even have to decide which of two possibilities are the customers and which are the servers. A classic example here is the so-called "machine-repairman" problem that will be discussed in Section 7.11. You may have a choice that is governed more by your point of view than by the physical characteristics of the problem.

By convention, the classical notation for the number of servers in a queueing system is a lowercase c. In remembering this, it may be helpful to know that the notation originally referred to *channels* in the telephone systems that were the principal application of early queueing theory.

The arrival process describes how the customers appear for service. Generally speaking, it is specified by naming some kind of a stochastic process, along with whatever parameters and simplifying assumptions are appropriate. Almost always, but not necessarily, customers are assumed to arrive one at a time. Usually, the arrival process is taken to be at least a renewal process, which means that the times between arrivals are independent random variables; and furthermore the process is assumed stationary, which means that the parameters do not vary over time. Nonstationary processes would be useful to model, for example, rush-hour traffic or slow periods, but the additional complications in the analysis and parameter estimation are sufficiently troublesome to cause one to avoid them whenever possible.

The simplest arrival process to deal with is a Poisson process, in which the times between arrivals are independent negative exponentially distributed random variables. The same process, viewed another way, gives a Poisson distributed random variable for the number of customers arriving in any time interval (hence the name for the process). Aside from several mathematical properties that make the Poisson arrival process convenient to use in queueing models, the data requirements are absolutely minimal. A single parameter, the arrival rate, is enough to describe the entire process.

Especially when the arrival process is Poisson, but usually even in other cases, the arrival rate is conventionally represented by the Greek lowercase lambda (λ). If one were estimating this parameter from statistical data, it would generally be preferable to deal with the reciprocal of lambda, which would be the mean time between arrivals. That is, you would collect a number of interarrival times and average them to get the mean time between arrivals; the mean arrival rate would be one divided by this value. For example, a 5-minute average interarrival time would correspond to an arrival rate of 1/5 arrivals per minute. Of course, you may use any convenient time unit (seconds, minutes, hours, etc.), as long as you are careful to maintain consistency throughout your model.

This service process is often regarded as being similar to the arrival process, but there are important differences. Strictly speaking, it is not a stochastic process, because it is interrupted by the idle periods when there are no customers present. However, it is usually described by specifying a distribution or distribution family for the successive service times, along with the assumption that these are independent of one another and also independent of the arrival process. Sometimes the parameters of the distribution may reflect a dependence on the number of customers already present. For example, the servers may work faster when the queue grows long. The easiest service time distribution to work with is the negative exponential distribution, which is fully characterized by a single parameter, the mean service rate (or its reciprocal, the mean service time). The usual notation for a service rate is a Greek lowercase mu (μ).

A queueing system consists of one or more servers, an arrival process, and a service process, along with some additional assumptions about how the system works. The word "queue" is sometimes used to describe the whole system, but we will be careful to reserve the use of this word for just that part of the system that holds the excess customers who cannot gain immediate access to a server. Thus, the number of customers in the system at any given time will equal the number of customers in the queue plus the number of customers in service. Of course, these numbers will vary over time as customers come and go, so they are formally stochastic processes.

If there is a limited amount of waiting room, or for any other reason customers are prohibited from joining a queue once it reaches a certain length, we say that the queue or system has finite capacity. Although you may be tempted to regard every real-life system as having finite capacity (simply because nothing in real life is infinite), you should realize that specifying a capacity adds a complication to the analysis. We do so, therefore, only when we have reason to believe that the limited capacity is a significant factor in determining the behavior of the system. A capital N is the usual notation for the capacity of the system; the capacity of the queue would be $N-c$.

The phrase "queue discipline" refers to the rule by which the next customer to be served is selected. The most common queue discipline is "first come-first served," often abbreviated FCFS. In some inventory applications, the same rule is called "first in-first out," or FIFO. In most cases where the customers are human, a policy of fairness dictates that customers should be served in the order of their arrival, which is equivalent to a FCFS discipline. However, there are exceptions to the general rule, and there can even be cases where other queue disciplines make more sense. "Last come-first served," abbreviated LCFS, and "random service selection," abbreviated RSS, are other simple rules. More complicated ones involve various kinds of priorities or classifications.

The population of customers from which the arrivals come is usually assumed to be infinite, so that the arrival process is unaffected by the number of customers already in the system. Sometimes, though, it is useful to be able to limit the number of potential arrivals to some finite value. For example, a maintenance facility may service only 20 trucks in a fleet; if they are all in the shop, there are no additional arrivals possible. This is somewhat different from a limited queue capacity, as we shall see. At any rate, the phrase "finite source population" refers to the case of a limited number of potential customers.

By now, you must have realized that many variations in queueing models are possible. The assumptions can combine in so many ways that it can be a challenge just to keep straight which particular combination is under discussion. Fortunately, this problem was recognized long ago, and a standard notation was developed. It is called Kendall's notation, after the man who proposed an early version of the classification scheme. Briefly, the idea is to specify a particular queueing model by a code consisting of six terms separated by slashes. The first term is an alphabetic code for the type of arrival process assumed. The second term is a similar code for the service process. The most commonly used letter codes are

M for negative exponential interarrival or service times.
D for deterministic interarrival or service times.
E_k for Erlang-k interarrival or service times.
GI for general independent interarrival times.
G for general service times.

The third, fourth, and fifth terms are either numbers or variables for the number of servers, capacity of the system, and size of the source population, respectively. The final term is an alphabetic code for the queue discipline, such as

FCFS or LCFS. The last two or three terms may be omitted if one intends them to assume their usual values of infinite capacity, infinite source population, and first come-first served discipline. The first three terms are always given.

To give a couple of examples, the designation $M/D/2/N$ would indicate a Poisson arrival process, deterministic service times, two servers, and a system capacity of N; and $GI/E_3/c/10/10/\text{LCFS}$ would indicate general independent interarrival times, Erlang-3 service times, c servers, a system capacity of 10, a finite source of size 10, and a last come-first served queue discipline. The simplest models, which will be explored in some depth, are the $M/M/1$, the $M/M/c$, the $M/M/c/c$, and the $M/M/\infty$.

7.4 PERFORMANCE MEASURES

Whatever the assumptions of a particular model may be, there are several commonly desired performance measures. Depending on the circumstances that motivated the construction of the model in the first place, one or several of these may stand out as being the key values and others may be of little concern. In different circumstances, the relative importance attached to the various measures may reverse. For example, in one instance the primary concern may be for the mean delay encountered by customers, and in another it may be for the utilization of the servers.

Although there are times when one would like more detailed information, the results that appear in this chapter are confined to steady-state performance. This term is often misunderstood. It does not mean that the queueing system has somehow settled down to a regular pattern of behavior. If we were observing the system in real time and gathering statistics, we would see perpetual fluctuations in the individual observations. For example, if we were to record the number of customers in the queue every 10 minutes, these values would continue to vary. However, if one were to calculate a running average of those numbers, the average would settle down to a consistent value after a sufficiently large number of observations. That is, further observations would have no significant effect on the statistic. In our analytical work, we are not calculating statistics from observations, but the concept is similar. We are determining values for performance measures that refer to expected behavior of the system over the long term.

Perhaps the most basic information we can have about the steady-state performance of a queueing system is the probability distribution of the number of customers present in the system. From this distribution, almost all of the other common measures can be computed. Later we will demonstrate methods for computing the distribution, but for now let us assume that it has already been determined, in order to show how it can be used. Regardless of what model we are dealing with (number of servers, etc.), let p_i represent the steady-state probability that i customers are present in the system. The index i will run from zero to whatever value corresponds to the capacity of the system, N or ∞.

A direct interpretation of p_i would answer the question "what is the probability of finding i customers in the system at some arbitrary time long after the last observation?" However, an indirect interpretation is often more useful. It can be proved that p_i is equal to the fraction of time, over the long run, that the system has exactly i customers present. For example, p_0 is the fraction of time the system is empty. Or, for another example, in a finite capacity system, p_N would be the fraction of time the system is full.

Recalling that c represents the number of servers, the sum $(p_c + p_{c+1} + p_{c+2} + \cdots)$ can be interpreted as the probability, or fraction of time, that all servers are busy. The same value could also be described as the probability of delay, or as the fraction of arriving customers who are forced to join the queue. Several other useful quantities can be constructed from subsets of the p_i distribution.

Two other distributions can be easily derived from the p_i one. The number of busy, or occupied, servers is a random variable which is closely related to the number of customers in the system. If b_i represents the probability that i servers are busy, then it is easy to see that for all $i<c$, $b_i = p_i$. That is, up until all servers are busy, having i servers busy is equivalent to having i customers in the system. The last possibility, b_c, is given by $b_c = p_c + p_{c+1} + \cdots$, since c or more customers in the system corresponds to having all servers busy.

Similar logic allows us to deduce the queue length distribution. If q_i represents the probability of finding i customers in the queue, then

$$q_0 = p_0 + \cdots + p_c$$
$$q_i = p_{c+i} \quad \text{for } i > 0$$

This derivation is easy to reproduce if you think of what values for the number in the queue correspond to what values for the number in the system.

Several mean values are so commonly computed that they have standard notation. The mean number of customers in the system (that is, the mean of the p_i distribution) is denoted by L. Of course, we can relate L to the p_i's by the basic definition of a mean,

$$L = \sum_i i p_i$$

but this is useful only after we know the p_i's, which will change form as the model assumptions vary. Hence, the formula for L in terms of the model parameters will change, depending upon the particular model in use, but the same letter is always used.

We will use B for the mean number of busy servers, from the b_i distribution, and L_q will represent the mean queue length, from the q_i distribution. U will stand for the average utilization of the servers. Depending on how the system works, the load may not be spread evenly over the servers, so the individual servers utilizations may vary. U is the average of these. The throughput, or output rate, or production rate of the queueing system, formally defined as the mean number of service completions per unit time, will be denoted by R.

The performance measures defined so far all relate to the number of customers somewhere in the system. Usually a customer is more interested in the time it takes to get through the queue or system. We use W to represent the steady-state mean time that a customer spends in the system, and W_q for the mean time in the queue.

This is quite a bit of notation to absorb all at once, but if you refer to Table 7.2, you will notice that there are mnemonic cues for each variable to help you remember. All of the capital letters refer to steady-state mean values; the lowercase letters indicate probabilities.

Table 7.2
SUMMARY OF PERFORMANCE MEASURE NOTATION

p_i	Probability *i* customers *p*resent
b_i	Probability *i* servers *b*usy
q_i	Probability *i* customers in *q*ueue
L	Mean *l*ength of system
B	Mean number of *b*usy servers
L_q	Mean *l*ength of *q*ueue
U	Mean *u*tilization
R	Mean throughput *r*ate
W	Mean *w*aiting time in system
W_q	Mean *w*aiting time in *q*ueue

7.5 RELATIONS AMONG THE PERFORMANCE MEASURES

As you might expect, there are many relationships among the performance measures. Some of these are quite obvious, and others are not so obvious. It is important to know these relationships, so that you can easily derive unknown values from known ones. Usually only one lengthy algebraic computation is enough to produce, for example, L, after which everything else is very easily produced using very minor transformations.

One of the simplest relations, which will hold for any queueing model possessing a steady-state solution, is

$$L = L_q + B$$

In words, the mean number in the system equals the mean number in the queue plus the mean number in service. A similar relation in terms of times is

$$W = W_q + (1/\mu)$$

where $(1/\mu)$ is the mean service time (whether or not the distribution is negative exponential). Both of these relations follow from the general property that the mean of a sum is the sum of the means. Another easy one is

$$U = \frac{B}{c}$$

which says that the average utilization equals the mean number of busy servers divided by the number of servers. If there is only one server, we can write

$$U = B = 1 - p_0$$

The throughput rate is only slightly more complicated. In steady-state, the mean number of service completions, or departures, per unit time must equal the mean number of arrivals per unit time (for, if not, the mean number in the system could not be a stable value). Hence, if the arrival rate is a constant λ and if every arrival is accepted, then

$$R = \lambda$$

However, finite capacity systems will not accept every arrival. Only those customers who arrive when the system is less than full are allowed into the system. In such cases, one must adjust the pure arrival rate (the attempts) downward to obtain the effective arrival rate. If arrivals "see" random states (a characteristic of all of the models developed in this chapter, but not all queueing models), the fraction of attempts that are successful is $1 - p_N$. Hence, for all of these finite capacity models

$$R = \lambda(1 - p_N)$$

It might be expected that mean waiting times are related to mean numbers of customers, in the sense that long lines would tend to imply long waiting times. Few people would guess, however, that there is a very simple direct relationship linking L, W, and R. The equation, known as Little's formula, is

$$L = RW$$

Once any two of these quantities are fixed, the third is determined. If any one is fixed, the other two must vary in direct proportion. In particular, in an infinite capacity system, where $R = \lambda$,

$$L = \lambda W$$

Incidentally, when you see Little's formula referenced in other books, it will most likely be stated in this latter form. However, experience has shown that students have difficulty remembering to modify λ in finite capacity cases, so in this book we have elected to sacrifice tradition for clarity.

Little's formula applies on a very broad scale to all kinds of problems. In fact, any system that transforms input to output over time, and possesses steady-state performance measures corresponding to L, R, and W will obey that law. At one extreme, the system could involve an arbitrarily complicated network of interacting processes; at the other, it might involve no queueing at all. The rule can be applied to any portion of a system, or to collections of systems. As an immediate example, we could restrict attention to just the queue portion of a queueing system to obtain the useful relation:

$$L_q = RW_q$$

To cite a more remote application, we could be dealing with an entire factory. Little's formula would state that the in-process inventory for the factory as a whole (L) equals the production rate of the factory (R) times the average flowtime of jobs through the factory (W). Alert readers will realize that Little's formula is a powerful tool.

7.6 MARKOVIAN QUEUEING MODELS

There is a class of queueing models for which detailed solutions are quite easy to obtain and furthermore allow a lot of variations. The class consists of those models which can be treated as special cases of the continuous time Markov process discussed in the last chapter. To maintain independence of the chapters, our treatment here will not assume that you have already studied Markov processes in detail. However, you will find that both your understanding and ability to develop new models will be greater if you take the time to learn that material.

To be a Markovian queueing system, both interarrival times and service times *must* be negative exponentially distributed. As already mentioned, this distribution family has the advantage of being fully characterized through only one parameter, which is conveniently related to the mean. Of course, that also implies that one has very limited control over the distribution; for example, one cannot separately adjust the mean and the variance.

One might well ask how reasonable or realistic are those assumptions. The answer is, usually, not very. Hardly ever will you encounter in real life a situation where the negative exponential assumptions can be rigorously defended. Actually, however, the proper question to ask is not how good the assumptions are, but how good the results are. In other words, we should be content to use the assumptions, even knowing that they are not exactly correct, if they enable us to obtain acceptable results. From this point of view, the Markovian assumptions seem better than you might expect. To be sure, there are instances where the use of the negative exponential distribution could produce very misleading results, and one should always be cautious in drawing conclusions in critical situations. Still, the simple Markovian models work surprisingly well in many practical situations.

When the queueing model is Markovian, you can set up and solve a set of linear equations whose solution provides the steady-state distribution of the number of customers in the system, that is, the p_i distribution. All of the other performance measures we have mentioned can then be computed from these. To be able to write down the proper equations, you must either understand the material presented in the last chapter, or learn the simplified method about to be presented here.

First, you must identify the possible states of the queueing system. Normally these will be integers corresponding to the possible numbers of customers present in the system (i.e., 0, 1, 2, . . . , N), although more complicated state descriptions may sometimes be necessary. Then, for each state, you must identify the possible transitions *out* of that state. For example, an arrival to the system would normally change the state from i to $i+1$ (unless i is already the maximum value in a finite capacity system), and a service completion would normally change it to $i-1$. Just think of all of the events that could occur when you are in a particular state, and trace the change of state that each would cause. Do not bother considering all of the ways to *enter* each state; these will take care of themselves.

The transition diagram provides a convenient visual portrayal of the structure of the states and transitions. Each state is represented by a small labeled circle, and each transition by an arrow, in the obvious way. Figure 7.6 shows a portion of a typical diagram.

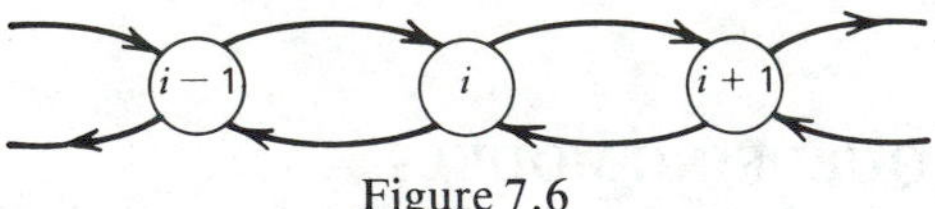

Figure 7.6
A portion of a transition diagram.

The next step, which is the only one posing even slight difficulties, is to figure out the appropriate transition rate to associate with each transition. We will be going through several cases individually. Once you get the idea, you will be able to handle all kinds of variations easily. In the very simplest case, the $M/M/1$ model, each arrival transition (i to $i+1$) is assigned the ordinary arrival rate, λ, and each service-completion transition (i to $i-1$) is assigned the service rate μ. So, for this particular model, the transition diagram looks like Fig. 7.7.

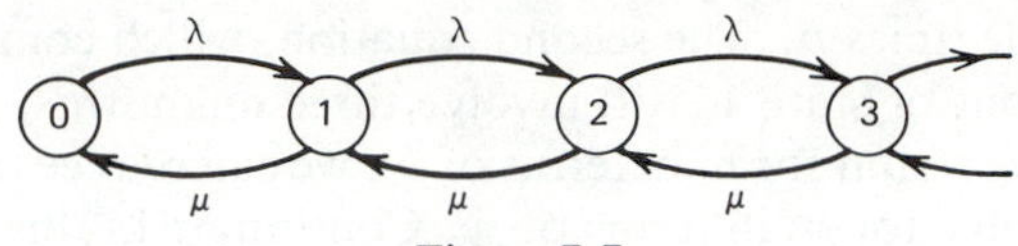

Figure 7.7
Transition diagram for $M/M/1$.

From the transition diagram, you can set up the steady-state equations for the p_i using the principle of *flow balance*. The logic of this principle is intuitively appealing, easy to remember, and simple enough to use confidently. Most importantly, it does yield the correct equations in every case. To state the principle, we must introduce the concept of "flow" along a transition arc. If an arc is from point i to point j, with an associated transition rate of r_{ij}, then the flow along the arc is $r_{ij}p_i$. When we first write this, of course, we do not know the value of p_i, so that variable will appear as an unknown in the equation. The principle of flow balance declares that, in steady-state, the total flow into every state must equal the total flow out. When the principle is applied to the states of the transition diagram, one by one, each state produces one equation, which is linear in the unknown p_i's. Collectively, the full set of equations determines the full p_i distribution, with one minor qualification.

Each equation will be linear, and there will be exactly as many equations as there are unknowns. The equations will be consistent, so there is sure to be a solution. The problem is that the equations will contain one dependency. That is, any one of the equations could be eliminated without loss of information. (This property follows from the basic properties of transition rates, and is more fully explained in Chapter 6.) However, there is one more equation that must be satisfied by the p_i values, and that is the requirement that the probabilities sum to one. This equation, called the normalizing equation, will be independent of all the others. Consequently the set of flow balance equations (less any one) together with the normalizing equation will jointly determine a unique solution. Moreover, you may be certain in advance that the solution will be of proper form for a probability distribution. In short, nothing can go wrong.

Frequently (virtually always in simple queueing models), the transition diagram is of the so-called *birth–death* form. This means that the states correspond to integers and that all transitions are to the state just above (a "birth") or to the state just below (a "death"). Whenever the number of customers in the system is an adequate state description and customers arrive and are serviced individually, the transition diagram will be of the birth–death form. A model in which, for example, customers are serviced in pairs would not have the birth–death form, because a single transition could change the state from i to $i-2$. All of the relations and methods discussed thus far would still apply; in particular, the flow balance equations would be valid. However, there is a definite advantage to dealing with the birth–death form when it comes to solving the equations.

The equations for a birth–death model have a characteristic form which makes them particularly easy to solve by hand. The first equation, which is constructed by examining the transitions in and out of state 0, will have only two unknowns, p_0 and p_1. This happens because state 0 is "connected" only to state 1. The coefficients of the unknowns will vary depending upon model assumptions, but it will always be possible to solve for p_1 in terms of p_0. That is, p_1 can be written as some multiple times p_0. The second equation, which comes from observing the flow into and out of state 1, will involve three unknowns, p_0, p_1, and p_2. By substituting our expression for p_1 in terms of p_0, we can reduce the unknowns to p_2 and p_0, and then solve for p_2 in terms of p_0. Continuing in this way, each successive equation introduces a new unknown that can be reduced to some multiple of p_0. Eventually, we have all of the unknowns expressed in terms of the one undetermined value of p_0.

The final step uses the normalizing equation to find the value of p_0. Since all of the probabilities must sum to one, and each probability is now expressed as a multiple of p_0, this one equation can be manipulated to yield an expression for p_0. If the model has an infinite number of states, there will be an infinite series to sum.

The description of the method of solution is perhaps more difficult to follow than the computations are to carry out. It really amounts to no more than simple algebraic substitution. Once you have gone through an example or two (referring, if necessary, to the paragraphs above), it will seem obvious how to proceed.

If the equations are not of the birth–death form, you may have difficulty solving them by hand. In such cases, you may be forced to use a computer program to get numerical solutions. It is possible that difficulties of another kind may appear in any such attempt, because queueing equations are quite often "ill-conditioned." Nothing in this book will demand sophisticated knowledge of numerical methods, but you should be aware of the possibility in more general cases.

7.7. THE *M*/*M*/1 MODEL

The easiest Markovian model to set up and solve is the single server, infinite capacity version. The bakery shop example used at the beginning of the chapter could serve as a concrete application of this model. Of course, we must assume that the arrivals occur according to a Poisson process (or, equivalently, that the interarrival times are independent and negative exponentially distributed). This assumption is quite reasonable for the bakery shop example. We must also assume that service times are negative exponentially distributed, which is *not* very likely to be satisfied by the type of service performed in a bakery shop. However, our objective at this time is more to develop the models than to select the best one, so we will proceed with the awareness that the assumptions are imperfect. Eventually, we will develop a more suitable model for the bakery shop.

In accordance with the usual notational conventions, we use λ for the mean arrival rate and μ for the mean service rate. These will be carried through the analysis in symbolic form, to permit the derivation of general formulas. However, if it helps to have specific numbers the first time through, you could imagine that customers arrive at the mean rate of 10 per hour ($\lambda = 10$) and the mean service rate is 20 per hour ($\mu = 20$). Such rates would be obtained from stopwatch type data if you observed an average time between arrivals of 6 minutes and an average service time of 3 minutes. Realistically, of course, one could not expect to deal with such nice round numbers.

The state of the queueing system will be the number of customers present, which will be indicated by an integer number from 0 to infinity. The assumption that customers arrive and are serviced individually gives the birth–death structure to the transition diagram. The arrival rate and service rate are associated with the appropriate transition arcs in the obvious way, yielding the transition diagram already shown in Fig. 7.7.

Applying the principle of flow balance to each state, starting with state 0, we get the steady-state equations:

$$\begin{aligned}\lambda p_0 &= \mu p_1\\(\lambda+\mu)p_1 &= \lambda p_0+\mu p_2\\(\lambda+\mu)p_2 &= \lambda p_1+\mu p_3\\(\lambda+\mu)p_3 &= \lambda p_2+\mu p_4\end{aligned}$$

and so on. These equations can be rearranged into the more common form, where the unknowns are aligned on the same side of the equality, to yield the equivalent:

$$\begin{array}{lllll}-\lambda p_0+ & \mu p_1 & & & =0\\ \lambda p_0- & (\lambda+\mu)p_1+ & \mu p_2 & & =0\\ & \lambda p_1- & (\lambda+\mu)p_2+ & \mu p_3 & =0\\ & \cdot & \cdot & \cdot & \cdot\\ & \cdot & \cdot & \cdot & \cdot\\ & \cdot & \cdot & \cdot & \cdot\end{array}$$

Notice that all but the first equation are covered by the general equation form:

$$\lambda p_{i-1}-(\lambda+\mu)p_i+\mu p_{i+1}=0$$

Using the solution strategy outlined above for birth–death type equations, we solve the first equation for p_1 in terms of p_0, giving

$$p_1=\left(\frac{\lambda}{\mu}\right)p_0$$

Substituting this in the second equation and solving for p_2 yields

$$p_2=\left(\frac{\lambda}{\mu}\right)^2 p_0$$

Continuing in this manner, we find that, in general,

$$p_i=\left(\frac{\lambda}{\mu}\right)^i p_0$$

To be certain that this form is, indeed, the solution for the entire infinite set of equations, it should be verified by substituting it into the general equation.

The remaining unknown, p_0, is evaluated by imposing the normalizing equation.

$$\begin{aligned}1&=p_0+p_1+p_2+\cdots\\1&=p_0+\left(\frac{\lambda}{\mu}\right)p_0+\left(\frac{\lambda}{\mu}\right)^2 p_0+\cdots\\1&=p_0\left[1+\left(\frac{\lambda}{\mu}\right)+\left(\frac{\lambda}{\mu}\right)^2+\cdots\right]\end{aligned}$$

The infinite series that appears within the brackets is a simple ratio series that will converge to the finite sum

$$\frac{1}{1-(\lambda/\mu)}$$

provided that the quantity (λ/μ) is less than 1. (The implications of this restriction will be explored more fully later.) Consequently, the solution for p_0 is

$$p_0 = 1 - \frac{\lambda}{\mu}$$

Putting this expression back into the solution for the general state probability produces the final solution

$$p_i = \left(\frac{\lambda}{\mu}\right)^i \left(1 - \frac{\lambda}{\mu}\right)$$

which holds for all i from 0 to infinity. This probability distribution happens to be of a common family—the geometric distribution.

Because the two parameters λ and μ appear only together as a ratio, it is conventional to replace them by the single parameter ρ (a lowercase Greek rho), where $\rho = \lambda/\mu$. The steady-state probability solution then can be written in the slightly simpler form

$$p_i = \rho^i(1-\rho) \qquad i = 1,2,\ldots\infty$$

provided that $\rho < 1$. The new parameter, ρ, is called the traffic intensity. It is, of course, the ratio of the arrival rate to the service rate, but there are also other ways to interpret it. Expressing it in the form

$$\rho = \frac{1/\mu}{1/\lambda}$$

shows that it is the ratio of the mean service time to the mean time between arrivals. It could also be thought of as the arrival rate, λ, times the mean service time, $1/\mu$, so one could interpret ρ as the mean number of arrivals during a period equal to the mean service time. All of these interpretations give some logical explanation for the requirement that appeared in the solution procedure that ρ be less than one. In simple words, the model will not give an answer if the customers are arriving faster, on the average, than they can be serviced.

Additional interpretations of ρ can be obtained from the steady-state solution. The term p_0 is the steady-state probability that the system is empty, which can also be thought of as the fraction of time (over the long run) that the server is idle. Hence, the fraction of time the server is busy, or the utilization, U, is $1 - p_0$. But since, for this model, $p_0 = 1 - \rho$, we have $U = \rho$. Similarly, we can argue that ρ is also equal to B, the mean number of busy servers, since there is only one. Once again, these interpretations suggest that the traffic intensity should approach, but not exceed, the value 1 to maximize the use of the server. However, when we have probed a little deeper into the behavior of the queue as a function of the traffic intensity, we will expose some dangers in drawing such a conclusion.

Since we have the probability distribution of the number of customers in the system, we can calculate the mean, L.

$$
\begin{aligned}
L &= \sum_{i=0}^{\infty} i p_i \\
&= 0p_0 + 1p_1 + 2p_2 + \cdots \\
&= (1-\rho)[\rho + 2\rho^2 + 3\rho^3 + \cdots] \\
&= \frac{\rho}{(1-\rho)}
\end{aligned}
$$

The infinite series converges to a finite sum only if $\rho < 1$, but this condition has already been assumed.

We can also obtain the queue length distribution, q_i, from the p_i:

$$
\begin{aligned}
q_0 &= p_0 + p_1 = (1-\rho) + \rho(1-\rho) = 1 - \rho^2 \\
q_i &= p_{i+1} = \rho^{i+1}(1-\rho) \qquad \text{for } i \geq 1
\end{aligned}
$$

and from this distribution, we can get the mean number of customers in the queue, L_q

$$
\begin{aligned}
L_q &= 0q_0 + 1q_1 + 2q_2 + \cdots \\
&= \frac{\rho^2}{1-\rho}
\end{aligned}
$$

The difference between L and L_q should be, and is, equal to B, the mean number of busy servers, which is equal to ρ.

Figure 7.8 graphs both L and L_q as a function of the traffic intensity (or utilization rate), ρ, over the full range of permissible values, $0 \leqq \rho < 1$. This figure reveals what happens as we attempt to obtain full utilization of the server—the expected number in the queue and in the system grows to infinity. It would be bad enough if the queue might occasionally grow to infinity, or if there were some probability that it might grow to infinity, but this result is even worse. It says that the *mean* queue length grows to infinity.

If this result seems so extraordinary that it casts doubt on the model, you must realize that there is a good reason why we do not witness infinite queues in everyday life. Whenever they start to grow too long, people intercede and alter the system. In other words, even if we manage to achieve $\rho = 1$, we will never see that system reach steady state. The model does correctly reflect what *would* happen *if* the system were allowed to operate.

Closer examination of the curves reveals even further cause for concern. In order to limit the mean queue length to a reasonably small size, say 5, the utilization must be a disappointingly low 0.85. That is, we must be prepared to tolerate 15% idleness of the server in order to avoid a queue which averages more than 5. Of course, even with an *average* queue length of 5, it may, occasionally, be considerably longer. Furthermore, the mean queue length is very sensitive to small changes in ρ in this range.

Thus the stochastic model has revealed that there is an unavoidable conflict between the desire to obtain full utilization of a server and the desire to keep the mean queue length short. The intuitive sense that the system is "in balance" when the arrival rate is equal to the service rate is clearly wrong. The ideal ratio of arrival rate to service rate is something less than one; its specific value depends on the relative costs of idleness versus congestion.

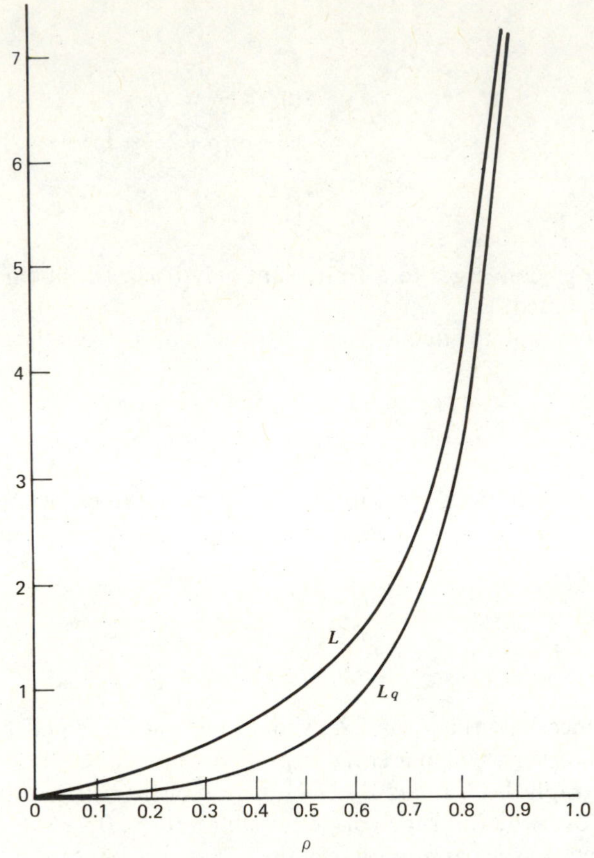

Figure 7.8
L and L_q as functions of ρ.

A misunderstanding of the queueing phenomenon has undoubtedly been the source of a good many disputes between labor and management, between customers and agents, and so on. Most people encountering a queue interpret the situation too narrowly. For example, a manager who sees that a worker is idle on one occasion may jump to the conclusion that the "feed" rate is unnecessarily small, and attempt to increase it. He may very well discover later that a very long queue had developed, and be forced to reduce it again. Over a period of time, as he persists in trying to achieve a "balance" between the feed rate and the worker's capability, and is consistently frustrated, he may reach the conclusion that the worker is deliberately thwarting his efforts. The worker, over the same period of time, may reach the conclusion that the manager is attempting to run a "sweatshop." In fact, both the occasional idleness and the occasional backlogs may be attributable solely to the *variability* in interarrival and/or service times.

7.8
LIMITED QUEUE CAPACITY

When waiting space is limited, a number of minor modifications must be made to the previously developed model. In terms of the queueing notation introduced earlier, we are now concerned with the $M/M/1/N$ system, in which N represents the maximum number of units or customers allowed in the system. Because the arrival process would ordinarily continue to generate arrivals even when

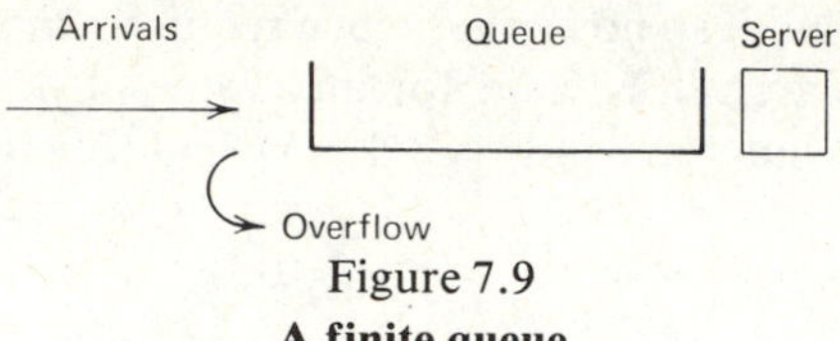

Figure 7.9
A finite queue.

the system was full, we must assume that these "blocked" arrivals are lost to the system.

A schematic version of the finite queue is shown in Fig. 7.9; the associated transition diagram is given in Fig. 7.10. The latter is identical to Fig. 7.7, except that it is finite.

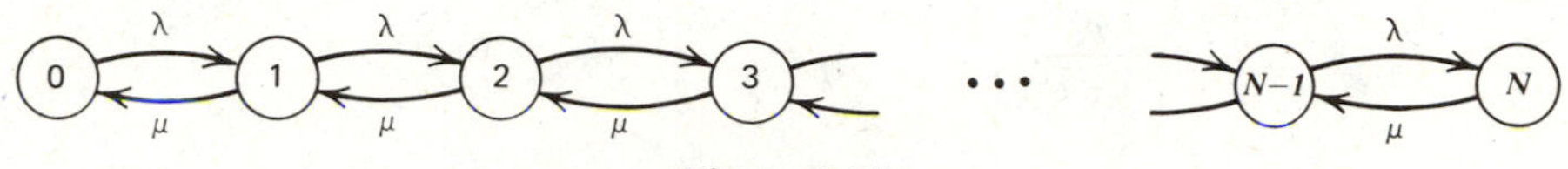

Figure 7.10
The transition diagram for a finite queue.

The steady-state equations are identical to those used previously, except that the last equation is now a special case:

$$0 = -\lambda p_0 + \mu p_1$$
$$0 = \lambda p_{j-1} - (\lambda + \mu) p_j + \mu p_{j+1} \qquad \text{for } 0 < j < N$$
$$0 = \lambda p_{N-1} - \mu p_N$$

The fact that the last equation is of an altered form does not influence the solution for the p_j in terms of p_0, because the equations contain a dependency. That is, we do not require the last equation. We obtain, as before,

$$p_j = \left(\frac{\lambda}{\mu}\right)^j p_0 \qquad j = 0, 1, \ldots, N$$

(It is easy to verify that this solution is consistent with the neglected last equation.) So far, then, the limitation on system capacity has not made any difference.

At this point, however, we must apply the normalizing equation to determine p_0.

$$1 = p_0 + p_1 + p_2 + \cdots + p_N$$

$$1 = p_0 + \frac{\lambda}{\mu} p_0 + \left(\frac{\lambda}{\mu}\right)^2 p_0 + \cdots + \left(\frac{1}{\mu}\right)^N p_0$$

$$1 = p_0 \left[1 + \frac{\lambda}{\mu} + \left(\frac{\lambda}{\mu}\right)^2 + \cdots + \left(\frac{\lambda}{\mu}\right)^N \right]$$

$$1 = p_0 \left[\frac{1 - \left(\frac{\lambda}{\mu}\right)^{N+1}}{1 - \frac{\lambda}{\mu}} \right]$$

$$p_0 = \frac{1 - \frac{\lambda}{\mu}}{1 - \left(\frac{\lambda}{\mu}\right)^{N+1}} = \frac{1 - \rho}{1 - \rho^{N+1}}$$

This time the series is finite, so no convergence requirement appears to restrict the value of ρ. The formula given is valid for any value of ρ, with the exception of $\rho=1$, in which case the sum is trivially $N+1$, so $p_0=1/N+1$. In summary,

$$p_j=\rho^j\left(\frac{1-\rho}{1-\rho^{N+1}}\right) \qquad j=0, 1, \ldots, N$$

unless $\rho=1$, in which case,

$$p_j=\frac{1}{N+1} \qquad j=0, 1, \ldots, N$$

The mean number in the system, L, can be obtained from the distribution, using well known finite-sum formulas

$$L=\frac{\rho}{1-\rho}\left(\frac{1-\rho^N}{1-\rho^{N+1}}\right)-\frac{N\rho^{N+1}}{1-\rho^{N+1}} \qquad \text{for } \rho\neq 1$$

$$=\frac{N}{2} \qquad \text{for } \rho=1$$

If the traffic intensity, ρ, is low and N is large, L will be close to $\rho/(1-\rho)$, as you would expect. If ρ is much greater than one, L approaches N. This, too, makes sense. A heavily overloaded system will tend to remain nearly full. This is one way to get good utilization of the server. Of course, when the arrival rate is much greater than the service rate, most arrivals will be blocked and therefore never be served by the system. Depending on the application, this loss may be anything from perfectly acceptable to intolerable.

Incidentally, the fraction of customers or items lost due to blocking is easily quantified. p_N gives the probability that the system is full.

$$p_N=\rho^N\left(\frac{1-\rho}{1-\rho^{N+1}}\right) \qquad \text{for } \rho\neq 1$$

$$=\frac{1}{N+1} \qquad \text{for } \rho=1$$

The same quantity can be interpreted as the fraction of time, over the long run, that an arriving customer will find the system full; hence, it is the fraction of arriving customers who are blocked. The *number* blocked per unit time would be this fraction times the number of arrivals per unit time, or $p_N\lambda$. This number may be useful in determining how many more customers would be served if the waiting space, N, were increased. One of the problems at the end of the chapter explores the idea further.

The case $\rho=1$ is interesting, in that it can provide additional insight into why a balance between the arrival rate and service rate is generally not desirable. When $\rho=1$, the distribution of the number in the system is uniform; that is, all states are equally likely. The mean would be "half full." For any finite N, items are being blocked. In an industrial setting where the items are workpieces that *must* be processed, the only recourse (short of unbalancing the arrival and service rates) is to increase the waiting room. But as N is increased, the states remain equally likely and the mean is still "half full." In the limit, as N goes to infinity, the probability of being in any particular state goes to zero and the mean goes to infinity (or half of infinity, which is the same thing). In the terminology of Markov processes used in chapter 6, the states remain recurrent, but become null.

7.9
MULTIPLE SERVERS

Another important and easily handled variation involves multiple servers drawing from a single queue, as depicted in Fig. 7.2. Such a system would be designated by $M/M/c/N$, where c is the number of servers. The system capacity N would certainly have to be at least c. One important special case occurs when $N = c$; that is, when the maximum number permitted in the system equals the number of servers. We will call this the "no waiting" case. Another special case occurs when $N = \infty$, or when there is infinite waiting room so that all arrivals may ultimately be served.

Assuming that all servers operate at the same mean rate, μ, there is no need to keep track of which servers are busy. It will be sufficient, in describing the state of the system, to indicate the number of customers in the system. Despite the fact that service may take place on a number of customers simultaneously, departures will occur one at a time. (The probability of two or more services being completed at exactly the same instant is zero.) Consequently it will still be appropriate to model this system as a birth-death process. What is different now is that the transition rates will be state dependent. If there is only one customer in the system, he will receive service from one of the servers at the rate μ. If there are two customers present, they will both receive service, *each* at the rate μ. The transition from state 2 to state 1 will occur as soon as *either* completes service, so the rate at which this occurs is twice as great as the rate would be if only one server were working, or 2μ. Similarly, the transition rate from 3 to 2 is 3μ, and so on. (If you find it difficult to think in terms of rates, you may wish to convince yourself that these rates are correct by translating them to mean times between transitions.) The transition rates corresponding to service completions continue to increase in this way until all c servers are occupied, after which they remain at the value $c\mu$. The transition diagram is shown in Fig. 7.11.

Figure 7.11
The transition diagram for multiple servers.

If we now restrict attention to the "no waiting" case, or the $M/M/c/c$ system, the transition diagram would extend only to the state c. The steady-state equations are:

$$\begin{aligned}
0 &= -\lambda p_0 + \mu p_1 \\
0 &= \lambda p_0 - (\lambda + \mu) p_1 + 2\mu p_2 \\
0 &= \lambda p_1 - (\lambda + 2\mu) p_2 + 3\mu p_3 \\
&\vdots \\
0 &= \lambda p_{c-1} - c\mu p_c
\end{aligned}$$

These can be solved in the usual way for birth–death equations to yield

$$p_j = \frac{1}{j!}\left(\frac{\lambda}{\mu}\right)^j p_0 \qquad \text{for } j = 0, 1, \ldots c$$

Defining $\rho = \lambda/\mu$, as in the $M/M/1$ queue,

$$p_j = \frac{\rho^j}{j!} p_0$$

Using the normalizing equation to evaluate p_0, we find

$$p_0 = \frac{1}{\sum_{i=0}^{c} \frac{\rho^i}{i!}}$$

The sum in the denominator does not simplify any further, but it might be helpful to observe that it is the first $c+1$ terms of the infinite series expression of e^ρ. If c is very large relative to ρ, so that the remaining terms in the infinite series are small, we can say that p_0 is approximately $e^{-\rho}$, and

$$p_j = \frac{\rho^j}{j!} e^{-\rho}$$

is, approximately, the distribution of the number of occupied servers. This distribution is, of course, Poisson. The approximation becomes exact as c goes to infinity. One way in which a real-life system could have an unlimited number of servers is for each arrival to serve himself.

More generally, when c is finite and the approximation is not appropriate, the steady-state distribution is

$$p_j = \frac{\frac{\rho^j}{j!}}{\sum_{i=0}^{c} \frac{\rho^i}{i!}}$$

and this distribution has been called a truncated Poisson distribution.

One term, in particular, is of special interest. The probability that all servers are occupied, or that the system is full, is given by

$$p_N = \frac{\frac{\rho^N}{N!}}{\sum_{i=0}^{c} \frac{\rho^i}{i!}}$$

This formula, known variously as *Erlang's lost call formula*, the *Erlang B*-formula, or the *first Erlang function*, has been used extensively in the design of telephone systems. Tables and graphs are readily available in, for example, Ref. 2.

The mean number of occupied servers or customers in the system is given by

$$L = \rho[1 - p_N]$$

This relation has a number of interesting verbal interpretations. The reader is encouraged to try to think of some.

Of course, all of these results were derived from a Markov model. In particular, service times were assumed to be negative exponentially distributed. Perhaps surprisingly, exactly the same results occur when the service time distribution is arbitrary. A. K. Erlang, who first obtained all of these results (for the negative-exponential service time case) as early as 1917, conjectured that they would hold for arbitrary service times. The conjecture was confirmed, however, only as recently as 1969 (13). As you might expect, the mathematics used is beyond the level of this text, but it is useful to know that the solution for the $M/G/c/c$ system (including the case $c=\infty$) is the same as for the $M/M/c/c$ system.

The solutions just discussed were for the "no waiting" case. Returning now to the more general case in which some, perhaps infinite, waiting capacity is available, the steady-state equations would be

$$\begin{aligned} 0 &= -\lambda p_0 + \mu p_1 \\ 0 &= \lambda p_{j-1} - (\lambda + j\mu)p_j + (j+1)\mu p_{j+1} && \text{for } j=1,\ldots,c-1 \\ 0 &= \lambda p_{j-1} - (\lambda + c\mu)p_j + c\mu p_{j+1} && \text{for } j=c,\ c+1,\ldots \end{aligned}$$

If the total system capacity, N, is finite, there would be one more equation for the last state.

$$0 = \lambda p_{N-1} - c\mu p_N$$

The solution of these equations offers no difficulties in principle, but the solutions do require some skill with summing of finite or infinite series. If you enjoy that sort of thing, there are many problems to try your hand at, just by varying c and N. If you would prefer to bypass the algebra, Table 7.3 is provided to cover the cases where $N=\infty$ and $c=1, 2, \ldots, 6$. The formulas are all in terms of a traffic intensity parameter ρ, which is defined by $\rho=\lambda/c\mu$ (notice the denominator includes the number of servers as well as the service rate), and in each case the convergence requirement $\rho<1$ is assumed. Formulas for p_0 and L_q are given, from which you can easily obtain other performance measures using the relations in Section 7.5.

Table 7.3
RESULTS FOR M/M/c

No. of Servers	p_0	L_q
1	$1-\rho$	$\dfrac{\rho^2}{1-\rho}$
2	$\dfrac{1-\rho}{1+\rho}$	$\dfrac{2\rho^3}{1-\rho^2}$
3	$\dfrac{2(1-\rho)}{2+4\rho+3\rho^2}$	$\dfrac{9\rho^4}{2+2\rho-\rho^2-3\rho^3}$
4	$\dfrac{3(1-\rho)}{3+9\rho+12\rho^2+8\rho^3}$	$\dfrac{32\rho^5}{3+6\rho+3\rho^2-4\rho^3-8\rho^4}$
5	$\dfrac{24(1-\rho)}{24+96\rho+180\rho^2+200\rho^3+124\rho^4}$	$\dfrac{625\rho^6}{24+72\rho+84\rho^2+20\rho^3-75\rho^4-125\rho^5}$
6	$\dfrac{5(1-\rho)}{5+25\rho+60\rho^2+90\rho^3+90\rho^4+54\rho^5}$	$\dfrac{324\rho^7}{5+20\rho+35\rho^2+30\rho^3-36\rho^5-54\rho^6}$

$\rho=\lambda/c\mu<1$

7.10 AN EXAMPLE

To illustrate several of the queueing models and at the same time demonstrate their usefulness in a realistic situation, consider the following scenario. It concerns a grain elevator at a small town in one of the plains states where wheat is the principal crop. During the harvest season, trucks loaded with wheat from the fields arrive at the elevator, where they must quickly deposit their loads and return to the fields for another. The check-in process involves weighing the truck, drawing samples for moisture and contamination tests, and a few other details, before the load can be dumped through a grate.

The farmers are *very* concerned about getting their crop off the field and into the elevator quickly. Once the wheat is ripe, it is highly vulnerable to rain or wind. Any delay could threaten a significant portion of a farmer's income for the whole year. Of course, all of the fields in a region ripen about the same time, so it is not surprising that a traffic problem could develop at the elevator.

To keep the numbers simple, let us suppose that the average interarrival time for trucks is 6.67 minutes and the average service time is 6 minutes. Then, using the standard formula for the $M/M/1$ model, we could expect an average time at the elevator (W) of one hour per truck, which matches the actual experience of the drivers. This delay is considered by the farmers to be intolerable.

At a meeting of the farmers' cooperative, three suggestions are put forth:

1. By adding sideboards to the trucks (to permit slightly larger loads) the average interarrival time could be lengthened to 10 minutes. Simultaneously, an extra worker at the check-in could shorten the average service time to 4 minutes. These relatively minor adjustments would cost an estimated $30,000.
2. Some farmers feel that, although the first proposal would help, it would not really solve the problem. They believe that more significant changes are warranted. The addition of another complete check-in station, including scale, test equipment, and personnel, would essentially double the service capacity. Arriving trucks would join a single line, and the truck at the front of the line would move to the first available of the two check-in points. This change would cost about $400,000.
3. A few farmers feel that the entire grain elevator facility should be duplicated on the other side of town. In addition to doubling the service capacity, this proposal would split the arrivals into equal halves, since each farmer would attend the nearer of the two elevators. The cost of this proposal would be close to $1,000,000.

In discussing the merits of the various proposals, the farmers find the low cost of the first proposal to be attractive, but many question the effectiveness of such minor changes in dealing with such a serious delay problem. The third proposal seems to offer an "ultimate fix," but the cost would be an enormous burden. The second proposal is favored by many who seek a compromise between cost and effectiveness. Basing your judgment on what has been presented thus far, which would you choose?

In the terminology of queueing theory, the first proposal involves changing the parameter values of the model, but no structural changes. The second proposal is for a two server, single queue system; and the third is for a two server, separated queue system. We will use the simple Markov models to perform our analysis, despite the fact that it is highly unlikely that service times would be negative exponentially distributed. We may excuse this model defect by pointing out that we are seeking only comparative performance, not absolute predictions.

The three proposals are represented by Table 7.4. The analysis, using results from Table 7.3 and the relations of Section 7.5, is straightforward, except possibly in the case of the third proposal. The important key to the third proposal is to realize that the system separates into two independent $M/M/1$ queueing systems and insofar as any individual truck is concerned, the service process is as if there

Table 7.4

Proposal	Model	Arrival Rate	Service Rate	Waiting Time (W)
Existing	$M/M/1$	9 per hr	10 per hr	60 min
1	$M/M/1$	6	15	6.67
2	$M/M/2$	9	10	7.52
3	$M/M/1$	4.5	10	10.91

were only one. Consequently, the analysis reduces to obtaining results for an ordinary $M/M/1$.

The results are startling, to say the least! The least expensive proposal turns out to yield the greatest benefits. It is unlikely that any of the farmers would realize, from intuition alone, what the analysis reveals. This example clearly illustrates why formal methods are needed when dealing with queueing phenomena. The human mind, which is naturally adept in many judgment situations, is just not very good at dealing with the kind of random variability that accounts for queues.

If you remain puzzled, or perhaps even skeptical, about the outcome of this example, you should satisfy yourself that you understand how each of the suggested changes actually affects the queue. The comparison of the original system, proposal 1, and proposal 3 can be simply related to traffic intensity.

The contrast between proposal 2 and proposal 3 is interesting because a very general conclusion can be drawn. The question might be posed as follows: given equivalent traffic conditions, is it better to have one line for two servers, or a separate line for each server? If you compare the formulas in Table 7.3, you will find that it is always better to have a single line. One might suspect that this conclusion is somehow related to the peculiarities of the Markovian assumptions, but even more general models support the same conclusion. Hence, we have discovered a general principle of queueing system design. Other things being equal, it is preferable to design a service system so that just one line feeds all of the servers and the customer at the head of the line is directed to the first available server. We call this concept the "principle of pooling," referring to the common pool of customers that feed the servers.

In recent years, some banks and airline ticket counters have adopted this principle. Perhaps you have had the experience of witnessing the change and seeing the interesting first-time reactions of customers to the unfamiliar new system. For the most part, however, the principle seems to be unknown to the average person. Certainly, one can find many opportunities to improve existing systems by applying the principle of pooling. As a design exercise, you may enjoy figuring out how to exploit the principle in supermarket checkouts. As you do so, be sure to think about side-benefits and possible disadvantages, as well as the pure queueing aspects.

7.11 FINITE SOURCES

The next variation to be developed models the situation that occurs when the potential arrivals form a fixed, finite population. Such a situation would occur if, for example, the server were an office copying machine which only a dozen peo-

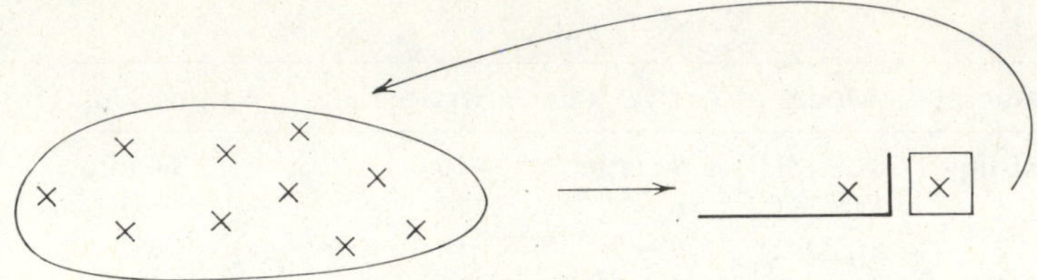

Figure 7.12
The finite source queue.

ple were authorized to use. Figure 7.12 depicts the situation. It is assumed that when a customer completes service, he returns to the source.

The queueing nomenclature for this model is $M/M/1/N/N$. The second N refers to the number of items in the source, or the maximum number of arrivals. The first N represents the capacity of the system. Although the capacity could conceivably differ from the maximum number of arrivals, there is no point in considering a greater capacity—it would never be used—and it will be obvious how to treat a lesser capacity.

The difference between this model and previous ones is in the arrival rates. It is intuitively clear that when most of the population are in the queue, the arrival rate should be lower than when most are in the source. Certainly, when *all* are in the queue, the arrival rate should be zero.

Let λ represent the mean arrival rate *per customer*. This would be statistically estimated by recording the times spent in the source by each customer, that is, the durations of the intervals beginning when a customer returns to the source after completing service and ending when the same customer reappears at the queue requiring another service. Assuming these times fit a negative exponential distribution with a common mean for all customers, λ would be the reciprocal of the average time. Note that λ *cannot* be estimated by recording times between arrivals to the queue, because these times will be dependent on the number already in the queue.

If there is only one customer in the source, and the other $N-1$ are in the system, the arrival rate will obviously be λ. If there are two in the source, the arrival rate will be twice as great, or 2λ. The logic here is similar to that used to obtain the service transition rates when there were multiple servers. Continuing in this manner, we obtain the transition diagram shown in Fig. 7.13. As in previous models, the state represents the number in the system (queue plus server), but it would also be possible in this case to use the number in the source as the indicator.

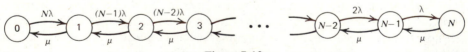

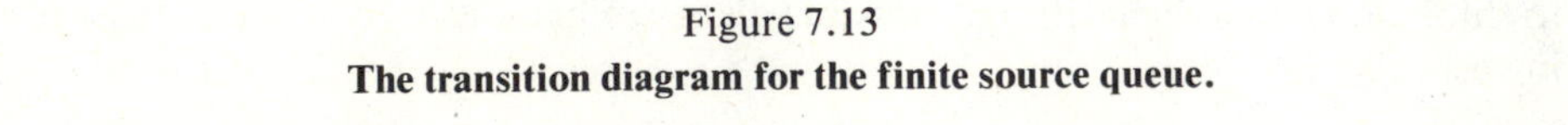

Figure 7.13
The transition diagram for the finite source queue.

The steady-state equations are

$$0 = -N\lambda p_0 + \mu p_1$$
$$0 = (N-j+1)p_{j-1} - [(N-j)\lambda + \mu]p_j + \mu p_{j+1} \qquad \text{for } j = 1, 2, \ldots, N-1$$
$$0 = \lambda p_{N-1} - \mu p_N$$

The general solution, obtained in the usual way, is

$$p_j = \frac{N!}{(N-j)!}\left(\frac{\lambda}{\mu}\right)^j p_0$$

The expressions for p_0, L, and L_q are obtainable in a straightforward manner, but do not reduce to simple formulas that could provide additional insight.

The finite source arrival rates can be combined with the multiple server service rates to provide a useful model which is commonly called the "machine interference model" or the "N machine, R repairmen" model. In this application, the N machines form the source population, and the repairmen are the servers. The breakdown of a machine constitutes an "arrival" (not to imply, necessarily, that broken machines are physically moved to a repair shop). A repair restores the machine to operation and, in that sense, returns it to the source. Formally the model is $M/M/R/N/N$.

The problem usually posed is, "given the failure and repair rates, λ and μ, and a fixed number of machines, N, what is the optimal number of repairmen, R, to service the machines?" The method of solution is, of course, to model the various alternatives and to select that one which achieves the most advantageous trade-off between the cost of repairmen and the cost of idle machines. One of the problems at the end of the chapter explores this optimization problem more explicitly.

7.12 QUEUE DISCIPLINES

It has been natural to assume throughout previous sections that the queues operated according to the first-come, first-served (FCFS) rule. Most systems serving human customers are at least intended to operate by this rule, simply because anything else would be perceived of as "unfair." Occasionally, however, practical constraints prevent strict adherence to a FCFS rule. Or sometimes the arrivals are not human customers for whom the order of service is an issue. For example, they might be workpieces which are to be operated on by a machine. In such cases there is no obvious incentive to hold to the FCFS discipline. In fact, if the workpieces are stacked on one another, the last-come, first-served (LCFS) discipline is a natural one. If they are spread out in no particular arrangement, the random selection for service (RSS) discipline may be appropriate.

At first glance it may appear that differences of this kind would immediately invalidate all previous results; however, such is not the case. A bit of reflection about the logic that went into the construction of the transition diagrams will convince the reader that the steady-state distribution of the number in the system is the same for any queue discipline which does not depend on service times. That is, the server could choose the next customer arbitrarily, as long as his choice is not biased by a prediction of the customer's service times, and the transition rates would be the same. The corresponding steady-state equations, and therefore the solution, would also remain the same. Since the distributions are the same, the mean (as well as other moments) of the number in the system or of the number in the queue would be the same. In other words, the L's and the L_q's that have been calculated for the FCFS cases are also applicable for the LCFS and RSS cases.

Furthermore, because the $L = \lambda W$ and $L_q = \lambda W_q$ relations apply for any queue discipline, the mean waiting times in the system and in the queue are the same for all three cases. On the average, it will take just as long to pass through a first-come, first-served system as it will through a last-come, first-served system—a fact that is certainly not intuitively obvious.

What measures *are* affected by the queue discipline? The primary one is the distribution of waiting times. Just because the mean waiting times are unaffected by the discipline, there is no cause to infer that, say, variances are also unaffected. In fact, it can be shown that, among the three disciplines mentioned so far, the FCFS rule will produce the smallest variance in waiting times, the LCFS rule will produce the largest, and the RSS rule will produce an intermediate value. In a sense, the risk of a very long wait is minimized if customers are served in the order of their arrival. By the same token, the chance of a very short wait is also minimized by the same policy. Thus the policy has the effect of tending to equalize waiting times, relative to other policies. It is in this sense, but in this sense only, that the first-come, first-served policy is "fairer" than other methods. Despite virtually universal acceptance of and insistence on the FCFS rule when a queue consists of human beings, there is no quantitative indication that *average* waiting times are any better under this rule.

One way to reduce average waiting times for *selected* customers is to use a priority (PR) discipline. There are many possible priority schemes, at least some of which can be modeled as Markov systems. The interested reader should consult a specialized textbook on queueing theory, such as Refs. 2, 5, 7, and 11.

Some of the more difficult priority schemes to model happen to yield some of the most useful results. Recall that the arguments leading to the conclusion that mean waiting times were unaffected by the queue discipline contained the qualification, "provided that the discipline is independent of service times." It may be correctly inferred that queue disciplines that *are* dependent on service times *will* influence the mean waiting times, W and W_q. This fact, in turn, suggests that wise selection of a service-time dependent queue discipline can *reduce* mean waiting time. Providing "express" service for customers whose predicted service times are short would be one way to implement such an idea.

In many industrial applications of queueing theory, the arrivals are not customers but jobs. Processing (service) times, or at least estimates of them, are available for use in scheduling. It is possible to imagine many seemingly reasonable scheduling concepts, such as "give highest priority to the job with the greatest work content," or "give highest priority to the job with the nearest due date." Out of all of these rules-of-thumb, one has been shown to produce surprisingly good results (i.e., small mean waiting times) under widely varying circumstances. This rule, which is briefly stated as the shortest processing time (SPT) rule, attributes highest priority to jobs requiring the least amount of time to complete. The rule can be applied strictly—even to the extent of interrupting or "preempting" a job being serviced whenever a job which could be finished in less time comes along—or only weakly. A method that separated arrivals into long and short jobs and then gave priority to the short ones would implement the SPT rule in a weak way. Even this method can achieve dramatic reductions in mean waiting time. In fact, almost any way of implementing the SPT idea will perform with surprising efficacy. For a more complete discussion, demonstrating comparative results for various queueing disciplines, see Ref. 1, Chapter 8.

7.13 NON-MARKOVIAN QUEUES

Aside from a few more general results which have been cited without proof, all of the queueing theory in this chapter has been based on Markov models. A remarkably rich variety of useful models was obtainable merely through suitably adjusting the transition rates. In each case, however, the probability distributions involved were negative exponential. This assumption, although certainly convenient, is not always realistic. Service time distributions, in particular, frequently lack the "forgetfulness" property that is characteristic of the negative exponential family. Thus there is a practical need for models that do not rely on strict Markov assumptions.

The Erlang family of distributions, a subclass of the Gamma family (see Appendix B), is representable in terms of Markov processes if a certain trick is employed. The trick is based on the fact that an Erlang-k distributed random variable is equivalent to the sum of k negative exponentially distributed random variables. To show how the trick would work in a specific example, suppose that we wish to model an $M/E_3/1$ system. Each service time, which is actually Erlang-3 distributed, can then be thought of as consisting of three negative exponentially distributed "stages." A supplementary variable will have to be incorporated into the state definition to keep track of which stage the service process is in; but once this is done, the process becomes Markovian. The states would be designated as follows:

"0" means the system is empty
"(i, j)" means the system contains i customers and the server is currently in stage j

The transition diagram, for the $M/E_3/1$ system, would look like Fig. 7.14. In this diagram, each arrow directed to the right represents a possible "arrival" transition, and each one directed down or to the upper left represents a "service stage" completion. The times between any such transitions are, of course, negative exponential. Although the resulting process is not of the birth-death type, as were most of the queueing models, it *is* a continuous-time Markov process.

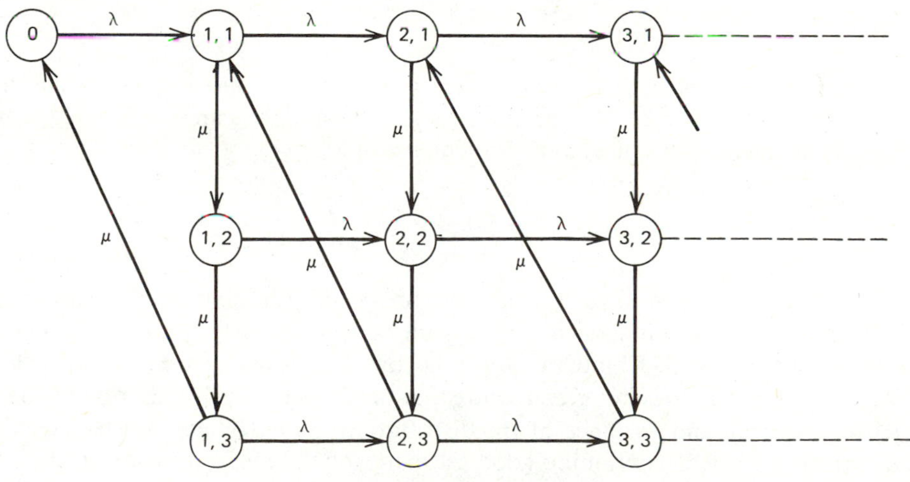

Figure 7.14
The transition diagram for $M/E_3/1$.

The same idea can be exploited to model systems with Erlang-distributed interarrival times. With two supplementary variables, it is even possible to model the system in which both the arrival process and service process are of the Erlang type. Although no difficulties in principle occur, in practice the steady-state solution is tedious to obtain and the expressions are rather nasty. Solutions for the $M/E_k/1$ and $E_k/M/1$ systems are derived in Ref. 11, pp. 164–169.

Another device is commonly used when either the service times or interarrival times are arbitrarily distributed but the other is negative exponentially distributed. The idea is known as "embedding a Markov chain," and is generally attributed to D. G. Kendall. The basic concept is to view the process only at selected moments, ignoring the dynamic behavior of the process at intermittent times. The moments at which the process is viewed—the embedding points—are chosen so that the discrete-time Markov assumption will hold. That is, given the state at one of these points, enough information is available to predict (probabilistically) the state at the next point, and any additional information is superfluous.

In the case of the $M/G/1$ system, for example, we may embed a Markov chain at the moments of service completions. At these times, and at these times only, it will be sufficient to know the number in the system to predict the future. At any other times, one would also have to know something like "elapsed service time on the customer currently in service." Because the arrival process is assumed to be negative exponential, it does not matter how long it has been since the last arrival. The same idea can be used to model the $GI/M/1$ queue, embedding at the moment when arrivals occur. It does not work for the $GI/G/1$ queue because the only possible embedding points at those moments at which an arrival and a service completion exactly coincide, but these moments are so rare that they may be considered useless for modeling purposes.

The embedded Markov chain method leads to a very useful and important result for the $M/G/1$ queue. It turns out that, for this system, the four measures L, L_q, W, and W_q depend on no more than the mean and variance of the service time distribution. The formula for L, known as the Pollaczek-Khintchine formula, is

$$L = \frac{2\rho - \rho^2 + \lambda^2 V(S)}{2(1-\rho)}$$

where $\rho = \lambda/\mu$, or the arrival rate times the mean service time, and $V(S)$ = the variance of the service time. The formulas for L_q, W, and W_q can be obtained from this one by the methods of Section 7.5. For example,

$$L_q = L - \rho = \frac{\rho^2 + \lambda^2 V(S)}{2(1-\rho)}$$

It is convenient to have such a powerful result. As mentioned earlier, it is frequently the case that service time distributions are clearly not negative exponential. The Pollaczek-Khintchine formula covers arbitrary service time distributions. Moreover, it does not even require that the form of the distribution be identified; the mean and variance of the distribution are sufficient. Of course, the statistical problem of obtaining good estimates of the mean and variance from empirical data is much easier than the problem of fitting an appropriate distribution to the same data. Because L, L_q, W, and W_q are independent of any queue discipline which does not depend on service times, the Pollaczek-Khintchine formula is not limited to the FCFS case. The restrictions to remember are that it applies only to single server, infinite capacity systems with Poisson arrivals.

One quick application of the formula yields an interesting insight. Consider the $M/D/1$ system, in which service times are constant. For this system, $V(S)=0$, so

$$L_q = \frac{\rho^2}{2(1-\rho)}$$

Recalling the analogous result for the $M/M/1$ system,

$$L_q = \frac{\rho^2}{(1-\rho)}$$

we see that the average number in the queue would be reduced by exactly one half if the variability were eliminated from the service times. In a sense, half of the queue can be attributed to service-time variability. The other half can be charged to arrival-time variability.

7.14 NETWORKS OF QUEUES

In many, if not most, of the practical situations you will encounter, the queues will occur not in isolation but as part of an organized system. For example, a factory would ordinarily contain dozens of queues, linked together by the logical sequence of the production process. Such systems often exhibit complicated behavior resulting not only from the direct interaction of arrival and service processes, but such additional factors as branching, merging, and looping of the traffic streams.

Perhaps the most obvious approach to dealing with networks of queues is to separate them into subsystems, each of which has only one queue, and then analyze each subsystem individually. This would permit use of the rich variety of models for single queues, while immediately extending the range of applications to systems of arbitrary size. Unfortunately, there are two problems with this approach: (1) it does not always work, in the sense of yielding valid results, and (2) even when it is technically correct, it neglects the interactions among the queues, which are often the most critical aspect of network behavior.

There is a class of queueing networks for which the decomposition strategy works very effectively. Provided that all of the required conditions are satisfied, each queue and its associated servers will act, and can be modeled, exactly as if it were an independent Markovian queue of the $M/M/c$ type. (Strictly speaking, the internal traffic streams are not always Poisson processes in these networks, but the results come out the same regardless.) The required conditions are:

1. All external arrivals to the network occur in independent Poisson streams. There may be several different streams entering at different points.
2. All service times are negative exponentially distributed with rates that depend at most upon the local queue.
3. All queues have unlimited capacity. Blocking or overflow are not permitted.
4. Any branching of the internal traffic streams is probabilistic, with probabilities that are independent of everything except the position in the network. In other words, after completing service at subsystem i, a customer would go next to subsystem j with probability p_{ij}. (One can think of the sequence of stations visited as a discrete-time Markov chain with one absorbing state.)

Any network of queues satisfying these conditions is said to be an "open network having a product form solution." The "open" part of the description refers to the fact that arrivals from the outside are accepted, and the product form description refers to the idea that the network factors into independent subsystems. The analysis of each subsystem amounts to straightforward application of the *M/M/c* results. The only step that may not be obvious is the calculation of the net arrival rate to each subsystem, since it comprises (possibly) a direct external arrival stream and several internal traffic streams coming from other subsystems. In the most general case, where each of k subsystems has its own external arrival stream (at rate λ_i) and every subsystem is connected to every other, the effective arrival rates e_i are given by the solution to the so-called "traffic equations":

$$
\begin{aligned}
p_{11}e_1 + p_{21}e_2 + \cdots + p_{k1}e_k + \lambda_1 &= e_1 \\
p_{12}e_1 + p_{22}e_2 + \cdots + p_{k2}e_k + \lambda_2 &= e_2 \\
&\vdots \\
p_{1k}e_1 + p_{2k}e_2 + \cdots + p_{kk}e_k + \lambda_k &= e_k
\end{aligned}
$$

The matrix of coefficients is the transient submatrix of an absorbing Markov chain (where absorption is equivalent to the departure of a customer from the network), and will be nonsingular in all practical situations. Hence, there is always a unique solution.

There is another class of computationally manageable networks of queues called "closed networks having a product form solution." Conceptually, these networks differ from the open type by maintaining a captured population of customers; there are neither arrivals nor departures. Although this class is somewhat more difficult to handle, it has the distinct advantage of representing some dependency among the subsystems. That is, the closed network acts as a coherent system, not just a collection of unrelated components. For this reason, the class is very useful in applications, even when the open network would seem the more natural choice. One can always define a dummy server to act as the "outside" generator of arrivals to the rest of the system to convert a superficially open network into a closed one.

The assumptions required for a closed network to be of product form (which in this case means that the network "almost factors") are the same as those required for the open network, excluding the first. However, the results for closed networks are generally less sensitive to deviations from the assumptions than are the results for open networks. This insensitivity property provides another reason to employ the closed network methods in practical applications, since the conditions will rarely be met very precisely.

As mentioned, one must pay a computational penalty to deal with the closed network case, but the methods are very effective at solving even large networks. Because the methods are iterative in nature, a computer is required. A full description of the theory and computational methods for closed networks can be found in Kleinrock (7).

7.15 CONCLUSIONS

It would be nice to be able to claim that most important queueing models have now been covered, and that, having completed this chapter, you should be equipped to handle the majority of situations you are likely to encounter. Regrettably, such is not the case. Real-life problems typically involve complexities not even touched on here. Often, the only practical analysis tool turns out to be simulation. On the other hand, there are many insights to be gained into even very complicated situations from studying the simpler models, and one should certainly attempt to develop this kind of understanding before resorting to simulation.

Recommended Readings

Hopefully, this introduction has provided some motivation to study the subject more thoroughly. Any of the textbooks (2), (3), (4), (5), (7), or (14) would provide a good entry into the standard methods of queueing theory. For an emphasis on applications, complete with detailed case studies, see (8) or (9).

REFERENCES

1. Conway, R. W., W. L. Maxwell, and L. W. Miller, *Theory of Scheduling*, Addison-Wesley, Reading, Mass., 1967.
2. Cooper, R. B., *Introduction to Queueing Theory*, 2nd ed., Elsevier North-Holland, New York, 1981.
3. Cox, D. R., and W. L. Smith, *Queues*, Wiley, New York, 1961.
4. Giffin, W. C., *Queueing: Basic Theory and Applications*, Grid, Columbus, Ohio, 1978.
5. Gross, D., and C. M. Harris, *Fundamentals of Queueing Theory*, Wiley, New York, 1974.
6. Khintchine, A. Y., *Mathematical Methods in the Theory of Queueing*, Hafner Publishing Co., New York, 1960.
7. Kleinrock, L., *Queueing Systems, Vol. 1,* Wiley, New York, 1975.
8. Kleinrock, L., *Queueing Systems, Vol. II*, Wiley, New York, 1976.
9. Lee, A. M., *Applied Queueing Theory*, St. Martin's Press, New York, 1966.
10. Newell, G. F., *Applications of Queueing Theory*, Chapman and Hall, London, 1971.
11. Saaty, T. L., *Elements of Queueing Theory*, McGraw-Hill, New York, 1961.
12. Stidham, S., Jr., "A Last Word on $L = \lambda W$," *Operations Research, 22*(1974), 417–421.
13. Takacs, L., "On Erlang's Formula," *Am. Math. Stat., 40*(1969), 71–78.
14. White, J. A., J. W. Schmidt, and G. K. Bennett, *Analysis of Queueing Systems,* Academic Press, New York, 1975.

EXERCISES

1. What is meant by *queue discipline*?
2. Describe Kendall's notation for identifying queueing models with examples.
3. What is Little's formula and what is its use?
4. Describe some of the performance measures used in analyzing queues.
5. What is a Markovian queueing model?
6. What is the meaning of a finite source population?
7. Describe the principle of *flow balance* and its uses.
8. What is a transition diagram and what does an arc in a transition diagram represent?

9. Set up the transition diagrams for the following queueing system models:
   (a) $M/M/2/4$
   (b) $M/M/3/3$
   (c) $M/M/1/3/3$

10. Write and solve the steady-state equations for the above models.

11. Using the "stair-step" graphical conventions of Section 7.2,
    (a) Identify the graphical correspondence to an idle server.
    (b) Figure out how the graph would be affected by multiple servers, and by limited waiting room. Redraw the block graph under these changed conditions.
    (c) Let $D(n)$ represent the delay experienced by the nth customer, $S(n)$ represent the service time of the nth customer, and $T(n)$ represent the time between the nth and $(n-1)$th arrival. Verify, using the graph, that $D(n) = \max[0, D(n-1) + S(n-1) - T(n)]$.

12. The following questions make use of the relations in Section 7.5. No particular assumptions are made about the distribution of interarrival or service times.
    (a) If a factory maintains an average in-process inventory equivalent to 300 work orders, or jobs, and an average job spends 6 weeks in the factory, what is the production rate of the factory in units of jobs per year?
    (b) If an average of 3000 people per day pass through an airport lobby, and the average time spent in the lobby is an hour and a half, what is the average number of people occupying the lobby?
    (c) If a queueing system has three servers, the average number of customers in the system is 6.4, and the average number in the queue is 4.0, what is the average utilization of each server?

13. The text provides formulas for L for the $M/M/1$ and $M/M/1/N$ queueing systems. Find the formulas for W and W_q for both of these systems. Plot the results as functions of ρ to see how the mean waiting times behave as ρ approaches 1.

14. To model a single server, infinite capacity queueing system in which customers are "discouraged" from entering when the queue grows long, let the arrival rate be dependent on the state in the following way: when there are i customers in the system, let the arrival rate be $\lambda/i+1$. Make the usual Markov and stationarity assumptions.
    (a) Show the steady-state equations and solve them if you can. Be sure to note any convergence requirement.
    (b) Explain in words what the parameter λ represents in this model, and how it could be measured. (Note: discouraged customers are never seen.)
    (c) Show an expression for W_q, the steady-state mean time spent in the queue. Be as precise as you can.

15. Consider a barbershop with two chairs, two barbers, and no room for customers to wait. Say that the state of the system is the number of customers in the shop: 0, 1, or 2. If there is an empty chair when a customer arrives, he enters the shop and his haircut begins. If both chairs are occupied when he arrives, he does not enter the shop. As soon as a customer's haircut is completed, he leaves the shop instantaneously. The barbers do not assist one another when there is only one customer in the shop. On the average a customer arrives every 10 minutes, and each haircut takes an average of 15 minutes.
    (a) Set up the transition diagram and steady-state equations.
    (b) Solve the steady-state equations for the distribution of the number of customers in the shop.
    (c) Calculate the expected number of busy barbers.
    (d) Calculate the expected number of customers turned away per hour.

(e) Notice in the above problem that some potential customers are turned away. If the shop had another barber, it might be able to profit from additional paying customers. On the other hand, the extra barber would have to be paid. Describe how you might determine whether it would be worthwhile for the shop to hire another barber.

16. A parking lot for a small shopping center has spaces for 100 cars. Assume that cars arrive according to a Poisson process, and that the durations of shopping trips are negative exponentially distributed. Also assume that a car will not wait for a parking space if the lot is full.

(a) Identify an appropriate queueing model.
(b) Write the steady-state equations and give the solution.
(c) Interpret, in this context, the steady-state probablities, L, W, L_q, and W_q. That is, explain what these mean in terms of the parking lot.
(d) Explain how you would determine whether spaces for more cars would be worth having.
(e) Explain what, if anything, could be done to model the situation if the shopping-trip durations were not negative exponential.

17. A job shop has four numerically controlled machine tools that are capable of operating on their own (i.e., without a human operator) once they are set up with the proper cutting tools and all adjustments are made. Each setup requires the skills of an experienced machinist, and the time needed to complete a setup is negative exponentially distributed with a mean of 1/2 hour. When the setup is complete, the machinist just pushes a button, and the machine requires no further attention until it has finished its lot size and is ready for another setup. The lotsize production times are negative exponentially distributed with a mean of 1 hour. The question is, "how many machinists should there be to tend the machines?" At opposite extremes, there could be one machinist tending all four machines, or there could be one machinist for each of the machines. The optimal number to have obviously depends on a trade-off between the cost of machinists and the cost of idle machines. Of course, machinists are paid the same regardless of how much work they do, but each machine incurs idle-time costs only when it is idle.

(a) Assume that the cost of a machinist (including fringe benefits, overhead, and the like) is \$20 per hour, and that the cost of an idle machine (including lost revenues and interest on the capital invested in the machine) is \$60 per hour of idleness. Using the finite source model suggested in Section 7.11 for each of the possible numbers of machinists, 1, 2, 3, or 4, evaluate the total cost for each alternative and select the optimal number of machinists.
(b) Assume that the cost of a machinist is fixed at some arbitrary value (such as \$20 per hour), and let the cost of an idle machine be expressed as some multiple, k, of that value. Generalize the results of part (a), by finding the ranges over which k may vary to produce the same optimal number of machinists. In other words, for what values of k would one machinist be optimal, for what values would two be optimal, and so on.

18. Consider an airport with two runways, one of which is used solely for take-offs and the other solely for landings. Assume that a plane, whether landing or taking off, will occupy a runway for an average of 2 minutes, where by "occupy" we mean "prevent use by any other plane." Delay on the ground, although unpleasant for passengers, does not pose a safety hazard; delay in the air, on the other hand, is of serious concern. Suppose that F.A.A. regulations specify that the mean delay in the air (i.e., W_q) must not exceed 10 minutes. Assume that planes arrive according to a Poisson process, and construct a model from the information given above to answer the following questions.

(a) What is the maximum tolerable load on the airport in terms of the mean number of planes that can arrive per hour?

(b) Would it make much difference if you were told that the standard deviation of the time that a plane will occupy a runway is 1 minute?

(c) What practical suggestion can you make as to what might be done to increase the load capacity of the airport, short of building new runways?

19. An executive must establish a secretarial staff. He has been allocated enough money to hire either two "class A" secretaries or three "class B" secretaries. In this company, secretaries are classified and paid according to their work efficiency. One "class A" secretary can complete the same job in an average of 2/3 the time that a "class B" secretary can, and is paid accordingly. Thus, at first glance, the two alternatives seem equal. Develop a Markovian queueing model (where the secretaries are the servers and the jobs submitted to them form a single queue with a FCFS discipline) to evaluate which, if either, alternative is preferable to the other in terms of keeping the backlog of work to a minimum.

20. A new hospital is to be built in a particular location. Restricting attention to just one section of the hospital, the maternity wing, imagine that the following information has been obtained:

 1. For the population that the hospital is intended to serve, we may expect an average of 12 deliveries per day. (This estimate is based on the national birth rate, but is confirmed by local data.)
 2. Recovery rooms are no problem. If necessary, rooms outside the maternity wing can be used. Delivery rooms are also not a problem because they are used only briefly for the actual delivery. The potential "bottleneck" is the availability of labor rooms.
 3. The average time that a labor room will be occupied by a patient is 3.5 hours, but of course there is considerable variability in this time. Preparation of a room for the next patient can be accomplished in half an hour, if necessary.
 4. If all labor rooms are occupied, an arriving maternity patient will be directed to another hospital. This is done very reluctantly, but state laws necessitate such action.

 (a) Provide a formula for the fraction of maternity patients who are turned away, as a function of the number of labor rooms designed into the hospital.

 (b) Provide a formula for the occupancy rate of the labor rooms, that is, the long-term average fraction occupied. (This figure is to be multiplied by the room rate to predict annual income from the labor rooms.)

 (c) Formulate a reasonable optimization problem, the solution to which would indicate the appropriate number of labor rooms to provide. (Introduce whatever cost parameters and the like that you feel are needed.)

 (d) Suppose that additional data were to reveal that the standard deviation of the labor room occupancy time is 1 hour. What effect would this information have on your analysis?

21. There is a problem with excessive delays at a self-service copying machine. It is a good machine, but recent increases in demand have strained its ability to keep up. The "rated" speed of the machine is 15 copies per minute, but this does not allow for the time the customer spends aligning the material, checking the darkness, and so on. A time study reveals that the actual service times of customers, allowing for all of the variations that occur, are negative exponentially distributed with a mean of two minutes. Of this, a quarter (or 1/2 minute) is "machine time" and the rest is "human time."

 The company representative offers two proposals. One is to replace the machine with a more expensive, faster machine which has a rated capacity of 45 copies per minute. The other is to introduce another machine which is identical to the original one. The cost of the present machine is, say, k_1 dollars per hour. The faster machine would cost k_2 dollars per hour. The two-machine system would cost $2k_1$. Waiting time is valued at k_3 dollars per hour. The demand rate is 24 customers per hour.

Using Markovian queueing models, show how to evaluate the proposals to find the most cost effective alternative.

22. At a particular exit ramp of a turnpike, there is a single toll booth. It was originally thought that there would be insufficient traffic at this point to justify more than one booth, but the traffic has increased recently. There are now (during rush hour, which is the period of concern) an average of 210 cars per hours, and the average service time is 15 seconds. Two proposals have been made. The first is to add another toll booth identical to the first (assume the traffic would split evenly and at random between the two). The second is to add an automatic (i.e., unmanned) booth that accepts exact change only. That booth would have an average service time of only 5 seconds, but only a third of the arrivals (selected at random) would have exact change. The performance measure of concern is the average delay incurred by the people who exit at this ramp.
    (a) Using an appropriate queueing model, estimate the delay under the present system, the first proposal, and the second proposal.
    (b) Recognizing that the service times are probably not negative exponential but are certainly not constant either, consider whether the results of part (a) should be modified.
    (c) Taking into account any other factors that may have been neglected, what is your recommendation?

CHAPTER 8
INVENTORY MODELS

8.1
INTRODUCTION

Inventory theory has evolved through several stages since it began in the 1920s. At first, it had very simple models that used only a few parameters to capture the key factors. Later these models were embellished to include more details by adding more parameters, but ignored variability and uncertainty. Gradually, probabalistic models were developed in the 1950s to capture the effects of unpredictable demand and lead times.

All of these models suffered from one limitation—they dealt with only one product at a time. The real life inventory facing many people was to manage an enormous variety of items, with enough interrelations among them to pose a management problem. Consequently, a separate data processing oriented subject called inventory control or inventory management evolved. Here the major concern was for organizing and maintaining records rather than optimizing performance. At first, the procedures were manual (card files and the like), but they quickly became computerized as the hardware permitted.

In the early 1970s, a growing recognition of the deficiencies of both approaches fostered a virtual revolution in industrial practice. A technique called Materials Requirements Planning (MRP) came into use. Later, this technique was renamed Manufacturing Resource Planning, without changing the initials. The central idea of MRP is to time the production or acquisition of batches of components so that they are available as they are needed in assemblies. The earliest versions assumed deterministic demand for end products, and would simply work backwards to calculate quantities and timing for the subassemblies and base components. Gradual refinements have extended the programs to make them more realistic. Any contemporary book on operations management should contain an explanation of MRP.

In the late 1970s and early 1980s, another movement emerged to revolutionize manufacturing practice. Stimulated by higher interest costs (which penalize excessive inventories) and by the example of a few very successful Japanese companies, the concept of just-in-time (JIT) production became popular. These shifting philosophies of inventory management indicate that the subject is still alive and growing.

In this chapter we shall explore only the most basic of the classical inventory models. The point of the study is to reveal the most fundamental aspects of inventory systems, to support an understanding of the relationships among the key variables. This is an important point, well worth emphasizing. The models are *not* intended to provide numbers or formulas that can be applied directly. Although an occasional situation may come along with just the right combination of conditions to justify the use of one of these models, such an occurrence will be rare. But you will frequently use the understanding of inventory behavior that you gain from these simple models.

It is worthwhile to attempt to achieve a very broad perspective of the problems that are addressed in inventory theory. In addition to the obvious applications to stocks of physical goods—lightbulbs, toothpaste, raw materials to be used in some production process, and the like—there exist many less obvious opportunities to use the models developed in this chapter. For example, the number of engineers employed by a company or the number of students enrolled in a college can be regarded as inventories. Various modes of natural attrition would comprise what we have called the demand process, and hiring or recruitment would constitute replenishment. The amount of equity capital available for corporate growth can be regarded as inventory. As it is used up, it must be replenished through the issuance of new stocks or bonds. Sometimes it is useful to think not of the physical items but of the space they occupy as the inventory. For example, the space available for new books in a library can be thought of as an inventory. As it is consumed, it must be replenished. These examples only begin to indicate the wealth of opportunity for application of the models and insights obtainable from the study of inventory theory. Figure 8.1 represents an inventory system in its broadest sense.

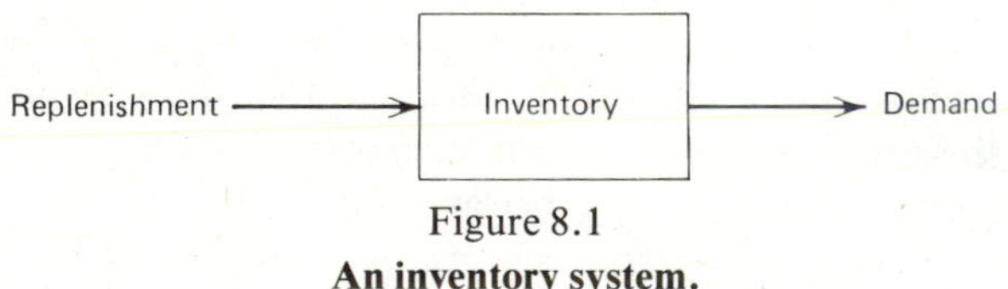

Figure 8.1
An inventory system.

I. DETERMINISTIC MODELS

Among the reasons for holding inventory is to avoid the time, cost, nuisance, and so on, of constant replenishment. On the other hand, to replenish only infrequently would imply large inventories. It is apparent that some sort of trade-off is involved, and the models of this section make this trade-off explicit. These are not the only justifications for inventory control because, of course, there are other reasons as well for holding inventory.

A common feature of these models is that they assume demand to be completely predictable. In those situations where it is *not* (perhaps a majority), they fall far short of representing proper trade-offs. The models should not be criticized on this account—they simply are not intended to represent those situations. At the same time, one should be aware of this limitation. Probabalistic models will be dealt with later in the chapter.

8.2 THE CLASSICAL ECONOMIC ORDER QUANTITY

We begin with the simplest possible version. We assume a single commodity. We penalize frequent replenishment by saying that a setup or ordering charge of $\$a$ is incurred each time an order is placed. We penalize excessive inventory by saying that each unit will cost $\$h$ to store for 1 unit of time (typically, the time unit is 1 year, but may be anything provided that other time units are consistent.) The "driving force" of the system—the process for which the inventory is held—is a steady demand for d units per unit time. All demand must be met without delay. To keep things simple, we assume that replenishment occurs instantaneously whenever an order for replenishment is made (*zero lead time*).

We now give a verbal argument for how the system should operate. Starting at any time with an inventory of y, the inventory will decrease at the rate d until it reaches zero or an order is placed. If it reaches zero, an order must be placed immediately to satisfy all demand without delay.

Would there be any reason to order before inventory reaches zero? No, because that would imply that the remaining items were held in inventory (thereby accumulating carrying costs) unnecessarily. If they were needed at all, they could have been ordered in the next shipment and the carrying cost would have been saved. Figure 8.2, graphically illustrates the pattern of inventory. We call the order quantity for the first order Q_1. By a repetition of the same argument, we can be sure that inventory will again fall to zero (at the same rate, of course) before the second order, of size Q_2, is placed.

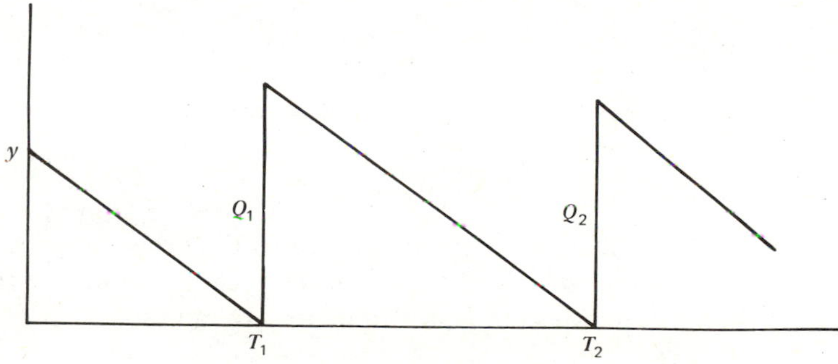

Figure 8.2
The inventory pattern.

Will Q_2 be larger than, smaller than, or equal to Q_1? Without even knowing what determines Q_1 and Q_2, we may argue that they must be equal. Q_1 is determined at time T_1 by taking into account costs and revenues for (possibly) the entire future of the process. But because the future is infinite, and the process is not changing in time, the entire future looks the same at time T_2 as it did at time T_1. Therefore, *however* Q_1 is determined, Q_2 will be determined in the same way and will therefore have the same value. By induction, all order sizes are equal, and the inventory level will behave as in Fig. 8.3. This "sawtooth" pattern is typical of inventory models.

Since the order size is always the same, we will use Q (without a subscript) to denote the common value, and use T to denote the common length of time

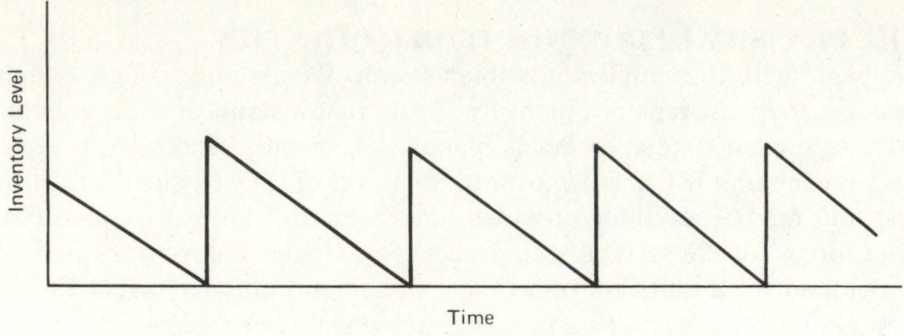

Figure 8.3
The inventory pattern.

between orders. T is just the length of time required to deplete Q units at the rate d, so

$$T = \frac{Q}{d}$$

The only remaining problem is to determine Q. There are a number of ways to set up a cost function to show how costs depend on Q, some of which are fruitless. One approach that will not work is to try to write the sum of all future costs. Because the time horizon is infinite, the sum of all future costs will be infinite no matter what Q is. An alternative would be to "discount" future costs, but since this would involve an additional parameter, it will not be pursued here. Another alternative would be to minimize the cost *per cycle*. This makes some sense, since these cycles are all the same and if you minimize the cost of one, you have minimized them all.

Following up on this idea, we could write

$$\text{cost/cycle} = \text{order cost} + \text{inventory holding cost}$$

$$\text{cost/cycle} = a + \left(\frac{Q}{2}\right)(h)\left(\frac{Q}{d}\right)$$

Here $Q/2$ is the average inventory over the length of the cycle, h is the cost per unit per unit of time, and Q/d is the duration of the cycle. To minimize this function, we differentiate with respect to Q, set equal to zero, and solve for Q.

$$\frac{d}{dQ}(\text{cost/cycle}) = \frac{Qh}{d} = 0$$

$$Q = 0$$

This apparent oddity for a solution is not the result of improper mathematics. The minimum of the cost function written does, indeed, occur at $Q = 0$. The intuitive reason is that we lower carrying costs by reducing Q and shortening the length of the cycle. The setup or order charge remains, of course, and as we shorten the cycle, we incur more and more of these charges over the same period of time. The problem is that we do not really want to minimize the cost *per cycle*, but rather the cost over some fixed interval of time.

Beginning again, we write:

$$\frac{\text{cost}}{\text{unit time}} = \left(\frac{\text{order cost}}{\text{cycle}}\right)\left(\frac{\text{no. of cycles}}{\text{unit time}}\right) + (\text{average inventory})\left(\frac{\text{holding cost}}{\text{unit time}}\right)$$

$$= a\frac{d}{Q} + \frac{Q}{2}h$$

This expression can be justified on its own, or it may be obtained from the cost/cycle by multiplying by the number of cycles per unit time, which is d/Q.

We are interested in this expression as a function of Q. In particular, we wish to select Q so as to minimize the cost rate. (See Fig. 8.4.) The first term behaves like the reciprocal function, decreasing monotonically to zero; the second term is linear in Q, increasing monotonically from zero. Hence, the sum will possess a unique minimum and that minimum will occur where the derivative vanishes. Setting the derivative to zero,

$$\frac{dC}{dQ} = -\frac{ad}{Q^2} + \frac{h}{2} = 0$$

and solving for Q, the optimum value of Q,

$$Q = \sqrt{\frac{2ad}{h}}$$

This is the famous economic order quantity, or EOQ formula. It is also known as the *Wilson-Harris formula*, *the economic lot size*, the "*square-root*" law, and other names.

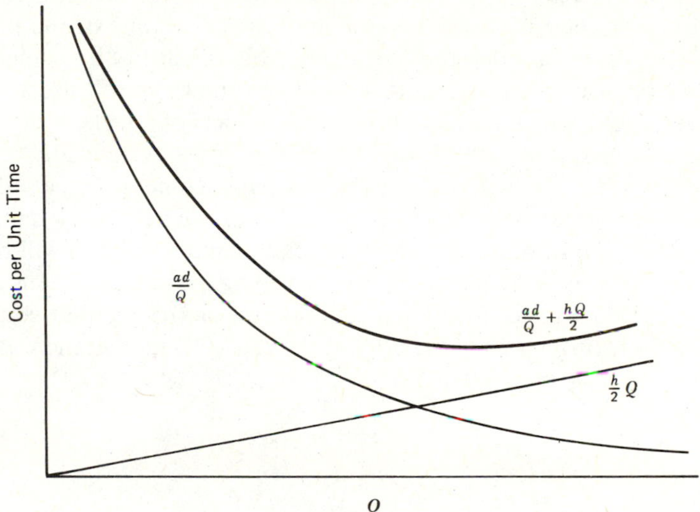

Figure 8.4
The cost rate as a function of Q.

Since $T = Q/d$, the optimal time between replenishments, or the duration of a cycle, is

$$T = \sqrt{\frac{2a}{dh}}$$

and by substituting the optimal Q in the cost function, we find that the minimum cost per unit time is $C = \sqrt{2adh}$. This expression, of course, neglects the cost of the goods. If the latter is desired, it is $d \times$ (cost per item).

Close examination of the EOQ formula reveals that the optimal-order quantity behaves about the way you would expect. Since a is in the numerator, the optimal-order quantity increases when the ordering cost does. Similarly, since h is in the denominator, the optimal-order quantity decreases when the holding cost increases. If the demand rate, d, increases then so does Q. Thus it is not surprising that the formula contains the ratio ad/h. (Incidentally, it should not be necessary to memorize this form, since a little brief reflection along the lines described should be sufficient to reconstruct it.) The only possibly surprising parts to the formula are the 2 in the numerator and the square-root sign. However, even the square-root sign could have been predicted if one considered the dimensions of ad/h:

$$\frac{ad}{h}=\frac{(\text{units/time})(\$)}{(\$/\text{units})/\text{time}}=\frac{\text{units}\cdot\$}{\text{time}}\cdot\frac{\text{units}\cdot\text{time}}{\$}=(\text{units})^2$$

Thus, the square-root sign must appear to make the dimension of ad/h consistent with that of Q. The 2 in the numerator originated from the fact that the average inventory is $Q/2$. A simple mnemonic way to account for it is to say that "the numerator is twice as significant as the denominator because it has two parameters in it."

8.3 A NUMERICAL EXAMPLE

One can hardly begin to appreciate the generality of the model just developed until one has seen the variety of possible applications. Some of the problems at the end of the chapter are designed to suggest this versatility by using situations that do not obviously involve inventory at all. For now, we will use a rather prosaic, typical inventory situation to illustrate the model concretely.

Consider a newspaper publishing concern that must periodically replenish its supply of paper stock. We will suppose that the paper comes in large rolls and that the printers use it up at the rate of 32 rolls per week. The cost of replenishment (which includes the cost of bookkeeping, trucking, and handling) will be taken to be $25 plus the cost of the paper. The cost of keeping the paper on hand, including rent for the space occupied, insurance, and interest on the capital tied up, will be $1 per roll per week. In terms of the notation of the EOQ model, then,

$$d=32$$
$$a=25$$
$$h=1$$

Consequently, the optimal number of rolls to order at a time would be

$$Q=\sqrt{\frac{2(32)(25)}{1}}=40\text{ rolls}$$

The time between orders would be

$$T=\frac{Q}{d}=\frac{40}{32}=1.25\text{ weeks}$$

and the cost of operating the system would be $C=\$40$ per week.

8.4 SENSITIVITY ANALYSIS

One of the important characteristics of the EOQ model is its "robustness." It tends to give reasonably good results even when parameter values are in error. To see why this is so, imagine that one of the values in the example just given were in error by as much as 100%. Say, for example, that the order cost a were really \$50 instead of \$25. Then the correct value of Q should be

$$Q = \frac{\sqrt{(2)(32)(50)}}{1} = \frac{\sqrt{(2)(2)(32)(25)}}{1}$$
$$= (\sqrt{2})40$$

or about 1.41 times as great as the answer previously obtained. In other words, an error of 100% in the input produced only a 41% error in the result. Similar errors in d or h give similar results, as the readers may verify for themselves. Overestimation is no more serious than underestimation. The cost function is also similarly insensitive to parameter errors. It is, of course, the square-root form that provides this very desirable quality.

The conclusion, then, is that the EOQ model may be applied with some confidence even in those situations that do not permit much confidence in the parameter values. This is perhaps the predominant case. Holding costs are very difficult to isolate from fixed overhead costs; some contributions to order costs may be hidden, and so on. It is reassuring to know that the parameter values do not need to be known very precisely to get reasonably good (i.e., close to optimum) results.

On the other hand, the above argument provides no assurance whatsoever that the EOQ formula will give good results when the *assumptions* of the model are not met. This point has been frequently overlooked, or misunderstood, with potentially disastrous consequences. Sometimes students remember that the EOQ formula is "robust" and use that fact to justify its use when, for example, demand is not really predictable. Later, when a probabilistic demand model is developed, an example will be given to illustrate the perils of using the EOQ model for that case.

8.5 NONZERO LEAD TIME

If an order for replenishment must be placed some fixed time in advance—that is, if there is a delay between placement of the order and receipt of the goods—then it is only necessary to anticipate sufficiently far in advance when the inventory will be exhausted and to place the order at the time which is such that the goods will arrive exactly when the inventory runs out. Since (in this model) the demand rate is fixed and known, this presents no problem.

If the lead time is denoted by L, then the quantity dL will be consumed during the lead time. Hence, the order for replenishment should be placed when the stock falls to the level dL. (See Fig. 8.5.)

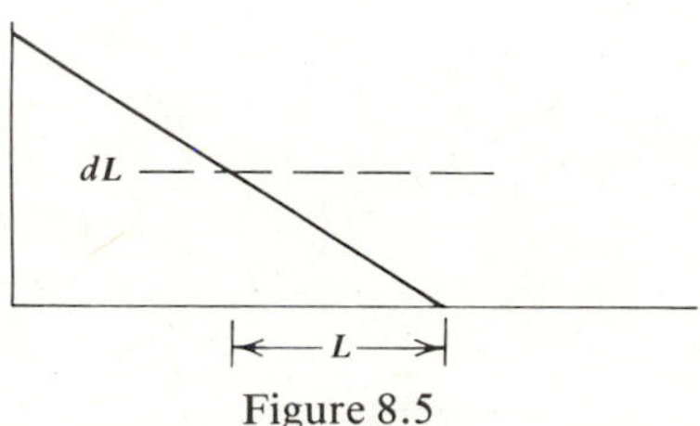

Figure 8.5
Lead-time demand.

8.6 THE EOQ WITH SHORTAGES ALLOWED

If shortages are not prohibited, but merely penalized, it is possible that the optimal policy is deliberately to run out and accumulate back orders before replenishment occurs. Intuitively, the idea is that the cost of running out may be sufficiently small relative to the cost of holding inventory that the cost trade-off results in doing both. The pattern of inventory would still be "saw-toothed" (see Fig. 8.6), but would drop below the zero level. Here, negative inventory represents goods that are "sold" but are not "delivered." In addition to the notation already used, we need

b = backorder penalty cost, proportional to both the number of backorders and time (like h)

S = inventory level just after replenishment (a portion of Q)

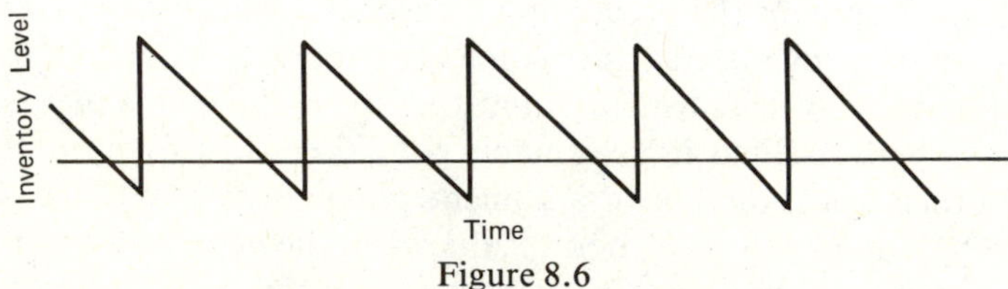

Figure 8.6
Inventory pattern with shortages.

Figure 8.7 may help to identify the terms in the cost function. The cost per unit time is as follows:

$$\frac{\text{cost}}{\text{unit time}} = \left(\frac{\text{order cost}}{\text{cycle}}\right)\left(\frac{\text{no. cycles}}{\text{unit time}}\right) + \left(\frac{\text{holding cost}}{\text{unit time}}\right)(\text{ave. inventory})$$

$$+ \left(\frac{\text{backorder cost}}{\text{unit time}}\right)(\text{ave. backorder}) = a\left(\frac{d}{Q}\right) + h\left(\frac{S^2}{2Q}\right) + b\left(\frac{(Q-S)^2}{2Q}\right)$$

Taking the partial derivatives with respect to Q and S, setting equal to zero, and solving yields the following results:

$$Q = \sqrt{\frac{2ad}{h}\left(\frac{h+b}{b}\right)}$$

$$S = \sqrt{\frac{2ad}{h}\left(\frac{b}{h+b}\right)}$$

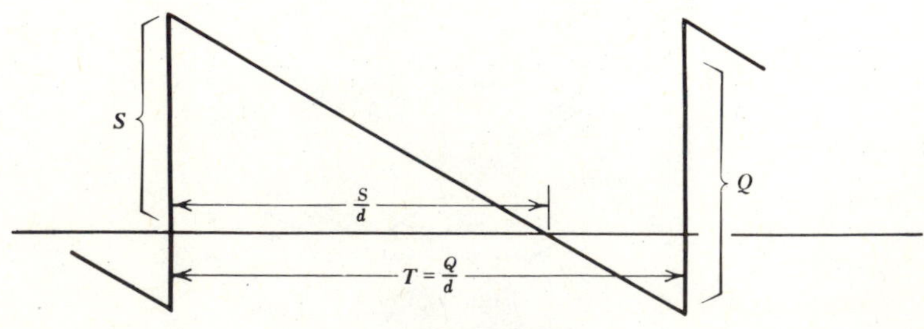

Figure 8.7
The terms identified.

Resubstituting, the minimum cost per unit time will be

$$C=\sqrt{2adh\left(\frac{b}{h+b}\right)}$$

which is smaller by the factor $\sqrt{b/h+b}$ than the optimal cost when back orders are prohibited.

8.7 THE PRODUCTION LOT-SIZE MODEL

Sometimes the replenishment of stock occurs gradually, rather than all at once. Such would be the case if the item is produced internally instead of purchased from an outside supplier. Because this result is more aptly described as a lot size than an order quantity, the model is referred to as the *production lot size*, or PLS, model.

The modification required in the EOQ model is relatively minor. Assuming no shortages are permitted, the inventory pattern would be as shown in Fig. 8.8. Production occurs during the intervals shown as darkened line segments and labeled T_p. Of course, usage also occurs during these intervals, so the inventory never reaches a level equal to the production lot size. Let p denote the production rate, which must be greater than d, the usage rate, in order for the model to make any sense. Starting from an inventory level of zero, production begins and demand continues, so the net rate of increase is $p-d$. This rate of increase continues for the time T_p, which is the time required to produce the total lot size, Q. But $T_p=Q/p$, so the maximum level of inventory reached is

$$T_p(p-d)=Q\left(1-\frac{d}{p}\right)$$

Once the maximum inventory level is established, it is a straightforward matter to formulate the average inventory expression and, thereby, the cost function. After differentiating with respect to Q and solving, the optimal PLS turns out to be

$$Q=\sqrt{\frac{2ad}{h}\left(\frac{p}{p-d}\right)}$$

If, as would often be the case, p is much greater than d, this formula would produce a result not significantly different from that produced by the ordinary EOQ formula.

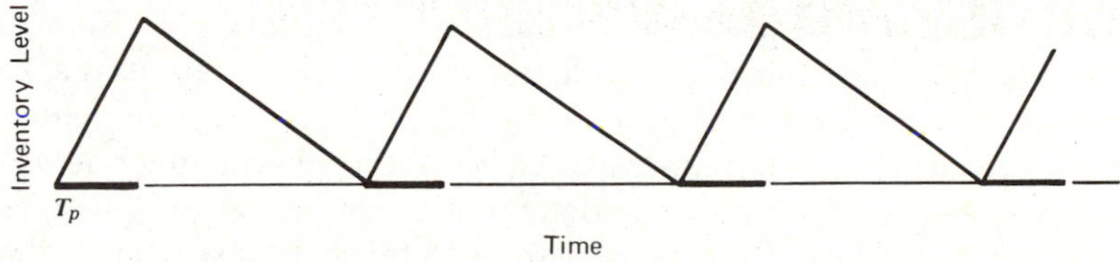

Figure 8.8
The PLS inventory pattern.

8.8 OTHER DETERMINISTIC INVENTORY MODELS

Hansmann (5) gives a variation of the PLS model in which the holding cost is proportional to production costs that are in turn a function of the lot size Q. He also considers a model in which holding cost is a step function related to Q, representing the situation in which additional space must be rented in fixed increments. A third variation makes the demand rate price dependent. Starr and Miller (8) have an interesting analysis of when items should be aggregated and ordered as a group. Buffa (1) addresses the problem of coordinating production-lot sizes for multiple items requiring the same production facilities.

Having seen classical optimization methods used to advantage in determining optimal inventory policies, the reader may already have wondered whether there is not some way to make use of linear programming for the same general purpose. Of course, there is. Example 3.2-4 on p. 87 suggests one such approach. Although it was convenient in that example to use wording indicating that replenishment of stock occurred through production, it could just as well be purchased externally. In fact, when the structure of the linear programming problem is that of a transportation problem and, further, when the costs are such that the "below diagonal" costs in the transportation tableau are all prohibitively large (as they are in Example 3.2-4 and would be whenever back orders are prohibited), it is possible to obtain the optimal solution by hand in a single pass; that is, no iterations will be required. When applicable, this model could be implemented on quite a large scale even without the aid of a computer. The method is described and proved optimal in Johnson (6).

Linear programming models enjoy considerable versatility. They can readily accommodate multiple items, changing demand rates (to represent, say, seasonal variations), and several kinds of constraints. Because large linear programming problems can be solved economically, there is little resistance to embellishing the model with additional aspects of the real-world problem. Of course, the solution will not appear as a closed-form expression, as it did in the models we have seen in this chapter.

There is one significant variation that linear programming models will *not* readily accept. A setup (or order) cost is a fixed charge that is incurred whenever production (or a purchase) occurs but is not incurred in any period for which the production or order quantity is zero. To incorporate such a cost would destroy the linearity of the objective function, and thereby remove the problem from the realm of linear programming. Dynamic programming, described in Chapter 10, is well suited to such problems; Section 10.9 contains an example.

II. PROBABILISTIC MODELS

The principle objective of the remainder of this chapter is to reveal, through a sequence of models, the significance of uncertainty in inventory decisions. There will be no attempt to be complete in cataloging the possible variations that can occur, nor even to be very rigorous in deriving the models that are presented. The mathematics required to incorporate aspects of uncertainty is much more difficult than that required for any of the deterministic models already treated. The temptation to dismiss those aspects—perhaps with the loose justification that variations will average out in the long run anyway—is strong. The major point of this section is that it is perilous to do so.

The first probabilistic model to be considered introduces the uncertainty aspects while dropping the dynamic aspects. The resulting model has a number of direct applications, but also serves as a transition to the more important and more complicated model subsequently considered.

8.9
THE NEWSBOY PROBLEM: A SINGLE PERIOD MODEL

One class of inventory problems requires that the order quantity decision be made only once for the entire demand process. That is, the opportunity to replenish the stock as it becomes depleted does not occur. Given that the total demand over the period in question is uncertain, the dilemma is to order *enough*, so that the full potential for profit may be realized, but not *too much*, so as to avoid losses on the excess. A street corner newsboy faces this dilemma. Although his problem recurs daily, each day's paper is unique and cannot be sold the following day. Hence, his decision on how many to buy may be made without regard to any day but the current one. In fact, since it will not matter when within the day he sells each paper, but only how many are sold by the end of the day, all aspects having to do with the passage of time may be safely neglected.

The model we develop to solve the problem will be oriented toward the newsboy's problem, but there are many inventory problems of a similar nature. Just to list a few suggestive examples, consider how to determine:

1. How many hot dogs to have on hand to sell at a particular ball game.
2. How many Christmas trees to stock for the Christmas season.
3. How much fresh bread to bake for a given day.
4. How many Easter lilies to plant for next spring.
5. How many swimsuits of a currently fashionable style to produce.

The major distinguishing characteristics of the newsboy-type problem are that the order quantity decision is a "one shot" affair and that, despite uncertainty in the demand, a proper trade-off must be achieved between the consequences of too much and too little.

Let c denote the cost per item and s the selling price per item. Then $s-c$ will represent the profit per item actually sold. Let v be the salvage value per item. v may be zero if excess items are utterly worthless, or even negative if it actually costs something to dispose of them. In any case, v is assumed to be less than c, for if it is not, one could profit even from the excess and would therefore desire unlimited quantities. The amount $c-v$ could be called the potential loss per unsold item.

As in earlier models, the order quantity will be represented by Q, and the fixed cost of placing an order by a. With negligible extra effort, we can include a provision for goodwill lost when customers' demands are not met, so let p represent the "lost sales penalty." The units of p will be dollars per item. The demand, being uncertain, will be described by a random variable D whose probability distribution is specified by $p_D(x)$. We may think of $p_D(x)$ as the probability that total demand will equal x.

To aid in formulating an expression for profit in terms of Q, temporarily imagine that x is fixed. There will be two expressions for profit depending on whether x, the amount demanded, is less than or greater than Q, the amount available. In the former case, all x units will be sold for s dollars each, the remainder $Q-x$ will be sold for v dollars each, and no sales will be lost. The cost of ordering the Q items will be $a+cQ$. This leads to the profit expression:

$$P(Q \mid x) = sx + v(Q-x) - a - cQ \qquad \text{if } x \leq Q$$

On the other hand, if demand exceeds supply, all Q items will be sold for s dollars each, there will be no excess items to dispose of at the salvage rate, but there will be a cost of p for each of the $(x-Q)$ demands not met. The cost of ordering the goods would remain the same as before. Consequently,

$$P(Q \mid x) = sQ - p(x-Q) - a - cQ \qquad \text{if } x > Q$$

Now, to get an overall expression for total *expected* profit, each possible profit, as given by the above two expressions, must be weighted by the probability of occurrence and the results summed. The new expression is

$$EP(Q) = \sum_{x=0}^{Q} (sx + vQ - vx - a - cQ)p_D(x) + \sum_{x=Q+1}^{\infty} (sQ - px + pQ - a - cQ)p_D(x)$$

which simplifies slightly to

$$EP(Q) = \sum_{x=0}^{Q} [(s-v)x + vQ]p_D(x) + \sum_{x=Q+1}^{\infty} [(s+p)Q - px]p_D(x) - a - cQ$$

The problem now remaining is to find the value of Q that maximizes this expression.

Although it has been logical up to this point to treat x and Q as integer valued, it will be easier to obtain and to express the solution if they are treated as continuous. Specifically, we would like to be able to differentiate the expected profit function to obtain the maximum. To do so, the sums must be replaced by integrals and $p_D(x)$ by $f(x)$, which must be interpreted as a density function. The revised expression is

$$EP(Q) = \int_0^Q [(s-v)x + vQ]f(x)dx + \int_Q^{\infty} [(s+p)Q - px]f(x)dx - a - cQ$$

This is now a continuous function of Q, having a shape something like that of Fig. 8.9.

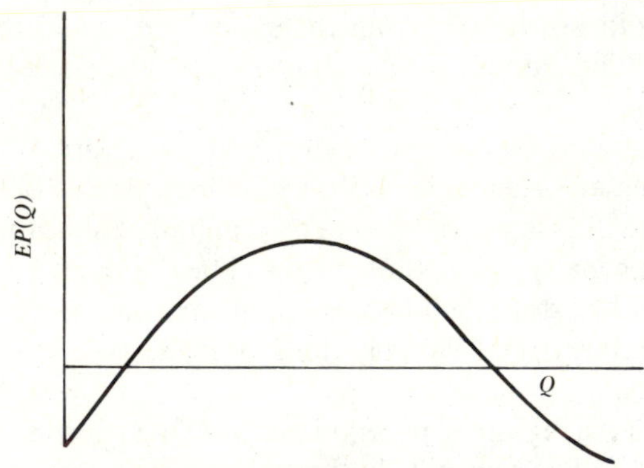

Figure 8.9
The expected profit as a function of Q.

To find the maximum will necessitate taking the derivatives of integral terms involving Q in the limits of integration. There is no difficulty posed by this; the formula is well known as *Leibniz's* rule. Because it may not be familiar to some readers, the rule is given here.

If

$$g(u) = \int_{a(u)}^{b(u)} h(u, x)dx$$

then

$$\frac{dg}{du} = \int_{a(u)}^{b(u)} \frac{dh}{du}dx + h[u, b(u)]\frac{db}{du} - h[u, a(u)]\frac{da}{du}$$

Application of Leibniz's rule to $EP(Q)$ gives for the derivative

$$\frac{dEP(Q)}{dQ} = v\int_0^Q f(x)dx + (s+p)\int_Q^\infty f(x)dx - c$$

Because $f(x)$ is a density function, the two integrals sum to 1. Setting the derivative equal to zero and simplifying,

$$0 = v\left[1 - \int_Q^\infty f(x)dx\right] + (s+p)\int_Q^\infty f(x)dx - c$$

$$c - v = (s+p-v)\int_Q^\infty f(x)dx$$

$$\int_Q^\infty f(x)dx = \frac{c-v}{s+p-v}$$

This is about as close as we can come to a closed-form expression for Q, unless we have further information about the nature of $f(x)$. It is not very satisfying as a solution, but it does specify a value for Q. The left-hand side can be interpreted as the probability that demand will exceed Q. In words, then, Q must be selected so that the probability of running out is equal to the value given by the right-hand side.

A solution of this kind is not uncommon among inventory models. We shall encounter it again in the next section. Whenever Q is determined by requiring the complementary cumulative distribution of demand to equal a certain expression, such as above, the solution is called a *critical ratio* policy.

Before illustrating the model with numerical values, let us devote some thought to the meaning of the critical ratio expression in this case. The numerator, $c - v$, is just what we earlier identified as the potential loss per unsold item. The denominator, $s + p - v$, can be more easily interpreted by adding and subtracting c in the appropriate places to yield the equivalent $(s-c)+p+(c-v)$. It can now be seen to equal the sum of the potential profit per item sold, the potential loss per unmet demand, and the potential loss per unsold item. The reader may want to reflect on what happens to the critical ratio, and hence to Q, as these unit profits or losses increase or decrease one at a time. Such reflection will be of aid in understanding the nature of the critical ratio solution, as well as in assuring that the results make sense.

Incidentally, if you are bothered by the continuity assumption that was made for mathematical convenience, you may be assured that the final result in the discrete case is completely analogous. That is Q is selected so that

$$\sum_{x=Q}^{\infty} p_D(x) = \frac{c-v}{s+p-v}$$

(or, if exact equality cannot be achieved, Q is the largest value such that the left-hand side barely exceeds the right). Details of the discrete case treatment may be found in Hadley and Whitin (4, p. 298).

To illustrate the model, suppose that the newsboy pays 8 cents per paper and sells them for 15 cents each. If he has any left, he can return them for 1 cent credit each. Lost sales involve no direct cost, so $p=0$. Assume that demand is normally distributed with a mean of 150 papers and a standard deviation of 25. According to the model, Q should be selected so that

$$\int_{Q}^{\infty} f(x)dx = \frac{c-v}{s+p-v} = \frac{8-1}{15+0-1} = 0.5$$

In other words, Q should be chosen so that one-half of the area under the density function should lie to the right of Q. Since the normal distribution is symmetric, this would imply $Q=150$. In this particular case, it is optimal to order a quantity equal to the mean demand.

But now suppose that the newsboy finds an alternative buyer who is willing to pay 5 cents per paper late in the day. That is, suppose that v increases from 1 to 5 cents. Then the critical ratio becomes

$$\frac{8-5}{15+0-5} = 0.3$$

and the optimal Q increases to where only three-tenths of the area under the normal density curve lies to the right. From normal tables, the implied Q is $150+(0.52)(25)=163$.

Hadley and Whitin (4) have several numerical examples worked out (pp. 299–303). They also show how to extend the model to cover multiple items subject to a common constraint, using dynamic programming (pp. 304–307).

The reader may have observed that the fixed cost of ordering, a, does not appear in the critical ratio and will therefore not affect the value of Q. However, there is a situation in which this cost can play an important role. If a is large enough, the optimal profit may be negative, in which case it may be best not to engage in the business at all. Suppose, for example, that the newsboy has fixed daily costs of \$10, independent of how many papers he orders or sells. In the first case calculated above, according to which he should order 150 papers, his expected total profit would be -90 cents. (This is, of course, found by substituting $Q=150$, along with the other parameter values, into the expected profit function.) In other words, if he engages in selling papers at all the *best* he can do is to lose an average of 90 cents a day. Clearly, he would be better off in some other business. In the second case, when the salvage value is much better and the optimal Q is 163, one would expect that his expected profits would be higher. They are indeed, but not enough to justify staying in business. They turn out to be -37 cents per day.

8.10
A LOT SIZE, REORDER POINT MODEL

Having, to this point, examined the deterministic economic lot-size model (with variations) and a probabilistic model in which time plays no role, the objective of this section is to incorporate both probabilistic and time aspects into a single model. The result will share characteristics with each of the aforementioned models, but will obviously be more complicated than either.

To begin, we assume that the status of the inventory is known at all times. In some real-life situations, this assumption would not hold. The kind of inventory system in which inventory is "taken" and therefore is of known quantity only at periodic intervals is called a *periodic review* system. In the current discussion, the system is one of *continuous review*.

There will be a lead time for replenishment, denoted by L. Although it will initially be assumed that L is fixed, we will eventually want to consider the effect of making L a random variable. Even when there is no uncertainty in L, the fact that demand during L is unpredictable implies that the order for replenishment must be initiated in anticipation of the *possibility* of running out. The policy assumed for this model is that the replenishment order will be initiated the moment that the inventory level falls to a certain value called the *reorder point*, denoted r. The value of r is one of the quantities to be determined.

The other decision variable is Q, the order quantity. Because orders always occur when the inventory level is r, and everything else is the same, there is no reason to vary the order quantity from one time to the next.

Regardless of what L, r, and Q may be, there will be a positive probability that the inventory will be depleted before the lead time expires. Consequently, what happens when this occurs must be specified. There are two obvious choices: demands that occur when the stock is out are accumulated as back orders and fulfilled when the order arrives, or such demands are simply never fulfilled. The former case is called the *back-orders case*; the latter is referred to as the *lost sales case*. Both are treated here. In either case, p will be used as the penalty cost per item. That is, if the inventory is exhausted, each unit demanded up until the stock is replenished will cost p dollars. Notice that this penalty is independent of the duration of the stockout period.

In keeping with previous models, the fixed cost of placing an order will be denoted by a, and the holding cost per item per unit time by h. The demand rate will be denoted by d, although this must now be understood to be an expected value of a random variable. The units of d are items per unit time, usually taken to be per year. The probability distribution of demand during a lead time L is given by the density function $f(x)$. The mean of this probability distribution will be represented by μ.

Although it is not apparent in the notation, there is an implicit relation among d, L, and μ. If, for example, d is 600 units per year, and the lead time L is 2 months, or $\frac{1}{6}$ year, then μ would have to be 100 units. More generally, the relation $dL = \mu$ must hold; for if it did not, the expected demand during a lead time would be either greater or less than at other times. Since the three parameters are all related, any one could be expressed in terms of the other two, and we could get by with one less symbol. However, there is little incentive to eliminate any one of the three when you consider their separate, individually useful interpretations. Consequently all three will be retained.

A graph of one possible pattern of the varying inventory level over time is shown in Fig. 8.10. This particular graph would apply to the back-orders case, since the inventory level does go negative at one point. Notice that while L and Q are the same for each cycle, the cycles are not of equal duration nor are the inventory levels just after replenishment (i.e., the peaks) of equal height. For contrast, Fig. 8.11 shows a lost-sales case, in which the inventory can never go below zero. Notice that relative to the back-orders case, the peaks will be higher by an amount equal to the lost sales. In other words, the full order quantity Q goes into inven-

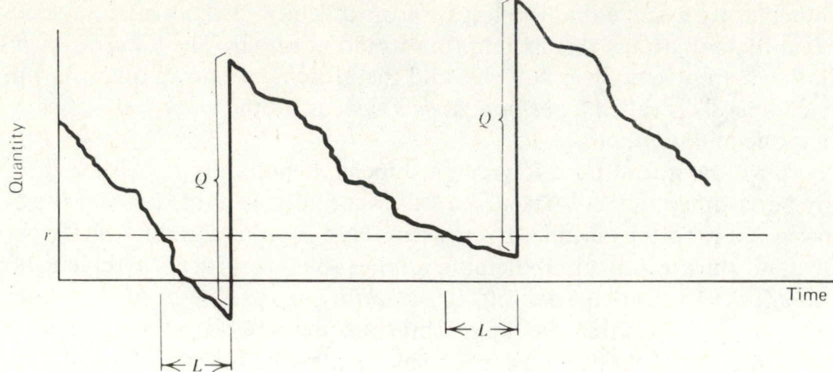

Figure 8.10
Inventory pattern in backorders case.

tory in the lost-sales case, whereas some may be allocated to filling prior demands in the back-orders case. Other factors being equal, then, the average inventory level will be slightly higher in the lost-sales case than in the back-orders case. This observation will be recalled when formulating an expression for the average inventory.

The first step in the formulation of the expected annual-cost function is to identify the separate components of cost. Accordingly, let

$$\text{EAC}(Q, r) = \text{OC} + \text{SC} + \text{HC}$$

where EAC(Q, r) stands for Expected Annual Cost, as a function of the decision variables Q and r, OC stands for Ordering Cost, SC for Stockout Cost, and HC for Holding Cost. The three terms on the right can now be developed separately.

The ordering cost is quite easy to obtain. It is a, the cost per order, times the expected number of cycles per year. Since d is the expected demand per year and Q is the amount sold per cycle, the expected number of cycles per year is d/Q, provided that all demand is met. So, at least in the back-orders case,

$$\text{OC} = a\frac{d}{Q}$$

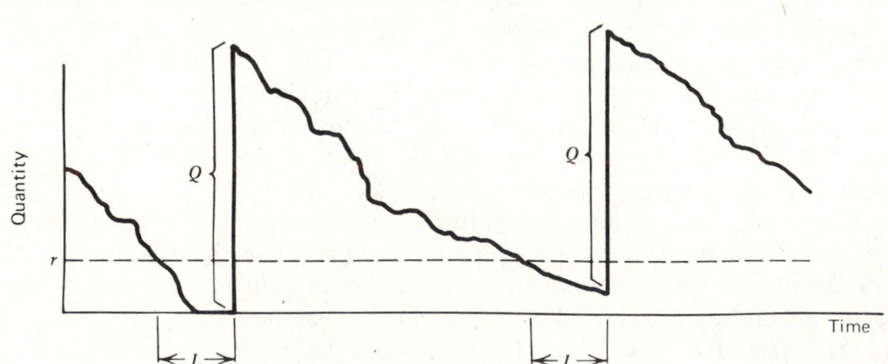

Figure 8.11
Inventory pattern in lost sales case.

In the lost-sales case, this expression is not exactly correct. If we assume, however, that p is sufficiently large to ensure that the policy ultimately used will result in few lost sales, the same expression will suffice as a close approximation to the expected ordering cost even in the lost-sales case.

The expected stockout cost, in either the back-orders or lost-sales case, will be p, the cost per back order or lost sale, times the expected number of back orders or lost sales per cycle, times the average number of cycles per year. The average number of cycles per year is d/Q, by the argument just given, so the only remaining problem is to determine the expected number of back orders or lost sales per cycle. Of course, it is only during a lead time that a stockout situation might occur. It will occur if the demand during that lead time exceeds r, which is the quantity on hand when the lead time begins. So if x represents the demand during a lead time, then the number of back orders or lost sales is

$$\begin{array}{ll} 0 & \text{if } x \leqq r \\ x-r & \text{if } x > r \end{array}$$

To get the expected value of this function of x, we must weight the values by their probability of occurrence and sum. Let $B(r)$ denote the expected number of back orders or lost sales per cycle. Then

$$B(r) = \int_0^r 0 f(x)dx + \int_r^\infty (x-r)f(x)dx = \int_r^\infty (x-r)f(x)dx$$

Further comments will be made later on the subject of how to compute $B(r)$, but for now it will suffice to represent the total expected stockout cost per year as

$$\text{SC} = p\frac{d}{Q}B(r)$$

It remains to express the expected annual holding cost. The argument to be employed now is far from rigorous, but it does yield the correct expression and gives a plausible explanation of where the terms come from. A properly cautious development can be found in Hadley and Whitin (4, pp. 175–188). The expected holding cost per year will be h, the holding cost per item per year, times the average inventory over a year. But the average inventory over a year will be the same as the average inventory over a "typical" cycle, where by a typical cycle we mean one whose behavior is identical to the expected behavior of all cycles. That is, it begins at an inventory level equal to the expected level at the beginning of all cycles, lasts a duration equal to the expected duration, and so on. We cannot hope to observe such a cycle in reality; it is merely a conceptual device to aid in formulating the expression for the average inventory level. The device "works" only because expected values, or averages, behave pretty much as intuition would suggest.

Imagine, then, this typical cycle, and consider the back-orders case. The cycle begins when an order arrives. Our only definite information about the inventory level is that it was r when this arriving order was initiated. Since that time, for the period L, demand has continued in its random fashion. The expected demand during L is μ, so the expected inventory level just before the order arrives is $r-\mu$ (which may conceivably be negative), and the expected inventory level at the start of our cycle is $Q+r-\mu$. The demand process then depletes the inventory, another replenishment order is placed, and the cycle ends with an expected inventory level of $r-\mu$ (see Fig. 8.12). The average inventory over this cycle is midway between $Q+r-\mu$ and $r-\mu$, or

$$\frac{Q+r-\mu+r-\mu}{2}=\frac{Q}{2}+r-\mu$$

By this argument, the expected annual holding cost expression is

$$\text{HC}=h\left(\frac{Q}{2}+r-\mu\right)$$

But this argument was for the back-orders case. It is apparent from Figs. 8.11 and 8.12 that, other things being equal, the average inventory in the lost-sales case will be slightly higher than in the back orders case. The expected inventory level just before the order arrives is not quite $r-\mu$, because now the inventory is not allowed to go negative. To see what it is instead, let x be the demand during the lead time L. Then the inventory level is

$$\begin{array}{ll} r-x & \text{if } x \leqq r \\ 0 & \text{if } x > r \end{array}$$

Taking the expected value of this function of x gives for the expected inventory level

$$\begin{aligned}\int_0^r (r-x)f(x)dx+\int_r^\infty 0f(x)dx &= \int_0^r (r-x)f(x)dx \\ &= \int_0^\infty (r-x)f(x)dx-\int_r^\infty (r-x)f(x)dx \\ &= r\int_0^\infty f(x)dx-\int_0^\infty xf(x)dx+\int_r^\infty (x-r)f(x)dx \\ &= r-\mu+B(r)\end{aligned}$$

In words, the expected inventory level just before replenishment is greater by an amount equal to the expected number of lost sales. The average inventory can be found as in the back-orders case. It works out to give for the expected annual holding cost in the lost-sales case

$$\text{HC}=h\left(\frac{Q}{2}+r-\mu+B(r)\right)$$

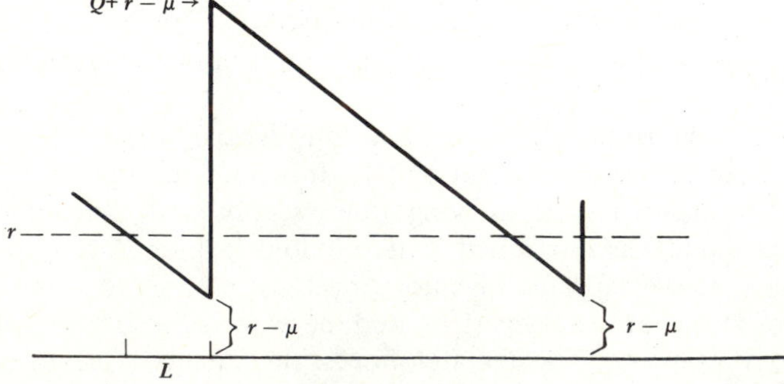

Figure 8.12
The typical cycle.

The total expected annual cost is therefore

$$\text{EAC}(Q, r) = \frac{ad}{Q} + \frac{pd}{Q}B(r) + h\left(\frac{Q}{2} + r - \mu\right)$$

in the back-orders case, and the same expression plus $hB(r)$ in the lost-sales case. This is the function of Q and r which is to be minimized; that is, the order quantity and reorder point that yield the smallest expected annual cost have to be found. The function happens to be convex. Barring extraordinary circumstances, the minimum will occur for values of Q and r that are nonnegative and finite, and will occur at that point where the partial derivatives with respect to Q and r both vanish.

So, taking the derivative with respect to Q,

$$\frac{\delta \text{EAC}(Q, r)}{\delta Q} = -\frac{ad}{Q^2} - \frac{pdB(r)}{Q^2} + \frac{h}{2}$$

setting it equal to zero and solving for Q, we find

$$Q = \sqrt{\frac{2d[a + pB(r)]}{h}}$$

Note the similarity of this expression to the classical, deterministic economic-order quantity formula. It appears that the optimal-order quantity when demand is probabilistic is the same as it would be in the deterministic case if a were increased by the amount $pB(r)$. Since p is the cost per back order or lost sale and $B(r)$ is the expected number of back orders or lost sales per cycle, $pB(r)$ is the expected cost attributable to stockouts per cycle. If you think about it, the addition of this quantity to a makes some sense. Of course, r and therefore $B(r)$ are still unknown.

Returning, then, to the expected annual-cost function and taking the partial derivative with respect to r, we get

$$\frac{\delta \text{EAC}(Q, r)}{\delta r} = h + \frac{pd}{Q}\frac{dB(r)}{dr}$$

in the back-orders case, or

$$\frac{\delta \text{EAC}(Q, r)}{\delta r} = h + \left(\frac{pd}{Q} + h\right)\frac{dB(r)}{dr}$$

in the lost-sales case. Recalling that

$$B(r) = \int_r^\infty (x - r)f(x)dx$$

and making use of Leibniz' rule to differentiate the integral (as in the newsboy problem),

$$\frac{dB(r)}{dr} = -\int_r^\infty f(x)dx$$

Substituting this back into the partial derivative, setting it equal to zero, and solving, we get

$$\int_r^\infty f(x)dx = \frac{hQ}{pd}$$

in the back-orders case, or

$$\int_{r}^{\infty} f(x)dx = \frac{hQ}{pd + hQ}$$

in the lost-sales case.

As was true in the newsboy problem, the best expression for r that we can obtain is a critical ratio solution. The left-hand side is the complementary cumulative distribution of lead-time demand evaluated at r, or the probability that the demand during L exceeds r. In words, then, the reorder point r must be set high enough so that the probability of running out is just equal to the critical ratio. It would be worthwhile for the reader to verify to his own satisfaction that the critical ratios make sense. For example, if p is very large, the critical ratio will tend to be small, which will in turn force the reorder point to be large enough to provide a generous margin of safety against stockouts.

Although we now have expressions for both Q and r, the problem is not solved because the evaluation of each of the expressions requires a knowledge of the value of the other. To get Q we need r, and to get r we need Q. Fortunately, there is an iterative approach that works out quite nicely in practice. It proceeds as follows:

1. As an initial, temporary value, let $B(r) = 0$, and solve for $Q_1 = \sqrt{2d[a + pB(r)]/h}$. This would amount to solving for the deterministic optimal-order quantity.
2. Using Q_1, find r_1 from the critical ratio rule:

$$\int_{r_1}^{\infty} f(x)dx = \frac{hQ}{pd} \qquad \text{(back-orders case)}$$

$$= \frac{hQ}{pd + hQ} \qquad \text{(lost-sales case)}$$

3. Using r_1, evaluate

$$B(r_1) = \int_{r_1}^{\infty} (x - r_1) f(x)dx$$

4. Using $B(r_1)$, solve for Q_2 as in step 1.
5. Using Q_2, find r_2 as in step 2.
6. Continue in this manner until no change occurs in the values of Q_i and r_i. The final values obtained are the optimal-order quantity and reorder point.

In practice, convergence is usually rapid. Often Q_2 and r_2 will be so close to optimal that further iterations are not worth the trouble.

Still, compared to the immediately calculable formula available for the deterministic case, the solution procedure required for this model is cumbersome. Although it would pose no difficulty for implementation as a computer algorithm, it would not seem realistically practical for use as a hand method in a day-to-day real-world setting. On the other hand, it is not inconceivable that an abbreviated procedure, simply and carefully explained with appropriate tables provided, could be made operational as an everyday decision tool. Here the solution is to be employed more as a pedagogical tool, to help in reaching a deeper understanding of the significance of uncertainty in demand to inventory systems. Once we know how important a factor it is, we can decide intelligently how to (or indeed, whether to) include it in our considerations. It may be, at least under certain conditions, that the error introduced by ignoring demand uncertainty is so minor that the considerable effort necessary to include it is unwarranted.

Because we do not have a closed-form solution, but only an iterative approach for obtaining it, a direct evaluation is difficult. For the most part, we will resort to numerical examples to show what difference it makes to include uncertainty. First, however, it is necessary to devote some attention to the evaluation of the integral involved in $B(r)$.

Recall that

$$B(r)=\int_r^\infty (x-r)f(x)dx$$

where $f(x)$ is the lead-time demand density function. When $f(x)$ is complicated, as it may reasonably be expected to be most of the time, the integration may not be trivial. Perhaps the most important special case occurs when the lead-time demand distribution is normal, in which case

$$B(r)=\int_r^\infty (x-r)\frac{1}{\sigma\sqrt{2\pi}}\exp\left[-\frac{1}{2}\left(\frac{x-\mu}{\sigma}\right)^2\right]dx$$

Fortunately this rather imposing expression can be translated to an expression involving only the standard normal density and cumulative distribution functions, which are readily available in tables. After some preliminary manipulation,

$$B(r)=\int_r^\infty [(x-\mu)+(\mu-r)]\frac{1}{\sigma\sqrt{2\pi}}\exp\left[-\frac{1}{2}\left(\frac{x-\mu}{\sigma}\right)^2\right]dx$$

$$=\int_r^\infty \left[\frac{x-\mu}{\sigma}+\frac{\mu-r}{\sigma}\right]\frac{1}{\sqrt{2\pi}}\exp\left[-\frac{1}{2}\left(\frac{x-\mu}{\sigma}\right)^2\right]dx$$

a change of variable from x to $y=(x-\mu)/\sigma$ produces

$$B(r)=\int_{(r-\mu)/\sigma}^\infty y\frac{1}{\sqrt{2\pi}}e^{-(1/2)y^2}\sigma dy+\int_{(r-\mu)/\sigma}^\infty \left(\frac{\mu-r}{\sigma}\right)\frac{1}{\sqrt{2\pi}}e^{-(1/2)y^2}\sigma dy$$

$$=\sigma\int_{(r-\mu)/\sigma}^\infty y\frac{1}{\sqrt{2\pi}}e^{-(1/2)y^2}dy+(\mu-r)\int_{(r-\mu)/\sigma}^\infty \frac{1}{\sqrt{2\pi}}e^{-(1/2)y^2}dy$$

Now the first integral is easy to evaluate.

$$\int_{(r-\mu)/\sigma}^\infty y\frac{1}{\sqrt{2\pi}}e^{-(1/2)y^2}dy=\frac{1}{\sqrt{2\pi}}\int_{(r-\mu)/\sigma}^\infty ye^{-(1/2)y^2}dy$$

$$=\frac{1}{\sqrt{2\pi}}\left[-e^{-(1/2)y^2}\Big|_{(r-\mu)/\sigma}^\infty\right]$$

$$=\frac{1}{\sqrt{2\pi}}\exp\left[-\frac{1}{2}\left(\frac{r-\mu}{\sigma}\right)^2\right]$$

$$=f\left(\frac{r-\mu}{\sigma}\right)$$

where $f(x)$ is the standard normal density function. The second integral is even easier; it is already tabulated as the complementary cumulative of the standard normal. In summary,

$$B(r)=\sigma f\left(\frac{r-\mu}{\sigma}\right)+(\mu-r)G\left(\frac{r-\mu}{\sigma}\right)$$

where $G(x)$ is the complementary cumulative distribution function for the standard normal. (Just to be sure there is no confusion, the lowercase f refers to the height of the density function, and the uppercase G refers to the area under the right-hand tail of the density function.) Of course, this expression is valid only when the lead-time-demand distribution is normal. If it is, say, gamma distributed, some other method of evaluating $B(r)$ must be found.

The assumption that lead-time demand is in fact normal is frequently justified by the Central Limit Theorem. The total demand can be written as a sum of the demands of individual customers, so if there are enough of them, the total should approach a normally distributed random variable. There are some weaknesses in this argument, but it provides some reassurance that the normality assumption is not completely arbitrary, in many real-world problems.

8.11 SOME NUMERICAL EXAMPLES

For the first illustration of the lot-size, reorder-point model, suppose that a large department store chain sells blank recording tape under its own brand name. The tape itself is identical to that of a well-known recording tape manufacturer, and is in fact supplied to the chain by that manufacturer. Because of the special labeling and packaging, however, there is a lead time of 5.2 weeks, or 1/10 year. Assume that the demand during a period of this length is normally distributed with a mean of 1000 tapes and a standard deviation of 250. By implication, the mean annual demand must be 10,000 tapes. The cost of paper work and handling associated with placing an order is \$100, and the holding cost is 15 cents per tape per year. Although we shall want to see what happens in the lost-sales case as well, assume for now that back orders are taken. The penalty is assumed to be \$1.00 per tape back ordered.

According to the iterative procedure, we first find

$$Q_1 = \sqrt{\frac{2da}{h}} = \sqrt{\frac{2(10{,}000)(100)}{0.15}} = 3651.5$$

This would, of course, suffice as the economic-order quantity if demand were deterministic. Proceeding, we next find

$$\int_{r_1}^{\infty} f(x)dx = \frac{Q_1 h}{pd} = 0.055$$

where $f(x)$ is normal with mean 1000 and standard deviation 250. Standardizing to permit use of the Standard Normal tables,

$$\int_{(r_1-1000)/250}^{\infty} f(x)dx = 0.055$$

we find

$$\frac{r_1 - 1000}{250} = 1.60$$

$$r_1 = 1400$$

Now, to find $B(r_1)$, the expected number of back orders if the reorder point is 1400, we have recourse to the formula

$$B(r) = \sigma f\left(\frac{r-\mu}{\sigma}\right) + (\mu - r)G\left(\frac{r-\mu}{\sigma}\right)$$

which in this case is

$$\begin{aligned} B(1400) &= 250f(1.60) + (1000 - 1400)G(1.60) \\ &= 250(0.111) - 400(0.055) \\ &= 5.822 \end{aligned}$$

This is an accurate result, in the sense that if the reorder point is 1400, then the expected number of back orders per cycle will be 5.82. Of course, the reorder point has not yet achieved is final value.

Beginning the second iteration, we find that the revised-order quantity is

$$\begin{aligned} Q_2 &= \sqrt{\frac{2d[a + pB(r_1)]}{h}} \\ &= \sqrt{\frac{2(10{,}000)[100 + 5.822]}{0.15}} \\ &= 3756.27 \end{aligned}$$

which is slightly larger than the previous value. The critical ratio increases just a little

$$\int_{r_1}^{\infty} f(x)dx = \frac{Q_2 h}{pd} = 0.05634$$

so the reorder point is only slightly lower

$$\frac{r_2 - 1000}{250} = 1.586$$

$$r_2 = 1396.5$$

With a lower reorder point, the expected number of back orders will increase

$$\begin{aligned} B(1397.5) &= 250f(1.586) - 397.5G(1.586) \\ &= 250(0.113) - 397.5(0.056) \\ &= 5.99 \end{aligned}$$

But this time when we reevaluate Q,

$$Q_3 = 3759.86$$

we find that the change in the order quantity is only about 1 unit (versus a change of about 100 units previously). In fact, the change in the next reorder point is not detectable:

$$r_3 = 1396.5$$

At this point, convergence may be considered complete. Since, from a practical standpoint, the order quantity and reorder point must be integers, there is no sense in trying to achieve accuracy beyond the first decimal figure.

The computations are summarized in Table 8.1.

Table 8.1
SOLVING FOR Q AND r

Iteration Number	Q	Critical Ratio	r	$B(r)$
1	3651.5	0.055	1400	5.82
2	3756.3	0.0563	1396.5	5.91
3	3757.9	0.0563	1396.5	5.91

From this table, it is easy to see the direction and magnitudes of the changes that occur from one iteration to the next. It is apparent that the first value for r was close to optimal, but was just a bit too large. One might be content to use this value, particularly since it happens to be a nice round number. The second value for Q appears to be close to optimal; further iterations increase it only slightly. As a practical procedure for hand computation, then, one might suggest finding r only once and Q only twice. Because the convergence of both Q and r to their respective optimal values is monotonic, we may be confident that r_1 is above its optimum and Q_2 is beneath its optimum. Consequently, the reorder point and order quantity could be rounded down and up, respectively, to operationally convenient values. In this case, $r = 1400$ and $Q = 3760$ would be good, realistic answers.

It would be a good idea for the student to attempt to duplicate the above calculations. A certain degree of delicacy is required to avoid computational errors. The expression for $B(r)$ is particularly sensitive to errors that may be introduced almost casually when r is rounded off or a value is interpolated from the normal tables.

The calculations for the lost-sales version of the same problem are, of course, very similar; only the expression for the critical ratio is different. Table 8.2 summarizes the values obtained.

Table 8.2
LOST-SALES CASE

Iteration Number	Q	Critical Ratio	r	$B(r)$
1	3651.5	0.052	1407.5	5.31
2	3747.2	0.053	1402.5	5.67
3	3753.5	0.053	1402.5	5.67

A careful examination of the formulas will reveal that, other things being equal, the order quantity must be lower and the reorder point higher in the lost-sales case as compared with the back-orders case. On the other hand, as the tables for this example show, the differences may be so small as to be negligible. Inasmuch as many real-life situations cannot be classified as either pure back-orders or pure lost-sales cases, but lie somewhere between the two, it is reassuring to know that these variations are not very important. Since we have made this point, our further examination of the results will be limited to the back-orders case.

The minimum cost for the example has not yet been exhibited. To obtain it requires only that the optimal values of Q and r be resubstituted into the cost function.

$$\text{EAC}(Q, r) = \frac{(100)(10{,}000)}{Q} + \frac{(1)(10{,}000)B(r)}{Q} + (0.15)\left[\frac{Q}{2} + r - 1000\right]$$

$$\begin{aligned}\text{EAC}(3757.9,\ 1397.5) &= \$266.10 + 15.73 + 341.31\\ &= \$623.15\end{aligned}$$

Just to verify the intuitive notion that it "shouldn't make much difference" if rounded-off values are used instead of the optimal ones, the cost expression can be evaluated at $Q = 3760$ and $r = 1400$:

$$\begin{aligned}\text{EAC}(3760,\ 1400) &= \$265.96 + 15.48 + 342.00\\ &= \$623.44\end{aligned}$$

As expected, the difference is not worth being concerned about.

An interesting cost comparison can be made between the above expected annual cost and the annual cost that would be incurred if the demand were deterministic. For the latter, the optimal-order quantity would be the first Q_1 found, or

$$Q = 3651.5$$

and the (deterministic) cost associated with operating the system would be \$547.72. It is apparent that a substantial portion of the expected annual cost is attributable solely to the fact that demand is uncertain. In other words, uncertainty is expensive. As the reader may verify, the difference between the two costs is accentuated as σ, a measure of the unpredictability of demand, increases. It is also a fact that (as one would expect) the difference diminishes to zero as σ does.

An even more dramatic comparison can be made between the expected annual cost using the optimal values for the probabilistic model and the expected annual cost using the values that would be obtained from the deterministic model. In other words, what would the penalty be if one used the simple model when, in fact, the more complicated one were appropriate? If uncertainty in the demand is ignored, the optimal Q would be 3651.5 and the r would be 1000. Using these values in the expected annual-cost function from the probabilistic model,

$$\text{EAC}(3651.5, 1000) = \$820.86$$

Compared to the minimum cost of \$623.15, this cost represents a serious departure from the optimum.

8.12 VARIABLE LEAD TIMES

We have seen that the reorder point (and hence the cost) is quite sensitive to the variance of the lead-time demand. This was all done under the assumption that the lead time was itself known with certainty. That is, all variability in lead-time demand was due to the demand process, and none to the supply process. It would be more realistic, surely, to allow for variability in both.

There is not too much additional explanation required in the way of theory. Let $y(x \mid t)$ be the conditional density function of demand, given that the lead time is t. Let $g(t)$ be the density function for the lead time t. Then $f(x)$, the lead-time-demand density function (as used in the previous models) is just the marginal density

$$f(x) = \int_0^\infty y(x \mid t)g(t)dt$$

Hence, if $y(x \mid t)$ and $g(t)$ are known separately, they can—in principle, at least—be combined to give the $f(x)$ used before.

From the practical standpoint, there are not too many cases for which the $f(x)$ will be a familiar, tabulated distribution. One case that can be handled is that in which $y(x \mid t)$ is Poisson and $g(t)$ is of the gamma family. In this case, $f(x)$ turns out to be negative binomial [see Hadley and Whitin (4, p. 117)].

Because the payoff, in terms of added insight, to be obtained from pursuing the general case is meager, an alternative course will now be taken. Suppose that, whatever distributions are involved in $y(x \mid t)$ and $g(t)$, the resulting $f(x)$ is normal. If the demand quantities are sufficiently large and consist of a sufficient number of individual orders, the Central Limit Theorem can be used to justify this assumption. But if $f(x)$ is normal, only the first two moments are required.

We are interested now in how these two moments are affected by the variability in lead time.

Let $E(Y)$ and $V(Y)$ represent the mean and the variance, respectively, of the demand per day. Similarly, let $E(L)$ and $V(L)$ represent the mean and variance of the lead time in days. Finally, let μ and σ^2 denote the mean and variance of the lead-time demand, which are the only parameters required when $f(x)$ is normal. Then using the relation above it can be shown that

$$\mu = E(Y)E(L)$$

which is exactly what one would expect. For example, if the expected demand per day is 50 units and the expected lead time is 16 days, then the expected lead-time demand would be 800 units. However, it can also be shown [Feller (2, p. 164)] that

$$\sigma^2 = E(L)V(Y) + E(Y)^2 V(L)$$

which is perhaps more than one would expect. The first term is not surprising; it is just what you would get if the lead time were fixed. For example, if the variance of the demand per day were 25, then over a 16-day lead time the variances for each day would add up to a total lead-time demand variance of 400. The squaring of $E(Y)$ in the second term, however, tends to accentuate the effect of $V(L)$, the lead-time variance. If the lead time is not fixed, but has a variance of 4, then the total lead-time demand variance would be

$$\begin{aligned} \sigma^2 &= (16)(25) + (50)^2(4) \\ &= 400 + 10{,}000 \\ &= 10{,}400 \end{aligned}$$

It can be seen that the second term has contributed much more to the total than the first term. The implication is that variability in lead times can have a surprisingly significant impact on lead-time demand variability. The latter, it will be recalled, had a surprisingly significant impact on cost.

To get a better feel for the magnitude of the variability generated when both demand and lead time are random variables, it would be preferable to compare standard deviations. Table 8.3 does so for the numerical examples just presented.

Table 8.3
FIXED AND RANDOM LEAD TIMES

	Fixed Lead Time	Random Lead Time
$E(Y)$, mean demand per day	50	50
$\sqrt{V(Y)}$, stan. dev. of same	5	5
$E(L)$, mean lead time	16	16
$\sqrt{V(L)}$, stan. dev. of same	0	2
μ, mean lead-time demand	800	800
σ, stan. dev. of same	20	102

Notice in the second column that the standard deviations are small relative to the means for each of the two random variables considered separately. If one did not know better, it would be easy to assume that the combined effect would be small, and therefore to neglect to consider lead-time variability. But as the table shows, the standard deviation of lead-time demand is more than 5 times as large when lead-time variability is included than it is when this variability is neglected.

In more physical terms, the risk of running out of stock is greatly increased when one is unsure of how long it will be until replenishment occurs. The reorder point and (to a lesser extent) the order quantity should reflect this greater risk. The consequences of failing to account for this source of variability can be severe.

8.13 THE IMPORTANCE OF SELECTING THE RIGHT MODEL

Imagine the following scenario, which has doubtlessly occurred in real life many times. A young engineer, fresh out of college and armed with a whole arsenal of textbook techniques, joins an old, established company. He immediately notices that order quantities and reorder points are determined solely by the subjective judgements of an "old hand" who never even graduated from high school. Eager to make an impression and confident that he can improve on the status quo, the engineer proposes the institution of a modern, computerized system of inventory control based on the (deterministic) EOQ formula. In selling the proposal to his superiors, he goes so far as to predict the savings that could be expected. Taking the one item used in the numerical example given in Section 8.12 as representative, he might predict an annual cost, for this one item, of $547.22. Compared with an experienced cost of, say, $750.00 for the same item under the existing system, the proposal seems quite attractive. The young engineer is patted on the back, and his system is implemented.

About a year later, when the new costs are accumulated, it is discovered that the actual cost under the new system has been, for the one item, $900.00. Similarly disappointing costs are experienced for the other items. As soon as his superiors figure out that the computer is not to blame, that it has operated as it was supposed to, the young engineer is fired. The responsibility for inventory control is restored to the "old hand," who emerges as the hero and only benefactor of the whole chain of events. For ever after, the company management is suspicious of bright young engineers bearing mathematical models. For his own part, the young engineer has "matured" as a result of his tragic experience; he will never again place his faith in models. Thus the "young" engineer becomes an "old" engineer.

The story is a sad one. It is sadder yet because no one in the story ever detected the true culprit. The unrecognized villain in the story is variability or, more precisely, the neglect of it. There was nothing inherently wrong with the model used; the flaw was in using it inappropriately. The EOQ was never intended to apply to situations in which demand is other than completely predictable.

If the young engineer had used the model of Section 8.11 for the item in question, he would have obtained a different order size and reorder point and a less optimistic predicted cost of $623.44. Of course, when uncertainty is present, there is risk associated with any course of action, so he should have been careful to explain that the predicted cost was only an expected value. By exploring the sensitivity of the cost function to various possible demand patterns, he could even place a sort of confidence interval on his prediction. With the right model, the chances are that he *could* improve upon the existing system.

To be sure, it is not always so. The "old hand," with years of experience to develop a feel for the trade-offs involved, is sometimes hard to surpass. An ordinary individual, given no special training but plenty of time to learn from his mistakes, can evolve into a very sophisticated self-adaptive control system. The most serious objection to becoming dependent on such a person is that he cannot be replaced. If his only method depends on an unquantifiable "sixth sense" that cannot be expressed or conveyed to anyone else, his successor will have to repeat the same mistakes to develop the same ability. One reason to turn to more objective methods of decision making is to avoid some of the inevitable disruptions that occur when personnel change.

The correct moral to draw from the story is that mathematical models in general, or inventory models in particular, *can* contribute to impressive improvements in systems, but there is no guarantee that they will do so. A perfectly good model incompetently applied can lead to grossly misleading conclusions. In any case, a model should not be considered an adequate substitute for good judgement.

8.14 CONCLUSIONS

At the conclusion of the last chapter, it was pointed out that it would be a serious mistake to believe that queueing models exist to cover most real-life situations. The same statement must be made with regard to inventory models. It is easy to find real-world problems for which no appropriate model has yet been developed. For example, situations in which the stocks of tens of thousands of individual, but interrelated, parts must be controlled are commonplace. Often the interrelations are so complicated that mere record keeping is a massive task requiring sophisticated information organizing systems. The models of this chapter—indeed, nearly all of the inventory models that appear in the literature—treat only the simplest kinds of interactions. As a consequence, inventory theory has been severely criticized by many practitioners faced with real-world problems; some have dismissed it as wholly useless. But to draw such a conclusion would be just as wrong as to believe that it can solve all problems.

What inventory models *can* do is to provide quantitative decision-making aids in a limited number of cases, and valuable qualitative insights in many other cases. Often, the understanding of the real-world system that is revealed through the construction of a model would not be obtained even from years of direct experience with the system.

Recommended Readings

One of the best resources with which to continue your study of inventory theory would be the book by Hadley and Whitin (4). Although it is old now, it is still considered the classic presentation of the theoretical view of inventory. A somewhat different perspective is offered by, for example, Greene (3) or Plossl and Wight (7). Buffa (1), Hanssman (5), and Starr and Miller (8) are examples of a number of available books that attempt to straddle the "theory versus practice" issue.

REFERENCES

1. Buffa, E. S., and W. H. Taubert, *Production-Inventory Systems: Planning and Control*, Irwin, Homewood, Ill., 1972.
2. Feller, William, *An Introduction to Probability Theory and Its Applications*, Vol. 11, Wiley, New York, 1966.
3. Greene, J. H., *Production and Inventory Control Handbook*, McGraw-Hill, New York, 1970.
4. Hadley, G., and T. M. Whitin, *Analysis of Inventory Systems*, Prentice-Hall, Englewood Cliffs, N. J., 1963.

5. Hanssman, F., *Operations Research in Production and Inventory Control*, Wiley, New York, 1962.

6. Johnson, S. M., "Sequential Production Planning over Time at Minimum Cost," *Management Science*, Vol. 3, 435–437, 1957.

7. Plossl, G. W., and Oliver W. Wight, *Material Requirements Planning by Computer*, American Production and Inventory Control Society, 1971.

8. Starr, Martin K., and David W. Miller, *Inventory Control: Theory and Practice*, Prentice-Hall, Englewood Cliffs, N. J., 1962.

EXERCISES

1. What is lead time and why is it important?

2. What is the unit for the holding cost parameter?

3. What is a newsboy problem?

4. Define reorder point. How is it related to lead time demand?

5. Explain the difference between a *periodic review* system and a *continuous review* system. Is one better than the other? Why or why not?

6. Is it possible to have a negative inventory level? If so, explain why.

7. A cafeteria uses up paper napkins at the rate of 12 boxes per week. They are so bulky that it is a nuisance to store many boxes. For accounting purposes, the company figures that the space costs 20 cents per box per week. The cost of placing an order and the handling that is involved when it arrives, regardless of the size of the order, is $10.00. How frequently should the orders be placed, and for how many boxes?

8. A company that has a definite plan for expansion will need to hire and train 60 new engineers per year. The cost of running a training program is $20,000, independent of the number of trainees. These engineers earn annual salaries of $30,000, so the company prefers not to hire and train them before they are needed. On the other hand, they *must* be available when they are needed; and since it costs so much to train them, they will be trained in advance of the time they are needed and in groups. They receive full salary while in the pool of trained, but not yet needed, engineers. What should the training-group size be, how often will the six-week training sessions be held, and how much will this program cost the company?

9. A pizza delivery man makes regular runs to a particular dormitory each evening. He hates to make a separate trip for each order because of the time it would take, the gas for the truck, and so on. On the other hand, he cannot deliver all at once because the orders come in at various times and the customers want quick service. He figures the orders (from the same dorm) come in at a rate of eight per hour in the evening. The cost of a trip is considered to be 60 cents. A pizza sells for $2.00. The cost of delaying a pizza beyond the normal preparation time, which the customer expects, is estimated to be 1 cent per pizza per minute. The cost has something to do with the reduction of future business from dissatisfied customers. Treating the prepared pizzas as inventory, and the delay cost as a holding cost, determine the optimal number of pizzas to deliver in each trip. What is the average delay? What is the maximum delay?

10. A company retains its cash reserves primarily in the form of short-term certificates of deposit which earn at the rate of 8%. Periodically, however, withdrawals are made to meet payroll and other cash requirements. These outflows occur through a checking account that earns no interest. The transfer of funds from the certificates of deposit involve penalties and service charges amounting to $150 each time a transfer is made. If the outflow of cash from the checking account is $300 per day, how often should the transfers be made? Suppose that an option is available to borrow cash from the bank at an interest rate of 0.033 cents per dollar per day (or 12% per year). It is suggested

that it may sometimes be cheaper to meet immediate cash requirements by borrowing, rather than by transferring funds from the certificates of deposit. Would it ever be advantageous to do so? Explain why it would or would not. If so, how often would transfers from the certificates of deposit occur?

11. Formulate the production-lot size cost function and verify that the solution given in Section 8.7 is correct. Also verify the comment that if p is much greater than d, one might as well use the ordinary EOQ, by calculating the ratio of the PLS to the EOQ when $p = 10d$.

12. A drugstore orders copies of a popular monthly magazine at a cost of 75 cents per copy. At the end of the month, when the next issue is delivered, any leftover copies can be returned to the distributor for a credit of 25 cents. The cover price of the magazine is \$1.50. Monthly demand for the magazine can be regarded as normally distributed with a mean of 50 copies and a standard deviation of 15. How many copies should the store order?

13. A certain vending machine dispenses sandwiches. Each morning, fresh sandwiches are put in, and the excess, if any, from the previous day are removed. Sandwiches cost the vendor 35 cents, and are sold for 85 cents. Day-old sandwiches are sold for 12 cents apiece to a skid-row soup kitchen. Assuming that daily demand is Poisson distributed with a mean of 25, determine the number of sandwiches to put into the machine each day.

14. In Section 8.9, the solution to the newsboy problem was expressed as a critical ratio policy. Suppose that demand is uniformly distributed over the range $[a, b]$. That is,

$$f(x) = \frac{1}{b-a} \qquad \text{for } a \leq x \leq b$$
$$= 0 \qquad \text{elsewhere}$$

Derive an explicit, closed-form expression for Q.

15. In planning a conference to be held next year, the Operations Research Society of America must decide on an appropriate number of hotel rooms to reserve for its members. We cannot be sure how many rooms will be required, but judging from previous conferences, the distribution is approximately normal with a mean of 1000 and a standard deviation of 200. Rooms at the conference hotel normally rent for \$35 per night, but have been offered to ORSA for \$25. If ORSA reserves too many rooms for its members, the excess rooms may be released to the hotel for possible use by other guests, but a charge of \$8 will be assessed for each such room. On the other hand, if not enough rooms are reserved, some members will at least have to pay the higher rate, and may even be forced to stay at another hotel. The inconvenience to a member who does not obtain one of the prereserved rooms is valued at \$7 (this is in addition to the higher room cost). How many rooms should ORSA reserve? *Hint*: Although ORSA, as an organization, does not profit, one can think of the collective benefits received by the members as profit.

16. In Section 8.10, a method was given for evaluating $B(r)$ when lead-time demand is normally distributed. Derive a method for evaluating $B(r)$ when lead-time demand is Poisson distributed, using tables of the complementary cumulative Poisson distribution.

17. Determine Q, r, and EAC (optimal order size, reorder point, and expected cost per year) under the assumptions:
    (a) Deterministic demand at the rate 2000 per year; lead time is $\frac{1}{10}$ year; and no back orders are permitted.
    (b) Same, but back orders are permitted, with penalty cost of \$160 per back order (independent of time).

(c) Probabilistic demand; normally distributed (mean = 200, standard deviation = 40) during fixed lead time of $\frac{1}{10}$ year (so mean annual demand is 2000); back orders permitted with penalty cost of \$160 per back order.

(d) Same as part (c) but back orders are not permitted; lost sales penalty if \$160 per lost sale. In each case, the holding cost is \$16 per item per year, and the ordering cost is \$4000.

18. Solve for the optimal reorder point and order quantity using the data of the example in Section 8.11 but substituting for the lead-time demand distribution the data of Table 8.3. That is, find r and Q when lead-time demand is normal with $\mu = 800$ and $\sigma = 20$, then resolve with $\mu = 800$ and $\sigma = 102$. Substitute the optimal values into the cost function to contrast the cost of fixed versus random lead time.

CHAPTER 9
SIMULATION

I. BASIC CONCEPTS

9.1 INTRODUCTION

Simulation analysis is a natural and logical extension to the analytical and mathematical models inherent in operations research. The preceding eight chapters of this book have been concerned with formulating decision models that not only closely approximate the real-world environment but also produce answers through standard numerical and/or mathematical manipulations of the resulting equations. It is evident that there are many situations that cannot be represented mathematically because of the stochastic nature of the problem, the complexity of problem formulation, or the interactions needed to adequately describe the problem under study. For many situations defying mathematical formulation, simulation is the only tool that might be used to obtain relevant answers.

The word simulation has been used rather loosely in the preceding discussion. Therefore, before we proceed it will be necessary to develop a working definition of this term. The following definition has been adopted from Naylor et al. (41).

Simulation is a numerical technique for conducting experiments on a digital computer, which involves logical and mathematical relationships that interact to describe the behavior and structure of a complex real-world system over extended periods of time.

Simulation has often been described as the process of creating the essence of reality without ever actually attaining that reality itself. Within the context of this chapter, simulation will involve the construction, experimentation, and manipulation of a complex model on a digital computer. The techniques that will be described have also been implemented on modern microcomputers, and elements of our discussion will directly address those possibilities.

Although simulation is sometimes viewed as a ''method of last resort,'' often to be employed when all else fails, recent advances in simulation methodologies, software availability, and technical developments have made simulation one of the most widely used and accepted tools in systems analysis and operations research. Most recent surveys on the use of operations research methodologies place simulation analysis either first or second to linear programming applications. In addition to the reasons previously stated, Naylor (57) has suggested that simulation analysis might be appropriate for the following reasons.

1. Simulation makes it possible to study and experiment with the complex internal interactions of a given system whether it be a firm, an industry, an economy, or some subsystem of one of them.
2. Through simulation, one can study the effects of certain information, organizational, and environmental changes of the operation of a system by making alterations in the model of the system and by observing the effects of these alterations on the system's behavior.
3. A detailed observation of the system being simulated may lead to a better understanding of the system and to suggestions for improving it, which otherwise would be unobtainable.
4. Simulation can be used as a pedagogical device for teaching both students and practitioners basic skills in theoretical analysis, statistical analysis, and decision making.
5. The experience of designing a computer simulation model may be more valuable than the actual simulation itself. The knowledge obtained in designing a simulation study frequently suggests changes in the system being simulated. The effects of these changes can then be tested via simulation before implementing them on the actual system.
6. Simulation of complex systems can yield valuable insight into which variables are more important than the others in the system and how these variables interact.
7. Simulation can be used to experiment with new situations about which we have little or no information, so as to prepare for what may happen.
8. Simulation can serve as a "preservice test" to try out new policies and decision rules for operating a system, before running the risk of experimenting on the real system.
9. For certain types of stochastic problems the sequence of events may be of particular importance. Information about expected values and moments may not be sufficient to describe the process. In these cases, simulation methods may be the only satisfactory way of providing the required information.
10. Simulation analysis can be performed to verify analytical solutions.
11. Simulation enables one to study dynamic systems in either real time, compressed time, or expanded time.
12. When new elements are introduced into a system, simulation can be used to anticipate bottlenecks and other problems that may arise in the behavior of the system.

Simulation analysis always begins with a *model* of the system that needs to be studied. In digital computer simulation this step is accomplished through the construction of a computer program that "describes" the system under study to the appropriate computer configuration. This representation might be in the form of a FORTRAN program, graphical interactive displays, or a complex simulation language such as SIMSCRIPT, SLAM II, or GPSS. Once this step has been completed, the model of the system is acted on and the results of these actions are observed over long simulated periods of time. In essence, the experimenter is acting on the created model rather than the actual system itself. Although simulation analysis is usually performed on a microprocessor or mainframe computer, such a representation is not always needed to study complex problems in operations research. Examples in this chapter clearly illustrate this premise through "hand" simulations. Regardless of the mode used to perform simulation analysis, a great deal of experience is desirable to adequately exploit the real powers of simulation. This background is often best gained through modeling experience, enabling a simulation analyst to create unique skills in this area. For this reason, simulation modeling is often more of an "art" than a science. This art is best cultivated rather than taught, although the basic tools and modeling logic can be gained through diligent study of simulation methodologies. Because art is obtained via experience, our main focus in this chapter will be the scientific and mathematical basis of systems simulation.

9.2 THE PHILOSOPHY, DEVELOPMENT AND IMPLEMENTATION OF SIMULATION MODELING

Simulation is one of the easiest tools of management science to use, but probably one of the hardest to apply properly and perhaps the most difficult from which to draw accurate conclusions. With the widespread use of powerful microcomputers, simulation is readily available to most managers and engineers engaged in operations research activities. With a reasonably knowledgeable programmer, a powerful general purpose simulation language can be obtained for less than $1000 that provides desk-top analysis capabilities for many real-world problems. However, the skills required to develop and to operate an effective simulation model are substantial. The variability or dispersion of simulation results is a significant problem and may require long and complex simulation analysis to draw meaningful conclusions from the simulation.

It is the purpose of this chapter to examine the process of simulation and the necessary tools to perform such analysis. A special emphasis will be placed on the problems associated with the mechanics of simulation modeling itself.

The Simulation Process

It is appropriate to examine the entire process by which simulation analysis is planned and performed. The design of the simulation model itself is a critical portion of any study, but it is not the only one with which the user must be concerned. The activities shown in Figs. 9.1, 9.2, and 9.3 are the major ones in any simulation study. Figure 9.1 covers the presimulation tasks whereas Figs. 9.2 and 9.3 are the actual simulation activities, broken down into design and operational groups of activities.

Presimulation Activities

In any general view of situations in which systems simulation is used, the first activity is a recognition of the problem. This recognition leads directly to the study and analysis of the system itself and culminates in the establishment of an objective directed toward solving the problem. Typical objectives could be categorized as follows:

Characterization of system performance.

- Selection of operation parameters for an existing system.
- Selection of operation parameters for a proposed system.
- Exploration of system behavior.
- Modification of an existing system.
- Design of a new system.

At this stage in the process, the user must evaluate the different tools or techniques available relative to the modelling objective and the system with which one is dealing. If the problem is one of establishing optimal parameters and the system fits or can be made to fit one of the available techniques previously discussed in this text, then this is obviously the most appropriate technique to use. The user could also develop a unique mathematical model if none of the textbook methods apply to the problem. These models are generally developed along two main lines of thought; *deductive* models and *algorithmic* models. Economic order quantities

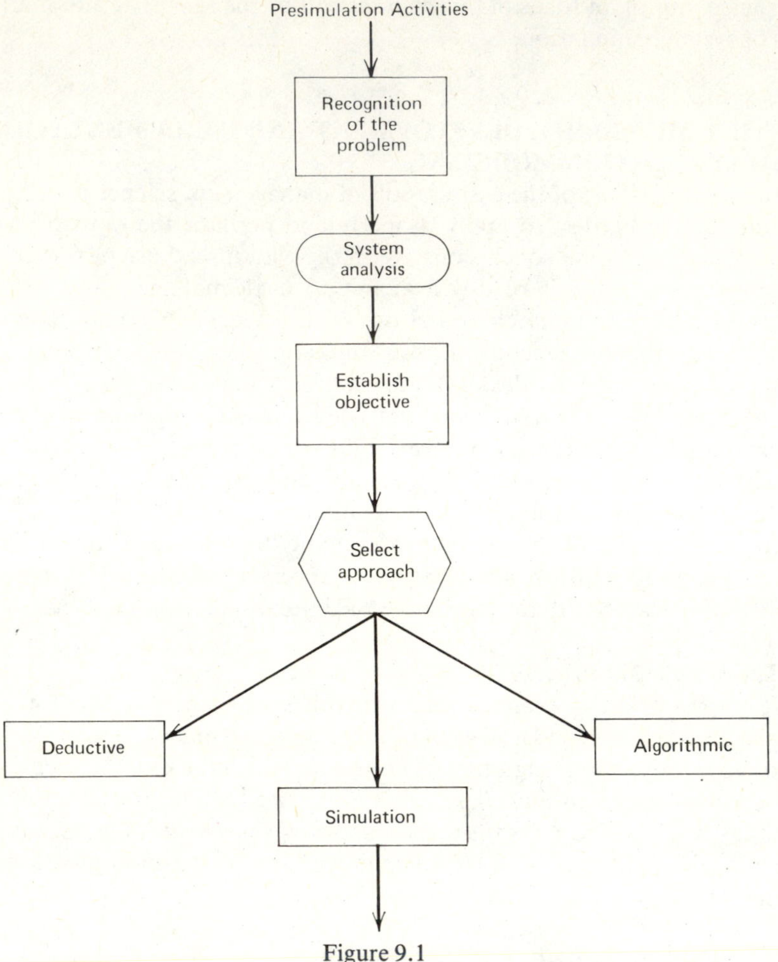

Figure 9.1

(EOQ) in inventory control techniques are examples of deductive solution methods. Algorithmic techniques are inductive, iterative techniques for developing numerical solutions to specific problems. Linear programming is an example of a technique in which an algorithm, usually the simplex algorithm, leads to an optimal solution. If the problem is inherently an optimization problem, and can be cast in a framework for which there is such an algorithm, then that approach might be preferred over simulation. Still another alternative is to experiment with the operating system itself. There are techniques, such as evolutionary operation (EVOP), that provide a means of systematically evaluating changes in the operating parameters of the system. This approach should be appropriate in dealing with systems that are highly responsive to parameter changes.

Simulation is the appropriate technique where it is not feasible to experiment on the system itself or where direct analytic techniques are not available. In the first case, it may be too expensive to experiment with the existing system. Such experimentation might change the operating characteristics of the system itself, and thus the risk that such experimentation might harm the system's performance is so great that the possibility is discarded. It is also possible that the system is just not available for experimentation. Alternatively, if the problem is to design an in-

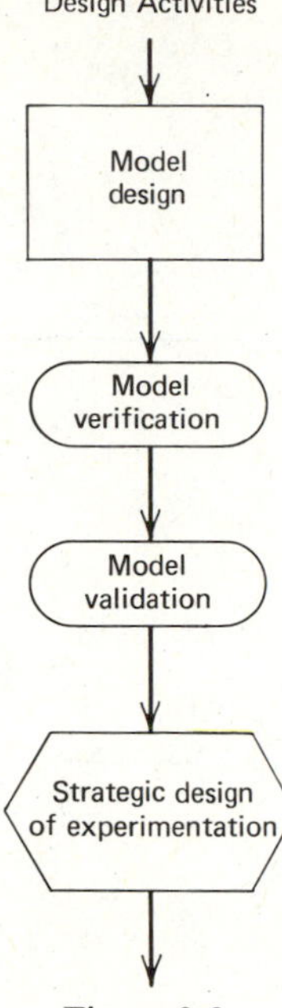

Figure 9.2

ventory control system for a planned facility, the system does not even exist prior to the time it is needed. In the second case, the complexity of many production systems precludes the application of analytic techniques, either deductive or algorithmic. If the system contains many stochastic elements of a complex nature, the resulting model probably resists analytic treatment. The two situations just discussed, where experimentation on the real-world system is not possible and where appropriate analytical techniques are not available, are the general *raison d'être* behind any simulation study. This rationale suggests the following definition of simulation:

> A technique of problem solving based upon experimentation performed on a model of the real-world system.

Developmental Activities

The first developmental activity is the *design and implementation* of the simulation model. The specific tasks in this activity will not be discussed in detail in the remaining sections of this chapter. However, an excellent discussion of this phase is found in many simulation textbooks. Assuming for the time being that these tasks are completed, the next activity is *verification* of the model. A verified model is one that has been proved to behave as its designer intended. This is an important activity in that without satisfactory and explicit verification, it is possible to have a model that appears to work satisfactorily but that gives answers that are actually erroneous. Fishman and Kiviat (9) suggest techniques for verification, including statistical methods, that go far beyond the usual practices of simple comparative analysis and manual checking of calculations.

It is also important to both understand and exercise methods of *model validation*. Validation is one of the most critical activities performed in any simulation

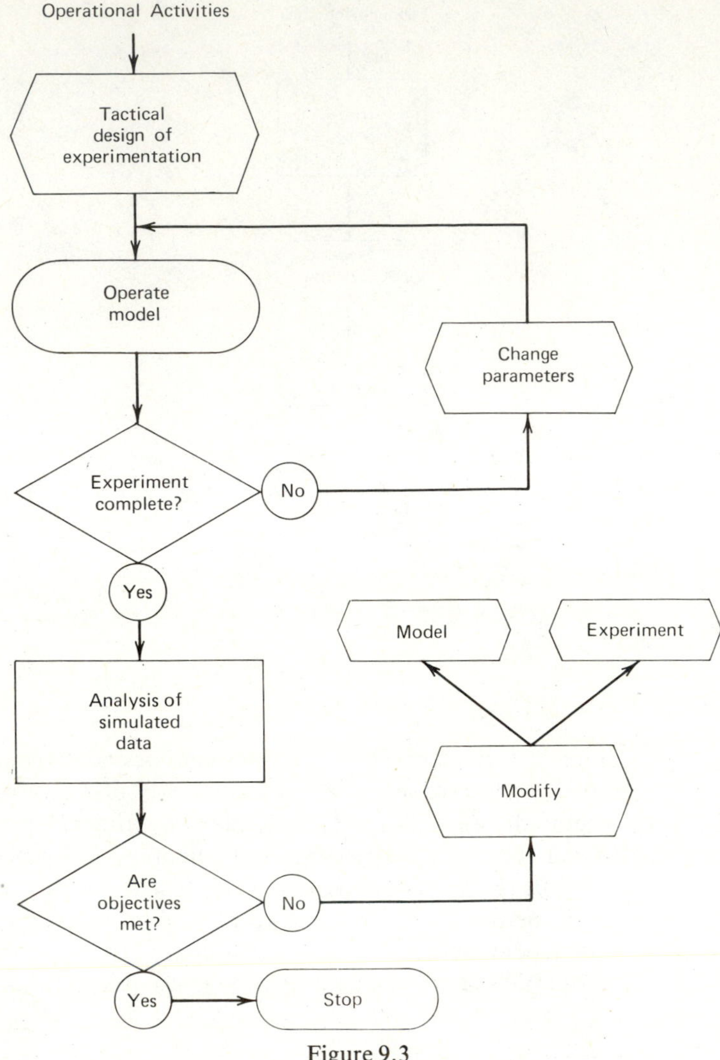

Figure 9.3

study. It is also one of the most difficult to accomplish satisfactorily. A validated model is one that has been proved to be a reasonable abstraction of the real-world system it is intended to represent. The usual approach to validation is to run the model with historical data and compare model results with actual system results for the same historical period. Such comparisons usually are not valid because of the fact that the model may be experimental or predictive in nature. It is also difficult to make a statistical comparison of results in many situations where such a comparison is appropriate, because of the requirement that equilibrium be reached before results are measured. It may take considerable time to reach equilibrium with the computer simulation model while the real-world system might never exist in a state of equilibrium—thereby severely complicating the comparison.

Strategic design refers to the activity of designing and planning the experimentation to be done with the simulation model and is the next activity. It includes specification of information to be determined and the accuracy of that information. The two classes of experimentation, as in the establishment of an objective, deal with the exploration of system behavior and/or optimization of system parameters. Exploration of system behavior is undertaken in an attempt to explain the relationship between results and the controllable parameters of the simulation. Optimization is performed to find the combination of parameter levels that minimize or maximize the results of the simulation. Experimental designs, like full factorial, fractional factorial, and the like are appropriate for exploration experimentation. For optimization, optimum seeking techniques are available, and though in many instances they cannot guarantee a global optimum, they can give good practical results. The article by Hunter and Naylor (22) is a particularly good outline of experimental design for simulation studies, and Schmidt and Taylor (49) have a good discussion on simulation optimization techniques.

Operational Activities

At this stage in the simulation process the model has been designed, implemented, studied, and its use has been planned. The remaining activities are to carry out the actual stimulation experimentation. This must include tactical design of the experiments that are to be performed as the first activity. Conway (7) describes this activity as determining how many simulation runs are to be executed and how data are to be collected from each run. This includes establishing initial conditions for variables in the models and estimating parameters so that the simulated system will reach a state of equilibrium as soon as possible. The user must also determine how equilibrium will be recognized so that data can be gathered without being biased by transients from start up of the run. Other considerations are the sample size required for the data to be gathered and the techniques to be used to compare alternative systems if this is the study objective. In this latter case, the user will be interested in relative results from the simulation runs. He or she can apply methods such as using the same sequence of random numbers for each run, which will reduce the residual variation between sets of results and thus will permit a reduction in length of the simulation runs themselves.

The simulation runs themselves can be represented as a loop, as shown in Fig. 9.3. In this loop, the model is run for the specific time, parameters are changed, and the model is run again. This is repeated until the specific experimentation has been completed. This is followed by an analysis activity in which the simulated data are processed and statistics are developed. Techniques such as regression analysis and analysis of variance are widely used methods of interpreting these data with respect to the original objective.

If the objective has been met, the simulation study is completed at this point. However, because simulation is a trial-and-error process, it will often be that the objective has not been satisfied, which leaves two general alternatives available. The first is to modify the model so that it will facilitate discrimination among simulated systems and then to rerun the experiment. The second alternative is to use the original model but to alter the design of the experimentation, using new search techniques or more powerful experimental designs.

9.3 DESIGN OF SIMULATION MODELS

A well-designed model is a key to meaningful simulation results. If the model builder overlooks the requisites of simulation that have previously been discussed, the simulation study may fall far short of accomplishing its objectives, or may indicate a false achievement when there is none. Likewise, if the model fails to take into consideration the important aspects of the real-world system, then the model cannot be validated. If such a model is used without validation, predictions of system behavior based on the model may be erroneous. The discussion of model design presented here will consider two groupings of design consideration: simulation technique related design considerations and practical restrictions due to computer software availability.

Simulation and the Time Advance Procedures

Time control is generally divided into two classes, though Nance (58) advances the idea of a continuous time-flow mechanism. The two typical types are *uniform time flow* and *variable time flow.* With uniform time flow, the model is advanced and processed through each and every time period simulated at fixed steps or intervals. Variable, or next-event, time-flow mechanism causes time to be incremented between only those periods that have events occurring. The model is processed only at these times, and performance figures are adjusted to account for the periods skipped. If the system has events that occur very frequently and with predescribed regularity (as in forecasting), then uniform time flow might be the best choice. However, if the events are irregular, then the time periods skipped will reduce computer running time and will justify the additional programming generally required for variable time flow.

Simulation is a useful tool because of its ability to handle complex systems that require the modeling of interacting stochastic variates. These are the means for modeling empirical or theoretical distributions of real-world parameters. The stochastic treatment of relevant events greatly increases the realism and applicability of the simulation model. These stochastic variates are generated by manipulation of pseudorandom numbers calculated by the computer. Most computer facilities include a generator of this kind in their software libraries, or can easily develop one. However, both random-number generation and stochastic-variate generation will be discussed in some detail as they are fundamental to any simulation analysis. Variance reduction methods, the next of the technique-related topics, will not be discussed in any detail, even though this can be an important means of reducing the time and cost of using simulation. These methods are generally applied to the scheme in which the simulation model utilizes random numbers. The result is that the variance of response variables is reduced substantially, and this can be used as a basis for reducing the number of iterations required for developing results at the desired level of confidence.

The topics already discussed are often to some degree dependent on the simulation language chosen for the study. If a simulation language is utilized, the time-flow mechanism and often methods of random-number generation and stochastic-variate generation are already specified within the confines of the language. Some of the general-purpose simulation languages being used today include the following:

SIMSCRIPT 2.5	SIMAN
SLAM II	MAST
DYNAMO	SEE-WHY
GPSS V	GEMS II
SIMULA	MAP I
SOL	IDSS 2·0

The detailed discussions of these languages can be found in the technical literature and can be helpful when one is selecting a language to use. A brief discussion of selected simulation languages can be found in Section 9.18.

Conclusions

Simulation is a useful and appropriate management science technique for use in the analysis of complex problems. The technique permits the testing and evaluation of current, proposed, or conceptualized systems without risk to current system performance or the need for real-world experimentation. It also permits the study of complex problems where direct analytic solution is not possible. Simulation is also an easy technique in which to develop proficiency, although it often requires a wide range of skills. The remainder of this chapter will be devoted to the task of developing these skills.

II. EXAMPLES OF SIMULATION MODELING

9.4 PERFORMANCE OF A BASEBALL HITTER

To illustrate the fundamental concepts of simulation analysis, let us consider the following simple example. Mighty Mantle is the top hitter for the Houston Blastros. Mighty is red hot, and has connected for 24 hits in 48 times at bat. His hits include 12 singles, 6 doubles, 2 triples, and 4 home runs. The Blastros wish to offer an incentive contract stipulating that if Mighty Mantle gets 20 total bases in his next 25 times at bat, he will receive a $1000 cash bonus. Can we expect Mighty Mantle to collect his bonus?

To simulate the performance of Mighty Mantle in his next 25 at-bats, we must describe statistically his behavior at the plate. There are several ways to do this, but the following method will be used for illustrative purposes. Since there have been 24 hits in 48 times at bat, there is a probability of .50 that Mighty will get a hit on any one trip to the plate. Given that a hit is obtained, there is a 50% chance that it will be a single, a 40% chance that it will be a double, 8.33% that it will be a triple, and a 16.37% chance that it will be a home run. Each of these events must be represented by a random phenomenon. Since the event "Get a Hit" is only one of two equally likely outcomes, a fair coin could be flipped to decide if a hit is made. If a hit is made, the following distribution describes whether it is a single, double, triple, or home run (Table 9.1).

Table 9.1
DISTRIBUTION OF HITS

Hit (x)	P_x	P_x	Dice Outcome
Single	1/2	18/36	7, 6, 5, 4
Double	1/4	9/36	8, 9
Triple	1/12	3/36	2, 3
Home run	1/6	6/36	10, 11, 12

Behavior of the random variable "hit," can be associated with the throw of dice. If we consider the sum of digits on the upturned faces, there are exactly 12 mutually exclusive and independent outcomes when a pair of dice is thrown. The student should be able to verify that the following density function corresponds to those outcomes.

X (sum of faces)	2	3	4	5	6	7	8	9	10	11	12
P_x (probability)	1/36	2/36	3/36	4/36	5/36	6/36	5/36	4/36	3/36	2/36	1/36

Association of the probability of various outcomes with the type of hit yields the relationship shown in Table 9.1. The simulation experiment is defined as follows. Flip a coin: if the outcome is a head, the batter has a hit; if a tail, the batter is out. If a hit is indicated, roll a pair of dice. If the sum of the upturned faces is a 4, 5, 6, or 7, record a single (1 base). If the sum is an 8 or 9, record a double (2 bases). If a 1 or 2, record a triple (3 bases), and if 10, 11, or 12, record a home run (4 bases). Terminate the experiment when 20 total bases have been accumulated. The results of the experiment are shown in Table 9.2.

Table 9.2
SIMULATION OF MIGHTY MANTLE

Trial	Flip	Dice	Hit	Bases	Total Bases
1	T	—			
2	H	7	Single	1	1
3	H	5	Single	1	2
4	T	—			
5	T	—			
6	T	—			
7	H	7	Single	1	3
8	H	6	Single	1	4
9	T	—			
10	H	3	Triple	3	7
11	T	—			
12	T	—			
13	H	11	Home run	4	11
14	H	6	Single	1	12
15	H	9	Double	2	14
16	T	—			
17	H	8	Double	2	16
18	H	4	Single	1	17
19	T	—			
20	T	—			
21	H	6	Single	1	18
22	H	7	Single	1	19
23	H	10	Home run	4	23

The simulation experiment resulted in a total of 23 total bases accumulated in 23 times at bat, Hence, the bonus is won. This experimental result can be compared to a theoretical result. The student will be asked to verify that based on available data (past performance), the expected number of bats needed to accumulate 20 total bases is 20.86 (21). The probability of achieving 20 total bases in 25 at-bats is approximately 0.552. Hence, the experimental result appears quite plausible. However, several factors should be noted. First, we must assume that the coin is perfectly "fair" and that the dice are perfectly "balanced." In addition, the flips and rolls must be executed with no known bias. Second, even under these ideal conditions, if the experiment is repeated, the results will not be the same. Simulation analysis is a "snapshot" of reality—it is a sampling procedure that predicts a single outcome from a stochastic environment. Note that if we repeated this experiment thousands of time, and then averaged our results, we would anticipate an experimental result that is very close to the theoretical result. The first observation is a basis for the need to obtain random numbers as "pure" as possible, and the second observation provides the necessity to execute simulation models on a digital computer over long (simulated) periods of time. The next example more clearly points out the need for simulation analysis to proceed until steady-state results are achieved, unless the analyst is interested in only transient results, or the simulation model represents a system that may never stabilize. The necessity for large quantities of random numbers is more fully explored in a later section of this chapter.

9.5 SIMULATION OF A TOOL CRIB

To illustrate the concept of simulation, and to demonstrate how meaningful properties of operational behavior can be calculated, let us consider a simple tool crib scenario. Mechanics arrive at a tool crib supply office in a random fashion to replace worn out tools. Tools are replenished on a first-come, first-serve basis by a single tool crib attendant. For simplicity, assume that the customer arrival times and service times are known for a stream of 20 customers. These are given in Table 9.3. The student should recognize this problem as a simple single-server

Table 9.3
CUSTOMER ARRIVAL AND SERVICE TIMES

Customer Number	Time of Arrival (in minutes)	Service Time (in minutes)
1	3.2	3.7
2	10.9	3.8
3	13.0	4.2
4	14.2	3.2
5	17.0	2.2
6	19.2	4.3
7	20.5	2.6
8	27.3	2.0
9	32.0	2.4
10	35.0	3.5
11	38.2	3.0
12	45.3	2.4
13	49.8	3.2
14	52.7	3.7
15	54.8	3.8
16	56.5	3.2
17	61.3	4.3
18	67.1	2.6
19	71.6	3.5
20	75.0	4.0

Table 9.4
SIMULATION OF A TOOL CRIB

Event Time	Customer Number	Event Type	Arrival Time (1)	Start Service Time (2)	Departure Time (3)	Time in System (3)–(1)	Number in Queue	Server Status	Server Idle Time
0.0	—	—	—	—	—	—	—	Idle	
3.2	1	Arrive	3.2	3.2	6.9	3.7	0	Busy	3.2
6.9	1	Depart	—				0	I	
10.9	2	A	10.9	10.9	14.7	3.8	0	B	4.0
13.0	3	A	13.0	14.7	18.9	5.9	1	B	
14.2	4	A	14.2	18.9	22.1	7.9	2	B	
14.7	2	D					1	B	
17.0	5	A	17.0	20.5	22.7	5.7	2	B	
18.9	3	D					1	B	
19.2	6	A	19.2	22.7	27.0	7.8	2	B	
20.5	7	A	20.5	27.0	29.6	9.1	3	B	
22.1	4	D					2	B	
22.7	5	D					1	B	
27.0	6	D					0	B	
27.3	8	A	27.3	29.6	31.6	4.3	1	B	
29.6	7	D					0	B	
31.6	8	D					0	I	
32.0	9	A	32.0	32.0	34.4	2.4	0	B	0.4
34.4	9	D					0	I	
35.0	10	A	35.0	35.0	38.5	3.5	0	B	0.6
38.2	11	A	38.2	38.5	41.5	3.3	1	B	
38.5	10	D					0	B	
41.5	11	D					0	I	
45.3	12	A	45.3	45.3	47.7	2.4	0	B	3.8
47.7	12	D					0	I	
49.8	13	A	49.8	49.8	53.0	3.2	0	B	2.1
52.7	14	A	52.7	53.0	56.7	4.0	1	B	
53.0	13	D					0	B	
54.8	15	A	54.8	56.7	60.5	5.7	1	B	
56.5	16	A	56.5	60.5	63.7	7.2	2	B	
56.7	14	D					1	B	
60.5	15	D					0	B	
61.3	17	A	61.3	63.7	68.0	6.7	1	B	
63.7	16	D					0	B	
67.1	18	A	67.1	68.0	70.6	3.5	1	B	
68.0	17	D					0	B	
70.6	18	D					0	I	
71.6	19	A	71.6	71.6	75.1	3.5	0	B	1.0
75.0	20	A	75.0	75.1	79.1	4.1	1	B	
75.1	19	D					0	B	
79.1	20	D					0	I	

queueing problem with known arrival and service characteristics. We wish to analyze the characteristics of the tool crib to determine: (a) the average time in the system, (b) the expected waiting line, and (c) the server utilization.

To proceed, we should recognize that the service system is composed of mechanics, who arrive for service, and a tool crib attendant, who provides service. The service mechanism (attendant) is either busy or idle, and an arrival (mechanic) is either waiting for service or being served. The condition of these entities determines the status of the system. Reflecting on these facts, we should realize that *system status* is only changed by an *arrival* or a *departure*. There are no other *status disturbing events* in this model. Simulation analysis is always concerned with status disturbing events, since only at these points in time will relevant statistics change. In this example, the point in time at which status changes (arrival or departure) will be called an *event*. Simulated time will advance from event to event, and relevant system statistics will be updated at each event as necessary. This is called *next event simulation*, and this type of analysis is the most common for system simulation models. Table 9.4 is used to record all data of interest to this model. Relevant to the model objectives, only three sets of statistics need to be maintained: (1) time in system per customer; (2) the length of the queue, and therefore the time the queue was that length; and (3) server status at each event and, hence, the length of time the server was idle. Figure 9.4 is a histogram of the server's idle time and the queue lengths at status disturbing events. Either Table 9.4 or Fig. 9.4 can now be used to calculate meaningful system behavioral statistics.

1. Average time in the system:

$$\frac{\sum_{i=1}^{20}\left[\text{depart time }(i) - \text{arrival time }(i)\right]}{20}$$

$$= \frac{97.7}{20}$$

$$= 4.89 \text{ min}$$

2. Expected queue length:

$$\frac{\sum_{i}\left[\text{queue length } i * \text{time interval for queue length } i\right]}{\text{total simulation time}}$$

$$= (1\times 1.2 + 2\times 0.5 + 1\times 2.3 + 2\times 1.9 + 1\times 0.3 + 2\times 1.3 + 3\times 1.6 + 2\times 0.6 + 1\times 4.3 + 1\times 2.3 + 1\times 0.3 + 1\times 0.3 + 1\times 1.7 + 2\times 0.2 + 1\times 3.7 + 1\times 2.4 + 1\times 0.9 + 1\times 0.1)/79.1$$

$$= \frac{33.6}{79.1}$$

$$= 0.42$$

3. Server utilization:

$$1 - \frac{\sum(\text{server idle time})}{\text{total simulation time}}$$

$$= 1 - \frac{3.2 + 4.0 + 0.4 + 0.6 + 3.8 + 2.1 + 1.0}{79.1}$$

$$= 0.81$$

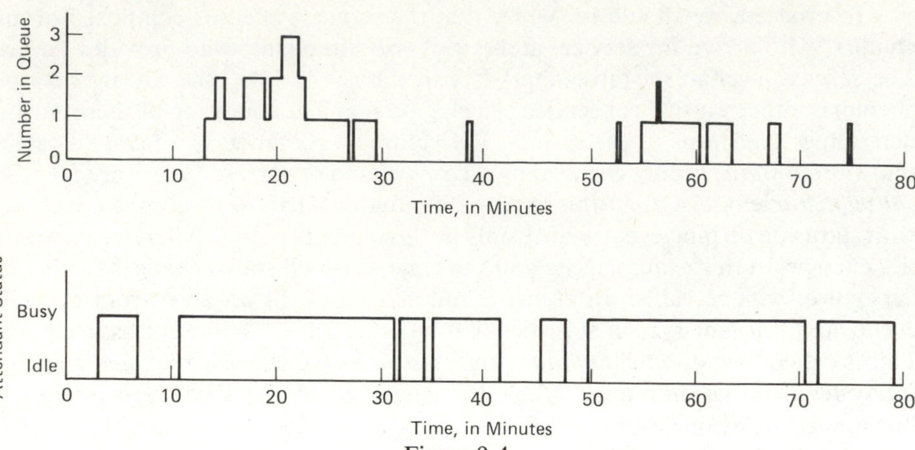

Figure 9.4

Histograms of attendant status and number in queue.

Several points should be made relevant to this analysis. First, it is unlikely that the analyst would have data predicting the arrival times and service times *a priori*. In actuality, these data would have been historical, to be recovered from standard times, a data base, or obtained from a work sampling study. The data used in this example are best interpreted to represent sample system behavior. Second, the statistics that we have calculated (e.g., time in system, expected queue length, and server utilization) are based on a finite set of data. In reality, much more data would be required to accurately predict system performance. Hence, the results obtained are best regarded as *transient*. As more data are used, and the time horizon of 79.1 min is extended, sample statistics would stabilize to relatively constant values. At this point, the results would represent *steady state* solutions. How much data are needed to obtain steady state, and how long one should simulate, is a matter of great statistical concern which, in general, lies beyond the scope of this book. However, we shall briefly discuss this issue again later in this chapter. Third, it is unlikely that this type of data would be used to perform a simulation analysis. It is common practice to reduce large amounts of data to a known statistical density function, and then to use mathematical techniques to generate statistical deviates. These techniques are discussed in a later section of this chapter. The next example illustrates a more realistic approach to dealing with a finite set of real-world data.

9.6 PRODUCTION LINE MAINTENANCE [Schmidt (59)]

Five production lines are to be maintained by one repair crew. When one of the lines fails it is repaired unless the repair crew is occupied on another line, which case it must wait for service. The repair crew services the lines in the order in which they fail. The production system operates three shifts per day, 5 days per week.

The time elapsed between start-up of a line and its failure has been observed to vary from one run to another in an unpredictable or random manner. Similar variation has been observed for the time to repair a line when it fails. The observed frequency distribution of failure and service times for line number 1 are given in Table 9.5 and graphically in Figs. 9.5 and 9.6. These data are based on an

Table 9.5
FREQUENCY DISTRIBUTION OF FAILURE AND SERVICE TIMES

Failure Time		Service Time	
Time Interval (weeks)	Observed Frequency	Time Interval (weeks)	Observed Frequency
0.00–0.02	35	0.000–0.002	35
0.02–0.04	20	0.002–0.004	23
0.04–0.06	16	0.004–0.006	28
0.06–0.08	20	0.006–0.008	13
0.08–0.10	17	0.008–0.010	16
0.10–0.12	16	0.010–0.012	13
0.12–0.14	8	0.012–0.014	6
0.14–0.16	6	0.014–0.016	11
0.16–0.18	5	0.016–0.018	3
0.18–0.20	2	0.018–0.020	3
0.20–0.22	5	0.020–0.022	5
0.22–0.24	8	0.022–0.024	4
0.24–0.26	3	0.024–0.026	1
0.26–0.28	5	0.026–0.028	3
0.28–0.30	1	0.028–0.030	2
0.30–0.32	1	0.030–0.032	2
0.32–0.34	1	0.032–0.034	2
0.34–0.36	1	0.034–0.036	2
0.36–0.38	1	0.036–0.044	0
0.38–0.40	0	0.044–0.046	1
0.40–0.42	2	0.046–0.074	0
0.42–0.44	0	0.074–0.076	1
0.44–0.46	1		
Total	174	Total	174

observation period of 20 weeks. Data of a similar nature were collected on the remaining four lines during the same period, but to avoid unnecessary complications in the model development they will not be presented.

In this system, once the times of failure and repair are known the status of all of the lines and the repair crew can be determined. Therefore, if the analyst can predict these times in some way he or she can write a computer simulation program that will reproduce the characteristics of the system. Failure time and service time are random variables and cannot be predicted with certainty, however, and the simulation analyst therefore attempts to generate these times in a manner that will reproduce the variability observed in the past on the assumption that similar variability can be anticipated in the future. However, this does not imply that behavior realized in the past will be reproduced identically in the future. The latter point is important. Twenty-weeks' data have been collected on the system considered here. If the system were observed for an additional 20 weeks, one would not expect an identical repetition of the first 20 weeks. Assuming that the system does not change radically from the first 20 weeks to the second, however, one would expect to find similar distributions of service and failure times for the two periods. That is, one would not expect radical departures from Figs. 9.5 and 9.6 for the

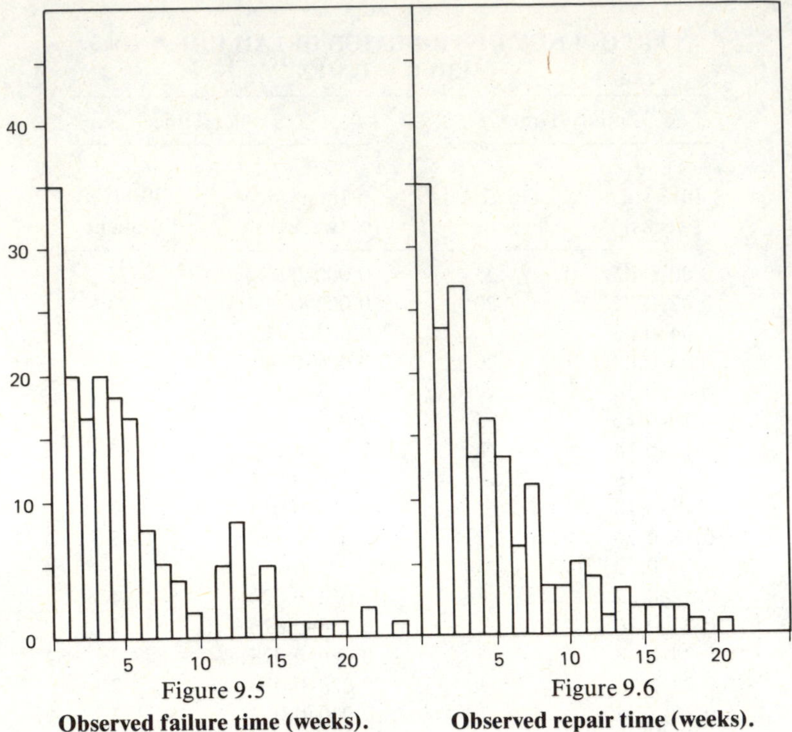

Figure 9.5
Observed failure time (weeks).

Figure 9.6
Observed repair time (weeks).

second 20 weeks. Therefore, for this problem the analyst would attempt to reproduce the variability of service time and failure time without reproducing identical values of these variables.

The variability of a random variable is generally represented by *frequency distributions* like those given in Table 9.5 and Figs. 9.5 and 9.6, or by a relative frequency distribution. A relative frequency distribution is the same as a frequency distribution except that the frequency counts in each interval or category are divided by the total number of observations.

An elaboration of the methods for generating random variables will be discussed later in this chapter. The essential characteristic of a realistic random variable generator is that it will reproduce the variability observed in the random variable studied.

The first step in developing a random variable or process generator is the calculation of the *cumulative distribution function* of the random variable. Assume that the range of values for the random variable has been broken into n intervals and that f_i is the frequency count for the ith interval. Then the cumulative frequency up to and including the jth interval, F_j, is given by

$$F_j = \frac{1}{n} \sum_{i=1}^{j} f_i$$

To illustrate, the calculation of the cumulative frequency distribution is summarized in Table 9.6 or the service-time data given in Table 9.5.

The next step in generating a random variable is to select or generate a random number. A random number will be defined as a random variable that is uniformly distributed over the interval 0 to 1. That is, each number between 0 and 1

Table 9.6
DEVELOPMENT OF THE RELATIVE CUMULATIVE DISTRIBUTION OF SERVICE TIME

Time Interval (Weeks)	Observed Frequency Distribution	Relative Frequency Distribution	Relative Cumulative Frequency Distribution
0.000–0.002	35	0.2011	0.2011
0.002–0.004	23	0.1322	0.3333
0.004–0.006	28	0.1609	0.4942
0.006–0.008	13	0.0747	0.5689
0.008–0.010	16	0.0920	0.6609
0.010–0.012	13	0.0747	0.7356
0.012–0.014	6	0.0345	0.7701
0.014–0.016	11	0.0632	0.8333
0.016–0.018	3	0.0173	0.8506
0.018–0.020	3	0.0173	0.8679
0.020–0.022	5	0.0287	0.8966
0.022–0.024	4	0.0230	0.9196
0.024–0.026	1	0.0057	0.9253
0.026–0.028	3	0.0173	0.9426
0.028–0.030	2	0.0115	0.9541
0.030–0.032	2	0.0115	0.9656
0.032–0.034	2	0.0115	0.9771
0.034–0.036	2	0.0115	0.9886
0.036–0.044	0	0.0000	0.9886
0.044–0.046	1	0.0057	0.9943
0.046–0.074	0	0.0000	0.9943
0.074–0.076	1	0.0057	1.0000

has an equal and independent chance of occurring. Let RN be the random number selected and let x_i be the upper limit of the ith interval. If

$$F_{i-1} < RN \leqq F_i$$

then the value of the random variable will be defined (arbitrarily) as x_i.

In summary, random times to failure and line repair times can be generated through the use of the cumulative distribution functions associated with these random events. Random deviates are obtained through the use of *random numbers* distributed uniformly on the interval 0–1. Generation of random numbers is discussed in great detail in Section 9.8. For the present time, assume that an ample supply of these random numbers is available for use.

Returning to the illustration in Table 9.6, if the random numbers, RN, are uniformly distributed between 0 and 1, then we would expect 20.11% of these numbers to be less than 0.2011, and 20.11% of the service times generated would have a value of 0.002. We would expect 13.22% of the random numbers to fall between 0.2011 and 0.3333 leading to a value of service time of 0.004. Continuing in this manner, we can readily see that the generative process given here will lead, in the long run, to proportions of times in each interval that correspond to the proportions observed. Although the use of the upper bound for each interval will cre-

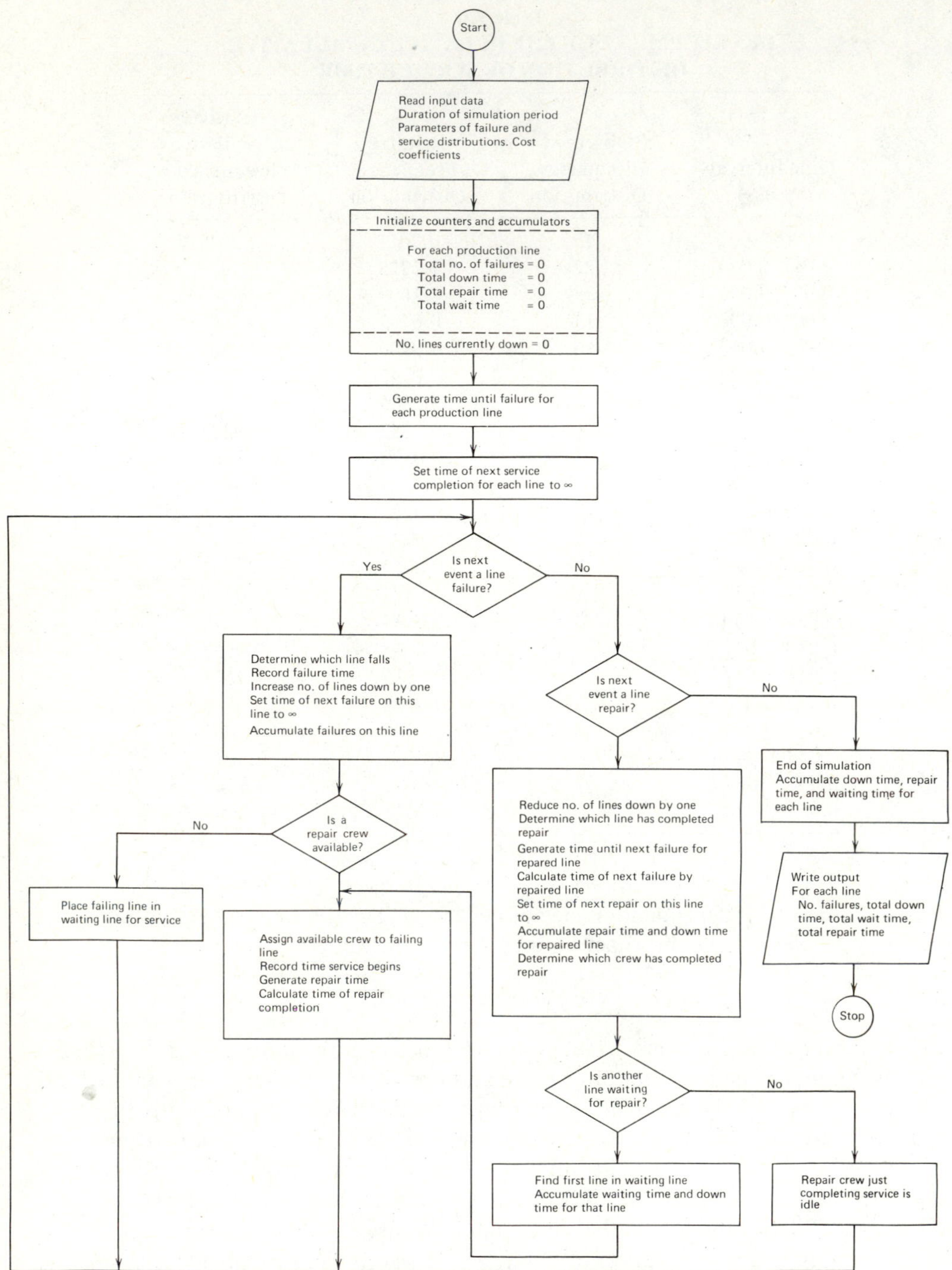

Figure 9.7
Flowchart for the maintenance simulator.

ate a slight bias with regard to the continuous time scale, this is used only for illustrative purposes and could be corrected through a continuous approximation.

To illustrate the simulation of this system, a 5.5 hour period beginning at midnight will be simulated. The logic of the simulation model is summarized in

Fig. 9.7. The first step is to generate the time of the first failure for each line. These times are given as follows:

First failure, line 1 = 2:17
First failure, line 2 = 4:12
First failure, line 3 = 10:24
First failure, line 4 = 3:08
First failure, line 5 = 3:21

The first event that alters the operating status of the system is a failure on the first line at 2:17. As defined above, a repair of line one should commence. Therefore, a service time for line 1 is generated and is 27 minutes. This creates the next status-changing event, which occurs at 2:44. Since the line will fail again, a new failure time is generated for line 1 and is 30 hours and 15 minutes. Therefore, line 1 fails again at 8:59 the next day.

The next event is a failure of line 4 at 3:08. Since no other lines are already down, a service time for line 4 is generated and is 32 minutes. Thus line 4 is up again at 3:40. Before line 4 is repaired, line 5 fails at 3:21 and is the next status-changing event. However, since the repair crew is busy on line 4, line 5 must wait for service. The next event is the repair of line 4 at 3:40, at which time repairs commence on line 5. Placing line 4 in service requires the generation of its next failure time and is 31 hours 12 minutes, or at 10:52 the next day. Repair time for line 5 is 22 minutes and repairs are completed at 4:02 and is the next event. At 4:02 a new failure time is generated for line 5 and is 5 minutes or occurs at 4:07. Since all lines are operating at 4:02, the next status-changing event is a line failure. The next line to fail is 5 at 4:07. Repairs for line 5 commence at 4:07, since no other lines are down at this time. Repair time on line 5 is generated and is 18 minutes; line 5 going into operation again at 4:25. The next event which occurs is a failure of line 2 at 4:12. Since the repair crew will be occupied on line 5 until 4:25, line 2 must wait for repairs until this time. The next event in the simulation is a repair of line 5 at 4:25 requiring generation of a new failure time for line 5 and a service time for line 2, since repairs for line 2 start on completion of the repairs for line 5. Generated failure time for line 5 is 58 minutes and service time for line 2 is 19 minutes. Therefore, line 2 starts operating at 4:44. At 4:44, a new failure time is generated for line 2 and is 9 minutes. Hence, line 2 fails at 4:53 and is the next event. Service on line 2 starts at 4:53, since line 2 is the only line down at that time. Generated service time for line 2 is 83 minutes. Therefore, line 2 is placed in operation at 6:16. At 5:23 line 5 fails but must wait for service, since the repair crew is occupied on line 2 at this time.

The above sequence of events is summarized in Table 9.7.

Although an extensive simulation analysis of this system by hand could become tedious if continued for long periods of study, using Fig. 9.7 and the given data, we could easily construct a computer program and conduct extended analysis. A computer program was utilized for this particular example, and the system was studied for 20 weeks of operation. These results are summarized in Table 9.8.

Input-Output Analysis

Although an extensive analysis of this output will not be performed at this time, it is evident that line 4 is the most reliable and that line 5 is down more than any other. In addition, the repair crew utilization is given by $11.04/20.0 = 0.552$ while there are repairs in queue $(8.54)(100)/20.0 = 42.7\%$ of the time. These fig-

Table 9.7

FIVE AND ONE-HALF-HOUR PERIOD OF SIMULATION FOR FIVE PRODUCTION LINES

Time of Change in System Status Hr:min	Event Altering System Status	Status							Cumulative Time	
		Line 1	Line 2	Line 3	Line 4	Line 5	Crew	Lines Down	Repair (min)	Waiting (min)
Midnight	None	Operating	Operating	Operating	Operating	Operating	Idle	0	0	0
2:17	Line 1 fails	Down (In service)					Line 1	1	0	0
2:44	Line 1 repaired	Operating					Idle	0	27	0
3:08	Line 4 fails				Down (In service)		Line 4	1	27	0
3:21	Line 5 fails					Down (Waiting)	Line 4	2	40	0
3:40	Line 4 repaired				Operating	Down (In service)	Line 5	1	59	19
4:02	Line 5 repaired					Operating	Idle	0	81	19
4:07	Line 5 fails					Down (In service)	Line 5	1	81	19
4:12	Line 2 fails		Down (Waiting)				Line 5	2	86	19
4:25	Line 5 repaired		Down (In service)			Operating	Line 2	1	99	32
4:44	Line 2 repaired		Operating				Idle	0	118	32
4:53	Line 2 fails		Down (In service)				Line 2	1	118	32
5:23	Line 5 fails					Down (Waiting)	Line 2	2	148	32

Table 9.8
RESULTS OF 20 WEEKS SIMULATION OF FIVE PRODUCTION LINES

Line Number	Number of Failures	Total Down Time (weeks)	Total Repair Time (weeks)	Total Waiting Time (weeks)
1	177	2.46	1.82	0.64
2	694	5.90	3.58	2.32
3	354	2.84	1.16	1.70
4	115	2.11	1.76	0.35
5	1093	6.25	2.72	3.53

ures indicate poor utilization of crew repair availability. Preventative maintenance procedures, increased crew sizes, and other system characteristics could easily be studied through the use of a more extensive simulation model.

The primary objective in presenting the previous production line model is to demonstrate how real-world data can be translated into probability density functions, and subsequently used to generate random events that represent behavior of the real world. We hope that in the complexity of model development and the subsequent tabular analysis of system behavior the student did not overlook the important task of model verification or model validation. Recall that model verification is the determination of whether a given model operates exactly as the model builder intended. Although one might choose to expand the simulation logic represented by the flow chart in Figure 9.7, or the analysis represented in Table 9.8, there can be little doubt that the model is indeed performing as described in the model definition phase. As we previously indicated, model validation is a much more difficult task, which poses the question: Does the model adequately represent the behavior of the real world? For this particular example, let us examine Table 9.8 in an attempt to validate model performance. Note that Table 9.8 exhibits a considerable spread of failures between production lines. In particular, the rate of failure for line 5 is more than nine times that of line 4. One should ask: Should we expect this result to occur under the given model assumptions? Consider the methodology that was used to generate production line failures. Note that within the boundaries of the previous problem description, a frequency distribution of time to failure was given in Table 9.5. Although not directly stated, it was tacitly assumed that this failure time distribution was used to generate failures for all five production lines. If this was indeed true, one should suspect that since all five production lines fail according to the same failure rate distribution, and that they are serviced according to the same service distribution, the relative number of failures through time should be approximately equal. Clearly, these results cannot be justified under current modeling assumptions. One might ask what is causing this apparent discrepancy in simulation results. Without further information, it is virtually impossible to predict exactly why this unexpected result occurred. However, one might conjecture about several things that might cause these unexpected results. First, one might examine the random number generator used in generating the full 20 weeks of tabular data. It is possible, but not likely, that for some unknown reason the random number generator got caught in a cycle continually generating short time-to-failures for line 5 and then generating long time-to-failures for lines 1 and 4. However, although this is a possibility, it does not completely describe the behavior of the number of failures. A second and more plausible observation is that the results presented in Table 9.8 did not actually come from the single distribution of time-to-failure nor time-to-repair. Indeed, these are exactly the conditions that generated the results of Table 9.8. Although they admittedly do not correspond to the example as previously presented, we have done this to point out an important fact relative to the acceptance of simulation results. This important fact is as follows. Do not blindly accept the results of a simulation experiment as being fact simply because its results are generated by a rigorous process or by a computer. One should always closely examine the totality of the simulation output to determine if the simulation results are realistic relevant to both the assumptions used in defining the model and relative to the methodologies that were used to generate results. It is indeed this type of questioning that defines the process of model validation. In conclusion, one might ask if *any* results would be valid if a single distribution of time-to-failure and a single distribution of time-to-repair are used in the simulation analysis. The answer to this question can only be given provided that the system analyst defines the exact reason for performing the simulation experiment. If, in fact, the simulation experiment was performed to determine the operational characteristics and utilization of the repair crew, then a single distribution for a time-to-repair and a time-to-failure will not affect the validity of the simulation results. However, if one is interested in not only the behavior of the repair crew but also in the individual performances of each of the five production lines, then it will be necessary to collect field data that is related to each of the five production lines, and to have multiple time-to-failure distributions and multiple time-to-repair distributions, each corresponding to a single line. Clearly, one must be aware of the desired simulation results while defining both the methodology and the logic used in performing a simulation experiment.

The analysis performed through simulation is obviously dependent on the problem to be resolved. A simulation model may be used simply to estimate the values of certain measures of system performance, or might be utilized to design and implement complex system procedures. In the example cited above, the simulation model was used to estimate the frequency of line failure, waiting time, and down time for the five production lines considered. Simulation may also be used to determine the values of certain decision variables that minimize the cost of operation of the system. As noted above, the number of repair crews used in the maintenance problem might be varied to determine the number of crews that will minimize the total cost of maintenance per week. Simulation is also useful in determining the sensitivity of certain measures of system performance to changes in constants or variables of the system. In the preceding example the mean time until line failure might be a function of the age of the line. Thus the analyst might increase mean time until failure by constant increments to determine the changes in system performance that can be anticipated in the future.

It is evident from this illustrative hand example that:

1. Large quantities of random numbers are required to generate system status-changing events.
2. Refinements of the random deviate generation scheme would be desirable to more closely approximate the stochastic nature of the system under study. To deal with these two problems, the next section is devoted to the generation of random numbers, and the succeeding section is devoted to the advanced techniques useful in generating a wide variety of random deviates.

III. PSEUDO-RANDOM NUMBERS

GENERATION OF RANDOM DEVIATES

In nearly all simulation experiments there exists a need for generating random statistical deviates from a certain distribution. The distribution will be the one that adequately describes and represents the physical process involved at that point in the experiment. During an actual simulation experiment, this process of generating a random deviate from a particular distribution may have to be done many times for many distributions, depending on the complexity of the model being investigated by the experiment.

This section is concerned with the techniques of generating the required statistical deviates on a digital computer.

The general process for generating a random deviate from a specific distribution will nearly always follow this pattern:

1. Generate a random number from the uniform distribution.
2. Perform a mathematical transformation of the uniform random number or numbers which produces a random deviate from the desired distribution.
3. Use the transformed deviate in the experiment as required.

These three steps are then repeated many times for the chosen distribution (and other distributions as required) as the experiment proceeds in time. Of course, it should be realized that at times the simulation model will require variates directly from the random-number generator, in which case step 2 will not be necessary.

Since the generation of a random number is the beginning for the generation of deviates from the more complicated distributions, the uniform distribution on the interval [0, 1] will be discussed first and then followed by discussions of (1) the techniques for transforming the uniform variate into a variate from a more complicated distribution, and (2) the more complicated distributions themselves.

9.7 THE UNIFORM DISTRIBUTION AND ITS IMPORTANCE TO SIMULATION

The uniform distribution over the interval [0, 1] is given by

$$f(x_0) = 1 \qquad \text{for } 0 \leqq x_0 \leqq 1 \tag{9.1}$$

$$F(x_0) = x_0 \qquad \text{for } 0 \leqq x_0 \leqq 1 \tag{9.2}$$

where the probability density function $f(x_0)$ = probability $(X = x_0)$ and the cumulative distribution function $F(x_0)$ = probability $(X \leqq x_0)$. The uniform distribution will be used as described above in deriving statistical generators for other probability distributions throughout this chapter. The term *uniform distribution* will refer to the above distribution over the interval [0, 1], and uniform random numbers or uniform random variates will refer to numbers generated from a uniform distribution over this range. Uniform deviates over a general range are easily generated through a scale transformation.

The probability density function and the cumulative distribution function are shown in Figures 9.8 and 9.9, respectively.

The importance of the uniform distribution stems from its use as the foundation in generating random deviates (variates) from more complicated distributions required in simulation experiments. As a simulation experiment proceeds in time, uniformly distributed random numbers are repeatedly being generated, giv-

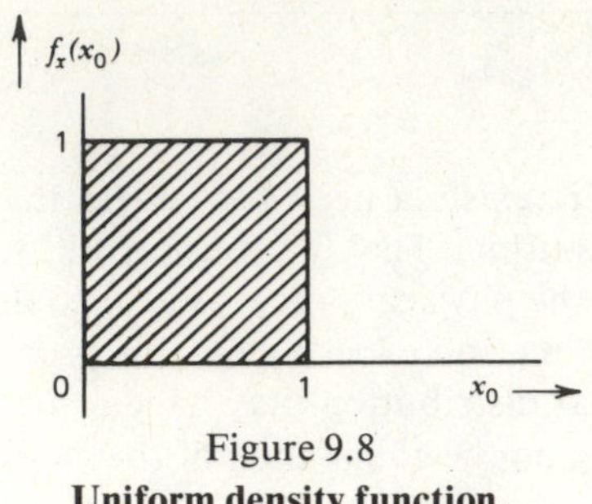

Figure 9.8
Uniform density function.

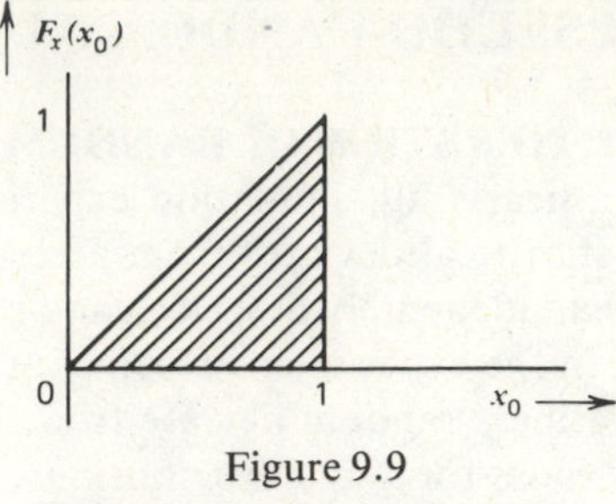

Figure 9.9
Uniform distribution function.

ing rise to various combinatorial operations on these deviates to produce random variates from any other statistical distribution required by the simulation.

Tocher (52) has suggested three modes of providing uniform random numbers on digital computers. These include (1) external provision, (2) internal generation by a random physical process, and (3) internal generation of sequences of digits by a recursive equation. The first method involves the recording of random-number tables (47, Rand Tables) on magnetic tape for input into a digital computer and then treating these random numbers as data for the problem of interest. This method is now of only historical significance. The second method uses an outside physical process whose random results can be read into the computer as a sequence of digits. Among the external processes that have been used are the random decay of radioactive material and the thermal noise in an electric valve circuit. The major objections to this method are that the results are not reproducible and that the random physical processes may develop a defect in their randomness. To our knowledge, this procedure has never been used. The third alternative, and the one that is most acceptable for digital computer simulations, involves the generation of "pseudorandom numbers" by an algorithmic-recursive-type equation. Being algorithmic and recursive indicates that the results of the previous calculations will be used in determining the next calculation. The ith term is used in a formula to calculate the $(i+1)$st term; the $(i+1)$st term is used to calculate the $(i+2)$nd term, and so on. Any series of numbers created by this method can never be truly random but, for all practical purposes, formulas have been developed that prove highly satisfactory. To determine whether a given sequence is satisfactory, various statistical tests have been designed to test the properties of the variates in question. We discuss these tests after we describe the current and historical methods used to generate these sequences of uniform deviates.

9.8 GENERATION OF RANDOM NUMBERS

Producing a number by some algorithm takes away some of the randomness that it should possess. A truly "random" number or occurrence can be produced only by some physical phenomenon, such as white noise. For this reason, numbers produced by algorithms are correctly referred to as pseudorandom numbers. Understanding this, pseudorandom numbers shall from this point be referred to as random numbers.

Properties of Uniformly Distributed Numbers

A random number generator should have the following properties:

1. The numbers generated should have as nearly as possible a uniform distribution.
2. The generator should be fast.

3. The generator program should not require large amounts of core.
4. The generator should have a long period (i.e., it should produce a large sequence of numbers before the sequence begins to cycle).
5. The generator should be able to produce a different set of random numbers to to reproduce a series of numbers depending on where it starts the sequence.
6. The method should not degenerate to repeatedly produce a constant value.
7. The generator should not produce a zero.

Several methods will now be examined that have been developed to generate sequences of pseudorandom numbers. The first three methods (Midsquare, Midproduct, and Fibonacci) are of historical significance and have detrimental, limiting characteristics. The last technique is the most popular in use today and is referred to as the congruential methods.

Midsquare Technique

We select a four-digit integer seed to initialize the generator. Our first random number is obtained from the seed in the following manner: The seed is squared and all digits except the middle four are ignored. The result is then normalized to give the first random number; this number is subsequently used as the new seed. Pseudorandom numbers can be generated in this manner, each time using the previous random number as the new seed.

The Midsquare technique is never used today. The method has a tendency to degenerate rapidly. If the number zero is ever generated, all subsequent numbers generated will also have a zero value unless steps are provided to handle this case. Furthermore, the method is slow, since many multiplications and divisions are required to access the middle digits in a fixed-word binary computer.

Midproduct Technique

This method is similar to the Midsquare technique except that a successive number is obtained by multiplying the current number by a constant, K, and taking the middle digits. The formula is

$$S_{n+1} = K(S_n)$$

The Midproduct technique has the following properties:

1. A longer period than the Midsquare technique.
2. More uniformly distributed than the Midsquare technique.
3. The method tends to degenerate.

Fibonacci Method

This generator is based on the Fibonacci sequence and is represented by

$$X_{n+1} = (X_n + X_{n-1}) \text{ Modulo } M$$

The method usually produces a period of length greater than M; however, the pseudorandom numbers obtained by using the Fibonacci method fail to pass the tests for randomness. Consequently the method does not give satisfactory results.

9.9 THE LOGIC IN GENERATING UNIFORM RANDOM VARIATES VIA A CONGRUENTIAL METHOD

Congruential methods of random number generation can be explained in the following manner. Consider this relation:

$$r_{i+1} = [Ar_i + C] \text{ Modulo } M$$

This formula can be expressed in words as follows. Choose a positive number A and a positive number C. Let r_i be an initial random number chosen ahead of time. r_i is called the *random number seed.* Multiply r_i by A, add C to this total, and divide the result by M. Keep the *remainder* and call this r_{i+1}. Repeat the procedure. If $C=0$, then the procedure is called a *multiplicative congruential generator*. If $C\neq 0$, then this is called a *mixed generator*. Surprisingly, mixed generators have not performed well in practice. Hence, we shall only consider the multiplicative congruential generator.

$$r_{i+1} = [Ar_i] \text{ Modulo } M \qquad i=1, 2, 3, \ldots$$

Let us consider a simple example with $r_1 = 5$, $A = 5$, and $M = 17$. The following sequence of pseudorandom numbers is produced:

$$5, 8, 6, 13, 14, 2, 10, 16, 12, 9, 11, 4, 3, 15, 7, 1, 5, \ldots$$

The sequence begins to repeat after 16 random numbers. Note that every number between 1 and 16 is in the sequence, but neither zero nor 17 ever appear, since the number zero would repeat forever. This particular choice of A, r, and X_i yields a *period* of 16. The numbers "appear" to occur at random, but they obviously repeat this sequence forever. The period of a random number generator and the "randomness" are obviously a function of the parameters A, M, and r_1. Number theory has been examined to suggest better choices for A, M, and r_1. A logical choice for M would be the largest number that can exist, since division by that number would always yield a value between zero and one. For a binary computer, the range of positive integers are 2^B, where B is this number of maximum digits per word.

For example, for a 32-bit machine the choice for M would be 2^{32}. Positive integers therefore range from 1 to $2^{32}-1$. The parameter A should be selected so that it is at least 5 digits, odd, and relative prime to M. Furthermore, it should not contain long sequences of zeros or ones. For a 32-bit machine, a good choice for A is 3^{19}; and this is also a good choice for r_1. Following this procedure for a 32-bit machine, numbers between 1 and $2^{32}-1$ will be produced. However, note that an anomaly occurs. Since all numbers are odd and relatively prime, *even* remainders can never occur! Nevertheless, mathematical tests have accepted this generator as a statistically sound choice—the numbers appear random. The numbers $A=3^{19}$ and $M=2^{32}-1$ are IBM choice for use on an IBM 360. Their choice of a seed is 65539, which produces a period of 2^{29}. The reader is encouraged to explore this logic through direct experimentation on either a mainframe or microcomputer using the appropriate value for B in 2^B-1.

9.10
TESTING A UNIFORM RANDOM-NUMBER GENERATOR

Since a random deviate from a certain distribution is created by performing a transformation on a uniform variate, the main emphasis on statistical testing should concern the ability of a random-number generator to accurately generate sequences of numbers uniformly distributed on the (0,1) interval. It is assumed that the transformations involved in transforming deviates of the uniform distribution to deviates of the particular distribution desired are mathematically correct. Hence, if anything is statistically wrong with the final distribution, it will be because of the deficiencies in the original random-number generator. Therefore, the following statistical tests should only be performed on the sequence of numbers generated by the uniform generator.

1. *The Frequency Test*. A goodness-of-fit procedure, such as the Chi Square Test or the Kolmogorov-Smirnov Test, is used to determine the uniformity of digits generated.
2. *The Gap Test*. Gap tests statistically analyze the number of digits that occur between the generation of a specified digit.
3. *The Runs Test*. The runs test (a) checks for sequences of digits that successively fall below (above) 0.50 or (b) checks to see if unusual runs up or down occur in successive numbers.
4. *The Poker Test*. The poker test examines individual digits to see if a single number or combinations of numbers form pairs, threes, and so forth, in successive sets of five random numbers.
5. *Serial Tests*. Serial tests are used to check the randomness of successive numbers in a sequence.
6. *Product Tests*. Product tests measure the independence (correlation) between the sequences of random numbers.
7. *Tests for Autocorrelation*. Autocorrelation tests examine the dependency of a particular number or another number in a sequence. These tests attempt to determine whether there are patterns or relationships among sets of random numbers.
8. *The Maximum Test*. The maximum test examines sequences of numbers to detect abnormally high or low numbers.

Students interested in more detail should refer to Appendix C, where the first four tests are given with numerical examples.

IV. TECHNIQUES FOR GENERATING RANDOM DEVIATES

9.11 THE INVERSE TRANSFORMATION METHOD

The inverse transformation technique deals with the cumulative distribution function, $F(x)$, of the distribution to be simulated. Since $F(x)$ is defined over the interval (0, 1), we can generate a uniform random variate R [also defined over the interval (0, 1)] and set $F(x)=R$. Then x is uniquely determined by the relation $F(x)=R$ and $x=F^{-1}(R)$ is the variate desired from the given distribution. The difficulty with this technique lies in finding the inverse transformation such that $F^{-1}(R)=x$. If this inverse function can be established, we only need generate various uniform random numbers and perform the inverse function on them to obtain random deviates from the distribution desired.

EXAMPLE 9.11-1

Generate a random variate from

$$f(x)=\begin{cases} 3x^2 & 0\leqq x\leqq 1 \\ 0 & \text{Elsewhere} \end{cases}$$

then

$$F(x)=\int_{t=0}^{x} 3t^2 dt$$

Hence,

$$F(x)=x^3$$

Since $F(x)$ is the cumulative distribution function of $f(x)$, we can replace $F(x)$ by a random number R. The inverse transformation is, therefore,

$$x=F^{-1}(R)=(R)^{1/3}$$

Hence, to generate a random deviate from the probability density $f(x)=3x^2$, a random deviate from the uniform distribution is generated (call it R) and the desired deviate $x=(R)^{1/3}$.

The problem with the inverse transformation technique is that for many distributions, the inverse function, $F^{-1}(R)$, does not exist or it is so complicated as to be impractical. When this is the case, either approximations or other techniques must be used.

The Exponential Distribution

In simulation experiments, the exponential distribution is used to describe the time interval between occurrences of like events, which are often arrivals in queueing problems. If the probability that an event will occur in a small-time interval is very small, and if the occurrence of this event is independent of the occurrence of other events, then the time interval between occurrence of these events is exponentially distributed and the process is a Poisson process. In addition to queueing processes, the exponential distribution is also used to describe component failure rates in reliability analysis.

The probability density function and cumulative distribution function are given as

$$f(x)=\lambda e^{-\lambda x} \qquad \text{for } x \geqq 0$$
$$F(x)=1-e^{-\lambda x} \qquad \text{for } x \geqq 0$$

To generate a random deviate from the exponential distribution, the inverse transform technique is easily applied. We have $F(x)=1-e^{-\lambda x}$; so we generate a uniform random deviate R and let

$$R=F(x)=1-e^{-\lambda x}$$

To find the inverse transform:

$$R=1-e^{-\lambda x}$$

or

$$1-R=e^{-\lambda x}$$

but $1-R$ is also from the uniform distribution, so let

$$R=e^{-\lambda x}$$

or

$$x=-\frac{\ln R}{\lambda}$$

Hence

$$F^{-1}(R)=-\frac{\ln R}{\lambda}=x$$

The procedure is to generate uniform random deviates and apply $F^{-1}(R)$ to generate exponential variates.

Weibull Distribution

The Weibull distribution is a family of density functions widely used in describing failure rate characteristics in reliability analysis.

The density function is

$$f(x) = \alpha\beta x^{\beta-1} e^{-\alpha x^{\beta}} \qquad x \geq 0$$

for $x > 0$, $\alpha > 0$, $\beta > 0$, the Weibull density function generates a family of probability density curves as α and β change their values.

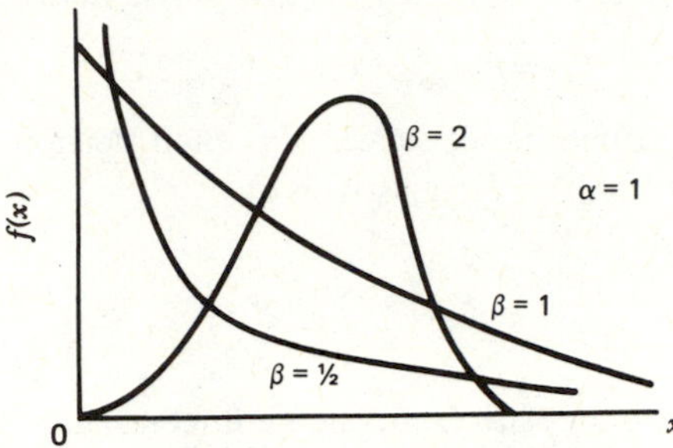

To generate a Weibull random deviate, the inverse transform technique can be used.

EXAMPLE 9.11-2

Define the cumulative distribution function.

$$F(x) = \alpha\beta \int_{t=0}^{x} t^{\beta-1} e^{-\alpha t^{\beta}} dt$$

let $y = \alpha t^{\beta}$

$$dy = \alpha\beta t^{\beta-1} dt$$

then

$$F(x) = \alpha\beta \int_{0}^{\alpha x^{\beta}} t^{\beta-1} e^{-y} \frac{dy}{\alpha\beta t^{\beta-1}}$$

Hence,

$$F(x) = 1 - e^{-\alpha x^{\beta}}$$

We must now find the inverse transform $F^{-1}(R)$.

$$F(x) = R = 1 - e^{-\alpha x^{\beta}}$$ where R is from the uniform distribution

$$1 - R = e^{-\alpha x^{\beta}}$$ but $1 - R$ is also from the uniform distribution, so

$$R = e^{-\alpha x^{\beta}}$$

and

$$x = \left[-\frac{1}{\alpha} \ln R \right]^{1/\beta}$$

Hence, to generate a series of deviates from the Weibull distribution, we only need to generate random deviates from the uniform distribution and apply the inverse transform:

$$F^{-1}(R) = \left[-\frac{1}{\alpha} \ln R \right]^{1/\beta} = x$$

The Geometric Distribution

A random variable X, defined as the number of failures in a sequence of Bernoulli trials before the first success occurs, is known as a geometric random variable. This distribution is related to the binomial distribution and has been used in the area of quality control and for lag distributions in econometric models.

The probability density function for the geometric distribution is

$$f(x) = pq^x \qquad x = 0, 1, 2, \ldots$$

By definition, p is the probability of success for each Bernoulli trial and $q = 1 - p$. The cumulative distribution function is given by

$$F(x) = \sum_{k=0}^{x} pq^k$$

To generate a random deviate from the geometric distribution, we make use of the fact that

$$1 - F(x) = q^{x+1}$$

and that $[1 - F(x)]/q$ has unit range. It will be left as an exercise for the student to show that these relationships are correct.

Once the above is accepted, let

$$R = q^x \qquad \text{(inverse technique)}$$

$$\ln R = x \ln q$$

Since x must be an integer, choose x such that x is the largest integer that satisfies

$$x \leq \ln R / \ln q$$

9.12 THE REJECTION TECHNIQUE

The rejection technique consists of drawing a random value from an appropriate distribution and subjecting it to a test to determine whether or not it will be accepted for use. To illustrate, let $f(x)$ be a frequency density function such that

$$f(x) = 0 \qquad \text{for } a > x > b$$

and

$$0 \leq f(x) \leq M \qquad \text{for } a \leq x \leq b$$

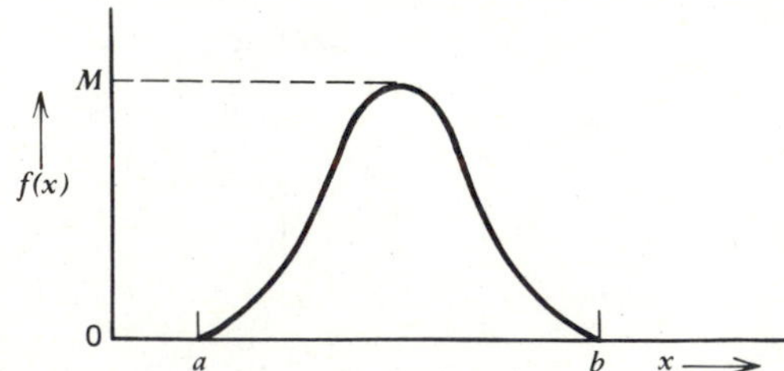

To carry out the method:

1. Generate two uniform random variates, R_1 and R_2.
2. Form the prospective random variate from $f(x)$; $x = a + (b - a)R_1$.

3. Test to see if $R_2 \leqq f(a+(b-a)R_1)/M$.
4. If the inequality holds, then accept $x=a+(b-a)R_1$ as the variate generated from $f(x)$.
5. If the inequality is violated, generate two new random numbers and try again. Note that M is simply the *mode* of $f(x)$.

The theory behind this method is based on the fact that the probability of R_2 being less than or equal to $(1/M)f(x)$ is $(1/M)f(x)$. Hence, if the prospective candidate x is chosen at random according to $x=a+(b-a)R_1$ and is rejected if $R_2>(1/M)f(x)$, then the probability density function of the accepted x's will be exactly $f(x)$. Note that if no rejection were used, that is, just step 2 is executed, then x would be distributed uniformly between a and $a+b$. With the added rejection test, however, we are picking only those x's conforming to the original distribution $f(x)$.

EXAMPLE 9.12-1

Generate a random deviate from:

$$f(x)=3x^2 \qquad 0\leq x\leq 1$$

then $M=3$, $a=0$, $b=1$

1. Generate R_1 and let $x=a+(b-a)R_1=R_1$.
2. Generate R_2 and compare it with
3. If $R_2 \leqq R_1^2$.
   then accept $x=R_1$ as a variate from $f(x)$
4. If $R_2>R_1^2$, reject $x=R_1$ and repeat steps 1–3.

Tocher (52) has shown that the expected number of trials before a successful random deviate from $f(x)$ is generated is M. This suggests that the method may be quite inefficient for certain density functions.

The Beta Distribution

Consider the beta distribution defined by the following density function.

$$f(x)=\frac{\Gamma(\alpha+\beta)}{\Gamma(\alpha)\Gamma(\beta)}x^{\alpha-1}(1-x)^{\beta-1} \qquad 1\geqq x\geqq 0$$

where α and β are the distribution parameters and

$$\Gamma(\phi)=\int_0^\infty x^{\phi-1}e^{-x}dx$$

The beta distribution is often used in describing the distribution of percentages, and has been extensively utilized in a related form in the analysis of PERT/CPM project scheduling problems. The particular shape of $f(x)$ is determined by α and β. (See Fig. 9.10.)

First determine the value of M.

$$M=\text{mode}=\left.\frac{df(x)}{dx}\right|_{x=0}$$

Hence

$$M=\frac{\alpha-1}{\alpha+\beta-2}$$

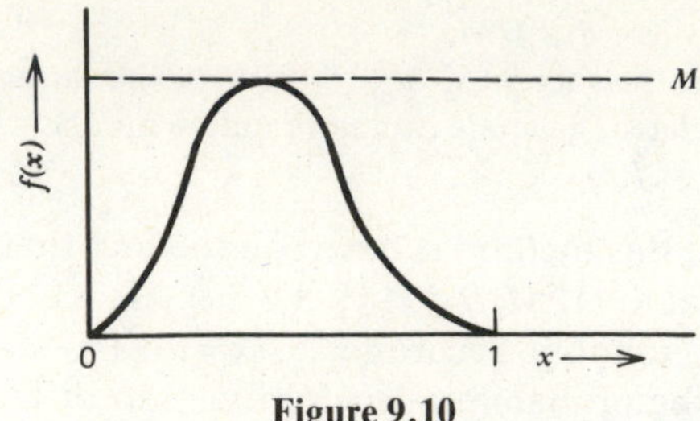

Figure 9.10

Note that M is only defined for certain values of α and β. The procedure is as follows:

1. Choose (or determine) α and β.
2. Calculate $M = (\alpha - 1)/(\alpha + \beta - 2)$.
3. Proceed with the normal rejection scheme.
4. Generate two random numbers; $R1$ and $R2$.
5. If $R2^*M \leqq f(x \equiv R2) = \dfrac{\Gamma(\alpha+\beta)}{\Gamma(\alpha)\Gamma(\beta)} R2^{\alpha-1}(1 - R2)^{\beta-1}$

   Then Deviate $= R2$

   $\left\{ \begin{array}{l} \text{if } R2^*M > f(x \equiv R2) \\ \text{Then go to step 4 and start again.} \end{array} \right.$

The Gamma Distribution

One of the most flexible distributions in engineering applications is the gamma density function. If a random variable is nonnegative and follows a unimodal distribution, the chances are good that a member of the gamma distribution can adequately describe the variable in question.

The gamma distribution is described by

$$f(x) = \frac{\beta^\alpha}{\Gamma(\alpha)} x^{\alpha-1} e^{-x/\beta} \qquad x \geq 0$$

$$\Gamma(x) = \int_0^\infty x^{\alpha-1} e^{-x} dx$$

where $\alpha > 0$, $\beta > 0$, and x is nonnegative. As α increases, the gamma distribution approaches a normal distribution asymptotically. If $\alpha = 1$, the gamma distribution is identical to the exponential distribution already discussed. A technique will now be discussed that generates general gamma variates. However, more efficient generation schemes for specialized related gamma forms will be given later in this section.

A Sampling Rejection Technique for Gamma Variates (Johnk's Method).

THEOREM 1

The sum of two gamma random variables with parameters $G_1(\alpha_1, \beta)$ and $G_2(\alpha_2, \beta)$ is also gamma with parameters $G(\alpha_1 + \alpha_2, \beta)$.

THEOREM 2

If U_1 and U_2 are continuous uniform random variables described by

$$f(U_i) = 1 \qquad i = 1, 2 \qquad 1 \geqq U_i \geqq 0$$

and

$$x=U_1^{1/A} \qquad y=U_2^{1/B}$$

and if it is true that

$$x+y \leqq 1$$

then

$$z=\frac{x}{x+y}$$

is Beta distributed with distribution function:

$$f(z)=\frac{\Gamma(A+B)}{\Gamma(A)\Gamma(B)} z^{A-1}(1-z)^{B-1} \qquad 1 \geqq z \geqq 0.$$

Note that this is a second method for generating Beta variates.

THEOREM 3

If X is a random variable gamma distributed with $\alpha=n+k$, $\beta=1$; and if Y is beta distributed with parameters $A=n$ and $B=k$; then $Z=XY$ is gamma distributed with $\alpha=n$, $\beta=1$.

Theorems 1, 2, and 3 are well known, and can be verified through the use of simple linear transformations and convolution theory.

The entire Johnk procedure depends on the ability to produce a random Beta deviate using Theorem 2, which is a rejection technique. In general, we can state:

Lemma 1

Given a Beta distribution function with parameters A and B, if we define $x=u_1^{1/A}$ and $y=u_2^{1/B}$, then the probability that $x+y \leqq 1$ is given by

$$Pr\{x+y \leqq 1\}=\frac{AB\Gamma(A)\Gamma(B)}{\Gamma(A+B)(A+B)}$$

If we define $P=\{\Gamma(A,B)[AB \mid (A+B)]\}^{-1}$ then an expression is obtained for the expected number of rejections in obtaining one Beta variate. Various combinations of A and B are given below:

B \ A	1	3	5
1	1	4	6
3	4	30	56
5	6	56	252

A general conclusion is that if Johnk's technique is used in Beta generations *per se*, the time required to obtain a statistical deviate could prove prohibitive except for selected combinations of A and B. At this point let us consider the use of this technique to generate gamma variates via Theorem 3. Suppose that we desire to generate a gamma-distributed random variable with noninteger shape parameter, α. Now define $n=\alpha-(\alpha)$, where (α) is the largest truncated integer, and similarly define $k=1-\alpha+(\alpha)$. By Theorem 3 let $A=n$ and $B=k$. Under these rules, it is necessary that $\alpha \equiv n+k \equiv 1$. It follows that if a Beta variate is generated following the parameters A and B, and multiplied by a gamma variate with parameters $\alpha=1$ and $\beta=1$, then a gamma variate with $\alpha=n$ and $\beta=1$ will be produced. Two facts are now significant and should be noted. A gamma distribution with parameters $\alpha=1$, $\beta=1$ is given by

$$f(x)=e^{x} \qquad 0 \leqq x \leq \infty$$

from which a variate is easily produced using an inverse transform. Since $A=n$ and $B=k$, A and B will always be less than one. By Lemma 1, the expected number of rejections in the beta generation phase would be

$$R=[AB\Gamma(A)\Gamma(B)]^{-1}=[(1-B)B\Gamma(1-B)\Gamma(B)]^{-1} \approx \left[\frac{(1-B)B\Pi}{\sin n\Pi}\right]^{-1}$$

Hence, the expected number of rejections is approximated by $1.33 \geqq R \geqq 1$.

The fundamental ingredients have now been established to construct a random deviate generator for a gamma density with arbitrary noninteger shape parameter α and scale parameter β. The algorithm is as follows:

Define

(i) α is a noninteger shape parameter.

(ii) $\alpha_1=(\alpha)$ is the truncated integer root of α.

(iii) u_i is the ith random number $1 \geqq u_i \geqq 0$.

1. Let $x=-\ln \prod_{i=1}^{\alpha_1} u_i$
2. Set $A=\alpha-\alpha_1$; $B=1-A$
   a. Set $j=1$.
   b. Generate a random number, u_j, and set $y_1=(u_j)^{1/A}$.
   c. Generate a random number, u_{j+1}, and set $y_2=(u_{j+1})^{1/B}$.
   d. If $y_1+y_2 \leqq 1$, go to (f).
   e. Set $j=j+2$; go to (b).
   f. Let $z=y_1/(y_1+y_2)$ so that z is a beta variable with parameters A and B by Theorem 2.
3. Generate a random number, u_N, and let $Q=-\ln(u_N)$.
4. The desired deviate is $D=(x+zQ)\beta$ by Theorem 1 and Theorem 3.

9.13 THE COMPOSITION METHOD

In this technique $f(x)$, the probability density function of the distribution to be simulated, is expressed as a probability mixture of properly selected density functions. This procedure is based on the definition of conditional probability or the law of compound probabilities.

Mathematically let $g(x \mid y)$ be a family of one-parameter density functions where y is the parameter identifying a unique $g(x)$. If a value of y is now drawn from a cumulative distribution function $H(y)$ and then if x is sampled from the $g(x)$ for that chosen y, the density function for x will be

$$f(x)=\int_{-\infty}^{\infty} g(x \mid y)dH(y)$$

By using this principle, more complicated distributions can be generated from simpler distributions that are themselves easily generated by the inverse transform technique or the rejection technique.

EXAMPLE 9.13-1

Taken from Butler (5). Generate a random variate from

$$f(x) = n\int_1^\infty y^{-n}e^{-xy}dy$$

let

$$dH(y) = n\,dy/y^{n+1} \qquad 1 < y < \infty,\ n \geqq 1$$

and

$$g(x) = ye^{-yx}$$

A variate is now drawn from a density function whose cumulative distribution function is $H(y)$. Once this y is selected, it determines a particular $g(x) = ye^{-yx}$. The desired variate from $f(x)$ is then simply a generated variate from $g(x) = ye^{-yx}$. To carry out the above instructions, generate two uniform variates R_1 and R_2 and let

$$S_1 = R_1^{-1/n}$$

$$x = \frac{1}{S_1}\log R_2$$

then x is the desired variate from

$$f(x) = n\int_1^\infty y^{-n}e^{-yx}dy$$

This technique is appropriate when generating deviates of "higher" type distributions from simpler ones. The difficulty lies in identifying the $H(y)$ and $g(x \mid y)$ needed to produce a given $f(x)$ under the relationship:

$$f(x) = \int_{-\infty}^{\infty} g(x \mid y)dH(y)$$

Fortunately, mathematical statisticians have provided us with several useful relationships called "convolutions," which can be used in generating certain random deviates. The following examples illustrate the procedure.

The Poisson Distribution

If the time intervals between like events are exponentially distributed, the number of events occurring in a unit interval of time has the Poisson distribution. Applications of the Poisson random variable include areas such as inventory control, queueing theory, quality control, traffic flow, and many other areas of management science.

The probability density function for the Poisson distribution is given by

$$f(x) = \frac{\lambda^x e^{-\lambda}}{x!} \qquad x = 0, 1, 2, \ldots, \infty$$

where λ is the expected number of occurrences per unit time. This implies that the time between events is exponentially distributed with mean $1/\lambda$.

We can use this relationship between the exponential and Poisson distributions to generate deviates from the Poisson distribution. A Poisson deviate x can be defined in the following manner:

$$\sum_{i=1}^{x} y_i \leqq 1 \leqq \sum_{i=1}^{x+1} y_i$$

where $y_1, y_2, \ldots, y_{x+1}$ are random deviates from the exponential distribution having mean $1/\lambda$ and are generated by (inverse transform technique)

$$y_i = -\frac{1}{\lambda} \ln R_i$$

where R_i is from the uniform distribution. In summary, the cumulative sums are generated until the inequality holds. When this occurs, x is the Poisson random deviate desired.

Another form of this same procedure is to define the Poisson deviate x when

$$\sum_{i=1}^{x} y_i \leqq \lambda \leqq \sum_{i=1}^{x+1} y_i$$

where y_i are again deviates from the exponential distribution, but with mean 1, that is,

$$y_i = -\ln R_i$$

The two techniques are essentially the same, but the first inequality seems to be more in agreement with the definition of the exponential distribution, since the y_i's actually have the true mean $1/\lambda$.

The Erlang Distribution

The Erlang distribution is a form of the gamma distribution with k equal to a positive integer. Mathematical statisticians have proved that this distribution is just the sum of k exponential variables, each with expected value $1/k$.

Hence, to generate an Erlang deviate we only need to sum k exponential deviates, each with expected value $1/k$. Therefore, the Erlang variate x is expressed as

$$x = \sum_{i=1}^{k} y_i = -\frac{1}{\alpha} \sum_{i=1}^{k} \ln R_i$$

where y_i is an exponential deviate generated by the inverse transform technique and R_i is a random number from the uniform distribution.

The Binomial Distribution

A random variable x defined as the number of successful events in a sequence of n independent Bernoulli trials, each with probability of success p, is known as a binomial random variable. The binomial distribution is one of the most important statistical distributions used in the areas of statistical sampling and quality control.

The probability density function for the binomial distribution is given as

$$f(x) = \binom{n}{x} p^x q^{n-x} \qquad x = 0, 1, \ldots, n$$

where p = probability of success per trial
$q = 1 - p$
n = number of trials
x = number of successes, an integer.

To generate a binomial deviate with parameters p and n the procedure is as follows:

1. Generate n uniform random deviates.
2. Count the number of uniform variates less than or equal to p.
3. The number found under step 2 is equal to the value of the binomial variate.

This procedure can then be repeated as many times as desired to generate other binomial deviates.

Another procedure involves using the normal distribution as an approximation to the binomial for cases where $n \geqq 20$ and $np \leqq 10$. Since the binomial variate is an integer, the normal variate used as an approximation must be rounded to the closest integer value. This method is faster but is only an approximation.

9.14 MATHEMATICAL DERIVATION TECHNIQUE

Under this technique, various functional relationships are used in order that certain complicated probability density functions can be represented in a simpler form from which deviates can easily be generated by the more common techniques. The normal distribution is one of the most widely used in applied statistics and has two parameters; μ = expected value, and σ^2 = variance. The cumulative distribution function of a normal random variable x is given by

$$F(x) = \int_{-\infty}^{x} \frac{1}{\sigma\sqrt{2\pi}} e^{-\frac{1}{2}[(x-\mu)/\sigma]^2} dx \qquad -\infty \leqq x \leqq \infty$$

This function cannot be evaluated analytically; hence, many of the methods for generating normal deviates use approximation methods.

If the parameters μ and σ^2 have values of zero and 1, respectively, the distribution is known as the standard normal density with

$$f(z) = \frac{1}{\sqrt{2\pi}} e^{-z^2/2} \qquad -\infty \leqq z \leqq \infty$$

Any normal variate x with parameters μ and σ^2 can be converted to a standard normal variate z by using

$$z = (x - \mu)/\sigma$$

Conversely, to convert a standard normal deviate z to a normal deviate with parameters μ and σ^2, use

$$x = \sigma z + \mu$$

Hence, to generate a normal deviate x with parameters μ and σ^2, we only need to generate a standard normal deviate and apply the preceding equation. Many methods for generating standard normal deviates exist. Three examples will be given. The first is a mathematical derivation technique (Box and Muller), which is explained below. The second and third will be explained in the section on approximation techniques (one technique attributed to Kahn and one to Hastings).

The Box and Muller Technique for Generating Normal Deviates [Box and Muller (4)]

The joint probability density function for two independent standard normal deviates x_1 and x_2 is

$$f(x_1, x_2) = \frac{1}{2\pi} e^{-1/2(x_1^2 + x_2^2)}$$

Consider the substitutions

$$x_1 = r \cos \theta$$
$$x_2 = r \sin \theta$$

then

$$f(r, \theta)\, dr\, d\theta = \frac{1}{2\pi} e^{-r^2} r\, dr\, d\theta$$

so that θ has a uniform distribution over the interval $(0, 2\pi)$ and r^2 has an exponential distribution. Hence, we generate two uniform random deviates R_1 and R_2 and let

$$r^2 = -2\ln R_1 \longleftrightarrow \text{(exponential generation}$$
$$r = (-2\ln R_1)^{1/2} \qquad \text{by inverse transform technique)}$$
$$\theta = 2\pi R_2 \longleftrightarrow \text{(scaled uniform generation)}$$

Then we have as the two desired normal deviates

$$x_1 = [2\ln(1/R_1)]^{1/2}\cos(2\pi R_2)$$
$$x_2 = [2\ln(1/R_1)]^{1/2}\sin(2\pi R_2)$$

The Box and Muller procedure generates two exact deviates from the standard normal distribution (usually requiring approximation methods) by using a mathematical derivation (polar coordinates). Analytical manipulations reduce the procedure to the generation of an exponential deviate and a uniform deviate, and a combination of these results according to the original substitutions.

Another application of this technique is given by the *log normal* distribution. Since the log normal deviate x is related to the normal deviate y by the mathematical relationship $x = \ln y$, we need only to generate a normal deviate y and use the mathematical relationship

$$x = \exp\left\{ 1/2 \ln\left(\frac{\mu^4}{\sigma^2 + \mu^2} \right) + \left[-2 \ln\left(\frac{\sigma^2 + \mu^2}{\mu^2} \right) \ln (RN) \right]^{1/2} \cos (2\pi RN) \right\}$$

to find a deviate x from the log normal distribution. (μ and σ^2 are the mean and variance of the desired log normal density function.)

9.15 APPROXIMATION TECHNIQUES

The approximation methods are all related to the inverse transform technique and are used when exact methods are either too complicated or are, in fact, impossible. With approximation methods, one of three different quantities is approximated. These are

1. The probability density function of the distribution to be simulated.
2. The cumulative distribution function of the distribution to be simulated.

3. The inverse transform to the cumulative distribution function of the distribution to be simulated.

Approximation methods are used when the cumulative distribution function cannot be obtained from the given probability density function, or when the inverse transform cannot be obtained even if $F(x)$ is available. When this is true, an approximation for the probability density function must be made that facilitates the derivation of the cumulative distribution function, or an approximation must be made on $F(x)$ directly. Once the cumulative distribution function has been found to the approximated probability density function, the inverse transform can often then be applied, and deviates are generated in the manner previously discussed.

Examples of Approximation Techniques

Suppose we wish to generate a deviate from the standard normal density function given by

$$f(x) = \frac{1}{\sqrt{2\pi}} e^{-x^2/2} \qquad \infty \geqq x \geqq -\infty$$

An approximation has been given by Kahn:

$$e^{-x^2/2} \approx \frac{2e^{-kx}}{(1+e^{-kx})^2}$$

for

$$x > 0 \qquad \text{and} \qquad k = \sqrt{8/\pi}$$

The cumulative distribution function to the approximation is easily calculated as

$$F(x) = \left[\frac{2}{1+e^{-kx}}\right] - 1$$

The inverse to this approximation cumulative distribution function is then

$$x = \frac{1}{k} \ln \left[\frac{1+R_1}{1-R_1}\right]$$

A random sign is then attached to this variate where R_1 is a deviate from the uniform distribution. Kahn's approximation derives a normal random deviate directly from an approximation to the density function, $f(x)$. The second method is to obtain an approximate inverse transform directly from this cumulative distribution function. When this occurs, an approximation for the cumulative distribution function is made that facilitates the derivation of the inverse transform. Once the inverse transform is found, deviates are again generated according to the inverse transform technique previously defined. Since this approximation method is not used as often as the following one, no examples will be given for this method.

Under the third approximation method, an approximation of the inverse transform itself is found and is applied directly to generate deviates from the distribution in question. One form often used for approximating the inverse function is

$$F(x) = a + bx + cx^2 + \alpha(1-x)^2 \ln (x) + Bx^2[\ln (1-x)]$$

a, b, c, α, and B are constants that give rise to the best approximation. From the cumulative distribution function, it is possible to construct a table giving the hypothetical frequencies for the various values of the variate from 0 to 1. The method of least squares can then be used to pass the curve through the series of points in the table. Tocher (52) illustrates that for the normal distribution with mean μ and standard deviation σ, this technique will yield the following:

$$a=(16834\mu-13452.96\sigma)\times 2^{-11}$$
$$b=(26953.865\sigma)\times 2^{-11}$$
$$c=0$$
$$\alpha=(-3772.769\sigma)\times 2^{-11}$$
$$B=(3772.769\sigma)\times 2^{-11}$$

Many other forms are used to approximate the inverse transform. One method attributed to Hastings for generating standard normal deviates is (52):

let

$$y=[-2\ln(1-R)]^{\frac{1}{2}}$$

then

$$x=F^{-1}(R)=y-\frac{a_0+a_1y}{b_0+b_1y+b_2y^2}$$

where x is the desired standard normal deviate and R is from the uniform distribution. The approximated inverse transform is specified by

$$a_0=2.30753$$
$$a_1=0.27061$$
$$b_0=1.0$$
$$b_1=0.99229$$
$$b_2=0.04481$$

9.16 SPECIAL PROBABILITY DISTRIBUTIONS

The Chi-Square Distribution

The Chi-square distribution is a gamma distribution with $\alpha=1/2$ and $k=\nu/2$, and therefore, has expected value equal to $(1/\alpha)k=2k=\nu$ (where ν is called the degree of freedom). The chi-square distribution is important for its use in testing of hypotheses regarding goodness of fit.

There are two cases to consider when generating chi-square deviates as the sum of k exponential deviates (gamma generation).

Case 1. If the value of ν is even, then k is an integer and the gamma density function reduces to an Erlang-θ with $\theta=\nu$. In this case, a Chi-square density is a special gamma density (Erlang). Hence, the following formula is appropriate:

$$x=-\frac{1}{\alpha}\sum_{i=1}^{k}\ln R_i$$

for the chi-square deviates.

Case 2. If the value of ν is odd then $k=\nu-1$ and

$$x=-\frac{1}{\alpha}\sum_{i=1}^{k}\ln R_i+z^2$$

is used for the chi-square deviate where z^2 is the square of a standard normal variate whose generation was discussed previously. R_i is again a uniform random deviate.

Another technique for generating a chi-square variate makes use of the fact that a chi-square variable with ν degrees of freedom can be represented as the sum of the squares of ν standard normal deviates. Using this gives

$$x = \sum_{i=1}^{\nu} z_i^2$$

as a chi-square variate where the z_i are deviates generated from the standard normal distribution.

The Student's *t* Distribution

The t random variable is defined as the ratio of a standard normal variable to the square root of a chi-square variable divided by its degrees of freedom. The t distribution is used in testing statistical hypotheses.

Mathematically, the t random variable is given as

$$x = \frac{z}{\sqrt{\dfrac{x_1}{\nu}}}$$

where z is a standard normal variable and x_1 is a chi-square variable with ν degrees of freedom.

A deviate from the t distribution is given by

$$x = \frac{z_1}{\left(\sum_{i=2}^{\nu+1} \dfrac{z_i^2}{\nu}\right)^{1/2}}$$

where the z's are standard normal variates.

The *F* Distribution

If x_1 and x_2 are chi-square random variables with degrees of freedom of a and b, respectively, then

$$x = \frac{x_1/a}{x_2/b}$$

is an F random variable with degrees of freedom a and b.

To generate an F variate, we only need to use

$$x = \frac{\dfrac{\sum_{i=1}^{a} z_i^2}{a}}{\dfrac{\sum_{i=a+1}^{a+b} z_i^2}{b}} = \frac{b \sum_{i=1}^{a} z_i^2}{a \sum_{i=a+1}^{a+b} z_i^2}$$

where z_i are standard normal deviates used to produce a chi-square variate as shown in the section on the chi-square distribution.

V. SIMULATION LANGUAGES

9.17
AN OVERVIEW

Let us suppose that a simulation model is to be constructed to study a large industrial complex. Further suppose that there are many production lines similar to those studied in Section 9.6 with a large number of repair crews servicing the plant. Imbedded within this framework are various subsets of activities that also interact with the production lines: machine centers, work forces, inventory systems, rework operations, and so on. It is obvious that if a simulation model is to be constructed and used to study proposed changes within this environment, there are many interrelated events that will have to be processed as the simulation progresses through time. These events will have to be processed in proper order as they occur through time, and any related changes in the system due to event status will have to be properly executed. In addition, each event might be stochastic in nature such that many different probability density functions must be used along with literally thousands of random numbers. Furthermore, to assess the impact of selected operational changes, sequencing rules, various queue lengths or queue disciplines, and possible additional changes in the system, there will have to be a wide variety of statistical arrays maintained and continuously updated.

Although all of the above requirements can be handled through proper statement by statement computer coding, it would be extremely convenient and efficient to have many of these tasks "preprogrammed" and packaged in such a way as to be easily accessable, versatile, and reliable for a wide variety of simulation applications.

A *simulation language* is the vehicle through which these required simulation tasks are supplied to potential systems analysts. Proper use of simulation languages can be crucial to the success or failure of a proposed simulation project. Proper language selection can determine the economic feasibility of a simulation study, and might result in manpower savings by a factor of 10 or more. Although simulation languages differ in their construction, logic, ease of usage, accessability, and flexibility, each attempts to supply to a potential user the following standard capabilities:

1. Structured data input.
2. A predetermined time-flow mechanism.
3. Echo checks and/or error checks for program inputs and logic structure.
4. Random-number generation routine(s).
5. A variety (capability) of random deviate generators.
6. A *clock routine* that automatically stores, sequences, and chronologically selects simulated events through time and maintains model equilibrium.
7. Automatic statistical collection (generation) functions.
8. Standard simulation output of relevant data and simulation statistics.
9. Ease of usage.
10. Proper documentation and instructions for any user.

Although most simulation languages attempt to provide the above standard capabilities, there are many ways in which one simulation language might differ from another. Some of these differences are as follows:

1. Mode and nature of data entry.
2. Degree of documentation.

3. Procedures for obtaining random numbers and generating random deviates.
4. Base code from which the language is constructed (FORTRAN, COBOL, SIMSCRIPT, PL-I, and so on).
5. Time-flow mechanisms (uniform or variable time flow).
6. Ease of usage and difficulty to learn.
7. Initialization of the program.
8. Methods of collecting, collating, and analyzing data.
9. Format and extent of output reports.
10. Basic intent of the language.
11. Primary event classification (events discrete in time or continuous through time).

The advantage in using a preconstructed simulation language is best explained by considering the actual process of translating an abstract or real-world problem into a simulation program. In general, the analyst has the objectives of the study and the scope of the model well defined. He or she will usually be proficient in at least one computer programming language (usually FORTRAN or COBOL). The next step in the modeling process would be a clear definition of the problem either symbolically (flow charts), descriptively (English, Spanish, etc.), or conceptually (prototype models, scaled figures, etc.). As soon as the problem to be studied is clearly defined and categorized, the analyst is ready to proceed with the construction of a simulation program. In the process of model construction, the program will have to be checked, debugged, and probably changed many times. If the analyst cannot program, but must rely on an outside programming source, serious communication problems may arise. If the analyst happens to know a common language such as FORTRAN, and does his or her own programming, the communication problems will be reduced; however, this does not eliminate the efforts required to prepare, test, and debug a simulation program after the basic situation is understood.

Simulation analysts had to face these problems repeatedly in the early years of simulation application. It was the repetition of these common processes that led to the development of simulation languages in the early 1960s by groups of people who understood the simulation process. The major advantage in using a simulation language is the savings in time and effort required to structure and debug the total simulation model. Although there will undoubtedly be a process of trial, error, and debugging, the total required for these functions is in general drastically reduced. Another primary advantage is that once the model has been debugged and constructed, it is generally easier to modify and perform experiments on a program constructed from a simulation language than one constructed from a multipurpose language such as FORTRAN.

In addition to the aforementioned benefits in terms of increased user flexibility, savings in total programming and debug time, and standardization of programming procedures, a rather unique change in modeling philosophy occurs simply because the modeler is using a dedicated and specifically designed simulation language. Many simulation languages are actually programming languages in their own right. That is, such a simulation language can be used for many purposes and has a vocabulary unique to that particular framework. If such a simulation language is repeatedly used for modeling purposes, the analyst begins to structure his or her thinking and the model-building process within the confines of that particular language. If a simulation language begins to be used in this manner, it actually becomes an aid in the problem-solving process and model formulation. This change in modeling philosophy often reduces or eliminates a critical step in the problem-solving process—the task of converting conceptualization into computer code. This step is absorbed directly into the problem-solving process and becomes as natural as the verbal description of a problem. The analyst begins to "think" as a simulator.

9.18 COMPARISON OF SELECTED EXISTING SIMULATION LANGUAGES

All formal simulation languages have characteristics. Some are better suited to certain classes of problems than others. The relative importance and usefulness of a simulation language to a potential user is a function of the user's interests and modeling intent. In addition, pedagogical considerations can be quite diverse from one modeler to another. The interests of a learner might be different from those of an expert, the choice from a systems analyst's viewpoint might be quite different from that of higher management, and the language needed by a chemical engineer for process simulation might differ from that of an industrial engineer in a piece-part manufacturing plant. All users, however, are generally interested in the following common considerations [Emshoff and Sisson (8)].

- A language that facilitates model construction and formulation.
- A language that is easy to learn and use.
- A language that provides adequate debugging and error diagnostics.
- A language that can be used on a wide range of problems.

The first criterion relates to the moral, physical, and mental structure of the problem analyst. The statements or commands used in the modeling process should agree with the general conceptual framework under which the analyst might normally operate. The second and third criteria deal with the completeness, integrity, and physical structure of the language itself. These functions are normally embedded in the language structure and cannot be easily changed, if at all. The last criteron deals with the flexibility, completeness, and inherent purpose of the language itself. Several simulation languages will now be described in general terms.

GPSS (General Purpose Systems Simulator)

GPSS is a simulation language originally developed by G. Gordon for the IBM Corporation in the early 1960s (15).

Three major versions of GPSS are now commonly used. (1) GPSS/V, (2) GPSS/H, and (3) GPSS P/C. GPSS/V is a descendent of GPSS/GPSS III and is recognized as the *de facto* standard version of GPSS. GPSS/H is a compiler and run-time support package for GPSS, and is usually upwards compatible to GPSS/V. GPSS P/C is a microcomputer implementation of GPSS/V with some extensions on the IBM PC or PC-XT. GPSS in all forms represents a transaction-flow orientation conceptualized via block diagrams that resemble process flow charts but possess special attributes, logic, and control characteristics. GPSS constructs a logical model of the system under study through the use of block commands. There are more than 40 different specialized blocks in GPSS. These commands perform specific functions that are unique to the language. Time is advanced in fixed units as transactions flow through a specified sequence of block commands. Transactions might possess certain attributes that can be used to make logical decisions at chosen block commands. Each block type might have names, symbols, or numbers associated with it, and each block consumes a specified amount of time to process a transaction. Block types can handle one item (facilities) or multiple items simultaneously (store). For example,

Typical block types	**Originate:**	Creates transactions.
	Queue:	Creates storage space and maintains certain queueing statistics.
	Split:	Creates multiple transactions from one transaction.
	Hold:	Stores a transaction for a designated amount of time.

GPSS III is particularly effective for studying inventory and queuing type problems. It provides a fixed format output, and collects predesignated statistics automatically. GPSS III advances time in fixed increments with an integer clock, but the time-flow mechanism uses next event time advance logic.

SIMSCRIPT 2.5

The SIMSCRIPT simulation language was developed at the Rand Corporation by H. M. Markowitz et al. in the early 1960s (31). SIMSCRIPT is a complete programming language. Although it was originally designed for simulation analysis, it can be used as a general purpose programming language. It requires a special compiler, and is available only on certain computer systems. SIMSCRIPT views the world as one in which the status is unchanged except at certain points in time called event times. A system is described in terms of the status-changing events, and the components of which the system is composed are called *entities*. Properties or characteristics that are associated with entities are called *attributes*, and a group or groups of entities are called sets. The logic of the simulator is constructed through a series of user-constructed statements similar in nature to FORTRAN IV, but all statements are in SIMSCRIPT. These statements perform designated functions, but must be strung together in such a manner as to create the correct program sequencing. Events are normally of two types: those that are generated internally to the simulator (endogenous events) and those that are created outside the simulation framework (exogenous events). Each event desired for the simulation model requires the construction of a separate small event subroutine. The initial conditions, elements of the system, and other required input data are initially entered in the program through definition cards. SIMSCRIPT provides standard error diagnostics and has a fixed format output for common statistics. SIMSCRIPT 2.5 is a variable time event simulator, or a next event simulator.

Although SIMSCRIPT II is currently considered to be the SIMSCRIPT standard, C.A.C.I. has recently announced a microcomputer version called PC SIMSCRIPT II.5. It is totally upwards compatible, operating on the IBM P/C, PC-XT, and PC-AT and is supported by SIMLAB, a model development environment.

MAP/1

MAP/1 is a simulation-based Modeling and Analysis Program (MAP) used to design and evaluate batch manufacturing systems. It provides capabilities for modeling and analyzing the various components of batch manufacturing systems, offering engineers an efficient means of understanding system behavior and improving system productivity. Using MAP/1, in-depth simulation analyses can be performed to investigate alternative system configurations and operating procedures.

MAP/1 was designed to model and to analyze batch manufacturing systems in which individual parts are processed discretely at work stations and are transported in quantities greater than or equal to one. MAP/1 can be used to analyze job shops, flow shops, assembly lines, and computerized manufacturing systems with robotic and FMS components.

A batch manufacturing system is an integrated set of machines, personnel, materials, and transportation equipment which produces parts. There is a construct in MAP/1 to represent each of these major components of batch manufacturing systems.

Stations are defined as places where processing occurs. Each group of like machines is represented as a station. Stations do not always include machines. For example, an inspection station may involve a purely manual process.

Transporters are devices that move parts between stations. They may be devices like forklifts, conveyors, or cranes.

Personnel are the people who perform processing at stations and operate transporters. In MAP/1, personnel are categorized by the types of tasks they perform.

Batch manufacturing systems turn arriving raw materials into finished parts by processing the materials at the various stations in the system. Each type of part has an operational flow by which it is made. Parts are characterized by the sequence of processing steps that produce them. A unique processing sequence may be specified for each part.

Once these basic components of the batch manufacturing system have been described, additional information about scheduling, shift operations, and station and transporter breakdowns is specified. This is followed by simulation specific information such as the initial conditions for the model, the simulation start time, the simulation finish time, and the number of simulation runs to be made.

MAP/1 provides a small set of easily learned constructs to represent the major structural components of the system being modeled. The user simply combines the structural components in the correct order to represent his or her system and then parameterizes each component.

MAP /1 is an easily learned simulation language that produces output reports that support the analysis of batch manufacturing systems. These reports include system trace, throughput reports, utilization reports, inventory reports, and timing reports.

MAP/1 is only available in a mainframe version, but it is supported by TESS. TESS is a simulation software support system with graphical input/output capabilities, data base management, and other technical support.

SIMULA

SIMULA is an old simulation language, which was developed by Dahl and Nygaard (64) specifically for the Univac division of the Sperry Rand Corporation. SIMULA is among a class of simulation languages that translates chosen syntactic structure into an ALGOL base. PROSIM also translates into an ALGOL base [Emshoff and Sisson (8)]. One of the unique features of SIMULA is the capability to create, destroy, and modify existing and new processes created by collections of SIMULA control statements. SIMULA creates a common data file that is accessable by all processes. SIMULA also deals with activities that can be created or destroyed through structured groups of SIMULA statements and commands. A transaction can be either created or destroyed by processes; however, the procedure must be constructed by the user for each individual case. Program logic is controlled by a master clock routine, and the SIMULA language provides preprogrammed logic to connect all of the components of the model. As in SIMSCRIPT 2.5 and GPSS, SIMULA provides a random number generator, several random variable generation schemes, a fixed format output, and error-checking devices. SIMULA also allows the user to program special capabilities into the language if so desired.

DYNAMO

The simulation language DYNAMO was created by P. Fox and A. L. Pugh at the Massachusetts Institute of Technology in the early 1960s (46). DYNAMO was created for the specific purpose of studying systems that can be described by a set of finite difference equations. The operating framework from which DYNAMO was derived was Industrial Dynamics, a language created by J. W. Forrester some years earlier to study individual companies or a consortium of industrial firms in response to alternative courses of action and management policies [Forrester (12)]. DYNAMO essentially attempts to discretize inherently continuous relationships and operating characteristics through fixed-time advance mechanisms. In doing so, continuously time-varying processes ideally solved via an analog computer can be discretely approximated on a digital computer. If the time advance is fine enough, the continuous time domain is accurately approximated. All DYNAMO models depend on the transfer of information and entities described in terms of rates of flow. Certain decision functions need to be created to describe how these rates of flow actually effect the system under study. The commands used in DYNAMO are very similar to FORTRAN-type statements. However, DYNAMO creates structured levels of modeling variables that can be used to describe a wide variety of process relationships. In general, DYNAMO will operate on practically any digital computer with three tape drives and one data channel. Documentation is readily available [Pugh (46)]. DYNAMO has to date been most effectively used in econometric modeling [Naylor et al. (41)], and simulation of industrial complexes [Pugh (46)] along with urban, social, and world-system models.

There are two other languages specifically constructed to deal with simulation models involving differential or difference equations. Those languages are:

1. CSMP (International Business Machines, 1967). Primarily created to solve engineering design problems.
2. CONRAD (GEC/AE I Automation, Inc., in England, 1970). A general language used for continuous processes; it is still in the last stages of development.

SLAM II

Developed by Pritsker (62), SLAM II is a FORTRAN-based simulation language that allows alternative approaches to modeling. It supports three different modeling viewpoints in a single, integrated framework. SLAM II permits network, discrete event, and continuous modeling perspectives, or any combination of the three, to be used in developing a single simulation model.

Network Modeling. A set of specialized node and branch symbols comprise the network modeling component of SLAM II. These symbols represent elements of processes such as equipment, storage areas, and decision points. The modeling task consists of incorporating these symbols into a network diagram that represents the system pictorially. The entities in the system (representing, for example, parts and information) flow through the network model.

Nodes in SLAM II are processing points that may cause an arriving entity to wait in a queue, change variable values, collect statistics, or free a resource. Other nodes provide entry or exit points in the network. Branches (or activities) represent the routing of entities between nodes. Routing may be deterministic, probabilistic, or based on status variables. Time delays occur on branches representing processing time, travel times, or waiting times. The SLAM II processor performs a simulation analysis of the network and provides estimates of system performance on standard summary reports that can be complemented with user-defined output.

Event Modeling. In the event-oriented part of SLAM II, the modeler uses FORTRAN subroutines to define the events that may occur and the mathematical and logical relationships describing the system changes that each type of event will produce.

The discrete event orientation in SLAM II requires that the user write a subroutine to establish the initial conditions for the simulation and event routines to specify the changes in system status for each event. If necessary, specialized outputs of system variables not included in the standard SLAM II reports can easily be obtained.

SLAM II provides a set of standard subprograms that perform common simulation functions such as scheduling events, file storage and retrieval, collecting and accessing statistics, and generating random samples. The executive program (part of the SLAM II processor) controls the simulation by advancing time and initiating calls to the event subroutines. Hence, SLAM II completely relieves the modeler of the task of sequencing events to occur chronologically.

Continuous Modeling. A continuous model is coded in SLAM II by specifying the algebraic, differential, or difference equations that describe the dynamic behavior of the system. The modeler codes these state variable equations in FORTRAN by employing a set of special SLAM-defined storage arrays. When the model includes differential equations, SLAM II automatically integrates them to calculate the values of the state variables within an accuracy prescribed by the modeler.

Technical Description. SLAM II consists of 135 subprograms with over 10,000 lines of source code. It is written in ANSI Standard FORTRAN '66, so that a standard FORTRAN compiler is the only prerequisite software required to use SLAM II. SLAM II is available in both a mainframe version and a 16-bit microcomputer version. Graphical input and graphical/animated output is supported by TESS.

SIMAN

Developed by Pegden (63), SIMAN is a simulation language designed around a logical modeling framework in which the simulation problem is decomposed into a "model" component and an "experiment" component. The model describes the elements of the system such as machines, storage points, and workpieces, and their interrelationships. The experiment specifies the experimental conditions under which the model is to be exercised. This includes elements such as the initial conditions for the run, machine capacities, and the type of statistics to be recorded. Since the experimental conditions are specified external to the model description, they can be changed without affecting the basic model definition.

SIMAN uses a block diagram modeling approach with the flexibility of the discrete event and continuous modeling orientations.

Block diagrams are the primary means of modeling discrete systems in SIMAN. These are linear top-down flowgraphs that depict the flow of entities through the system. The block diagram is constructed as a sequence of blocks whose shapes indicate their general function. The sequencing of blocks is shown with arrows that control the flow of entities from block to block.

There are 10 basic block types in SIMAN, and these basic block types perform 40 different elemental functions. SIMAN allows a modeler to code FORTRAN subroutines, which call event subroutines within the SIMAN Subprogram Library to perform standard functions such as file manipulations, event scheduling, and random sampling.

Continuous systems are modeled in SIMAN by coding the equations of the model in FORTRAN. SIMAN automatically integrates any differential equations in the model to compute the response of the system over simulated time. The system response variables can then be plotted by using commands in the SIMAN Output Processor.

SIMAN is available in both mainframe and microcomputer versions. SIMAN is supported by two specialized programs called BLOCKS and PLAYBACK. BLOCKS is for input graphics, and PLAYBACK is for output graphics. Animated output is available through a supported product called CINEMA.

9.19 THE MICROCOMPUTER REVOLUTION IN SIMULATION APPLICATIONS

In recent years, the methodologies of systems simulation have become increasingly important tools in manufacturing systems analysis and in all other major areas of engineering. The main driving force behind this upsurgence has been the development and successful application of major simulation languages like those described in the preceding discussion. However, without a doubt, the most important factors leading to the acceptance of modern systems simulation analysis techniques have been the dual development of extremely powerful and flexible microcomputers and the corresponding development of microcomputer simulation languages. It is not unlikely that in the near future the equivalent of a VAX or IBM mainframe machine may be on every student's desk or within close proximity to any engineer. Today, 32-bit microcomputers with 40-megabyte hard disks are not difficult to find in modern manufacturing complexes. Perhaps the most exciting development with regard to simulation has been the graphics and animation capabilities that have become available as graphics software and microcomputer hardware mature into mainstream products. In our previous discussion, we noted that several major simulation languages now have available microcomputer versions of the mainframe language that are totally upwards compatible and perfectly capable of modeling and analyzing relatively large systems. It is conjectured that within the next 5 to 10 years incredible simulation animation and modeling systems will be released to both enhance and astound any engineer interested in performing simulation analysis. For the present, however, one should carefully

Table 9.9
SIMULATION LANGUAGES

Simulation Language	Language Base	Event Monitors	Arithmetic	Modeling Base
GEMS	FORTRAN	Discrete	Real	Network
GPSS/H	GPSS	Discrete	Integer	Block diagram
GPSS/PC	GPSS	Discrete	Integer	Block diagram
IDSS 2.0	FORTRAN/Machine[1]	Combined	Real	Network
INTERACTIVE	PASCAL	Discrete	Real	Network
MAP-1	FORTRAN '66/'77	Discrete	Real	Statements
MAST	FORTRAN	Discrete	Real	—
MICRONET	FORTH	Discrete	Integer	Network
PC SIMSCRIPT II.5	SIMSCRIPT	Discrete	Real	Event
PSIM	PASCAL	Discrete	Real	Event
Q-GERTS	FORTRAN	Discrete	Real	Network
SEE-WHY	FORTRAN	Discrete	Real	Event
SIMAN	FORTRAN '77	Combined	Real	Block diagram
SIMAN P/C for micro	FORTRAN '77	Combined	Real	Block diagram
SIMSCRIPT II.5	SIMSCRIPT	Discrete	Real	Event
SLAM II[1]	FORTRAN '77	Combined	Real	Network
SLAM II PC	FORTRAN '77	Combined	Real	Network
INSIGHT	FORTRAN	Combined	Real	Network

[a]Lodestone II, Inc., 3833 Texas Avenue, Suite 460, Bryan, TX 77802 (409) 846-4171
[b]Wolverine Software Corporation, 7630 Little River Turnpike, Suite 208, Annadale, VA 22003
[c]Minuteman Software, P.O. Box 171, Stow, MA 01775 (617) 897-5662
[d]Pritsker and Associates, Inc., P.O. Box 2413, West Lafayette, IN 47906 (317) 463-5557
[e]Micro Simulation, 50 Milk Street, Suite 1500, Boston, MA 02109 (617) 451-8448
[f]CMS Research, Inc., 945 Bavarian Court, Oshkosh, WI
[g]C.A.C.I., 12011 San Vincente Blvd., Los Angeles, CA 90049 (213) 476-6511
[h]ISTEL, Inc., 83 Cambridge Street, Burlington, MA 01803 (617) 272-7333
[i]Systems Modeling Corp., Calder Sq., P.O. Box 10074, State College, PA 16805 (814) 238-5919
[j]SYSTECH Inc., P.O. Box 509203, Indianapolis, IN 46250

choose both the mainframe and microcomputer capabilities that are already developed for current use. The following are several important capabilities that one should consider when choosing microcomputer-based simulation systems.

1. Operating systems required: APPLE DOS, P/C DOS, UNIX, etc.
2. Core requirements to execute and load the simulation language: 64K, 256K, 512K, etc.
3. The base language used: FORTRAN, BASIC, FORTH, C, PL-1, etc.
4. Mainframe compatability: upwards compatability, data base management, transferability, networking, etc.
5. Graphics support: Plot 10 compatability, standardized interfaces, portability, etc.
6. Interfaced to support programs: LOTUS 1-2-3, SUPERCALC, FRAMEWORK, TOPVIEW, etc.
7. Database support systems: MICRORIM, D-BASE, R-BASE, ORACLE, etc.

Although it is beyond the scope of this introductory chapter to explore microcomputer capabilities, requirements, and simulation availabilities, Table 9.9 provides a short survey of the available simulation languages now running on microcomputers and their corresponding mainframe versions. Since the microcomputer field is evolving so rapidly, and simulation support tools are now under development, it is likely that this information will be obsolete in a short period of time

Table 9.9 (Continued)

Hardware	Graphic Input	Application Areas	Information
[10]	No	Manufacturing systems, general purpose	a
[6]	No	General purpose	b
IBM PC	No	General purpose	c
CAD-LINK[9]	Yes	General purpose, man. sys.	d
Apple, IBM PC	No	General purpose	e
[7]	No	Batch manufacturing systems	d
IBM PC	No	Design of FMS	f
Apple, IBM PC	No	General purpose	d
IBM PC-XT-AT	No	General purpose	g
Apple, IBM PC	No	General purpose	e
[7]	No	General purpose, man. sys.	d
Cromemco or System[3]	Yes	Manufacturing systems, general purpose	h
[8]	No	General purpose, man. sys.	i
IBM PC	Yes[4]	General purpose, man. sys.	i
IBM PC	No	General purpose	g
[10]	Yes[5]	General purpose	d
IBM PC	No	General purpose	d
[8]	Yes	Medical systems, general purpose	j

[1]IDSS 2.0 must be run on a VAX 11/780, VAX 11/750, or CAD-LINK microprocessor.
[2]Software that will run on an IBM PC may run on IBM compatibles.
[3]The necessary hardware to run SEE-WHY can be purchased with the software.
[4]BLOCKS, available from Systems Modeling Corp., can be used to input graphically the block diagram.
[5]TESS (The Extended Simulation System), available from Pritsker and Associates, Inc., can be used to graphically input the network.
[6]IBM 370, 43XX, 30XX, AHMDAHL, and VAX; also IBM XT/370 and MICROVAX.
[7]32-bit mainframe with FORTRAN COMPILERS I.
[8]Both IBM P/C AND 32 bit mainframe versions requiring FORTRAN COMPILERS.
[9]Designed for VAX 11/780 or similar with plot 10 compatibility.
[10]32-bit mainframe, T/I Professional, and IBM P/C.

In addition to the simulation languages that we have previously discussed and that appear in Table 9.9, there exist several other important languages that either serve a special purpose or have been limited to a small interested group of users. In this short introductory section, it is impossible to discuss all the simulation languages. Interested readers should refer to the selected references at the end of this chapter and the current literature for further information.

VI. ADVANCED CONCEPTS IN SIMULATION ANALYSIS

Although it is beyond the scope of this chapter to explore in detail the total spectrum associated with simulation analysis, it should be noted that the actual construction, debugging, and production of a workable simulation program might very well be only a starting point for a more comprehensive simulation analysis. Several advanced areas of digital simulation analysis are as follows:

1. Design of computer simulation experiments.
2. Variance reduction techniques.
3. Statistical analysis of simulation output.
4. Optimization of simulation parameters.

9.20 DESIGN OF SIMULATION EXPERIMENTS

Once a simulation model has been constructed and meaningful system statistics are being generated, the experimenter may be interested in learning more about the underlying structure or properties of the system being studied. In particular, it might be desirable to quantify the effects of deliberately changing relevant factors over a given region of interest. Through an investigation of simulation output it is possible to create a *response surface* in which the various factors involved are designated at different levels of operation. To characterize these surfaces, *experimental designs* are often used to analyze, quantify, and predict the effects of response-surface change. Experimental designs such as *full factorials, latin squares, fractional designs,* and *rotatable designs* have been successfully employed with digital simulation experiments [Hunter and Naylor (22)]. Through the use of *linear and curvilinear* regression, the results of these experiments have been reduced to formulas that can *predict* system behavior over a wide range of changes. The primary purpose of experimental designs is to determine *which variables* are most important, *how* these variables influence the response of our simulation model, and *why* certain results occur as they do in the simulation experiment. Experimental designs are expected to play an increasingly important role in future analysis of digital simulation output.

9.21 VARIANCE REDUCTION TECHNIQUES

Simulation experiments are constructed to gain meaningful information about certain aspects of the system under study. For example, in simulating a single-channel queueing problem one might be primarily interested in the expected waiting line or expected queue that develops during the servicing period. Normally, one would observe the system at intervals of change, update all relevant queueing statistics, and print out desired information after a period of simulated time. The entire simulation might yield a result for *expected queue length*. Now, since the simulation itself is a random process, the output is by definition a *random variable*. Being a random variable, simulation output inherently contains statistical fluctuation and great care should be exercised in using these results. In words, we wish our output to be as accurate as possible considering statistical variation. A common way to reduce this measure of uncertainty is to *replicate* the experiment and *average* the simulation results. From elementary statistics we know that for a random variable x and an estimator $\bar{x}$ the following is true:

$$E(\bar{x}) = E(x)$$
$$VAR(\bar{x}) = \frac{VAR(x)}{n}$$

Hence, we reduce the variance of our estimator by a factor of $(1/n)$ by doing n times as much work. Since variance is obviously related to the amount of work involved, we can define efficiency as:

$$\text{efficiency} = 1/(\text{variance})(\text{work})$$

A *variance reduction* technique is therefore one that reduces the inherent random deviation in our statistical output. Obviously, we would do well to seek methods that will reduce variance by factors proportionately more than the work involved to accomplish this reduction.

There have been several statistically-based techniques which help to accomplish this goal. An excellent survey is given by Naylor (41,57); while Hammersley and Handscomb (19) provide a theoretical basis for variance reduction techniques. Numerical examples are given for a few common techniques in Hillier and Liebermann (60). Some of these techniques are listed below.

1. Use of expected values.
2. Stratified sampling.
3. Importance sampling.
4. Control variates.
5. Antithetic variates.
6. Quasirandom numbers.

Practical applications of these techniques are very few in number, and the investigation of their use and efficiency is a topic for further research.

9.22 STATISTICAL ANALYSIS OF SIMULATION OUTPUT

Closely related to the problems of variance reduction and simulation efficiency are the problems of relevant statistical analysis of simulation output.

A requirement exists for the development of a statistical technique or set of techniques to (1) validate, (2) authenticate, and (3) collate time series data generated by digital simulation models. *Validation* is an implication that the model adequately represents the "real-world" situation; *authentication* denotes the establishment of a measure of confidence in a single set of model results; *collation* connotes a critical comparison of two or more sets of model output. The process for validating, authenticating, and collating digital simulation models is designated as *verification*. This section addresses itself to this requirement for a subset of time-series data generated by simulation models.

Four main types of statistical functions have evolved for describing random data: (1) mean-square values, (2) probability density functions, (3) autocorrelation functions, and (4) power spectral density functions. In the case of describing joint properties of random data, three types of statistical functions have been developed to satisfy this descriptive requirement: (1) joint probability density functions, (2) cross-correlation functions, and (3) cross-spectral density functions for the amplitude, time, and frequency domains, respectively. Each of these areas are subject to certain weaknesses and strengths. Fishman and Kiviat (9) recently suggested spectral analysis as a superior technique for evaluating autocorrelated simulation results. Spectral analysis has been employed successfully in the study of the time-dependent nature of physical processes and economic processes (57).

The initial requirement for either collation or validation of simulation results is the examination of the data for *stationarity, normality,* and *randomness*. It is assumed that *autocorrelation* exists and consequently, standard statistical techniques for analysis of simulation data cannot be accurately utilized. The assumption of autocorrelation is well founded, for it is well known that most digital simulation data is highly autocorrelated. The purpose of a statistical analysis of simulation output would be to remove the effects of autocorrelation from this data, and in the process detect other abnormalities that are not readily evident, such as influential noise and periodicities. Once these effects have been removed, the data should then exhibit the properties of randomness and independence necessary for conventional statistical analysis, and provide a more reliable basis for sound operating decisions.

Not only does the effect of autocorrelation bias most simulation results, but the problem is confounded by the interaction with initial conditions, modes of sampling, and external problems due to faulty deviate generators. The exploration and analysis of these problem areas is a source of current research. Fishman (10) has published an excellent text that explores the mathematical and practical aspects of the problem. Duket (61) has also investigated this area and concludes that the currently suggested techniques for the analysis of simulation output may be deficient. Certainly the problem is a highly complex one and will require much further research.

9.23 OPTIMIZATION OF SIMULATION PARAMETERS

Although simulation is primarily a tool for systems *analysis*, there is no reason why one should not employ it for systems *optimization* when the model is appropriate. For example, instead of determining the expected queue length of a servicing facility and analyzing the effects of such a queue, one might be interested in determining the *optimum* queue length with respect to minimized system costs. Optimization of simulation experiments can be achieved through an interface with the tools of linear and nonlinear programming. In addition, statistical techniques such as response surface analysis have been used within the "spirit" of simulation analysis for long periods of time. Direct search techniques that require only a "black box" response can be used to guide a simulation to an optimum response with respect to the chosen decision variables. Some of these optimization techniques are discussed in Chapter 11 (Golden Section Search, Hooke-Jeeves Search); others are available in books in the reference list. The primary problem appears to be in the interface between sound simulation technology and mathematical optimization. However, with the increasing popularity of the "team approach" to problem solving these difficulties are rapidly being resolved.

9.24 SUMMARY AND CONCLUSIONS

This chapter presents some of the major aspects of digital simulation analysis. Once viewed as a plaything for idle computer programmers, simulation now plays a major role in the solution of real-world problems from all phases of operations research. Fundamental to the development of any simulation model is the generation of random or "pseudorandom numbers"; and subsequently the generation and use of statistical random deviates. Sections 9.7 through 9.16 dealt in some detail with these two problems. Since a sound random number is an absolute necessity, general tests for determining the validity of a random number generator were explained in Section 9.10.

Simulation languages are now taking a central role in determining the scope and applicability of simulation modeling. Part V surveyed the major simulation languages in such a way as to impart general knowledge about their characteristics to the casual reader. Although simulation languages will continue to play a central role in large-scale systems simulation, there may exist problems that can be adequately formulated and solved by hand or on one of the ever increasing "home computers" currently available on the market today. Part II illustrated the logic and procedures of simulation analysis using three fundamental problems.

Simulation analysis provides a means by which the systems analyst can experiment in a representative problem area without having to deal directly with the real-world system itself. Since simulation is a sampling procedure used to produce relevant statistical results about complex problems, the most important thing about the simulation process is the production, collation, and interpretation of output data. Some of the ways in which this problem might be addressed were discussed in Section 9.21.

Regardless of the position which one wishes to take regarding the scope, purpose, and applicability of operations research, one thing is clear—simulation analysis now plays a central role in the study and solution of complex engineering problems, and it is a tool with which every systems analyst should be familiar. For additional readings, the reader should refer to any of the simulation texts given under "Selected Reference Texts."

SELECTED REFERENCE TEXTS

Banks, J. and J. S. Carson III, *Discrete Event System Simulation*, Prentice-Hall, Englewood Cliffs, N. J., 1984.

Bratley, P., B. L. Fox, and L. E. Schrage, *A Guide to Simulation*, Springer-Verlag, New York, 1984.

Fishman, G. S., *Principles of Discrete Event Simulation*, Wiley, New York, 1978.

Law, A. M., and W. D. Kelton, *Simulation Modeling and Analysis*, McGraw-Hill, New York, 1982.

Pegden, C., *SIMAN, Systems Modeling Corporation*, P. O. Box 10074, State College, Pa., 1984.

Pritsker, A. A. B., *Introduction to Simulation and SLAM II*, 2nd ed., Halstead Press, New York, 1984.

REFERENCES

1. Aitchison, J., and J. A. C. Brown, *The Lognormal Distribution*, Cambridge University Press, Cambridge, 1957.
2. Allard, J. L., R. A. Dobell, and T. E. Hull, "Mixed Congruential Random Number Generators for Decimal Machines," *Journal of the ACM, 10*(2), 131-132 (April 1963).
3. Berman, M. G., "Generating Random Variates from Gamma Distributions with Non-Integer Shape Parameters," The Rand Corporation, R-641-PR, Santa Monica, Calif., November, 1970.
4. Box, G. E. P., and M. E. Muller, "A Note on the Generation of Normal Deviates," *Annals of Mathematical Statistics*, XXIX, 610–611 (1958).
5. Butler, J. W., "Machine Sampling from Given Probability Distributions," in *Symposium on Monte Carlo Methods*, edited by H. A. Meyer, Wiley, New York, 1956.
6. Chorafas, D. N., *Systems and Simulation*, Academic Press, New York, 1965.
7. Conway, R. W., "Some Tactical Problems in Digital Simulation," *Manag. Sci., 10*(1), 47–61 (October 1963).
8. Emshoff, J. R., and R. L. Sisson, *Design and Use of Computer Simulation Models,* MacMillan, New York, 1970.
9. Fishman, G. S., *Concepts and Methods in Discrete Event Digital Simulation*, Wiley, New York, 1973.
10. Fishman, G. S., and P. J. Kiviat, "Digital Computer Simulation: Statistical Considerations," Memorandum RM-5387-PR, The Rand Corporation, Santa Monica, Calif., November 1967.
11. Fishman, G. S., and P. J. Kiviat, "The Analysis of Simulation-Generated Time Series," *Manag. Sci., 13*, 525–557 (March 1967).

12. Forrester, Jay, *Industrial Dynamics*, MIT Press, Cambridge, Mass., 1961.
13. Gaver, D. P., "Statistical Methods for Improving Simulation Efficiency," Management Sciences Research Report No. 169, Graduate School of Industrial Administration, Carnegie-Mellon University, Pittsburgh, Pa. (August 1969).
14. Golden, D. G., and J. D. Schoeffler, "*GSL-A* Combined Continuous and Discrete Simulation Language," *Simulation, 20*, 1–8 (January 1973).
15. Gordon, G., *System Simulation*, Prentice-Hall, Englewood Cliffs, N. J., 1969.
16. Greenberger, M., "Method in Randomness," *Communications of the ACM, VIII* (3), 177–179 (1965).
17. Grosenbaugh, L. R., "More on Fortran Random Number Generators," *Communications of the ACM, 12*(11), 639 (November 1969).
18. Gruenberger, F. J., *Problems for Computer Solution*, Wiley, New York, 1965.
19. Hammersly, J. M., and D. C. Handscomb, *Monte Carlo Methods*, Wiley, New York, 1964.
20. *Handbook of Mathematical Functions,* edited by M. Abramowitz and I. A. Stegun, Applied Mathematics Series 55, Department of Commerce, Government Printing Office, Washington, D.C., p. 950, 1967.
21. Hull, T. E., and A. R. Dobell, "Random Number Generators," *Society for Industrial and Applied Mathematics, 4*(3), 320 (July 1962).
22. Hunter, J. S., and T. H. Naylor, "Experimental Designs for Computer Simulation Experiments," *Manag. Sci., 16*(7), 422–434 (March 1970).
23. Hurst, N. R., "GASP IV: A Combined Continuous/Discrete FORTRAN Based Simulation Language," unpublished Ph.D. Thesis, Purdue University, Lafayette, Ind., 1973.
24. IBM Corporation, General Purpose Simulation System/360 OS and DO5 Version 2 User's Manual, SH20-0694-0, White Plains, N. Y., 1969.
25. Kiviat, P. J., R. Villanueva, and H. Markowitz, *The SIMSCRIPT II Programming Language*, Prentice-Hall, Englewood Cliffs, N.J., 1969.
26. Kleine, H., "A Survey of Users' Views of Discrete Simulation Languages," *Simulation, 14*(5), 225–229 (May 1970).
27. Knuth, D. E., *The Art of Computer Programming*, Vol 2, Addision Wesley, Reading, Mass., 1968.
28. Kruskal, B., "Extremely Portable Random Number Generator," *Communications of the ACM, 12*(2), 93–94 (February 1969).
29. Larson, H. J., *Introduction to Probability Theory and Statistical Inference,* Wiley, New York, 1969.
30. Maclaren, M. D., and G. Marsaglia, "Uniform Random Number Generators," *Journal of the ACM, 12*(1), 83–89 (January 1965).
31. Markowitz, H. M., H. W. Karr, and B. Hausner, *SIMSCRIPT: A Simulation Programming Language,* Prentice-Hall, Englewood Cliffs, N.J., 1963.
32. Marsaglia, G., and T. A. Bray, "One-Line Random Number Generators and Their Use in Combinations," *Communications of the ACM, 1*(11), 757–759 (November 1968).
33. Marsaglia, G., and M. D. MacLaren, "A Fast Procedure for Generating Normal Random Variables," *Communications of the ACM, VII*, 4–10 (1964).
34. Marsaglia, G., and M. D. MacLaren, "Uniform Random Number Generators," *Journal of the Association for Computing Machinery, XII*, 83–89 (1965).
35. Martin, F. F., *Computer Modelling and Simulation*, Wiley, New York, 1968.
36. McMillan, C., and R. F. Gonzales, *Systems Analysis: A Computer Approach to Decision Models*, Richard D. Irwin, Inc., Homewood, Ill., 1965.

37. Meier, R. C., "The Application of Optimum-Seeking Techniques to Simulation Studies: A Preliminary Evaluation," *Journal of Financial and Quantitative Analysis, 2*(1), 31–51 (March 1967).

38. Meier, R. C., W. T. Newell, and H. L. Pazer, *Simulation in Business and Economics*, Prentice-Hall, Englewood Cliffs, N.J., 1969.

39. Mihram, G. A., "On Antithetic Variates," *Proceedings of the 1973 Summer Computer Simulation Conference*, Montreal, 91–95, July 17–19, 1973.

40. Mize, J. H., and J. C. Cox, *Essentials of Simulation*, Prentice-Hall, Englewood Cliffs, N.J., 1968.

41. Naylor, T. H., J. L. Balintfy, D. S. Burdick, and Kong Chu, *Computer Simulation Techniques*, Wiley, New York, 1966.

42. Phillips, D. T., "Applied Goodness of Fit Testing," AIIE Monograph Series, AIIE-OR-72-1, Atlanta, Ga., 1972.

43. Phillips, D. T., and C. S. Beightler, "Procedures for Generating Gamma Variates with Non-Integer Parameter Sets," *J. Stat. Comput. Simul., 1*, 197–208 (1972).

44. Pritsker, A. A. B., and P. J. Kiviat, *Simulation with GASP II*, Prentice-Hall, Englewood Cliffs, N.J., 1969.

45. Pritsker, A. A. B., *The GASP IV User's Manual*, Pritsker and associates, W. Lafayette, Ind., 1973.

46. Pugh, A. L. III, DYNAMO II User's Manual, The M.I.T. Press, Cambridge, Mass., 1970.

47. Rand Corporation, *A Million Random Digits with 1,000,000 Normal Deviates,* Free Press, New York, 1955.

48. Schriber, T., *A GPSS Primer* (preliminary printing), Ulrich's Books, Ann Arbor, Mich., 1972.

49. Schmidt, J. W., and R. E. Taylor, *Simulation and Analysis of Industrial Systems,* Richard D. Irwin, Inc., Homewood, Ill., 1970.

50. SCi Simulation Software Committee, "The SCi Continuous System Simulation Language (CSSL)," *Simulation, 9*, 281–303 (December 1967).

51. Tocher, K. D., "The Application of Automatic Computers to Sampling Experiments," *Journal of the Royal Statistical Society, B16*, 39–61 (1954).

52. Tocher, K. D., *The Art of Simulation*, Van Nostrand, Princeton, N.J., 1963.

53. Tocher, K. D., "Review of Computer Simulation Languages," *Oper. Res. Quart., 16*, 189–217 (June 1965).

54. Tramposch, H., and H. A. Jones, Jr., "Impact Problems Efficiently Solved with 1130 CSMP," *Simulation, 14*, 73–79 (February 1970).

55. Weibull, W., "A Statistical Distribution of Wide Applicability," *Journal of Applied Mechanics, XVIII*, 293–297 (1951).

56. Wilson, Benjamin, *Integral Calculus*, London, 1891.

57. Naylor, T. H., *Computer Simulation Experiments with Models of Economic Systems*, Wiley, New York, 1971.

58. Nance, R. E., "On Time Flow Mechanisms for Discrete System Simulation," Computer Science Center, Southern Methodist University, Dallas, Tex., 1969.

59. Schmidt, J. W., "Fundamentals of Simulation," *Proceedings 1974 Systems Engineering Conference,* Minneapolis, Minn. (November 1974).

60. Hillier, F. S., and G. J. Lieberman, *Introduction to Operations Research*, 2nd ed., Holden-Day, San Francisco, 1974.

61. Duket, S. D., *Simulation Output Analysis*, unpublished Masters Thesis, Purdue University, Lafayette, Ind., 1974.

62. Pritsker, A. A. B., *Introduction to Simulation and SLAM II,* 2nd ed., Halstead Press, New York, 1984.

63. Pegden, C., *SIMAN*, Systems Modeling Corporation, P. O. Box 10074, State College, Pa., 1984.

64. Dahl, O. J., and K. Nygaard, "SIMULA-An Algol-based Simulation Language," *Comm. of the ACM, 9*, 671–678 (September 1966).

EXERCISES

1. Define the term "simulation" as used in this chapter.
2. List 10 reasons why simulation analysis is appropriate for many real-world problems.
3. List five disadvantages in using simulation analysis.
4. Explain the terms "compressed time" and "expanded time" related to simulation analysis.
5. Why is modern simulation analysis more of an "art" than "science"? How does this fact relate to the scientific content of Section III?
6. Explain the difference in *deductive* and *algorithmic* models. Give two examples of each type.
7. What is the difference in model *verification* and model *validation*?
8. Explain the difference in *uniform time* advance simulation procedures and *variable time* advance procedures. When would one be preferred over the other? Which procedure do you think is faster? Explain.
9. Explain why a simulation experiment must usually be executed over long periods of simulated time.
10. What is simulation bias? Give an example. What is a transient analysis? Give an example. Can you think of a real-world system that is always transient?
11. What is a *random number generator*? Why is this term a misnomer? What are the desirable properties of random number generators?
12. Describe in words what is an inverse transform as related to random deviate generation.
13. Why are *rejection techniques* usually inefficient?
14. Explain the logic of composition techniques. Give three good examples.
15. List 10 major advantages to using a simulation language.
16. Describe the difference in next event simulation modeling and continuous simulation modeling.
17. List five reasons why one would want to use a microcomputer-based simulation language.
18. What is *variance reduction*? Why is this extremely important to simulation analysis?
19. What is autocorrelation? How does it effect simulation analysis?
20. Describe how one would "optimize" the use of simulation. Why is this such a difficult task? Why might an "optimized simulation model" not truly yield an optimal solution?
21. The Neetee-Eatee Hamburger Joint specializes in soybean burgers. However, soyburgers are not the only commodity sold. Customers may also buy cokes, fishburgers,

and so on. Customers arrive according to the following interarrival rate distribution over the lunch period from 11:00 AM to 1:00 PM:

Time Between Arrivals (minutes)	Probability
3	0.30
5	0.20
6	0.15
8	0.20
10	0.15

People who want burgers for lunch usually arrive in groups. The following distribution of arrivals has been observed over a period of past history:

Number of People	Probability
1	0.40
2	0.30
3	0.20
4	0.10

Each individual customer orders between none and two hamburgers according to the following:

Burgers per Person	Probability
0	0.20
1	0.65
2	0.15

Due to fluctuations in burger-eating per person, the average stay per person is a random variable given by the following:

Length of Stay (minutes)	Probability
10	0.10
15	0.40
20	0.30
25	0.20

If a group enters the restaurant, the time in the system for the group is determined by the longest length of stay for an individual in that group.

Using random number tables, simulate behavior of the hamburger joint for a period of 6 hours. Determine answers to the following questions:

1. How many burgers would have to be made on the average on an hourly basis?
2. What is the expected stay in the dining area per group of people?

22. Simulate a multiple-channel service system with three parallel channels, equal selection probabilities, exponential time between arrivals, and exponential service times. Assume an infinite queue length.

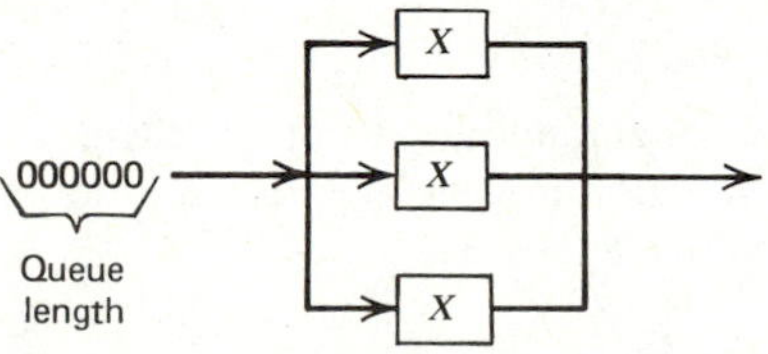

Service Times

Server 1: 6.2 minutes per unit
Server 2: 5.1 minutes per unit
Server 3: 4.8 minutes per unit

Interarrival Time: 3.5 minutes per unit

Determine the percent utilization of each service facility for a period of simulated time equal to 3 hours. Compare this estimate to the exact answer using the appropriate formula from Chapter 7.

23. A conveyor-serviced production system consists of two operators in series performing essentially the same operation. A schematic of this operation is as follows:

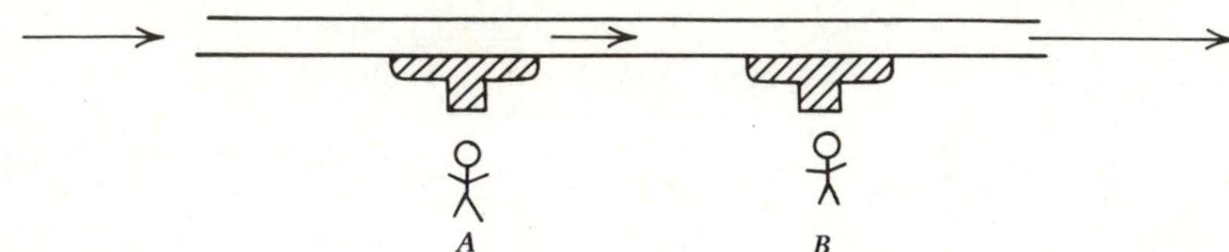

Items arrive via a continuous belt conveyor according to an exponential density function with mean equal to 2.8 minutes per item. Server A and Server B service these items according to a normal density function with the following parameters.

Server	μ	σ^2
A	5.2	1.0
B	5.0	1.0

Server A has in-process storage for two items, and Server B has in-process storage for one item (in addition to item being serviced). Simulate the behavior of this system from a completely idle state for a period of 300 time units and estimate the following:

(a) Percent utilization of Server A.
(b) Percent utilization of Server B.
(c) Number of balkers per 100 time units.

24. List five desirable properties that a "good" random number generator should possess.

25. Explain the difference in "random numbers" and "pseudorandom numbers."

26. Demonstrate the procedure of random deviate generation using the midsquare technique starting with the number 3264. What is the period for this sequence of random numbers?

27. Repeat Exercise 26 using the same random-number seed for the midproduct technique. Use $k = 47$.

28. Using the table of random numbers given in Appendix C, run the following random number tests.

(a) The frequency test.
(b) The gap test.
(c) The poker test.
(d) The runs test.

29. Using the Box-Muller normal random generation scheme for standard normal variates ($\mu = \theta$; $\sigma^2 = 1$), write a function to generate normal random deviates from the general normal density function ($\mu = \theta$; $\sigma^2 = \varphi$).

30. Use the fundamental theory and logic of the *rejection technique* to estimate the area under the following curve.

$$Y = 0.65 \sin(x) \qquad \pi \geqq x \geqq 0$$

Use 100 points to perform the required calculations.

31. Develop a random deviate generation scheme for the following probability density function.

$$f(x) = \begin{cases} x^2/2 & 1 \geqq x \geqq 0 \\ 2.3e^{-x} & \infty \geqq x \geqq 1 \end{cases}$$

32. The following probability density function is known as the *Raleigh* distribution.

$$f(x) = \begin{cases} x/\sigma^2 \exp\left(-\dfrac{x^2}{2\sigma^2}\right) & x \geqq 0 \\ & \sigma > 0 \\ 0 & \text{elsewhere} \end{cases}$$

Develop a random deviate generator for this density function.

33. The *triangular* density function is a continuous density that follows this general form:

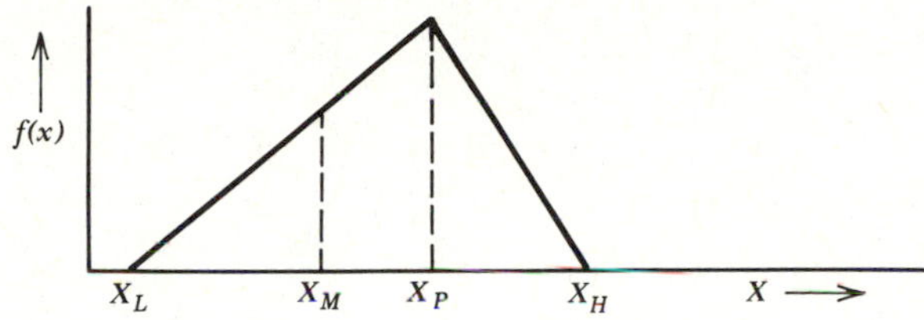

where X_L = minimum value
X_H = maximum value
X_P = most likely value
X_M = mean value

The triangular density function is given with respect to the position of X_P, such that:

$$f(x) = \begin{cases} \dfrac{2(X - X_L)}{(X_H - X_L)(X_P - X_L)} & X \leqq X_P \\ \dfrac{2(X_H - X)}{(X_H - X_L)(X_H - X_P)} & X > X_P \end{cases}$$

Develop a random deviate generator for the triangular density function.

34. Structure a new random deviate generation scheme for the standard normal density function using the *rejection technique*. Comment on your new method. Would it be efficient; if not, why?

35. According to the test, a Poisson deviate with parameter λ can be generated by finding the number of exponential random deviates used to satisfy the following inequality.

$$\sum_{i=1}^{x} y_i \leqq 1 \leqq \sum_{i=1}^{x+1} y_i$$

where $y_i = 1/\lambda \ln(R_i)$
R_i = the ith random number in a sequence of random numbers.
Verify this fact.

CHAPTER 10
DYNAMIC PROGRAMMING

I. BASIC CONCEPTS

10.1
INTRODUCTION

In most operations research problems the objective is to find the optimal (max or min) values of the "decision variables," that is, those variables that can change or be controlled within the problem structure. Usually, these variables are dealt with *simultaneously* or collectively. Each of us, however, has been faced with problems in which it might be possible to break our decisions up into smaller components or parts (decomposition) and then recombine our previous decisions in some form or another to obtain the desired answer (composition). This approach is called *multistage problem solving*, and dynamic programming is a systematic technique for reaching an answer in problems of this nature. Many techniques are found in this book for solving various optimization problems. Numerous algorithms have been developed to solve both the linear and nonlinear objective functions that are subject to various constraint configurations. One might think that all of these procedures could be classified as those dealing with either linear or nonlinear functions, but dynamic programming cannot be uniquely classified in either category. Properly applied, dynamic programming cuts across all fields of mathematical programming. Important applications have surfaced in inventory control theory, network flows, job-shop scheduling, production control, integer programming, and many other areas. Dynamic programming has also proved useful in solving problems relevant to all fields of engineering. Although it might seem that we are about to produce the panacea of algorithmic procedures, one should be forewarned that this is not true. Like all operations research techniques, dynamic programming has its limitations and weaknesses. However, when applicable the technique can be spectacular with regard to its computational savings.

10.2
HISTORICAL BACKGROUND

The founding father of dynamic programming, and the man primarily responsible for the development of dynamic programming, is Richard Bellman.

Bellman first developed the concepts of dynamic programming in the late 1940s and early 1950s while working as a researcher at The Rand Corporation.

Bellman wrote several papers about his work and, in 1957, a book, *Dynamic Programming* (5). Since that time this book has been a continuing source of unique applications and problem-solving logic for numerous engineering problems. A second book by Bellman appeared in 1961 (6), and in 1962 (8) a third book was produced in collaboration with S. E. Dreyfus, a Rand colleague. As Bellman and his associates began to popularize the techniques and methodologies of dynamic programming, important contributions were made by other authors. Aris (1, 2) authored two books on dynamic programming, the first in 1961 and the second in 1964. In late 1964, Aris, Nemhauser, and Wilde (3) developed a generalized theory for dealing with branched, cyclic, and looping multistaged systems. L. G. Mitten (20, 21, 22) also contributed greatly during this period and supplied many of the underlying ideas for important future development. Mitten (20), Denardo (11), and Dreyfus (12) all independently contributed to the mathematical properties of the dynamic programming approach.

In 1966 G. L. Nemhauser produced an excellent text dealing with the applications of dynamic programming, *Introduction to Dynamic Programming* (24). Other references include the book by Beightler, Phillips, and Wilde, *Foundations of Optimization* (4), and one by D. J. White, *Dynamic Programming* (38). A complete treatment of nonserial dynamic programming can be found in the text by Bertele and Brioschi, *Nonserial Dynamic Programming* (9). A survey paper by Thomas also provides a good overview of the field (36). Morin (32), and Morin and Esogbue (31), have provided computationally efficient algorithms for dynamic programming utilizing a unique state reduction algorithm based on what they call the "embedded state space" approach. Computer programs and a general theory for closed form solutions to nonserial dynamic programming problems have been developed by Pope, Curry, and Phillips (25). Other important references are given in the bibliography at the end of this chapter.

II. THE DEVELOPMENT OF DYNAMIC PROGRAMMING

10.3 MATHEMATICAL DESCRIPTION

Suppose that we are faced with a problem in which there are known input parameters and we desire to use these parameters (max or min our problem) in an optimal fashion. Denote the point at which we make a decision as a *stage*, and denote our input parameters as the *state*. Let the decision itself be governed by some sort of equation or rule, called a *transformation*. Pictorially,

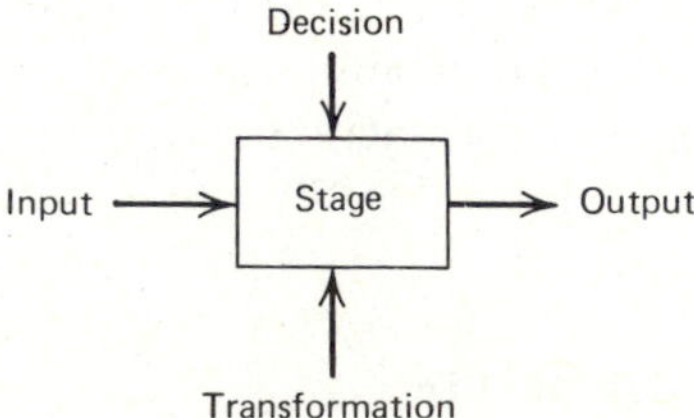

Let us imagine that at each *stage* we are forced to make a *decision*. Every decision that we make has a relative worth, or benefit (good or bad!) reflected by a decision benefit equation. Let this equation be represented or denoted as a *return*

function, since for every set of decisions we make, we get a return on each decision. This *return function* will, in general, depend on both the *state variable* and the *decision* made at stage *n*. An *optimal* decision at stage *n* would be that decision which yields a maximum (or minimum) return for a given value of the state variable, S_n.

Functionally, for a single stage we have

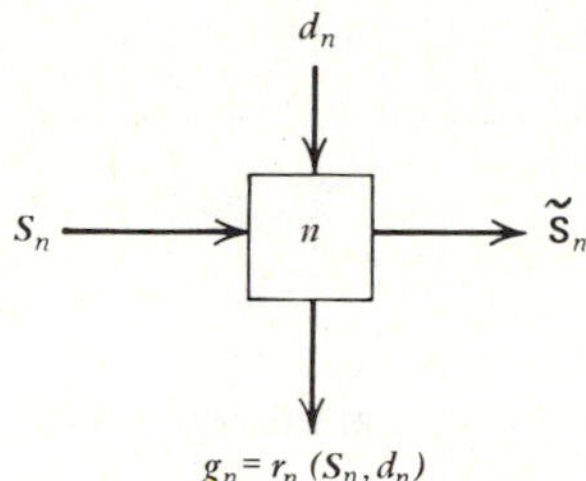

where

S_n = input state $\qquad$ $\tilde{S}_n$ = output state

n = stage number $\qquad$ g_n = return function = $r_n(S_n, d_n)$

d_n = decision

It may be beneficial to clarify these definitions with a representative example. Consider a manager (the decision maker) at a particular point in time (the stage). She has at her disposal a certain amount of money to invest (state variable) in one of 10 possible projects (decision variables). Each project would yield a return on her investment (return function). If one assumes that the possible investment alternatives are described by incrementally increasing costs, then the possible investment decisions obviously depend on the amount of investment money available and the project chosen. Hence, as previously stated, each possible return is a function of both the *state variable* and the feasible *decision variable*.

Suppose that we are faced with a number of decision points (stages) related in some manner by a *transition function*:

$$S_{n-1} = S_n \circledast d_n$$

[output of stage *n*] = [input to stage *n*] ⊛ [decision made at stage *n*]

Examples of stage transformations are given by

⊛	Transition function
+	$S_{n-1} = S_n + d_n$
−	$S_{n-1} = S_n - d_n$
×	$S_{n-1} = S_n \cdot d_n$
±√	$S_{n-1} = S_n \pm \sqrt{d_n}$

The units of S_n, d_n and S_{n-1} must be homogeneous. These units might be dollars, machines, or any other designation. These units are determined by the particular problem being solved, as will be illustrated in the examples in this chapter.

Further suppose that there are exactly N stages at which a decision is to be made. These N stages are all linked by the transition function(s) previously described. Functionally,

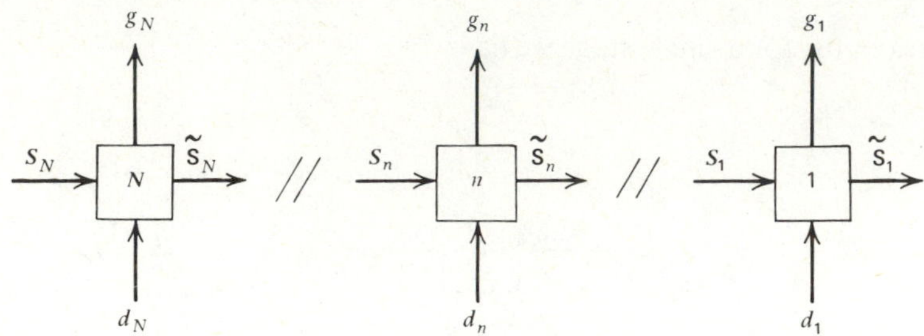

An N-stage multistage system.

Since a state variable is both the output from one stage and an input to another, it is sometimes represented by more than one symbol, namely,

$$\tilde{S}_{i+1} = S_i \qquad i = 1, 2, \ldots, (N-1)$$

Note also that the stages are numbered in an *opposite direction* to the flow of information. The usefulness (or necessity) of such a convention will be made clearer as dynamic programming is explored.

In dynamic programming, one solves multivariable optimization problems *sequentially*, or one *stage* at a time. Hence, it will be necessary to keep track of all the returns accumulated in our decision process as we proceed from stage to stage. Denote by $f_n(S_n, d_n)$ the *accumulated* total return calculated over n-stages, given a particular state variable, S_n. Similarly, denote by $f_n^*(S_n)$ the *optimal* n-stage total return for a particular input state, S_n. That is, a particular value of S_n might give rise to many possible decisions, d_n, among which is a decision, d_n^*, which gives rise to an optimal n-stage total return $[f_n^*(S_n)]$.

Since $f_n^*(S_n)$ consists of accumulated optimal returns, it can be written as

$$\begin{aligned} f_n^*(S_n) &= \operatorname*{opt}_{d_n, d_{n-1}, \ldots, d_1} \{g_n \circledast g_{n-1} \circledast \cdots \circledast g_1\} \\ &= \operatorname*{opt}_{d_n, d_{n-1}, \ldots, d_1} \{r_n(d_n, S_n) \circledast r_{n-1}(d_{n-1}, S_{n-1}) \circledast \cdots \\ &\qquad \circledast r_1(d_1, S_1)\} \end{aligned}$$

In this general formulation $\circledast$ represents any operand dictated within the context of the problem at hand and, in addition, might change from one stage to the next. For example, $\circledast$ might represent addition, subtraction, division, or multiplication ($+$, $-$, $\div$, $\times$, respectively).

Now, suppose that we are dealing with a minimization problem, possessing additive transitions and additive returns. This implies that the optimization problem is represented as follows:

$$\begin{aligned} f_n^*(S_n) = \min_{d_n, d_{n-1}, \ldots, d_1} &\{r_n(d_n, S_n) + r_{n-1}(d_{n-1}, S_{n-1}) + \cdots \\ &+ r_1(d_1, S_1)\} \end{aligned}$$

How would one actually obtain $f_n^*(S_n)$ using dynamic programming?

10.4 DEVELOPING AN OPTIMAL DECISION POLICY

If our multistage system actually looks like the one just illustrated, then we can notice some interesting characteristics:

1. There are exactly N points at which a decision must be made.
2. If we *start* at stage 1, then nothing affects an optimal decision except the knowledge of the *state* of the system at stage 1 and the choice of our *decision variable*.
3. Stage 2 only effects the decision at stage 1; the choice we make at stage 2 is governed only by the *state* of the system at stage 2 and the restrictions on our decision variable. . . . etc. to stage N

To begin, suppose that we *knew* the optimal *policy* (the set of decisions that would lead to an optimal value of our *return function*) for every possible *state* at *stage 1*. It is true that if we make an optimal decision at stage 1 corresponding to a given state S_1, this decision is unaffected by whatever occurs at stages 2, 3, . . . , N (since they *precede* stage 1). Now suppose that stages 1 and 2 are connected by the following relation (transition function):

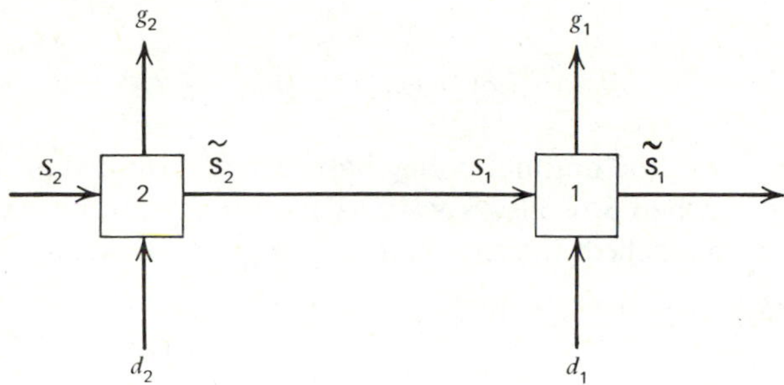

Transition function.

It is obvious from previous discussions that the one-stage return is given by

$$g_1 = r_1(S_1, d_1)$$

and the *optimal* one-stage return is found by searching over all possible decision variables (defined by a particular state variable). Hence,

$$f_1^*(S_1) = \underset{d_1}{\text{opt}}\,\{r_1(S_1, d_1)\}$$

Now note that the range of d_1 is *determined* by S_1, but S_1 is determined by what has happened in the previous stage. Specifically, define $S_1 = S_2 - d_2$ for this example.* Consider now the total optimal two-stage return:

$$f_2^*(S_2) = \underset{d_2}{\text{opt}}\,\{r_2(S_2, d_2) + f_1^*(S_1)\}$$

or

$$f_2^*(S_2) = \underset{d_2}{\text{opt}}\,\{r_2(S_2, d_2) + f_1^*(S_2 - d_2)\}$$

since

$$S_1 = S_2 - d_2$$

*There are many other possible relations, this particular one is considered only for illustrative purposes. Others will be dealt with later.

An interesting fact is now observed, namely that $f_2^*(S_2)$ is only a function of S_2 and d_2 (provided that $f_1^*(S_1)$ is known for all possible values of S_1)! By continuing the above logic recursively, it is clear that for a general N-stage system one could write

$$f_N^*(S_N) = \underset{d_N}{\text{opt}} \{r_N(d_N, S_N) + f_{N-1}^*(S_{N-1})\}$$

The entire procedure now reveals itself through the above equation. If one is given the input state S_N, it is possible to recover $f_N^*(S_N)$ *provided* one has available $f_{N-1}^*(S_{N-1})$. Proceeding further, $f_{N-1}^*(S_{N-1})$ can only be determined by knowing $f_{N-2}^*(S_{N-2})$, etc., until one finally needs $f_1^*(S_1)$. (But this is where the entire procedure started!)

From the above logic, the computational procedure is to determine $f_1^*(S_1)$ for all possible values of S_1. Hence, $f_1^*(S_1)$ is defined for any state variable at stage 1. Knowing $f_1^*(S_1)$, $f_2^*(S_2)$ is now determined for all possible values of S_2 through the following recursion,

$$f_2^*(S_2) = \underset{d_2}{\text{opt}} \{r_2(S_2, d_2) + f_1^*(S_1)]$$

or

$$f_2^*(S_2) = \underset{d_2}{\text{opt}} \{r_2(S_2, d_2) + f_1^*(S_2 - d_2)\}$$

Notice that once the optimal policy has been determined at stage 2, for any incoming state S_2, then S_1 is also known. This process can be repeated until the Nth stage has been reached. In general, the recursive relationship is given by

$$f_n^*(S_n) = \underset{d_n}{\text{opt}} \{r_n(S_n, d_n) + f_{n-1}^*(S_n - d_n)\} \qquad n = 1, 2, \ldots, N \quad (\text{where } f_0^*(S_0) \equiv 0)$$

Note that this process can be carried on through any number of stages, N, and that at any stage $1 \leqq n \leqq N$ an *optimal policy* for the n stage problem is readily available for *any* input, S_n. Note also that computationally the process was started at stage 1 and proceeded from right to left to stage N. In certain problems, it may be more desirable to start at stage N and proceed to stage 1 in the same fashion. The latter procedure is called *forward analysis* or *forward recursion* whereas the former is called *backward analysis* or *backward recursion*. Both procedures are essentially the same mathematically, but computationally one may be much easier than the other. The choice is largely dependent on the ability or ingenuity of the researcher, and the form of the problem at hand. In addition, the mathematical formulations of the transition function and the return function will vary from problem to problem. In general, the recursive relationship is given by

$$f_n^*(S_n) = \underset{d_n}{\text{opt}} \{r_n(S_n, d_n) \circledast f_{n-1}^*(S_{n-1})\}$$

$$S_{n-1} = S_n \circledast d_n$$

Note that this gives rise to multiplicative, divisional, and other relationships from stage to stage. The key step in solving the general dynamic programming problem is the decomposition of problem into N separate optimizations involving $f_1^*(S_1)$, $f_2^*(S_2), \ldots, f_N^*(S_N)$. This general decomposition is always possible for additive ($\pm$) returns. For the general case, the properties of *separability* and *additivity* must hold. These concepts are beyond the scope of this introductory chapter but the interested reader can see references [11] and [12] for details. Finally, the states of the system, the decisions, and the return function are generally known from the original problem formulation. Because each problem has a unique transition function, and the choice of states change from problem to problem, dynamic programming is largely an *art*, not primarily a *science*. The more exposure you have to dynamic programming problems, the better your understanding of them. Hence, we will look at a number of different problems.

10.5 DYNAMIC PROGRAMMING IN PERSPECTIVE

Dynamic programming is a mathematical technique dealing with the optimization of multistage processes. The basic concept is contained within the "principle of optimality":

> The optimal set of decisions in a multistage decision process problem has the property that whatever the initial stage, state, and decisions are, the remaining decisions must constitute an optimal sequence of decisions for the remaining problem, with the stage and state resulting from the first decision (or occurring naturally) considered as initial conditions.

Basic Features of a Dynamic Programming Problem

1. In dynamic programming problems, decisions regarding a certain problem are typically optimized at subsequent *stages* rather than simultaneously. This implies that if a program is to be solved using dynamic programming, it must be separated into N subproblems.
2. Dynamic programming deals with problems in which choices, or *decisions*, are to be made at each *stage*. The set of all possible choices is reflected and/or governed by the *state* at each stage.
3. Associated with each decision at every stage is a *return function* that evaluates the choice made at each decision in terms of the contribution that the decision can make to the overall objective (maximization or minimization).
4. Each stage N the *total* decision process is related to its adjoining stages by a quantitative relationship called a *transition function*. This transition function can either reflect discrete quantities or continuous quantities depending on the nature of the problem.
5. Given the current state, an optimal policy for the *remaining stages* in terms of a *possible input state* is independent of the policy adopted in previous stages.
6. The solution procedure always proceeds by finding the optimal policy for each possible input *state* at the present stage.
7. A recursive relationship is always used to relate the optimal policy at stage n to the $(n-1)$ stages that follow. This relationship is given by

$$f_n^*(S_n) = \underset{d_n}{\text{opt}}\,\{r_n(d_n) \circledast f_{n-1}^*(S_n \circledast d_n)\}$$

   Here the symbol $\circledast$ denotes any mathematical relationship between S_n and d_n; including addition, subtraction, multiplication, and root operations.
8. By using this recursive relation, the solution procedure moves from stage to stage—each time finding an optimal policy for each *state* at that *stage*—until the optimal policy for the last stage is found. Once the N-stage optimal policy has been discovered, the N-component decision vector can be recovered by tracing back through the N-stage transition functions.

III. ILLUSTRATIVE EXAMPLES

10.6 A PROBLEM IN OIL TRANSPORT TECHNOLOGY

The concepts and computational procedures involved in solving dynamic programming problems are best illustrated through the use of an example. Consider the following problem.

The Black Gold Petroleum Company has recently found large deposits of oil on the North Slope of Alaska. To develop this field, a transportation network must be developed from the North Slope to one of 10 possible shipping points in the United States. The total transmission line will require eight pumping stations between a North Slope ground oil storage plant and the shipping points. A number of sites are possible for each substation, but not every site in one area can be reached from every site in the next area for a number of reasons. These include (1) geographic inaccessibility due to terrain features such as mountains or lakes, (2) the inability to purchase lease right-of-ways, and (3) restricted wildlife areas. Associated with each pair of pumping stations is the cost of constructing the connecting transmission line. The construction cost between any two points is not a constant, but varies from a given origin to a given destination. The problem is to determine a feasible pumping configuration for the crude oil, while minimizing the attendant manufacturing costs. The following diagram represents the problem with accessible sites connected by arcs, labeled with the cost of constructing those arcs (Figure 10.1).

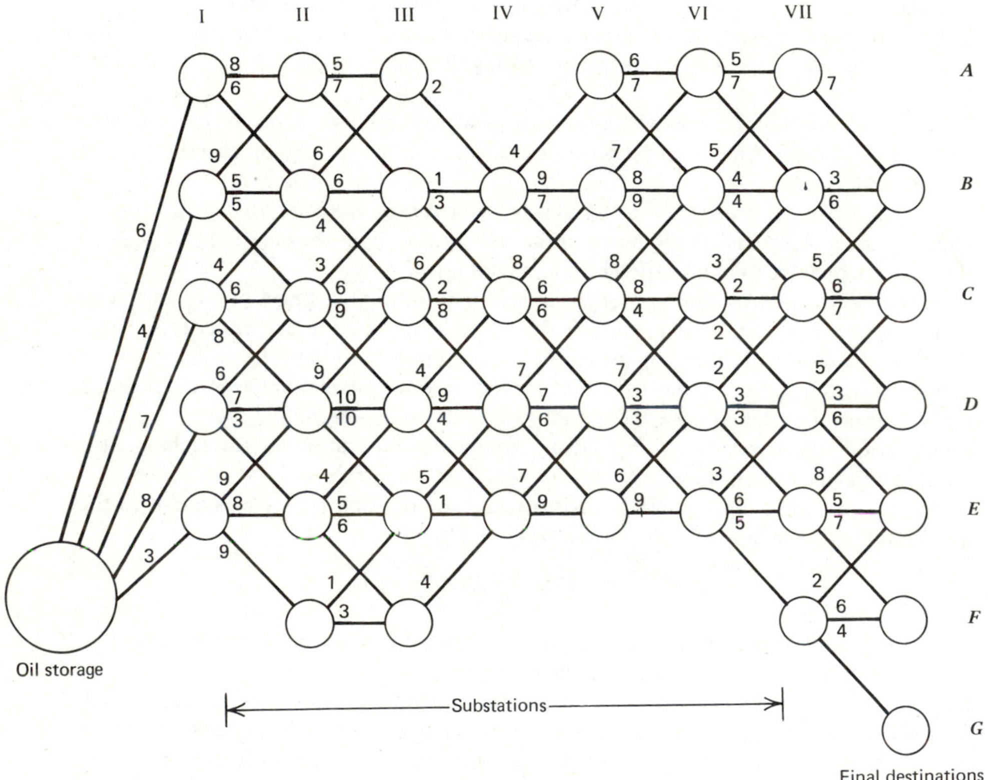

Figure 10.1
The distribution network.

Solution Procedure

The typical analyst might view this problem as one of a finite set of feasible shipping paths. Following this line of reasoning, one might determine that there are more than 100 feasible paths to consider. Each would have to be evaluated to guarantee an optimal solution. Furthermore, as the network grows linearly in size, the number of feasible alternatives to be examined grows at a combinatorial rate. Discarding this line of reasoning, one might decide to be "greedy" about the whole matter and develop a solution by choosing the path of least resistance (least-cost path) starting at the oil storage location and proceeding from left to right in Fig. 10.1. The following sequence of oil flow is then generated.* The

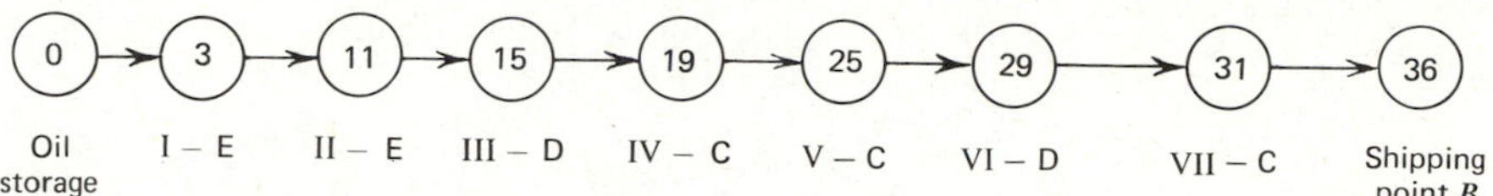

numbers that appear in each node represent the total *accumulated* cost incurred in reaching that particular node. A quick check reveals that this sequence cannot be the optimal (least-cost) solution, since by violating our selection rule at position VI-D, we obtain a solution costing only 35 units, namely,

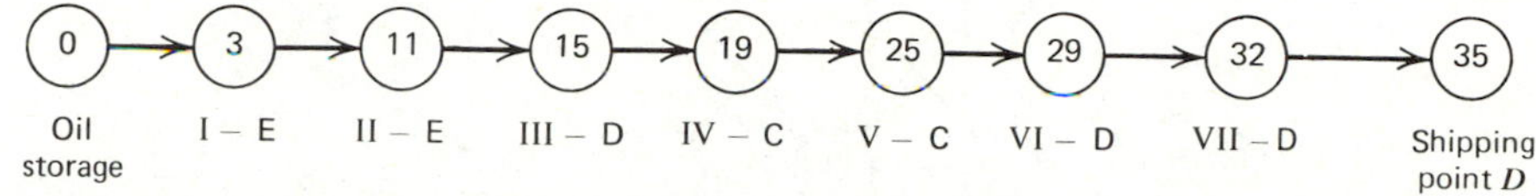

Suppose that we consider an entirely different approach to the problem solution. Let us assume that we have transported oil to some station in region VII. Further assume that we have no idea *how* we got there or what it has cost us to get there. Our primary concern is to pump the oil from whichever station we are at in region VII to a final shipping point at minimum cost. If we adopt the notation that the total cost of proceeding from the current point to a final shipping point will be written inside the node, we obtain the following:

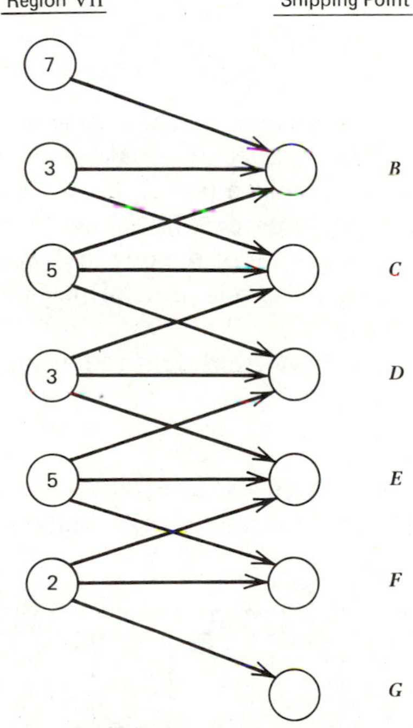

The following observation should now be obvious—*Regardless of where we might be in region VII, we have a minimum cost policy that leads us to our final destination*. Now suppose that we repeat this logic by assuming that the oil has

*Note that this is not the only sequence one can generate under this criterion, since at node IV-C one has two choices of where to go.

been pumped to region VI and we wish to proceed to a final destination at minimum cost. Again, we do not know *where* we are, but minimum costs can be calculated for every pumping station. These are given below.

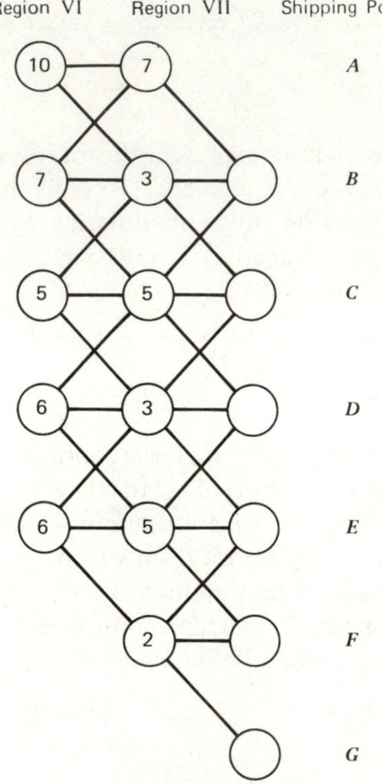

In examining the logic of this approach, a fundamental idea is readily conceived that is directly related to our decision-making process.

If a decision is made at any given point, this decision must be optimal with respect to all other decisions that necessarily follow. In other words, all decisions that follow are a direct consequence of the one that was just made, and are dictated by the present decision. This basic principle has already been introduced as the *Principle of Optimality*.

Hence, proceeding recursively from right to left until the oil storage node is reached, Fig. 10.2 is produced.

Interpretation of Results

From Fig. 10.2, there are seven possible solutions, each of which yield a minimum cost of 30 units. The optimal solution can easily be illustrated by a tree diagram.

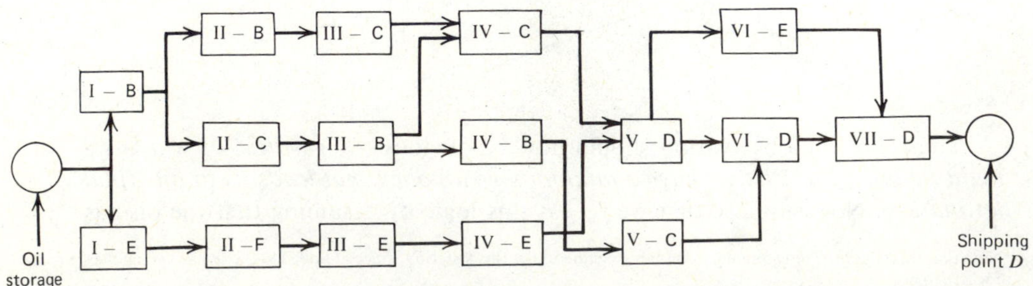

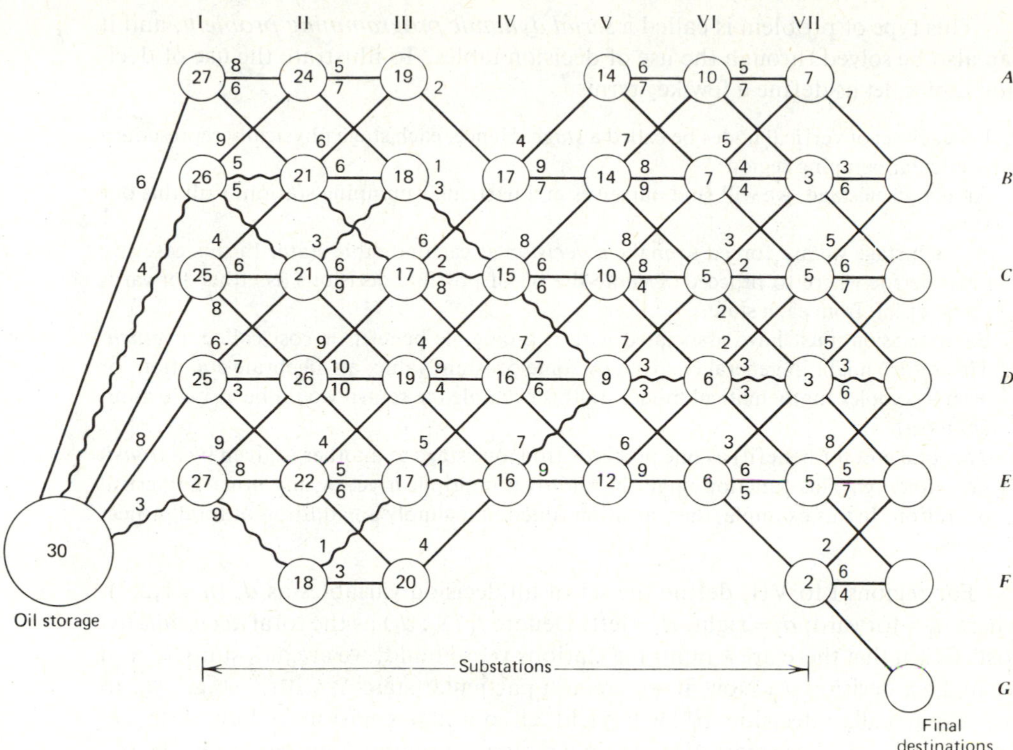

Figure 10.2
Final analysis.

From these results several interesting observations can be made.

1. Multiple optimal solutions have been obtained, not just a single solution. This is an extremely desirable result, since if there exist alternative solutions, they will be uncovered in the course of the problem solution.
2. The total number of paths evaluated in the problem solution increased in a linear fashion, proportional only to the number of pumping stations in each vertical region. Hence, the original combinatorial problem has been dramatically reduced.
3. From any vertical region, the *optimal* policy is given from any pumping station within that region to a final shipping point. This could be extremely important as the project progresses.

For example, suppose that pumping stations have been constructed up to the station at node IV-C. Now suppose that we are unable to obtain building rights from node IV-C to node V-D. The new optimal policy is easily recovered, since we know the optimal completion costs from any station in region V to final shipping. The optimal policy is to build link nodes IV-C to V-C and continue along the following path.

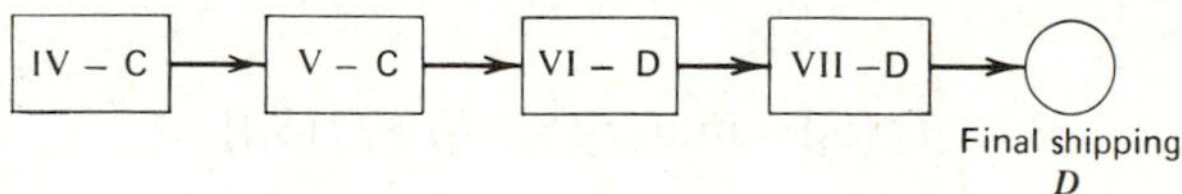

The new optimal cost is 31 units. This type of suboptimal evaluation is almost always available in dynamic programming results.

This type of problem is called a *serial dynamic programming problem*, and it can also be solved through the use of decision tables. To illustrate the use of decision tables, let us define a few key terms.

1. Let each set of vertical nodes be called a *stage*. Hence, each stage physically represents a particular pumping region.
2. At any given stage, we will find ourselves at a particular pumping station. Call this our *state*.
3. At each stage we are forced to make a *decision* at each possible state. In this case, we must decide where to proceed. Denote the set of possible decisions as either forward, right, or left from each state.
4. Each decision must have associated with it a tangible benefit or cost called a *return*. This return might in general be given by some constant value, an algebraic equation, or even a complex mathematical model. In this example the return would be a single number (cost).
5. The total cost (or benefit) as one proceeds from one stage to another is given by a *transition function*. This function might be additive, multiplicative, or any other numerical operation. In this example, the transition function is simply an addition of total project costs.

For regions I to VII, define the set of all decision variables as d_n $(n=1,2,3)$, where $d_1=$ forward; $d_2=$ right; $d_3=$ left. Denote $f_k(S_k, d_k)$ as the total *accumulated* cost. Given that there are k pumping stations yet to build, we are at a state S_k, and we make a decision d_k. Now, if we are at a particular state S_k with k stages yet to go, a particular decision d_n^* will yield an optimal decision policy, $f_k^*(S_k) = \underset{d_n}{\text{opt}}\ f_n(S_n, d_n)$. If there exist N stages in a dynamic programming problem, the objective is to find a policy set $f_1^*(S_1), f_2^*(S_2), \ldots, f_N^*(S_N)$. Consider a decision table for region VII. We will call this stage 1, since it is the first decision point, and enter $r_1(S_1, d_1)$ in the body of Table 10.1.

Table 10.1
STAGE 1 (REGION VII)

S_1 \ d_1	R	L	F	d_1^*	$f_1^*(S_1)$
A	7	—	—	R	7
B	6	—	3	F	3
C	7	5	6	L	5
D	6	5	3	F	3
E	7	8	5	F	5
F	4	2	6	L	2

Note that for any possible state we might be in, an optimal decision policy and return is given by d_1^* and $f_1^*(S_1)$, respectively. Proceeding to stage 2 (region VI) we obtain Table 10.2. Note that the entries in the table for stage 2 contain the cumulative costs of progressing from region VI to a final destination. In mathematical terms,

$$f_2^*(S_2) = \min_{d_2} \{r_2(S_2, d_2) + f_1^*(S_1)\}$$

It is important to note that $f_1^*(S_1)$ is *only* a function of the state variable S_1, and S_1 is uniquely determined once a particular decision, d_2, is known at stage 2. Proceeding recursively, Tables 10.3 to 10.7 are generated in a similar fashion.

Table 10.2
STAGE 2 (REGION VI)

S_2 \ d_2	R	L	F	d_2^*	$f_2^*(S_2)$
A	10	—	12	R	10
B	9	12[a]	7	F	7
C	5	6	7	R	5
D	8	7	6	F	6
E	7	6	11	L	6

[a]If one is in state B at stage 2 and decides to go left, this will cost 5 units. This decision puts one at state A in stage 1. From the decision table at stage 1, state A, the optimal policy costs f^* $(A) = 7$ units. Hence, the total cost is 12 units.

Table 10.3
STAGE 3 (REGION V)

S_3 \ d_3	R	L	F	d_3^*	$f_3^*(S_3)$
A	14	—	16	R	14
B	14	17	15	R	14
C	10	15	13	R	10
D	9	12	9	R,F	9
E	—	12	15	L	12

$$f_3^*(S_3) = \min_{d_3}\{r_3(S_3, d_3) + f_2^*(S_2)\}$$

Table 10.4
STAGE 4 (REGION IV)

S_4 \ d_4	R	L	F	d_4^*	$f_4^*(S_4)$
B	17	18	23	R	17
C	15	22	16	R	15
D	18	17	16	F	16
E	—	16	21	L	16

$$f_4^*(S_4) = \min_{d_4}\{r_4(S_4, d_4) + f_3^*(S_3)\}$$

Table 10.5
STAGE 5 (REGION III)

S_5 \ d_5	R	L	F	d_5^*	$f_5^*(S_5)$
A	19	—	—	R	19
B	18	—	18	R,F	18
C	24	23	17	F	17
D	20	19	25	L	19
E	—	21	17	F	17
F	—	20	—	L	20

$$f_5^*(S_5) = \min_{d_5}\{r_5(S_5, d_5) + f_4^*(S_4)\}$$

Table 10.6
STAGE 6 (REGION II)

S_6 \ d_6	R	L	F	d_6^*	$f_6^*(S_6)$
A	25	—	24	F	24
B	21	25	24	R	21
C	28	21	23	L	21
D	27	26	29	L	26
E	26	23	22	F	22
F	—	18	23	L	18

$$f_6^*(S_6) = \min_{d_6}\{r_6(S_6, d_6) + f_5^*(S_5)\}$$

Table 10.7
STAGE 7 (REGION I)

S_7 \ d_7	R	L	F	d_7^*	$f_7^*(S_7)$
A	27	—	32	R	27
B	26	33	26	R,F	26
C	34	25	27	L	25
D	25	27	33	R	25
E	27	35	30	R	27

$$f_7^*(S_7) = \min_{d_7}\{r_7(S_7, d_7) + f_6^*(S_6)\}$$

The optimal solution is given by computing the minimum cost routes from the oil storage depot to each possible state in stage 1. The results are given in Table 10.8.

Table 10.8
STAGE 8

S_8 \ d_8	A	B	C	D	E	d_8^*	$f_8^*(S_8)$
Oil storage	33	30	32	33	30	B,E	30

At this point, our recursive analysis is complete and from Table 10.8 an optimal cost route has been found that costs 30 units. However, we do not know what this route will be until we retrace an optimal path successively through stages 8, 7, . . . 1. This is easily done, since we know an optimal decision for every possible input state at each subsequent stage. At stage 8, the optimal policy is to proceed to state A. Looking at stage 7, the optimal policy starting in state A is to proceed right ($d_7^* = R$). Moving to the right from stage 7 results in a move to state B at stage 6. Proceeding in this manner, three optimal solutions are generated that correspond to those obtained previously.

Finally, note that since there is only one fixed state for the last stage (*first* decision point), the analysis becomes trivial at this stage. Dynamic programming problems with a fixed initial state are usually called *initial value* problems. Such problems are, in general, easier to solve by using computations that run against the sequence of physical decisions, as this example illustrates. Such an analysis has previously been defined as *backward recursion*.

In some instances, the separation of the decision space into N stages is best solved by addressing each stage in natural order. This approach is called *forward recursion* and will not be demonstrated in this chapter. Forward recursion is most useful in *nonserial dynamic programming*. The interested reader should see Nemhauser (24) or Beightler, Phillips, and Wilde (4).

10.7 A FACILITIES SELECTION PROBLEM

The Shop-Right Corporation in College Station, Texas is interested in expanding its marketing efforts to other parts of the state. A pilot survey of possible expansion sites has identified three location alternatives within a 100 mile radius of the current corporate marketing area. Table 10.9 summarizes potential revenues and authorized construction costs for four different building sizes that are being considered at each construction site.

Table 10.9
BUILDING SIZES, REVENUES, AND COSTS

Site	B_1		B_2		B_3		B_4		Do Not Build	
Size	R_1	C_1	R_2	C_2	R_3	C_3	R_4	C_4	R_5	C_5
I	.50	1	.65	2	.80	3	1.4	5	0	0
II	.62	2	.78	5	.96	6	1.8	8	0	0
III	.71	4	1.2	7	1.6	9	2.0	11	0	0

Costs and potential revenues are expressed in millions of dollars. Both the costs and revenues are nonlinear, monotonically increasing functions due to economies of scale in building procedures and market influence as a function of size.

If the Shop-Right Corporation can only afford to invest $21 million in the total expansion project, what locations should be selected and how large a store should be constructed at each new location?

The solution of this problem must yield two independent decisions: where to build the store(s) and how large. The problem is easily solved by dynamic programming, once this problem structure is identified. The parameters to be specified are related to the definition of *stages, state variables, decision variables, the return function*, and *the transition function*.

Stages. The stages for this problem are identified by considering what primarily influences and controls the decision logic. The definition of stages is the element in dynamic programming that transforms the problem from a multivariable optimization problem to an equivalent series of single variable optimization problems. Clearly, if a site is not chosen for any construction, then the decision of building size is not relevant. Hence, the elements that control the entire decision process are building sites. This defines a three-stage dynamic programming problem.

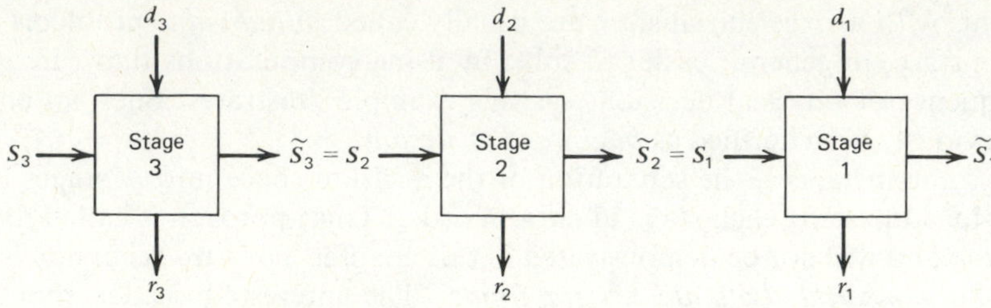

State Variables. The state variable in a dynamic programming problem is that element in the problem statement that directly controls the decision variable selection process. The state variable usually represents the limited resources that constrain the domain of allowable decisions. In this problem, the limiting resource is money available for construction. Given an infinite supply of money (assuming the construction of a building at each site is economically attractive), one would simply construct the largest store possible at each site. Hence, a proper definition of the state variable for this problem would be the amount of money remaining to be spent at each site as the decision process unfolds. Note that initially, the entire $21 million is available, but as decisions to spend money at each site are considered sequentially, the amount of money subsequently available changes as a function of the decision space.

Decision Variables. Although there are two decisions to be made (where to build and what size to build), the decision of *where* to build is automatically determined by the proper (optimal) sequence of decisions that specify what size building to construct. If the decision is "not to build," that site is eliminated. Hence, the only decision variable that needs to be considered is what size building to construct.

Return Function. The return function should reflect the intrinsic worth of making a decision at any given stage. Hence, the return function at each stage is simply the profit achieved by constructing a certain size of store. Note that in this problem statement, the profit is given as a single number. However, the profit could be calculated by a complex mathematical function or even by a comprehensive simulation model as a function of the decision.

Transition Function. The transition function must represent the change in available capital as the decision process proceeds from stage to stage. Hence, this transition function relates the amount of dollars originally available at stage n to the amount of dollars available at stage $(n-1)$ after a decision d_n has been made at stage n. This is mathematically expressed as follows:

$$S_{n-1} = S_n - C_n$$

where S_{n-1} = amount of capital available for construction at stage $n-1$

S_n = amount of capital originally available for construction at stage n

C_n = cost of building (directly related to d_n)

The Recursive Formula. The dynamic programming recursive formula is as follows:

$$f_n^*(S_n) = \max_{4 \geq d_n \geq 0} \{f_n(S_n, d_n) + f_{n-1}^*(S_{n-1})\} \qquad n = 1, 2, 3, 4$$

where:

$$S_{n-1} = S_n - C_n$$
$$f_n(S_n, d_n) = r_n(S_n, d_n) = R_n$$

Note that the decision to not build at site n is represented by $d_n = 0$. The return function represents the profit realized by building a store of size d_n, which is determined by the amount of capital, S_n, available at the stage where a decision is made; $R_n = r_n(S_n, d_n)$. The entire problem solution can be carried out through decision tables similar to those presented in the previous example.

We will begin by referring to site 1 as stage I (Table 10.10). Clearly, stage I could represent any site, since the order of examination is not relevant to the total decision process. It is important to realize that stage I represents the last decision point in the sequence of 3. Hence, the amount of money available to spend is unknown, since stages II and III are assumed to have been previously examined. However, we are assured that $21 \geq S_1 \geq 0$.

Table 10.10
STAGE I (SITE 1)

$$f_{\text{I}}^*(S_1) = \max_{4 \geq d_1 \geq 0} \{r_1(S_1, d_1)\}$$

where: $r_1(S_1, d_1) = R_1$

S_1 \ d_1	0	1	2	3	4	d_{I}^*	$f_{\text{I}}^*(S_1)$
$21 \geq S_1 \geq 5$	0	.50	.65	.80	1.4	4	1.4
$4 \geq S_1 \geq 3$	0	.50	.65	.80	—	3	.80
2	0	.50	.65	—	—	2	.65
1	0	.50	—	—	—	1	.50
0	0	0	0	0	0	0	0

Note that although we do not know what S_1 actually is at this stage, an optimal decision d_{I}^* and return $f_{\text{I}}^*(S_1)$ has been determined for all possible inputs. Aiso note that the state variable vector has been reduced for convenience from 21 rows to only 5 rows. This simplification occurs because the one stage returns for all decisions and states over this range are all identical. The student should verify that an expanded table using all 21 rows can be simplified to this format (see Table 10.10).

The next set of calculations are shown in the stage II analysis (see Table 10.11). d_2 might assume one or more of the values depending on the available capital (S_2). For example, if there is $S_2 = \$20$ million available, then any building size is possible. If we choose $d_2 = 2$, this yields a profit of $R_2 = 0.780$, requiring an expenditure of $C_2 = \$5$ million. This leaves \$15 million for stage I. That is: $r_2(S_2, d_2) = R_2 = 0.78$ and $S_1 = S_2 - d_2 = 20 - 5 = 15$. From Table 10.10, $f_{\text{I}}^*(S_1 = 15) = 1.4$. Hence, $f_{\text{2}}^*(S_2 = 20) = 2.18$.

To further illustrate these results, suppose that there is \$9 million available to spend at stage II ($S_2 = 9$). Any one of the four building sizes can be constructed with this amount of money. Assume that one decides to build size 2 ($d_2 = 2$). The revenue from this decision, which is an obvious function of the available capital S_2, is $r_2(S_2, d_2) = R_2 = 0.78$. This expenditure of \$5 million in construction costs

Table 10.11
STAGE II (SITE 2)

$$f_2^*(S_2) = \max_{4 \geq d_2 \geq 0} \{r_2(S_2, d_2) + f_1^*(S_1)\}$$

$$S_1 = S_2 - C_2$$
where: $r_2(S_2, d_2) = R_2$

S_2 \ d_2	0	1	2	3	4	d_2^*	$f_2^*(S_2)$
$21 \geq S_2 \geq 13$	1.4	2.02	2.18	2.36	3.2	4	3.20
12	1.4	2.02	2.18	2.36	3.2	4	2.60
11	1.4	2.02	2.18	2.36	3.2	4	2.60
10	1.4	2.02	2.18	2.6	2.45	3	2.45
9	1.4	2.02	1.58	1.61	2.30	4	2.30
8	1.4	2.02	1.58	1.61	1.80	1	2.02
7	1.4	2.02	1.43	1.61	—	1	2.02
6	1.4	1.42	1.28	1.46	—	3	1.46
5	1.4	1.42	.78	.96	—	1	1.42
4	.80	1.27	—	—	—	1	1.27
3	.80	1.12	—	—	—	1	1.12
2	.65	.62	—	—	—	0	.65
1	.50	—	—	—	—	0	.50
0	0	—	—	—	—	0	0

leaves \$4 million for stage 1: $S_1 = S_2 - C_2 = 9 - 5 = 4$. Hence, $f_1^*(S_1) = 4 = 0.80$ and $r_2(S_2,\ d_2) = 0.78 + 0.80 = 1.58$. Over all possible decisions related to $S_2 = 9$, $f_2^*(S_2 = 9) = 2.30$ when $d_2^* = 4$.

The final set of calculations is shown in stage III (Table 10.12). Stage III is the last in a sequence of three stagewise decisions. However, stage III actually represents the first physical decision point. Since site 3 is actually the first decision point, we know that exactly \$21 million is available for use ($S_3 = 21$). Note that this was not true for stages 1 and 2, since the expenditures at site 3 were unknown. For example, if the first option was chosen ($d_3 = 1$), then this would cost $C_1 =$ \$4 million and yield a return of $r_3(S_3,\ d_3) = 0.71$. This would leave $S_2 = S_3 - C_3 = 21 - 4 =$ \$17 million for stage II. From stage II, we already have calculated that $f_2^*(S_2 = 17) = 3.2$. Hence, $r_3(S_3,\ d_3) = 3.2 + 0.71 = 3.91$. Over all possible decisions at stage III, $f_3^*(S_3 = 21) = 4.4$ when $d_3^* = 2$.

Table 10.12
STAGE III (SITE 3)

$$f_3^*(S_3) = \max_{4 \geq d_3 \geq 0} \{r_3(S_3, d_3) + f_2^*(S_2)\}$$

$$S_2 = S_3 - C_3$$
where: $r_3(S_3, d_3) = R_3$

S_3 \ d_3	0	1	2	3	4	d_3^*	$f_3^*(S_3)$
21	3.2	3.9	4.4	4.2	4.3	2	4.4

The problem solution is now known. The Shop-Right Corporation will realize a profit of \$4.4 million per year. The construction decisions can be recovered by tracing the expenditure/decision process successively from stage III to stage I. In stage III, the optimal decision is to choose $d_3^* = 2$ at a cost of \$7 million. This leaves \$14 million for stage II (SITE 2). From stage II, if $S_2 = 14$ the optimal decision is to invest in building size $d_2^* = 4$ at a cost of \$8 million, leaves \$6 million for stage I. From stage I, $d_1^* = 4$ is optimal at a cost of \$5 million. The optimal policy is given in Table 10.13 for a total cost of \$20 million. Notice that we did not spend all of the available capital.

Table 10.13
OPTIMAL POLICY

Site	Building Size	Profit	Cost
III	2	1.20	7
II	4	1.80	8
I	4	1.40	5
Totals:		\$4.40 million	\$20 million

Now, assume that due to other investment opportunities, the corporation decides to invest only \$15 million. What is the new optimal policy? This solution can be obtained by creating only one additional row in the stage III calculations. The new solution is generated in Table 10.14.

Table 10.14
NEW STAGE 3 (SITE 3)

$$f_3^*(S_3) = \max_{4 \geq d_3 \geq 0} \quad r_3(S_3, d_3) + f_2^*(S_3 - C_3)$$

where: $\quad r_3(S_3, d_3) = R_3$

S_3 \ d_3	0	1	2	3	4	d_3^*	$f_3^*(S_3)$
21	3.2	3.91	4.4	4.2	4.3	2	4.4
15	3.2	3.31	3.22	3.06	3.27	1	3.31

The new solution uses all of the \$15 million, and is shown in Table 10.15.

Table 10.15
NEW SOLUTION

Site	Building Size	Profit	Cost
III	1	0.71	4
II	4	1.80	8
I	3	0.80	3
Totals:		\$3.31 million	\$15 million

Those type of calculations provide a form of sensitivity analysis unique to dynamic programming. Such evaluations of alternate optimal policies as a function of available capital provide powerful tools for decision-making alternatives.

10.8 THE OPTIMAL CUTTING STOCK PROBLEM

The Optimal Order paper company has received orders for four different groups of publications. The following orders have been placed.

8 rolls of 2 ft paper at \$2.50 per roll
6 rolls of 2½ ft paper at \$3.10 per roll
5 rolls of 4 ft paper at \$5.25 per roll
4 rolls of 3 ft paper at \$4.40 per roll

Due to heavy demand on the printing process, the paper company only has 13 ft of paper from which to fill these orders. If partial orders can be filled, which orders and how many of each should be filled to maximize total profits?

This problem is actually representative of a larger class of mathematical programming problems known as "knapsack problems" or the "fly-away kit problems." Mathematically, this problem can be represented as an all-integer-linear programming problem in the following manner:

$$\max_{x \epsilon S} \{x_1 P_1 + x_2 P_2 + \cdots + x_n P_n\}$$

$$\text{s.t.} \quad x_1 Q_1 + x_2 Q_2 + \cdots + x_n Q_n \leqq U$$

$$x_1, x_2, \ldots, x_n \text{ integer}$$

$$x_i \geqq 0 \qquad \text{all } i$$

where P_i = profit/cost per unit, Q_i = consumption per unit, $i = 1, 2, \ldots, n$

U = upper limit on consumption

This class of problems will be explored in greater detail in Chapter 11, but we shall now show that this problem can be solved by using dynamic programming. To do so, we must define the problem *stages*, *decision variables*, *state variables*, *return function*, and *transition function*. For this particular problem, we establish the following definitions.

Stages. The stages for this problem will be different orders. Since four orders are under consideration, a four-stage dynamic programming problem must be analyzed. Schematically, it can be represented by the following stage diagram.

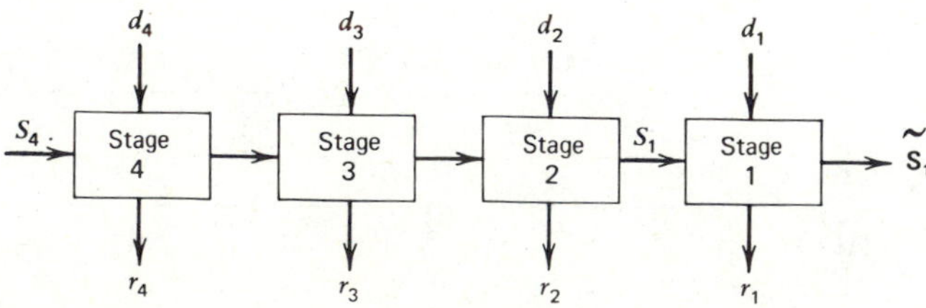

State Variables. The choice of a state variable is probably not immediately obvious for this problem. In general, the definition of the state variables in a particular dynamic programming problem will depend on the foresight and ingenuity of a given problem solver. For this example, a proper choice would be to define the state variable at stage n to be the *remaining* feet of paper left for the order being processed at stage n and *all* remaining stages.

Decision Variables. The decision variables will be how many rolls of paper to manufacture at each stage. A lower bound for the decision variable at each stage is obviously zero, and an upper bound is given by the number of rolls physically possible under the availability constraint. These upper bounds are given by $[F_o/L_n]$, where F_o is the feet of available paper, L_n is the desired length of the nth order, and [ ] denotes the largest-integer portion of the ratio F_o/L_n. However, at stage n there is always exactly S_n feet of paper left for orders n, $(n-1)$, . . . , 1, hence, at each stage d_n will be given by $d_n \leqq [S_n/L_n]$, with $d_n \leqq [13/L_n]$ only when $S_n = F_o = 13$ ft.

Return Function. The return function at the nth stage will be the additional revenue gained by making d_n rolls of the nth order.

Transition Function. The transition function should reflect the amount of available paper left to use at the nth stage. Hence, the transition function at the nth stage is the amount of paper left for consumption at stages $(n-1)$, $(n-2)$, . . . , 1. This is given by

$$S_{n-1} = S_n - d_n L_n \qquad n = 2,3,4.$$

Note that when $n = 1$

$$S_0 = S_1 - d_1 L_1$$

Since S_0 physically represents the amount of paper unused in the entire process, this should be as close to zero as possible. Hence, d_1 is uniquely determined by its maximum allowable value, given by $d_1 = [S_1/L_1]$, and thus will be the fractional part of this ratio.

The dynamic programming problem is therefore given by the following expression at the nth stage:

$$f_n^*(S_n) = \max_{0 \leqq d_n \leqq \left[\frac{S_n}{L_n}\right]} \{r_n(S_n, d_n) + f_{n-1}^*(S_{n-1})\}$$

where
$$S_{n-1} = S_n - d_n L_n$$

and
$$f_0^*(S_0) \equiv 0$$

$$f_n(S_n, d_n) = r_n d_n + f_{n-1}^*(S_n - d_n L_n)$$
$$n = 1,2,3,4$$

Since there is no reason to order our stages in any particular manner, we may associate any paper order with any stage. For illustrative purposes, let

stage 1 ⇒ 2.5 ft paper,
stage 2 ⇒ 4 ft paper,
stage 3 ⇒ 3 ft paper, and
stage 4 ⇒ 2 ft paper.

The solution is given in Tables 10.16 to 10.19.

Table 10.16
STAGE 1

$$f_1^*(S_1, d_1) = \{3.10d_1\}$$

$$f_1^*(S_1) = \max_{5 \geqq d_1 \geqq 0} \{3.10d_1\}$$

d_1 \ S_1	0	1	2	3	4	5	6	7	8	9	10	11	12	13
0	0	0	0	0	0	0	0	0	0	0	0	0	0	0
1	—	—	—	3.10	3.10	→	→	→	→	→	→	→	→	→
2	—	—	—	—	—	6.20	6.20	→	→	→	→	→	→	→
3	—	—	—	—	—	—	—	—	9.30	9.30	→	→	→	→
4	—	—	—	—	—	—	—	—	—	—	12.40	12.40	→	→
5	—	—	—	—	—	—	—	—	—	—	—	—	—	15.50
$f_1^*(S_1)$	0	0	0	3.10	3.10	6.20	6.20	6.20	9.30	9.30	12.40	12.40	12.40	15.50
d_1^*	0	0	0	1	1	2	2	2	3	3	4	4	4	5

Table 10.17
STAGE 2

$$f_2^*(S_2) = \max_{3 \geqq d_2 \geqq 0} \{5.25d_2 + f_1^*(S_2 - 4d_2)\}$$

d_2 \ S_2	0	1	2	3	4	5	6	7	8	9	10	11	12	13
0	0	0	0	3.10	3.10	6.20	6.20	6.20	9.30	9.30	12.40	12.40	12.40	15.50
1	—	—	—	—	5.25	5.25	5.25	8.35	8.35	11.45	11.45	11.45	14.55	14.55
2	—	—	—	—	—	—	—	—	10.50	10.50	10.50	13.60	13.60	16.70
3	—	—	—	—	—	—	—	—	—	—	—	—	15.75	15.75
$f_2^*(S_2)$	0	0	0	3.10	5.25	6.20	6.20	8.35	10.50	11.45	12.40	13.60	15.75	16.70
d_2^*	0	0	0	0	1	0	0	1	2	1	0	2	3	2

Table 10.18
STAGE 3

$$f_3^*(S_3) = \max_{4 \geqq d_3 \geqq 0} \{4.40d_3 + f_3^*(S_3 - 3d_3)\}$$

d_3 \ S_3	0	1	2	3	4	5	6	7	8	9	10	11	12	13
0	0	0	0	3.10	5.25	6.20	6.20	8.35	10.50	11.45	12.40	13.60	15.75	16.70
1	—	—	—	4.40	4.40	4.40	7.50	9.65	10.60	10.60	12.75	14.90	15.85	16.80
2	—	—	—	—	—	—	8.80	8.80	8.80	11.90	14.05	15.0	15.0	17.15
3	—	—	—	—	—	—	—	—	—	13.20	13.20	13.20	16.30	18.45
4	—	—	—	—	—	—	—	—	—	—	—	—	17.60	17.60
$f_3^*(S_3)$	0	0	0	4.40	5.25	6.20	8.80	9.65	10.60	13.20	14.05	15.0	17.60	18.45
d_3^*	0	0	0	1	0	0	2	1	1	3	2	2	4	3

Table 10.19
STAGE 4

$$f_4^*(S_4) = \max_{6 \geqq d_4 \geqq 0} \{2.50d_4 + f_3^*(S_4 - 2d_4)\}$$

S_4 \ d_4	0	1	2	3	4	5	6	$f_4^*(S_4)$	d_4^*
13	18.45	17.50	18.20	17.15	16.20	16.90	15.0	18.45	0

The optimal solution results in a total profit of $f_4^*(S_4 = 13) = \$18.45$. The optimal policy is recovered by working backwards through stages 4, 3, 2, and 1. The results are summarized in Table 10.20.

Table 10.20
OPTIMAL POLICY

Stage n	S_n	Optimal Policy	Transition to Stage $(n-1)$
4	$S_4 = 13$	$d_4^* = 0$	$S_3 = S_4 - 2d_4^* = 13$
3	$S_3 = 13$	$d_3^* = 3$	$S_2 = S_3 - 3d_3^* = 4$
2	$S_2 = 4$	$d_2^* = 1$	$S_1 = S_2 - 4d_2 = 0$
1	$S_1 = 0$	$d_1^* = 0$	$S_0 = 0$

Since S_4 is known (specified) to be $S_4 = 13$, this is again an *initial value* dynamic programming problem. Note that all available paper was used since $S_0 \equiv 0$ in the final analysis. Now suppose you are told that due to tattered edges, only 12 ft of paper will be available to fill all orders. How is the analysis changed? In a general mathematical programming approach, it would probably require a re-solution of the entire problem. However, note that in the dynamic programming approach, all that is changed is the value of S_4 (from $S_4 = 13$ to $S_4 = 12$). But by the principle of optimality, if S_4 changes to a lower value, all subsequent analysis at stages 3, 2, and 1 will still be valid, since they have been computed for all possible input values of S_3, S_2, and S_1, respectively. (Note this will not be true for *all* values if S_4 is made larger.) Using this basic premise, Table 10.21 yields optimal solutions for the range of input values $S_4 = 13$, 12, 11, and 10.

$$f_4^*(S_4) = \max_{0 \leqq d_4 \leqq \left[\frac{S_4}{L_4}\right]} \{r_4(S_4, d_4) + f_3^*(S_3)\}$$

where $S_3 = S_4 - 2d_4$

Table 10.21
OPTIMAL SOLUTIONS

d_4 \ S_4	13	12	11	10
0	18.45	17.60	15.0	14.05
1	17.50	15.70	15.70	13.10
2	18.20	15.60	14.65	13.80
3	17.15	16.30	13.70	12.95
4	16.20	15.25	14.40	10.0
5	16.90	12.50	12.50	12.50
6	15.0	15.0	—	—
$f_4^*(S_4)$	18.45	17.60	15.70	14.05
d_4^*	0	0	1	0

Input State S_4	d_4^*	d_3^*	d_2^*	d_1^*	$f_4^*(S_4)$
13	0	3	1	0	\$18.45
12	0	4	0	0	\$17.60
11	1	3	0	0	\$15.70
10	0	2	1	0	\$14.05

Now suppose that $S_4 = 13$ and you receive a directive that at least one roll of 2-ft paper *must* be shipped. Since this type of order corresponds to stage 4, we are really saying that $d_1 \geqq 1$. Since the optimal solution already derived violates this constraint, what further calculations are necessary to solve this problem? The answer is *None*! Since we already have optimal policies for any set of input state variables, stage 4 is examined for the best policy. From stage 4 calculations, it is clear that the best policy is to make exactly two rolls of 2-ft paper ($d_1^* = 2$). This will lead to the optimal policy shown in Table 10.22.

Table 10.22

Stage n	d_n^*
4	2
3	3
2	0
1	0
$f_4^*(S = 13) = \$18.20$	

Note that this solution is counter to engineering intuition, since logic would dictate that if we are forced to make a 2-ft roll, we should only make one. However, this decision would result in a profit of \$17.50. Following the optimal policy, this new restriction has cost us only \$0.25.

10.9
A PROBLEM IN INVENTORY CONTROL

The "Dry Feet" shoe store sells rubber shoes for protective use in snow. Past experience has indicated that the selling season is only 6 months long, and lasts from October 1 through March 31. The sales division has forecast the following demands for next year.

Month	Demand
October	40
November	20
December	30
January	40
February	30
March	20

All shoes sold by this store are purchased from outside sources. The following information is known about this particular shoe.

Purchasing Conditions. The unit purchasing cost is $4 per pair; however, the supplier will only sell in lots of 10, 20, 30, 40, or 50 pair. Any orders for more than 50 or less than 10 will not be accepted.

Quantity Discounts. The following quantity discounts apply on lot size orders.

Lot Size	Discount (%)
10	4
20	5
30	10
40	20
50	25

Ordering Costs. For each order placed, the store incurs a fixed cost of $2. In addition, the supplier charges an average amount of $8 per order to cover transportation costs, insurance, packaging, and so forth, irrespective of the amount ordered.

Storage Limitations. Due to large in-process inventories, the store will carry no more than 40 pair of shoes in inventory at the end of any one month. Carrying charges are $0.20 per pair per month, based on the end-of-month inventory. Since the sale of shoes for snow is highly seasonable and subject to design changes, it is desired to have both incoming and outgoing seasonal inventory at zero.

Assuming that demand occurs at a constant rate throughout each month and that the holding cost is based on the end of the month inventory, find an ordering policy which will minimize total seasonal costs.

Define the following relevant terms.

Stages. Each month of the 6-month ordering cycle will constitute a single stage. The state diagram will be as shown below.

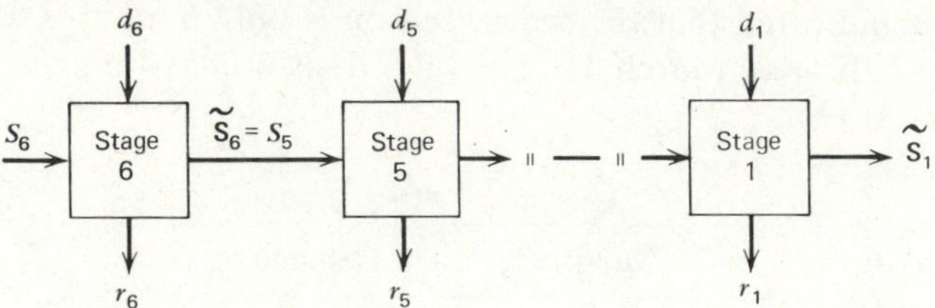

State Variables. At the nth stage, the state variable should be defined as the amount of entering inventory, given that there are n months remaining in the present selling period.

Decision Variables. At the nth stage, the decision to be made is how many pairs of shoes should be ordered to satisfy the demand during stages n, $(n-1)$, . . . , 1.

Transition Function. The transition function must relate the state variable at stage n to the state variable at stage $(n-1)$. This function is given by

$$S_{n-1} = S_n + d_n - D_n \qquad n = 1,2,\ldots,6$$

where $S_0 = \tilde{S}_1 \equiv 0$

$S_6 \equiv 0$

$S_n =$ state variable at nth stage

$d_n =$ decision variable at nth stage

$D_n =$ demand at nth stage

Note that $(S_n + d_n - D_n)$ will be the quantity of items for which a holding cost of $h_n = \$0.20$, $n = 1,2,\ldots,6$ per unit will be incurred.

Return Function. The return function at each stage should reflect the total cost resulting from the particular decision made at that stage. The return function at the nth stage is given by

$$r_n(d_n, S_n) = \phi(d_n) + h_n(S_n + d_n - D_n) \qquad n = 1,2,\ldots,6$$

where $\phi(d_n) =$ order cost function at nth stage. In this problem $\phi(d_n)$; $n = 1,2,\ldots,6$ is composed of a fixed cost of \$10 per order, plus a unit cost that depends on the number of units ordered.

$h_n =$ holding cost per unit per month. This cost is the same for all stages and is equal to \$0.20 per pair; $h_n = \$0.20$, $n = 1,2,\ldots,6$.

The dynamic programming formulation is, therefore, represented in the following mathematical form:

$$f_n^*(S_n) = \min_{d_n} \{\phi(d_n) + h_n(S_n + d_n - D_n) + f_{n-1}^*(S_{n-1})\} \qquad n = 1,2,\ldots,6$$

where $f_0^*(S_0) \equiv 0$

$$S_{n-1} = S_n + d_n - D_n \qquad n = 1,2,\ldots,6$$
$$S_0 \equiv 0$$
$$S_6 \equiv 0$$

Solution Technique

In computing this equation recursively, it will be convenient to refer to the following cost data.

Units Ordered	$\phi(d_n)$	Comment (% discount)
10	48	4
20	86	5
30	118	10
40	138	20
50	160	25

Stage 1 (March). Since it is desired to reduce all inventory to zero by the end of March, $S_0 \equiv 0$. Demand for stage 1 is 20 units, so it follows that S_1 will be either 0, 10, or 20 units, and $d_1^* = D_1 - S_1$.

$$f_1^*(S_1) = \min_{d_1} \{\phi(d_1)\}$$

S_1	d_1^*	$f_1^*(S_1)$
0	20	86
10	10	48
20	0	0

Stage 2 (February)

$$f_2^*(S_2) = \min_{d_2} \{\phi(d_2) + 0.20(S_2 + d_2 - 30) + f_1^*(S_1)\}$$

where $S_1 = S_2 + d_2 - 30$

S_2 \ d_2	0	10	20	30	40	50	d_2^*	$f_2^*(S_2)$
0	—	—	—	204	188	164	50	164
10	—	—	172	168	142	—	40	142
20	—	134	136	122	—	—	30	122
30	86	98	90	—	—	—	0	86
40	50	52	—	—	—	—	0	50

Stage 3 (January)

$$f_3^*(S_3) = \min_{d_3} \{\phi(d_3) + 0.20(S_3 + d_3 - 40) + f_2^*(S_2)\}$$

where $S_2 = S_3 + d_3 - 40$

S_3 \ d_3	0	10	20	30	40	50	d_3^*	$f_3^*(S_3)$
0	—	—	—	—	302	304	40	302
10	—	—	—	282	282	286	30,40	282
20	—	—	250	262	264	252	20	250
30	—	212	230	244	230	218	10	218
40	164	192	212	210	196	—	0	164

Stage 4 (December)

$$f_4^*(S_4) = \min_{d_4} \{\phi(d_4) + 0.20(S_4 + d_4 - 30) + f_3^*(S_3)\}$$

where $S_3 = S_4 + d_4 - 30$

S_4 \ d_4	0	10	20	30	40	50	d_4^*	$f_4^*(S_4)$
0	—	—	—	420	422	414	50	414
10	—	—	388	402	392	384	50	384
20	—	350	370	372	362	332	50	332
30	302	332	340	342	310	—	0	302
40	284	302	310	290	—	—	0	284

Stage 5 (November)

$$f_5^*(S_5) = \min_{d_5} \{\phi(d_5) + 0.20(S_5 + d_5 - 20) + f_4^*(S_4)\}$$

where $S_4 = S_5 + d_5 - 20$

S_5 \ d_5	0	10	20	30	40	50	d_2^*	$f_5^*(S_5)$
0	—	—	500	504	474	468	50	468
10	—	462	472	454	446	452	40	446
20	414	434	422	426	430	—	0	414
30	386	384	394	410	—	—	10	384
40	336	356	378	—	—	—	0	336

Stage 6 (October)

$$f_6^*(S_6) = \min_{d_6} \{\phi(d_6) + 0.20(S_6 + d_6 - 40) + f_5^*(S_5)\}$$

where $S_5 = S_6 + d_6 - 40$

Since the seasonal nature of the problem dictates a zero ending inventory, there will be no inventory carried from stage 6. Hence,

S_6 \ d_6	0	10	20	30	40	50	d_6^*	$f_6^*(S_6=0)$
0	—	—	—	—	606	608	40	606

From stage 6, the optimal policy is easily recovered from the transition functions and is given by

$$d_6^* = 40 \qquad d_3^* = 40$$
$$d_5^* = 50 \qquad d_2^* = 50$$
$$d_4^* = 0 \qquad d_1^* = 0$$

Total cost $= 138 + \{160 + 0.20(30)\} + 0 + 138 + \{160 + .2(20)\}$

Total cost $= 606 = f_6^*(S_6 = 0)$

IV. CONTINUOUS STATE DYNAMIC PROGRAMMING

10.10 INTRODUCTION

In every problem presented thus far, a tabluar decision table has been required listing all possible combinations of the decision variable and the state variable at each stage. This solution technique has been necessitated by the fact that we are dealing with integer valued components at each stage. In many problems, this restriction can be removed so that both the decision variables and the state variables are free to take on any value in the allowable range of values (the feasible alternatives). Such problems will be called *continuous state* dynamic programming problems. Computationally, the removal of the finite valued state and decision variable assumption allows us to use all of the tools of continuous mathematical programming, including derivatives and direct search techniques. The following example illustrates the computational procedure.

10.11 A NONLINEAR PROGRAMMING PROBLEM

Maximize: $$\sum_{i=1}^{3} r_i$$

Subject to: $$S_{i-1} = 3S_i - d_i \qquad i = 1,2,3$$
$$S_i \geqq d_i \geqq 0$$

where

$$r_1 = 3d_1$$
$$r_2 = 2d_2$$
$$r_3 = d_3^2$$

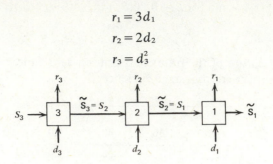

r_3 r_2 r_1

$S_3 \rightarrow$ [3] $\tilde{S}_3 = S_2 \rightarrow$ [2] $\tilde{S}_2 = S_1 \rightarrow$ [1] $\rightarrow \tilde{S}_1$

d_3 d_2 d_1

The problem is one of maximizing the set of three stage returns, where the return at a given stage is a function of the decision made at that stage. The solution is as follows:

Stage 1

$$\max_{S_1 \geqq d_1 \geqq 0} \{r_1\} = \max_{S_1 \geqq d_1 \geqq 0} \{3d_1\}$$

Since d_1 can assume any value on the range $S_1 \geqq d_1 \geqq 0$, it is obvious that d_1^* (the optimal value of d_1) should be as large as possible. Therefore

$$d_1^* = S_1$$
$$f_1^*(S_1) = 3S_1$$

Stage 2

$$\max_{S_2 \geqq d_2 \geqq 0} \{r_2 + f_1^*(S_1)\} = \max_{S_2 \geqq d_2 \geqq 0} \{2d_2 + 9S_2 - 3d_2\}$$
$$= \max_{S_2 \geqq d_2 \geqq 0} \{9S_2 - d_2\}$$
$$\therefore\ d_2^* = 0$$
$$f_2^*(S_2) = 9S_2$$

Stage 3

$$\max_{S_3 \geqq d_3 \geqq 0} \{r_3 + f_2^*(S_2)\} = \max_{S_3 \geqq d_3 \geqq 0} \{d_3^2 + 27S_3 - 9d_3\}$$

The objective at stage 3 is to maximize the function $f = d_3^2 - 9d_3 + 27S_3$. This is

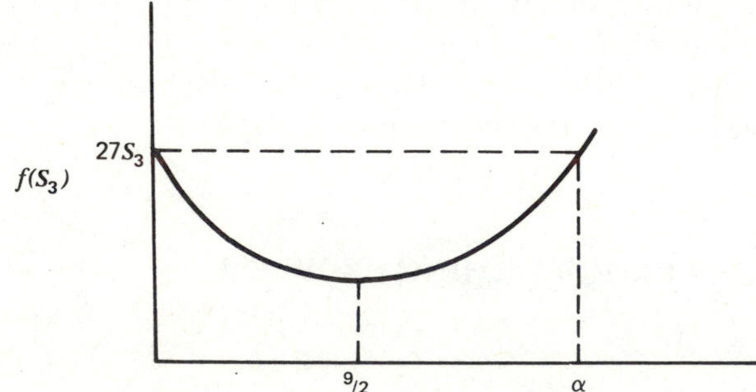

Figure 10.3

a convex function in d_3 as shown in Fig. 10.3. From Fig. 10.3 it is obvious that the optimal decision policy would be

$$\begin{array}{llll} \text{if:} & S_3 < \alpha & \text{then:} & d_3^* = 0 \\ & S_3 > \alpha & & d_3^* = S_3 \\ & S_3 = \alpha & & d_3^* = 0 \quad \text{or} \quad S_3 \end{array}$$

The point α is easily found, since

$$27S_3 = \alpha^2 - 9\alpha + 27S_3$$
$$\Rightarrow \alpha = 9$$

Hence,

$$f_3^*(S_3) = \begin{cases} 27S_3 & \text{for } S_3 \leqq 9 \\ S_3^2 + 18S_3 & \text{for } S_3 > 9 \end{cases}$$

The optimal decision policy is now available for any input state S_3. Solution values are given in Table 10.23 for selected inputs.

Table 10.23
OPTIMAL SOLUTIONS

S_3	d_1^*	d_2^*	d_3^*	Optimal Return $f_3^*(S_3)$
3	27	0	0	81.0
6	54	0	0	162
9	81(54)	0	0(9)	243
12	72	0	12	360

10.12 A PROBLEM IN MUTUAL FUND INVESTMENT STRATEGIES*

Continuous-Variable, Initial Value Problem

Suppose you are considering investment in two mutual funds. You have $10,000 to invest right now and will be able to invest an additional $1000 per year for each of the next 4 years. At the beginning of each investment period, you must decide how much of your available capital to invest in each fund. Once invested, the money cannot be withdrawn until the end of the 5-year period. The investments will earn money in two different ways: (1) Each fund has a long-term dividend potential realized as a percent return per year on accumulated capital, and the value of any investment left in the fund is expected to increase at this growth rate; (2) each fund also has a short-term interest-dividend rate, and any investment in some period will return cash to you at the end of the period at the particular rate of interest. This cash is available for reinvestment. Any money not invested in one of these funds earns you nothing.

What you need is a "5-year plan" for investment, the goal being to maximize total investment returns at the end of the fifth year.

*This problem was adapted from one presented by Wilde and Beightler (29) originally conceived by Bellman (5).

Fund	Short-Term Rates (i) 1	2	3	4	5	Long Term Dividend (I)
A	0.020	0.0225	0.0225	0.025	0.025	0.04
B	0.060	0.0475	0.050	0.040	0.040	0.03

Use backward recursion and consider the last year first:

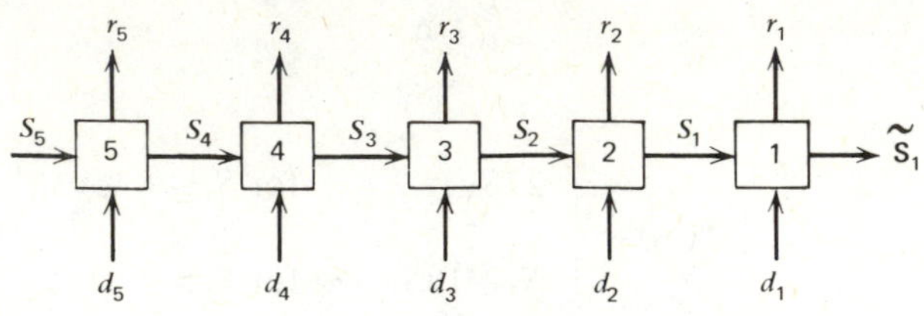

Define Stage: each investment period

State: S_n = amount of capital available for investment at beginning of year $(6-n)$

Decision: d_n = amount of capital to invest in fund A at beginning of year $(6-n)$

$\therefore (S_n - d_n)$ = amount of money for fund B

Return

r_n = future value of long-term earnings for stages 5, 4, 3, 2.

r_1 = present value of all earnings for stage 1.

Transition Function

$$S_{n-1} = i_A d_n + i_B(S_n - d_n) + 1000$$
$$= 1000 + i_B \cdot S_n + d_n(i_A - i_B) \qquad \text{for } n = 1,2,3,4$$

and $S_5 = 10{,}000$.

The goal is to

$$\max R = \sum_{n=1}^{5} r_n$$

where

$$r_n = (1 + I_A)^n \cdot d_n + (1 + I_B)^n(S_n - d_n)$$
$$= d_n[(1 + I_A)^n - (1 + I_B)^n] + S_n(1 + I_B)_n \qquad \text{for } n = 2, 3, 4, 5$$

and

$$r_1 = (1 + I_A)^1 d_1 + (1 + I_B)^1(S_1 - d_1) + i_A d_1 + i_B(S_1 - d_1)$$
$$= d_1[I_A - I_B + i_A - i_B] + S_1[1 + I_B + i_B]$$

Stage 1

$$f_1^*(S_1) = \max_{0 \leq d_1 \leq S_1} \{d_1(.04 - .03 + .025 - .040) + S_1(1 + .03 + .040)\}$$
$$= \max_{d_1} \{1.070S_1 - .005d_1\}$$

The optimal decision is therefore to make d_1 as small as possible. Hence,

$$d_1^* = 0$$
$$f_1^*(S_1) = 1.070S_1$$

Stage 2

$$f_2^*(S_2) = \max_{0 \leqq d_2 \leqq S_2} \{r_2 + f_1^*(S_1)\}$$
$$= \max_{d_2} \{d_2(1.04^2 - 1.03^2) + S_2(1.03)^2 + 1.07(1000 + .04S_2 - .015d_2)\}$$
$$= \max_{d_2} \{1070 + 1.1037S_2 + .0046d_2\}$$

Hence,

$$f_2^*(S_2) = 1070 + 1.108S_2 \qquad \text{and} \qquad d_2^* = S_2.$$

Stage 3

$$f_3^*(S_3) = \max_{0 \leqq d_3 \leqq S_3} \{r_3 + f_2^*(S_2)\}$$
$$= \max_{d_3} \{d_3(1.04^3 - 1.03^3) + S_3(1.03)^3 + 1070 + 1.108(1000 + .05S_3 - .0275d_3)\}$$
$$= \max_{d_3} \{2178 + 1.1481S_3 + .0018d_3\}$$

Hence,

$$f_3^*(S_3) = 2178 + 1.15S_3 \qquad \text{and} \qquad d_3^* = S_3$$

Stage 4

$$f_4^*(S_4) = \max_{0 \leqq d_4 \leqq S_4} \{r_4 + f_3^*(S_3)\}$$
$$= \max_{d_4} \{d_4(1.04^4 - 1.03^4) + S_4(1.03)^4 + 2178 + 1.15(1000 + .045S_4 - .025d_4)\}$$
$$= \max_{d_4} \{3328 + 1.1772S_4 + .0156d_4\}$$

Hence,

$$f_4^*(S_4) = 3328 + 1.193S_4 \qquad \text{for} \qquad d_4^* = S_4$$

Stage 5

$$f_5^*(S_5) = \max_{0 \leqq d_5 \leqq S_5} \{r_5 + f_4^*(S_4)\}$$

But

$$S_5 = 10{,}000$$

Therefore,

$$f_5^*(S_5) = \max_{d_5} \{d_5(1.04^5 - 1.03^5) + 10{,}000(1.03)^5 + 3328 + 1.193(1000 + .05(10{,}000) - .04d_5)\}$$

or

$$f_5^*(S_5) = \max_{d_5} \{16{,}711 + .0097d_5\}$$

Hence,

$$f_5^*(S_5) = 16{,}808 \qquad \text{and} \qquad d_5^* = S_5 = 10{,}000$$

Summary

Beginning of Year	Investment in Fund: A	B
1	10,000	0
2	All available funds	0
3	All available funds	0
4	All available funds	0
5	0	All available funds

Optimal total return $= f_5^*(S_5 = \$10{,}000) = \$16{,}808$ at the end of the fifth year.

V. MULTIPLE STATE VARIABLES

10.13 THE "CURSE OF DIMENSIONALITY"

In all the example problems we have solved thus far, each was characterized by the presence of a single state variable at every stage in the dynamic programming formulation. Computationally, the presence of multiple state variables creates major difficulties in the solution of dynamic programming problems. Consider a 10-stage problem with a single state variable composed of five components at each stage. If we can make a feasible decision at every stage over all five components, the solution procedure will require at most $(5 \times 5 \times 10) = 250$ calculations to be made in the problem solution. Now consider the addition of another state variable at each stage consisting of five components. Each stage might now require $(5 \times 5 \times 5)$ calculations, while the entire program could require as many as $(5 \times 5 \times 5 \times 10) = 1250$ separate calculations in the decision/state variable matrices. In addition, to compute an optimal solution to the complete problem, it is necessary to retain an optimal solution to *every* state variable *combination* at each stage. If one considers a problem with five state variables, each with 10 components, the amount of information required for both to be computed and stored would be astronomical. This dramatic increase in the amount of work (and storage) required to reach an optimal solution has been called the "curse of dimensionality" by Bellman (5).

The difficulties encountered in multiple-state dynamic programming will be illustrated through the solution of the following nonlinear, integer programming problem.

10.14 A NONLINEAR, INTEGER PROGRAMMING PROBLEM

$$\begin{aligned} \max \quad & 13x_1 - 5x_2^2 + 30.2x_2 - x_1^2 + 10x_3 - 2.5x_3^2 \\ \text{s.t.} \quad & 2x_1 + 4x_2 + 5x_3 \leqq 10 \\ & x_1 + x_2 + x_3 \leqq 5 \\ & x_1, x_2, x_3 \geqq 0 \qquad \text{and integer} \end{aligned}$$

Define the *decision variables* as $d_1=x_1$, $d_2=x_2$ and $d_3=x_3$, and let the *stages* correspond to the variables x_i, $i=1,2,3$.

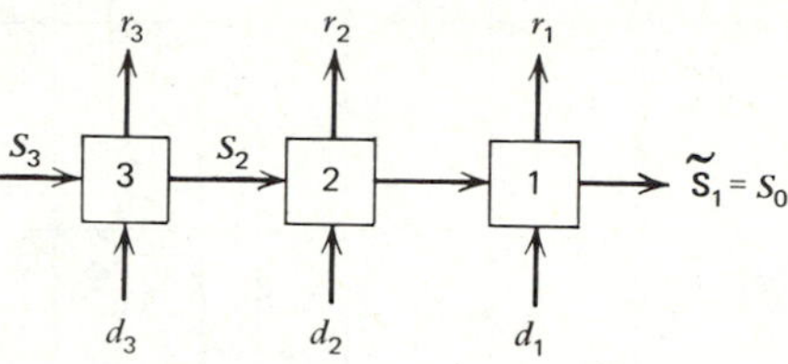

Since the present problem has two equality constraints, there will be two state variables to search over. Denote these two state variables at the nth stage as S_n and y_n, $n=1,2,\ldots,N$, such that

$$S_{n-1}=S_n-\alpha_n d_n \qquad n=1,2,\ldots,N$$
$$y_{n-1}=y_n-\nu_n d_n$$

The optimization problem now becomes

$$\begin{aligned}\max \quad & 13d_1-5d_2^2+30.2d_2-d_1^2+10d_3-2.5d_3^2\\ \text{s.t.} \quad & 2d_1+4d_2+5d_3\leqq 10\\ & d_1+d_2+d_3\leqq 5\\ & d_1,\ d_2\geq 0\\ & d_1,\ d_2=0,\ 1,\ 2,\ldots\end{aligned}$$

Stage 1

$$f^*(S_1,y_1)=\max_{0\leq d_1\leq \min\left[\frac{10}{2},\,5\right]}\{13d_1-d_1^2\}$$

Since
$$\frac{\partial f(S_1,y_1)}{\partial d_1}=13-2d_1=0$$

This implies that $d_1=6.5$, and the second derivative verifies that this is a maximum. However, note from the bounds on the decision variable that $5\geqq d_1^*\geqq 0$, so that the optimum solution would be to make d_1 as large as possible. Hence,

$$d_1^*=\min \mathrm{I}\begin{Bmatrix}S_1/2\\ y_1\end{Bmatrix}$$

where $I=$ integer values of $\dfrac{S_1}{2}$ and y_1

The solution to the stage 1 problem is given by Table 10.24 for all feasible combinations of S_1 and y_1.

Stage 2

$$f_2^*(S_2,y_2)=\max_{0\leq d_2\leq\left[\frac{10}{3},\,5\right]}\{30.2d_2-5d_2^2+f_1^*(S_1,y_1)\}$$
$$\text{s.t.}\quad S_1=S_2-4d_2$$
$$y_1=y_2-d_2$$

Table 10.24
TWO-TABLE VARIABLE COMPUTATIONS

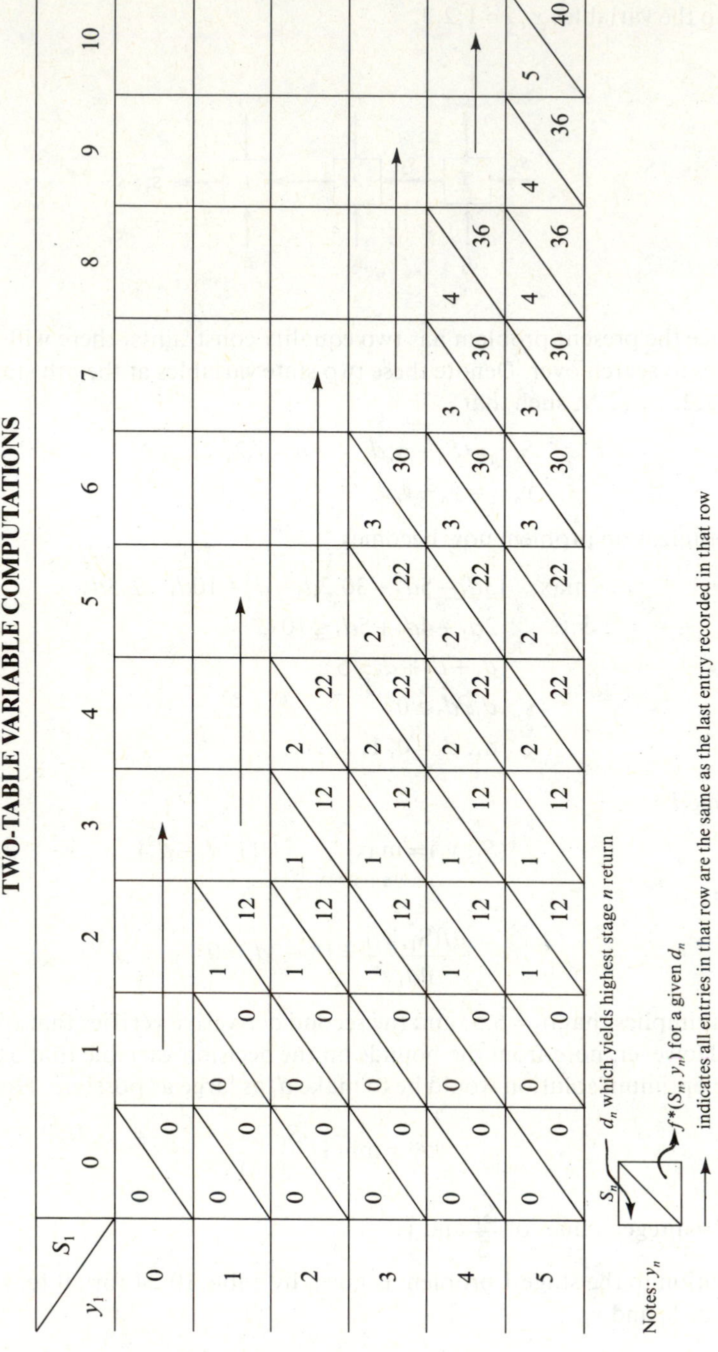

y_1 \ S_1	0	1	2	3	4	5	6	7	8	9	10
0	0 / 0	→	→	→							
1	0 / 0	0 / 0	1 / 12	→	→	→					
2	0 / 0	0 / 0	1 / 12	1 / 12	2 / 22	→	→	→			
3	0 / 0	0 / 0	1 / 12	1 / 12	2 / 22	2 / 22	3 / 30	→	→	→	
4	0 / 0	0 / 0	1 / 12	1 / 12	2 / 22	2 / 22	3 / 30	3 / 30	4 / 36	→	→
5	0 / 0	0 / 0	1 / 12	1 / 12	2 / 22	2 / 22	3 / 30	3 / 30	4 / 36	4 / 36	5 / 40

Notes: y_n / S_n — d_n which yields highest stage n return

$f^*(S_n, y_n)$ for a given d_n

→ indicates all entries in that row are the same as the last entry recorded in that row

Since

$$f_2(S_2, y_2) = 30.2d_2 - 5d_2^2$$

$$\frac{\partial f_2}{\partial d_2} = 30.2 - 10d_2 = 0$$

Therefore,

$$d_2 = 3 \text{ [a maximum point since } \frac{\partial^2 f_2}{\partial f_2^2} < 0\text{]}$$

However, d_2 cannot assume a value of 3 since if d_1 is zero, $4d_2 \leqq 10$ and max $\{d_2\} = 2$. In addition, it may not be optimal to choose d_2 as large as possible because of the variable dependencies. Hence, feasible combinations are evaluated at stage 2 by taking into account the stage 1 returns (see Table 10.25).

Stage 3

$$f_3^*(S_3 - y_3) = \max_{0 \leqq d_3 \leqq \left[\frac{10}{5}, 5\right]} \{10d_3 - 2.5d_3^2 + f_2^*(S_2, d_2)\}$$

$$\text{s.t.} \quad S_2 = S_3 - 5d_3$$

$$y_2 = y_3 - d_3$$

But at stage 3, the state variables are known to be $S_3 = 10$ and $y_3 = 5$. Therefore, the stage constraints become

$$S_2 = 10 - 5d_3 \qquad \text{and} \qquad y_2 = 5 - d_3$$

Since

$$f_3(S_3, y_3) = 10d_3 - 2.5d_3^2$$

$$\frac{\partial f_3}{\partial d_3} = 10 - 5d_3 = 0$$

which implies $d_3 = 2$ [a maximum stationary point]

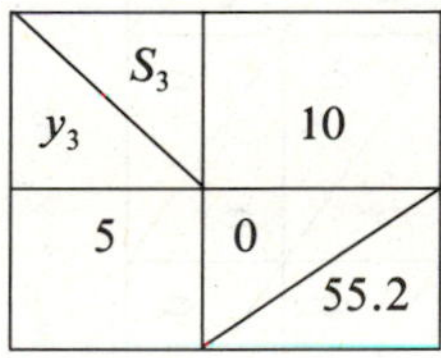

[Note that $d_3^* = 0$ was chosen, as usual, from the entire set of feasible d_3]

Hence,

$$f_3^*(S_3 = 10, y = 5) = 55.20$$

Using the stage transition functions, it is easily found that the optimal solution is given by

$$d_3^* = x_3^* = 0$$

$$d_2^* = x_2^* = 1$$

$$d_1^* = x_1^* = 3$$

Although it was unnecessary to compute optimal returns for all possible state variable combinations (due to the decision restrictions), it is obvious that a great many more computations are necessary to solve a two-state variable problem than a one-state variable problem. In general, a dynamic programming formulation

Table 10.25
STAGE TWO CALCULATIONS

y_2 \ S_2	0	1	2	3	4	5	6	7	8	9	10
0	0 / 0	→	→	→							
1	0 / 0	0 / 0	0 / 12	0 / 12	1 / 25.2	→	→	→			
2	0 / 0	0 / 0	0 / 12	0 / 12	1 / 25.2	1 / 25.2	1 / 37.2	1 / 37.2	2 / 40.4	2 / 40.4	2 / 52.4
3	0 / 0	0 / 0	0 / 12	0 / 12	1 / 25.2	1 / 25.2	1 / 37.2	1 / 37.2	2 / 47.2	2 / 40.4	2 / 52.4
4	0 / 0	0 / 0	0 / 12	0 / 12	1 / 25.2	1 / 25.2	1 / 37.2	1 / 37.2	1 / 47.2	2 / 40.4	1 / 52.2
5	0 / 0	0 / 0	0 / 12	0 / 12	1 / 25.2	1 / 25.2	1 / 37.2	1 / 37.2	1 / 47.2	2 / 40.4	1 / 55.2

becomes quite unattractive when the number of state variables involved are three or more.

10.15
ELIMINATION OF STATE VARIABLES

The elimination of state variables remains a source of current research for those interested in the practical application of large dynamic programming problems. Currently three methods are suggested for combatting the ''curse of dimensionality.'' Bellman and Dreyfus (8), Nemhauser (24), and later Phillips (34) suggest using Lagrangian multipliers to reduce state space dimensionality. The procedure is to form a new Lagrangian function in which Lagrangian multipliers are introduced to eliminate all constraints but one from the problem formulation. The effect is to reduce the M-dimensional state space (M constraints) and the N-dimensional decision space (N decision variables) to a one-dimensional state space and $N+M-1$ dimensional decision space (N original decision variables and $M-1$ Lagrangian multipliers). The solution procedure will involve a search for the optimal values of the $M-1$ Lagrangian multipliers, with each stage of the search requiring the solution of a serial dynamic programming problem.

The second approach involves the introduction of a surrogate constraint composed of M surrogate multipliers and N decision variables resulting in an $M+N$ decision space with a single-state space. Greenberg and Pierskalla (15) have suggested this approach, and show that the surrogate algorithm will always produce a duality gap at most as large as the associated Lagrangian gap. To date, this procedure has not been formally exploited to the authors' knowledge.

A third approach has recently been suggested by Morin and Esogbue (31) involving what they call the ''embedded state variable'' algorithm. Morin and Marsten (32) have reported solutions to nonlinear knapsack-type problems with 10 state variables. Although a complete discussion of these techniques falls well beyond the scope of this introductory article, more detailed treatments are provided in the references discussed in this section. We now turn our attention to the final major topic to be discussed—stochastic dynamic programming.

VI. STOCHASTIC SYSTEMS

10.16
STOCHASTIC DYNAMIC PROGRAMMING—A BRIEF OVERVIEW

In addition to serial and nonserial structures, multistage systems can be further divided into two large classes: deterministic systems and stochastic systems. In deterministic systems it is assumed that all the variables are known or can be determined exactly. In other words, there are no random variables. Deterministic systems can be used to describe real world problems where all the variables are known exactly or where the variations of the variables are small enough that the error caused by using their expected values as their exact values is acceptable.

Real world problems which contain random variables with variances large enough that an unacceptable error is produced by replacing them with their expected values should be described by stochastic systems.

Stochastic Serial Systems

In the deterministic models previously discussed, it was assumed that there are no random variables. Since there are no random variables, if the input state and the decision of a stage are known, then the return and output state of the stage can be determined exactly. For stochastic systems this is no longer true. If the return functions are stochastic, that is, $r_n = r_n(S_n, d_n, K_n)$, and the transition functions are deterministic, that is, $S_n = t_n(S_n, d_n)$, then the output states can be determined exactly; however, the returns cannot. If both the return functions and the transition functions are stochastic, then neither can be determined exactly.

Once either of the functions are made stochastic, then some decision must be made as to what the objective is. Here, we assume that the objective is to optimize the expected value of the return function.

An elaborate theory justifies the use of an expected value criterion in a stochastic environment where the process is to evolve only once. This theory is based on the determination of a "utility," which is a function associating real numbers with consequences by means of a person's personal evaluation of preferences among gambles (see Savage, 35).

The approach used to solve stochastic systems with stochastic transition functions depends on whether there is any feedback concerning the stages' input states from the system. If there is no feedback from the system to the decision maker, then the decision set must be obtained *a priori* and dynamic programming is not generally applicable. However, if there is feedback from the system in that the input state of a stage is known before the stage's decision is made, then dynamic programming is generally applicable, since dynamic programming determines the optimal decision at each stage as a function of the input state to the stage. Dreyfus (12, 13) points out that the optimal decision set obtained using feedback performs at least as well as the *a priori* optimal decision set obtained without using feedback, on the average. This is because the *a priori* decision set obtained without using feedback can be considered as a subset of all the possible decision sets using feedback.

It makes no difference in the optimal decision set or the optimal return for deterministic problems whether feedback is used or not. Since the transition functions are deterministic, the same information as was obtained from feedback (stage's input state) can be determined exactly from the transition function. This is also true for stochastic systems where the transition functions are deterministic.

If the objective is to maximize the expected value of the sum of the stage's returns and both the return functions and the transition functions are stochastic, then the same basic set of recursive equations previously developed in this chapter can be reformulated to obtain solutions.

Let $r_n(S_n, d_n, D_n)$ equal the return from stage n as a function of the input state S_n, the decision d_n, and the random variable K_n. And let $f_n(S_n)$ equal the conditional expected value of the return from the stages 1 through n given that the system is in state S_n at the beginning of stage n and that an optimal policy is followed. Then

$$f_1(S_1) = \max_{d_1} \{E[r_1(S_1, d_1, K_1)]\}$$

and

$$F_n(S_n) = \max_{d_n} \{E[r_n(S_n, d_n, K_n) + f_{n-1}(S_{n-1})]\}, \qquad \text{for } n = 2, \ldots, N$$

E is the expected value operator. Substituting for S_{n-1} from the transition function $S_{n-1} = t_n(S_n, d_n, K_n)$ yields

$$f_n(S_n) = \max_{d_n} \{E[r_n(S_n, d_n, K_n) + f_{n-1}(t_n(S_n, d_n, K_n))]\}, \qquad \text{for } 2, \ldots, N$$

Although the random variables in the transition functions do not greatly affect the solution of the initial-value serial problems (except for the additional computations required for the expected values), this is not true for a final-value problem or a two-point boundary problem, (Beightler, Phillips and Wilde (4)). Since the output state can no longer be determined exactly from the input state and the decision, specifying the value the final state variable will take is no longer possible. This means that advanced techniques such as state inversion and decision inversion are no longer possible.

Stochastic Systems

The solution of generalized stochastic dynamic programming problems remains a source of current research. However, specialized algorithms are available for expected value problems. Since this area constitutes a broad and complex area of multistaged optimization, the reader is referred to Nemhauser (24) and Beightler, Phillips, and Wilde (4) for a comprehensive treatment of this area.

10.17 SUMMARY AND CONCLUSIONS

This chapter explains the optimization technique known as dynamic programming. Dynamic programming is not a mathematical algorithm, but a *solution procedure* that has been effectively applied to a wide variety of operations research problems. The successful application of dynamic programming is often based on the *art* of modeling rather than the mathematical aspects of problem solution. This chapter developed the fundamental approach to problem solving via dynamic programming, while cultivating the art through a series of representative example problems. It is important to place the entire contents of this chapter into its proper perspective. Although dynamic programming is an extremely powerful optimization technique when it is applied to a wide range of problems, it is not the panacea of all solution procedures. Quite to the contrary, dynamic programming is only one of many operations research techniques that should be included in the engineers' code of problem-solving capabilities. This chapter illustrates the solution procedure for a wide range of engineering problems, each of which exhibits unique characteristics. In general, any optimization problem that can be decomposed by the principle of optimality is a candidate for efficient solution via dynamic programming. If the problem can also be formulated with three or less state variables, spectacular computational savings can often be realized. It is hoped that the material in this chapter will serve to facilitate increased applications of this powerful technique.

REFERENCES

1. Aris, R., *The Optimal Design of Chemical Reactors*, Academic, New York, 1961.
2. Aris, R., *Discrete Dynamic Programming*, Blaisdell Publishing Co., New York, 1964.
3. Aris, R., G. Nemhauser, and D. J. Wilde, "Optimization of Multi-stage Cycle and Branching Systems by Serial Procedures," *A.I.Ch.E. Journal*, *10*, 913–919 (1964).
4. Beightler, C. S., D. T. Phillips, and D. J. Wilde, *Foundations of Optimization*, 2nd ed., Prentice-Hall, Englewood Cliffs, N.J., 1976.
5. Bellman, R., *Dynamic Programming*, Princeton University Press, Princeton, N.J., 1957.
6. Bellman, R., *Adaptive Control Processes: A Guided Tour*, Princeton University Press, Princeton, N.J., 1961.

7. Bellman, R., "Some Problems in the Theory of Dynamic Programming," *Econometrica*, *22*, 37–48 (1954).
8. Bellman, R., and S. E. Dreyfus, *Applied Dynamic Programming*, Princeton University Press, Princeton, N.J., 1962.
9. Bertele, W., and F. Brioschi, *Nonserial Dynamic Programming*, Academic Press, New York, 1972.
10. Crisp, R. M. and C. S. Beightler, "Closed-Form Solutions to Certain Linear Allocation Problems," *AIIE Transactions*, 323–327 (December 1969).
11. Denardo, E. V., and L. G. Mitten, "Elements of Sequential Decision Processes," *A.I.I.E. Journal*, *XVIII* (1), 106–112 (1967).
12. Dreyfus, S. E., *Dynamic Programming and the Calculus of Variations*, Academic, New York, 1965.
13. Dreyfus, S. E., "Computational Aspects of Dynamic Programming," *Operations Research*, *5*, 409–415 (1957).
14. Gibson, D., "Dynamic Programming: A Process Design Tool," *Industrial Engineering*, 25–31 (December 1969).
15. Greenberg, H. J., and W. P. Pierskalla, "Surrogate mathematical programming," *Oper. Res. 18*, 924–939 (1970).
16. Hadley, G., *Nonlinear and Dynamic Programming*, Addison-Wesley, Reading, Mass., 1964.
17. Hastings, N. A. J., *Dynamic Programming With Management Applications*, Crane, Russak, and Co., New York, 1973.
18. Kaufmann, A., *Graphs, Dynamic Programming, and Finite Games*, Academic, New York, 1967.
19. Kaufmann, A., and R. Cruon, *Dynamic Programming*, Academic, New York, 1967.
20. Mitten L. G., "Composition Principles for Synthesis of Optimal Multistage Processes," *Operations Research*, *12*, 610–619 (1964).
21. Mitten, L. G., and G. L. Nemhauser, "Multistage Optimization," *Chemical Engineering Progress*, *59*, 52–60 (1963).
22. Mitten, L. G., and G. L. Nemhauser, "Optimization of Multistage Separation Processes by Dynamic Programming," *Canadian Journal of Chemical Engineering*, *41*, 187–194 (1963).
23. Nemhauser, G. L., "Decomposition of Linear Programs by Dynamic Programming," *Naval Research Logistics Quarterly*, *11*, 191–196 (1964).
24. Nemhauser, G. L., *Introduction to Dynamic Programming*, Wiley, New York, 1966.
25. Pope, D. N., G. L. Curry, and D. T. Phillips, "Closed Form Solutions to Nonserial, Nonconvex Quadratic Programming Problems Using Dynamic Programming," *Journal of Mathematical Analysis and Applications*, *86*, 628–647 (1982).
26. Roberts, S. M., *Dynamic Programming in Chemical Engineering and Process Control*, Academic, New York, 1964.
27. Saaty, T. L., *Mathematical Methods of Operations Research*, McGraw-Hill, New York, 1959.
28. Wald, A., *Statistical Decision Functions*, Wiley, New York, 1950.
29. Wilde, D. J., and C. S. Beightler, *Foundations of Optimization*, Prentice-Hall, Englewood Cliffs, N.J., 1967.
30. Meier, W. L., and C. S. Beightler, "Branch compression and absorption in nonserial multistage systems," *J. Math. Anal. Appl. 21* (1968).
31. Morin, T. L., and A. M. O. Esogbue, "The embedded space approach to reducing dimensionality in dynamic programs of higher dimensions," *J. Math. Anal. Appl. 48*(3) (December 1974).

32. Morin, T. L., and R. E. Marsten, "An algorithm for nonlinear knapsack problems," *Manage. Sci. 22*(10) (June 1976).

33. Parker, M. W., *The Analysis of Multistage Serial Forms*, unpublished Doctoral Dissertation, University of Arkansas, 1969.

34. Phillips, D. T., A. Ravindran, and J. J. Solberg, *Operations Research; Principles and Practice*, Wiley, New York, 1976.

35. Savage, L. J., *The Foundations of Statistics*, Wiley, New York, 1954.

36. Thomas, M. E., "A survey of the state of art in dynamic programming," *AIIE Trans. 8*(1) (March 1976).

37. Von Neumann, J., and O. Morgenstern, *Theory of Games and Economic Behavior*, Princeton University Press, Princeton, New Jersey, 1947.

38. White, D. J. *Dynamic Programming*, Holden-Day, San Francisco, 1969.

39. Wilde, D. J., and C. S. Beightler, *Foundations of Optimization*, Prentice-Hall, Englewood Cliffs, New Jersey, 1967, First Edition.

EXERCISES

1. Is dynamic programming a linear programming recursive procedure or a nonlinear programming procedure?

2. Define in words the following concepts:
   (a) Stages.
   (b) States.
   (c) Transformation function.
   (d) Return function.
   (e) State variable.
   (f) Decision variable.
   (g) Transition function.

3. Explain the differences in *forward recursion* and *backward recursion*.

4. Why is dynamic programming more of an art than a science?

5. In your own words, state the "principle of optimality."

6. In the example of Section 10.7, assume that because of political considerations building size 1 must be chosen for site 3. What is the new optimal policy? How would the problem solution procedure be effected if building size 1 was dictated at site 1?

7. Explain why the restriction of integrality in the decision space makes a dynamic programming problem easier to solve?

8. Explain the effect of adding a second constraint to a dynamic programming problem. Describe a recursive procedure to solve such a program.

9. List 10 types of problems that can be formulated and solved with dynamic programming. What restrictions are placed on each problem formulation?

10. Assume that the solution space for an equivalent formulation to the knapsack problem of Section 10.8 is continuous. Set up and explain the necessary recursive relationships and transformations.

11. An electric utility is building a large generating station and a regional power distribution network to serve a large number of customers scattered over a wide area. Because of the large area to be served, they decide to build 10 substations to handle the distribution of power to specific localities. The problem is where to locate the substations along the transmission line so as to minimize the cost of building high-tension lines, assuming that local distribution costs are roughly the same no matter where the substa-

tion is built in a given local area. A number of sites are possible for each substation, but not every site in one area can be reached from every site in the next area for a number of reasons including (1) geographic inaccessibility due to terrain features such as lakes and mountains, and (2) the inability to purchase or lease right-of-way for the lines. The following diagram represents the problem with accessible sites connected by arcs labeled with the cost of connecting them (in appropriate units).

(a) Determine the minimum cost power distribution plan.

(b) Assume due to political considerations, you must end at C. Determine the new minimum-cost solution.

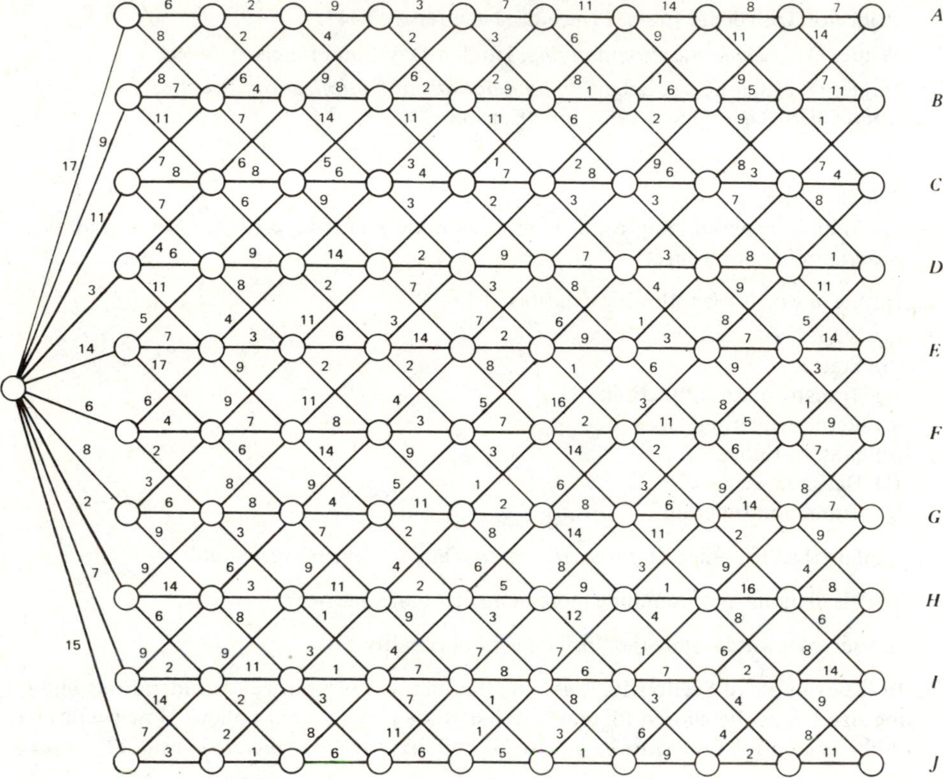

12. A company is planning their buying policy for a 7-month period. It manufactures crankcases for a large auto manufacturer and buys its raw steel in lot sizes of 500, 1000, and 1500 lb. However, quantity discount prices lot sizes at

500 lb, $3000
1000 lb, $5000
1500 lb, $7500

Anticipated usage over the 7-month period is as follows:

months	1	2	3	4	5	6	7
usage	700	1100	800	400	900	1200	600

An inventory holding charge of one-half dollar per pound is incurred per month, based on the end-of-month inventory.

The company's storage space will only handle 600 lb at a time. There is no steel in storage at the start of the planning period.

Minimize the total cost.

13. A group of businessmen are considering the establishment of a gambling casino. The businessmen have already purchased a building with a floor space limitation of 25 sq yd. These men are considering using four types of gambling. They include blackjack, poker, craps, and roulette. The estimated value per table and space required is given as follows:

	Game	Space Required Per Table, S_i	Profit Added per Table: First Table	Second Table	Third Table	Fourth Table
$i=1$	Blackjack	4	10	7	4	1
$i=2$	Poker	5	9	9	8	8
$i=3$	Craps	6	11	10	9	8
$i=4$	Roulette	3	8	6	4	2

It is important to note that the investors have realized that their marginal returns decrease for each game as more tables are added. This indicates that the clientele may fill one roulette table all the time but that a second table might be idle part of the time. Thus we have the varying values for each item (game). How many tables should be installed for each game to maximize profits?

14. The Davis Car Rental Agency has four cars available at Central Headquarters. There are requests from six marketing outlets for one car apiece. Based on customer satisfaction, mileage and transportation costs, the following cost matrix has been constructed for car delivery.

Cars \ Market	1	2	3	4	5	6
1	7	12	9	15	8	14
2	5	10	5	12	6	13
3	8	10	7	16	7	12
4	9	11	8	14	7	11

(a) Set up this problem as a mathematical programming problem. Can you identify this particular type of problem?

(b) Solve this problem using dynamic programming.

15. Platinum has been discovered in two different countries. The mines are named A and B. Each dollar spent mining platinum in mine A (at the beginning of the year) yields at the end of that year and each succeeding year, 1 lb of platinum and \$30 in backing capital, which is generously "loaned" to the mine by another government. Each dollar spent mining platinum in mine B, in a similar fashion, yields 0.5 lb of platinum and \$100 in backing capital. The governments of both countries have just been overthrown and the new ruler (of both countries) would like to know the best way to allocate his available money (\$$S_3$) so as to maximize the amount of platinum produced during the next 3 years.

16. A rancher wishes to build up his herd of cattle but has an initial capital of only \$5000. The rancher can purchase two strains of cattle, A or B. For each \$1000 invested in cattle A one obtains each year two calves and \$500 in operating capital. Cattle type B yields three calves on the average and \$200 capital. How should money be allocated over the next 4 years in order to build up the herd maximally? How many cattle will the

rancher have at the end of 4 years? *Note: Assume that this problem can be worked as a continuous dynamic programming problem, and round the optimal solution.*

17. A wine distiller produces wine in a process that consists of n stages. At each stage one faces the decision of whether to sell the yield from that stage or purify it further. Revenue derived from a vat of wine sold after the nth stage is $250n$. The cost of passing through stage n is $4n^2$.

    (a) Through how many stages should the wine be passed in order to maximize profit?
    (b) What is the maximum profit?

    Important: The number of stages must be treated as a discrete variable.

18. Solve the following problem using dynamic programming.

$$\begin{aligned} \text{Minimize:} \quad & f(x) = 2x_1^2 - 3x_2 - 4x_3^2 \\ \text{Subject to:} \quad & 3x_1 + 2x_2 + 6x_3 \geqq 16 \\ & x_2 \leqq 4 \\ & x_3 \leqq 5 \\ & x_1, x_2, x_3 \geqq 0 \end{aligned}$$

19. Solve the following problem using dynamic programming.

$$\begin{aligned} \text{Maximize:} \quad & f(x) = 2x_1^3 - 3x_1 + x_2^2 - 4x_2 + x_3 \\ \text{Subject to:} \quad & 4x_1 + 2x_2 + 3x_3 \leqq 15.3 \\ & x_1, x_2, x_3 \geqq 0 \end{aligned}$$

20. Re-solve Exercise 19 under the assumption that x_1, x_2, and x_3 can only take on integer values.

21. Solve the following problem using dynamic programming.

$$\begin{aligned} \text{Minimize:} \quad & f(x) = x_1(2x_1 - 4) + 3x_2(x_2 - 1) + x_3^2 - 7x_3 + 7 \\ \text{Subject to:} \quad & 2x_1 + 3x_2 + 2x_3 \geqq 15 \\ & x_1, x_2, x_3 \quad \text{integer} \\ & x_1, x_2, x_3 \geqq 0 \end{aligned}$$

22. Solve the following problem using dynamic programming.

$$\begin{aligned} \text{Maximize:} \quad & f(x) = 2x_1 - x_1^2 + x_2 \\ \text{Subject to:} \quad & 2x_1^2 + 3x_2^2 \leqq 6 \\ & x_1, x_2 \geqq 0 \end{aligned}$$

23. Develop a set of recursive equations to solve the following class of dynamic programming problems.

Minimize:

$$\sum_{n=1}^{N} a_i x_i^{b_i}$$

Subject to:

$$\prod_{n=1}^{N} c_i x_i^{d_i} \geqq k$$

$$x_i \geqq 0 \qquad i = 1, 2, \ldots, N$$

24. Using the results of Exercise 23, solve the following problem.

$$\begin{aligned} \text{Minimize:} \quad & 3x_1^2 + 4x_2^2 + x_3^2 \\ \text{Subject to:} \quad & x_1 x_2 x_3 \geqq 9 \\ & x_1, x_2, x_3 \geqq 0 \end{aligned}$$

25. A government analyst has identified five critical subsystems that are instrumental to the mission effectiveness of a new fighter aircraft that has been designed by the Aero-Astro aircraft company. It has been determined that the overall effectiveness of the aircraft is equal to the product of the individual effectiveness of each of the five subsystems.

Although numerous bids have been received, Aero-Astro has identified two options for each subsystem that meet the design specifications. Only $100k are available for these subsystems.

What is the maximum effectiveness that can be obtained for this amount?

TABLE OF AVAILABLE OPTIONS

	UHF Radio	TACAN	WPNS Delivery	Radar	Navigation System
Option A	Eff = 0.95	0.90	0.95	0.90	0.95
	Cost = 20*k*	15*k*	30*k*	40*k*	20*k*
Option B	Eff = 0.90	0.85	0.90	0.85	0.8
	Cost = 15*k*	10*k*	25*k*	30*k*	10*k*

CHAPTER 11

NONLINEAR PROGRAMMING

I. BASIC CONCEPTS

11.1 INTRODUCTION

In Chapter 2 the fundamentals of linear programming were explored in great detail. As the terms indicate, all functional forms dealt with in that chapter were linear in nature. In several instances, one might have wondered about the validity of such assumptions. Indeed in many situations the assumption of linearity as applied to a real-world process might be questionable. In recent years, there has been a great deal of research applied to the solution of what we will call "nonlinear programming problems." Methods that are derived to solve the broad set of problems that make up this functional classification will be referred to as nonlinear programming algorithms. A major disadvantage in studying the field of nonlinear programming is the wide variety of techniques that are presently used to attack nonlinear programming problems. A nonlinear programming problem is characterized by terms or groups of terms that involve intrinsically nonlinear functions. For example: sin (x), cos (y), $e^{x_1+x_2}$, ln (x_3), etc. Nonlinearities also arise as a result of interactions between two or more variables, such as: x_1x_2, x_1 ln (x_2), $x_2^{x_3}$, and so on. In studying linear programming solution techniques, there was a basic underlying structure that was exploited in solving those problems. Primarily this structure dictated that an optimal solution could be found by (cleverly) solving sets of linear equations. It was also known that an optimal solution would always be found at an extreme point of the feasible solution space. In solving nonlinear programming problems, an optimal solution might be found at an extreme point, a point *interior* to the feasible region, or at a point of discontinuity. In addition, algorithmic techniques might involve the solution of simultaneous linear equations, simultaneous nonlinear equations, or both. Many algorithmic procedures have been suggested for solving nonlinear programming problems; however, only a small subset of all procedures have actually proved useful in solving real-world problems. Indeed, some of the more successful techniques resort to approximation techniques using linear programming subproblems.

This chapter defines the basic characteristics of nonlinear programming, and explains several useful algorithms employed in solving nonlinear programming problems. The algorithms explained in this chapter are chosen for three basic reasons: first, they are widely used in practice and have met with a measurable degree of success; second, each technique chosen illustrates a fundamental approach to the solution of nonlinear programming problems; and third, they have all been programmed for the digital computer and are available without great difficulty. Finally, in attempting to convey the algorithmic methodology and underlying theory of nonlinear programming, the major concepts are conveyed at a fundamental level for greater understanding of the student. In practice, there are many complex modifications to the basic underlying concepts presented here. These modifications are best classified as advanced nonlinear programming and will not be explained in this text.

Problem Definition. Define the following following generalized mathematical program:

Minimize: $f(\mathbf{x}) \quad \mathbf{x} \in E^n$

Subject to:

$$H_j(\mathbf{x}) = 0 \qquad j = 1, 2, \ldots, M$$
$$G_k(\mathbf{x}) \leqq 0 \qquad k = 1, 2, \ldots, \bar{M}$$
$$\mathbf{x} = (x_1, x_2, \ldots, x_N)$$

Without loss of generality, the following definitions will be directed at minimizing solutions to the above problem.

Note that there are N decision variables defined by M equality constraints and $\bar{M}$ inequality constraints. If both the objective function and all constraints are *linear*, then we have a linear programming problem. If any component of $f(\mathbf{x})$, $H_j(\mathbf{x})$, or $G_k(\mathbf{x})$ contains nonlinear functions, then we have a nonlinear programming problem. All mathematical programs can be expressed in this form. Nonnegativity conditions on any of the solution variables are implicitly defined as part of the $\bar{M}$ inequality constraint set. Any solution vector $\mathbf{x}$ that satisfies all sets of M equality constraints and $\bar{M}$ inequality constraints is called an *admissible* or *feasible* solution. A particular set of solution variables that yield a minimizing value for $f(\mathbf{x})$ is called the *optimizing solution vector*. Such a solution vector will be denoted by $\mathbf{x}^*$ and need not necessarily be unique. Indeed, many nonlinear programming problems possess multiple solution vectors that yield the same optimal value for $f(\mathbf{x})$. The question of whether or not a nonlinear programming problem possesses a finite optimal solution at all depends on whether or not the problem is *bounded* in the solution variables. Determining if a (unique) optimal solution does exist to a bounded nonlinear programming problem is dependent on the shape of the objective function and the constraint set.

Consider a sequence of points $x_1 > x_2 > x_3 > \cdots > x_n$. If $f(x_1) > f(x_2) > \cdots > f(x_n)$, then the function is said to be monotonically increasing. If $f(x_1) < f(x_2) < \cdots < f(x_n)$, then the function is said to be monotonically decreasing. If $f(x_n) \geqq f(x_{n+1})$ or $f(x_n) \leqq f(x_{n+1})$, then the function is monotonically nonincreasing and monotonically nondecreasing, respectively. If over a given region a function increases (decreases) to a certain point and then decreases (increases) monotonically, the function is said to be *unimodal*. A unimodal function has only one peak (valley). Functions with two or more "peaks" are called *multimodal*.

A saddle point is one that *appears* to be an optimal solution from a local viewpoint but is in fact inferior to some other point in the solution space. Since there might be many such points within a general nonlinear programming formu-

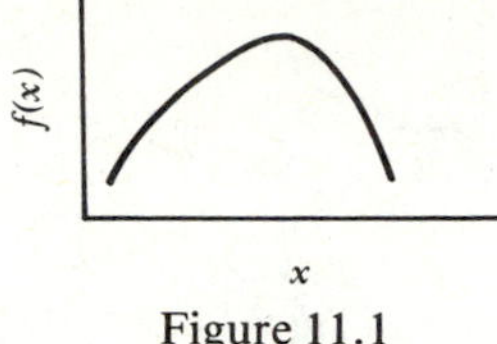

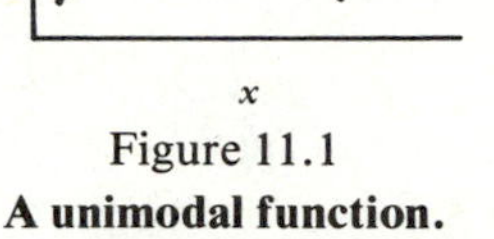

Figure 11.1
A unimodal function.

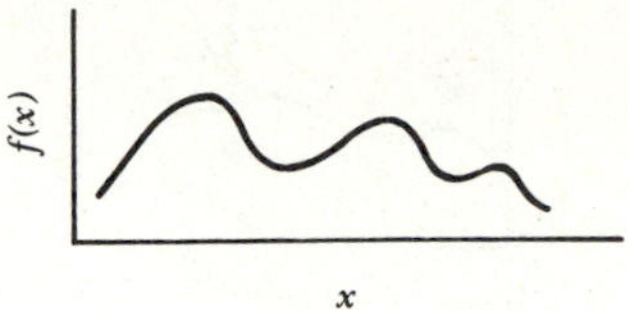

Figure 11.2
A multimodal function.

lation, these points are called *stationary points* or in this case *local minima*. (See Figs. 11.1–11.3.) Stationary points might exist at a true minimum, a point of inflection, a saddle point, or even a maximum! The best optimizing solution is called a *global minimum*. Most nonlinear programming algorithms yield solutions that are *local minima*. This is caused primarily by algorithmic procedures that depend on the local properties of a nonlinear programming problem. Fortunately many real-world formulations possess only a global minimum. However, when this is not the case, alternate local minima solutions must be examined to determine the best. Intimate knowledge of the problem at hand or a different set of starting procedures often help to find the global minimum. In certain cases, *any* solution identified as locally optimal is indeed a *global* optimal solution. Since such cases greatly enhance the power of nonlinear programming algorithmic procedures, it would be worth our time to examine those cases in which this property is true. In general, functions can be classified as (a) continuous, (b) discontinuous, or (c) discrete (see Figs. 11.4–11.6).

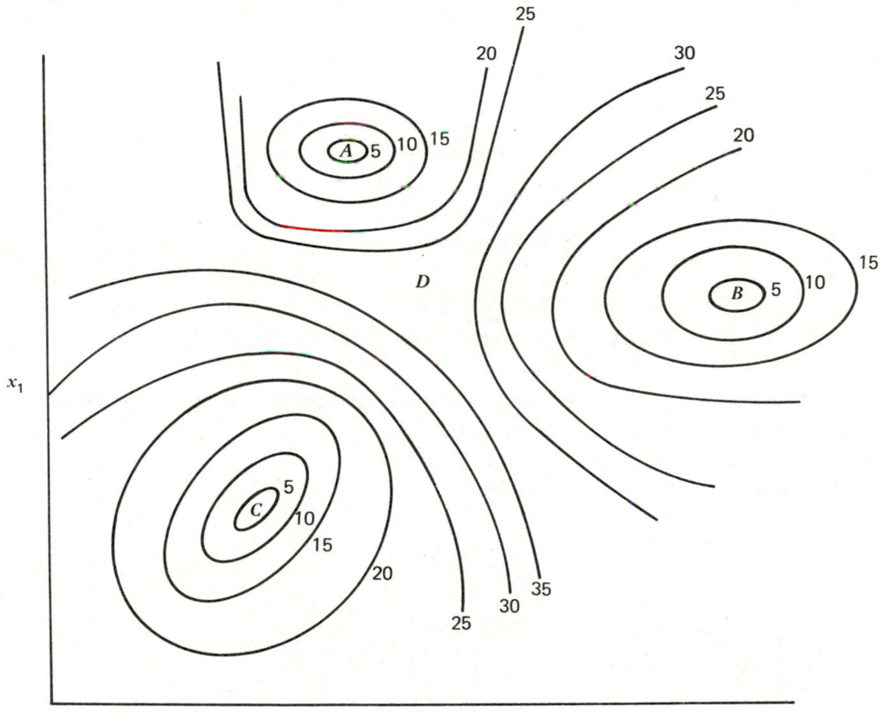

Figure 11.3
A multimodal function with a saddle point.

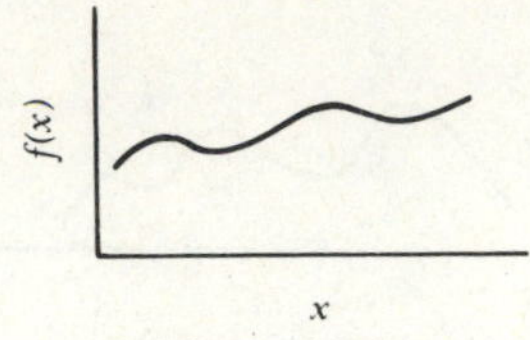

Figure 11.4
A continuous function.

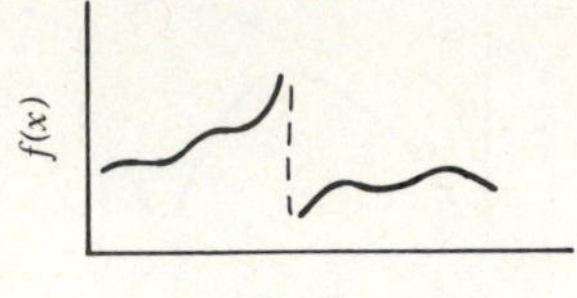

Figure 11.5
A discontinuous function.

A special type of function that we will find extremely beneficial is a *concave* or *convex* function. Mathematically a *convex* function is defined as follows:

Definition of a Convex Function. Given any two points in n-dimensional space $(\mathbf{x}_1, \mathbf{x}_2)$, if the following inequality holds at all pairs of points the function is said to be *convex.*

$$f[(1-\theta)\mathbf{x}_1+\theta\mathbf{x}_2] \leqq [(1-\theta)f(\mathbf{x}_1)+\theta f(\mathbf{x}_2)] \qquad 1 \geqq \theta \geqq 0$$

A concave function is similarly defined with the inequality reversed. Although this definition will prove useful mathematically, what does it mean? Geometrically it means that if a function is convex (concave) and if a line is drawn between any two points on the surface of the function, the line segment joining these two points will lie entirely above (below) that function.

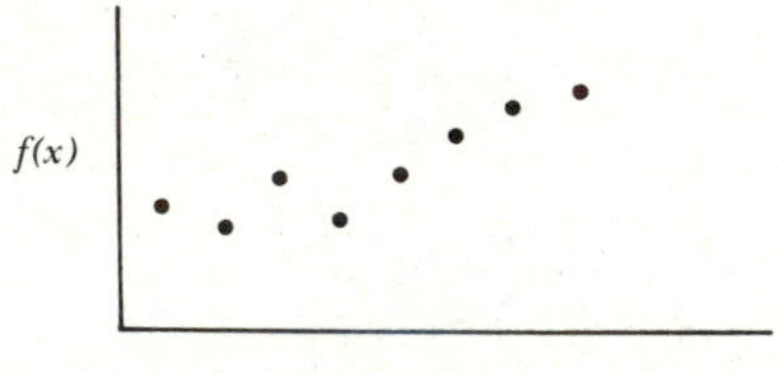

Figure 11.6
A discrete function.

Note from Figs. 11.7 to 11.12 several important facts:

1. The definition of a convex (concave) function is not dependent on the definition of a function being continuous (discontinuous).
2. A function can be concave over one region and convex over another.
3. A linear function is *both* concave and convex.

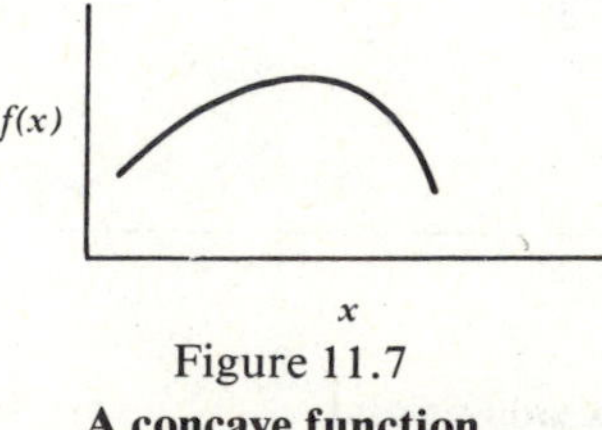

Figure 11.7
A concave function.

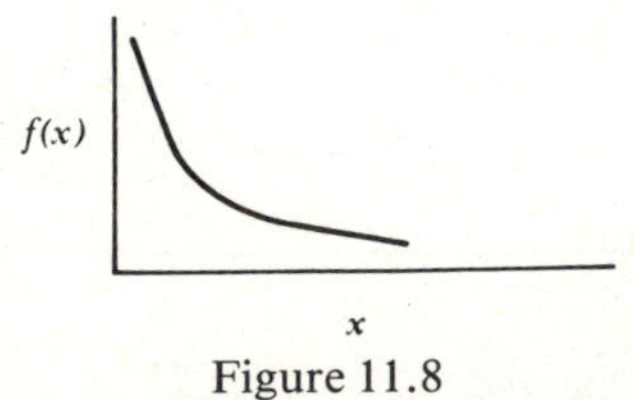

Figure 11.8
A convex function.

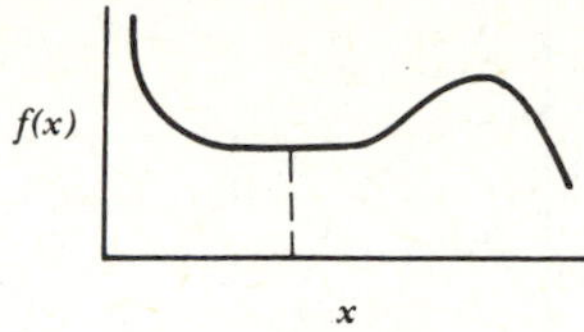

Figure 11.9
A function neither concave nor convex.

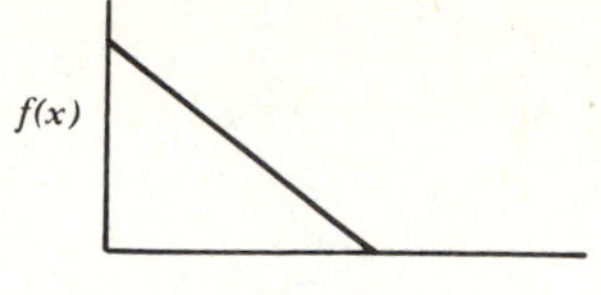

Figure 11.10
A function which is both concave and convex.

The Search for Optimal Solutions

In solving nonlinear programming problems it might appear that we are searching for the proverbial "needle in a haystack." With an infinite number of solution vectors and the possibilities of multiple optima, it might appear to be hopeless. Fortunately, there are several fundamental theorems that can be utilized to guide our search even in the face of such complexities. Moreover, if such conditions as convexity (concavity) are met, the characterization of the (unknown) optimal solution becomes relatively well defined. However, lest the reader become relatively secure at this point, let us state that although the characterization of the problem becomes easier, the actual solution procedure might be extremely complex.

If we are dealing with bounded continuous functions, a theorem by Weierstrass guarantees us that a maximum or minimum will always exist, either at a point interior to the boundaries of feasible solution variables or at the boundary itself. This is intuitively clear, since a bounded function must always possess a maximum or minimum value somewhere within the region of interest. If the function is *continuous* over the domain of interest, stationary points can be located through the use of differential calculus provided all derivatives can be found. The calculus tells us that a stationary point will exist in the interior or at a boundary if the partial derivatives of an unconstrained function vanish (become zero) at a particular solution vector. The constrained case is similarly treated through the use of constrained derivatives and will be discussed later in this chapter. In addition, if there are discontinuities in our function, then one (or more) of these derivatives might fail to exist. These points would then need to be considered separately. Finally an optimal solution might exist at a boundary point defined by the constraints on the problem. For example, this is the case in linear programming, which is a special (completely degenerate) case of nonlinear programming.

In summary, if we are to devise a procedure for solving nonlinear programming problems, we need to examine the following three candidates:

1. All points at which the continuous first derivatives are all zero.
2. All points interior to the region at which discontinuities exist for the first derivatives.
3. Points on the boundaries of the solution space.

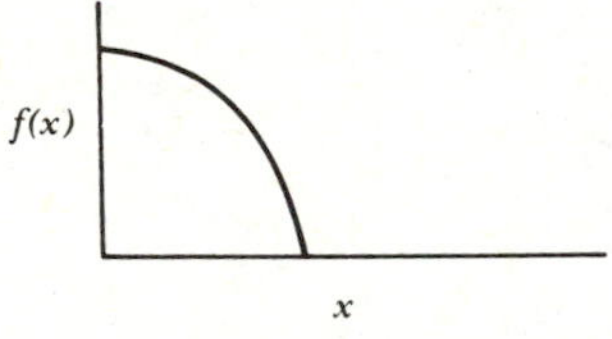

Figure 11.11
A monotonic concave function.

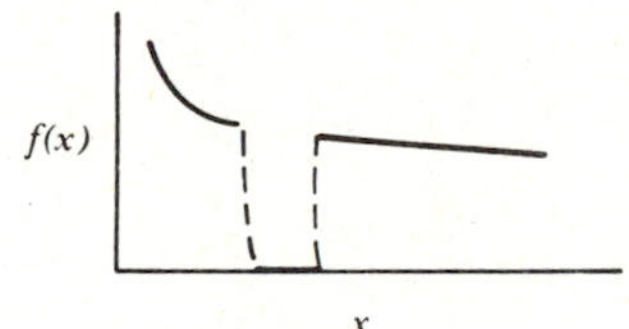

Figure 11.12
A discontinuous convex function.

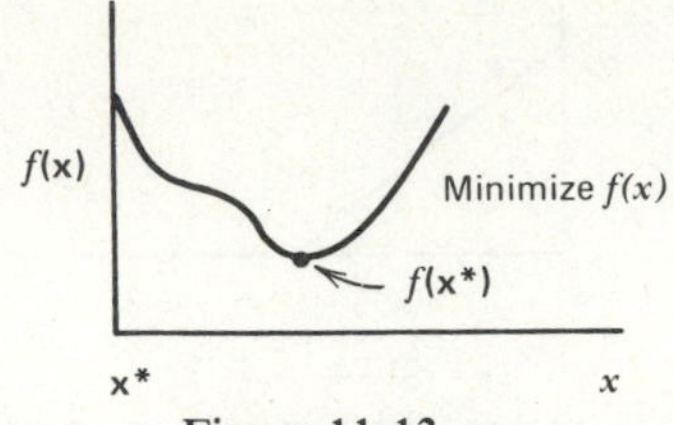

Figure 11.13
An interior solution to an unconstrained minimization problem.

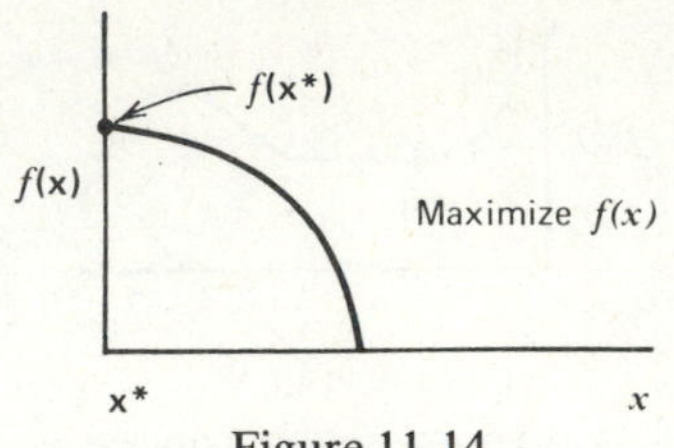

Figure 11.14
A boundary solution to an unconstrained maximization problem.

The Effects of Concavity/Convexity on the Search for an Optimum

Case A. Maximum or Minimum—Unconstrained. If the nonlinear programming problem consists of only an objective function, $f(\mathbf{x})$, and if the objective function is convex (concave), then a unique (single) optimum solution will be found at a point (a) interior to the feasible region where all derivatives vanish or (b) at a boundary point (see Figs. 11.13 and 11.14).

Case B. Maximization—Constrained. If the nonlinear programming problem consists of both an objective function and constraints, the uniqueness of an optimal solution depends on the nature of both the objective function and the constraint set. If the objective function is *concave*, and the constraint set forms a *convex* region, there will be only one maximizing solution to the problem. Hence any stationary point must be a *global maximum* solution.

Case C. Minimization—Constrained. If the nonlinear programming problem consists of both an objective function and constraints, and if the objective function is *convex* and the constraint set also forms a *convex* region, then any stationary point will be a *global minimizing* solution.

Case D. Minimizing (Maximizing) a Concave (Convex) Function. If one is minimizing (maximizing) a concave (convex) function, then the optimal solution will only be found at one of the extreme points of the constraint set. For example, if one is *minimizing* the function given in Fig. 11.15, the points A and B are all that need to be examined. The problem in performing the required search is that for most nonlinear programming problems—especially real-world formulations—this set of (possible) solution points might be extremely large.

Case E. The Linear Function. A linear function forms a class of optimization problems all to itself. By previous definition, a linear function is both convex and concave. Thus if the solution space is convex, then a solution

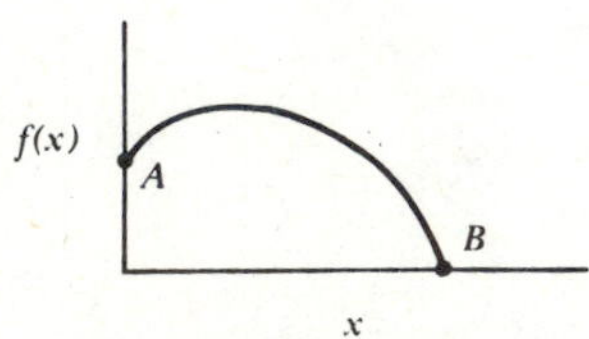

Figure 11.15
Minimizing a concave function.

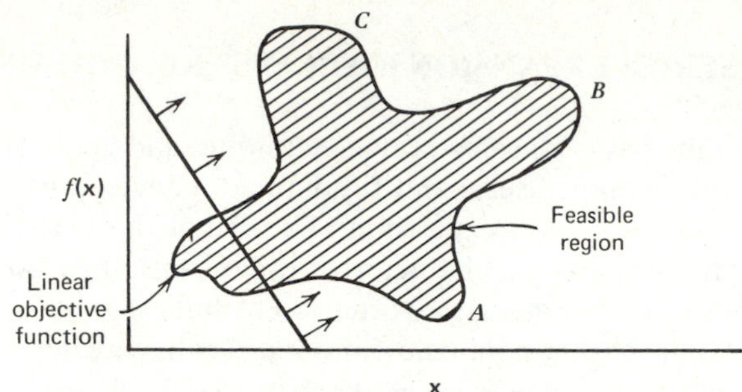

Figure 11.16
Linear optimization with a nonconvex solution space.

will always be found at a boundary. This observation is precisely what gave rise to the simplex algorithm of linear programming.

Case F. Nonconvex Regions. If the constraint set forms a nonconvex solution space, then any algorithmic procedure based on local properties of the nonlinear programming problem might produce a local stationary point that may be neither globally maximum nor minimum. This observation is valid even if the objective function is linear in nature. Figure 11.16 illustrates this theory since points A, B, and C are all locally optimal to a maximizing procedure, with point B being the global maximizing solution.

The Quest for Optimality

Having identified some of the properties of nonlinear programming, we are now in a position to present some solution algorithms useful in solving nonlinear programming problems. As previously mentioned, the process of algorithmic development is still a very active research activity. Efficient algorithms have been derived for certain classes of problems, but it is doubtful that an algorithm will ever be developed that is universally accepted as a comprehensive solution technique. Hence, unlike linear programming, there does not exist a *basic* underlying structure or concept such as the simplex algorithm.

Although the general problem remains extremely complex, the optimal solution will, in general, be found at an interior point at which first derivatives vanish, a boundary point, or a discontinuity. Except in certain special cases, the key as to where the optimal solution must be found will remain unknown. Hence, a comprehensive search must consider both interior points and boundary conditions. When possible, analytical techniques will be employed to guide our search. The most useful (and common) techniques will be first and second derivatives or a Taylor's series approximation. In the cases where analytical techniques cannot be used, such as in a "black box" response, a pattern or numerical search will have to be conducted. The problems of discontinuities can only be handled by examining the objective function value at each point of discontinuity, and retaining the maximum (minimum) solution. In general, this chapter will deal with what we will call "well-behaved" functions; that is, those that do not have any discontinuities or singularity points such as outward pointing cusps. These methods should be sufficient to attack many real-world applications. The remaining cases will be left to further study from the rapidly expanding technical literature and texts on nonlinear programming.

11.2
TAYLOR'S SERIES EXPANSIONS; NECESSARY AND SUFFICIENCY CONDITIONS

In developing the basic concepts of the techniques and goals of nonlinear programming, perhaps the most useful and fundamental development is the use and understanding of Taylor's series expansion. The use of this series surfaces in the recognition of optimal solutions, characterization of stationary points, and in linearizing (gradient) and quadratic approximation techniques.

Suppose that an arbitrary nonlinear function $f(x)$ having derivatives of all orders can be represented by a power series expansion of the form:

$$f(x) = C_0 + C_1(x-a) + C_2(x-a)^2 + - + C_n(x-a)^n \tag{11.1}$$

THEOREM 1

A power series may be differentiated term by term within its interval of convergence.

By this theorem;

$$f'(x) = C_1 + 2C_2(x-a) + 3C_3(x-a)^2 + - + nC_n(x-a)^{n-1}$$
$$f''(x) = 2!C_2 + 3\cdot 2C_3(x-a) + - + n(n-1)C_n(x-a)^{n-2}$$
$$f'''(x) = 3!C_3 + - - - + n(n-1)(n-2)C_n(x-a)^{n-3}$$

and, in general,

$$f^{(n-1)}(x) = (n-1)!C_{n-1} + n!C_n(x-a) + - + \tag{11.2}$$

If we substitute $x=a$ in these equations and also in $f(x)$, we can solve for the C's in the following manner:

$$\text{If} \qquad x=a: \qquad f(x) = f(a) = C_0$$

Hence,

$$\begin{array}{ll} f'(a) = C_1 & C_1 = f'(a) \\ f''(a) = 2!C_2 & C_2 = f''(a)/2! \\ f'''(a) = 3!C_3 & C_3 = f'''(a)/3! \\ \vdots & \vdots \\ f^{(n-1)}(a) = (n-1)!C_{n-1} & C_{n-1} = f^{(n-1)}(a)/(n-1)! \end{array}$$

Thus $f(x)$ can be represented in a power series involving the differentials of $f(x)$ itself, namely,

$$f(x) = f(a) + f'(a)(x-a) + \frac{f''(a)}{2!}(x-a)^2 + - + \frac{f^{(n-1)}(a)}{(n-1)!}(x-a)^{n-1} \tag{11.3}$$

This is called *Taylor's series*.*

Note if we let $x = a + h$ we get a useful form of this series

$$f(a+h) = f(a) + f'(a)h + \frac{f''(a)}{2!}h^2 + - + \tag{11.4}$$

*Brook Taylor (1685–1731).

and if we let $a=0$, we obtain

$$f(h)=f(x)=\sum_{n=0}^{\infty}\frac{f^n(0)}{n!}x^n \tag{11.5}$$

This is known as *Maclaurin's series*.

Unconstrainted Optimization and Its Relationship to Taylor's Series

Recall that for Taylor's series,

$$f(x)=f(a+h)=f(a)+f'(a)h+\frac{f''(a)}{2!}h^2+-+ \tag{11.6}$$

Now suppose we consider a small deviation, Δx, away from a point x_0 to a point x; if we associate Δx with h and x_0 with a;

$$\begin{aligned} f(x)&=f(x_0+\Delta x)=f(x_0)+f'(x_0)\Delta x+\frac{f''(x_0)}{2!}[\Delta x]^2+-+ \\ f(x)&=f(x_0)+\left(\frac{df}{dx}\right)_{x_0}\Delta x+\frac{1}{2}\left(\frac{d^2f}{dx^2}\right)_{x_0}\Delta x^2+-+ \end{aligned} \tag{11.7}$$

This equation says that if we are given the value of a function and all of its derivatives at some point x_0, the value of that function can be determined at any other nearby point. The accuracy of such a determination is dependent on the number of terms needed to adequately represent an arbitrary function. Two forms of Eq. 11.7 will be subsequently explored in great detail, those being the linear approximation (first two terms) and the quadratic approximation (first three terms).

EXAMPLE 11.2-1

Given $f(x)=4-7x+x^2$

(a) Calculate the value of this function at $x_0=2$.

(b) Find the value of $f(x)$ at $x=4$ by using (a) and the derivatives of $f(x)$.

(a) $f(x_0)\,|_{x_0=2}=4-7(2)+4=-6$

(b) $f'(x)=2x-7 \qquad \Delta x=2$

$f''(x)=2$

$f'''(x)=0$

Hence,

$$\begin{aligned} f(x=4)&=f(x_0)+f'(x_0)(2)+\frac{f''(x_0)}{2!}(2)^2 \\ &=-6+(-3)(2)+\frac{(2)(2)}{2} \end{aligned}$$

or

$$f(x)=-8$$

The true value of the objective function is given by

$$f(x)\,|_{x=4}=4-28+16=-8$$

Note that since the function we wish to approximate is actually a quadratic function, the first three terms of Eq. 11.7 gave an *exact* answer.

EXAMPLE 11.2-2

Calculate the value of $f(x)=2x^3-3x^2+x-4$ at $x=7$ starting from the point $x_0=3$

$$f(x_0)\,|_{x_0=3}=26$$

From Eq. 11.7,

$$f(x)=f(x_0)+\left(\frac{df}{dx}\right)_{x_0}\Delta x+\frac{1}{2}\left(\frac{d^2f}{dx^2}\right)\Delta x^2+\cdots+$$

Define

$$x_0=3$$
$$\Delta x=4$$

and

$$\frac{df}{dx}=6x^2-6x+1$$
$$\frac{d^2f}{dx^2}=12x-6$$

Hence,

$$\left(\frac{df}{dx}\right)_{x_0}=37 \qquad \Delta x=4$$
$$\left(\frac{d^2f}{df^2}\right)_{x_0}=30$$

Therefore,

$$f(x)=26+(37)(4)+\tfrac{1}{2}(30)(16)$$
$$f(x)=414$$

The true value of $f(x)$ at $x=7$ is given by

$$f(x)=542$$

Note that the Taylor's series approximation underestimated the correct answer. The reason for this lies in the fact that the true expression is cubic, and we have used a quadratic approximation. The reader will be asked to show in Exercise 13 that a cubic approximation will yield the exact answer.

Conditions for a Local Minimum. Suppose that we are given a point x^* suspected to be a local minimum. How do we check to see if this hypothesis is true?

Consider an arbitrary deviation from x^* given by ϵ. This gives a *neighborhood* surrounding x^*; $x^*\pm\epsilon$. Denote this neighborhood by δ. Therefore, x will be contained in δ if

$$0\leqq |x-x^*|\leqq\epsilon$$

Now let

$$x=x^*+\Delta x$$

where

$$-\epsilon\leqq\Delta x\leqq\epsilon$$

x^* is a *local* minimum if

$$f(x^*) \leqq f(x)$$

Hence, for $x \in \delta$, if $x = x^* + \Delta x$; then using Eq. 11.7,

$$f(x) = f(x^*) + \left(\frac{df}{dx}\right)^* \Delta x + \left(\frac{d^2f}{dx^2}\right)^* \frac{\Delta x^2}{2!} + + - + \tag{11.8}$$

The expressions $(df/dx)^*$ and $(d^2f/dx^2)^*$ represent the first and second derivatives of $f(x)$ evaluated at the point x^*. Suppose we now define

$$\delta x = f(x) - f(x^*) = \left(\frac{df}{dx}\right)^* \Delta x + \frac{1}{2!}\left(\frac{d^2f}{dx^2}\right)^* \Delta x^2 \tag{11.9}$$

where we ignore terms of higher orders.

Since we have assumed x^* to be a local minimum, then $\delta x = f(x) - f(x^*)$ must be positive for all perturbations $x = x^* \pm \epsilon$ in the neighborhood of x^*. Note at this point that Δx can either be + or − in the first term $[(df/dx)^* \Delta x]$; however, by the above arguments,

$$\left(\frac{df}{dx}\right)^* \Delta x + \frac{1}{2!}\left(\frac{d^2f}{dx^2}\right)^* \Delta x^2 \geqq 0 \tag{11.10}$$

In general, if Δx is very small, then Δx^2 will be even smaller and the first term will completely dominate the sign of the whole series (unless, of course, it is zero—this will be discussed later). Hence, there are three possible cases:

Case I. If $(df/dx)^* > 0$, then Δx must also be > 0 for x^* to be a local minimum.
Case II. If $(df/dx)^* < 0$, then Δx must also be < 0 for x^* to be a local minimum.
Case III. If $(df/dx)^* \equiv 0$, then Δx can be anything and the sign of the series is dominated by the second term, that is,

$$\frac{1}{2!}\left(\frac{d^2f}{dx^2}\right)^* \Delta x^2$$

Since $1/2!$ and Δx^2 are always positive regardless of Δx, then it is necessary that $(d^2f/dx^2)^* > 0$.

Now, suppose that *both* $(df/dx)^* = 0$ and $(d^2f/dx^2)^* = 0$. In this case, no information is obtained from either the first or second derivatives. When this situation occurs, one is forced to obtain information from higher-order derivatives.

From Eq. 11.7, note that the next term in the Taylor's series expansion contains Δx^3. This implies that by the same arguments previously applied to the first derivative term, the third term *must* also vanish if the point is either a minimum or a maximum. This is true because if it *did not* vanish, the value of the function could change signs with arbitrary changes in Δx. Such a situation exists in Figs. 11.17 and 11.18. Note from Figs. 11.17 and 11.18 that if the third-order terms fail to vanish, then the point is characterized as a *saddle point* or a *point of inflection*. Therefore, if the third-order terms do vanish, then the nature of the stationary point is determined by the sign of the fourth-order terms. That is,

If: $$\left(\frac{d^4f}{dx^4}\right)^* > 0 \text{ the stationary point is a minimum} \tag{11.11}$$

$$\left(\frac{d^4f}{dx^4}\right)^* < 0 \text{ the stationary point is a maximum} \tag{11.12}$$

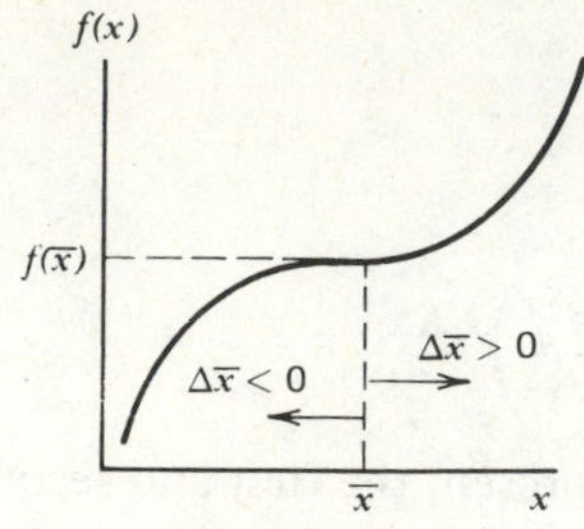

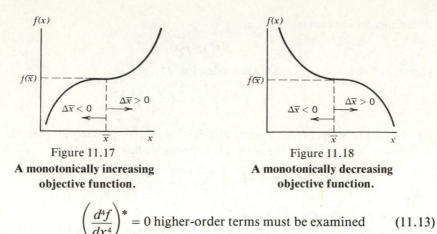

Figure 11.17
A monotonically increasing objective function.

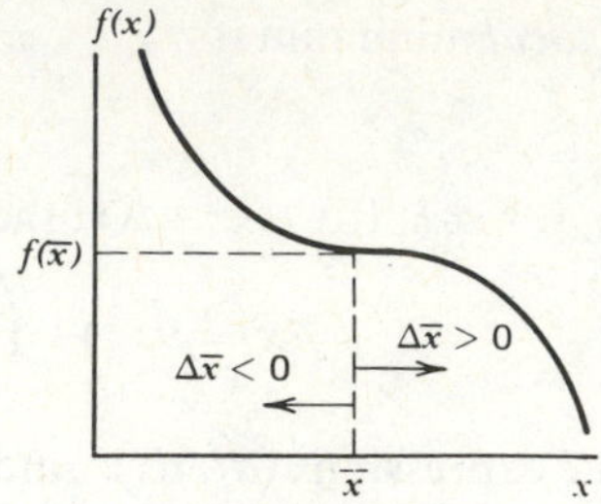

Figure 11.18
A monotonically decreasing objective function.

$$\left(\frac{d^4f}{dx^4}\right)^* = 0 \text{ higher-order terms must be examined} \tag{11.13}$$

In continuing those arguments, it should now be clear that the first *nonvanishing derivative* must be *even*, and the sign of that derivative evaluated at the stationary point must be *positive* for a minimum and *negative* for a *maximum*. If the first nonvanishing derivative is *odd*, then the point is neither a maximum nor a minimum, but a point of inflection.

We can now summarize all this into compact rules that can easily be followed for finding or identifying minimum and maximum points for an unconstrained function that yields continuous derivatives.

Necessary Conditions for a Local Minimum.

$$\left(\frac{df}{dx}\right)^* = 0 \qquad \text{and} \qquad \left(\frac{d^2f}{dx^2}\right)^* \text{must be} \geqq 0 \tag{11.14}$$

Sufficient Conditions for a Local Minimum.

$$\left(\frac{df}{dx}\right)^* = 0 \qquad \text{and} \qquad \left(\frac{d^2f}{dx^2}\right)^* > 0 \tag{11.15}$$

The preceding arguments can easily be applied to *maximizing* stationary points so that

$$\left(\frac{df}{dx}\right)^* = 0 \qquad \text{and} \qquad \left(\frac{d^2f}{dx^2}\right)^* \leqq 0 \tag{11.16}$$

are *necessary conditions* and

$$\left(\frac{df}{dx}\right)^* = 0 \qquad \text{and} \qquad \left(\frac{d^2f}{dx^2}\right)^* < 0 \tag{11.17}$$

are *sufficient conditions.*

Note that *all points* for which this is true must be minima (maxima), but there may exist some points for which this may not be true!

EXAMPLE 11.2-3

Minimize: $\qquad f(x) = x_1^3 - 3x_1 + 6$

The necessary condition for a minimizing solution is that the first derivative vanish. Hence,

$$\frac{df(x)}{dx_1} = 3x_1^2 - 3 = 0$$

Therefore,

$$x_1^* = +1 \qquad \text{or} \qquad x_1^* = -1$$

Employing the second-derivative test in one dimension,

$$\left(\frac{d^2f}{dx_1^2}\right)^* = 6x_1$$

Case I. At the point $x_1^* = 1.0$, the second derivative is positive. Therefore, this solution is a minimizing stationary point.

Case II. At the point $x_1^* = -1.0$, the second derivative is negative. Therefore, this solution is a maximizing point.

Multivariable Unconstrained Optimization

Define Taylor's series for an n-dimensional expansion as follows:

Let

$$\begin{aligned} x_1 &= x_1^* + \Delta x_1 \\ x_2 &= x_2^* + \Delta x_2 \\ &\vdots \\ x_n &= x_n^* + \Delta x_n \end{aligned} \qquad \mathbf{x} = \begin{bmatrix} x_1 \\ x_2 \\ \vdots \\ x_n \end{bmatrix} \qquad \mathbf{x}^* = \begin{bmatrix} x_1^* \\ x_2^* \\ \vdots \\ x_n^* \end{bmatrix} \qquad \Delta\mathbf{x} = \begin{bmatrix} \Delta x_1 \\ \Delta x_2 \\ \vdots \\ \Delta x_n \end{bmatrix}$$

Hence,

$$\mathbf{x} = \mathbf{x}^* + \Delta\mathbf{x}$$

Assuming that the function $f(\mathbf{x})$ is continuous and possesses all nth-order partial derivatives, the following series expansion corresponds to Eq. 11.3:

$$\begin{aligned} f(\mathbf{x}) &= f(\mathbf{x}^* + \Delta\mathbf{x}) \\ &= f(\mathbf{x}^*) + \sum_{i=1}^{n}\left(\frac{\partial f}{\partial x_i}\right)^* \Delta x_i + \frac{1}{2!}\sum_{i=1}^{n}\sum_{j=1}^{n}\left(\frac{\partial f^2}{\partial x_i\,\partial x_j}\right)^* \Delta x_i \Delta x_j \\ &\quad + \frac{1}{3!}\sum_{i=1}^{n}\sum_{j=1}^{n}\sum_{k=1}^{n}\left(\frac{\partial f^3}{\partial x_i\,\partial x_j\,\partial x_k}\right)^* \Delta x_i \Delta x_j \Delta x_k + - + \end{aligned} \tag{11.18}$$

For clarity, let us develop the concepts of necessary and sufficiency conditions for only two dimensions.

Taylor's Series Development in Two Dimensions. Consider the two-dimensional vector $\mathbf{x}$ in the neighborhood of $\mathbf{x}^*$. Define

$$\mathbf{x} = \begin{bmatrix} x_1 \\ x_2 \end{bmatrix} \qquad \mathbf{x}^* = \begin{bmatrix} x_1^* \\ x_2^* \end{bmatrix} \qquad \Delta\mathbf{x} = \begin{bmatrix} \Delta x_1 \\ \Delta x_2 \end{bmatrix}$$

By Taylor's series;

$$\begin{aligned} f(\mathbf{x}) &= f(\mathbf{x}^* + \Delta\mathbf{x}) \\ &= f(\mathbf{x}^*) + \left(\frac{\partial f(x)}{\partial x_1}\right)^* \Delta x_1 + \left(\frac{\partial f(x)}{\partial x_2}\right)^* \Delta x_2 \\ &\quad + \frac{1}{2!}\left(\frac{\partial f^2(x)}{\partial x_1^2}\right)^* \Delta x_1^2 + \frac{1}{2!}\left(\frac{\partial f^2(x)}{\partial x_2^2}\right)^* \Delta x_2^2 \\ &\quad + (2)\frac{1}{2!}\left(\frac{\partial f(x)}{\partial x_1\,\partial x_2}\right)^* \Delta x_1 \Delta x_2 + - - + \end{aligned} \tag{11.19}$$

Now define ∇f as follows:

$$\nabla f = \left(\frac{\partial f}{\partial x_1}, \frac{\partial f}{\partial x_2}\right)$$

and

$$H\langle \mathbf{x} \rangle = \begin{bmatrix} \dfrac{\partial^2 f}{\partial x_1^2} & \dfrac{\partial^2 f}{\partial x_1\,\partial x_2} \\[2ex] \dfrac{\partial^2 f}{\partial x_2\,\partial x_1} & \dfrac{\partial^2 f}{\partial x_2^2} \end{bmatrix} \tag{11.20}$$

Using the above definitions, we can write:

$$f(\mathbf{x}) = f(\mathbf{x}^*) + [\nabla f]^* \Delta\mathbf{x} + \tfrac{1}{2}\Delta\mathbf{x}^T [H^*(\mathbf{x})]\Delta\mathbf{x} + - + \tag{11.21}$$

The quantities $[\nabla f]^*$ and $[H^*(\mathbf{x})]$ are the *gradient vector* and the *Hessian matrix* evaluated at the point $\mathbf{x}^*$. These two expressions will play an increasingly important role in future developments and should be carefully noted. For simplicity, we will often call ∇f the *gradient* and, $H(\mathbf{x})$ the *Hessian*.

EXAMPLE 11.2-4

Suppose $f(\mathbf{x}) = x_1^2 + x_1x_2 + x_2^2$. Evaluated at $\mathbf{x}^* = [x_1, x_2] = [2, 3]$

$$f(\mathbf{x}) = 19$$

What is $f(\mathbf{x})$ at $\mathbf{x} = [3, 5]$? If

$$\mathbf{x}^* = [2, 3] \qquad \Delta\mathbf{x} = [1, 2]$$

then

$$\frac{\partial f}{\partial x_1} = 2x_1 + x_2 \qquad \frac{\partial f}{\partial x_2} = 2x_2 + x_1$$

$$\frac{\partial^2 f}{\partial x_1\,\partial x_2} = 1 \qquad \frac{\partial^2 f}{\partial x_2^2} = 2 \qquad \frac{\partial^2 f}{\partial x_1^2} = 2$$

and

$$f(\mathbf{x}) = f(\mathbf{x}^*) + \left(\frac{\partial f}{\partial x_1}\right)^* \Delta x_1 + \left(\frac{\partial f}{\partial x_2}\right)^* \Delta x_2 + \tfrac{1}{2}\Delta\mathbf{x}^T [H^*\langle\mathbf{x}\rangle]\Delta\mathbf{x} \tag{11.22}$$

Equation 11.22 represents a second-order Taylor's series expansion of $f(\mathbf{x})$. Using the vectors $\mathbf{x}^*$ and $\mathbf{x}$, one obtains

$$f(\mathbf{x}) = 19 + (7)(1) + 8(2) + \tfrac{1}{2}(2)(1) + \tfrac{1}{2}(2)(4) + \tfrac{1}{2}(1)(2)(2)$$
$$f(\mathbf{x}) = 49$$

Check: $\qquad f(\mathbf{x}) = [3]^2 + (3)(5) + [5]^2 = 49$

Note that since $f(\mathbf{x})$ was actually a quadratic function, we exactly predicted the value of the function.

Sufficiency Conditions for a Local Minimum (Maximum). From previous discussions, we are sure that for a local minimum or maximum:

1. For a local minimum to exist, all local perturbations in the present solution vector must produce a *positive* change in the objective function.

2. For a local maximum to exist, all local perturbations in the present solution vector must produce a *negative* change in the objective function.

How can we guarantee that we have a local maximum or minimum? Since $\Delta\mathbf{x}$ is very small, then all first-order terms will dominate the sign of the entire Taylor's series. Hence, a necessary condition for a local *minimum* to exist is that the *gradient* vanish.

$$(\nabla f)^* \equiv 0$$

In addition, for a local minimum to exist it is *sufficient* to show that ∂f is positive for all points within the neighborhood of $\mathbf{x}^*$. Since the *gradient* must vanish, then the *Hessian* now controls the sign of the entire series. At this point, note that since we are dealing with real numbers the *Hessian* can be uniquely specified in terms of the controlling sign by means of the following five conditions.

The Hessian matrix $H\langle\mathbf{x}\rangle$ *is said to be*

1. *Positive definite* if for all $\Delta\mathbf{x}$

$$(\Delta\mathbf{x})^T H^*\langle\mathbf{x}\rangle(\Delta\mathbf{x}) > 0 \tag{11.23}$$

2. *Negative definite* if for all $\Delta\mathbf{x}$

$$(\Delta\mathbf{x})^T H^*\langle\mathbf{x}\rangle(\Delta\mathbf{x}) < 0 \tag{11.24}$$

3. *Indefinite* if for all $\Delta\mathbf{x}$

$(\Delta\mathbf{x})^T H^*\langle\mathbf{x}\rangle(\Delta\mathbf{x})$ can take both positive and negative values for all changes in the neighborhood. (11.25)

4. *Positive semidefinite* if for all $\Delta\mathbf{x}$

$$(\Delta\mathbf{x})^T H^*\langle\mathbf{x}\rangle(\Delta\mathbf{x}) \geqq 0 \tag{11.26}$$

5. *Negative semidefinite* if for all $\Delta\mathbf{x}$

$$(\Delta\mathbf{x})^T H^*\langle\mathbf{x}\rangle(\Delta\mathbf{x}) \leqq 0 \tag{11.27}$$

It then follows by examining the Taylor's series expansion that the following rules can be used to characterize a stationary point.

Sufficient Conditions for a Local Minimum at $\mathbf{x}^*$. (11.28)
1. $(\nabla f)^* \equiv 0$
2. $|H^*\langle\mathbf{x}\rangle|$ is *positive definite*

Sufficient Conditions for a Local Maximum at $\boldsymbol{x}^*$. (11.29)
1. $(\nabla f)^* \equiv 0$
2. $|H^*\langle\mathbf{x}\rangle|$ is *negative definite*

Sufficient Conditions for a Saddle Point. (11.30)
1. $(\nabla f)^* \equiv 0$.
2. $|H^*\langle\mathbf{x}\rangle|$ is *indefinite*

If $(\nabla f)^* \equiv 0$ and $|H^*\langle\mathbf{x}\rangle|$ is any semidefinite form, then higher-order terms must be examined to determine the nature of the solution vector. Note that the nature of the Hessian matrix is easily determined from the functional second derivatives. For the reader who is unfamiliar with these tests, Appendix A has been included to facilitate the above rules.

Checking a Stationary Point for the n-Dimensional Unconstrained Optimization Problem. When dealing with functions of dimensions higher than two, it is convenient to define a procedure that is useful in identifying a stationary point (one at which the *Gradient* vanishes). To this end, define the following n-dimensional matrix of second *partial derivatives*.

$$H^*(\mathbf{x}) = \begin{bmatrix} \left(\frac{\partial^2 f}{\partial x_1^2}\right)^* & \left(\frac{\partial^2 f}{\partial x_1, \partial x_2}\right)^* & \cdots & \left(\frac{\partial^2 f}{\partial x_n, \partial x_1}\right)^* \\ \left(\frac{\partial^2 f}{\partial x_2, \partial x_1}\right)^* & \left(\frac{\partial^2 f}{\partial x_2^2}\right)^* & \cdots & \left(\frac{\partial^2 f}{\partial x_n, \partial x_2}\right)^* \\ \cdot & \cdot & & \cdot \\ \cdot & \cdot & & \cdot \\ \cdot & \cdot & & \cdot \\ \left(\frac{\partial^2 f}{\partial x_n, \partial x_1}\right)^* & \left(\frac{\partial^2 f}{\partial x_n, \partial x_2}\right)^* & \cdots & \left(\frac{\partial^2 f}{\partial x_n, \partial x_n}\right)^* \end{bmatrix} \tag{11.31}$$

For this matrix, there exists exactly n-determinants formed from the single element in the upper-left-hand corner, successively through the entire matrix moving from upper left to lower right. Designate these determinates by $D_1, D_2, \ldots, D_n$. The following tests are valid:

1. For a stationary point to be a *minimum*, it is sufficient that $D_1, D_2, \ldots, D_n$ all be *positive* quantities.
2. For a stationary point to be a *maximum*, it is sufficient that all *even* determinants are *positive* and all *odd* determinants are *negative*.

$$D_j < 0 \qquad j = 1,3,5,\ldots$$
$$D_j > 0 \qquad j = 2,4,6,\ldots$$

If these conditions are not exactly satisfied, then the point *may* or *may not* be an optimal solution. In this case, higher-order tests must be employed or all stationary points examined.

The Special Case of $n=2$ (*Two Dimensions*). In the special case where there are only two decision variables, the above conditions reduce to a convenient test for a stationary point. Note that Eq. 11.31 reduces to the following sufficient conditions for a minimum:

$$\left(\frac{\partial^2 f}{\partial x_1^2}\right)^* > 0 \tag{11.32}$$

$$\left[\left(\frac{\partial^2 f}{\partial x_1^2}\right)^* \left(\frac{\partial^2 f}{\partial x_2^2}\right)^* - \left(\frac{\partial^2 f}{\partial x_1\, \partial x_2}\right)^{2*}\right] > 0 \tag{11.33}$$

In layman's terms, these conditions are sufficient because they guarantee that the function is convex over the solution space. In technical terms, these conditions must be met to guarantee a positive definite Hessian matrix. These conditions are sufficient for a maximizing point provided that > can be replaced with < in both instances. Weaker conditions can be stated in the following manner, if

$$\left(\frac{\partial^2 f}{\partial x_1^2}\right)^* \geqq 0 \tag{11.34}$$

$$\left[\left(\frac{\partial^2 f}{\partial x_1^2}\right)^* \left(\frac{\partial^2 f}{\partial x_2^2}\right)^* - \left(\frac{\partial^2 f}{\partial x_1\, \partial x_2}\right)^{2*}\right] \geqq 0 \tag{11.35}$$

Then the above conditions are necessary for a local minimum but are not sufficient.

EXAMPLE 11.2-5

Minimize: $f(\mathbf{x}) = x_1^2 + x_2^2 - 2x_1 + x_1x_2 + 1$

Is the point $\mathbf{x} = (x_1, x_2) = (\frac{4}{3}, -\frac{2}{3})$ a local minimum? From the above,

$$\frac{\partial f}{\partial x_1} = 2x_1 - 2 + x_2 \qquad \frac{\partial^2 f}{\partial x_1^2} = 2$$

$$\frac{\partial f}{\partial x_2} = 2x_2 + x_1 \qquad \frac{\partial^2 f}{\partial x_2^2} = 2$$

$$\frac{\partial^2 f}{\partial x_1\,\partial x_2} = 1$$

From Eqs. 11.32 and 11.33, using the point $x_1 = \frac{4}{3}$, $x_2 = -\frac{2}{3}$, one obtains

$$\left[\left(\frac{\partial^2 f}{\partial x_1^2}\right)^* \left(\frac{\partial^2 f}{\partial x_2^2}\right)^* - \left(\frac{\partial^2 f}{\partial x_1\,\partial x_2}\right)^{2*}\right] > 0$$

which yields

$$(2)(2) - (1) = 3 > 0$$

and

$$\left(\frac{\partial^2 f}{\partial x_1^2}\right)^* = 2 > 0$$

Therefore, the point $x_1^* = \frac{4}{3}$ and $x_2^* = -\frac{2}{3}$ is a minimizing solution.

EXAMPLE 11.2-6

Consider the following objective function.

Minimize: $f(\mathbf{x}) = x_2^2 + 3x_1^6 + 5x_2^4$

From $f(\mathbf{x})$, one obtains

$$\frac{\partial f}{\partial x_1} = 18x_1^5 \qquad \frac{\partial f}{\partial x_2} = 2x_2 + 20x_2^3 \qquad \frac{\partial^2 f}{\partial x_1\,\partial x_2} = 0$$

$$\frac{\partial^2 f}{\partial x_1^2} = 90x_1^4 \qquad \frac{\partial^2 f}{\partial x_2^2} = 60x_2 + 2$$

From the first partial derivatives, it is evident that the point $x_1^* = 0$ and $x_2^* = 0$ is a stationary point. From the second partial derivatives, and the results of Eqs. 11.32 and 11.33,

$$\left[\left(\frac{\partial^2 f}{\partial x_1^2}\right)^* \left(\frac{\partial^2 f}{\partial x_2^2}\right)^* - \left(\frac{\partial^2 f}{\partial x_1\,\partial x_2}\right)^{2*}\right] = (0)(2) - 0 = 0 \qquad (11.36)$$

Note that Eq. 11.36 satisfies the necessary conditions for a local minimum given by Eqs. 11.34 and 11.35. However, the point cannot be identified without reference to higher-order tests. The student will be asked to show in Exercise 14 that this point is indeed a global minimum solution.

II. UNCONSTRAINED OPTIMIZATION

11.3
FIBONACCI AND GOLDEN SECTION SEARCH

Fibonacci Search

Fibonacci search is a univariate search technique that can be used to find the maximum (minimum) of an arbitrary unimodal, univariate objective function. The name Fibonacci search has been attributed to this technique because of the search procedure's dependency on a numerical sequence called Fibonacci numbers. Consider the following recursive relationship which generates an infinite series of numbers:

$$X_n = X_{n-1} + X_{n-2} \qquad n = 2,3, \ldots \tag{11.37}$$

Define

$$X_0 = 1 \qquad \text{and} \qquad X_1 = 1 \tag{11.38}$$

The above equation generates a series of numbers that are known as Fibonacci numbers (Table 11.1).

Table 11.1
THE FIBONACCI SEQUENCE

Identifier	Sequence	Fibonacci Number
F_0	0	1
F_1	1	1
F_2	2	2
F_3	3	3
F_4	4	5
F_5	5	8
F_6	6	13
F_7	7	21
F_8	8	34
F_9	9	55
F_{10}	10	89
F_{11}	11	144
F_{12}	12	233
F_{13}	13	377
F_{14}	14	610
F_{15}	15	987
Etc.	Etc.	Etc.

Although the name Fibonacci might be new to some readers, the Fibonacci sequence of numbers has a long historical background. The derivation of this series goes back to Leonardo of Pisa, who was known as Fibonacci. The use of these numbers has appeared in diverse areas such as the reproduction of rabbits and the mathematical structure of pineapple scales! (30)

The Fibonacci search technique is a sequential search technique that successfully reduces the interval in which the maximum (minimum) of an arbitrary nonlinear function must lie. To apply this technique, the assumption of unimodality must be invoked or the technique may locate a stationary point or completely fail. If the assumption of unimodality holds, it can be shown that the Fibonacci search technique is an optimal search technique in the minimax sense. That is, compared to known univariate search techniques, in a sequence of N functional evaluations it will yield the minimum/maximum interval of uncertainty. The interval of uncertainty is defined as the interval in which the optimum solution is known to exist. The Fibonacci search technique can be derived from geometrical considerations, provided that the goal of achieving a minimax search strategy is followed. A lucid derivation is presented by Converse (5), or Beightler, Phillips, and Wilde (2).

Suppose that a unimodal objective function is known to possess a maximum over a range of values from some point A to some point B (Fig. 11.19): Fibonacci search is an efficient search technique that can be used to find either the maximum or minimum of a nonlinear, unconstrained objective function in one variable. The function must be bounded and unimodal to guarantee convergence to a global solution.

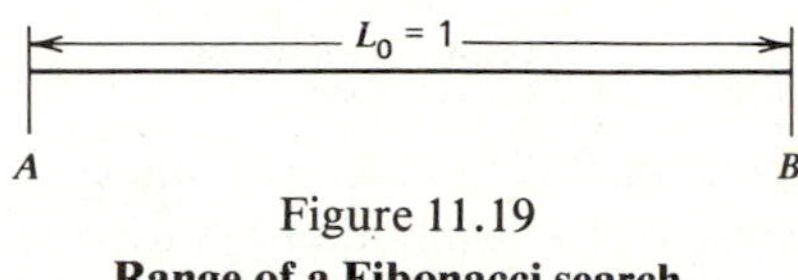

Figure 11.19
Range of a Fibonacci search.

Define the initial interval of search to be of length L_0. This interval is called the initial interval of uncertainty, since we are uncertain where the optimum lies but are assured that it must be between points A and B. Assume that our goal is to reduce the initial interval of uncertainty to some finite length L_N using exactly N functional evaluations. Define the following variables relevant to our subsequent discussion.

Let L_n = length of the interval of uncertainty after n functional evaluations

X_n = value of the variable X for which we seek an optimal value after N functional evaluations; $n = 1,2, \ldots N$.

f_n = value of the objective function using X_n; $n = 1,2, \ldots N$

ϵ = the minimum separation allowed between any two points over the interval L_0.

The value of ϵ represents the *resolution* that can be obtained experimentally between the points X_n and X_{n-1}. The resolution is actually a statement of how close one point X_n can be to another point X_{n-1} and still distinguish between the change in $f(x)$. The basic logic of our search will be to eliminate regions in which the optimum cannot lie based on strategically placed points of functional evaluation.

Consider the following example in which an (unknown) function $f(x)$ is to be successively evaluated over the interval from $X = 5.0$ to 15.0 to determine the *maximum* value of $f(x)$ over that interval. Rather arbitrarily, assume that a functional evaluation at point $X_1 = 12$ yields $f(x_1) = 10.7$ (Figure 11.20).

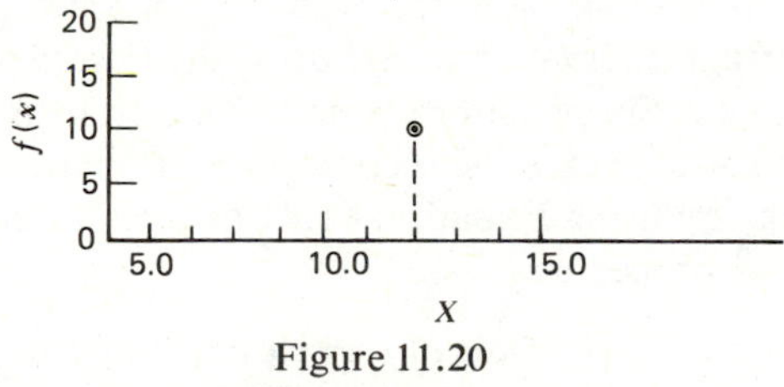

Figure 11.20
First evaluation.

Note we do this by using our previous definitions, $L_0 = 10$ and $L_1 = 10$. In other words, only one functional evaluation yields no knowledge of where the optimal (maximum) solution lies. Assume a second arbitrary evaluation yields $f(x_2) = 5.0$ at $x_2 = 9$ (Figure 11.21).

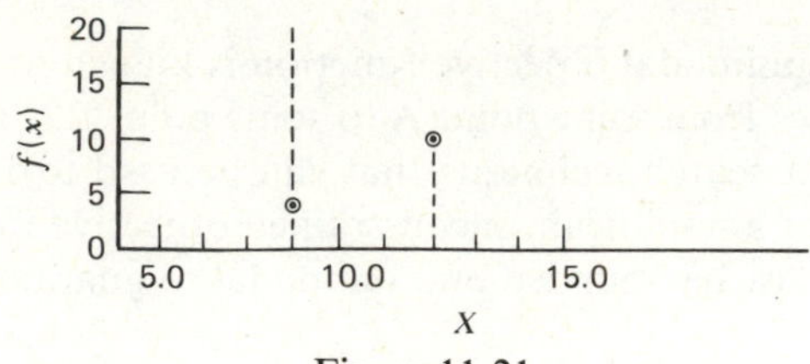

Figure 11.21
Second evaluation.

At this evaluation, we can conclude that the optimal solution *cannot lie* between $X = 5.0$ and $X = 9$, because of the unimodality assumption. Hence, in two functional evaluations we reduce our interval of uncertainty from $L_0 = L_1 = 10.0$ to $L_2 = 6$. Note that an optimally efficient search procedure would be one that results in the *minimum interval of uncertainty* after N functional evaluations. Fibonacci search is optimum in a *minimax* sense: it minimizes the maximum interval of uncertainty after N functional evaluations. The following procedure is optimal in a minimax sense and is called a *Fibonacci search technique*.

One can show by rigorous proof that the length of the final interval of uncertainty is given by the following equation [Phillips, Ravindran, and Solberg (20) or Converse (5)]:

$$L_N = \frac{L_0}{F_N} + \frac{\epsilon F_{N-2}}{F_N} \tag{11.39}$$

Note that the final interval of uncertainty is a function of the number of experimental evaluations (N), the allowable resolution (ϵ), and the initial search interval (L_0).

One can also prove that the length of the interval of uncertainty after the first two functional evaluations is given by the following relationship:

$$L_2 = \frac{1}{F_N}[L_0 F_{N-1} + \epsilon(-1)^N] \tag{11.40}$$

Finally, one can prove that the following equation is valid throughout the search procedure:

$$L_n = L_{n-2} - L_{n-1} \qquad n = 3, 4, \ldots, N \tag{11.41}$$

The last two equations provide the means to execute a Fibonacci search. Since the minimax search procedure calls for symmetrical placement of experimental points within the current interval of uncertainty, the first two points can be determined by calculating L_2. Simply subtract this value from B to locate one point, and add this value to A to locate the second point. Subsequent points can be located in the same way by using Equation 11.41 as the search progresses. The following steps define the procedure:

Step I. Define the end points of the search, A and B.

Step II. Define the number of functional evaluations, N, that are to be used in the search.

Step III. Define the minimum resolution parameter, ϵ.

Step IV. Define the initial interval and first interval of uncertainty as $(B-A)$.

Therefore, $L_0 = L_1 = (B-A)$ (11.42)

Step V. Define the *second* interval of uncertainty as follows:

$$L_2 = \frac{1}{F_N}[L_0 F_{N-1} + \epsilon(-1)^N] \tag{11.43}$$

Where F_N and F_{N-1} are *Fibonacci numbers.*

Step VI. Locate the first two functional evaluations at the two symmetric points X_1 and X_2, defined as follows:

$$X_1 = A + L_2 \tag{11.44}$$
$$X_2 = B - L_2$$

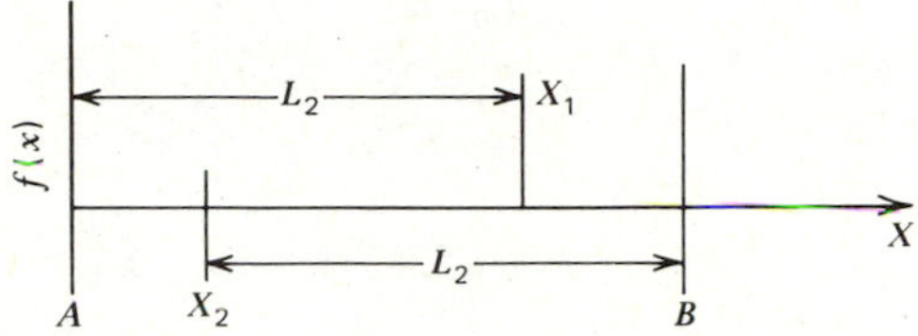

Step VII. Calculate $f(X_1)$ and $f(X_2)$, and eliminate the interval in which the optimum cannot lie.

Step VIII. Use the relationship $L_n = L_{n-2} - L_{n-1}$ to locate subsequent points of evaluation within the remaining interval of uncertainty.

Continue to repeat Steps VII and VIII until N functional evaluations have been executed. The final solution can either be an average of the last two points evaluated (X_N and X_{N-1}) or the best (max/min) functional evaluation.

Summary

All that is needed to begin a Fibonacci search are (1) The minimum separation (resolution) between any two functional evaluations (ϵ), (2) the number of experiments that will be run (N), and (3) the bounding values of B and A. Fibonacci search can be used with either univariate maximization or minimization problems.

EXAMPLE 11.3-1

Maximize the function $f(x) = -3X^2 + 21.6X + 1.0$, with a minimum resolution of 0.50 over six functional evaluations. The optimal value of $f(x)$ is assumed to lie in the range $25 \geqq X \geqq 0$.

Solution

From Eq. 11.40:

$$L_2 = \frac{1}{F_N}[L_0 F_{N-1} + \epsilon(-1)^N] \tag{11.45}$$
$$= \frac{1}{13}[25(8) + 0.50]$$

$$L_2 = 15.4231 \tag{11.46}$$

The first two functional evaluations will be conducted over the range $25 \geqq X \geqq 0$, symmetrical within this interval. Therefore,

$$X_1 = 15.4231 \tag{11.47}$$

$$X_2 = 9.5769 \tag{11.48}$$

$$f(X_1) = -379.477 \tag{11.49}$$

$$f(X_2) = -67.233 \tag{11.50}$$

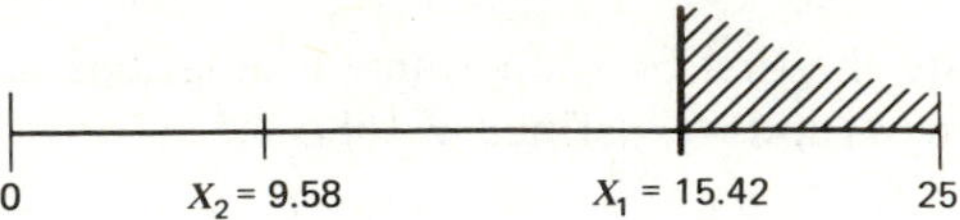

Hence, the region to the right of $X_1 = 15.42$ can be eliminated. Note that

$$L_0 = 25 \tag{11.51}$$

$$L_1 = 25 \tag{11.52}$$

and

$$L_2 = 15.4231 \tag{11.53}$$

Thus

$$L_3 = L_1 - L_2 = 25 - 15.4231 = 9.5769 \tag{11.54}$$

Symmetrically within the present interval of uncertainty, the two new points would be $X_3 = 9.5769$ and $X_4 = 5.8462$. Note that one of the new functional evaluations corresponds to one of the old functional evaluations. As previously noted, this will always occur in a Fibonacci search. Table 11.2 shows the progression through the six functional evaluations.

Table 11.2
FIBONACCI SEARCH

Functional Evaluations (n)	Interval of Uncertainty	X_{n-1}	$f(X_{n-1})$	X_n	$f(X_n)$
2	$15.4231 \geqq X \geqq 0$,[15.4231]	9.5769	−67.233	15.4231	−379.477
3	$9.5769 \geqq X \geqq 0$,[9.5769]	5.8462	24.744	9.5769	−67.233
4	$5.8462 \geqq X \geqq 0$,[5.8462]	3.731	39.83	5.8462	24.744
5	$5.8462 \geqq X \geqq 2.115$,[3.731]	2.115	32.26	3.731	39.83
6	$4.2304 \geqq X \geqq 2.115$,[2.115]	3.731	39.83	4.2304	38.688

At the sixth functional evaluation, the interval of uncertainty is established as

$$I_6 = 2.115 \tag{11.55}$$

The best estimate of the optimal solution is given by

$$X_5^* = 3.731 \tag{11.56}$$

Hence,

$$f(X_5^*) = 39.83 \tag{11.57}$$

The resolution is

$$\epsilon = 4.2304 - 3.731 \tag{11.58}$$
$$\Rightarrow \epsilon = 0.4994$$

Note that $\epsilon \leq 0.50$. The optimal solution is actually $X^* = 3.60$, and will be approached through further searching, provided that the ϵ criterion is relaxed.

Golden Section Search

In performing a Fibonacci search, the two primary drawbacks are the *a priori* specification of the resolution factor (ϵ) and the number of experiments to be performed (N). It is obvious that if the search is functioning properly, the successive experiments will gradually reduce the interval of uncertainty. From Eq. 11.39;

$$\lim_{\substack{\epsilon \to 0 \\ N \to \infty}} \{L_N\} = \lim_{\substack{\epsilon \to 0 \\ N \to \infty}} \left\{ \frac{L_0}{F_N} + \frac{\epsilon F_{N-2}}{F_N} \right\} = 0 \tag{11.59}$$

This simply says that the final interval of uncertainty will converge to zero as the number of functional evaluations increase to infinity, provided that ϵ is allowed to be infinitely small.

From Eq. 11.40;

$$\lim_{\substack{N \to \infty \\ \epsilon \to 0}} \{L_2\} = \lim_{\substack{N \to \infty \\ \epsilon \to 0}} \left\{ \frac{1}{F_N}(L_0 F_{N-1} + \epsilon(-1)^N \right\} = L_0 \left[\frac{F_{N-1}}{F_N} \right] \tag{11.60}$$

One can show that in the limit the ratio of F_{N-1}/F_N goes to 0.618. The reader can empirically verify this by extending Table 11.1 and calculating the ratios.

This is known as the *golden ratio* or *golden section*, and has been used quite extensively in architectural applications. If we apply these results, it is immediately obvious that $L_2 = 0.618\ L_0 = 0.618\ L_1$. this is all we need to start a modified Fibonacci search. This modified version is known as *golden section search* [Wilde (30)]. Termination criteria can be based on a number of physical or mathematical considerations [Beveridge and Schecter (4) and Converse (5)]. For most practical purposes, the search can terminate when (1) the functional evaluations X_n and X_{n-1} become arbitrarily close, or (2) changes in the objective function, $f(X)$, become negligible. In comparison to the Fibonacci search procedure the golden section search is less efficient. This is quite logical, since the golden section search was derived from the Fibonacci search. The primary difference lies in the goal of minimizing the maximum interval of uncertainty as the search proceeds. Since Eq. 11.60 is not dependent on resolution considerations nor on the number of functional evaluations, the minimax principle is lost for early searches. However, part of this efficiency is regained as the number of search points increase. In practice, the golden section search is often used because it requires less information to implement each search, and is by construction self-starting.

The general procedure is similar to the Fibonacci search and proceeds as follows:

Step I. Define the initial interval of uncertainty as $L_0 = B - A$, where B is the upper bound of the search and A is the lower bound.

Step II. Determine the first two functional evaluations at points X_1 and X_2 defined by

$$X_1 = A + 0.618(B - A)$$
$$X_2 = B - 0.618(B - A)$$

Step III. Eliminate the appropriate region in which the optimum cannot lie.

Step IV. Determine the region of uncertainty defined by

$$L_{j+1} = L_{j-1} - L_j \qquad j = 2, 3, \ldots$$

Where

$$L_0 = B - A$$
$$L_1 = B - A$$
$$L_2 = X_1 - A$$

or:

$$L_2 = B - X_2$$

depending on the region eliminated at Step III.

Step V. Establish a new functional evaluation using the result of Step IV; Evaluate $f(x)$ at this point, and then go to Step III. Repeat this procedure until a specified convergence criteria is satisfied.

EXAMPLE 11.3-2

Minimize: $f(x) = x^4 - 15x^3 + 72x^2 - 1135x$

Terminate the search when

$$|f(X_n) - f(X_{n-1})| \leqq 0.50$$

The initial range of X is $1 \leqq X \leqq 15$.

Solution

The first two points are placed symmetrically within the interval $1 \leqq X \leqq 15$. The golden section ratio places these points at

$$X_1 = 1 + 0.618(15 - 1) = 9.652$$

and

$$X_2 = 15 - 0.618(15 - 1) = 6.348$$

Hence,

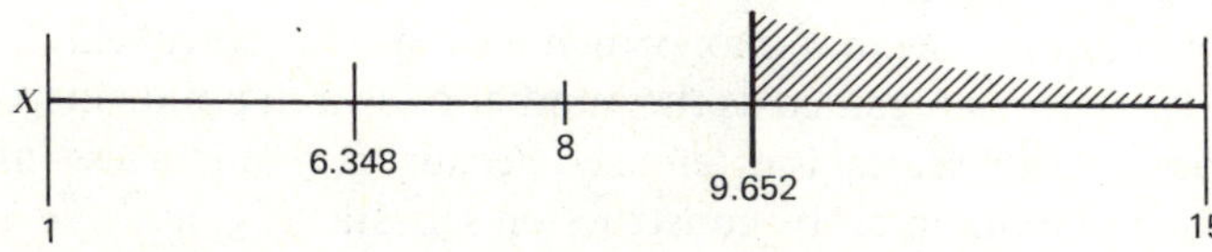

$$f(X_1) = 595.70$$
$$f(X_2) = -168.82$$

Therefore, the region to the right of $X=9.652$ can be eliminated, and the interval of uncertainty after two functional evaluations is given by

$$9.652 \geqq X \geqq 1$$

From this point on, the search procedure is exactly the same as a Fibonacci search. Table 11.3 shows the progression of the golden section search through 10 iterations.

Table 11.3
GOLDEN SECTION SEARCH

Functional Evaluations (n)	X_{n-1} (right)	$f(X_{n-1})$	X_n (left)	$f(X_n)$	Interval of Uncertainty	Length
2	9.652	595.70	6.346	−168.80	$9.652 \geqq X \geqq 1$	8.652
3	6.346	−168.80	4.304	−100.06	$9.652 \geqq X \geqq 4.304$	5.348
4	7.609	−114.64	6.346	−168.80	$7.609 \geqq X \geqq 4.304$	3.305
5	6.346	−168.80	5.566	−147.61	$7.609 \geqq X \geqq 5.566$	2.043
6	6.828	−166.42	6.346	−168.80	$6.828 \geqq X \geqq 5.566$	1.262
7	6.346	−168.80	6.048	−163.25	$6.828 \geqq X \geqq 6.048$	0.780
8	6.530	−169.83	6.346	−168.80	$6.828 \geqq X \geqq 6.346$	0.482
9	6.643	−169.34	6.530	−169.83	$6.643 \geqq X \geqq 6.346$	0.297

At iteration number 9, note that

$$f(X_9) = -169.34 \quad \text{and} \quad f(X_8) = -169.83$$

Hence,

$$|f(X_9) - f(X_8)| = 0.49$$

Since termination criteria are satisfied, the golden section search will stop at this point. The best answer is given by

$$X^* = 6.643$$
$$f(X^*) = -169.34$$

11.4 THE HOOKE AND JEEVES SEARCH ALGORITHM

The Fibonacci and golden section search algorithms are very effective in dealing with univariate nonlinear functions that are assumed to be unimodal. In practice, these searches can often be utilized over any bounded single variable search, but if the function is not unimodal, then global optimization is not guaranteed. In general, univariate search can be used in multivariable optimization through successive perturbations of each decision variable. The procedure for an N-variable optimization problem would be to fix $(N-1)$ variables at a chosen value, and to search over the Nth decision variable until a maximizing (minimizing) solution is found with respect to that one variable. The procedure is then repeated by choosing one of the original fixed $(N-1)$ variables as a decision variable and finding a new optimal solution. The procedure is repeated until no change in any one variable will bring about an improvement in the current value of the objective function. This approach is called *sectioning*, and is perhaps the simplest of the multivariable optimization techniques. The sectioning approach works very well provided there is no strong interaction between the decision variables. Once strong interactions enter the optimization problem, there are long ridges and steep valleys often formed on the response surface, and the procedure tends to oscillate wildly as the one-dimensional searches are performed. Conceptually visualize a mountain climber trying to negotiate a steep ridge running from southwest to northeast who is restricted to small movements in the north-south directions only. Such would be the effect of search along coordinate axes over a function that runs diagonal to those axes. It is obvious that a better approach would be one that retains the simplicity of a coordinate axis search, but that provides an opportunity to change distance, direction, or a combination of both. The method of Hooke and Jeeves accomplishes those objectives. Without loss of generality, the method of Hooke and Jeeves will now be explained with reference to function minimization.

The Hooke-Jeeves algorithm consists of two distinct phases. The first is an *exploratory search* phase that serves to establish a direction of improvement, and a second is a *pattern move* that extracts the current solution vector to another point in the solution space. Using function minimization for illustrative purposes, the algorithm proceeds as follows: First, an initial solution vector is chosen $\mathbf{x}^{(0)} = (x_1^{(0)}, x_2^{(0)}, \ldots, x_n^{(0)})$. The initial value of the objective function is given by $f(\mathbf{x}^{(0)})$. Label this point *Set 1*. An initial *exploratory search* is now conducted about this point to find a direction of objective function improvement. Define a *perturbation* vector $\mathbf{P} = (\Delta x_1, \Delta x_2, \ldots, \Delta x_n)$ that will be used to systematically change the current solution vector. Choosing each variable in turn, an objective function evaluation is made at $x_k^{(0)} \pm \Delta x_k$; $k = 1, 2, \ldots, n$. In particular, suppose that $f(\mathbf{x})$ is evaluated at $\mathbf{x}_1^{(0)} = (x_1^{(0)} + \Delta x_1, x_2^{(0)}, \ldots, x_n^{(0)})$. If an improvement is found in $f(\mathbf{x})$ at $f(\mathbf{x}_1^{(0)})$ namely $f(\mathbf{x}_1^{(0)}) < f(\mathbf{x}^{(0)})$, then the current value of the objective function is updated to $f(\mathbf{x}_1^{(0)})$. If this move fails to improve the objective function, then the vector $\mathbf{x}_1^{(0)} = (x_1^{(0)} - \Delta x_1, x_2^{(0)}, \ldots, x_n^{(0)})$ is tried. This procedure is followed for each decision variable in turn, until the last decision variable has been changed. The final solution vector is accepted as a point in space which indicates a direction of objective function improvement. Call this point $\mathbf{x}^{(1)}$ and label it as *Base 1*. The *pattern move* phase is now implemented and consists of moving from $\mathbf{x}^{(0)}$ through $\mathbf{x}^{(1)}$ to a new point $\mathbf{x}^{(2)}$ defined by

$$\mathbf{x}^{(2)} = \mathbf{x}^{(0)} + 2(\mathbf{x}^{(1)} - \mathbf{x}^{(0)})$$

or

$$\mathbf{x}^{(2)} = 2\mathbf{x}^{(1)} - \mathbf{x}^{(0)} \tag{11.61}$$

Call this point *Base 2*.

The point $\mathbf{x}^{(2)}$ is not immediately accepted. Before a decision is made to change the current accepted solution to *Base 2*, another *exploratory search* is conducted about *Base 2*. Performing this search as was done previously, a new point $\mathbf{x}^{(3)}$ will be established. At this time, a comparison is made between $f(\mathbf{x}^{(3)})$ and the *Base 1* solution vector. If $f(\mathbf{x}^{(3)}) < f(\mathbf{x}^{(1)})$, then $\mathbf{x}^{(3)}$ is accepted as the new solution and labeled *Base 1*. The point from which additional moves will now be made is updated to $\mathbf{x}^{(1)}$. Hence, $\mathbf{x}^{(1)}$ is now labeled *Set 1*. We are now ready to make another pattern move from point $\mathbf{x}^{(1)}$ (Set 1) through point $\mathbf{x}^{(3)}$ (Base 1) to a point $\mathbf{x}^{(4)}$ (Base 2). Exploratory searches will now be conducted about *Base 2* to determine if the pattern move was a success. This sequence of moves is repeated until an exploratory search about the point *Base 2* fails to yield objective function improvement. If this occurs, the pattern search is said to be a *failure*. When this occurs, the solution vector at *Base 1* is returned to the original status of *Set 1*, and the procedure begins anew around the point *Set 1* as if it were the initial solution vector. If an exploratory search about *Set 1* fails to yield an improved solution vector, then the change vector $\mathbf{P} = (\Delta x_1, \Delta x_2, \ldots, \Delta x_n)$ should be reduced to $\mathbf{P} = (\Delta x_1/2, \Delta x_2/2, \ldots, \Delta x_n/2)$ and another *exploratory search* conducted. When every component of $\mathbf{P}$ becomes less than a predetermined increment, the process terminates and *Set 1* is acepted as the optimal solution. Himmelblau (13) has suggested refined rules for termination, but our simple procedure will be used for illustrative purposes.

In general, after the *Initial exploratory search*, a point $\mathbf{x}^{(k)}$ is labeled *Set 1*. A point $\mathbf{x}^{(k+2)}$ is labeled *Base 1*. A projection is made from $\mathbf{x}^{(k)}$ through $\mathbf{x}^{(k+2)}$ to a point $\mathbf{x}^{(k+3)}$, labeled *Base 2*. If an exploratory search about point $\mathbf{x}^{(k+3)}$ is successful, then point $\mathbf{x}^{(k+1)}$ is accepted as *Set 1*, point $\mathbf{x}^{(k+3)}$ is accepted as *Base 1*, and the process repeated. If an exploratory search about *Base 2* results in failure, then *Base 1* is treated as if it were the initial solution vector, relabeled *Set 1*, and the entire procedure started anew.

EXAMPLE 11.4-1

Minimize: $f(\mathbf{x}) = 3x_1^2 + x_2^2 - 12x_1 - 8x_2$

Suppose that we start from the point $\mathbf{x}^0 = (1, 1)$ with an initial change vector $\mathbf{P} = (0.50, 0.50)$. The value of $f(\mathbf{x})$ at the point $\mathbf{x}^0$ is given by $f(\mathbf{x}^0) = -10.0$.

Exploratory search from $\mathbf{x}^0 = (1, 1)$; Set $1 = (1, 1)$

$$x_1^{(1)} = 1 + 0.50 = 1.50 \qquad f(1.5, 1) = -18.25 \text{ (success)}$$
$$x_2^{(1)} = 1 + 0.50 = 1.50 \qquad f(1.5, 1.5) = -21.0 \text{ (success)}$$

The exploratory search is successful; hence,

$$\mathbf{x}^{(1)} = (1.5, 1.5); \qquad f(\mathbf{x}^{(1)}) = -21.0$$
$$\text{Set } 1 = (1, 1) = \mathbf{x}^{(0)}$$
$$\text{Base } 1 = (1.5, 1.5) = \mathbf{x}^{(1)}$$

A *pattern move* is now employed.

$$\mathbf{x}^{(2)} = 2\mathbf{x}^{(1)} - \mathbf{x}^{(0)} = \text{base } 1 - \text{set } 1$$

Hence,

$$\mathbf{x}^{(2)} = (2.0, 2.0) \qquad f(\mathbf{x}^{(2)}) = -24.0$$
$$\text{Base } 2 = (2.0, 2.0)$$

The success or failure of the pattern move is now determined through a second exploratory search about Base 2. Functional comparisons will be based on $f(\mathbf{x}^{(2)}) = -24.0$.

Exploratory search from $\mathbf{x}^{(2)} = (2.0, 2.0)$

$$x_1^{(3)} = 2.0 + 0.50 = 2.5 \qquad f(2.5, 2) = -23.25 \text{ (failure)}$$
$$x_1^{(2)} = 2.0 - 0.50 = 1.5 \qquad f(1.5, 2) = -23.25 \text{ (failure)}$$
$$x_2^{(3)} = 2.0 + 0.50 = 2.5 \qquad f(2.0, 2.5) = -25.75 \text{ (success)}$$

Hence,

$$\mathbf{x}^{(3)} = (2.0, 2.5) \quad \text{and} \quad f(\mathbf{x}^{(3)}) = -25.75$$

At this point, Base 1 is accepted as the best possible solution to date. An algorithmic labeling update now occurs before proceeding.

$$\text{Set } 1 = (1.5, 1.5) = \mathbf{x}^{(1)}$$
$$\text{Base } 1 = (2.0, 2.5) = \mathbf{x}^{(3)}$$

A *pattern move* is now employed along the direction of function minimization. Since this direction is actually along the same line of improvement as before, the pattern move results in an *acceleration* in that direction.

$$\mathbf{x}^{(4)} = 2\mathbf{x}^{(3)} - \mathbf{x}^{(1)} = \text{Base } 1 - \text{Set } 1$$

Hence,

$$\mathbf{x}^{(4)} = (2.5, 3.5) \qquad f(\mathbf{x}^{(4)}) = -27$$

Again $f(\mathbf{x}^{(4)})$ is not immediately accepted or rejected, but an *exploratory search* is conducted about $\mathbf{x}^{(4)}$.

$$x_1^{(5)} = 2.5 + 0.50 = 3.0 \qquad f(3.0, 3.5) = -24.75 \text{ (failure)}$$
$$x_1^{(5)} = 2.5 - 2.0 \;= 2.0 \qquad f(2.0, 3.5) = -27.75 \text{ (success)}$$
$$x_2^{(5)} = 3.5 + 0.50 = 4.0 \qquad f(2.0, 4.0) = -28.0 \text{ (success)}$$

At this point, Base 1 = (2.0, 2.5) is accepted as the best solution to date, and the following labels updated:

$$\text{Set } 1 = \mathbf{x}^{(3)} = (2.0, 2.5)$$
$$\text{Base } 1 = \mathbf{x}^{(5)} = (2.0, 4.0)$$

A *pattern move* now takes place from Set 1 through Base 1.

$$\mathbf{x}^{(6)} = 2\mathbf{x}^{(5)} - \mathbf{x}^{(3)} = \text{Base } 1 - \text{Set } 1$$

Hence,

$$\mathbf{x}^{(6)} = (2.0, 5.5) \quad \text{and} \quad f(\mathbf{x}^{(6)}) = -25.75$$

An exploratory search is compared to $f(\mathbf{x}^{(5)}) = -28.0$.

$$x_1^{(7)} = 2.0 + 0.50 = 2.5 \qquad f(2.5, 5.5) = -25 \text{ (failure)}$$
$$x_1^{(7)} = 2.0 - 0.50 = 1.5 \qquad f(1.5, 5.5) = -25 \text{ (failure)}$$
$$x_2^{(7)} = 5.5 + 0.5 = 6.0 \qquad f(2.0, 6.0) = -24 \text{ (failure)}$$
$$x_2^{(7)} = 5.5 - 0.5 = 5.0 \qquad f(2.0, 5.0) = -27 \text{ (failure)}$$

At this point, the pattern search is deemed a failure. The algorithm now returns to point $\mathbf{x}^{(5)}$ and is begun anew as if $\mathbf{x}^{(5)}$ were $\mathbf{x}^{(0)}$ initially. If a pattern search about the point $\mathbf{x}^{(5)}$ is successful, then a pattern move will be made in the direction of improvement. If the pattern search about point $\mathbf{x}^{(5)}$ fails, then the step size vector $\mathbf{P} = (\Delta x_1, \Delta x_2, \ldots, \Delta x_n)$ is changed to $\mathbf{P} = (\Delta x_1/2, \Delta x_2/2, \ldots, \Delta x_n/2)$, and the process begins again. In this case, the procedure will terminate at point $\mathbf{x}^{(5)}$, since the (global) minimizing solution is $\mathbf{x}^* = (2.0, 4.0)$. The sequence of moves illustrating Example 11.4-1 are given in Fig. 11.22.

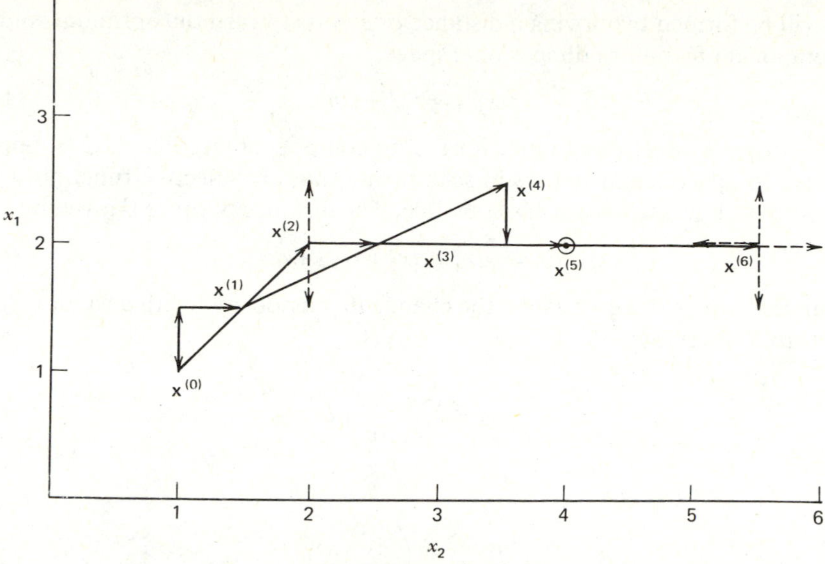

Figure 11.22
A Hooke and Jeeves Search.

11.5 GRADIENT PROJECTION

Undoubtedly the most widely used component in nonlinear programming is the *gradient* of a function. To characterize and define the nature of a functional gradient, consider the point $\mathbf{x}_j$ in Figure 11.23.

Suppose that we are at the point $\mathbf{x}_j$ and we seek to find the (unknown) maximum value of $f(\mathbf{x})$, which occurs at point A. Suppose further that the only properties which we can utilize in our search for the optimum are local values dependent on the coordinates of the current solution vector. Utilizing only this local information, suppose we wish to proceed from the current point $\mathbf{x}_j$ to a new point $\mathbf{x}_{j+1}$ in such a way that we approach the optimum at the fastest possible rate. In particular, suppose that we wish to move a distance s from a point $\mathbf{x}_j$ to a new point $\mathbf{x}_{j+1}$.

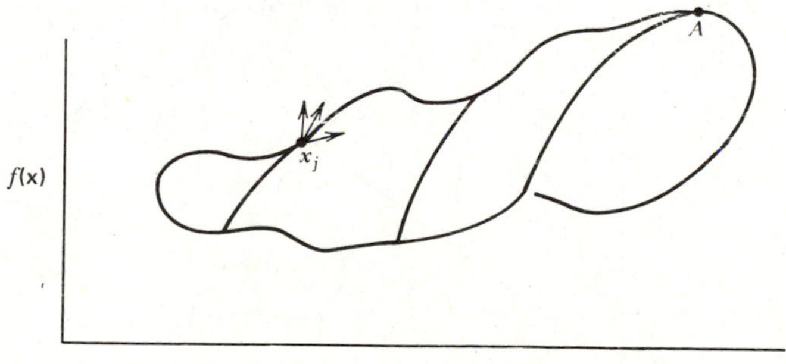

Figure 11.23
A multidimensional nonlinear function.

$\mathbf{x}_{j+1}$ will be formed by moving a distance of s units toward the optimum solution. In component form for i dimensional space;

$$x_{j+1}^{(i)} = x_j^{(i)} + sm_i \tag{11.62}$$

where m_i is the direction of move for the ith component (see Fig. 11.24). Suppose we wish to take a small step ds in such a way that an objective function $y = f(\mathbf{x})$ increases or decreases as much as possible. The distance of move is given by

$$ds = \sqrt{dx_1^2 + dx_2^2 + \cdots + dx_n^2} \tag{11.63}$$

Assuming y to be differentiable, the change in y associated with a set of displacements dx_i is given by:

$$dy = \sum_{i=1}^{n} \left(\frac{\partial y}{\partial x_i} \right) dx_i \tag{11.64}$$

or

$$\frac{dy}{ds} = \sum_{i=1}^{n} \left(\frac{\partial y}{\partial x_i} \right) \frac{dx_i}{ds} \tag{11.65}$$

Now a particular set of displacements will make dy/ds as large or small as possible. This is the direction of steepest ascent or descent. Viewed as an optimization problem, we wish to maximize or minimize Eq. 11.65 subject to Eq. 11.63.

Maximize/Minimize:
$$\frac{dy}{ds} = \sum_{i=1}^{n} \left(\frac{\partial y}{\partial x_i} \right) \frac{dx_i}{ds} \tag{11.65}$$

Subject to:
$$ds = \sqrt{\sum_{i=1}^{n} dx_i^2}$$

Hence, forming the Lagrangian function (see Section 11.6)

Maximize/Minimize:

$$\sum_{i=1}^{n} \left(\frac{\partial y}{\partial x_i} \right) \frac{dx_i}{ds} - \lambda \left[1 - \sum_{i=1}^{n} \left(\frac{dx_i}{dx} \right)^2 \right]$$

Differentiating with respect to dx_i/ds,

$$\frac{\partial y}{\partial x_i} - 2\lambda \left(\frac{dx_i}{ds} \right) = 0 \qquad i = 1, 2, \ldots, n \tag{11.66}$$

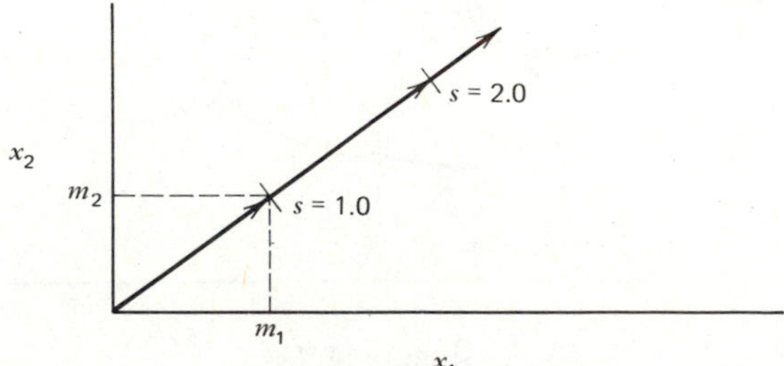

Figure 11.24
A two-dimensional move.

and with respect to the Lagrangian multiplier, λ,

$$\sum_{i=1}^{n}\left(\frac{dx_i}{ds}\right)^2=1$$

Therefore

$$\frac{1}{4\lambda^2}\sum_{i=1}^{n}\left(\frac{\partial y}{\partial x_i}\right)^2=1$$

or

$$2\lambda=\pm\sqrt{\sum_{i=1}^{n}\left(\frac{\partial y}{\partial x_i}\right)^2}$$

In parametric form, this says for the ith component,

$$x_{j+1}^{(i)}=x_j^{(i)}+\left[\frac{\partial y}{\partial x_i}\cdot\frac{1}{2\lambda}\right]s=x_j^{(i)}+m_i s$$

Hence, using Eq. 11.66, a move giving the greatest *increase* in y is given by

$$m_i=\frac{\dfrac{\partial y}{\partial x_i}}{\sqrt{\sum_{i=1}^{n}\left(\dfrac{\partial y}{\partial x_i}\right)^2}}\qquad i=1,2,\ldots,n$$

and that giving the maximum *decrease* is

$$m_i=\frac{-\dfrac{\partial y}{\partial x_i}}{\sqrt{\sum_{i=1}^{n}\left(\dfrac{\partial y}{\partial x_i}\right)^2}}\qquad i=1,2,\ldots,n$$

Note that the numerator is the gradient, while the denominator is a normalizing factor.

EXAMPLE 11.5-1

A problem of steepest descent. Let the function be $y(\mathbf{x})=(4-x_1)^2+x_2^2$ and our objective is to find the minimum point given an initial solution of (0, 0).

That is,

$$\mathbf{x}^{(1)}=(0,0)$$
$$y(\mathbf{x})=(4-x_1)^2+x_2^2$$

The direction of steepest descent for the ith component is given by

$$m_i=\frac{-\dfrac{\partial y_i}{\partial x_i}}{\sqrt{\sum_{j=1}^{n}\left(\dfrac{\partial y}{\partial x_i}\right)^2}}$$

For the given problem

$$\frac{\partial y}{\partial x_1} = -8 + 2x_1 \qquad \frac{\partial y}{\partial x_2} = 2x_2$$

$$\left.\frac{\partial y}{\partial x_1}\right|_{(0,0)} = -8 \qquad \left.\frac{\partial y}{\partial x_2}\right|_{(0,0)} = 0$$

$$m_1 = \frac{8}{\sqrt{64}} = 1 \qquad m_2 = \frac{0}{\sqrt{64}} = 0$$

$$\therefore \quad x_1^{(2)} = 0 + s(1) = s \qquad x_2^{(2)} = 0 + s(0) = 0$$

$\therefore \quad y = (4 - s)^2$ is to be minimized with respect to s

$$\frac{dy}{ds} = -8 + 2s = 0; \qquad \Rightarrow s = 4$$

$$\therefore \quad x_1^{(2)} = 4 \qquad x_2^{(2)} = 0$$

$$\mathbf{x}^{(2)} = (4, 0)$$

$y \mid_{\mathbf{x}^{(2)}} = 0$. Hence, $\mathbf{x}^{(2)}$ is a stationary point

Evaluating the Hessian matrix proves $\mathbf{x}^{(2)}$ to be the minimum point. The point (4, 0) is the nucleus of a family of circles. Note we were able to obtain the optimal solution in one step. This is because of the special nature of the objective function (see Fig. 11.25).

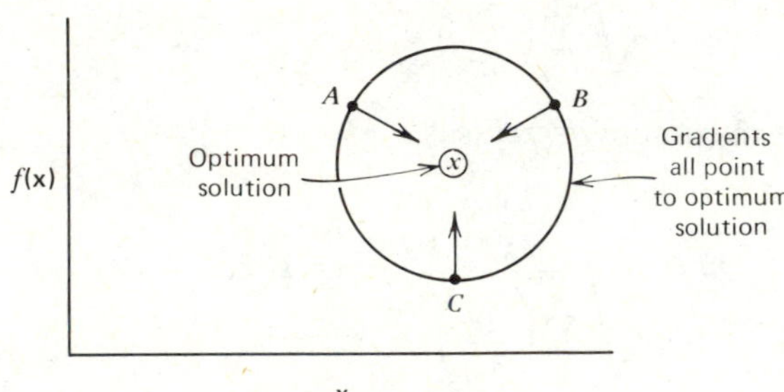

Figure 11.25
Gradient projection with circular contours.

EXAMPLE 11.5-2
Elliptical contours

Minimize: $\qquad y = 3x_1^2 + 4x_2^2$

The direction of steepest descent is given by

$$m_i = \frac{-\dfrac{\partial y}{\partial x_1}}{\sqrt{\displaystyle\sum_{i=1}^{n}\left(\frac{\partial y}{\partial x_i}\right)^2}}$$

Let us start at the point (1, 1).

It takes three iterations to reach the optimal solution, and the appropriate values are given in Table 11.4.

Table 11.4
SOLUTION OF EXAMPLE 11.5-2

Iteration Number	m_1	m_2	Optimal s	x_1	x_2
0	—	—	—	1	1
1	−0.6	−0.8	1.3736	0.1458	−0.0989
2	−0.8	0.6	0.2	0.0171	−0.0211
3	−0.518	0.8525	0.199	0	0

EXAMPLE 11.5-3

Consider the following problem

$$\text{Minimize:} \qquad y(\mathbf{x}) = 2(x_1 + x_2)^2 + 2(x_1^2 + x_2^2)$$

Pass 1. Let us start at the point $\mathbf{x}_0^{(1)} = \{5, 2\}$. Hence,

$$y_1(\mathbf{x}) = 156.0$$

$$\frac{\partial y}{\partial x_1} = 4(x_1 + x_2) + 4x_1 \qquad \frac{\partial y}{\partial x_2} = 4(x_1 + x_2) + 4x_2$$

Evaluated at $\mathbf{x}_0^{(1)}$ we obtain

$$\left(\frac{\partial y}{\partial x_1}\right)\Bigg|_{x_0^{(1)}} = 48 \qquad \left(\frac{\partial y}{\partial x_2}\right)\Bigg|_{x_0^{(1)}} = 36$$

The direction of steepest descent is given by

$$m_1^{(1)} = \frac{-48}{\sqrt{48^2 + 36^2}} = \frac{-48}{60} = -0.8$$

$$m_2^{(1)} = \frac{-36}{\sqrt{48^2 + 36^2}} = \frac{-36}{60} = -0.6$$

Then, the position of a new point closer to the optimal value of $y(\mathbf{x})$ is

$$x_1^{(2)} = x_1^{(1)} + sm_1^{(1)} = 5 - 0.8s$$
$$x_2^{(2)} = x_2^{(1)} + sm_2^{(1)} = 2 - 0.6s$$

The optimal value of s is that which will minimize the value of $y(\mathbf{x})$ in the direction of steepest descent. Namely the value which minimizes

$$\bar{y}(\mathbf{x}) = 2[7 - 1.4s]^2 + 2[(5 - 0.8s)^2 + (2 - 0.6s)^2]$$

Using any one-dimensional search technique or a simple first derivative, we find $s = 5.07$ yielding $\bar{y}(\mathbf{x}) = 3.97$. Thus the new base point is given by

$$x_1^{(2)} = 5 - 0.8(5.07) = 0.944$$
$$x_2^{(2)} = 2 - 0.6(5.07) = -1.042$$

Hence, $y_2(\mathbf{x}) = 3.977$ (note the significant decrease in the objective function).

Pass 2. Evaluating the partial derivatives at the new base point:

$$\left(\frac{\partial y}{\partial x_1}\right) = 3.384 \qquad \left(\frac{\partial y}{\partial x_2}\right) = -4.56$$

The direction of steepest descent:

$$m_1^{(2)} = \frac{-3.384}{\sqrt{(3.384)^2 + (-4.56)^2}} = \frac{-3.384}{5.678} = -0.5959$$

$$m_2^{(2)} = \frac{4.56}{\sqrt{(3.384)^2 + (-4.56)^2}} = \frac{4.56}{5.678} = 0.8030$$

Then the position of a new point closer to the optimum value of $y(\mathbf{x})$ is

$$x_1^{(3)} = x_1^{(2)} + sm_1^{(2)} = 0.944 - 0.5959s$$

$$x_2^{(3)} = x_2^{(2)} + sm_2^{(2)} = -1.042 + 0.8030s$$

Again s is defined by minimizing

$$y(\mathbf{x}) = 2[-0.098 + 0.2071s]^2 + 2[(0.944 - 0.5959s)^2 + (-1.042 + 0.8030s)^2]$$

which, again using the Fibonacci search, yields $s = 1.36$ from which $\bar{y}(\mathbf{x}) = 0.11$

$$\begin{aligned} x_1^{(3)} &= 0.944 - 0.5959(1.36) = 0.1336 \\ x_2^{(3)} &= -1.042 + 0.8030(1.36) = 0.0501 \end{aligned} \Rightarrow y_3(\mathbf{x}) = 0.11$$

Pass 3. Evaluating the partial derivatives at the new base point:

$$\left(\frac{\partial y}{\partial x_1}\right) = 1.2692 \qquad \left(\frac{\partial y}{\partial x_2}\right) = 0.9352$$

The direction of steepest descent is given by

$$m_1^{(3)} = \frac{-1.2692}{\sqrt{(1.2962)^2 + (0.9352)^2}} = \frac{-1.2692}{1.5765} = -0.8051$$

$$m_2^{(3)} = \frac{0.9352}{1.5765} = -0.5932$$

$$x_1^{(4)} = x_1^{(3)} + sm_1^{(3)} = 0.1336 - 0.8051s = 0.0289$$

$$x_2^{(4)} = x_2^{(3)} + sm_2^{(3)} = 0.0501 - 0.5932s = -0.027$$

$$y(\mathbf{x}) = 2[0.1837 - 1.3983s]^2 + 2[(0.1336 - 0.8051s)^2 + (0.0501 - 0.5932s)^2]$$

Again, using a one-dimensional search, we get $s = 0.13 \Rightarrow \bar{y}(\mathbf{x}) = 0.00$ to two decimal places. The procedure will terminate at this solution vector. The (accepted) optimal solution is $\mathbf{x}^* = (0.03, -0.03)$. The true optimal solution is $\mathbf{x}^* = (0.0, 0.0)$. This can easily be verified through further iterations.

Scaling and Oscillation

The difficulties encountered in solving the last problem arose because of the nonlinearities and elongated contours of the objective function. Strong interactions between solution variables also create great difficulties by forming long ridges or valleys that the gradient is not suited to negotiate. The problems associated with small oscillating movements through a long narrow valley or similar geometric configurations can sometimes be overcome by an appropriate scaling of the objective function. Consider the following problem.

$$\text{Minimize:} \qquad y(\mathbf{x}) = 100(x_2 - x_1^2)^2 + (1 - x_1)^2$$

This particular problem is called "Rosenbrock's function," and it has been constructed to challenge the gradient procedure. The student will be asked to solve the problem in Exercise 11.21 to appreciate its difficulty. However, consider the following transformation:

Let $$z_1 = 10(x_2 - x_1^2); \qquad z_2 = 1 - x_1$$

Hence, we now wish to solve

Minimize: $$y(\mathbf{z}) = z_1^2 + z_2^2$$

$$\left(\frac{\partial y}{\partial z_1}\right) = 2z_1 \qquad \left(\frac{\partial y}{\partial z_2}\right) = 2z_2$$

If $\mathbf{x}_0 = \{2, 2\}$; then $\mathbf{z}_0 = \{-20, -1\}$. Hence,

$$\left.\left(\frac{\partial y}{\partial z_1}\right)\right|_{z_0} = -40$$

$$\left.\left(\frac{\partial y}{\partial z_2}\right)\right|_{z_0} = -2$$

$$m_1^{(1)} = \frac{40}{\sqrt{(-40)^2 + (-2)^2}} \qquad m_2 = \frac{+2}{\sqrt{(-40)^2 + (-2)^2}}$$

$$\Rightarrow m_1^{(1)} \simeq 1.0 \qquad m_2^{(1)} \simeq 0.050$$

Therefore,

$$z_1^{(2)} = z_1^{(1)} + sm_1^{(1)}$$
$$z_2^{(2)} = z_2^{(1)} + sm_1^{(1)}$$
$$\Rightarrow z_1^{(2)} = -20 + s$$
$$z_2^{(2)} = -1 + 0.05s$$

The optimal s is given by

Minimize: $$y = (-20 + s)^2 + (0.050s - 1)^2$$

which is minimized by $s = 20$. Therefore,

$$z_1^{(2)} = -20 + 20 = 0$$
$$z_2^{(2)} = -1 + 1 = 0$$

At this point we note that the gradient will be zero, and therefore we have reached an optimal solution in only one step; $z_1^* = 0$; $z_2^* = 0$. Translated into our original problem, this yields $x_1^* = x_2^* = 1.0$, which is a global minimum to Rosenbrock's function.

A Gradient Based Method with Second Derivatives

To place the discussion that follows in its proper perspective, recall that the optimization methods of this section use a procedure by which the ith point is updated to the $(i+1)$st point for all n variables according to the following relation:

$$x_{j+1}^{(i)} = x_j^{(i)} + m_i s \qquad j = 1, 2, \ldots, n \tag{11.67}$$

The direction of search is the gradient of $f(\mathbf{x})$.

$$m_i = \nabla f(\mathbf{x}) = \left[\frac{\partial f(\mathbf{x})}{\partial x_j}\right] \qquad j = 1, 2, \ldots, n$$

By convention, we actually use the *normalized gradient*, which is simply a linear transformation of $\nabla f(\mathbf{x})$.

$$m_i = \frac{\dfrac{\partial f}{\partial x_j}}{\sqrt{\displaystyle\sum_{j=1}^{n}\left(\frac{\partial f}{\partial x_j}\right)^2}} \qquad j=1, 2, \ldots, n$$

In the discussion that follows, we simply refer to the gradient of $f(\mathbf{x})$ as $\nabla f(\mathbf{x})$, and computationally we will use the unscaled form of the above equation.

We have observed that the gradient is the "best local" direction of minimization (maximization) of a function $f(\mathbf{x})$ at a point $\mathbf{x}=\bar{\mathbf{x}}$. The negative gradient points in the direction of local maximum decrease, and actually points directly to the global minimum of a convex circular function (Example 11.5-1). However, since the gradient represents a linear function, it is usually *not* the *best* direction for general nonlinear functions. Since a gradient search strategy actually uses successive linear approximations, it is logical to consider the use of higher order derivatives to obtain more accurate information.

Consider once again the Taylor's series expansion of a nonlinear objective function $f(\mathbf{x})$. Recall from Eq. 11.18 that the value of a function at some point $\mathbf{x}$ can be approximated from the function and its value at the point $\mathbf{x}_0$.

$$f(\mathbf{x}) = f(\mathbf{x}_0) + \left(\frac{df}{dx}\right)_{\mathbf{x}_0} \Delta\mathbf{x} + \tfrac{1}{2}\Delta\mathbf{x}\left(\frac{d^2f}{dx_2}\right)_{\mathbf{x}_0} \Delta\mathbf{x} + - +$$

The term $(df/dx)\mathbf{x}_0$ is the gradient of $f(\mathbf{x})$ at the point $\mathbf{x}=\mathbf{x}_0$: $\nabla f(\mathbf{x}_0)$. The term $(d^2f/dx^2)_{\mathbf{x}_0}$ is the *hessian* of $f(\mathbf{x})$ at the point $\mathbf{x}_0$: $H(\mathbf{x}) = \nabla^2 f(\mathbf{x}_0)$. Hence, if we drop all terms of order 3 and higher, we obtain the following *quadratic approximation*:

$$\hat{f}(\mathbf{x}) = f(\mathbf{x}_0) + \nabla f(\mathbf{x}_0)\,\Delta\mathbf{x} + \tfrac{1}{2}\Delta\mathbf{x}\nabla^2 f(\mathbf{x}_0)\,\Delta\mathbf{x}$$

Note that $\hat{f}(\mathbf{x})$ is an *approximating function* to $f(\mathbf{x})$, since the terms involving powers of $\mathbf{x}$ three or higher are not present. Now let us assume that we wish to use this approximating function to project an optimal value of $f(\mathbf{x})$. In other words, let us determine a search direction m_i in Eq. 11.67. A logical procedure would be to simply proceed as before, and use the *gradient* of the approximating function $\hat{f}(\mathbf{x})$. The next projected point will be where the gradient of $\hat{f}(\mathbf{x})$ is zero.

Therefore

$$\nabla\hat{f}\,(\mathbf{x}) = \nabla f\,(\mathbf{x}_0) + \nabla^2 f(\mathbf{x}_0)\,\Delta\mathbf{x} \equiv 0$$

or

$$\Delta\mathbf{x} = -[\nabla^2 f(\mathbf{x}_0)]^{-1}\,\nabla f(\mathbf{x}_0)$$

This scheme gives rise to a higher order procedure that iteratively projects to an optimum solution by using the following scheme:

$$x_{j+1}^{(i)} = x_j^{(i)} - S[\nabla^2 f(\mathbf{x}_0)]^{-1}\,\nabla f(\mathbf{x}_0)$$

This "prescription" for optimization is commonly known as *Newton's method*. We demonstrate this procedure through the following example.

EXAMPLE 11.5-4 NEWTON'S METHOD

Consider once again the function of Example 11.5-3.

Minimize: $\quad f(\mathbf{x}) = 2(x_1+x_2)^2 + 2(x_1^2+x_2^2)$

The *gradient* of $f(\mathbf{x})$ is as follows:

$$\nabla f(\mathbf{x}) = \left[\frac{\partial f}{\partial x_1}, \frac{\partial f}{\partial x_2}\right]$$

Therefore,

$$\nabla f(\mathbf{x}) = [4(x_1 + x_2) + 4x_1, 4(x_1 + x_2) + 4x_2]$$

The *hessian* of $f(\mathbf{x})$ is as follows:

$$H(\mathbf{x}) = \nabla^2 f(\mathbf{x}) = \begin{bmatrix} \dfrac{\partial f^2}{\partial x_1^2} & \dfrac{\partial^2 f}{\partial x_1 \partial x_2} \\ \dfrac{\partial^2 f}{\partial x_2 \partial x_1} & \dfrac{\partial^2 f}{\partial x_2^2} \end{bmatrix}$$

Therefore,

$$\nabla^2 f(\mathbf{x}) = \begin{bmatrix} 8 & 4 \\ 4 & 8 \end{bmatrix}$$

As before, let us start our search at the point $\mathbf{x} = \mathbf{x}_0^{(1)} = \{5, 2\}$. Evaluated at $\mathbf{x}_0^{(1)}$ we obtain

$$\nabla f(\mathbf{x}_0) = [48, 36]$$

$$\nabla^2 f(\mathbf{x}_0) = \begin{bmatrix} 8 & 4 \\ 4 & 8 \end{bmatrix}$$

The two new (projected points) are given by

$$x_1^{(1)} = 5 - S\begin{bmatrix} 8 & 4 \\ 4 & 8 \end{bmatrix}^{-1} \begin{bmatrix} 48 \\ 36 \end{bmatrix}$$

$$x_2^{(1)} = 2 - S\begin{bmatrix} 8 & 4 \\ 4 & 8 \end{bmatrix}^{-1} \begin{bmatrix} 48 \\ 36 \end{bmatrix}$$

or

$$x_1^{(1)} = 5 - S \begin{bmatrix} \dfrac{1}{6} & -\dfrac{1}{12} \\ -\dfrac{1}{12} & \dfrac{1}{6} \end{bmatrix} \begin{bmatrix} 48 \\ 36 \end{bmatrix}$$

$$x_2^{(1)} = 2 - S \begin{bmatrix} \dfrac{1}{6} & -\dfrac{1}{12} \\ -\dfrac{1}{12} & \dfrac{1}{6} \end{bmatrix} \begin{bmatrix} 48 \\ 36 \end{bmatrix}$$

Therefore,

$$x_1^{(1)} = 5 - 5S$$

$$x_2^{(1)} = 2 - 2S$$

The optimal search parameter S can be found as in Example 11.5-3.

Minimize:

$$f(s) = 2(7 - 7S)^2 + 2[(5 - 5S)^2 + (2 - 2S)^2]$$

Clearly, $f(s)$ is minimized by $S=1$.
The new (projected) optimal solution is now given by

$$x_{\text{new}} = (x_1^{(1)}, x_2^{(1)}) = (0, 0)$$

Referring again to Example 11.5-3 or by direct verification, one can easily verify that this is the optimal solution:

$$x_1^* = 0$$
$$x_2^* = 0$$
$$f(\mathbf{x}^*) = 0$$

Note that although the first order procedure of Section 11.5, as used in Example 11.5-3, failed to completely converge in three iterations, Newton's method converged in one iteration. This is to be expected, since Newton's method uses second order information, $\nabla^2 f(\mathbf{x})$, and the function to be minimized is of second order. If Newton's method had been applied to a function involving terms of order 3 or higher, a series of iterations would have been necessary. The student should verify this by solving exercise 20 at the end of this chapter.

III. CONSTRAINED OPTIMIZATION PROBLEMS; EQUALITY CONSTRAINTS

11.6
LAGRANGE MULTIPLIERS

In most engineering problems—especially real-world formulations—the object is to optimize (maximize or minimize) a criterion (or objective) function subject to several constraints. Conceptually the introduction of side conditions that must be satisfied in an optimization problem presents no real difficulties; the feasible region of solutions is simply bounded or "constrained" by these side conditions. It seems that such a reduction of the feasible solution space would be highly beneficial; mathematically, however, it sometimes has devastating effects on rational solution techniques. For example, it can be relatively easy to solve a simple nonlinear optimization problem that is unconstrained, but if a few nonlinear *constraints* are imposed on the problem there are very few known solution techniques available. Certainly there are relatively few efficient solution techniques.

The mathematical technique of *Lagrange multipliers* has been developed to convert constrained optimization problems into unconstrained optimization problems. Of course, this can only be accomplished by creating a new problem of higher dimensions as we shall subsequently make evident.

Consider the problem of maximizing a continuous and differentiable function $y_0 = f(x_1, x_2, \ldots, x_n)$ subject to the constraint $g(x_1, x_2, \ldots, x_n) = \alpha$ where $g(\mathbf{x})$ is also continuous and differentiable. The above conditions suggest that we could choose the variable x_n in the constraint and express it in terms of the remaining $(n-1)$ variables such that

$$x_n = H(x_1, x_2, \ldots, x_{n-1})$$

We could then substitute this into the objective function to obtain

$$y_0 = \hat{f}[x_1, x_2, \ldots, x_{n-1}, H(x_1, x_2, \ldots, x_{n-1})]$$

In this form, classical methods can be employed, since the function is unconstrained. A necessary condition for extreme points (maximum or minimum) is that all the first derivatives vanish:

$$\frac{\partial y_0}{\partial x_j} = 0 \qquad j = 1, 2, \ldots, (n-1)$$

which yields by the chain rule

$$\frac{\partial y_0}{\partial x_j} = \frac{\partial \hat{f}}{\partial x_j} + \frac{\partial \hat{f}}{\partial x_n} \cdot \frac{\partial H}{\partial x_j} \qquad j = 1, 2, \ldots, (n-1)$$

However from $g(x_1, x_2, \ldots, x_n) = \alpha$ we see that

$$\frac{\partial g}{\partial x_j} + \frac{\partial g}{\partial x_n}\frac{\partial H}{\partial x_j} = 0 \qquad j = 1, 2, \ldots, (n-1)$$

$$\Rightarrow \frac{\partial H}{\partial x_j} = -\frac{\partial g}{\partial x_j} \bigg/ \frac{\partial g}{\partial x_n} \qquad \text{if} \qquad \frac{\partial g}{\partial x_n} \neq 0 \qquad j = 1, 2, \ldots, (n-1)$$

Therefore,

$$\frac{\partial y_0}{\partial x_j} = \frac{\partial \hat{f}}{\partial x_j} - \left[\frac{\partial \hat{f}}{\partial x_n} \cdot \frac{\partial g}{\partial x_j} \bigg/ \frac{\partial g}{\partial x_n} \right] = 0 \qquad j = 1, 2, \ldots, (n-1)$$

If the solution vector obtained is the maximizing vector, then $x_1^*, x_2^*, \ldots, x_n^*$ are the maximizing values. Denote

$$\lambda = \frac{\partial \hat{f}}{\partial x_n} \bigg/ \frac{\partial g}{\partial x_n}$$

so that

$$\frac{\partial \hat{f}}{\partial x_j} - \lambda \frac{\partial g}{\partial x_j} = 0 \qquad j = 1, 2, \ldots, n$$

and the enforced condition that

$$g(x_1, x_2, \ldots, x_n) = \alpha$$

Notice we now have $(n+1)$ equations in $(n+1)$ unknowns. These conditions are *necessary* for an optimum, provided that not all the derivatives $\partial g/\partial x_j$ vanish at $x_1^*, x_2^*, \ldots, x_n^*$. These necessary conditions can be obtained quite easily in practice, since we observe that we can write the following relation

$$y_0 = f(x_1, x_2, \ldots x_n) - \lambda[g(x_1, x_2, \ldots, x_n) - \alpha]$$

and then note that

$$\frac{\partial y_0}{\partial x_j} = \frac{\partial f}{\partial x_j} - \lambda \frac{\partial g}{\partial x_j} = 0$$

$$\frac{\partial y_0}{\partial \lambda} = g(x_1, x_2, \ldots, x_n) - \alpha = 0$$

which are exactly the same conditions previously derived for optimality.

EXAMPLE 11.6-1

Maximize: $\quad f(\mathbf{x}) = 3x_1^2 + x_2^2 + 2x_1x_2 + 6x_1 + 2x_2$

Subject to: $\quad 2x_1 - x_2 = 4$

Forming the *Lagrangian Function* one obtains

$$L(x_1, x_2, \lambda) = 3x_1^2 + x_2^2 + 2x_1x_2 + 6x_1 + 2x_2 - \lambda[2x_1 - x_2 - 4]$$

$$\frac{\partial L}{\partial x_1} = 6x_1 + 2x_2 + 6 - 2\lambda = 0$$

$$\frac{\partial L}{\partial x_2} = 2x_2 + 2x_1 + 2 + \lambda = 0$$

$$\frac{\partial L}{\partial \lambda} = 2x_1 - x_2 - 4 = 0$$

We now have three equations in three unknowns that we can solve simultaneously; and so, one obtains

$$x_1^* = \frac{7}{11}$$

$$x_2^* = -\frac{30}{11}$$

$$\lambda^* = \frac{24}{11}$$

The original objective function yields the value: $f(x_1^*, x_2^*) = 85.7$.

Although we have found a stationary point for this maximization problem, there is actually no guarantee that this particular solution vector is the one we seek. In fact, any solution vector obtained by this constrained optimization technique might be a maximum, minimum, or a saddle point. For this particular problem, we can check our solution by referring to the results of Section 11.1. Recall that if the objective function is *concave*, and the constraints form a *convex* set, then the solution will be a global maximum. Using Appendix A we know that if the Hessian matrix of a function is negative definite, then the function is concave. From the objective function,

$$H(\mathbf{x}) = \begin{bmatrix} \dfrac{\partial^2 f}{\partial x_1^2} & \dfrac{\partial^2 f}{\partial x_1 \partial x_2} \\ \dfrac{\partial^2 f}{\partial x_2 \partial x_1} & \dfrac{\partial^2 f}{\partial x_2^2} \end{bmatrix}$$

Hence,

$$H(\mathbf{x}) = \begin{bmatrix} 6 & 2 \\ 2 & 2 \end{bmatrix}$$

From Appendix A, since the matrix is symmetric and all principal diagonals are positive, the Hessian matrix is found to be *positive definite*. The objective function is thus *convex*, not concave. In addition, since the constraint is linear, it is convex and forms a convex set. Now recall from Section 11.1 that if a convex function is maximized, the solution will be found at an extreme point. In this example, the Lagrangian has located only a stationary point. The moral is to always check the nature of a Lagrangian stationary point.

Lagrangian Optimization; Everett's Method [Everett (10)]

Consider an arbitrary objective function $f(\mathbf{x})$, subject to K equality constraints of the form $g_i(\mathbf{x}) = b_i$, $i = 1, 2, K$, where $\mathbf{x}$ is a vector of j components,

$j \geqq K$. For simplicity, denote $H_i(\mathbf{x}) \equiv g_i(\mathbf{x})$ (note that we have deleted the resource specification). Now suppose we introduce K *nonnegative* ($\lambda_i \geq 0$) real numbers, along with the assumption that there is a vector $\mathbf{x}^*$, which *maximizes* $f(\mathbf{x})$

Maximize:

$$f(\mathbf{x}) - \sum_{i=1}^{K} \lambda_i H_i(\mathbf{x}) \qquad \mathbf{x}^* \in S$$

over all $\mathbf{x} \in S$. It is not necessary that this vector be unique, only that there is some $\mathbf{x}^*$ which maximizes the above function. We shall call this function the *Everett Lagrangian*. The above assumptions imply that

$$f(\mathbf{x}^*) - \sum_{i=1}^{K} \lambda_i H_i(\mathbf{x}^*) \geqq f(\mathbf{x}) - \sum_{i=1}^{K} \lambda_i H_i(\mathbf{x})$$

Hence,

$$f(\mathbf{x}^*) \geqq f(\mathbf{x}) + \sum_{i=1}^{K} \lambda_i [H_i(\mathbf{x}^*) - H_i(\mathbf{x})] \qquad \text{for all } \mathbf{x} \in S$$

Now since $\mathbf{x}^*$ is a maximizing vector, it must be true for all feasible values in the neighborhood of $\mathbf{x}^*$ that

$$\sum_{i=1}^{K} \lambda_i [H_i(\mathbf{x}^*) - H_i(\mathbf{x})]$$

be nonnegative; therefore, it must be true that $f(\mathbf{x}^*) \geqq f(\mathbf{x})$. This leads us to the main theorem:

THEOREM 1

1. λ_i, $i = 1, 2, \ldots, K$ are nonnegative real numbers.
2. If $\mathbf{x}^* \in S$ maximizes the function $f(\mathbf{x}) - \sum_{i=1}^{K} \lambda_i H_i(\mathbf{x}) \qquad \mathbf{x} \in S$ then
3. $\mathbf{x}^*$ maximizes $f(\mathbf{x})$ for all feasible $\mathbf{x}$.

What does this tell us? It says that for *any* choice of nonnegative λ_i, $i = 1, 2, \ldots, K$ that if an unconstrained maximum to the new Lagrangian function $f(\mathbf{x}) - \sum_{i=1}^{K} \lambda_i H_i(\mathbf{x})$ can be found, then this solution is also a valid solution to an original constrained optimization problem, but valid in the sense that the problem we have solved by choosing a set of nonnegative λ_i may not be the problem that we wish to solve! In fact, corresponding to the optimal solution of the Lagrangian after choosing a set of λ's is a set of constraints that use only a specified amount of resources—the amount required to produce the present solution. In other words, we have solved a constrained problem but perhaps not the problem we wanted to solve. The problem that we wish to solve is one in which all resources (the $\mathbf{b}$ vector) are completely used. Now it is true that if we choose another set of λ's we will solve another problem—the trick is to choose this new set of λ's in such a way as to approach our original constraint set. One possible way is to exhaustively search all possible combinations until the right one is chosen, but in most cases one can tell how to change the multipliers by simply observing the changes in the resource vector.

The procedure is as follows:

1. Choose a set of nonnegative multiplers and transform the constrained problem to an unconstrained one.
2. Find a vector $\mathbf{x}^*$ that maximizes this new function.
3. Use this maximizing vector $\mathbf{x}^*$ to calculate the values for each constraint $H_i(\mathbf{x}^*)$ $i=1, 2, \ldots, K$.
4. If $H_i(\mathbf{x}^*) \equiv b_i$ all i then an optimal solution has been found. If $H_i(\mathbf{x}^*) \neq b_i$ then go to 1 and repeat the procedure.

Putting the entire procedure into perspective, at each iteration we always solve *some* optimization problem. However, the problem that we have solved may not force $g(\mathbf{x}) - b_i = 0$. If we have not solved the desired problem, then we should choose a new set of $\{\lambda_i\}$ in such a way as to approach the solution we seek.

For completeness, it should be noted that there may not be a set of λ's that will generate $H_i(\mathbf{x}^*) \equiv b_i$. These conditions are known as *gaps*, and they are a subject of considerable interest. It has been noted, however, that these are relatively rare in real-world problems. The above procedure will always converge to an optimal solution in a finite number of steps provided that there are no gaps at the optimal solution and a solution exists. Hence, the real value of this technique is that it can be used on (1) nondifferentiable functions and (2) special nondifferentiable functions such as integer programming problems.

EXAMPLE 11.6-2

An astronaut's water container is to be stored in a space capsule wall. The container is made in the form of a sphere surmounted by a cone, the base of which is equal to the radius of the sphere. If the radius of the sphere is restricted to exactly 6 ft and a surface area of 450 ft² is all that is allowed, find the dimensions x_1 and x_2 such that the volume of the container is a maximum.

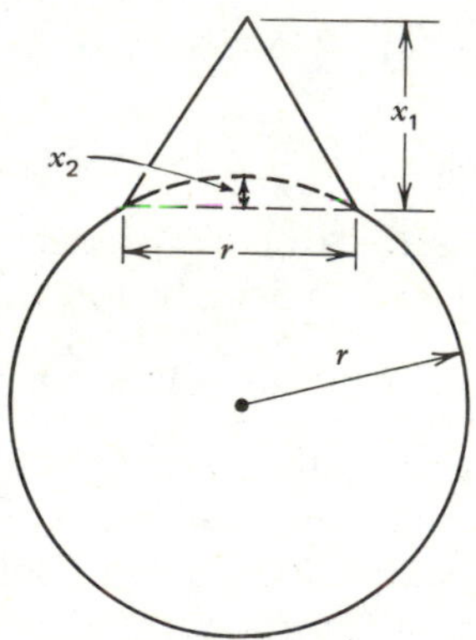

Volume of the conical top $= \frac{1}{3}\pi \frac{r^2}{4} x_1 = \frac{\pi r^2}{12} x_1$

Volume of the cut sphere $= \left[\frac{4}{3}\pi r^3 - \frac{1}{6}\pi x_2 \left(\frac{3r^2}{4} + x_2^2 \right) \right]$

Volume of the capsule $= \frac{\pi}{12} r^2 x_1 + \frac{4}{3}\pi r^3 - \frac{\pi}{6} x_2 \left(\frac{3r^2}{4} + x_2^2 \right)$

Surface area of cone $= \frac{\pi r}{2}\left[\sqrt{\frac{r^2}{4}+x_1^2}\right]$

Surface area of spherical portion $= \left[4\pi r^2 - \pi\left(\frac{r^2}{4}+x_2^2\right)\right]$

Total surface area $= \frac{\pi r}{2}\left[\sqrt{\frac{r^2}{4}+x_1^2}\right] + 4\pi r^2 - \pi\left(\frac{r^2}{4}+x_2^2\right)$

From the above information, the following problem can be constructed:

Maximize:

$$f(\mathbf{x}) = \frac{\pi}{12}r^2 x_1 + \tfrac{4}{3}\pi r^3 - \frac{\pi}{6}x_2\left(\frac{3r^2}{4}+x_2^2\right)$$

Subject to:

$$\frac{\pi r}{2}\left[\sqrt{\frac{r^2}{4}+x_1^2}\right] + 4\pi r^2 - \pi\left(\frac{r^2}{4}+x_2^2\right) = 450$$

or

Maximize:

$$f(\mathbf{x}) = \frac{\pi}{12}(36)x_1 + \tfrac{4}{3}\pi(216) - \frac{\pi}{6}x_2(\tfrac{3}{4}r^2 + x_2^2)$$

Subject to:

$$\frac{\pi(6)}{2}[\sqrt{\tfrac{36}{4} + x_1^2}] + 4\pi(36) - \pi(\tfrac{36}{4} + x_2^2) = 450$$

or

Maximize: $f(\mathbf{x}) = 9.43x_1 - 14.14x_2 - 0.52x_2^3 + 905.143$

Subject to: $9.43\sqrt{x_1^2+9} - 3.14x_2^2 = 25.714$

The Lagrangian function is given by

$$L(\lambda_1, x_1, x_2) = 9.43x_1 - 14.14x_2 - 0.52x_2^3 - \lambda(9.43\sqrt{x_1^2+9} - 3.14x_2^2)$$

The problem is now one of solving a two-variable, unconstrained maximization problem. This maximization can be carried out in any convenient manner. The Lagrangian technique suggested by Everett will now be applied.

Phase 1. A common starting point is to choose $\lambda = 0$ and solve the original unconstrained problem for the solution variables. If this solution (fortuitously) satisfies the problem constraints as equalities, the problem is solved. If not, then a nonzero value of λ should be chosen and the problem should be solved again. For $\lambda = 0$, the solution is clearly $x_1^* = \infty$. This is logical because it will create an infinite container.

Phase 2. Assume $\lambda = 3$.

$$L(x_1, x_2) = 9.43x_1 - 14.14x_2 - 0.52x_2^3 - 3(9.43\sqrt{x_1^2+9} - 3.14x_2^2)$$

Using the calculus,

$$\frac{\partial L(x_1, x_2)}{\partial x_1} = 9.43 - 3(9.43)(x_1^2+9)^{-1/2}x_1 = 0$$

$$\frac{\partial L(x_1, x_2)}{\partial x_2} = -14.14 - 1.56x_2^2 + 6(3.14)x_2 = 0$$

Note that both equations are simply quadratic functions. Solving, one obtains

$$x_1^* = 1.06$$
$$x_2^* = 0.80$$

The constraint yields

$$9.43\sqrt{10.125} - 3.14(0.8)^2 \equiv 27.99$$

Note that we have found the solution to a *particular* optimization problem. Specifically, the problem where the right-hand side of the constraint is exactly 27.99. However, this is not the problem we wish to solve. To solve the problem as stated, it is necessary to modify λ and try again. Hence, we shall *decrease* the value of λ to 2.0 in an attempt to decrease the available resource. Table 11.5 gives the result of Phase 3. Note that the constraint evaluated at $\lambda = 2.0$ is equal to 26.94, which is closer to the value we seek. Further iterations are also shown.

Table 11.5
EVERETT'S METHOD

Phase	λ	x_1^*	x_2^*	Constraint (Right-hand side)
1	0	∞	—	∞
2	3.0	1.06	0.80	27.99
3	2.0	1.73	1.35	26.94
4	1.5	2.68	2.79	13.52

Note that since at Phase 4 the resource requirement is too small, we shall increase the value of λ (we have passed the value we seek), continuing in this fashion as shown in Table 11.6.

Table 11.6
SOLUTION OF EXAMPLE 11.6-2

Phase	λ	x_1^*	x_2^*	Constraint (Right-hand side)
5	1.9	1.86	1.47	26.48
6	1.8	2.00	1.61	25.89

We are sufficiently close to the answer to terminate. The optimal solution is

$$x_1^* = 2.0 \text{ ft}$$
$$x_2^* = 1.61 \text{ ft}$$

11.7 EQUALITY CONSTRAINED OPTIMIZATION: CONSTRAINED DERIVATIVES

Wilde and Beightler (29) have devised a direct method for handling nonlinear optimization problems subject to equality constraints.

Consider the problem of finding a stationary point of an arbitrary function $f(\mathbf{x})$ subject to M differentiable constraints:

$$g_j(\mathbf{x}) = 0 \qquad j = 1, 2, \ldots, M$$

where

$$\mathbf{x} = (x_1, x_2, \ldots, x_N)$$

We desire to find an expression for the first partial derivatives of $f(\mathbf{x})$ at all the points that satisfy $g_j(\mathbf{x}) = 0$.

By Taylor's Theorem, for the points $(\mathbf{x} + \Delta\mathbf{x})$ in the neighborhood of $\mathbf{x}$, we have

$$f(\mathbf{x} + \nabla\mathbf{x}) - f(\mathbf{x}) = \nabla f(\mathbf{x}) \cdot \Delta\mathbf{x} + \text{higher order terms}$$

and

$$g_j(\mathbf{x} + \Delta\mathbf{x}) - g_j(\mathbf{x}) = \nabla g_j(\mathbf{x}) \cdot \Delta\mathbf{x} + \text{higher order terms}$$

(Note that $\nabla f(\mathbf{x})$ and $\nabla g_j(\mathbf{x})$ are gradient vectors as defined in Section 11.2.)

$$\text{as } \Delta x_j \to 0, \qquad \text{then higher order terms rapidly vanish}$$

Therefore, to a first-order approximation,

$$\partial f(\mathbf{x}) = \nabla f(\mathbf{x}) \cdot \partial\mathbf{x}$$

and

$$\partial g_j(\mathbf{x}) = \nabla g_j(\mathbf{x}) \cdot \partial\mathbf{x} \qquad j = 1, 2, \ldots, M$$

Since $g_j(\mathbf{x}) = 0$, then $\partial g_j(\mathbf{x}) = 0$ so we obtain at a stationary point

$$\partial f(\mathbf{x}) = \nabla f(\mathbf{x}) \cdot \partial\mathbf{x} \tag{11.68}$$

$$0 = \nabla g_j(\mathbf{x}) \cdot \partial\mathbf{x} \qquad j = 1, 2, \ldots, M \tag{11.69}$$

At $\mathbf{x}$ the partial derivatives become known constants so that Eqs. 11.68 and 11.69 give rise to $(M+1)$ equations in $(N+1)$ unknowns. That is, $\partial f(\mathbf{x})$ and the N components of the vector $\partial\mathbf{x}$. Three cases may exist:

1. $M = N$
2. $M > N$
3. $M < N$

Cases 1 and 2 are of no importance, since if $M = N$ the solution would be $\partial\mathbf{x} = 0$, and if $M > N$, then $(M - N)$ equations are redundant. Hence, we will concentrate on Case 3.

Partition $\mathbf{x}$ into two mutually exclusive sets, $\mathbf{Y}$ and $\mathbf{Z}$, such that $\mathbf{x} = (\mathbf{Y}, \mathbf{Z})$. Designate the vector $\mathbf{Y}$ as a set of M-independent or *decision* variables, and $\mathbf{Z}$ as a set of $(N - M)$-dependent or *state* variables. For simplicity, suppose that we define a partition of the gradient vector $f(\mathbf{x})$ into two parts defined by $\mathbf{Y}$ and $\mathbf{Z}$. The gradient vectors can now be rewritten as

$$\nabla f(\mathbf{x}) = \nabla f(\mathbf{Y}, \mathbf{Z}) = [\nabla f(\mathbf{Y}), \nabla f(\mathbf{Z})]$$

and

$$\nabla g_j(\mathbf{x}) = \nabla g_j(\mathbf{Y}, \mathbf{Z}) = [\nabla g_j(\mathbf{Y}), \nabla g_j(\mathbf{Z})] \qquad j = 1, 2, \ldots, M$$

Furthermore, define a nonsingular Jacobian matrix **J** by

$$\mathbf{J} = \nabla g(\mathbf{Y}) = \begin{bmatrix} \nabla g_1(\mathbf{Y}) \\ \nabla g_2(\mathbf{Y}) \\ \cdot \\ \cdot \\ \cdot \\ \nabla g_M(\mathbf{Y}) \end{bmatrix}$$

and a control matrix **C** by

$$\mathbf{C} = \nabla g(\mathbf{Z}) = \begin{bmatrix} \nabla g_1(\mathbf{Z}) \\ \nabla g_2(\mathbf{Z}) \\ \cdot \\ \cdot \\ \cdot \\ \nabla g_M(\mathbf{Z}) \end{bmatrix}$$

Equations 11.68 and 11.69 can now be rewritten as

$$\partial f(\mathbf{Y}, \mathbf{Z}) = f(\mathbf{Y})\partial \mathbf{Y} + f(\mathbf{Z})\partial \mathbf{Z}$$
$$\mathbf{J}(\partial \mathbf{Y}) = -\mathbf{C}(\partial \mathbf{Z})$$

Since **J** is chosen nonsingular, we can immediately see that

$$\partial f(\mathbf{x}) = [-f(\mathbf{Y})\mathbf{J}^{-1}\mathbf{C} + f(\mathbf{Z})]\partial \mathbf{Z} \tag{11.70}$$

Note that Eq. 11.70 reflects how the rate of change in $f(\mathbf{x})$ behaves with respect to a rate of change in the decision vector. This is analogous to conventional first derivatives, but these rate of changes are *constrained*. The quantity in brackets will be denoted by ∇_{YZ} and is formally a *constrained derivative*. Wilde and Beightler have shown that whenever constrained derivatives are used, the classical theory of *interior optima* may be employed. Therefore, at the desired stationary point, $\mathbf{x}^*$, a necessary condition for optimality is that $\nabla_{YZ} \equiv 0$. Formally, a constrained derivative is the rate of change of the objective function with respect to a certain decision variable, while holding all other decision variables constant. The state variables continually adjust to maintain feasibility. Note that by the construction of ∇_{YZ} only *feasible*, not arbitrary perturbations, are allowed. The above procedure is a Jacobian method for obtaining the constrained derivative. For further use, it will be convenient to develop the above results in matrix notation.

If Z_i represents the ith element of **Z** and Y_i the ith element of **Y**, then we can express the components of Eq. 11.70 as follows:

$$\nabla f(\mathbf{Y}) = \left(\frac{\partial f}{\partial Y_1} \quad \frac{\partial f}{\partial Y_2} \quad \cdots \quad \frac{\partial f}{\partial Y_M} \right) \tag{11.71}$$

$$\nabla f(\mathbf{Z}) = \left(\frac{\partial f}{\partial Z_1} \quad \frac{\partial f}{\partial Z_2} \quad \cdots \quad \frac{\partial f}{\partial Z_{N-M}} \right) \tag{11.72}$$

$$\mathbf{C} = \begin{bmatrix} \frac{\partial g_1}{\partial Z_1} & \frac{\partial g_1}{\partial Z_2} & \cdots & \frac{\partial g_1}{\partial Z_{N-M}} \\ \cdot & & & \cdot \\ \cdot & & & \cdot \\ \cdot & & & \cdot \\ \frac{\partial g_M}{\partial Z_1} & \frac{\partial g_M}{\partial Z_2} & \cdots & \frac{\partial g_M}{\partial Z_{N-M}} \end{bmatrix} \tag{11.73}$$

and

$$\mathbf{J} = \begin{bmatrix} \frac{\partial g_1}{\partial Y_1} & \frac{\partial g_1}{\partial Y_2} & \cdots & \frac{\partial g_1}{\partial Y_M} \\ \cdot & & & \cdot \\ \cdot & & & \cdot \\ \cdot & & & \cdot \\ \frac{\partial g_M}{\partial Y_1} & \frac{\partial g_M}{\partial Y_2} & \cdots & \frac{\partial g_M}{\partial Y_M} \end{bmatrix} \tag{11.74}$$

We now have all the necessary ingredients to solve equality-constrained optimization problems.

EXAMPLE 11.7-1

Minimize: $f(\mathbf{x}) = 5x_1^2 + x_2^2 + 2x_1x_2$

Subject to: $g_1(\mathbf{x}) = x_1x_2 - 10$

Since there are two variables in the problem and only one constraint, one must designate one of the variables as a *decision* variable, and the other as a *state* variable.

Let: x_1 = state variable

x_2 = decision variable

Hence,

$$\mathbf{Z} = (x_1)$$
$$\mathbf{Y} = (x_2)$$
$$\nabla f(\mathbf{x}) = \nabla f(\mathbf{Z}) = (10x_1 + 2x_2)$$
$$\nabla f(\mathbf{Y}) = (\ 2x_2 + 2x_1)$$
$$\mathbf{J} = [\nabla g_1(\mathbf{Y})] = [x_1]$$

Hence,

$$\mathbf{J}^{-1} = [x_1^{-1}]$$
$$\mathbf{C} = [\nabla g_1(\mathbf{Z})] = [x_2]$$

From Eq. 11.70

$$\nabla_{YZ} = [-\nabla f(\mathbf{Y})\mathbf{J}^{-1}\mathbf{C} + \nabla f(\mathbf{Z})]$$

Therefore,

$$\nabla_{YZ} = (10x_1 + 2x_2) - (2x_2 + 2x_1)(x_1^{-1})(x_2)$$

or

$$\nabla_{YZ} = 10x_1 - 2x_2^2x_1^{-1}$$

A necessary condition for optimality is that ∇_{YZ} vanish. Therefore,

$$\nabla_{YZ} = 0 \Rightarrow 2x_2^2 - 10x_1^2 = 0$$

We now have two equations in two unknowns which must be satisfied:

$$x_2^2 - 5x_1^2 = 0$$

and

$$x_1x_2 - 10 = 0$$

Solving simultaneously one obtains,

$$x_2^4 = 500$$
$$\Rightarrow x_2^* = \pm 4.7287$$

and

$$x_1^* = \pm 2.1147$$

Hence, we have two possible stationary points

$$\mathbf{x}_1^* = (2.1147,\ 4.7287)$$
$$\mathbf{x}_2^* = (-2.1147,\ -4.7287)$$

The reader can verify by examining the Hessian matrices of the objective function and constraint that both these points are minimizing solution vectors.

11.8 PROJECTED GRADIENT METHODS WITH EQUALITY CONSTRAINTS

Consider the problem of maximizing/minimizing a nonlinear objective function subject to linear or nonlinear constraints.

$$\begin{aligned} &\text{Maximize/Minimize:} && y = f(\mathbf{x}) \\ &\text{Subject to:} && g_i(\mathbf{x}) = b_i \qquad i = 1, 2, \ldots, m \\ & && \mathbf{x} = (x_1, x_2, \ldots, x_n) \end{aligned}$$

The classical method of gradient optimization cannot be used in this problem, since variations in the vector **x** might lead to inadmissible regions violating one or more constraints. Thus any perturbation in the solution vector **x** must be made in such a way as to automatically satisfy the constraints. Therefore, for any small displacement dx_i the following must be satisfied at an optimal solution.

$$dg_k = \sum_{i=1}^{n} \left(\frac{\partial g_k}{\partial x_i}\right) dx_i = 0 \qquad k = 1, 2, \ldots, m$$

The search for an optimal value must proceed in a restricted manner satisfying the constraints. As before, let us seek feasible perturbations that will make the rate of change in the objective function the greatest. Along a particular direction this rate of change is given by

$$\frac{dy}{ds} = \sum_{i=1}^{n} \left(\frac{\partial y}{\partial x_i}\right) \frac{dx_i}{ds}$$

where we require:

$$ds = \sqrt{(dx_1)^2 + (dx_2)^2 + \cdots + (dx_n)^2}$$

while at the same time satisfying

$$\frac{dg_k}{ds} = \sum_{i=1}^{n} \left(\frac{\partial g_k}{\partial x_i}\right) \frac{dx_i}{ds} \equiv 0 \qquad k = 1, 2, \ldots, m$$

Formulated in this manner, the optimal rate of changes dx_i/ds are those that make the following Lagrangian function take on a stationary value.

$$F=\frac{dy}{ds}+\lambda_0\left[1-\sum_{i=1}^{n}\left(\frac{dx_i}{ds}\right)^2\right]+\sum_{k=1}^{m}\lambda_k\left[\sum_{i=1}^{n}\left(\frac{\partial g_k}{\partial x_i}\right)\frac{dx_i}{ds}\right]$$

The necessary conditions are given by

$$\frac{\partial F}{\partial\left[\dfrac{dx_i}{ds}\right]}=\frac{\partial y}{\partial x_i}-2\lambda_0\left(\frac{dx_i}{ds}\right)+\sum_{k=1}^{m}\lambda_k\left(\frac{\partial g_k}{\partial x_i}\right)=0 \qquad i=1,2,\ldots,n \tag{11.75}$$

$$\frac{\partial F}{\partial \lambda_0}=\left[1-\sum_{i=1}^{n}\left(\frac{dx_i}{ds}\right)^2\right]=0 \tag{11.76}$$

$$\frac{\partial F}{\partial \lambda_k}=\left[\sum_{i=1}^{n}\left(\frac{\partial g_k}{\partial x_i}\right)\frac{dx_i}{ds}\right]=0 \qquad k=1,2,\ldots,m \tag{11.77}$$

From Eq. 11.73 we obtain

$$\frac{dx_i}{ds}=\frac{1}{2\lambda_0}\left[\frac{\partial y}{\partial x_i}+\sum_{k=1}^{m}\lambda_k\left(\frac{\partial g_k}{\partial x_i}\right)\right] \qquad i=1,2,\ldots,n \tag{11.78}$$

Equation 11.78 is the rate of change in x_i, which results in the greatest change in $f(\mathbf{x})$ subject to the m equality constraints.

If we substitute Eq. 11.78 into Eq. 11.77 we obtain

$$\sum_{i=1}^{n}\left(\frac{\partial g_j}{\partial x_i}\right)\left[\frac{\partial y}{\partial x_i}+\sum_{k=1}^{m}\lambda_k\left(\frac{\partial g_k}{\partial x_i}\right)\right]\frac{1}{2\lambda_0}=0 \qquad j=1,2,\ldots,m \tag{11.79}$$

or

$$\sum_{i=1}^{n}\left(\frac{\partial g_j}{\partial x_i}\right)\left(\frac{\partial y}{\partial x_i}\right)=-\sum_{i=1}^{n}\frac{\partial g_j}{\partial x_i}\sum_{k=1}^{m}\lambda_k\left(\frac{\partial g_k}{\partial x_i}\right) \qquad j=1,2,\ldots,m \tag{11.80}$$

Now, if we substitute Eq. 11.78 into Eq. 11.76:

$$\sum_{i=1}^{n}\left[\frac{1}{2\lambda_0}\left(\frac{\partial y}{\partial x_i}\right)+\sum_{k=1}^{m}\lambda_k\left(\frac{\partial g_k}{\partial x_i}\right)\right]^2=1$$

or

$$\sum_{i=1}^{n}\left[\left(\frac{\partial y}{\partial x_i}\right)+\sum_{k=1}^{m}\lambda_k\left(\frac{\partial g_k}{\partial x_i}\right)\right]^2=4\lambda_0^2$$

From which we can obtain using Eq. 11.80

$$2\lambda_0=\pm\sqrt{\sum_{i=1}^{n}\left[\left(\frac{\partial y}{\partial x_i}\right)^2+\frac{\partial y}{\partial x_i}\sum_{k=1}^{m}\lambda_k\left(\frac{\partial g_k}{\partial x_i}\right)\right]^2} \tag{11.81}$$

It should also be noted that $2\lambda_0$ is essentially a *constrained derivative*. Hence, when λ_0 becomes very small we must be near an optimum solution. This will guide our actions as we converge to an optimum. As previously, the positive component of Eq. 11.81 represents the rate of maximum increase in the objective function, while the negative portion is the rate of maximum decrease.

As in the unconstrained case, our objective is to start with a feasible base point, $x_j^{(\delta)}$, and construct a new base point $x_j^{(\delta+1)}$ in such a manner that the objective function (1) increases/decreases at an optimum rate, and (2) satisfies all constraints. The new base point will, as before, be some point at distance s along solution vector components $(\Delta x_1, \Delta x_2, \ldots, \Delta x_2)$. Algebraically

$$x_j^{(\delta+1)} = x_j^{(\delta)} + sm_j \qquad j = 1, 2, \ldots, n \tag{11.82}$$

Using Eq. 11.75 this becomes

$$x_j^{(\delta+1)} = x_j^{(\delta)} + s\left\{\frac{1}{2\lambda_0}\left[\frac{\partial y}{\partial x_j} + \sum_{k=1}^{m} \lambda_k \frac{\partial g_k}{\partial x_j}\right]\right\} \qquad j = 1, 2, \ldots, n \tag{11.83}$$

By the previous derivation, this will be an improved solution vector for small movements, s. The computational scheme is as follows:

1. Enumerate the n partial derivatives of $y = f(x)$ and the $(m \cdot n)$ partial derivatives $\partial g_k/\partial x_i$ $k = 1, 2, \ldots, m$; $i = 1, 2, \ldots, n$.
2. Determine an initial feasible base point $x_j^{(1)}$.
3. Using Eq. 11.80, and the above results, calculate the m Lagrangian multipliers $\lambda_1, \lambda_2, \ldots, \lambda_m$.
4. Using the results of (3), calculate λ_0 from Eq. 11.81.
5. Calculate the new constrained base point $x_j^{(2)}$ using Eq. 11.83.
6. If the value of λ_0 at (5) becomes zero, then the last base point represents the optimal solution. Otherwise, return to (3) with the new base point and continue.

EXAMPLE 11.8-1

Minimize: $\quad f(\mathbf{x}) = (x_1 - 3)^2 + (x_2 - 4)^2$

Subject to: $\quad 2x_1 + x_2 = 3$

Phase 1

Step I. $\quad \dfrac{\partial f}{\partial x_1} = 2(x_1 - 3) \qquad \dfrac{\partial f}{\partial x_2} = 2(x_2 - 4)$

$$\frac{\partial g}{\partial x_1} = 2 \qquad \frac{\partial g}{\partial x_2} = 1$$

Step II. A feasible starting solution is obviously given by $\mathbf{x}_1 = (1, 1)$. Hence,

$x_1^{(1)} = 1.0$; $x_2^{(1)} = 1.0$; $f(\mathbf{x}_1) = 13$.

Step III. $\quad \left(\dfrac{\partial f}{\partial x_1}\right)_{\mathbf{x}_1} = -4 \qquad \left(\dfrac{\partial f}{\partial x_2}\right)_{\mathbf{x}_1} = -6$

$$\left(\frac{\partial g}{\partial x_1}\right)_{\mathbf{x}_1} = 2 \qquad \left(\frac{\partial g}{\partial x_2}\right)_{\mathbf{x}_1} = 1$$

Hence, from Eq. 11.80

$$(2)(-4) + (1)(-6) = -\{(2)[\lambda_1(2)] + (1)[\lambda_1(1)]\}$$

from which, $\lambda = 14/5$

Step IV. Using Eq. 11.81;

$$4\lambda_0^2 = \left\{(-4)^2 + (-4)\left(\frac{14}{5}\right)(2)\right\}^2 + \left\{(-6)^2 + (-6)\left(\frac{14}{5}\right)(1)\right\}^2 = 409.6$$

or $2\lambda_0 = -20.24$

Step V. $$x_1^{(2)} = 1 + s\left\{\frac{1}{-20.24}\left[-4 + \frac{14}{5}(2)\right]\right\}$$
$$x_2^{(2)} = 1 + s\left\{\frac{1}{-20.24}\left[-6 + \frac{14}{5}(1)\right]\right\}$$

Suppose that we arbitrarily choose a step size $s = 2.0$, then

$$x_1^{(2)} = 0.763 \qquad \mathbf{x}_2 = (0.763, 1.474)$$
$$x_2^{(2)} = 1.474$$

Note that at this new base point the objective function changes from $f(\mathbf{x}_1) = 13$ to $f(\mathbf{x}_2) = 11.381$. Note also that the new solution still lies along the equality constraint. Since $\lambda_0 \neq 0$ we should proceed to a new base point.

Phase 2

Step III. $$\left(\frac{\partial f}{\partial x_1}\right)_{\mathbf{x}_2} = -4.47$$
$$\left(\frac{\partial f}{\partial x_2}\right)_{\mathbf{x}_2} = -5.052$$
$$\left(\frac{\partial g_1}{\partial x_1}\right)_{\mathbf{x}_2} = 2.0$$
$$\left(\frac{\partial g_1}{\partial x_2}\right)_{\mathbf{x}_2} = 1.0$$

Hence, from Eq. 11.80

$$(2)(-4.47) + (1)(-5.052) = -\{(2)(\lambda_1)(2) + (1)(\lambda_1)(1)\}$$
$$\lambda_1 = \frac{14}{5}$$

Step IV. Using Eq. 11.81

$$4\lambda_0^2 = \left\{(-4.47)^2 + (-4.47)\left(\frac{14}{5}\right)(2)\right\}^2 + \left\{(-5.052)^2 + (-5.052)\left(\frac{14}{5}\right)(1)\right\}^2$$

Therefore, $2\lambda_0 = -12.43$

Step V. $$x_1^{(3)} = 0.763 + s\left\{\frac{1}{-12.43}\left[-4.47 + \frac{14}{5}(2)\right]\right\}$$
$$x_2^{(3)} = 1.474 + s\left\{\frac{1}{-12.43}\left[-5.052 + \frac{14}{5}(1)\right]\right\}$$

Therefore

$$x_1^{(3)} = 0.49 \qquad \mathbf{x} = (0.49, 2.02)$$
$$x_2^{(3)} = 2.02$$

The objective function is now $f(\mathbf{x}_3) = 10.25$. Since we are still making improvements we proceed further.

Phase 3

Step III. $\left(\dfrac{\partial f}{\partial x_1}\right)_{\mathbf{x}_3} = -5.02 \qquad \left(\dfrac{\partial f}{\partial x_2}\right)_{\mathbf{x}_3} = -3.96$

$\left(\dfrac{\partial g}{\partial x_1}\right)_{\mathbf{x}_3} = 2 \qquad \left(\dfrac{\partial g}{\partial x_2}\right)_{\mathbf{x}_3} = 1$

Using Eq. 11.80

$$(2)(-5.02) + (1)(-3.96) = -5\lambda_1$$

$$\lambda_1 = \frac{14}{5}$$

Step IV. Using Eq. 11.81

$$4\lambda_0^2 = \left\{(-5.02)^2 + (-5.02)\left(\frac{14}{5}\right)(2)\right\}^2 + \left\{(-3.96)^2 + (-3.96)\left(\frac{14}{5}\right)(1)\right\}^2$$

$2\lambda_0 = -5.44$

Step V. $x_1^{(4)} = 0.69 + s\left\{\dfrac{1}{-5.44}\left[-5.02 + \left(\dfrac{14}{5}\right)(2)\right]\right\}$

$x_2^{(4)} = 1.62 + s\left\{\dfrac{1}{-5.44}\left[-3.96 + \left(\dfrac{14}{5}\right)(1)\right]\right\}$

Therefore,

$$x_1^{(4)} = 0.37 \qquad x_2^{(4)} = 2.26 \qquad \mathbf{x}_4 = (0.37,\ 2.26)$$

The objective function is now $f(\mathbf{x}_4) = 9.95$

Phase 4

Step III. $\left(\dfrac{\partial f}{\partial x_1}\right)_{\mathbf{x}_4} = -5.26 \qquad \left(\dfrac{\partial f}{\partial x_2}\right)_{\mathbf{x}_4} = -3.48$

$\left(\dfrac{\partial g}{\partial x_1}\right)_{\mathbf{x}_4} = 2.0 \qquad \left(\dfrac{\partial g}{\partial x_2}\right)_{\mathbf{x}_4} = 1.0$

Using Eq. 11.80;

$$(2)(-5.26) + (1)(3.48) = -5\lambda_1$$

$$\lambda_1 = \frac{14}{5}$$

Step IV. Using Eq. 11.81;

$$4\lambda_0^2 = \left\{(-5.26)^2 + (-5.26)\left(\frac{14}{5}\right)(2)\right\}^2 + \left\{(-3.48)^2 + (-3.48)\left(\frac{14}{5}\right)(2)\right\}^2$$

$2\lambda_0 = -2.96$

Step V. $x_1^{(5)} = 0.37 + 5\left\{\frac{1}{-2.96}\left[-5.26 + \left(\frac{14}{5}\right)(2)\right]\right\}$

$$x_2^{(5)} = 2.26 + 5\left\{\frac{1}{-2.96}\left[3.48 + \left(\frac{14}{5}\right)(1)\right]\right\}$$

Therefore,

$$x_1^{(5)} = 0.025$$
$$x_2^{(5)} = 2.95$$

The objective function is now $f(\mathbf{x}_5) = 9.98$. At this point, note that we have not improved our objective function at the last move. The reason for this is that we have moved too far and passed the optimum. The procedure would now be to return to the last base point, reduce the step size, and proceed again. Further iterations will not be given since the procedure should be clear from the previous iterations.

IV. CONSTRAINED OPTIMIZATION PROBLEMS: INEQUALITY CONSTRAINTS

11.9 NONLINEAR OPTIMIZATION—THE KUHN-TUCKER CONDITIONS

In the preceding section, we found that Lagrangian multipliers could be utilized in solving equality constrained optimization problems. Kuhn and Tucker have extended this theory to include the general nonlinear programming problem with both equality and inequality constraints. Consider the following general nonlinear programming problem.

Program I

Minimize: $f(\mathbf{x})$

Subject to: $h_j(\mathbf{x}) = 0 \qquad j = 1, 2, \ldots, m$

$g_j(\mathbf{x}) \geqq 0 \qquad j = m+1, \ldots, p$

THEOREM 1

If $\mathbf{x}^*$ is a solution to Program I, and the functions $f(\mathbf{x})$, $h_j(\mathbf{x})$, and $g_j(\mathbf{x})$ are once differentiable, then there exists a set of vectors μ^* and λ^* such that $\mathbf{x}^*$, μ^*, and λ^* satisfy the following relations:

$$h_j(\mathbf{x}) = 0 \qquad j = 1, \ldots, m \tag{11.84}$$

$$g_j(\mathbf{x}) \geqq 0 \tag{11.85}$$

$$\mu_j[g_j(\mathbf{x})] = 0 \tag{11.86}$$

$$\mu_j \geqq 0 \tag{11.87}$$

$$\frac{\partial f(\mathbf{x})}{\partial x_k} + \sum_{j=1}^{m} \lambda_j \left[\frac{\partial h_j(\mathbf{x})}{\partial x_k}\right] - \sum_{j=m+1}^{P} \mu_j \left[\frac{\partial g_j(\mathbf{x})}{\partial x_k}\right] = 0 \qquad k = 1, 2, \ldots, N \tag{11.88}$$

Equations 11.84 through 11.88 represent relationships that when satisfied are necessary conditions for an optimal solution to Program I. These relationships are known as the *Kuhn-Tucker conditions* after the individuals who first derived them. Note that Eqs. 11.84 and 11.85 specify primal feasibility Equations 11.86 are complementary slackness conditions, analogous to those in linear programming; Eq. 11.87 contains nonnegative dual variables corresponding to the Lagrangian multipliers previously introduced in Section 11.6.

In most real-world formulations of nonlinear programming problems, there exist nonnegativity conditions on all solution variables. These are also accommodated in Program I, since they are of the same form as the stated inequality constraints. However, we will consider this case separately to clarify certain notational problems that commonly appear in the literature. Consider the following statement of Program I with nonnegativity restraints on the solution variables.

Program II

Minimize: $f(\mathbf{x})$

Subject to:

$$h_j(\mathbf{x}) = 0 \qquad j = 1, 2, \ldots, m$$

$$g_j(\mathbf{x}) \geqq 0 \qquad j = m+1, \ldots, p$$

$$x_k \geqq 0 \qquad k = 1, 2, \ldots, N$$

THEOREM 2

If $\mathbf{x}^*$ is a solution vector to Program II, and if $f(\mathbf{x})$, $h_j(\mathbf{x})$, and $g_j(\mathbf{x})$ are once differentiable functions, then there exists a set of vectors μ^*, λ^*, and $\mathbf{V}^*$ such that $\mathbf{x}^*$, μ^*, λ^*, and $\mathbf{V}^*$ satisfy the following relations.

$$h_j(\mathbf{x}) = 0 \qquad j = 1, 2, \ldots, m \tag{11.89}$$

$$g_j(\mathbf{x}) \geqq 0 \qquad j = m+1, \ldots, p \tag{11.90}$$

$$x_k \geqq 0 \qquad k = 1, 2, \ldots, N \tag{11.91}$$

$$\mu_j[g_j(\mathbf{x})] = 0 \qquad j = m+1, \ldots, p \tag{11.92}$$

$$\mu_j \geqq 0 \tag{11.93}$$

$$V_k[x_k] = 0 \qquad k = 1, 2, \ldots, N \tag{11.94}$$

$$V_k \geqq 0 \tag{11.95}$$

$$\frac{\partial f(\mathbf{x})}{\partial x_k} + \sum_{j=1}^{m} \lambda_j \left[\frac{\partial h_j(\mathbf{x})}{\partial x_k} \right] - \sum_{j=m+1}^{P} \mu_j \left[\frac{\partial g_j(\mathbf{x})}{\partial x_k} \right] - V_k = 0 \qquad k = 1, 2, \ldots, N \tag{11.96}$$

Note, however, that by Eqs. 11.96 and 11.94, these conditions imply that

$$\Delta L_k(\mathbf{x}, \mu, \lambda) = \frac{\partial f(\mathbf{x})}{\partial x_k} + \sum_{j=1}^{m} \lambda_j \left[\frac{\partial h_j(\mathbf{x})}{\partial x_k} \right] - \sum_{j=m+1}^{P} \mu_j \left[\frac{\partial g_j(\mathbf{x})}{\partial x_k} \right] \geqq 0 \qquad k = 1, 2, \ldots, N \tag{11.97}$$

and

$$x_k[\Delta L_k(\mathbf{x}, \mu, \lambda)] = 0 \qquad k = 1, 2, \ldots, N \tag{11.98}$$

the reader should compare these equations to the more general results of Theorem 1. Let us now consider a series of examples explaining these conditions.

Discussion of the Kuhn-Tucker Conditions
Consider the following problem:

Program III

Minimize:	$f(\mathbf{x})$	Note that $x_j \geqq 0$ is a *special* requirement and not a *general* one
Subject to:	$x_j \geqq 0$	

Now, if the constraints did not exist a necessary condition for an optimal solution would be

$$\left.\frac{\partial f(\mathbf{x})}{\partial x_j}\right|_{\mathbf{x}^*} = 0 \qquad \text{all } j$$

The following example clearly shows why this need not be the case in the present problem.

We see that in Case II the constrained minimum is obviously not the free minimum as in Case I. In Case II the free minimum occurs at a negative value of $\mathbf{x}$, which is infeasible. Clearly, $\mathbf{x} \equiv 0$ is the solution to this problem. It is also clear that regardless of the functional form of $f(\mathbf{x})$, one of the quantities $(\mathbf{x}, [dF(\mathbf{x})/d\mathbf{x}])$ will be zero. Note also from the pictures that $dF/d\mathbf{x} \geqq 0$. Thus, the optimality conditions are

a. $\mathbf{x} \geqq 0$

b. $x_j \cdot \dfrac{df(\mathbf{x})}{dx_j} = 0 \qquad \text{all } j$

c. $\dfrac{df(\mathbf{x})}{dx_j} \geqq 0 \qquad \text{all } j$

If $f(\mathbf{x})$ is *convex*, then these conditions are both *necessary and sufficient.*

Using our newly acquired knowledge, let us proceed to a more general case.

Program IV

Minimize:	$f(\mathbf{x})$	
Subject to:	$g_i(\mathbf{x}) \geqq 0$	$i = 1, 2, \ldots, m$
	$\mathbf{x} \geqq 0$	

A *partial Lagrangian function* would be

$$L(\mathbf{x}, \mu) = f(\mathbf{x}) - \sum_{i=1}^{m} \mu_i [g_i(\mathbf{x})]$$

$$\mathbf{x} \geqq 0$$

$$\mu \geqq 0$$

Since the partial Lagrangian is now of the same form as Problem II, we can apply those results directly.

A. $x_j \geqq 0 \qquad j=1, 2, \ldots, n$

B. $x_j\left[\frac{\partial L(\mathbf{x}, \mu)}{\partial x_j}\right]=0 \qquad \text{or} \qquad x_j\left[\frac{\partial f(\mathbf{x})}{\partial x_j}-\sum_{i=1}^{m}\mu_o\frac{\partial g_i(\mathbf{x})}{\partial x_j}\right]=0$

C. $\frac{\partial f(\mathbf{x})}{\partial x_j}-\sum_{i=1}^{m}\mu_i\frac{\partial g_i(\mathbf{x})}{\partial x_j}\geqq 0 \qquad j=1, 2, \ldots, n$

These conditions are completely analogous to Eqs. 11.91, 11.97, and 11.98 in Program II, but they are not enough to characterize a solution, since the variables μ_i have not been dealt with completely. However, we can repeat these arguments for the μ vector, provided that we treat μ_j as simply another independent variable. (This is actually what we do in Lagrangian optimization.)

The conditions are, therefore,

D. $\mu_i \geqq 0 \qquad i=1, 2, \ldots, m$

E. $\mu_i\left[\frac{\partial L(\mathbf{x}, \mu)}{\partial \mu_i}\right]=\mu_i g_i(\mathbf{x})=0 \qquad i=1, 2, \ldots, m \qquad (11.99)$

F. $\frac{\partial L(\mathbf{x}, \mu)}{\partial \mu_i}=g_i(\mathbf{x})\geqq 0$

It will be left as an exercise for the student to show that the most general problem, stated at the beginning of Section 11.9, can similarly be derived using the same logic.

Before using the Kuhn-Tucker conditions, one must assume that certain irregular conditions will not occur in the solution space—particularly at a stationary point. These irregular conditions are specified in what Kuhn and Tucker call the *constraint qualification*.

A rigorous discussion of the constraint qualification is beyond the scope of this text. However, it essentially guarantees that over the region of interest there will be no singularities or other anomalies, such as outward pointing cusps. For the interested reader, an excellent discussion of the constraint qualification is contained in Himmelblau (13). For completeness, the following theorem specifies the conditions under which the Kuhn-Tucker conditions are not only necessary but also sufficient.

The Kuhn-Tucker Sufficiency Theorem

Consider the nonlinear programming problem as stated in Program I. Let the objective function $f(\mathbf{x})$ be convex, $g_j(\mathbf{x})$ be concave for all $j=m+1, \ldots, p$; and $h_j(\mathbf{x})$ be linear for all $j=1, 2, \ldots, m$. If there exists a solution $(\mathbf{x}^*, \mu^*, \lambda^*)$ satisfying the Kuhn-Tucker conditions (Eqs. 11.84 through 11.88), then $\mathbf{x}^*$ is an optimal solution to Program I.

EXAMPLE 11.9-1

Consider the following nonlinear programming problem,

$$\begin{aligned} \text{Minimize:} \quad & f(\mathbf{x})=x_1^2-x_2 \\ \text{Subject to:} \quad & x_1+x_2= 6 \\ & x_1 \geqq 1 \\ & x_1^2+x_2^2\leqq 26 \end{aligned}$$

Following Eqs. 11.80 and 11.81, we find that

$$f(\mathbf{x}) = x_1^2 - x_2; \qquad \nabla f(\mathbf{x}) = (2x_1, -1)$$
$$h_1(\mathbf{x}) = x_1 + x_2 - 6; \qquad \nabla h_1(\mathbf{x}) = (1, 1)$$
$$g_1(\mathbf{x}) = x_1 - 1; \qquad \nabla g_1(\mathbf{x}) = (1, 0)$$
$$g_2(\mathbf{x}) = 26 - x_1^2 - x_2^2; \qquad \nabla g_2(\mathbf{x}) = (-2x_1, -x_2)$$

The Kuhn-Tucker conditions are given below:

$$x_1 + x_2 = 6$$
$$x_1 \geqq 1$$
$$x_1^2 + x_2^2 \leqq 26$$
$$\mu_1(x_1 - 1) = 0$$
$$\mu_2(x_1^2 + x_2^2 - 26) = 0$$
$$\mu_1 \geqq 0, \ \mu_2 \geqq 0$$
$$2x_1 + \lambda_1 - \mu_1 + 2\mu_2 x_1 = 0$$
$$-1 + \lambda_1 + 2\mu_2 x_2 = 0$$

The Hessian matrix of $f(\mathbf{x})$, denoted by $H(\mathbf{x})$, is given by the following:

$$H(\mathbf{x}) = \begin{bmatrix} 2 & 0 \\ 0 & 0 \end{bmatrix}$$

Since $H(\mathbf{x})$ is positive semidefinite, the objective funciton $f(\mathbf{x})$ is convex (see Appendix A). Similarly, we can show that $g_2(\mathbf{x})$ is concave since its Hessian matrix is negative definite. Since $h_1(\mathbf{x})$ and $g_1(\mathbf{x})$ are linear, the conditions of the Kuhn-Tucker sufficiency theorem are satisfied. Hence, if there exists a solution ($\mathbf{x}^*$, μ^*, λ^*) satisfying the Kuhn-Tucker conditions, then $\mathbf{x}^*$ is optimal to the nonlinear program. By inspection we find the following solution,

$$x_1^* = 1, \qquad x_2^* = 5, \qquad \lambda_1 = 0, \qquad \mu_1 = 2.2$$

and $\mu_2 = 0.1$. This solution satisfies the Kuhn-Tucker conditions. Hence, $x_1^* = 1$ and $x_2^* = 5$ is optimal and the minimum value of $f(\mathbf{x}) = -4.0$.

11.10 QUADRATIC PROGRAMMING

A quadratic programming problem will be formally defined in the following manner. ("Prime" denotes the transpose of a matrix or vector.)

Minimize: $$f(\mathbf{x}) = \mathbf{cx} + \mathbf{x}'\mathbf{Qx} \tag{11.100}$$

Subject to: $$\mathbf{Ax} \geqq \mathbf{b} \tag{11.101}$$
$$\mathbf{x} \geqq 0$$

$\mathbf{c} = (c_1, c_2, \ldots, c_n)$ $\qquad$ $\mathbf{b} = (b_1, b_2, \ldots, b_m)$

$\mathbf{x} = (x_1, x_2, \ldots, x_n)$

$$\mathbf{Q} = \begin{bmatrix} q_{11} & q_{12} & \cdots & q_{1n} \\ q_{21} & q_{22} & \cdots & q_{2n} \\ \cdot & \cdot & & \cdot \\ \cdot & \cdot & & \cdot \\ \cdot & \cdot & & \cdot \\ q_{n1} & q_{n2} & \cdots & q_{nn} \end{bmatrix} \qquad \mathbf{A} = \begin{bmatrix} a_{11} & a_{12} & \cdots & a_{1n} \\ a_{21} & a_{22} & \cdots & a_{2n} \\ \cdot & \cdot & & \cdot \\ \cdot & \cdot & & \cdot \\ \cdot & \cdot & & \cdot \\ a_{m1} & a_{m2} & \cdots & a_{mn} \end{bmatrix}$$

The objective function (Eq. 11.100) is usually designated as a *quadratic form*; hence, the name Quadratic Programming for this class of problems. Note that the region of feasible solutions is defined over a set of *linear* inequality constraints and a set of nonnegativity constraints on the solution variables.

Before we attempt to solve this nonlinear programming problem, let us define the conditions under which a problem solution might be obtained, and how such a solution might be characterized. Since the constraint set of Eqs. 11.101 and 11.102 is linear, it is convex. Thus if the objective function $f(\mathbf{x})$ is convex, we know that if a local minimizing solution exists to the quadratic programming problem it will also be a global minimizing solution. The quadratic function is convex provided that the matrix $\mathbf{Q}$ in Eq. 11.100 is positive definite or positive semidefinite.* To insure convergence to a global minimum, the elements of $\mathbf{Q}$ are required to satisfy the above conditions. The elements of $\mathbf{c}$, $\mathbf{A}$, and $\mathbf{b}$, however, are arbitrary. At the present time, there are no computationally efficient methods for finding a (global) optimal solution to the quadratic programming problem when $\mathbf{Q}$ is a general symmetric matrix. In the general case, any algorithm might converge to a local maximum, a local minimum, or even a saddle point. For the present time, we will consider only the case where $\mathbf{Q}$ is positive (semi)definite. In the next section, an algorithm will be presented that can be used to efficiently solve the quadratic programming problem.

Before proceeding, it would be most desirable for us to point out the major difference in finding a solution to a linear programming problem and a quadratic programming problem. In linear programming, if a feasible optimizing solution exists to a particular linear programming problem, this solution will always occur at a vertex of the feasible solution space. Furthermore, if the problem possesses N decision variables, there will be exactly M of, say, $M \leqq N$ constraint inequalities satisfied as strict equalities at a nondegenerate optimal solution. In the quadratic programming problem, this need not be the case. In fact, optimality may occur at the intersection of $n \leqq M$ inequalities holding at equality, or at a point strictly interior to all constraint equations. Consider Figs. 11.26 to 11.29 for a two dimensional quadratic programming problem.

Minimize: $$f(\mathbf{x}) = x_1^2 - 4x_1 + 4 + x_2^2 - 2x_2 + 1$$

Note that the following optimal solutions occur for Figs. 11.26 to 11.29. (The objective function of Fig. 11.26 applies to Figs. 11.27 to 11.29.)

Figure	x_1^*	x_2^*	Constraints	Binding Constraints
11.27	2.0	1.0	0	0
11.28	2.0	1.0	2	0
11.29	2.0	1.0	2	1
11.30	1/2	1/2	2	2

These considerations should be noted, for we will subsequently deal with them directly. For the present time, recall the following useful observation. If the objec-

*For illustrative purposes, the quadratic programming problem has been presented as a minimization problem. If the objective function is to be maximized, the same arguments apply. In that case, the requirement would be for Q to appear as a negative definite or negative semidefinite matrix. See Appendix A for details.

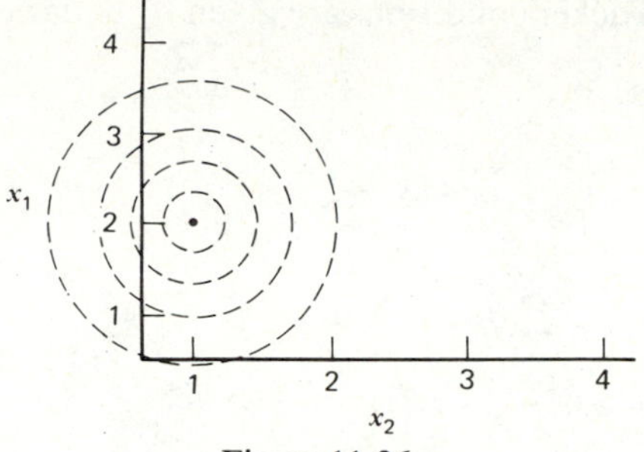

Figure 11.26
Circular contours.

tive function is convex and the constraints linear, then the problem is a *convex programming* problem. Hence, if the Kuhn-Tucker conditions of Section 11.9 are applied to this program, then any set of solution variables satisfying the Kuhn-Tucker conditions must be a global minimizing solution. Let us therefore develop an algorithm capable of finding a solution to the Kuhn-Tucker conditions generated from the quadratic programming problem of Eqs. 11.100 to 11.102.

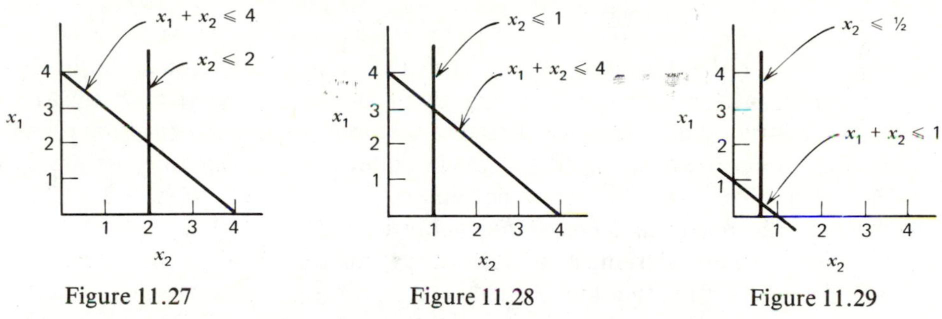

Figure 11.27 Figure 11.28 Figure 11.29

For clarity, consider the following algebraic representation of the above equations. (Without loss of generality, assume q is symmetric.)

$$\text{Minimize:} \qquad f(\mathbf{x}) = \sum_{j=1}^{n} c_j x_j + \sum_{j=1}^{n} \sum_{k=1}^{n} x_j q_{jk} x_k \tag{11.103}$$

$$\text{Subject to:} \quad g_i(\mathbf{x}) = \sum_{j=1}^{n} a_{ij} x_j \geqq b_i \qquad i = 1, 2, \ldots, m \tag{11.104}$$

$$\mathbf{x} \geqq 0$$

Comparing this quadratic programming problem to Program IV (Section 11.9)

$$g_i(\mathbf{x}) = \sum_{j=1}^{n} a_{ij} x_j - b_i \qquad i = 1, 2, \ldots, m$$

$$L(\mathbf{x}, \mu) = \sum_{j=1}^{n} c_j x_j + \sum_{j=1}^{n} \sum_{k=1}^{n} x_j q_{jk} x_k - \sum_{i=1}^{m} \mu_i \left[\sum_{j=1}^{n} (a_{ij} x_j - b_i) \right]$$

The associated Kuhn-Tucker conditions are given by equations A through F.

A. $\mathbf{x} \geqq 0$

$$\text{B.}\quad x_j\left[c_j + 2\sum_{k=1}^{n} q_{jk}x_k - \sum_{i=1}^{m} \mu_i a_{ij}\right] = 0 \tag{11.105}$$

$$\text{C.}\quad c_j + 2\sum_{k=1}^{n} q_{jk}x_k - \sum_{i=1}^{m} \mu_i a_{ij} \geqq 0 \qquad j = 1, 2, \ldots, n \tag{11.106}$$

D. $\mu_i \geqq 0$

$$\text{E.}\quad \mu_i\left[\sum_{j=1}^{n} a_{ij}x_j - b_i\right] = 0 \qquad i = 1, 2, \ldots, m \tag{11.107}$$

$$\text{F.}\quad \sum_{j=1}^{n} a_{ij}x_j - b_i \geqq 0 \qquad i = 1, 2, \ldots, m \tag{11.108}$$

Let the nonnegative variables v_j and s_i denote the bracketed quantities in Eqs. 11.105 and 11.107, respectively. This reduces Eqs. 11.105 and 11.107 to

$$\mu_i\left[\sum_{j=1}^{n} a_{ij}x_j - b_i\right] = \mu_i s_i = 0 \qquad i = 1, 2, \ldots, m \tag{11.109}$$

$$v_j[x_j] = 0 \qquad j = 1, 2, \ldots, n \tag{11.110}$$

Note from Eq. 11.109 that since $g_i(\mathbf{x})$ is simply the ith inequality constraint in the original problem formulation, this implies that either μ_i *or* $g_i(\mathbf{x})$ or both must be zero at optimality. This condition is known as *complementary slackness*, and has the following interpretation: If the Lagrangian multiplier for the ith constraint is positive, then that constraint must be binding at optimality ($g_i(\mathbf{x}) \equiv 0$). Conversely, if the ith constraint is *nonbinding* at optimality ($g_i(\mathbf{x}) > 0$) this implies that the Lagrangian multiplier must be *zero* at optimality. Finally, note that Eq. 11.108 is actually stating that the original m constraints must be satisfied once we find an (optimal) solution. In summary, the Kuhn-Tucker conditions are as follows:

$$2\sum_{k=1}^{n} q_{jk}x_k - \sum_{i=1}^{m} \mu_i a_{ij} - v_j = -c_j \qquad j = 1, 2, \ldots, n \tag{11.111}$$

$$\sum_{j=1}^{n} a_{ij}x_j - s_i = b_i \qquad i = 1, 2, \ldots, m \tag{11.112}$$

$$\mu_i s_i \equiv 0 \qquad i = 1, 2, \ldots, m \tag{11.113}$$

$$v_j x_j = 0 \qquad j = 1, 2, \ldots, n \tag{11.114}$$

$$\mathbf{x} \geqq 0 \qquad \mu \geqq 0 \qquad \mathbf{v} \geqq 0 \qquad \mathbf{s} \geqq 0$$

We now have a total of $(m + n)$ additional *nonlinear* equations which if observed would yield a stationary point to the original quadratic program. In this case, since $\mathbf{Q}$ is positive (semi)definite, the solution is guaranteed to be the global (optimum) solution. Wolfe (31) has exploited this structure to form a simplex-based algorithm to solve this problem. Note that if a feasible solution can be found to Eqs. 11.111 and 11.112, while maintaining the relations of Eqs. 11.113

and 11.114 the problem is solved. Solutions to the $(m+n)$ linear equations are easily obtained through a simplex operation on Eqs. 11.111 and 11.112 using a Phase I procedure (see Chapter 2). The conditions of Eqs. 11.113 and 11.114 can be maintained by noting that when a change of basis occurs, and for the present time assuming nondegeneracy, the following rules must be strictly satisfied:

1. If a decision variable x_j is a basic variable, the variable v_j may not be considered as a candidate for entering the basis, and vice versa.
2. If the variable μ_i is a basic variable, the variable s_i may not be considered as a candidate for entering the basis and vice versa.

Since this technique uses the basic simplex algorithm, it is called the Simplex Method for Quadratic Programming. A damaging characteristic of this procedure is that this method may fail to converge when the matrix of the quadratic form (Q) is positive semidefinite (instead of positive-definite). A more efficient and simple method for solving the Kuhn-Tucker conditions 11.111 through 11.114 has been developed and is known as the *complementary pivot method*. This is computationally more attractive than most of the methods available for solving quadratic programming problems when Q is positive semidefinite. This method has been developed as a general procedure for solving a special class of problems known as the *complementary problem* and is discussed in detail in Section 11.11.

11.11 COMPLEMENTARY PIVOT ALGORITHMS

Consider the general problem of finding a nonnegative solution to a system of equations of the following form:

Find vectors **x** and **z** such that

$$\mathbf{w} = \mathbf{M}\mathbf{z} + \mathbf{q} \tag{11.115}$$

$$\mathbf{w} \geqq 0, \qquad \mathbf{z} \geqq 0 \tag{11.116}$$

$$\mathbf{w}'\mathbf{z} = 0 \tag{11.117}$$

where **M** is an $(n \times n)$ square matrix and **w**, **z**, **q** are n-dimensional column vectors.

The above problem is known as a *complementary problem*. Note that there is no objective function to minimize or maximize in this formulation. Condition 11.115 represents a system of simultaneous linear equations; Condition 11.116 requires the solution to Condition 11.115 be nonnegative; Condition 11.117 implies $w_i z_i = 0$ for all $i = 1, 2, \ldots, n$ since $w_i, z_i \geqq 0$. Thus we have a single nonlinear constraint.

As an illustration, consider a problem with

$$\mathbf{M} = \begin{pmatrix} 1 & 2 & 3 \\ 4 & 5 & 6 \\ 6 & 7 & 8 \end{pmatrix}$$

and

$$\mathbf{q} = \begin{pmatrix} 2 \\ -5 \\ -3 \end{pmatrix}$$

The *complementary problem* is given by

Find $w_1, w_2, w_3, z_1, z_2, z_3$

Such that

$$
\begin{aligned}
w_1 &= z_1 + 2z_2 + 3z_3 + 2 \\
w_2 &= 4z_1 + 5z_2 + 6z_3 - 5 \\
w_3 &= 6z_1 + 7z_2 + 8z_3 - 3 \\
w_1, w_2, w_3, z_1, z_2, z_3 &\geqq 0 \\
w_1 z_1 + w_2 z_2 + w_3 z_3 &= 0
\end{aligned}
$$

Applications

The two important applications of the complementary problem are to solve linear and convex quadratic programming problems by converting them to an equivalent complementary problem.

Linear Programming. Consider the linear program:

Minimize: $\mathbf{c}'\mathbf{x}$

Subject to: $\mathbf{Ax} \geqq \mathbf{b}$

$\mathbf{x} \geqq 0$

where $\mathbf{A}$ is an $(m \times n)$ matrix, $\mathbf{c}$ and $\mathbf{x}$ are $(n \times 1)$ column vectors, and $\mathbf{b}$ is an $(m \times 1)$ column vector ("prime" denotes the transpose of a vector or matrix).

From duality theory, the dual of the above linear program is given by

Maximize: $\mathbf{b}'\mathbf{y}$

Subject to: $\mathbf{A}'\mathbf{y} \leqq \mathbf{c}$

$\mathbf{y} \geqq 0$

According to the complementary slackness theorem (see Section 4.2), if there exists feasible solutions to the primal and the dual problems satisfying the complementary slackness conditions, then the feasible solutions are in fact optimal to their respective problems. Thus to find an optimal solution to the linear program, we can solve the following problem.

Find vectors $\mathbf{x}$, $\mathbf{y}$, $\mathbf{u}$, $\mathbf{v}$ such that

$$\mathbf{v} = -\mathbf{A}'\mathbf{y} + \mathbf{c} \tag{11.118}$$

$$\mathbf{u} = \mathbf{Ax} - \mathbf{b} \tag{11.119}$$

$$\mathbf{x}, \mathbf{y}, \mathbf{u}, \mathbf{v} \geqq 0 \tag{11.120}$$

$$\mathbf{v}'\mathbf{x} + \mathbf{u}'\mathbf{y} = 0 \tag{11.121}$$

$\mathbf{u}$ and $\mathbf{v}$ denote the vectors of slack variables of the primal and dual problems, respectively.

Note that Equs. 11.118 through 11.120 represent the primal-dual feasibility while Eq. 11.121 represents the complementary slackness condition. Comparing Eqs. 11.118 through 11.121 to the complementary problem previously defined, we get the equivalent complementary problem for a linear program as

$$\mathbf{M} = \begin{pmatrix} \mathbf{0} & -\mathbf{A}' \\ \mathbf{A} & \mathbf{0} \end{pmatrix} \qquad \mathbf{w} = \begin{pmatrix} \mathbf{v} \\ \mathbf{u} \end{pmatrix} \qquad \mathbf{z} = \begin{pmatrix} \mathbf{x} \\ \mathbf{y} \end{pmatrix}$$

and

$$\mathbf{q} = \begin{pmatrix} \mathbf{c} \\ -\mathbf{b} \end{pmatrix}$$

It should be noted that **M** is a square, asymmetric, positive semidefinite matrix of order $(m+n)$.

Convex Quadratic Programming. Consider a convex quadratic programming problem of the form:

$$\begin{aligned} \text{Minimize:} \quad & f(\mathbf{x}) = \mathbf{c}'\mathbf{x} + \mathbf{x}'\mathbf{Q}\mathbf{x} \\ \text{Subject to:} \quad & \mathbf{A}\mathbf{x} \geqq \mathbf{b} \\ & \mathbf{x} \geqq 0 \end{aligned}$$

where: **Q** is an $(n \times n)$ matrix.

Assume **Q** is symmetric and is positive-definite or positive semidefinite.

In matrix notation, the Kuhn-Tucker optimality conditions to the above convex quadratic program can be written as follows. (See Eqs. 11.111 through 11.114 in Section 11.10.)

Find vectors $\mathbf{x}$, μ, $\mathbf{v}$, $\mathbf{s}$ such that

$$\mathbf{v} = 2\mathbf{Q}\mathbf{x} - \mathbf{A}'\mu + \mathbf{c} \tag{11.122}$$

$$\mathbf{s} = \mathbf{A}\mathbf{x} - \mathbf{b} \tag{11.123}$$

$$\mathbf{x}, \mu, \mathbf{v}, \mathbf{s} \geqq 0$$

$$\mathbf{v}'\mathbf{x} + \mathbf{s}'\mu = 0$$

comparing the above system of equations to the complementary problem, we note that

$$\mathbf{w} = \begin{pmatrix} \mathbf{v} \\ \mathbf{s} \end{pmatrix}, \quad \mathbf{z} = \begin{pmatrix} \mathbf{x} \\ \mu \end{pmatrix}, \quad \mathbf{M} = \begin{pmatrix} 2\mathbf{Q} & -\mathbf{A}' \\ \mathbf{A} & 0 \end{pmatrix}, \quad \text{and} \quad \mathbf{q} = \begin{pmatrix} \mathbf{c} \\ -\mathbf{b} \end{pmatrix}$$

Thus an optimal solution to the convex quadratic program may be obtained by solving the 3equivalent complementary problem shown above. It should be again noted that the matrix **M** is positive semidefinite, since **Q** is positive definite or positive semidefinite.

We shall illustrate the transformation of a linear and quadratic program to a complementary problem with the following examples.

EXAMPLE 11.11-1

Consider the linear program:

$$\begin{aligned} \text{Minimize:} \quad & f(\mathbf{x}) = 2x_1 + 3x_2 - x_3 \\ \text{Subject to:} \quad & 2x_1 - x_2 + 5x_3 \geqq 1 \\ & 3x_1 + x_2 - x_3 = 4 \\ & x_1 - x_2 + x_3 = 2 \\ & x_1, x_2, x_3 \geqq 0 \end{aligned}$$

Before the linear program can be converted to a complementary problem, the equality constraints have to be converted to inequalities as follows:

$$3x_1 + x_2 - x_3 \geqq 4$$

$$x_1 - x_2 + x_3 \geqq 2$$

$$(3x_1 + x_2 - x_3) + (x_1 - x_2 + x_3) \leqq 6$$

The last inequality simplifies to $4x_1 \leqq 6$. The linear program may thus be rewritten as

$$\begin{aligned} \text{Minimize:} \quad & f(\mathbf{x}) = 2x_1 + 3x_2 - x_3 \\ \text{Subject to:} \quad & 2x_1 - x_2 + 5x_3 \geqq 1 \\ & 3x_1 + x_2 - x_3 \geqq 4 \\ & x_1 - x_2 + x_3 \geqq 2 \\ & -4x_1 \geqq -6 \\ & x_1, x_2, x_3 \geqq 0 \end{aligned}$$

Comparing the above linear program to the notations previously introduced, we obtain:

$$\underset{(4\times 3)}{\mathbf{A}} = \begin{bmatrix} 2 & -1 & 5 \\ 3 & 1 & -1 \\ 1 & -1 & 1 \\ -4 & 0 & 0 \end{bmatrix}, \quad \mathbf{b} = \begin{pmatrix} 1 \\ 4 \\ 2 \\ -6 \end{pmatrix} \quad \text{and} \quad \mathbf{c} = \begin{pmatrix} 2 \\ 3 \\ -1 \end{pmatrix}$$

The equivalent complementary problem is given as:

$$\underset{(7\times 7)}{\mathbf{M}} = \begin{bmatrix} 0 & -\mathbf{A}' \\ \mathbf{A} & 0 \end{bmatrix} = \begin{bmatrix} 0 & 0 & 0 & -2 & -3 & -1 & 4 \\ 0 & 0 & 0 & 1 & -1 & 1 & 0 \\ 0 & 0 & 0 & -5 & 1 & -1 & 0 \\ 2 & -1 & 5 & 0 & 0 & 0 & 0 \\ 3 & 1 & -1 & 0 & 0 & 0 & 0 \\ 1 & -1 & 1 & 0 & 0 & 0 & 0 \\ -4 & 0 & 0 & 0 & 0 & 0 & 0 \end{bmatrix}$$

$$\underset{(7\times 1)}{\mathbf{q}} = \begin{bmatrix} \mathbf{c} \\ -\mathbf{b} \end{bmatrix} = \begin{bmatrix} 2 \\ 3 \\ -1 \\ -1 \\ -4 \\ -2 \\ 6 \end{bmatrix}$$

The values of z_1, z_2, and z_3 produce the solution to x_1, x_2, and x_3; whereas z_4, z_5, z_6 and z_7 correspond to the dual solution.

EXAMPLE 11.11-2

Consider a convex quadratic programming problem:

$$\begin{aligned} \text{Minimize:} \quad & f(\mathbf{x}) = -6x_1 + 2x_1^2 - 2x_1x_2 + 2x_2^2 \\ \text{Subject to:} \quad & -x_1 - x_2 \geqq -2 \\ & x_1, x_2 \geqq 0 \end{aligned}$$

For this problem

$$\mathbf{A} = (-1 \quad -1), \quad \mathbf{b} = (-2), \quad \mathbf{c} = \begin{pmatrix} -6 \\ 0 \end{pmatrix}, \quad \text{and} \quad \mathbf{Q} = \begin{bmatrix} 2 & -1 \\ -1 & 2 \end{bmatrix}$$

The equivalent complementary problem is given by

$$\underset{(3\times 3)}{\mathbf{M}} = \begin{pmatrix} \mathbf{Q}+\mathbf{Q}' & -\mathbf{A}' \\ \mathbf{A} & \mathbf{0} \end{pmatrix} = \begin{pmatrix} 4 & -2 & 1 \\ -2 & 4 & 1 \\ -1 & -1 & 0 \end{pmatrix}$$

and

$$\mathbf{q} = \begin{pmatrix} \mathbf{c} \\ -\mathbf{b} \end{pmatrix} = \begin{pmatrix} -6 \\ 0 \\ 2 \end{pmatrix}$$

The solution to z_1 and z_2 correspond to the optimal values of x_1 and x_2, since the matrix of the quadratic form (**Q**) is positive definite.

An Algorithm to Solve the Complementary Problem

Consider the complementary problem:

Find vectors **w** and **z** such that

$$\mathbf{w} = \mathbf{M}\mathbf{z} + \mathbf{q}$$
$$\mathbf{w}, \mathbf{z} \geqq 0$$
$$\mathbf{w}'\mathbf{z} = 0$$

Definitions

1. *Feasible Solution.* A nonnegative solution (**w**, **z**) to the system of the equation $\mathbf{w} = \mathbf{M}\mathbf{z} + \mathbf{q}$ is called a feasible solution to the complementary problem.
2. *Complementary Solution.* A feasible solution (**w**, **z**) to the complementary problem that also satisfies the complementarity condition $\mathbf{w}'\mathbf{z} = 0$ is called a complementary solution.

The condition $\mathbf{w}'\mathbf{z} = 0$ is equivalent to $w_i z_i = 0$ for all i. The variables w_i and z_i for each i is called a *complementary pair* of variables. Note that if the elements of the vector **q** are nonnegative, then there exists an obvious complementary solution given by $\mathbf{w} = \mathbf{q}$, $\mathbf{z} = 0$. Hence, the complementary problem is nontrivial only when at least one of the elements of **q** is negative. This means that the initial basic solution given by $\mathbf{w} = \mathbf{q}$, $\mathbf{z} = 0$ is *infeasible* to the complementary problem even though it satisfies the complementary condition $\mathbf{w}'\mathbf{z} = 0$.

At present there exists no general algorithm to solve all the complementary problems. When the matrix **M** satisfies certain special properties, a *Complementary Pivot Method* has been developed by Lemke (17) to determine the complementary solution if one exists. Specifically, the complementary pivot method is guaranteed to find a complementary solution when **M** satisfies any one of the following properties.

1. All the elements of **M** are positive.
2. The matrix **M** is positive definite.
3. All the principal determinants of the matrix **M** are positive.*

In other words, when **M** satisfies any of the above conditions, there always exists a complementary solution to the complementary problem irrespective of what values the elements of **q** assume.

We have previously stated that linear and convex quadratic programming problems give rise to a complementary problem where the matrix **M** is positive semidefinite. Under this case the complementary pivot method is guaranteed to terminate with a complementary solution only when a solution exists for that particular problem. In other words, it is possible for the complementary problem not to have a solution, since some linear and quadratic programs may not have optimal solutions.

*See Appendix A for the definition of principal determinants.

The Complementary Pivot Method

The complementary pivot method starts with the infeasible basic solution given by $\mathbf{w}=\mathbf{q}$; $\mathbf{z}=0$. To make the solution nonnegative, an artificial variable z_0 is added at a sufficiently positive value to each of the equations in the $\mathbf{w}-\mathbf{M}\mathbf{z}=\mathbf{q}$ system, so that the right-hand-side constants (q_i+z_0) become nonnegative. The value of z_0 will be the absolute value of the most negative q_i. We now have a basic solution given by

$$w_i=q_i+z_0, \qquad z_i=0 \qquad \text{for all } i=1, \ldots, n,$$

and

$$z_0=-\min_i(q_i)$$

Note that, even though this solution is nonnegative, satisfies the constraints and is complementary $(w_i z_i=0)$, it is not feasible because of the presence of the artificial variable z_0 at a positive value. We shall call such a solution an *almost complementary solution*.

The first step in the complementary pivot method is to find an almost complementary solution, by augmenting the original system of Eq. $(\mathbf{w}=\mathbf{M}\mathbf{z}+\mathbf{q})$ by an artificial variable z_0 as follows:

$$\begin{aligned} \mathbf{w}-\mathbf{M}\mathbf{z}-\mathbf{e}z_0&=\mathbf{q}\\ \mathbf{w}, \mathbf{z}, z_0&\geqq 0\\ \mathbf{w}'\mathbf{z}&=0 \end{aligned}$$

where

$$\underset{(n\times 1)}{\mathbf{e}}=(1, 1, \ldots, 1)'$$

Thus, the initial tableau becomes:

Basis	$w_1 \cdots w_s \cdots w_n$			$z_1 \cdots z_s \cdots z_n$			z_0	$\mathbf{q}$
w_1	1			$-m_{11}$	$-m_{1s}$	$-m_{1n}$	-1	q_1
w_s		1		$-m_{s1}$	$-m_{ss}$	$-m_{sn}$	-1	q_s
w_n			1	$-m_{n1}$	$-m_{ns}$	$-m_{nn}$	-1	q_n

where the m_{ij}'s are the elements of the $\mathbf{M}$ matrix.

Step I. To determine the initial almost complementary solution, the variable z_0 is brought into the basis replacing the basic variable with the most negative value. (Let $q_s=\min q_i<0$.) This implies that z_0 replaces w_s in the basis. Performing the necessary pivot operation yields the following tableau:

Tableau 1

Basis	$w_1 \cdots w_s \cdots w_n$			$z_1 \cdots z_s \cdots z_n$			z_0	$\mathbf{q}$
w_1	1	-1	0	m'_{11}	m'_{1s}	m'_{1n}	0	q'_1
z_0	0	-1	0	m'_{s1}	m'_{ss}	m'_{sn}	1	q'_s
w_n	0	-1	1	m'_{n1}	m'_{ns}	m'_{nn}	0	q'_n

where

$$q_s' = -q_s; \quad q_i' = q_i - q_s \qquad \text{for all } i \neq s;$$

$$m_{sj}' = \frac{-m_{sj}}{-1} = m_{sj} \qquad \text{for all } j = 1, \ldots, n;$$

$$m_{ij}' = -m_{ij} + m_{sj} \qquad \text{for all } j = 1, \ldots, n \qquad \text{and } i \neq s.$$

Note that

1. $q_i' \geqq 0 \quad i = 1, \ldots, n$
2. The basic solution $w_1 = q_1', \ldots, w_{s-1} = q_{s-1}', z_0 = q_s', w_{s+1}, = q_{s+1}', \ldots, w_n = q_n'$, and all other variables initially zero is an almost complementary solution.
3. The almost complementary solution becomes a complementary solution as soon as the value of z_0 is reduced to zero.

In essence the complementary pivot algorithm proceeds to find a sequence of almost complementary solutions (tableaus) until z_0 becomes zero. To do this, the basis changes must be done in such a way that the following conditions are met:

A. The complementarity between the variables must be maintained (i.e., $w_i z_i = 0$ for all $i = 1, \ldots, n$).
B. The basic solution remains nonnegative (i.e., the right-hand-side constants must be nonnegative in all tableaus).

Step II. In order to satisfy Condition A, we observe that the variables w_s and z_s are both out of the basis in Tableau 1. As long as either one of them is made basic, the complementarity between **w** and **z** variables will still be maintained. Since w_s just came out of the basis, the choice is naturally to bring z_s into the basis. Thus we have a simple rule for selecting the nonbasic variable to enter the basis in the next tableau. It is always the complement of the basic variable which just left the basis in the last tableau. This is called the *complementary rule*.

After selecting the variable to enter the basis, we have to determine the basic variable to leave. This is done by applying the *Minimum Ratio Test* similar to the one used in the simplex method so that Condition B is satisfied. Therefore to determine the variable to leave the basis, the following ratios are formed:

$$\frac{q_i'}{m_{is}'} \qquad \text{for those } i = 1, \ldots, n \text{ for which } m_{is}' > 0$$

Let

$$\frac{q_k'}{m_{ks}'} = \underset{m_{is}' > 0}{\text{Minimum}} \left(\frac{q_i'}{m_{is}'} \right)$$

This implies that the basic variable w_k leaves the basis, to be replaced by z_s. We now obtain the new tableau by performing the pivot operation with m_{ks}' as the pivot element.

Step III. Since w_k left the basis, the variable z_k is brought into the basis by the complementary rule, and the basis changes are continued as before until one of two things happens which indicates termination of the algorithm:

1. The minimum ratio is obtained in row s and z_0 leaves the basis. The resulting basic solution after performing the pivot operation is the complementary solution.
2. The minimum ratio test fails, since all the coefficients in the pivot column are nonpositive. This implies that there exists *no* solution to the complementary problem. In this case, we say that the complementary problem has a *Ray Solution* (17).

REMARKS

1. It has been shown that the complementary pivot method always terminates with a complementary solution in a finite number of steps whenever (a) all the elements of **M** are positive, or (b) **M** has positive principal determinants (includes the case where **M** is positive definite).

2. The most important application of complementary pivot theory is in solving linear and convex quadratic programming problems. We have seen that under these cases, **M** is a positive semidefinite matrix. It has been proved that whenever **M** is positive semidefinite, the algorithm will terminate with a complementary solution if one exists for that problem. In other words, termination 2 implies that the given linear program or quadratic program has no optimal solution.

For a proof of the above remarks the reader is referred to Cottle and Dantzig (7).

Let us illustrate the complementary pivot method for solving a convex quadratic progrma using Example 11.11-2.

Minimize: $f(\mathbf{x}) = -6x_1 + 2x_1^2 - 2x_1x_2 + 2x_2^2$

Subject to: $-x_1 - x_2 \geqq -2$

$x_1, x_2 \geqq 0$

The equivalent complementary problem is given below:

$$\underset{(3\times 3)}{\mathbf{M}} = \begin{pmatrix} 4 & -2 & 1 \\ -2 & 4 & 1 \\ -1 & -1 & 0 \end{pmatrix} \quad \text{and} \quad \mathbf{q} = \begin{pmatrix} -6 \\ 0 \\ 2 \end{pmatrix}$$

Since all the elements of **q** are not nonnegative, an artificial variable z_0 is added to every equation. The initial tableau is given below:

Tableau 1

Basis	w_1	w_2	w_3	z_1	z_2	z_3	z_0	**q**
w_1	1	0	0	−4	2	−1	(−1)	−6
w_2	0	1	0	2	−4	−1	−1	0
w_3	0	0	1	1	1	0	−1	2

The initial basic solution is $w_1 = -6$, $w_2 = 0$, $w_3 = 2$, $z_1 = z_2 = z_3 = z_0 = 0$. An almost complementary solution is obtained by replacing w_1 by z_0 as shown in Tableau 2. The almost complementary solution is given by $z_0 = 6$, $w_2 = 6$, $w_3 = 8$, $z_1 = z_2 = z_3 = w_1 = 0$. Since the complementary pair (w_1, z_1) is out of the basis, either w_1 or z_1 can be made a basic variable without affecting the complementarity between all pairs of variables ($w_i z_i = 0$). Since w_1 just left the basis, we bring z_1 into the basis. Applying the minimum ratio test, we obtain the ratios as (6/4, 6/6, 8/5). This implies that z_1 replaces w_2 in the basis. Tableau 3 gives the new almost complementary solution after the pivot operation. By applying the complementary rule, z_2 is selected as the next basic variable (w_2 just left the basis). The minimum ratio test determines w_3 as the basic variable to leave. The next almost complementary solution after the pivot operation is shown in Tableau 4.

Tableau 2

Basis	w_1	w_2	w_3	z_1	z_2	z_3	z_0	$\mathbf{q}$
z_0	−1	0	0	4	−2	1	1	6
w_2	−1	1	0	(6)	−6	0	0	6
w_3	−1	0	1	5	−1	1	0	8

Tableau 3

Basis	w_1	w_2	w_3	z_1	z_2	z_3	z_0	$\mathbf{q}$
z_0	$-\frac{1}{3}$	$-\frac{2}{3}$	0	0	2	1	1	2
z_1	$-\frac{1}{6}$	$\frac{1}{6}$	0	1	−1	0	0	1
w_3	$-\frac{1}{6}$	$-\frac{5}{6}$	1	0	(4)	1	0	3

By the complementary rule, z_3 becomes the next basic variable. Application of the minimum ratio test results in the replacement of z_0 from the basis. This implies that the next tableau will correspond to a complementary solution as shown in Tableau 5.

Tableau 4

Basis	w_1	w_2	w_3	z_1	z_2	z_3	z_0	$\mathbf{q}$
z_0	$-\frac{1}{4}$	$-\frac{1}{4}$	$-\frac{1}{2}$	0	0	$\left(\frac{1}{2}\right)$	1	$\frac{1}{2}$
z_1	$-\frac{5}{24}$	$-\frac{1}{24}$	$\frac{1}{4}$	1	0	$\frac{1}{4}$	0	$\frac{7}{4}$
z_2	$-\frac{1}{24}$	$-\frac{5}{24}$	$\frac{1}{4}$	0	1	$\frac{1}{4}$	0	$\frac{3}{4}$

The complementary solution is given by $z_1 = 3/2$, $z_2 = 1/2$, $z_3 = 1$, $w_1 = w_2 = w_3 = 0$.

Hence, the optimal solution to the given quadratic program becomes

$$x_1^* = \frac{3}{2}, \qquad x_2^* = \frac{1}{2}, \qquad \text{and} \qquad f(\mathbf{x}^*) = -\frac{11}{2}$$

Tableau 5

Basis	w_1	w_2	w_3	z_1	z_2	z_3	z_0	**q**
z_3	$-\frac{1}{2}$	$-\frac{1}{2}$	-1	0	0	1	2	1
z_1	$-\frac{1}{12}$	$-\frac{1}{12}$	$\frac{1}{2}$	1	0	0	$-\frac{1}{2}$	$\frac{3}{2}$
z_2	$-\frac{1}{12}$	$-\frac{1}{12}$	$\frac{1}{2}$	0	1	0	$-\frac{1}{2}$	$\frac{1}{2}$

Efficiency of the Complementary Pivot Method

A computer program to solve a complementary problem is given in Ravindran (21). In an experimental study conducted by Ravindran (22), the complementary pivot method has been tested against the simplex method for solving linear programming problems. The study reveals the superiority of the complementary pivot method over the simplex method in a number of randomly generated problems, both with regard to the number of iterations and computation time. A recent study by Ravindran and Lee (23) has shown the superiority of the complementary pivot method for solving convex quadratic programming problems as well.

11.12 SEPARABLE PROGRAMMING

Separable programming was first introduced by C. E. Miller in 1963. Beale (1) refers to separable programming as "probably the most useful nonlinear programming technique."

In separable programming, nonlinear programming problems are solved by approximating the nonlinear functions with piecewise linear functions and then solving the optimization problem through the use of a modified simplex algorithm of linear programming, or in special cases, the ordinary simplex algorithm.

A basic assumption in separable programming is that all functions in the problem be separable. Consider the following function of two variables:

$$f(x_1, x_2) = x_1^3 - 2x_1^2 + x_1 + x_2^4 - x_2 \tag{11.124}$$

This function is separable because it can be "separated" into two functions, each a function of one variable:

$$f(x_1, x_2) = f_1(x_1) + f_2(x_2) \tag{11.125}$$

where

$$f_1(x_1) = x_1^3 - 2x_1^2 + x_1 \tag{11.126}$$

and

$$f_2(x_2) = x_2^4 - x_2 \tag{11.127}$$

Hadley (12) shows how one can approximate a nonlinear separable function. Consider an arbitrary continuous function $f(x)$ of a single variable x, which is defined for all x, $0 \leqq x \leqq a$. This function might have the form illustrated in Fig. 11.30. Suppose we arbitrarily choose some points (refer to them as grid points)

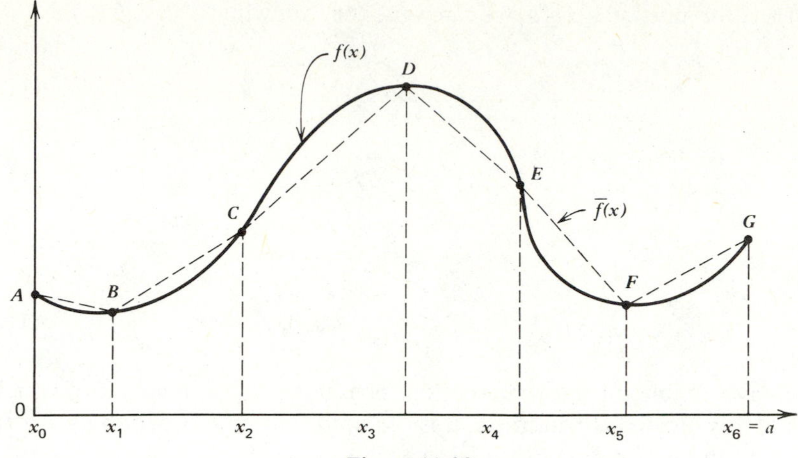

Figure 11.30
Approximation of a nonlinear function.

within the range of possible values for x, as shown in Fig. 11.30. Now, if for each x_k we compute $f_k = f(x_k)$ and connect the points (x_k, f_k) and (x_{k+1}, f_{k+1}) we will have formed our approximation function $\bar{f}(x)$, which is a piecewise linear function.

From Fig. 11.30 we can see that depending on the function being approximated and the distance between the grid points, the approximation is very close to the original function in some regions and in other regions it is not close at all. A better approximation can be achieved by adding more grid points to the grid and making the grid points closer together. The grid points do not have to be equally spaced.

To make use of our approximating function $\bar{f}(x)$ we must show how to express it analytically. Referring again to Fig. 11.30 when x is in the interval $x_k \leqq x \leqq x_{k+1}$, $f(x)$ is being approximated by $\bar{f}(x)$ where

$$\bar{f}(x) = f_k + \frac{f_{k+1} - f_k}{x_{k+1} - x_k}(x - x_k) \tag{11.128}$$

If x lies in the interval $x_k \leqq x \leqq x_{k+1}$ then it can be written

$$x = \lambda x_{k+1} + (1 - \lambda)x_k \tag{11.129}$$

for some λ, $0 \leqq \lambda \leqq 1$.

Solving Eq. 11.129 for $(x - x_k)$ we obtain

$$(x - x_k) = \lambda(x_{k+1} - x_k) \tag{11.130}$$

and substituting in Eq. 11.128 we have

$$\bar{f}(x) = \lambda f_{k+1} + (1 - \lambda)f_k \tag{11.131}$$

Now, letting $\lambda = \lambda_{k+1}$ and $1 - \lambda = \lambda_k$ we see that when $x_k \leqq x \leqq x_{k+1}$, there exist a unique λ_k and λ_{k+1} such that

$$x = \lambda_k x_k + \lambda_{k+1} x_{k+1} \tag{11.132}$$

$$\bar{f}(x) = \lambda_k f_k + \lambda_{k+1} f_{k+1} \tag{11.133}$$

$$\lambda_k + \lambda_{k+1} = 1 \tag{11.134}$$

where $\lambda_k, \lambda_{k+1} \geqq 0$.

In fact, for any x, $0 \leqq x \leqq a$, we can write the following:

$$x = \sum_{k=0}^{r} \lambda_k x_k \tag{11.135}$$

$$\bar{f}(x) = \sum_{k=0}^{r} \lambda_k f_k \tag{11.136}$$

where

$$\sum_{k=0}^{r} \lambda_k = 1, \qquad \lambda_k \geqq 0; \qquad k = 0, \ldots, r \tag{11.137}$$

and r is any suitable integer representing the number of segments into which the domain of x is divided. In addition, it is required that no more than two of the λ_k be positive, and if two are positive, say λ_k and λ_s, $s > k$, it must also be true that $s = k + 1$, or in other words, the λ's must be adjacent. These restrictions ensure that the approximating function will indeed be the dashed curve in Fig. 11.31.

Referring to Fig. 11.31 again, if λ_0 and λ_1 were positive and $\lambda_2, \ldots, \lambda_6$ were zero, then Eqs. 11.135 and 11.136 would give us a point on the line segment joining points A and B, a portion of our approximation function. However, if for instance λ_0 and λ_2 were allowed to be positive with $\lambda_1, \lambda_3, \ldots, \lambda_6$ equal to zero, we would then have a point on a line connecting A and C, which is not part of the approximation function, $\bar{f}(x)$; and if only λ_3 were positive, it must then equal 1, all other λ's would equal zero, and we would have point D, both on $f(x)$ and $\bar{f}(x)$.

McMillan (18) states that any continuous, nonlinear, and separable function $f(x_1, x_2, \ldots, x_n)$ can be approximated by a piecewise linear function and solved using a linear programming solution technique provided that the following prescription is applied.

$$f(\mathbf{x}) \approx \bar{f} = \sum_{k=0}^{r} \lambda_{k1} f_{k1} + \sum_{k=0}^{r} \lambda_{k2} f_{k2} + \cdots + \sum_{k=0}^{r} \lambda_{kn} f_{kn} \tag{11.138}$$

where

$$x_n = \sum_{k=0}^{r} \lambda_{kn} x_{kn} \qquad (r \text{ is defined as above}) \tag{11.139}$$

given that

1. $\sum_{k=0}^{r} \lambda_{kj} = 1; \qquad j = 1, \ldots, n$ (11.140)
2. $\lambda_{kj} \geqq 0; \qquad k = 0, \ldots, r, \qquad j = 1, \ldots, n$ (11.141)
3. No more than two of the λ's that are associated with any one variable j are greater than zero, and if two are greater than zero, they must be adjacent.

The following example illustrates the use of the above approximation method and the use of the simplex algorithm of linear programming (modified to maintain Restriction 3 above) to solve the following nonlinear programming problem:

Maximize: $f(\mathbf{x}) = 3x_1 + 2x_2$

Subject to: (1) $g(\mathbf{x}) = 4x_1^2 + x_2^2 \leqq 16$

(2) $x_1, x_2 \geqq 0$

Notice that x_1 and x_2 enter the problem linearly in the objective function. In Constraint 1, however, they are nonlinear in nature. Thus we must write both x_1 and x_2 in terms of the λ's. If a variable is linear throughout the entire problem it is not necessary to write it in terms of the λ's; it can be used as a variable itself. Both the objective function and constraint 1 are separable functions.

$$f = f_1(x_1) + f_2(x_2)$$

where

$$f_1(x_1) = 3x_1$$

and

$$f_2(x_2) = 2x_2$$

Also,

$$g = g_1(x_1) + g_2(x_2)$$

where

$$g_1(x_1) = 4x_1^2$$

and

$$g_2(x_2) = x_2^2$$

Now to approximate our problem by using piecewise linear functions of the form Eq. 11.138 we must first determine the domains of interest for the variables x_1 and x_2. From Constraints 1 and 2 the possible values for x_1 and x_2 are $0 \leqq x_1 \leqq 2$ and $0 \leqq x_2 \leqq 4$, respectively. Dividing the domains of interest for x_1 and x_2 arbitrarily into four segments each, we obtain the grid points shown in Table 11.6. The column labeled x_{k1} shows the grid points for variable x_1, and the column labeled x_{k2} shows the grid points for variable x_2.

The piecewise linear function to be used to approximate f is

$$\bar{f} = \sum_{k=0}^{4} \lambda_{k1} f_{k1} + \sum_{k=0}^{4} \lambda_{k2} f_{k2} \tag{11.142}$$

The approximation function for g is

$$\bar{g} = \sum_{k=0}^{4} \lambda_{k1} g_{k1} + \sum_{k=0}^{4} \lambda_{k2} g_{k2} \tag{11.143}$$

To evaluate Eqs. 11.142 and 11.143 we must compute f_{k1}, f_{k2}, g_{k1}, and g_{k2} at each of the grid points for x_1 and x_2. These values are given in Table 11.7.

Table 11.7
GRID POINTS AND VALUES OF FUNCTIONS AT GRID POINTS

K	x_{k1}	$f_{k1} = 3x_{k1}$	$g_{k1} = 4x_{k1}^2$	x_{k2}	$f_{k2} = 2x_{k1}$	$g_{k2} = x_{k2}^2$
0	0.0	0.0	0.0	0.0	0.0	0.0
1	0.5	1.5	1.0	1.0	2.0	1.0
2	1.0	3.0	4.0	2.0	4.0	4.0
3	1.5	4.5	9.0	3.0	6.0	9.0
4	2.0	6.0	16.0	4.0	8.0	16.0

We can now evaluate Eqs. 11.142 and 11.143 and write our approximation problem as

$$\text{Maximize: } \bar{f}=0.0\lambda_{01}+1.5\lambda_{11}+3.0\lambda_{21}+4.5\lambda_{31}+6.0\lambda_{41} +0.0\lambda_{02}+2.0\lambda_{12}+4.0\lambda_{22}+6.0\lambda_{32}+8.0\lambda_{42}$$

Subject to:

$$(1)\quad \bar{g}=0.0\lambda_{01}+1.0\lambda_{11}+4.0\lambda_{21}+9.0\lambda_{31}+16.0\lambda_{41} +0.0\lambda_{02}+1.0\lambda_{12}+4.0\lambda_{22}+9.0\lambda_{32}+16.0\lambda_{42}\leqq 16$$

To this problem we must add constraints of the form Eq. 11.140.

$$(2)\quad \lambda_{01}+\lambda_{11}+\lambda_{21}+\lambda_{31}+\lambda_{41}=1$$

$$(3)\quad \lambda_{02}+\lambda_{12}+\lambda_{22}+\lambda_{32}+\lambda_{42}=1$$

and the nonnegativity conditions of Eq. 11.141:

$$\lambda_{kj}\geqq 0; \qquad k=0,\ldots,4, \qquad j=1,2$$

Note that this approximating problem to our original nonlinear problem is linear. Thus, we can solve this problem using the simplex algorithm of linear programming if we modify it to ensure that in any basic solution no more than two of the λ's that are associated with either of the x_j variables are greater than zero and if two (rather than one) are greater than zero, then they must be adjacent. These restrictions will be incorporated into a "restricted basis rule" for the simplex algorithm.

Adding slack variables to convert Constraint 1 of the approximating problem to an equality, we have our first simplex tableau in Table 11.8.

Table 11.8
FIRST TABLEAU

c_j	0	1.5	3.0	4.5	6.0	0	2	4	6	8	0	
Basis	λ_{01}	λ_{11}	λ_{21}	λ_{31}	λ_{41}	λ_{02}	λ_{12}	λ_{22}	λ_{32}	λ_{42}	s	b
s	0.00	1.00	4.00	9.00	16.00	0.00	1.00	4.00	9.00	16.00	1.00	16.00
λ_{01}	1.00	1.00	1.00	1.00	1.00	0.00	0.00	0.00	0.00	0.00	0.00	1.00
λ_{02}	0.00	0.00	0.00	0.00	0.00	0.00	1.00	1.00	1.00	(1.00)	0.00	1.00
$\bar{c}$ Row	0.00	1.50	3.00	4.50	6.00	0.00	2.00	4.00	6.00	8.00	0.00	0.00

$\bar{f}=0.0$

The first tableau shows an initial basic feasible solution of: $\lambda_{01}=1.0$, $\lambda_{02}=1.0$, $s=16.0$, giving a value for the objective function of $\bar{f}=0.0$. The relative profit ($\bar{c}$) row indicates that the problem is not yet optimal and that bringing λ_{42} into the basis will make the largest contribution to the objective function. However, under the restricted basis rule we cannot bring λ_{42} into the basis unless it replaces λ_{02} in the basis. This is because we would have two nonadjacent λ's in the basis corresponding to the variable x_2. The minimum ratio rule of selecting the pivot row suggests that we can pivot either in row 1 or row 3. If we pivot in row 1, then λ_{02} will not leave the basis. We can pivot in row 3 to satisfy our restricted basis rule; the result of this pivot is shown in Table 11.9.

Checking the bottom row of the second tableau we see that the problem is not yet optimal. Bringing λ_{41} into the basis would make the largest contribution to the objective function; however, we would have to pivot in row 1 and this would vio-

Table 11.9
SECOND TABLEAU

Basis	λ_{01}	λ_{11}	λ_{21}	λ_{31}	λ_{41}	λ_{02}	λ_{12}	λ_{22}	λ_{32}	λ_{42}	s	b
s	0.00	(1.00)	4.00	9.00	16.00	−16.00	−15.00	−12.00	−7.00	0.00	1.00	0.00
λ_{01}	1.00	1.00	1.00	1.00	1.00	0.00	0.00	0.00	0.00	0.00	0.00	1.00
λ_{42}	0.00	0.00	0.00	0.00	0.00	1.00	1.00	1.00	1.00	0.00	0.00	1.00
$\bar{c}$ Row	0.00	1.60	3.00	4.50	6.00	−8.00	−6.00	−4.00	−2.00	0.00	0.00	8.00

$\bar{f}=8.0$

late the restricted basis rule because we would then have λ_{01} and λ_{41} in the basis. λ_{31} and λ_{21} cannot enter the basis for the same reason. Thus we finally pivot on the circled element. The result is shown in Table 11.10.

Table 11.10
THIRD TABLEAU

Basis	λ_{01}	λ_{11}	λ_{21}	λ_{31}	λ_{41}	λ_{02}	λ_{12}	λ_{22}	λ_{32}	λ_{42}	s	b
λ_{11}	0.00	1.00	4.00	9.00	16.00	−16.00	−15.00	−12.00	−7.00	0.00	1.00	0.00
λ_{01}	1.00	0.00	−3.00	−8.00	−15.00	16.00	15.00	12.00	(7.00)	0.00	−1.00	1.00
λ_{42}	0.00	0.00	0.00	0.00	0.00	1.00	1.00	1.00	1.00	1.00	0.00	1.00
$\bar{c}$ Row	0.00	0.00	−3.00	−9.00	−18.00	16.00	16.50	14.00	8.50	0.00	1.50	8.00

$\bar{f}=8.0$

Using the restricted basis rule again, we find that we can pivot on the circled element in the third tableau and it will improve the value of the objective function. Doing this, we obtain the tableau in Table 11.11.

Table 11.11
FOURTH TABLEAU

Basis	λ_{01}	λ_{11}	λ_{21}	λ_{31}	λ_{41}	λ_{02}	λ_{12}	λ_{22}	λ_{32}	λ_{42}	s	b
λ_{11}	1.00	1.00	(1.00)	1.00	1.00	0.00	0.00	0.00	0.00	0.00	0.00	1.00
λ_{32}	0.14	0.00	−0.43	−1.14	−2.14	2.28	2.14	1.71	1.00	0.00	−0.14	0.14
λ_{42}	−0.14	0.00	0.43	1.14	2.14	−1.28	−1.14	−0.71	0.00	1.00	0.14	0.86
$\bar{c}$ Row	−1.21	0.00	0.64	0.71	0.21	−3.43	−1.73	−0.57	0.00	0.00	−0.29	9.21

$\bar{f}=9.21$

Following this same procedure, after three more pivot operations we obtain the final tableau shown in Table 11.12.

Table 11.12
FINAL TABLEAU

Basis	λ_{01}	λ_{11}	λ_{21}	λ_{31}	λ_{41}	λ_{02}	λ_{12}	λ_{22}	λ_{32}	λ_{42}	s	b
λ_{21}	1.79	1.59	1.00	0.00	−1.39	1.79	1.59	1.00	0.00	−1.39	−0.20	0.40
λ_{32}	0.00	0.00	0.00	0.00	0.00	1.00	1.00	1.00	1.00	1.00	0.00	1.00
λ_{31}	−0.80	−0.60	0.00	1.00	2.40	−1.79	−1.59	−1.00	0.00	1.39	0.20	0.60
$\bar{c}$ Row	−1.79	−0.60	0.00	0.00	−0.60	−3.30	−1.59	−0.50	0.00	−0.10	−0.30	9.90

$\bar{f}=9.9$

Notice, by looking at the bottom row of the final tableau that further basis changes will not increase the value of the objective function. It is possible to reach a stage in the solution of such a problem where no further pivot operations are possible because of the restricted basis rule, but positive elements (for maximization problems) still appear in the bottom row of the tableau. Hadley (12) proves, however, that when such a stage is reached, as long as every λ vector that could enter the basis has a negative element in its bottom row (for maximization problems), then we have obtained a relative maximum of the approximating problem with respect to the original variables (x's).

Now from the final tableau we see that $\lambda_{21}=0.40$, $\lambda_{31}=0.60$, and $\lambda_{32}=1.00$. Note that this solution is consistent with conditions of Eqs. 11.140 and 11.141 and the restricted basis rule. Using Eq. 11.139 to recover the values of x_1 and x_2 we have

$$\begin{aligned}x_1 &= \lambda_{01}x_{01}+\lambda_{11}x_{11}+\lambda_{21}x_{21}+\lambda_{31}x_{31}+\lambda_{41}x_{41}\\ &= (0.00)(0.00)+(0.00)(0.50)+(0.40)(1.00)+(0.60)(1.50)\\ &\quad +(0.00)(2.00)=1.30\\ x_2 &= \lambda_{02}x_{02}+\lambda_{12}x_{12}+\lambda_{22}x_{22}+\lambda_{32}x_{32}+\lambda_{42}x_{42}\\ &= (0.00)(0.00)+(0.00)(1.00)+(0.00)(2.00)+(1.00)(3.00)\\ &\quad +(0.00)(4.00)=3.0\end{aligned}$$

Substituting in these values for x_1 and x_2 we get the value of the objective function:

$$f(\mathbf{x})=3(1.30)+2(3.00)=9.90$$

Notice that this value also appears in the lower right-hand corner of the final tableau in Table 11.12 as the value of the objective function of our approximating problem, $\bar{f}$. This illustrates that the two objective functions are approximately equal.

Figure 11.31 gives a graphical solution to the above problem to illustrate that the solution is approximately correct.

Thus we have seen that separable programming can be used to solve nonlinear programming problems with separable functions.

The following pertinent comments are made by Hadley on separable programming:

1. Adding more grid points to an approximation of a nonlinear function will improve the approximation but at the expense of increasing the number of variables in the approximating problem. This may, because of the restricted basic rule, considerably increase the number of iterations required in the simplex algorithm solution to the approximating problem.
2. If the set of feasible solutions to the original problem is convex, then any feasible solution to the approximating problem will also be a feasible solution to the original problem.
3. For more accurate solutions, it may be advantageous to solve a problem initially with a rather coarse grid (grid points widely spread) and then resolve it, using a finer grid only in the neighborhood of the solution to the first approximation. This would allow a closer approximation without increasing the number of grid points.
4. If we are maximizing, the objective function of the original problem is concave, and the constraint set is convex, then the same will be true for the approximating problem. The solution will be an approximation to a point at which the original problem assumes its global maximum. By solving a series of approximating problems, using finer and finer

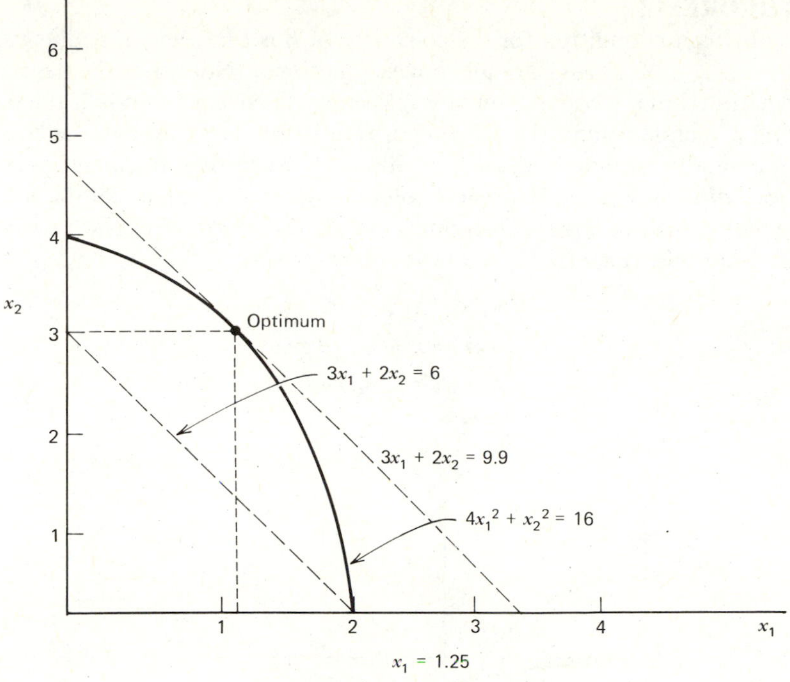

Figure 11.31
Separable programming example.

grids, the sequence of optimal values for the objective function of the approximating problem will approach, in the limit, the optimal value of the objective function for the original problem.

V. THE GENERAL NONLINEAR PROGRAMMING PROBLEM

11.13 NONLINEAR OBJECTIVE FUNCTION SUBJECT TO LINEAR OR NONLINEAR CONSTRAINTS: A CUTTING PLANE ALGORITHM

Consider the following general nonlinear programming problem:

Program I

Maximize: $f(\mathbf{x})$

Subject to: $h_i(\mathbf{x}) \leqq 0 \qquad i = 1, 2, \ldots, m$

Define the feasible region which contains all possible solution vectors by the set S.

$$S = \{\mathbf{x} \mid h_i(\mathbf{x}) \leqq 0\} \qquad i = 1, 2, \ldots, m$$

It will be assumed that $f(\mathbf{x})$ is a concave function, $h_i(\mathbf{x})$, $i = 1, 2, \ldots, m$ are linear or nonlinear convex functions, and that the region S forms a closed and bounded *convex set*. The reader not familiar with this concept is referred to Appendix A for definition and properties.

THEOREM 1

A sufficient condition for the convexity of S is that the constraint functions $h_j(\mathbf{x})$, $j=1, 2, \ldots, m+1$ are all convex functions. (Note that the negative of a concave function is a convex function.) Keeping Theorem 1 in mind, the basic underlying principle behind the Kelly algorithm is that even though the constraints are individually nonlinear, if they are all convex then they form a convex set. In addition, *any* convex set of feasible solutions (S for example) can be totally enclosed within a set of *linear* constraints (sometimes referred to as *half spaces*).

For example, consider the following constraint set.

$$h_1(\mathbf{x}) = x_1^2 + x_2^2 - 25 \leqq 0$$
$$h_2(\mathbf{x}) = x_1^2 + x_2 - 9 \leqq 0$$
$$h_3(\mathbf{x}) = -x_1 \leqq 0$$
$$h_4(\mathbf{x}) = -x_2 \leqq 0$$

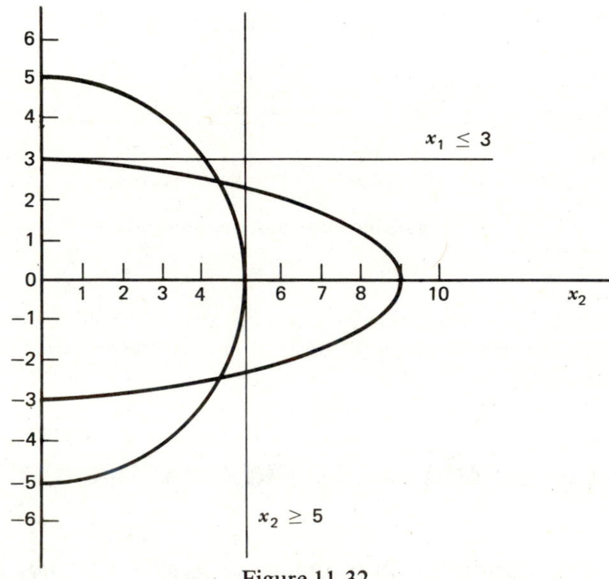

Figure 11.32
A constrained optimization problem.

It is clear from Fig. 11.32 that the feasible region defined by

$$K = \{(x_2, x_2) \mid x_1 \leqq 3; x_2 \leqq 5\}$$
$$x_1 \geqq 0; x_2 \geqq 0$$

contains all possible solution vectors defined by

$$\mathbf{S} = \{(x_1, x_2) \mid x_1^2 + x_2^2 \leqq 25; x_1^2 + x_2 \leqq 9\}$$
$$x_1 \geqq 0; x_2 \geqq 0$$

Of course, the set K would contain certain solution vectors that would not be defined by the set **S**, but every feasible solution to the original set **S** is also feasible to the (approximating) set K.

Kelly's Cutting Plane Algorithm

A method that can be employed to solve the above problem is a cutting plane algorithm developed by Kelly (15). The method is exceedingly simple in concept, and can be explained and implemented rather easily. The algorithm develops a series of ever-improving linear programming problems, whose solutions converge to the optimal solution of the original nonlinear programming problem. The procedure actually solves the following transformed problem.

Program II

Maximize: $Z = \mathbf{cx}$

Subject to: $h_i(\mathbf{x}) \leqq 0 \qquad i = 1, 2, \ldots, m$

Where: $Z = \mathbf{cx}$ is a linear function.

Program I can always be expressed in this form, since the following transformation is always valid.

Maximize: Z

Subject to: $h_i(\mathbf{x}) \leqq 0 \qquad i = 1, 2, \ldots, m$

$$h_{m+1}(\mathbf{x}) = Z - f(\mathbf{x}) \leqq 0$$

Hence, the remainder of this discussion will deal directly with an approximation to Program II. Before proceeding, let us note the following theorems which are useful in implementing the proposed algorithm.

Basic Logic of the Algorithm. The algorithm of Kelly replaces each nonlinear constraint $h_i(\mathbf{x})$ $i = 1, 2, \ldots, m+1$ by an approximating (linear) constraint $g_i(\mathbf{x})$, $i = 1, 2, \ldots, m+1$ such that the solution to the original problem is always contained in the approximating (linear programming) problem.

THEOREM 2

If $\mathbf{x}_0$ is an optimal solution to the approximating linear programming problem, and if the solution vector $\mathbf{x}_0$ also satisfies the (original) nonlinear constraint set, then $\mathbf{x}_0$ is also an optimal solution to the (original) nonlinear programming problem.

Consider the case where $\mathbf{x}_0$ is *not* feasible to the original nonlinear programming problem. If this occurs, then a new linear constraint is introduced that *eliminates* the previous linear programming optimal solution $\mathbf{x}_0$, but simultaneously preserves *all* feasible solutions (not yet attained) of the original nonlinear programming problem. Geometrically, this new constraint acts as a "cutting plane" that "cuts off" the previous linear programming solution. Thus the algorithm that we are describing is sometimes referred to as "Kelly's Cutting Plane" algorithm. We will continue to add "cutting plane" constraints until the optimal solution is reached.

THEOREM 3

To converge to a feasible, optimal solution, an optimal solution to the original nonlinear programming problem must be a *boundary point* solution, since all approximating functions are linear.

In summary, the algorithm selects a sequence of approximating solutions which are infeasible to the original problem that converges towards feasibility. The procedure is essentially for nonlinear constraints; any linear constraints in the original constraint set remain unchanged.

Determination of the Cutting Plane. Since the constraints $h_i(\mathbf{x})$, $i=1, 2, \ldots, m+1$ are convex and assumed differentiable, by a Taylor's Series expansion:

$$h_i(\mathbf{x}) - h_i(\bar{\mathbf{x}}) \geqq \nabla h_i(\bar{\mathbf{x}})(\mathbf{x} - \bar{\mathbf{x}}) \qquad \text{for any} \quad \bar{\mathbf{x}} \in E^n \tag{11.144}$$

When the optimal solution $\bar{\mathbf{x}}$ to the approximating problem does not satisfy the original constraint set, determine the constraint $h_K(\mathbf{x}) \leqq 0$ that is most violated. That is,

$$\operatorname*{Max}_j \{h_j(\bar{\mathbf{x}})\} = h_k(\bar{\mathbf{x}}) \qquad \text{for all } j \text{ for which } h_j(\bar{\mathbf{x}}) > 0$$

From Eq. 11.144, the new cutting plane constraint is given by the following:

$$g_K(\mathbf{x}) = h_k(\bar{\mathbf{x}}) + \nabla h_K(\bar{\mathbf{x}})(\mathbf{x} - \bar{\mathbf{x}}) \leqq 0 \tag{11.145}$$

Note that the new constraint must satisfy two conditions.

1. The current solution vector $\bar{\mathbf{x}}$ should violate the new constraint.
2. All feasible solutions to the original nonlinear constraint $h_K(\mathbf{x}) \leqq 0$ should satisfy the new constraint.

Proposition 1. For $\mathbf{x} = \bar{\mathbf{x}}$, the left-hand-side of Eq. 11.145 is $h_K(\bar{\mathbf{x}}) > 0$. Hence, Condition 1 is satisfied.

Proposition 2. Consider *any* $\mathbf{x}$ feasible to the original constraint $h_k(\mathbf{x})$. By the convexity assumption on $h_K(\mathbf{x})$,

$$h_K(\bar{\mathbf{x}}) + \nabla h_K(\bar{\mathbf{x}})(\mathbf{x} - \bar{\mathbf{x}}) \leqq h_K(\mathbf{x}) \leqq 0.$$

Hence any $\mathbf{x}$ feasible to the original constraint set also satisfy the approximating constraint.

Note that with the addition of the new constraint to the previous problem, the previous solution vector $\bar{\mathbf{x}}$ is now *infeasible* to the new problem. A new optimal solution is easily obtained through a *dual simplex* procedure (see Chapter 4).

Steps of the Algorithm.

Step I. Find an approximating linear constrained region that contains the (optimal) solution to the nonlinear programming problem to be solved. We can always choose large values of $x_j = \delta_j$, $j = 1, 2, \ldots, h$ such that s is enclosed in the set $\mathbf{k} = \{\mathbf{x} \mid x_j \leqq \delta_j\}$.

Step II. Solve the linear programming problem:

$$\begin{aligned} &\text{Maximize:} && \text{LP1} = \mathbf{cx} \\ &\text{Subject to:} && x_n \leqq \delta_n \qquad n = 1, 2, \ldots, n \end{aligned}$$

Let $\bar{\mathbf{x}}$ be optimal for LP-1.

Step III. Check whether $h_i(\bar{\mathbf{x}}) \leqq 0$, $i = 1, 2, \ldots, m+1$
If yes, terminate.
If no, go to Step IV.

Step IV. Find $h_K(\bar{\mathbf{x}}) = \max\limits_{j \in m+1} \{h_j(\bar{\mathbf{x}})\}\ h_j(\bar{\mathbf{x}}) > 0.$

Step V. Define a new constraint

$$g_{\text{new}}(\mathbf{x}) = h_K(\bar{\mathbf{x}}) + \nabla h_K(\mathbf{x} - \bar{\mathbf{x}}) \leqq 0$$

Step VI. Append this constraint to the previous LP problem and find a new optimal solution via a dual-simplex procedure. Call this the new $\bar{\mathbf{x}}$. Go to Step III.

Summary and Conclusions. Kelly's cutting plane method is widely used both as a direct solution technique, and as an auxiliary algorithm useful in implementing other related methods. There are several advantages and disadvantages associated with this technique.

Advantages

1. The technique uses a direct extension of the standard simplex algorithm (Dual Simplex). Therefore, it is very efficient for convex programming problems that are very nearly linear in structure.
2. There is relatively little work per algorithmic step. A simple dual simplex routine and a Taylor's series approximation is all that is normally required.
3. The algorithm is easy to implement and program, and it is computationally sound.

Disadvantages

1. The algorithm requires convexity to guarantee convergence; hence, it is not applicable to non-convex programming problems.
2. None of the intermediate solutions are feasible to the original problem.
3. The algorithm often exhibits slow convergence as the optimum is approached.
4. The size of the problem grows as more constraints are added, thus creating potential difficulties in computer core storage. Although this disadvantage might prove serious for larger problems, there has been some work done on schemes to drop from the solution procedure old, previously generated constraints.

EXAMPLE 11.13-1

Maximize: $f(\mathbf{x}) = x_1 + x_2$

Subject to:

$$\begin{aligned} h_1(\mathbf{x}) &= x_1^2 + x_2^2 - 25 \leqq 0 \\ h_2(\mathbf{x}) &= x_1^2 + x_2 - 9 \leqq 0 \\ x_1 &\geqq 0 \\ x_2 &\leqq 0 \end{aligned}$$

Step I. Suppose that the values of x_1 and x_2 are set to zero.

$$\text{Set } x_1 = 0 \qquad \max\{x_2\} = \max\{5, 9\} = 9$$
$$\text{Set } x_2 = 0 \qquad \max\{x_1\} = \max\{5, 3\} = 5$$

Hence, the feasible solution space is bounded by the half-space defined by

$$K = \{(x_1, x_2) \mid x_1 \leqq 5; \quad x_2 \leqq 9; \quad x_1, x_2 \geqq 0\}$$

Step II. Maximize: $z = x_1 + x_2$

Subject to:

$$\begin{aligned} g_1(\mathbf{x}) &= x_1 \leqq 5 \\ g_2(\mathbf{x}) &= x_2 \leqq 9 \\ x_1, x_2 &\geqq 0 \end{aligned}$$

$$x_1^{(0)} = 5.0, \ x_2^{(0)} = 9.0, \ Z^{(0)} = 14.0$$

Step III. Constraints are not satisfied

Step IV. Max $\{h_1(\mathbf{x}), h_2(\mathbf{x})\} = \text{Max}\{81, 25\} = h_1(\mathbf{x})$

Step V. Generate a cutting plane constraint

$$g_3(\mathbf{x}) = h_1(\mathbf{x}^{(0)}) + \nabla h_1(\mathbf{x}^{(0)})(\mathbf{x} - \mathbf{x}^{(0)}) \leqq 0$$

Since $\nabla h_1(\mathbf{x}^{(0)}) = (2x_1, 2x_2)$

$$g_3(\mathbf{x}) = 81 + (10, 18)\begin{pmatrix} x_1 \\ x_2 \end{pmatrix} - (10, 18)\begin{pmatrix} 5 \\ 9 \end{pmatrix}$$

$$\Rightarrow g_3(\mathbf{x}) = x_1 + 1.8x_2 \leqq 13.1$$

Step VI. The problem now becomes:

Maximize: $z = x_1 + x_2$

Subject to:

$$\begin{aligned} g_1(\mathbf{x}) &= x_1 \leqq 5 \\ g_2(\mathbf{x}) &= x_2 \leqq 9 \\ g_3(\mathbf{x}) &= x_1 + 1.8x_2 \leqq 13.1 \\ x_1, x_2 &\geqq 0 \end{aligned}$$

The reader will be asked to verify in Exercise 11.32 that the optimal solution to this problem is given by

$x_1^* = 5.0$

$x_2^* = 4.5$

$Z^* = 9.5$

$\mathbf{x}^{(1)} = (5.0, 4.5)$

Return to Step III

Step III. $h_1(\mathbf{x}^{(1)}) = 20.2$

$h_2(\mathbf{x}^{(1)}) = 20.5$

Both $h_1(\mathbf{x})$ and $h_2(\mathbf{x})$ are violated.

Step IV. Max $\{h_1(\mathbf{x}), h_2(\mathbf{x})\} = \max\{20.2, 20.5\} = h_2(\mathbf{x})$

Step V. Generate a new "cut"

$$g_4(\mathbf{x}) = h_2(\mathbf{x}^{(1)}) + \nabla h_2(\mathbf{x}^{(1)})(\mathbf{x} - \mathbf{x}^{(1)})$$

Since

$$\nabla h_2(\mathbf{x}) = (2x_1, 1)$$

$$g_4(\mathbf{x}) = 20.5 + (10, 1)\begin{pmatrix} x_1 \\ x_2 \end{pmatrix} - (10, 1)\begin{pmatrix} 5 \\ 4.5 \end{pmatrix} \leqq 0$$

$$\Rightarrow g_4(\mathbf{x}) = 10x_1 + x_2 \leqq 34$$

Step VI. The problem now becomes:

Maximize: $z = x_1 + x_2$

Subject to:

$$\begin{aligned} g_1(\mathbf{x}) &= x_1 \leqq 5 \\ g_2(\mathbf{x}) &= x_2 \leqq 9 \\ g_3(\mathbf{x}) &= x_1 + 1.8x_2 \leqq 13.1 \\ g_4(\mathbf{x}) &= 10x_1 + x_2 \leqq 34 \\ x_1, x_2 &\geqq 0 \end{aligned}$$

At this point, it should once again be emphasized that one would not completely resolve the above problem. Instead, $g_4(\mathbf{x})$ would be appended to the previous (optimal) linear programming tableau, and the new problem would be solved via a dual-simplex procedure.

The solution is given by

$$x_1^* = 2.83$$
$$x_2^* = 5.70$$
$$z^* = 8.53$$
$$\mathbf{x}^{(2)} = (2.83, 5.7)$$

Step III. $$h_1(\mathbf{x}^{(2)}) = 40.499: h_2(\mathbf{x}^{(2)}) = 13.71$$

Since both constraints are violated, the algorithm would add a new cutting-plane constraint and proceed as in previous iterations. Since the procedure should be clear at this point, further iterations will not be shown. Progression of the algorithm to the present solution vector is shown in Figure 11.33, and is evident that the optimal solution of $x_1^* = 2.114$ and $x_2^* = 4.531$ is being approached.

11.14 OPTIMIZATION BY GEOMETRIC PROGRAMMING†

The increased use of mathematical models in the analysis and optimization of industrial systems is one of the significant developments of modern engineering practice. Unfortunately, most real-life systems are such that models that describe them accurately usually prove too complex for solution by the algorithms available. This is especially true of problems in which the constraints are nonlinear or the objective function is of more than second degree.

In 1961, however, Zener (32) observed that a sum of component costs sometimes may be minimized almost by inspection when each cost depends on products of the design variables, each raised to arbitrary but known powers. Duffin (8) and Peterson (9) extended Zener's work, and Passy and Wilde (19) further generalized the method to include negative coefficients and reversed inequalities. Zener and Duffin called their method *geometric programming*, since it was based on a generalization of the arithmetic-geometric mean inequality. The basic theory and formal proofs for this new optimization technique can be found in (3) and (29).

Geometric Programming

Consider first a hypothetical, unconstrained problem wherein we want to minimize the total inventory and production cost associated with the manufacture of a certain product. Let this cost be denoted by y:

$$y = c_1 q^{-3} s^{-2} + c_2 q^3 s + c_3 q^{-3} s^3 \tag{11.146}$$

where

q = tons of product manufactured during a given period

s = fraction of production to be stored in inventory

c_1, c_2, and c_c are cost coefficients that vary with the time period and the production facilities.

†The authors would like to express their appreciation to Dr. C. S. Beightler, Dr. R. M. Crisp, Jr., and Dr. W. L. Meier for contributing the material which comprises this section.

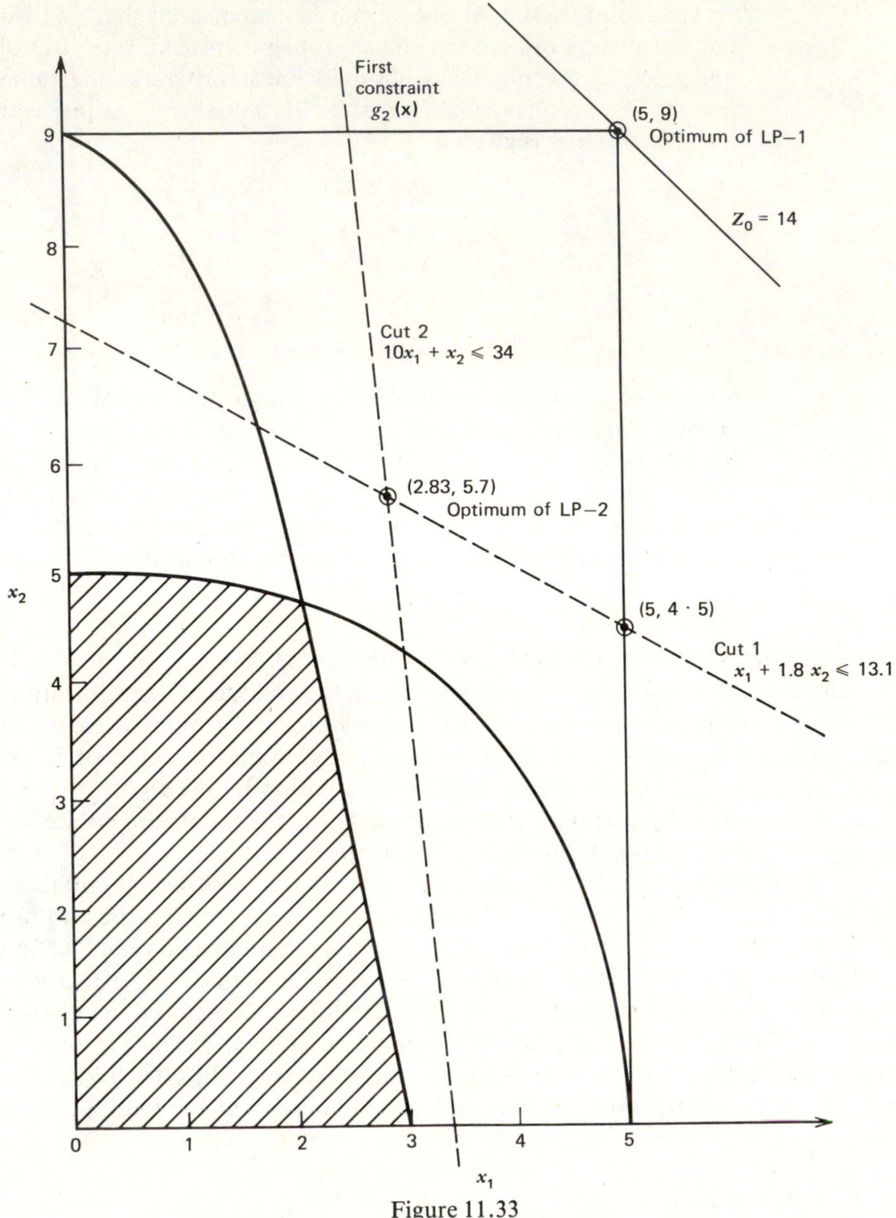

Figure 11.33
Cutting plane example.

The first term represents the setup and down-time costs; the second, the idle-time and fixed-production costs; the third, the inventory and handling costs.

Equation 11.146 is not a polynomial, since the exponents are not restricted to the positive integers. However, the c_j must be positive, and this requirement led Zener and Duffin to call such functions *posynomials*. The classical method for minimizing such a function involves setting to zero the first partial derivatives of y with respect to both q and s. This produces nonlinear simultaneous equations that, in general, are quite difficult to solve. In addition, the vanishing of these first derivatives is only a necessary condition for a minimum; sufficient conditions are more complicated, involving functions of higher order partial derivatives.

Consider now a different approach to the solution of this problem. A posynomial containing N variables and a total of T terms may be written as

$$y = \sum_{t=1}^{T} c_t \prod_{n=1}^{N} x_n^{a_{tn}} \tag{11.147}$$

As in the classical method, one sets the first partial derivatives to zero:

$$\left(\frac{\partial y}{\partial x_k}\right)^* = \frac{1}{x_k^*} \sum_{t=1}^{T} a_{tk} c_t \prod_{n=1}^{N} (x_n^*)^{a_{tn}} = 0; \qquad k = 1, \ldots, N \tag{11.148}$$

Instead of solving these nonlinear equations directly for the optimal values, x_k^*, Zener and Duffin made a substitution of variables based on a duality theory.

Define the optimal weights:

$$w_t \equiv \frac{c_t \prod_{n=1}^{N} (x_n^*)^{a_{tn}}}{y^*}; \; t = 1, \ldots, T \tag{11.149}$$

These weights must sum to unity:

$$\sum_{t=1}^{T} w_t = 1 \tag{11.150}$$

In addition (assuming that y^* and all of the x_k^* are nonzero), Eqs. 11.148 and 11.149 give

$$\sum_{t=1}^{T} a_{tk} w_t = 0; \qquad k = 1, \ldots, N \tag{11.151}$$

Notice that Eqs. 11.150 and 11.151 are *linear* in the weights, and do *not* depend on the cost coefficients c_t. For the example problem given by Eq. 11.146, these equations are simply

$$\begin{aligned} w_1 + w_2 + w_3 &= 1 \\ -3w_1 + 3w_2 - 3w_3 &= 0 \\ -2w_1 + w_2 + 3w_3 &= 0 \end{aligned}$$

from which it follows at once that

$$w_1 = 0.4, \qquad w_2 = 0.5, \qquad w_3 = 0.1.$$

Thus without knowing the optimal production rate or optimal inventory, it is nevertheless possible to state definitely that these optimal values must always be such as to make down-time and setup costs four times as large as the inventory and handling costs, and also such as to make the idle time and fixed production costs five times as large as the inventory and handling costs. Furthermore, this optimal distribution is totally unaffected by changes in the cost coefficients c_1, c_2, and c_3. This separation of technological effects, as reflected by the exponents a_{m}, from the economic effects, as measured by the coefficients c_t, is unique to the geometric programming algorithm.

Since the weights sum to unity,

$$y^* = \prod_{t=1}^{T} (y^*)^{w_t}$$

$$= \prod_{t=1}^{T} \left(c_t \prod_{n=1}^{N} (x_n^*)^{a_{tn}} / w_t \right)^{w_t}$$

$$= \prod_{t=1}^{T} \left(\frac{c_t}{w_t} \right)^{w_t} \prod_{t=1}^{T} \prod_{n=1}^{N} (x_n^*)^{a_{tn} w_t} \tag{11.152}$$

But, by Eq. 11.151,

$$\prod_{t=1}^{T} \prod_{n=1}^{N} (x_n^*)^{a_{tn} w_t} = \prod_{n=1}^{N} \prod_{t=1}^{T} (x_n^*)^{a_{tn} w_t}$$

$$= \prod_{n=1}^{N} (x_n^*)^{\sum_{t=1}^{T} a_{tn} w_t}$$

$$= 1$$

Therefore, Eq. 11.159 becomes

$$y^* = \prod_{t=1}^{T} \left(\frac{c_t}{w_t} \right)^{w_t} \tag{11.153}$$

and hence the minimal cost can be found, still without knowledge of the optimal values of the decision variables. In the illustrative example of Eq. 11.146, suppose that, for a particular time period and a given production facility, the unit costs are $c_1 = 60$, $c_2 = 50$, and $c_3 = 20$. Then from Eq. 11.153 the minimal cost is

$$y^* = \left(\frac{60}{0.4} \right)^{1.4} \left(\frac{50}{0.5} \right)^{0.5} \left(\frac{20}{0.1} \right)^{0.1}$$

$$= 125.8$$

If this is an acceptable cost, Eq. 11.149 can be used to find the optimal values of the decision variables

$$60q^{-3}s^{-2} = 0.4(125.8) = 50.32$$

$$20q^{-3}s^{3} = 0.1(125.8) = 12.58$$

so that

$$q^3 = 1.19s^{-2}$$

and

$$s^5 = 0.75, \qquad \text{or} \qquad s^* = 0.944$$

from which it follows that

$$q^3 = 1.34; \qquad q^* = 1.12 \text{ tons}$$

Notice that the only *nonlinear* equations encountered in this analysis are those of Equation 11.149, each of which contain only one term, and therefore *linear* in the logarithm of x_n.

The preceding example was special in that it had exactly one more term than variable; that is, $T=N+1$. Thus there were exactly as many linear equations (Eqs. 11.150 and 11.151) as there were variables, w_t, so that a unique set of weights satisfied these equations. When T is greater than $N+1$, further steps must be taken to find the optimum weights.

Before leaving this special situation, however, it will be interesting to examine a problem familiar to all readers: finding the economic lot size for a purchased product. A simple form of this problem may be stated as follows: For a given product, a wholesaler must decide what size lots he will put into stock periodically. The total variable cost associated with the purchase and storage of this product is

$$y=\tfrac{1}{2}Qh+ad/Q+\frac{(c+kQ)d}{Q} \tag{11.154}$$

where

$Q=$ lot size (pieces per order).
$h=$ carrying cost per piece per year
$d=$ annual requirements (pieces per year)
$a=$ cost of processing an order
$c, k=$ given constants.

The first term in this objective function represents the total carrying costs, the second term gives the total ordering costs, and the third term is the total purchasing costs. This last term includes quantity discounts, where $P+kQ$ is the price that the vendor charges for an order of Q pieces. Since Q is the only unknown, the constant term may be dropped, and Eq. 11.154 takes the form

$$y=c_1Q+c_2Q^{-1} \tag{11.155}$$

The optimal lot size, Q^*, can now be found directly by *inspection*, since Eqs. 11.150 and 11.151 are simply

$$w_1+w_2=1$$
$$w_1-w_2=0$$

Therefore, Q^* must be such as to make the two terms in Eq. 11.155 equal:

$$c_1Q^*=c_2Q^{*-1}$$

or

$$Q^*=\sqrt{\frac{c_2}{c_1}}$$

for any values of c_1 and c_2. From Eq. 11.154, $c_1=(1/2)I$ and $c_2=R(s+P)$, so that

$$Q^*=\sqrt{\frac{2d(a+c)}{h}}$$

Degrees of Difficulty

The difference between the number of variables and the number of independent linear equations is usually called the number of *degrees of freedom*. For Eqs. 11.151 and 11.152, this number is $T-(N+1)$. Zener and Duffin (9) suggest calling this quantity the number of *degrees of difficulty*, since they make the problem harder to solve.

Consider the objective function

$$y = 1000x_1 + 4 \times 10^9 x_1^{-1} x_2^{-1} + 2.5 \times 10^5 x_2 + 9000 x_1 x_2 \qquad (11.156)$$

This cost function is to be minimized by choosing positive values for x_1 and x_2. Using geometric programming, weights are introduced, and Eqs. 11.150 and 11.151 become

$$\begin{aligned} w_1 + w_2 + w_3 + w_4 &= 1 \\ w_1 - w_2 \qquad + w_4 &= 0 \\ - w_2 + w_3 + w_4 &= 0 \end{aligned}$$

Here, $N=2$, $T=4$, so that one degree of difficulty is present and the equations do not have a unique solution. There are, in fact, an infinity of solutions as can be seen by solving for the first three weights in terms of the fourth:

$$w_1 = \frac{1}{3}(2 - 2w_4) \qquad (11.157)$$

$$w_2 = \frac{1}{3}(1 + w_4) \qquad (11.158)$$

$$w_3 = \frac{1}{3}(1 - 2w_4) \qquad (11.159)$$

Any choice of w_4 will produce values for the other weights that will satisfy Eqs. 11.150 and 11.151. The problem is to select the optimal weights from among these infinite possibilities.

Although Eq. 11.153 holds for any number of degrees of difficulty, the minimum cost y^* cannot be computed until the optimal weights are known. If Eqs. 11.157, 11.158, and 11.159 are substituted into Eq. 11.153, the minimum cost will be given as a function, $d(w_4)$, of w_4 alone. Duffin has proved that $d(w_4)$, the *substituted dual function*, is *maximized* by the optimal weight, w_4^*, and that this maximum value is equal to the minimum cost y^*. Therefore,

$$d(w_4^*) = \max_{w_4>0} d(w_4) = y^* \qquad (11.160)$$

In general, he showed that a sufficient condition for y to be minimized is that the dual function defined by

$$d(w_1, \ldots, w_T) \equiv \prod_{t=1}^{T} \left(\frac{c_t}{w_t} \right)^{w_t} \qquad (11.161)$$

be maximum with respect to the w_t, subject to the conditions given by Eqs. 11.150 and 11.151.

The numerical value of w_4^* can be found most easily by setting to zero the first derivative of the logarithm of $d(w_4)$. When this is done (see 11.153), the optimal weight is found to be

$$w_4^* = 0.453$$

The other weights are found from Eqs. 11.157, 11.158, and 11.159:

$$w_1^* = w_3^* = 0.031; \qquad w_2^* = 0.484$$

and

$$y^* = \$12.6 \times 10^6$$

Although the weights are not independent of the cost coefficients c_t when there are degrees of difficulty, *bounds* on the weights can be found that do not depend on the cost coefficients. In this illustrative problem, for example, w_4 is bounded above by $\frac{1}{2}$, since, at that value, w_1 and w_3 would vanish. Hence, for *any* cost coefficients,

$$0 \leqq w_1 = w_3 \leqq \frac{1}{3}$$

$$\frac{1}{3} \leqq w_2 \leqq \frac{1}{2}$$

$$0 \leqq w_4 \leqq \frac{1}{2}$$

More degrees of difficulty, of course, complicate the problem still further. However, the following relationship,

$$y(x_1, \ldots, x_N) \geqq y^* = d^* \geqq d(w_1, \ldots, w_T), \tag{11.162}$$

can be used for obtaining quick estimates of the optimal value of the objective function. This is accomplished by neglecting $T-(N+1)$ of the terms in y_1, thereby producing a problem with no degrees of difficulty. This zero degree problem is very easy to solve for the weights, which are then substituted into the original dual function to obtain a lower bound on the true optimum cost. The corresponding x_n can then be used to construct an upper bound by substituting them into the objective function.

In the objective function given by Eq. 11.156, for example, if the third term is discarded ($w_3 = 0$), the resulting zero-degree problem satisfies Eqs. 11.150 and 11.151 when $w_1 = w_3 = 0$ and $w_2 = w_4 = \frac{1}{2}$. The dual function gives a lower bound as

$$d\left(0, \frac{1}{2}, 0, \frac{1}{2}\right) = \left(\frac{4 \times 10^9}{\frac{1}{2}}\right)^{\frac{1}{2}} \left(\frac{9000}{\frac{1}{2}}\right)^{\frac{1}{2}}$$

$$= 12.0 \times 10^6$$

Corresponding to these dual variables (weights) are the primal variables $x_1 = 411$ and $x_2 = 1.63$. Substitution of these values into the original objective function yields $y = 12.8 \times 10^6$. Hence, the use of Eq. 11.162 has produced tight bounds on the optimal value of the objective function:

$$12.0 \times 10^6 \leqq y^* \leqq 12.8 \times 10^6$$

Generalized Geometric Programming

Geometric programming has recently been generalized beyond the preceding description to allow the use of negative coefficients in both objective function and constraints, and also to permit reversed inequality constraints. The following is a summary of the most general formulation of the geometric programming problem. A complete discussion and derivation of these results may be found in (29).

Define the *generalized posynomial* $y_m(x)$ as

$$y_m \equiv \sum_{t=1}^{T_m} \sigma_{mt} c_{mt} \prod_{n=1}^{N} x_n^{a_{mtn}} \qquad m=0, 1, \ldots, M \tag{11.163}$$

where

$$\sigma_{mt} = \pm 1$$

is a signum function, and the coefficients c_{mt} are all positive.

The primal problem is to minimize

$$y \equiv y_o \tag{11.164}$$

subject to the constraints

$$y_m \leqq \sigma_m (\equiv \pm 1); \qquad m=1, \ldots, M \tag{11.165}$$

and

$$x_n > 0; \qquad n=1, \ldots, N \tag{11.166}$$

For the associated dual problem, consider a set of T variables ω satisfying a normality condition

$$\sum_{t=1}^{T_o} \sigma_{ot} \omega_{ot} = \sigma(\equiv \pm 1) \tag{11.167}$$

and N orthogonality conditions

$$\sum_{m=0}^{M} \sum_{t=1}^{T_m} \sigma_{mt} a_{mtn} \omega_{mt} = 0 \tag{11.168}$$

as well as T nonnegativity conditions

$$w_{mt} \geqq 0; \qquad m=0, 1, \ldots, M; \qquad t=1, \ldots, T_m \tag{11.169}$$

and M linear inequality constraints

$$\omega_{mo} \equiv \sigma_m \sum_{t=1}^{T_m} \sigma_{mt} \omega_{mt} \geqq 0; \qquad m=1, \ldots, M \tag{11.170}$$

where

$$T \equiv \sum_{m=0}^{M} T_m$$

From these variables, plus the c_{mt}, σ_{mt} and σ, the dual function can be formed as follows

$$d(\omega) \equiv \sigma \left[\prod_{m=0}^{M} \prod_{t=1}^{T_m} \left(\frac{c_{mt} \omega_{mo}}{\omega_{mt}} \right)^{\sigma_{mt} \omega_{mt}} \right]^{\sigma} \tag{11.171}$$

In this function, ω_{00} is defined to be equal to $+1$ and, in addition,

$$\lim_{\omega_{mt} \to 0} \left(\frac{c_{mt} \omega_{mo}}{\omega_{mt}} \right)^{\sigma_{mt} \omega_{mt}} = 1 \tag{11.172}$$

Then for every point x^o where y is locally minimum there exists a set of dual variables ω^o satisfying Equations 11.167 through 11.170 and such that

$$d(\omega^o)=y(x^o) \tag{11.173}$$

The dual function is stationary at ω^o with respect to all nonnegative ω_{mt}; in particular at the global minimum x^*, if it exists, the corresponding dual variables ω^* are such that

$$d(\omega^*)=y(x^*) \tag{11.174}$$

Once the dual variables ω and δ are known, the corresponding values of the primal variables x are found from the following relations:

$$c_{ot}\prod_{n=1}^{N} x_n^{a_{otn}} = \omega_{ot}\sigma y^o; \qquad t=1, \ldots, T_o \tag{11.175}$$

and

$$c_{mt}\prod_{n=1}^{N} x_n^{a_{mtn}} = \frac{\omega_{mt}}{\omega_{mo}}; \qquad t=1, \ldots, T_m$$

$$m=1, \ldots, M \tag{11.176}$$

From Eq. 11.175, it can be seen that σ will have the same sign as y^o. since there will always be more terms than variables x, N equations can be found that are solvable for the N primals. In addition, the solution of these equations is not difficult, since they are linear in $\log x_n$.

The following example will serve to illustrate the use of the above equations. Assume that it is desired to maximize the functional

$$y' = 5x_1^2 - x_2^2x_3^4$$

subject to the usual nonnegativity conditions and the inequality

$$-5x_1^2x_2^{-2} + 3x_2^{-1}x_3 \geqq 2$$

To get this into the proper form given by Eqs. 11.164 and 11.165, the negative of y' must be minimized, and all terms in the constraint multiplied by $-(1/2)$ to reverse its sense. The equivalent problem is to minimize $y(\equiv -y')$

$$y= -5x_1^2 + x_2^2x_3^4$$

subject to

$$\frac{5}{2}x_1^2x_2^{-2} - \frac{3}{2}x_2^{-1}x_3 \leqq -1$$

The dual variables must satisfy Eqs. 11.167 and 11.168:

$$\begin{aligned}
-\ \omega_{01} + \omega_{02} \qquad\qquad\quad &= \sigma \\
-2\omega_{01} \qquad + 2\omega_{11} \qquad &= 0 \\
2\omega_{02} - 2\omega_{11} + \omega_{12} &= 0 \\
4\omega_{02} \qquad\quad - \omega_{12} &= 0
\end{aligned}$$

The solution is $\sigma = -1$, $\omega_{01} = \frac{3}{2}$, $\omega_{02} = \frac{1}{2}$, $\omega_{11} = \frac{3}{2}$, and $\omega_{12} = 2$. By definition, $\omega_{oo} \equiv 1$ and, from Eq. 11.170, $\omega_{10} = \sigma_1(\omega_{11} - \omega_{12}) = \frac{1}{2}$.

The value of y at the stationary point x^o is given by $d(\omega^o)$:

$$y^o = -\left[\left(\frac{5\cdot 1}{\frac{3}{2}}\right)^{-3/2}\left(\frac{1\cdot 1}{\frac{1}{2}}\right)^{1/2}\cdot\left(\frac{\frac{5}{2}\cdot\frac{1}{2}}{\frac{3}{2}}\right)^{3/2}\left(\frac{\frac{3}{2}\cdot\frac{1}{2}}{2}\right)^{-2}\right]^{-1}$$
$$= -0.796$$

The optimal policy is found from Eqs. 11.175 and 11.176; the (absolute) value of the first term of the objective function is

$$5x_1^2 = \omega_{01}\sigma y^o = \frac{3}{2}(-1)(-0.796) = 1.19$$

so that

$$x_1^o = 0.488$$

The first term of the constraint is

$$\frac{5}{2}(0.488)^2 x_2^{-2} = \frac{\omega_{11}}{\omega_{10}} = \frac{\frac{3}{2}}{\frac{1}{2}} = 3$$

and thus

$$x_2^o = 0.445$$

The second term of the constraint gives

$$\frac{3}{2}(0.455)^{-1}x_3 = \frac{\omega_{12}}{\omega_{10}} = \frac{2}{0.5} = 4$$

whence

$$x_3^o = 1.19$$

Since there are zero degrees of difficulty, this is the only stationary point and, in fact, it is also the true minimum value of y under the inequality constraint. Hence, the true maximum of the original problem is $y'^* = 0.796$.

This example points up the power of geometric programming in the analysis of systems described by complex mathematical models. No other algorithm could solve such a highly nonlinear problem as this so easily. Also, since the dual variables depend only on the exponents and signs, and not on the magnitudes of the coefficients, a study of these invariants gives valuable insight into the economics of the process.

Discussion and Comments

The geometric programming algorithm in its present form can handle a large class of problems often found in practice without the necessity of using questionable linear or quadratic approximations. Thus highly nonlinear systems can be analyzed using this technique and accurate answers can be obtained. In addition, important invariance properties of the system are often discovered, giving optimal component proportions that are completely independent of fluctuating prices and unit charges. Geometric programming has great potential in engineering design and systems analysis; it is a technique that can be applied to a wide variety of practical situations. For a complete discussion of geometric programming, see Beightler and Phillips (3) or Beightler, Phillips and Wilde (2). A good discussion of the computational efficiency of geometric programming can be found in Reklaitis, Ravindran, and Ragsdell (24).

Recommended Readings

Nonlinear programming is a vast subject. Hopefully, this chapter has provided an introduction and some motivation to study the subject more thoroughly. One of the best resources with which to continue your study of nonlinear programming theory and applications would be the recent textbook by Reklaitis, Ravindran and Ragsdell (24). Other reference texts include Beightler, Phillips and Wilde (2), Cooper and Steinberg (6), Himmelblau (13), and Simmons (27). Chapter 12 of reference (24) contains the results of several experimental studies conducted to compare the efficiencies of different nonlinear programming algorithms. Practitioners interested in the strategies for optimization studies and engineering case studies should consult Chapters 13 and 14 of reference (24).

REFERENCES

1. Beale, E. M. L., P. J. Coen, and A. D. J. Flowerdew, "Separable Programming Applied to an Ore Purchasing Problem," *Journal of Applied Statistics* (July 1965).
2. Beightler, C. S., D. T. Phillips, and D. J. Wilde, *Foundations of Optimization*, Second Edition, Prentice-Hall, Englewood Cliffs, N.J., 1979.
3. Beightler, C. S., and D. T. Phillips, *Applied Geometric Programming*, Wiley, New York, 1976.
4. Beveridge, G. S., and R. S. Schecter, *Optimization: Theory and Practice*, McGraw-Hill, New York, 1970.
5. Converse, A. D., *Optimization*, Holt, Rinehart & Winston, New York, 1970.
6. Cooper, Leon, and D. Steinberg, *Introduction to Methods of Optimization*, W. B. Saunders, Philadelphia, 1970.
7. Cottle, R. W., and G. B. Dantzig, "Complementary Pivot Theory and Mathematical Programming," *Journal of Linear Algebra and Applications*, *1*, 105–125 (1968).
8. Duffin, R. J., and E. L. Peterson, "Constrained Minima Treated by Geometric Means," Westinghouse Scientific Paper 64-158-129-P3 (March 1974).
9. Duffin, R. J., E. L. Peterson, and C. Zener, *Geometric Programming*, Wiley, New York, 1966.
10. Everett, H., "Generalized Lagrangian Multipliers for Solving Problems of Optimal Allocation of Resources," *Operations Research*, *11*, 399–471 (1963).
11. Gottfried, B. S., and Joel Weisman, *Introduction to Optimization Theory*, Prentice-Hall, Englewood Cliffs, N.J., 1973.
12. Hadley, G., *Nonlinear and Dynamic Programming*, Addison-Wesley, Reading Mass., 1974.
13. Himmelblau, D. L. *Applied Nonlinear Programming*, McGraw-Hill, New York, 1972.
14. Hooke, R., and T. A. Jeeves, "Direct Search Solution of Numerical and Statistical Problems," *J. Assoc. of Comp. Machines*, *8*, 212–229 (1961).
15. Kelley, J. E., "The Cutting Plane Method for Solving Convex Programs," *J. Soc. Ind. Appl. Math*, *8*, 703–712 (1960).
16. Kunzi, H. P., and J. Krelle, *Nonlinear Programming*, Blaisdell Publishing Company, New York, 1966.
17. Lemke, C. E., "Bimatrix Equilibrium Points and Mathematical Programming," *Management Science*, *11*, 681–689 (1965).
18. McMillan, Claude, Jr., *Mathematical Programming*, Wiley, New York, 1970.
19. Passy, U., and Wilde, D. J., "Generalized Polynomial Optimization," *SIAM Journal on Applied Mathematics*, *15* (5) (September 1967).

20. Phillips, D. T., A. Ravindran, and J. Solberg, *Operations Research: Principles and Practice*, 1st ed., Wiley, New York, 1976.
21. Ravindran, A., "A Computer Routine for Solving Quadratic and Linear Programming Problems," *Communications of the Association for Computing Machines*, *15*, 818–820 (1972).
22. Ravindran, A., "A Comparison of Primal Simplex and Complementary Pivot Methods for Linear Programming," *Naval Research Logistics Quarterly*, *20*, 95–100 (1973).
23. Ravindran, A., and H. K. Lee, "Computer Experiments in Quadratic Programming Algorithms," *European J. of Operational Research*, *8*, 166–174 (1981).
24. Reklaitis, G. V., A. Ravindran, and K. M. Ragsdell, *Engineering Optimization: Methods and Applications*, Wiley, Interscience, New York, 1983.
25. Rosen, J. B., "The Gradient Projection Method for Nonlinear Programming, Part I, Linear Constraints," *J. Soc. Ind. Appl. Math.*, *8*, 181–217 (1960).
26. Saaty, T. L., and J. Bram, *Nonlinear Mathematics*, McGraw-Hill, New York, 1964.
27. Simmons, D. M., *Nonlinear Programming for Operations Research*, Prentice-Hall, Englewood Cliffs, N.J., 1975.
28. Sokolnikoff, I. S., and R. M. Redheffer, *Mathematics of Physics and Modern Engineering*, 2nd ed., McGraw-Hill, New York, 1966.
29. Wilde, D. J., and C. S. Beightler, *Foundations of Optimization*, Prentice-Hall, Englewood Cliffs, N.J., 1967.
30. Wilde, D. J., *Optimum Seeking Methods*, Prentice-Hall, Englewood Cliffs, N.J., 1964.
31. Wolfe, P., "The Simplex Method for Quadratic Programming," *Econometrica*, *27*, 382–398 (1959).
32. Zener, C., "A Mathematical Aid in Optimizing Engineering Designs," *Proceedings of the National Academy of Science*, *47*, (April 1961).

EXERCISES

1. Define the following terms:
   (a) Linear function
   (b) Nonlinear function
   (c) Convex set/nonconvex set
   (d) Gradient vector
   (e) Hessian matrix
   (f) Necessary conditions
   (g) Sufficiency conditions
2. A problem is described by a fourth-order equation, $f(x)$. Write out the Taylor's series expansion that will predict $f(x)$ at a point B given $f(x)$ at any point A.
3. Suppose a point satisfies the sufficiency conditions for a local minimum. How do you establish that it is a global minimum?
4. What are the primary uses of Kuhn-Tucker necessary and sufficient conditions?
5. Consider the minimization of $f(x)$ over a feasible region denoted by S. Suppose the gradient of $f(x) \neq 0$ for every $x \in S$. What can you say about the nature of optimal solution to this problem and why?
6. What is "restricted basis entry" and why is it necessary?
7. What is the purpose of adding a new constraint at each iteration of Kelley's algorithm?
8. Explain how linear constraints should be treated in the cutting plane method. Can linear equalities be accommodated?

9. Under what problem conditions is separable programming likely to be preferable to Kelley's cutting plane method?

10. Is Lemke's algorithm guaranteed to solve a quadratic programming problem always? Why or why not?

11. What is the difference between a "posynomial" and a "polynomial"?

12. What is a "ray solution" and what is its significance in the complementary pivot method?

13. Show that if a cubic approximation to $f(x) = 2x^3 - 3x^2 + x - 4$ is used from Eq. 11.7, the exact value of $f(x)$ at $x = 7$ can be predicted.

14. Show that the point $\mathbf{x} = (0, 0)$ is a global minimum solution to

$$f(\mathbf{x}) = x_2^2 + 3x_1^6 + 5x_2^4.$$

15. Minimize the following objective functions using a Golden section search. Use a resolution of $\epsilon = 0.10$.

(a) Minimize: $$f(x) = 3x^4 + (x-1)^2$$ $$4 \geqq x \geqq 0$$

(b) Minimize: $$f(x) = 4x \sin(x) \pi \geqq x \geqq 0$$

(c) Minimize: $$f(x) = 2(x-3)^2 + e^{0.5x^2}$$ $$100 \geqq x \geqq 0$$

16. An experimenter has obtained the following equation used to describe the trajectory of a space capsule:

$$f(x) = 4x^3 + 2x - 3x^2 + e^{x/2}$$

Use the Golden section search technique to find a root of the above equation. *Hint: A new problem needs to be formulated to facilitate solution.*

17. Solve the following optimization problems using a Hooke-Jeeves search. Start your search with unit moves along the coordinate axis and continue until the percent change in the objective function is 2 percent or less. Plot the trajectory of search.

(a) Minimize: $$f(\mathbf{x}) = 50 + (2.71 - x_1)^2 + (1 - x_2)^2$$
start from the point $x = (0.5, 0)$

(b) Maximize: $$f(\mathbf{x}) = \frac{1}{2x_1 + 6(x_2 + 1)^2}$$

start from the point $\mathbf{x} = (0, 0)$

18. Solve the following problem using gradient projection:

Minimize: $$f(\mathbf{x}) = 25(x_1 - 3x_2)^2 + (x_1 - 3)^2$$

19. Solve Exercise 18 using Newton's second derivative technique.

20. Solve the following problem using Newton's method.

Minimize: $$f(\mathbf{x}) = (3x_1 - 1)^3 + 4x_1x_2 + x_2^2$$

start the search from the point $\mathbf{x} = (1, 2)$

21. Search for the optimum solution to Rosenbrock's function using gradient projection. Proceed through only five steps, and graph the progress of the search toward the optimum solution of $\mathbf{x} = (1, 1)$. Start at the point $\mathbf{x} = (-1, 1)$.

22. Solve the following problems through the classical Lagrangian technique. (Do not search the Lagrangian multiplier.)

    (a) Minimize: $f(\mathbf{x}) = x_1^2 + x_2^2 - 4x_1 + 2x_2 + 5$

    Subject to: $g(\mathbf{x}) = x_1 + x_2 = 4$

    (b) Minimize: $f(\mathbf{x}) = (x_1 - 2)^2 + (x_2 - 1)^2$

    Subject to: $g(\mathbf{x}) = x_1 - 2x_2 + 1 = 0$

    (Himmelblau, 1972)

23. A problem that arises in oil-storage purchasing problems is that of determining how many gallons of each item to purchase for a minimum cost policy under restricted tank space requirements. The following formula describes system costs.

    Minimize:

$$f(\mathbf{x}) = \sum_{n=1}^{N} \left(\frac{\alpha_n \beta_n}{x_n} + \frac{\phi_n x_n}{2} \right)$$

where α_n = fixed cost for nth item
β_n = withdrawal rate per unit of time for nth item
ϕ_n = the holding cost per unit time for nth item.

The floor space constraint is given by

$$g(\mathbf{x}) = \sum_{n=1}^{N} f_n x_n \leq F$$

where f_n = space requirement for nth item
F = available space

Suppose that the following costs are determined:

Item (n)	α_n (\$)	β_n	ϕ_n (\$)	f_n (cubic feet)
1	9.60	3	0.47	1.4
2	4.27	5	0.26	2.62
3	6.42	4	0.61	1.71

(a) Verify that the positive optimal solution to the unconstrained problem is given by: $x = (11.07, 12.82, 9.176)$.

(b) If there is only 22 cubic feet of space available, solve the constrained problem as an equality constrained optimization problem using Everett's method of generalized Lagrangian multipliers.

24. Solve the following problem using constrained derivatives:

    Minimize: $f(\mathbf{x}) = 7x_1 - 6x_2 + 4x_3$

    Subject to: $x_1^2 + 2x_2^2 + 3x_3^2 = 1$

    $x_1 + 5x_2 - 3x_3 = 6$

25. Solve the following problem using constrained derivatives:

    Minimize: $f(\mathbf{x}) = 4x_1 - \frac{1}{2}x_2^2 - 12$

    Subject to: $x_1 x_2 = 4$

26. Solve the following problems using the projected gradient method:

(a) Minimize: $f(\mathbf{x}) = 4x_1 - x_2^2 - 6$

Subject to: $26 - x_1^2 - x_2^2 = 0$

(b) Minimize: $f(\mathbf{x}) = 25(x_1 - 3x_2)^2 + (x_1 - 3)^2$

Subject to: $x_1 + 2x_2 = 9$

27. Solve the following problem using only the Kuhn-Tucker conditions:

$$f(\mathbf{x}) = 100 - 1.2x_1 - 1.5x_2 + 0.3x_1^2 + 0.05x_2^2$$

Subject to:

$$g_1(\mathbf{x}) = x_1 + x_2 \geqq 35$$
$$g_2(\mathbf{x}) = x_1 \geqq 0$$
$$g_3(\mathbf{x}) = x_2 \geqq 0$$

28. Consider the following problem:

$$f(\mathbf{x}) = 100(x_2 - x_1^2)^2 + (1 - x_1)^2$$

Subject to:

$$g_1(\mathbf{x}) = x_1 + 1 \geqq 0$$
$$g_2(\mathbf{x}) = 1 - x_2 \geqq 0$$
$$g_3(\mathbf{x}) = 4x_2 - x_1 - 1 \geqq 0$$
$$g_4(\mathbf{x}) = 1 - 0.5x_1 - x_2 \geqq 0$$

What can you say about the following two solution vectors?

$$\mathbf{x} = (-1, 1) \qquad \text{and} \qquad \mathbf{x} = (-0.3773, 0.1557)$$

29. Solve the following quadratic programming problems.

(a) Minimize: $f(\mathbf{x}) = x_1^2 - x_1x_2 + 3x_2^2 - 4x_2 + 4$

Subject to: $x_1 + x_2 \leqq 1$

$x_1, x_2 \geqq 0$

(b) Minimize: $f(\mathbf{x}) = 2x_2^2 + 3x_1^2 + 3x_1x_2 - 25(x_1 + x_2)$

Subject to: $2x_1 + x_2 \leqq 5$

$x_1, x_2 \geqq 0$

30. Solve the following problem using separable programming.

Maximize: $f(\mathbf{x}) = x_1^2 - 2x_2^2 + 3x_3^2 + 4x_2 - 9x_3$

Subject to: $2x_1^2 + x_2 + x_3^2 \leqq 2.5$

$x_1, x_2 \geqq 0$

31. Show how the following problems can be solved using separable programming.

(a) Minimize: $f(\mathbf{x}) = x_1 + 3x_1x_2 + x_2x_3$

Subject to: $x_1x_2 + x_3 + x_2x_3 \geqq 4$

$x_1, x_2, x_3 \geqq 0$

(b) Minimize: $f(\mathbf{x}) = 2e^{x_1^2 + x_2^2} + (x_1 - x_2)^2$

Subject to: $x_1 + x_2 \geqq 5$

$x_1, x_2 \geqq 0$

32. Finish the example problem of Section 11.13 and verify that the optimal solution is given by: $\mathbf{x} = (5.0, 4.5)$ to an accuracy of ± 0.10.

33. Solve the following problem using Kelly's Cutting Plane algorithm.

Minimize: $f(\mathbf{x}) = x_1^2 + x_2$

Subject to: $x_1 + x_2 \geqq 4$

$x_1, x_2 \geqq 0$

34. Solve the following problem using Kelly's Cutting Plane algorithm.

Minimize: $f(\mathbf{x}) = 3x_1^2 + x_2^2$

Subject to:
$$h_1(\mathbf{x}) = 2x_1^2 + 3x_2^2 \leq 50$$
$$h_2(\mathbf{x}) = 4x_1^2 - x_2^2 \leq 50$$
$$x_1, x_2 \geq 0$$

35. Solve the following problems using geometric programming.

(a) Minimize: $f(\mathbf{x}) = 5x_1^2x_2^{-1}x_3 + 10x_1^{-3}x_2^2x_3^{-2}$

Subject to:
$$0.357x_1^{-1}x_3 + 0.625x_1^{-1}x_3^{-1} \leq 1$$
$$x_1, x_2, x_3 > 0$$

(b) Minimize: $f(\mathbf{x}) = x_1^{-1}x_2^{-1}x_3^{-1}$

Subject to:
$$2x_1 + x_2 + 3x_3 \leq 1$$
$$x_1, x_2, x_3 > 0$$

36. Find the dimensions of an open rectangular tank that give a minimum surface area if the capacity of the tank is equal to 1000 ft³. Solve this problem using:
(a) Lagrangian multipliers
(b) Constrained derivatives
(c) Projected gradient
(d) Geometric programming
(e) Newton's technique

Comment on the relative effectivness of the solution techniques.

37. A forest fire is burning down a narrow valley of width 2 miles at a velocity of 32 fpm. (Fig. 11.34). A fire can be contained by cutting a fire break through the forest across the width of the valley. A man can clear 2 ft of the fire break in a minute. It costs $20 to transport each man to the scene of fire and back, and each man is paid $6 per hour while there. The value of timber is $2000 per square mile. How many men should be sent to fight the fire so as to minimize the total costs?

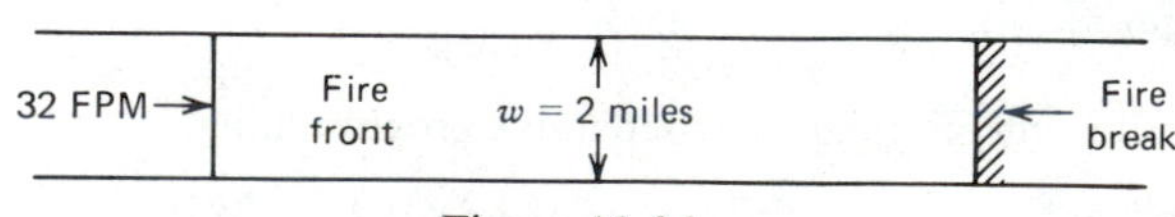

Figure 11.34
Problem 37 schematic.

38. From theoretical considerations it is believed that the dependent variable y is related to variable x via a two-parameter function:

$$y(x) = \frac{k_1 x}{1 + k_2 x}$$

The parameters k_1 and k_2 are to be determined by a least squares fit of the following experimental data:

x	y
1.0	1.05
2.0	1.25
3.0	1.55
4.0	1.59

Find k_1 and k_2, using (i) Pattern Search, (ii) Newton's method.

39. A scientist has observed a certain quantity Q as a function of a variable t. She has good reasons to believe that there is a physical law relating t and Q that takes the form:

$$Q(t) = a \sin t + b \tan t + c \qquad (11.177)$$

She wants to have the "best" possible idea of the value of the coefficients a, b, c using the results of her n experiments $(t_1, Q_1; t_2, Q_2; t_3, Q_3; \ldots; t_n, Q_n)$. Taking into account that her experiments are not perfect and also perhaps that Eq. (11.177) is not rigorous, she does not expect to find a perfect fit. So she defines an error term

$$e_i = Q_i - Q(t_i)$$

and has two different ideas of what "best" may mean:

1. Her first idea is to minimize

$$Z_1 = \sum_{i=1}^{n} |e_i|$$

2. Her second idea is to minimize

$$Z_2 = \sum_{i=1}^{n} e_i^2$$

She then discovers that for physical reasons the coefficients a, b, c must be nonnegative. Taking these new constraints into consideration, she minimizes again:

3.

$$Z_3 = \sum_{i=1}^{n} |e_i|$$

4.

$$Z_4 = \sum_{i=1}^{n} e_i^2$$

Show, with detailed justification, that she will have to solve, respectively,

(a) A linear program in case (1).
(b) A system of linear equations in case (2). [*Hint*: First show that in case (2), the problem reduces to minimizing an unconstrained convex function.]
(c) A linear program in case (3).
(d) A problem of the form

$$\omega = Mz + q \qquad \omega \geqslant 0 \qquad z \geqslant 0 \qquad \omega^T z = 0 \text{ in case (4).}$$

Specify what M, z, q, ω are. (*Hint*: Show first that the problem in this case reduces to a convex quadratic program.)
Note: Give all specifications you can on each problem: number of variables, number of constraints or size, etc.

40. A monopolist estimates the following price vs. demand relationships for four of his products:

Product 1: $x_1 = -1.5143p_1 + 2671$
Product 2: $x_2 = -0.0203p_2 + 135$
Product 3: $x_3 = -0.0136p_3 + 0.0015p_4 + 103$
Product 4: $x_4 = 0.0016p_3 - 0.0027p_4 + 19$

where x_j and p_j are the unknown demand and price for product j. The function to be maximized is $\Sigma_{j=1}^{4} p_j x_j$. The cost of production is fixed and hence can be discarded. Using the price-demand relationships, the objective function can be written as a quadratic function in either p_j or x_j. The products use two raw materials for production, and they give rise to the following capacity constraints:

$$0.026x_1 + 0.8x_2 + 0.306x_3 + 0.245x_4 \leqslant 121$$
$$0.086x_1 + 0.02x_2 + 0.297x_3 + 0.371x_4 \leqslant 250$$
$$x_1, \ldots, x_4 \geqslant 0$$
$$p_1, \ldots, p_4 \geqslant 0$$

Formulate a quadratic program to determine the optimal production levels that will maximize the total revenue $\Sigma p_j x_j$, and solve.

41. The electrical resistance network shown in Figure 11.35 is required to accommodate a current flow of 10 A from junction 1 to node 4. Given the resistance of each of the arcs, the problem is to determine the equilibrium distribution of current flows that will minimize the total power loss in the network.
    (a) Formulate the problem as a quadratic program.
    (b) Write down the Kuhn-Tucker conditions.
    (c) Using (b) and the fact that the optimal current flows have to be positive, determine the equilibrium distribution of current flows in the network.

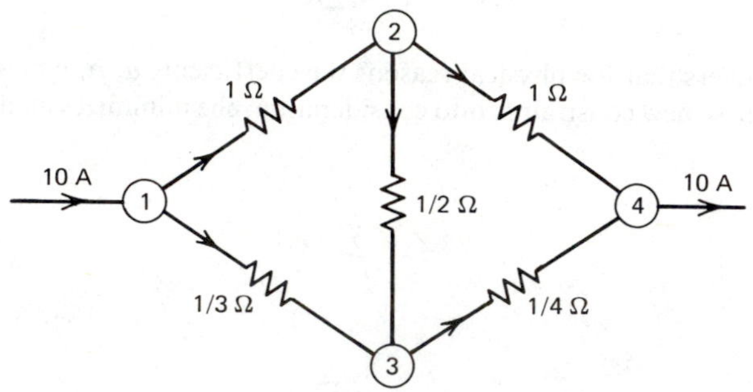

Figure 11.35
Electrical resistance network.

APPENDICES

APPENDIX A
REVIEW OF LINEAR ALGEBRA

A.1 SET THEORY

A *set* is a well-defined collection of things. By well defined we mean that given any object, it is possible to determine whether or not it belongs to the set.

The set $S=\{x \mid x \geqq 0\}$ defines the set of all nonnegative numbers. $x=2$ is an element of the set S, and is written as $2 \in S$ (2 belongs to S).

The *union* of two sets P and Q defines another set R so that $R=P \cup Q=\{x \mid x \in P$ or $x \in Q$, or both$\}$.

The *intersection* of two sets, written $P \cap Q$, defines a set $R=\{x \mid x \in P \text{ and } x \in Q\}$.

P is a *subset* of Q, written $P \subset Q$, if every element of P is in Q.

Disjoint sets have no elements in common. If P and Q are disjoint sets, then $x \in P$ implies $x \notin Q$ and vice versa.

The *empty* set, denoted by Φ, is a set with no elements in it.

A.2 VECTORS

A *vector* is an ordered set of real numbers. For instance, $\mathbf{a}=(a_1, a_2, \ldots, a_n)$ is a vector of n elements or components.

If $\mathbf{a}=(a_1, a_2, \ldots, a_n)$, and $\mathbf{b}=(b_1, b_2, \ldots, b_n)$, then

$$\mathbf{a}+\mathbf{b}=\mathbf{c}=(a_1+b_1, a_2+b_2, \ldots, a_n+b_n)$$
$$\mathbf{a}-\mathbf{b}=\mathbf{d}=(a_1-b_1, a_2-b_2, \ldots, a_n-b_n)$$

$\alpha\mathbf{a}=\mathbf{e}=(\alpha a_1, \alpha a_2, \ldots, \alpha a_n)$ for any scalar α positive or negative.

The vector $\mathbf{0}=(0, 0, \ldots, 0)$ is called the *null* vector.

The inner product of two vectors, written $\mathbf{a} \cdot \mathbf{b}$ or simply $\mathbf{ab}$ is a number given by

$$a_1b_1+a_2b_2+\cdots+a_nb_n$$

For example, if $\mathbf{a}=(1, 2, 3)$ and $\mathbf{b}=\begin{pmatrix}4\\5\\6\end{pmatrix}$, then $\mathbf{a} \cdot \mathbf{b}=4+10+18=32$.

A set of vectors $\mathbf{a}_1, \mathbf{a}_2, \ldots, \mathbf{a}_n$, is *linearly dependent* if there exists scalars $\alpha_1, \alpha_2, \ldots, \alpha_n$ not all zero, such that

$$\sum_{i=1}^{n} \alpha_i \mathbf{a}_i=\mathbf{0}$$

In this case, at least one vector can be written as a *linear combination* of the others. For example,

$$\mathbf{a}_1 = \lambda_2\mathbf{a}_2 + \lambda_3\mathbf{a}_3 + \cdots + \lambda_n\mathbf{a}_n$$

If a set of vectors is not dependent, then it must be *independent*.

A *vector space* is the set of all n-component vectors. This is generally called the *euclidean n*-space.

A set of vectors is said to *span* a vector space **V** if every vector in **V** can be expressed as a linear combination of the vectors in that set.

A *basis* for the vector space **V** is a set of linearly independent vectors that spans **V**.

A.3 MATRICES

A *matrix* **A** of size $(m \times n)$ is a rectangular array (table) of numbers with m rows and n columns.

Example A.3-1

$$\underset{(2\times 3)}{\mathbf{A}} = \begin{bmatrix} 1 & 2 & 3 \\ 4 & 5 & 6 \end{bmatrix}$$

is a matrix of two rows and three columns. The (i, j)th element of **A**, denoted by a_{ij}, is the element in the ith row and jth column of **A**. In Example A.3-1, $a_{12} = 2$ while $a_{23} = 6$. In general, a matrix of size $(m \times n)$ is written as,

$$\underset{(m\times n)}{\mathbf{A}} = [a_{ij}]$$

The elements a_{ij} for $i = j$ are called the *diagonal* elements; while the a_{ij} for $i \neq j$ are called the *off-diagonal* elements.

The elements of each column of a matrix define a vector called a *column vector*. Similarly, each row of a matrix defines a *row vector*. In Example A.3-1, the vectors $\mathbf{a}_1 = \begin{pmatrix} 1 \\ 4 \end{pmatrix}$, $\mathbf{a}_2 = \begin{pmatrix} 2 \\ 5 \end{pmatrix}$, and $\mathbf{a}_3 = \begin{pmatrix} 3 \\ 6 \end{pmatrix}$ are the column vectors of the matrix **A**; while the vectors $\mathbf{b}_1 = (1\ 2\ 3)$ and $\mathbf{b}_2 = (4\ 5\ 6)$ are the row vectors of **A**.

Thus a vector may be treated as a special matrix with just one row or one column. A matrix with an equal number of rows and columns is called a *square matrix*.

The transpose of a matrix $\mathbf{A} = [a_{ij}]$, denoted by $\mathbf{A}'$ or $\mathbf{A}^{\tau}$ is a matrix obtained by interchanging the rows and columns of **A**. In other words, $\mathbf{A}' = [a'_{ij}]$ where $a'_{ij} = a_{ji}$. The transpose of **A** defined in Example A.3-1 is given by

$$\underset{(3\times 2)}{\mathbf{A}'} = \begin{bmatrix} 1 & 4 \\ 2 & 5 \\ 3 & 6 \end{bmatrix}$$

The matrix **A** is said to be *symmetric* if $\mathbf{A}' = \mathbf{A}$. The *identity* matrix, denoted by **I**, is a square matrix whose diagonal elements are all one and the off-diagonal elements are all zero.

A matrix whose elements are all zero is called a *null matrix*.

Matrix Operations

The *sum* or *difference* of two matrices **A** and **B** is a matrix **C** (written $\mathbf{C} = \mathbf{A} \pm \mathbf{B}$) where the elements of **C** are given by

$$c_{ij} = a_{ij} \pm b_{ij}$$

For two matrices **A** and **B**, the product **AB** is defined if and only if the number of columns of **A** is equal to the number of rows of **B**. If **A** is an $(m \times n)$ matrix and **B** is an $(n \times r)$ matrix, then the product $\mathbf{AB} = \mathbf{C}$ is defined, whose size is $(m \times r)$. The (i, j)th element of **C** is given by

$$c_{ij} = \sum_{k=1}^{n} a_{ik} b_{kj}$$

Example A.3-2

$$\underset{(2\times 3)}{\mathbf{A}} = \begin{bmatrix} 1 & 2 & 3 \\ 4 & 5 & 6 \end{bmatrix}, \qquad \underset{(3\times 2)}{\mathbf{B}} = \begin{bmatrix} 1 & 2 \\ 3 & 4 \\ 5 & 6 \end{bmatrix}$$

$$\mathbf{AB} = \underset{(2\times 2)}{\mathbf{C}} = \begin{bmatrix} 22 & 28 \\ 49 & 64 \end{bmatrix}$$

Example A.3-3

Let $\mathbf{A} = \begin{bmatrix} 1 & 2 & 3 \\ 4 & 5 & 6 \end{bmatrix}$, $\mathbf{x} = \begin{bmatrix} 2 \\ 3 \\ 4 \end{bmatrix}$, and $\mathbf{y} = (2, 3)$. Then,

$$\mathbf{Ax} = \underset{(2\times 1)}{\mathbf{b}} = \begin{bmatrix} 20 \\ 47 \end{bmatrix},$$

while

$$\mathbf{yA} = \underset{(1\times 3)}{\mathbf{d}} = (14\ 19\ 24)$$

Note that **b** is a column vector, and **d** is a row vector. For any scalar α,

$$\alpha\mathbf{A} = [\alpha a_{ij}]$$

Matrix operations satisfy the following properties:

1. $(\mathbf{A}+\mathbf{B})+\mathbf{C} = \mathbf{A}+(\mathbf{B}+\mathbf{C})$
2. $\mathbf{A}+\mathbf{B} = \mathbf{B}+\mathbf{A}$
3. $(\mathbf{A}+\mathbf{B})\mathbf{C} = \mathbf{AC}+\mathbf{BC}$
4. $(\mathbf{AB})\mathbf{C} = \mathbf{A}(\mathbf{BC})$
5. $\mathbf{IA} = \mathbf{AI} = \mathbf{A}$
6. $(\mathbf{A}+\mathbf{B})' = \mathbf{A}'+\mathbf{B}'$
7. $(\mathbf{AB})' = \mathbf{B}'\mathbf{A}'$

In general $\mathbf{AB} \neq \mathbf{BA}$.

Determinant of a Square Matrix

The determinant of a square matrix **A**, denoted by $|\mathbf{A}|$, is a number obtained by certain operations on the elements of **A**. If **A** is a (2×2) matrix, then

$$|\mathbf{A}| = \begin{vmatrix} a_{11} & a_{12} \\ a_{21} & a_{22} \end{vmatrix} = a_{11}a_{22} - a_{12}a_{21}$$

If **A** is an $(n\times n)$ matrix, then

$$|\mathbf{A}| = \sum_{i=1}^{n} a_{i1}(-1)^{i+1}\,|\mathbf{M}_{i1}|$$

where $\mathbf{M}_{i1}$ is a submatrix obtained by deleting row i and column 1 of **A**. For example,

$$\text{if} \qquad \underset{(3\times 3)}{\mathbf{A}} = \begin{bmatrix} 1 & 2 & 3 \\ 4 & 5 & 6 \\ 7 & 8 & 9 \end{bmatrix}$$

then

$$|\mathbf{A}| = 1\begin{vmatrix} 5 & 6 \\ 8 & 9 \end{vmatrix} - 4\begin{vmatrix} 2 & 3 \\ 8 & 9 \end{vmatrix} + 7\begin{vmatrix} 2 & 3 \\ 5 & 6 \end{vmatrix}$$

$$= (45-48) - 4(18-24) + 7(12-15) = 0$$

A matrix is said to be *singular* if its determinant is equal to zero. If $|\mathbf{A}| \neq 0$, then $\mathbf{A}$ is called *nonsingular*.

Inverse of a Matrix

For a nonsingular square matrix $\mathbf{A}$, the inverse of $\mathbf{A}$ denoted by $\mathbf{A}^{-1}$, is a nonsingular square matrix such that

$$\mathbf{AA}^{-1} = \mathbf{A}^{-1}\mathbf{A} = \mathbf{I} \text{ (identity matrix)}$$

The inverse matrix ($\mathbf{A}^{-1}$) may be obtained by performing row operations on the original matrix ($\mathbf{A}$). The row operations consist of

1. Multiply or divide any row by a number.
2. Multiply any row by a number and add it to another row.

To find the inverse of a matrix $\mathbf{A}$, one starts by adjoining an identity matrix of similar size as $[\mathbf{A}]$. By a sequence of row operations $\mathbf{A}$ is reduced to $\mathbf{I}$. this will reduce the original $\mathbf{I}$ matrix to $\mathbf{A}^{-1}$ since

$$\mathbf{A}^{-1}[\mathbf{AI}] = [\mathbf{IA}^{-1}]$$

Example A.3-4

$$\mathbf{A} = \begin{bmatrix} 1 & 1 \\ 1 & -1 \end{bmatrix}$$

Since $|\mathbf{A}| = -2$, $\mathbf{A}$ is nonsingular, and hence $\mathbf{A}^{-1}$ exists. To compute $\mathbf{A}^{-1}$ start with the following matrix:

$$(\mathbf{AI}) = \left[\begin{array}{cc|cc} 1 & 1 & 1 & 0 \\ 1 & -1 & 0 & 1 \end{array}\right]$$

subtract row 1 from row 2:

$$\left[\begin{array}{cc|cc} 1 & 1 & 1 & 0 \\ 0 & -2 & -1 & 1 \end{array}\right]$$

Divide row 2 by -2

$$\left[\begin{array}{cc|cc} 1 & 1 & 1 & 0 \\ 0 & 1 & 1/2 & -1/2 \end{array}\right]$$

Subtract row 2 from row 1:

$$\left[\begin{array}{cc|cc} 1 & 0 & 1/2 & 1/2 \\ 0 & 1 & 1/2 & -1/2 \end{array}\right]$$

Thus

$$\mathbf{A}^{-1} = \begin{bmatrix} 1/2 & 1/2 \\ 1/2 & -1/2 \end{bmatrix}$$

Verify that $\mathbf{AA}^{-1} = \mathbf{A}^{-1}\mathbf{A} = \mathbf{I}$.

A.4 QUADRATIC FORMS

A function of n variables $f(x_1, x_2, \ldots, x_n)$ is called a *quadratic form* if

$$f(x_1, x_2, \ldots, x_n) = \sum_{i=1}^{n}\sum_{j=1}^{n} q_{ij}x_ix_j = \mathbf{x}'\mathbf{Qx}$$

where $\underset{(n\times n)}{\mathbf{Q}} = [q_{ij}]$ and $\mathbf{x}' = (x_1, x_2, \ldots, x_n)$. Without any loss of generality, $\mathbf{Q}$ can always be assumed symmetric. Otherwise $\mathbf{Q}$ may be replaced by the symmetric matrix $(\mathbf{Q}+\mathbf{Q}')/2$ without changing the value of the quadratic form.

Definitions

1. A matrix $\mathbf{Q}$ is *positive definite* if and only if the quadratic form $\mathbf{x}'\mathbf{Q}\mathbf{x}>0$ for all $\mathbf{x}\neq 0$. For example, $\mathbf{Q}=\begin{bmatrix} 2 & -1 \\ -1 & 2 \end{bmatrix}$ is positive definite.
2. A matrix $\mathbf{Q}$ is *positive semidefinite* if and only if the quadratic form $\mathbf{x}'\mathbf{Q}\mathbf{x}\geqq 0$ for all $\mathbf{x}$ and there exists an $\mathbf{x}\neq 0$ such that $\mathbf{x}'\mathbf{Q}\mathbf{x}=0$. For example, $\mathbf{Q}=\begin{bmatrix} 1 & -1 \\ -1 & 1 \end{bmatrix}$ is positive semidefinite.
3. A matrix $\mathbf{Q}$ is *negative definite* if and only if $-\mathbf{Q}$ is positive definite. In other words, $\mathbf{Q}$ is negative definite when $\mathbf{x}'\mathbf{Q}\mathbf{x}<0$ for all $\mathbf{x}\neq 0$. For example, $\mathbf{Q}=\begin{bmatrix} -2 & 1 \\ 1 & -3 \end{bmatrix}$ is negative definite.
4. A matrix $\mathbf{Q}$ is *negative semidefinite* if $-\mathbf{Q}$ is positive semidefinite. For example, $\mathbf{Q}=\begin{bmatrix} -1 & 1 \\ 1 & -1 \end{bmatrix}$ is negative semidefinite.
5. A matrix $\mathbf{Q}$ is *indefinite* if $\mathbf{x}'\mathbf{Q}\mathbf{x}$ is positive for some $\mathbf{x}$ and negative for some other $\mathbf{x}$. For example, $\mathbf{Q}=\begin{bmatrix} 1 & -1 \\ 1 & -2 \end{bmatrix}$ is indefinite.

Principal Minor

If $\mathbf{Q}$ is an $(n\times n)$ matrix, then the *principal minor* of order k is a submatrix of size $(k\times k)$ obtained by deleting any $(n-k)$ rows and their corresponding columns from the matrix $\mathbf{Q}$.

Example A.4-1

$$\mathbf{Q}=\begin{bmatrix} 1 & 2 & 3 \\ 4 & 5 & 6 \\ 7 & 8 & 9 \end{bmatrix}$$

Principal minors of order 1 are essentially the diagonal elements 1, 5, and 9. The principal minor of order 2 are the following (2×2) matrices:

$$\begin{bmatrix} 1 & 2 \\ 4 & 5 \end{bmatrix} \quad \begin{bmatrix} 1 & 3 \\ 7 & 9 \end{bmatrix}, \quad \text{and} \quad \begin{bmatrix} 5 & 6 \\ 8 & 9 \end{bmatrix}$$

The principal minor of order 3 is the matrix $\mathbf{Q}$ itself.

The determinant of a principal minor is called the *principal determinant*. For an $(n\times n)$ square matrix, there are in all 2^n-1 principal determinants.

Leading Principal Minor of order k of an $(n\times n)$ matrix is obtained by deleting the *last* $(n-k)$ rows and their corresponding columns. In Example A.4-1, the leading principal minor of order 1 is one (delete the last two rows and columns). The leading principal minor of order 2 is $\begin{bmatrix} 1 & 2 \\ 4 & 5 \end{bmatrix}$, while that of order 3, is the matrix $\mathbf{Q}$ itself. The number of leading principal determinants of an $(n\times n)$ matrix is n.

There are some easier tests to determine whether a given matrix is positive definite, negative definite, positive semidefinite, negative semidefinite or indefinite. *All these tests are valid only when the matrix is symmetric* (if the matrix $\mathbf{Q}$ is not symmetric, change $\mathbf{Q}$ to $(\mathbf{Q}+\mathbf{Q}')/2$ and then apply the tests).

Tests for Positive Definite Matrices

a. All diagonal elements must be positive.
b. All the *leading principal determinants* must be positive.

Tests for Positive Semi-Definite Matrices

a. All diagonal elements are nonnegative.
b. All the *principal determinants* are nonnegative.

1. To prove that a matrix is negative definite (negative semidefinite) test the negative of that matrix for positive definite (positive semidefinite).

2. A sufficient test for a matrix to be indefinite, is that at least two of its diagonal elements are of the opposite signs.

A.5 CONVEX AND CONCAVE FUNCTIONS

A function of n variables $f(x_1, x_2, \ldots, x_n)$ is said to be a *convex function* if and only if for any two points $\mathbf{x}^{(1)}$ and $\mathbf{x}^{(2)}$ and $0 \leqq \lambda \leqq 1$,

$$f[\lambda\mathbf{x}^{(1)} + (1-\lambda)\mathbf{x}^{(2)}] \leqq \lambda f(\mathbf{x}^{(1)}) + (1-\lambda)f(\mathbf{x}^{(2)})$$

A function $f(x_1, \ldots, x_n)$ is a concave function if and only if $-f(x_1, \ldots, x_n)$ is a convex function.

The gradient of a function $f(x_1, \ldots, x_n)$ is given by

$$\nabla\mathbf{f}(x_1, \ldots, x_n) = \left[\frac{\delta f}{\delta x_1}, \frac{\delta f}{\delta x_2}, \ldots, \frac{\delta f}{\delta x_n}\right]$$

The *hessian matrix* of a function $f(x_1, \ldots, x_n)$ is an $(n \times n)$ symmetric matrix given by

$$\mathbf{H}_f(x_1, \ldots, x_n) = \left[\frac{\delta^2 f}{\delta x_i \delta x_j}\right]$$

Test for Convexity of a Function

A function f is a convex function if the hessian matrix of f is positive definite or positive semidefinite for all values of $x_1, \ldots, x_n$.

Test for Concavity of a Function

A function f is concave if the hessian matrix of f is negative definite or negative semidefinite for all values of $x_1, \ldots, x_n$.

Example A.5-1

$$f(x_1, x_2, x_3) = 3x_1^2 + 2x_2^2 + x_3^2 - 2x_1x_2 - 2x_1x_3 + 2x_2x_3 - 6x_1 - 4x_2 - 2x_3$$

$$\nabla\mathbf{f}(x_1, x_2, x_3) = \begin{pmatrix} 6x_1 - 2x_2 - 2x_3 - 6 \\ 4x_2 - 2x_1 + 2x_3 - 4 \\ 2x_3 - 2x_1 + 2x_2 - 2 \end{pmatrix}$$

$$\mathbf{H}_f(x_1, x_2, x_3) = \begin{bmatrix} 6 & -2 & -2 \\ -2 & 4 & 2 \\ -2 & 2 & 2 \end{bmatrix}$$

To show that f is a convex function, we test $\mathbf{H}$ for positive definite or positive semidefinite property. Note that

1. $\mathbf{H}$ is symmetric
2. All diagonal elements are positive
3. The leading principal determinants are

$$|6| > 0, \qquad \begin{vmatrix} 6 & -2 \\ -2 & 4 \end{vmatrix} = 20 > 0, \qquad |H| = 16 > 0$$

Hence, $\mathbf{H}$ is a positive definite matrix, which implies f is a convex function. (As a matter of fact, when $\mathbf{H}_f$ is positive definite, f is said to be strictly convex with a unique minimum point.)

A.6 CONVEX SETS

A set $\mathbf{S}$ is said to be a convex set if for any two points in the set the line joining those two points is also in the set. Mathematically, $\mathbf{S}$ is a convex set if for any two vectors $\mathbf{X}^{(1)}$ and $\mathbf{X}^{(2)}$ in $\mathbf{S}$, the vector $\mathbf{X} = \lambda\mathbf{X}^{(1)} + (1-\lambda)\mathbf{X}^{(2)}$ is also in $\mathbf{S}$ for any number λ between 0 and 1.

Examples

Figures 1 and 2 represent convex sets, whereas Figure 3 is not a convex set.

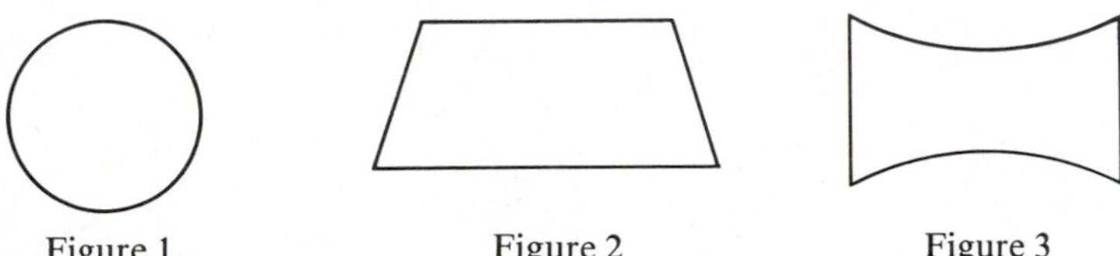

Figure 1 Figure 2 Figure 3

Theorem 1: The set of all feasible solutions to a linear programming problem is a convex set.

Theorem 2: The intersection of convex sets is a convex set.

Theorem 3: The union of convex sets is not necessarily a convex set.

Definition

A convex combination of vectors $\mathbf{X}^{(1)}, \mathbf{X}^{(2)}, \ldots, \mathbf{X}^{(k)}$ is a vector $\mathbf{X}$ such that

$$\mathbf{X} = \lambda_1\mathbf{X}^{(1)} + \lambda_2\mathbf{X}^{(2)} + \cdots + \lambda_k\mathbf{X}^{(k)}$$
$$\lambda_1 + \lambda_2 + \cdots + \lambda_k = 1$$
$$\lambda_i \geqq 0 \qquad \text{for } i = 1, 2, \ldots, k.$$

Extreme point or vertex of a convex set is a point in the set that cannot be expressed as the midpoint of any two points in the set.

For example, consider the convex set $\mathbf{S} = \{(X_1, X_2) \mid 0 \leqq X_1 \leqq 2,\ 0 \leqq X_2 \leqq 2\}$. This set has four extreme points given by (0, 0), (0, 2), (2, 0), and (2, 2).

A *hyperplane* is the set of all points $\mathbf{X}$ satisfying $\mathbf{CX} = Z$ for a given vector $\mathbf{C} \neq \mathbf{0}$ and scalar Z. The vector $\mathbf{C}$ is called the *normal* to the hyperplane. For example, $H = \{(X_1, X_2, X_3) \mid 2X_1 - 3X_2 + X_3 = 5\}$ is a hyperplane.

A *half space* is the set of all points $\mathbf{X}$ satisfying $\mathbf{CX} \leqq Z$ or $\mathbf{CX} \geqq Z$ for a given vector $\mathbf{C} \neq \mathbf{0}$ and scalar Z.

Theorem 4: A hyperplane is a convex set.

Theorem 5: A half space is a convex set.

Theorem 6: Every basic feasible solution to a linear-programming problem corresponds to an extreme point of the convex set of feasible solutions.

Theorem 7: If there exists an optimal solution to a linear program, then at least one of the extreme points of the convex set of feasible solutions will qualify to be an optimal solution.

Theorem 8: The set of all optimal solutions to a linear program forms a convex set.

Remarks

1. The fundamental theory of the simplex method is based on Theorems 1, 6, and 7.
2. Theorem 8 is extremely useful in practice. If we have two optimal solutions to a linear program, given by $\mathbf{X}^{(1)}$ and $\mathbf{X}^{(2)}$, then every solution X obtained by

$$\mathbf{X} = \lambda\mathbf{X}^{(1)} + (1 - \lambda)\mathbf{X}^{(2)} \qquad \text{for } 0 \leqq \lambda \leqq 1$$

is also an optimal solution. Thus a linear program may have (a) no optimal solution, (b) a unique optimal solution, or (c) an infinite number of optimal solutions.

REFERENCE

1. G. Hadley, *Linear Algebra*, Addison-Wesley, Reading, Mass., 1961.

APPENDIX B

PROBABILITY REVIEW

B.1 INTRODUCTION

Although it is likely that the student using this book will have had some previous exposure to probability theory, this appendix has been included to serve as a concise, convenient "refresher" summary of those probability concepts that are important in subsequent chapters. The student who is confident of his or her abilities in this area may elect to skim, or skip altogether, the entire appendix. On the other hand, the student who has had no previous exposure to the subject would be well advised to consult one of the introductory books listed at the end of the appendix.

I. BASIC DEFINITIONS

B.2 EXPERIMENTS, SAMPLE SPACES, AND EVENTS

An *experiment* is some well-understood procedure or process whose outcome can be observed but is not known with certainty in advance. The set of all possible outcomes is called the *sample space*. Whenever the sample space consists of a countable number of outcomes, it is said to be *discrete*; otherwise, it is *continuous*. An *event* is any subset of the sample space. When the result of the experiment becomes known, we would say that a specified event had *occurred* if the observed outcome is contained in the subset that is the event.

Of course, events are really no more than what you would ordinarily consider them to be. Often the most natural way to specify them is to describe them in words. However, the reason for defining them formally as sets is to establish a mathematical way to combine and manipulate them. The set theoretic union of two events produces another event. If $C = A \cup B$, we would say in words that event C had occurred if event A *or* event B occurred. Similarly, the intersection of two events corresponds to the word "and." The complement of any event is another event; we would say that $\bar{A}$ had occurred if A had not occurred. Two events are mutually exclusive if their intersection is the empty set, which can be thought of as the impossible event. In other words, two events are mutually exclusive if they could not both occur.

B.3 PROBABILITY

When the "probability of an event" is spoken of in everyday language, almost everyone has an idea of what is meant. It is fortunate that this is so, because it would be quite difficult to introduce the concept to someone who had never considered it before. There are at least three distinct ways to approach the subject, none of which is wholly satisfying.

The first to appear, historically, was the frequency concept. If an experiment were to be repeated many times, then the number of times that the event was observed to occur, divided by the number of times that the experiment was conducted, would approach a number that was defined to be the probability of the event. This definition proved to be somewhat limiting, however, because circumstances frequently prohibit the repetition of an experiment under precisely the same conditions, even conceptually.

To extend the notion of probability to a wider class of applications, the idea of "subjective" probabilities emerged. According to this idea, the probability of an event need not relate to the frequency with which it would occur in an infinite number of trials; it is just a measure of the degree of likelihood we believe the event to possess. Thus different people could attach different probabilities to the same event.

Most modern texts use a purely axiomatic definition. According to this notion, probabilities are just elements of an abstract mathematical system obeying certain axioms. This notion is at once the most powerful and the most devoid of real-world meaning. Of course, the axioms are not purely arbitrary; they were selected to be consistent with the earlier concepts of probabilities and to provide them with all of the properties everyone would agree they should have.

A probability measure is a function, $P(\ )$, mapping events onto real numbers, and satisfying

1. $0 \leqq P(A) \leqq 1$, for any event A.
2. $P(S) = 1$, where S is the whole sample space, or the "certain" event.
3. if $A_1, A_2, A_3 \cdots$ are a set of pairwise mutually exclusive events (finite or infinite in number), then $P(A_1 \cup A_2 \cup A_3 \cdots) = P(A_1) + P(A_2) + \cdots$.

Although probabilities have a number of other properties well worth mentioning, these three axioms are sufficient to derive the others.

It should be obvious that these three axioms are not enough to determine the probability of any event. For all but trivial sample spaces, there will exist an infinity of ways to assign probabilities to events while satisfying the three axioms. At this point, we are merely establishing properties or rules required of any way that might be chosen.

Some of the additional "basic laws" of probability (which can be proved from the above axioms) are

4. $P(\phi) = 0$, where ϕ is the empty set, or the impossible event.
5. $P(\bar{A}) = 1 - P(A)$.
6. $P(A \cup B) = P(A) + P(B) - P(A \cap B)$, for any two events, A and B.

Another "basic law" is, in reality, a definition of the *conditional probability* of an event, A, given that another event, B, has occurred. The notation for this conditional probability is $P(A \mid B)$, and the defining formula is

7. $P(A \mid B) = \dfrac{P(A \cap B)}{P(B)}$, provided $P(B) \neq 0$.

The notion of conditional probability conforms to the intuitive concept of altering our estimate of the likelihood of an event as we acquire additional information. That is, $P(A \mid B)$ is the new probability of A after we know that B has occurred. Because it is common in modeling applications to know $P(A \mid B)$ directly, but not to know $P(A \cap B)$, axiom 7 often appears in the equivalent form:

8. $P(A \cap B) = P(A \mid B)P(B)$.

Conditional probabilities are useful only when the events involved, A and B, have something to do with one another. If knowledge that B has occurred has no bearing on our estimate of the likelihood of A, we would say that the two events are *independent*, and write

9. $P(A \mid B) = P(A)$ if and only if A and B are independent.

This rule can be taken as the formal definition of independence. Combining axioms 8 and 9, it immediately follows that

10. $P(A \cap B) = P(A)P(B)$ if and only if A and B are independent.

Alternatively, axiom 10 could be taken as the definition of independence and axiom 9 would immediately follow.

A set of events $B_1, B_2, \ldots, B_n$ constitute a *partition* of the sample space S if they are mutually exclusive, that is,

$$B_i \cap B_j = \phi \text{ for all } i \neq j$$

and collectively exhaustive, that is,

$$B_1 \cup B_2 \cup \cdots B_n = S$$

In simple terms, a partition is just any way of grouping and listing all possible outcomes such that no outcome appears in more than one group. When the experiment is performed, one and only one of the B_i will occur. It is easy to prove that

11. $\sum_{i=1}^{n} P(B_i) = 1$, for any partition $B_1, B_2, \cdots, B_n$.

One of the most useful relationships in modeling applications is the following

12. $P(A) = \sum_{i=1}^{n} P(A \mid B_i)P(B_i)$, for any partition $B_1, B_2, \cdots, B_n$.

It is not difficult to prove this relation rigorously, using only the rules of set theory and probability. However, because it is more important to understand the relation at an intuitive level to know how and when to use it, a verbal argument will be given here. Suppose we want to know the probability of some event A. The situation is too complicated to know $P(A)$ directly, but we *could* state the probability of A if we knew which of a number of possible conditions held. That is, we know $P(A \mid B_i)$ for a set of mutually exclusive, collectively exhaustive conditions, B_i. It would make some sense to "average" these various possible values for the probability of A. But if the conditions B_i are not all equally likely, the various $P(A \mid B_i)$ should not be given equal weight in the average; each should be weighted according to the probability that the condition B_i does in fact hold, or $P(B_i)$. This logic produces formula 12.

B.4 RANDOM VARIABLES

Although events may be directly assigned probabilities, more commonly the events are first associated with real numbers that are then in turn associated with probabilities. For example, if the experiment involves observing the number of heads appearing when four coins are tossed, it would be natural to associate the possible outcomes with the integers 0, 1, 2, 3, and 4. These integers are not, in themselves, events, but the event corresponding to each value is evident. The function that assigns numbers to events is called a *random variable*. It is interesting to note that a random variable is, technically speaking, neither random nor a variable. It is conceptually convenient, however, to suppress all references to the real-world events and to regard a random variable as an ordinary variable whose value is randomly selected. In other words, once the random variable is well defined, we may speak of any value in the range of the random variable as if it were actually the event. It makes sense, thereby, to speak of the probability that a random variable, X, equals some particular number.

In most cases, the rule that provides the value in the range of the random variable to go with each real-world event is so obvious that no special attention need be given to it. It is important to realize, however, that *values* of random variables have probabilities associated with them only because the values correspond to *events* that deserve the probabilities directly.

If the values in the range of a random variable are integers (or, more precisely, a countable subset of the real numbers), the random variable is *discrete*. If the range consists of all values over an interval of the real numbers, the random variable is *continuous*.

A word of caution is in order with respect to the use of the word "random." Occasionally, particularly in statistical applications, the word carries the connotation of equal likelihood. For example, when we say "take a random sample" we mean (among other things) that each member of the sampled population should have an equal chance of being selected. In general, however, the word "random" does not carry any such connotation.

B.5 PROBABILITY DISTRIBUTIONS

Any rule that assigns probabilities to each of the possible values of a random variable is a *probability distribution*. The term is used somewhat loosely, because there are several different ways to specify such a rule. More precise terms are used when a particular form is intended.

For discrete random variables, the most obvious and commonly used method of specifying the rule is to indicate the probability for each value separately. The function $p(x)$, defined as

$$p(x) = P(X = x)$$

is called the *probability distribution function*. An alternative, equally sufficient method is to specify the *cumulative distribution function*, $F(x)$, defined as

$$F(x) = P(X \leqq x)$$

A third choice would be the complementary cumulative distribution function, $G(x)$, defined as

$$G(x) = P(X > x).$$

If any one of these is known, the others can be easily obtained in obvious ways. For example,

$$F(x) = \sum_{y=-\infty}^{x} p(y)$$

and

$$p(x) = F(x) - F(x-1)$$

For continuous random variables, the situation is somewhat complicated by the fact that range of possible values is uncountably infinite. It is not consistent with the axioms of probability to allow each individual value to have positive probability. In fact, with the possible exception of a countable number of points, each individual value must be assigned the probability zero! In contrast to the discrete case, a probability of zero does not imply that the corresponding event is impossible; it merely means that any one particular value is so unlikely, when considered next to the uncountably infinite set of alternatives, that the probability must be negligibly small. Consequently, it is fruitless to speak of the probabilities of particular values of random variables in the continuous case.

On the other hand, it makes perfect sense to speak of the probability that the value will fall within some interval. In particular, the cumulative distribution function, $F(x)$, is well defined by

$$F(x) = P(X \leqq x)$$

Because the probability that X will exactly equal x is zero,

$$P(X \leqq x) = P(X < x) + P(X = x) = P(X < x)$$

So $F(x)$ also describes $P(X<x)$. In other words, no distinction between strong and weak inequalities, or between open and closed intervals, need be made in the continuous case. Of course, the distinction must be scrupulously maintained in the discrete case.

From the definition, it is apparent that $F(x)$ must have the following properties

$$0 \leqq F(x) \leqq 1 \text{ for all } x$$

$$\lim_{x \to -\infty} F(x) = 0$$

$$\lim_{x \to \infty} F(x) = 1$$

$$F(y) \geqq F(x) \text{ for any } y > x$$

In words, the function $F(x)$ must be bounded between 0 and 1, must approach zero at the left extremity of its range and one at the right extremity, and must be monotonically nondecreasing. (Actually, the last three imply the first.) Conversely, *any* function having these properties will qualify as a cumulative distribution function for some continuous random variable.

Given the cumulative distribution function, one can easily express the probability that the random variable will assume a value within any specified region. For example,

$$P(a \leqq X \leqq b) = P(X \leqq b) - P(X < a) = F(b) - F(a)$$

The *complementary cumulative distribution function*, $G(x)$, defined by

$$G(x) = P(X > x)$$

or by

$$G(x) = 1 - F(x)$$

would also serve to describe fully the distribution.

The *probability density function*, $f(x)$, is a function that, when integrated between a and b, gives the probability that the random variable will assume a value between a and b. That is,

$$P(a \leq X \leq b) = \int_a^b f(x)dx$$

The relation between the density function and the distribution function is direct

$$F(x) = \int_{-\infty}^{x} f(y)dy$$

and

$$f(x) = \frac{d}{dx} F(x)$$

Although it may not seem to be the most natural way to describe a probability distribution, the density function is used more often than the cumulative or complementary cumulative distribution functions. In the case of a few distributions, only the density function can be expressed in closed form; the others must be expressed as integrals of the density function. It is important, therefore, that the student learn to think in terms of density functions. One of the first things to get straight is that the value of the density function at some point is *not* a probability. The only way to get a probability from a density function is to integrate it.

Any density function will have the properties

$$\int_{-\infty}^{\infty} f(x)dx = 1$$

$$f(x) \geqq 0 \qquad \text{for all } x$$

The first property is a direct consequence of the definition, but the second requires a brief argument. If $f(x)$ were negative at any point, then there would exist two points, a and b, such that the integral of $f(x)$ between a and b was negative. This would imply that

$$P(a \leqq X \leqq b) < 0$$

which is impossible. Therefore $f(x)$ must be nonnegative everywhere.

Any function, $f(x)$, having the two properties mentioned above will qualify as a probability density function for some continuous random variable. Notice, in particular, that there is no requirement that $f(x)$ be bounded above. The second property sometimes leads students to the mistaken presumption that $f(x) \leqq 1$. In fact, $f(x)$ can be much greater than 1 over a narrow range, provided only that the *integral* over any interval does not exceed 1. Notice also that there is no requirement that $f(x)$ be continuous. Functions that are discontinuous, or abruptly "jump" from one value to another, can be integrated without difficulty, provided only that the points of discontinuity are limited in number. The method, of course, is to separate the interval that you wish to integrate into a sequence of intervals, over each of which the density function is continuous.

Although, as already noted, it is important to keep in mind that $f(x)$ is not a probability, it is useful in many applications to be able to substitute something involving $f(x)$ into expressions as if it were a probability. A generally reliable device is to think of the notation $f(x)dx$ as representing the probability that the random variable equals x. The dx part of the expression can be regarded as an interval of infinitesimal width, so the product of $f(x)$ and dx is (roughly speaking) an area under the curve, or a probability. The presence of dx will indicate that an integration must be performed before an exact expression can be inferred.

Although the distinction between the discrete and continuous random variable cases is important, there are occasions when it is convenient tohave a unified terminology to cover both cases. The letters PDF may be used to refer to either the probability distribution function, in the discrete case, or the probability density function, in the continuous case. Similarly, the letters CDF will stand for the cumulative distribution function and CCDF for the complementary cumulative distribution function.

B.6 JOINT, MARGINAL, AND CONDITIONAL DISTRIBUTIONS

Whenever more than one random variable is involved in a single problem, there is a possibility that they are related. If so, it would not be sufficient to describe the probability distribution of each random variable in isolation; the relation between or among them must also be described. There are two methods in common use.

Suppose that two random variables, X and Y, are involved. The *joint cumulative distribution function*, or joint CDF, $F(x, y)$, is defined by

$$F(x, y) = P(X \leqq x,\ Y \leqq y)$$

In words, it is the probability that X takes on a value less than or equal to x and that Y takes on a value less than or equal to y. The same definition will suffice whether the random variables are both discrete, both continuous, or mixed. Conceptually the basic idea is to extend the notion of a cumulative distribution function to two dimensions. Obviously, the same basic idea can be used to extend the notion to higher dimensions.

If both X and Y are discrete, the *joint probability distribution function* is defined by

$$p(x, y) = P(X = x,\ Y = y)$$

If both are continuous, the *joint probability density function* is defined by

$$f(x, y) = \frac{\partial}{\partial x}\frac{\partial}{\partial y}F(x, y)$$

The latter must be integrated twice to obtain a probability. In particular,

$$P(a \leqq X \leqq b, c \leqq Y \leqq d) = \int_a^b \int_c^d f(x, y)dy\,dx$$

Each of these is just a two-dimensional extension of the appropriate function for single random variables, and can be extended to higher dimensions in the obvious way. The term "joint PDF" will describe either function.

Sometimes a joint PDF for two or more random variables is given, but one wants to know the PDF for just one of the random variables. That is, you might want to make a probability statement about, say, X, without regard to the value of Y. When both X and Y are discrete, the *marginal probability distribution* function of X is obtained from the joint PDF by

$$p(x) = \sum_{\text{all } y} p(x, y)$$

When both are continuous, the *marginal probability density function* of X is given by

$$f(x) = \int_{\text{all } y} f(x, y)dy$$

A marginal PDF is just an ordinary PDF, with all of the usual properties and interpretations. The word "marginal" merely conveys the information that it was obtained from a joint PDF.

By symmetry, the marginal PDF of Y is obtainable from the joint PDF of X and Y by summing or integrating over all values of X. If more than two random variables are involved in a joint PDF, the marginal PDF for any one can be found by summing or integrating over all values of all random variables other than the one whose marginal PDF is sought. Although it is not often used, the marginal CDF is, if anything, even easier to obtain from a joint CDF:

$$F(x) = \lim_{y \to \infty} F(x, y)$$

Dealing with the CDF also has the advantage of permitting a single expression to cover both the discrete and continuous cases.

Two random variables are *independent* if

$$F(x, y) = F(x)F(y) \qquad \text{(for all } x \text{ and } y)$$

or, in terms of PDF's,

$$p(x, y) = p(x)p(y) \qquad \text{(for all } x \text{ and } y)$$

for discrete random variables, and

$$f(x, y) = f(x)f(y) \qquad \text{(for all } x \text{ and } y)$$

for continuous random variables. When the random variables are independent, *but only then*, the joint distribution can be constructed from the marginals.

Independence of random variables is an extremely important concept. Not only must you know how to manipulate the functions in the presence or absence of the property, but also to judge whether in real-life situations the property can be reasonably assumed to hold. Because the mathematical definition may not be sufficiently revealing by itself to allow the student to grasp the concept at an intuitive level, a bit of further discussion seems warranted. When we say that the joint distribution can be obtained simply by multiplying the marginals, we are admitting that the joint distribution contains no more information than is already contained in the separate descriptions of the random variables. In other words, there is no need to account for the influence that one of the random variables might exert on another. This would be true if and only if no such influence exists. Although the definition of independence of random variables is very similar in appearance to the definition of independence of events, it is actually a much stronger requirement. In order for X and Y to be independent, it is necessary that *every* event associated with X be independent of *every* event associated with Y.

The method of expressing joint PDF's or CDF's is just one of the ways to describe a relationship between two random variables. The other method is based on the idea of fixing a value for one and describing the subsequent distribution for the other. If both are discrete, the *conditional probability distribution function* of X given Y is defined by

$$p(x \mid y) = P(X = x \mid Y = y)$$

In $p(x \mid y)$, x is the argument of the function, and y can be regarded as a parameter. In other words, we may insert various values of x into the function to get the probability that the random variable equals x, but this probability will be contingent upon the value of y. Through its definition as a conditional probability, the conditional PDF is easily related to the joint PDF by the expression

$$p(x \mid y) = \frac{p(x, y)}{p(y)}, \qquad \text{provided } p(y) \neq 0$$

An analogous function exists for continuous random variables, but cannot be defined directly in terms of a conditional probability. The *conditional probability density function* of X given Y is most simply defined in terms of the joint density function

$$f(x \mid y) = \frac{f(x, y)}{f(y)}, \qquad \text{provided } f(y) \neq 0$$

This function must be integrated with respect to x to yield a probability; the y simply acts as a parameter.

The conditional PDF of X given Y reduces to the marginal PDF if and only if X and Y are independent. Notationally,

$$p(x \mid y) = p(x) \qquad \text{for all } x, y$$

or

$$f(x \mid y) = f(x) \qquad \text{for all } x, y$$

if and only if X and Y are independent. These expressions are entirely consistent with our earlier discussion of independence. If knowledge of the value of Y contributes nothing to a probability statement involving X, it must be that X and Y are unrelated.

Whenever a conditional distribution and one marginal is given, the other marginal can be obtained. The procedure is first to obtain the joint distribution and then use that to get the desired marginal. In the discrete case, the expressions would be

$$p(x, y) = p(x \mid y)p(y)$$

$$p(x) = \sum_{\text{all } y} p(x, y)$$

Therefore,

$$p(x) = \sum_{\text{all } y} p(x \mid y)p(y)$$

The analogous formula in the continuous case would be

$$f(x) = \int_{\text{all } y} f(x \mid y)f(y)dy$$

Both of these expressions are very useful in modeling applications.

B.7 EXPECTATION

To describe a random variable completely requires a probability distribution in one of its various forms. If we were to require, however, a single number that best "summarized" the information contained in the distribution, we would almost certainly want to specify the "center" of the distribution. There are several ways to define "center," but the most useful is the expectation.

The *expectation* of a random variable X, denoted $E(X)$, is defined by

$$E(X) = \sum_{\text{all } x} xp(x)$$

when X is discrete, and

$$E(X) = \int_{\text{all } x} xf(x)dx$$

when X is continuous. The same quantity may be called the *expected value* of X (although this term is somewhat misleading), the *mean* of the distribution of X, or the *first moment* of the distribution of X. It should *not* be confused with an arithmetic average or a sample mean. The latter are statistical entities; we would compute them from data. An expectation is calculated from, and is an attribute of, a probability distribution. It can be regarded as a *weighted* average of the values of X, in which each possible value is weighted by the probability of its occurrence.

Although $E(X)$ is often called the expected value of X, one should be on guard against "expecting" $E(X)$ to occur as the value of X. Indeed, when X is discrete, $E(X)$ may not even be a *possible* value of X. It is true that if the experiment for which X is defined were to be repeated many times independently and the observed values of X were collected and averaged, then this average would be "close" to $E(X)$, in a certain probabilistic sense. However, this fact is a theorem of statistics (one form of the Law of Large Numbers), and has little significance for any single trial.

One of the reasons that the expectation is so useful as a measure of centrality is that it has a number of very convenient properties. For any random variable X and any constants a and b,

$$E(aX) = aE(X)$$

and

$$E(X + b) = E(X) + b$$

In words, both multiplicative and additive constants can be "pulled out" of the expectation. For any two random variables X and Y,

$$E(X + Y) = E(X) + E(Y)$$

In words, the expected value of a sum is the sum of the expected values. The same relation can be extended to sums of more than two random variables, and will hold whether or not the random variables are independent. Whenever X and Y are independent the expected value of the product will decompose; that is,

$$E(XY) = E(X)E(Y)$$

but this relation does not generally hold when the random variables are dependent.

Another convenience associated with using the expectation is the fact that the expectation of an arbitrary function of a random variable is easily expressed. Let $h(X)$ be any function of X. Then if X is discrete,

$$E(h(X)) = \sum_{\text{all } x} h(x)p(x)$$

and if X is continuous,

$$E(h(X)) = \int_{\text{all } x} h(x)f(x)dx$$

In other words, $h(x)$ merely replaces x in the definition of $E(X)$. These expressions are not a new definition, but are derived by considering a random variable $Y = h(X)$ and relating the distribution of Y to the distribution of X.

A concept used repeatedly in the next three chapters is that of *conditional expectation*. Formally, the *conditional expectation* of a random variable X given the value of a related random variable Y, is defined by

$$E(X \mid y) = \sum_{\text{all } x} xp(x \mid y)$$

or by

$$E(X \mid y) = \int_{\text{all } x} xf(x \mid y)dx$$

Usually, however, our use of the concept will be such that the conditional expectation may be known directly. For example, suppose that we are interested in an inventory problem and X represents the number of units of some goods sold during a specified period. If Y represents the number of customers who purchase some number of units during the period, and if the expectation of the number of units purchased is the same for each customer, say 3.6 units, then the conditional expectation of the number of units sold *given* that the number of customers is y would be $3.6y$, for any y. That is, we obtain

$$E(X \mid y) = 3.6y$$

without having to use the conditional probability distribution $p(X \mid y)$. The details of the logic involved are probably unnecessary, but, just to verify rigorously that the result is correct, we may argue as follows. The number of units sold, X, is the sum of the amounts sold to each individual customer. If the number of customers is specified to be y, then X will consist of the sum of y random variables. The expectation of a sum is the sum of the expectations; so if each of these is the same, namely 3.6, the sum of y of them is $3.6y$.

The conditional expectation of X given y can be combined with the distribution of Y to yield the unconditional expectation of X. In notation,

$$E(X) = \sum_{\text{all } y} E(X \mid y)p(y)$$

or

$$E(X) = \int_{\text{all } y} E(X \mid y)f(y)dy$$

A brief way to express both of these is

$$E(X) = E[E(X \mid y)]$$

but this form does not suggest how useful the relation is as a technique for formulating an expression for $E(X)$. The other forms suggest that the expectation of X can be thought of as a weighted average of the conditional expectations of X given y, taken over all possible conditions y, with each possible $E(X \mid y)$ weighted according to the probability of occurrence. Faced with the problem of expressing $E(X)$, then, one might try to find another random variable, Y, whose distribution is known or can be found, and which has the property that when the value of Y is specified, the expectation of X is easy to obtain. If such a random variable can be found, the expectation of X can be expressed easily. The logic is very similar to that used when expressing a marginal PDF in terms of a conditional PDF, as described in Section B.6.

B.8 VARIANCE AND OTHER MOMENTS

The nth *moment* of a random variable is defined as the expectation of the nth power of the random variable. Since X^n is just a special case of a function of X, the nth moment can be expressed as

$$E(X^n) = \sum_{\text{all } x} x^n p(x)$$

or

$$E(X^n) = \int_{\text{all } x} x^n f(x)dx$$

The first moment is, of course, the expectation. The nth *central moment* or the nth *moment about the mean* is defined as

$$E([X - E(X)]^n) = \sum_{\text{all } x} [x - E(X)]^n p(x)$$

or

$$= \int_{\text{all } x} [x - E(X)]^n f(x)dx$$

In words, it is the expectation of the nth power of the random variable after it has been "shifted" by subtracting the expectation.

After the expectation, the next most important single number used to summarize distributions is the second moment about the mean, more commonly known as the *variance*. Denoting the variance of X by $V(X)$,

$$V(X) = \sum_{\text{all } x} [x - E(X)]^2 p(x)$$

or

$$V(X) = \int_{\text{all } x} [x - E(X)]^2 f(x)dx$$

In both the discrete and continuous case, the variance can be shown to equal the second moment minus the expectation squared. That is,

$$V(X) = E(X^2) - E(X)^2$$

The variance, being defined as a weighted average of the squared deviations from the expectation, is a measure of the spread, or dispersion, of a probability distribution. One of the objections to its use for this purpose is that the units are not those of X, but of X^2. Use of the *standard deviation*, defined as the square root of the variance, overcomes this objection.

The properties of variances are not so obvious as those of expectations. Whereas the behavior of expectations conforms to what intuition would suggest, considerable care must be exercised in dealing with variances and standard deviations. The rules for dealing with multiplicative and additive constants are

$$V(aX) = a^2 V(X)$$

and

$$V(X + b) = V(X)$$

In words, a multiplicative constant can be "pulled out" of a variance, but must be squared; an additive constant can be "dropped out." When considering a sum of random variables, the variance of the sum will be the sum of the variances, *if* the random variables are independent. For two independent random variables X and Y,

$$V(X + Y) = V(X) + V(Y)$$

On the other hand, if the random variables are dependent, this relation will not generally hold. The correct expression for the general case requires another definition.

Given two random variables X and Y, the *covariance* of X and Y is defined by

$$\mathrm{COV}(X, Y) = E([X - E(X)][Y - E(Y)])$$

but this expression can be shown to equal

$$\mathrm{COV}(X, Y) = E(XY) - E(X)E(Y)$$

It will be recalled that when X and Y are independent, $E(XY) - E(X)E(Y)$, so the covariance of independent random variables is zero. The converse does not always hold true; that is, the mere knowledge that the covariance of random variables is zero would not be enough for one to conclude that they are independent. Indeed, examples can be constructed of dependent random variables for which the covariance equals zero. On the other hand, a nonzero covariance definitely implies a relationship between the random variables, so the covariance is used as a (somewhat imperfect) measure of the degree of dependence. Another, related, measure of dependence is the correlation coefficient between X and Y, usually denoted by ρ, which is defined as

$$\rho = \frac{\mathrm{COV}(X, Y)}{\sqrt{V(X)V(Y)}}$$

Returning to the variance of a sum of random variables, the general equation for two random variables is

$$V(X + Y) = V(X) + V(Y) + 2\mathrm{COV}(X, Y)$$

II. DISCRETE PROBABILITY DISTRIBUTIONS

B.9 THE DISCRETE UNIFORM DISTRIBUTION

When a random variable X has only a finite number of possible values, each of which can occur with equal likelihood, the distribution is called *discrete uniform*. Without serious loss of generality, we may assume that the range of X is $x = 1, 2, \ldots N$, in which case the probability distribution function is

$$p(x) = \frac{1}{N}, \; x = 1, 2, \ldots, N$$

When X has this range, the mean and variance are

$$E(X) = \frac{N+1}{2}$$

$$V(X) = \frac{N^2 - 1}{12}$$

Of course, a shift or scaling of the range of X will have a concomitant effect upon the PDF, mean, and variance. In any case, the PDF is just one divided by the total number of possible values, for each value, and the expectation falls at the midpoint of the range.

Although it has many uses, the discrete uniform distribution is not so important as it is frequently thought to be by beginners in probability. Elementary textbooks often give so much emphasis to combinatorial probability—using permutations and combinations to count the number of ways that events could occur and using these counts (together with the assumption of equal likelihood) to form probabilities—that it is easy to develop a concept of probability theory that is limited to this one special case. It is important to realize that the discrete uniform distribution is just one of many useful distributions.

B.10 THE BERNOULLI DISTRIBUTION

If a random variable must assume one of two values, 0 or 1, it is said to be a Bernoulli random variable. The corresponding experiment, which has only two possible outcomes, is called a Bernoulli trial. Usually the outcome that is mapped by the random variable onto the value "1" is called a "success;" the other is called a "failure." The distribution is given by

$$p(1) = p$$
$$p(0) = 1 - p$$

where p is the only parameter of the distribution, often referred to as the "probability of success."

The distribution may seem so trivial as to be undeserving of special attention. Although it is true that direct applications are limited, it turns out that a number of more important distributions can be derived from considering a sequence of independent Bernoulli trials.

B.11 THE BINOMIAL DISTRIBUTION

Let X be a discrete random variable defined over the range $x = 0,1,2, \ldots$. If

$$p(x) = \binom{n}{x} p^x (1-p)^{n-x}$$

then we say that X has a binomial distribution with parameters n and p, where n is a positive integer and $0 \leqq p \leqq 1$. The notation $\binom{n}{x}$ refers to the so-called "binomial coefficient" defined by $\binom{n}{x} = n!/x!(n-x)!$. Tables of binomial coefficents and the binomial distribution are readily available.

A binomially distributed random variable can usually be thought of as counting the number of successes in a sequence of n independent Bernoulli trials, where the probability of success on any trial is p.

The expectation, or mean number of successes, is np. The variance is $np(1-p)$.

B.12 THE POISSON DISTRIBUTION

Let X be a discrete variable defined over the range $x = 0, 1, 2, \ldots, \infty$. If

$$p(x) = \frac{\lambda^x e^{-\lambda}}{x!} \qquad x = 0, 1, 2, \ldots$$

then we say that X has a Poisson distribution with parameter λ, where λ must be positive. The Poisson distribution has a number of convenient properties that contribute to its usefulness in modeling. The expectation and variance are equal to one another, and are given simply by the parameter of the distribution:

$$E(X) = V(X) = \lambda$$

The distribution is reproductive; that is, the sum of Poisson distributed random variables will be another Poisson distributed random variable. The parameter of the sum random variable will be just the sum of the parameters of the constituent random variables.

One of the common usages of the Poisson distribution is an approximation to the binomial distribution when the number of trials (n) becomes large while the probability of occurrence (p) becomes small. All that is required for the approximation is to give the two distributions the same expectation. That is, let $\lambda = np$.

Another common use of the Poisson distribution is to describe the number of events occurring within some period of time. In this context, it is the usual practice to use λt as the parameter of the distribution, where t is interpreted as the length of the period and λ is now the mean "rate" at which events occur. The "Poisson process" and its properties will be discussed in some detail in the next chapter.

B.13 THE GEOMETRIC DISTRIBUTION

There are two common versions of the geometric distribution. If X is defined over the range $x = 1, 2, \ldots, \infty$ and has the PDF

$$p(x) = p(1-p)^{x-1} \qquad x = 1, 2, \ldots, \infty$$

where $0 \leqq p \leqq 1$, we would say that X has the geometric distribution beginning at 1. If it is defined over the range $x = 0, 1, \ldots, \infty$, and

$$p(x) = p(1-p)^{x} \qquad x = 0, 1, \ldots, \infty$$

we would say that X has the geometric distribution beginning at 0. It is apparent that one version is just a shifted version of the other, and that other shifts could be made without altering the form of the distribution. Both of these versions appear in applications and are easily confused.

The expectations and variance for the geometric distribution beginning at 1 are, respectively,

$$E(X) = \frac{1}{p}$$

$$V(X) = \frac{1-p}{p^2}$$

When the distribution begins at 0, the variance is the same, but the expectation is $(1-p)/p$.

A possible interpretation of X, when it begins at 1, is as the number of trials, in a sequence of independent Bernoulli trials, that will occur before the first success is observed. More precisely, it is the number of the trial on which the first success occurs. If X begins at zero, it could be thought of as counting the number of failures before the first success. In either case, X counts trials, so the geometric distribution is often regarded as a "waiting time" distribution. One should not confuse this interpretation of the geometric distribution with that of the binomial distribution. The latter fixes the number of trials and counts successes.

B.14 THE NEGATIVE BINOMIAL DISTRIBUTION

Let X be a discrete random variable defined over the range $x = r, r+1, \ldots, \infty$. We would say that X follows a negative binomial distribution if

$$p(x) = \binom{x-1}{r-1} p^r (1-p)^{x-r}, \; x = r, r+1, \ldots, \infty$$

where r is an integer $\geqq 1$, and $0 \leqq p \leqq 1$. Another name for the same distribution is the Pascal distribution. When $r = 1$, the distribution reduces to the geometric. The expectation and variance are

$$E(X) = \frac{r}{p}$$

$$V(X) = \frac{r(1-p)}{p^2}$$

The explanation for this distribution just extends that of the geometric. X represents the number of the trial, in a sequence of independent Bernoulli trials, on which the rth success occurs. Thus the negative binomial distribution is another waiting-time distribution. Thinking of X in this way suggests that the waiting time for the rth success ought to be the sum of r waiting times for the one success. Because the trials are independent, this logic is valid. It is a fact that the sum of r geometrically distributed random variables will yield a random variable whose distribution is negative bionomial with parameter r.

Sometimes the negative binomial distribution is used without any waiting-time interpretation, but simply because the parameters can be adjusted so as to fit a set of data. In this case, it may be desirable to have the range of X begin at zero, rather than r. If so, the appropriate PDF would be

$$p(x) = \binom{r+x-1}{x} p^r(1-p)^x \qquad \text{for } x = 0, 1, \ldots, \infty$$

The variance would be the same, but the expectation would be $r(1-p)/p$.

Table B.1 summarizes the properties of common discrete probability distributions.

III. CONTINUOUS PROBABILITY DISTRIBUTIONS

B.15 THE CONTINUOUS UNIFORM DISTRIBUTION

When a continuous random variable X is restricted to a finite range, $a \leqq x \leqq b$, and is such that "no value is any more likely than any other," then X would be appropriately described by the continuous uniform distribution. It is the obvious analog of the discrete uniform distribution, which restricted the random variable to a finite number of equally likely values. The description "no value more likely than any other" is somewhat loose, because, of course, the probability of any one value for a continuous random variable is zero. A better, although less intuitive, description would be "the probability that x falls within any interval in the range of X depends only on the width of the interval and not on its location."

In any case, the distribution, is rigorously defined by its probability density function

$$f(x) = \frac{1}{b-a} \qquad a \leqq x \leqq b$$

The expectation is at the midpoint of the range,

$$E(X) = \frac{a+b}{2}$$

and the variance is

$$V(X) = \frac{(b-a)^2}{12}$$

B.16 THE NORMAL DISTRIBUTION

Easily the most important continuous probability distribution, the normal distribution has been useful in countless applications involving every conceivable discipline. The usefulness is due in part to the fact that the distribution has a number of properties that make it easy to deal with mathematically. More importantly, however, the distribution happens to describe quite accurately the random variables associated with a wide variety of experiments.

Table B.1
DISCRETE PROBABILITY DISTRIBUTIONS

Name	Range	Parameters	PDF	Expectation	Variance
Discrete uniform	$x=1,2,\ldots,N$	$N=1,2,\ldots$	$p(x)=\frac{1}{N}$	$\frac{N+1}{2}$	$\frac{N^2-1}{12}$
Bernoulli	$x=0,1$	$0\leq p\leq 1$	$p(0)=1-p$ $p(1)=p$	p	$p(1-p)$
Binomial	$x=0,1,\ldots,n$	$n=1,2,\ldots$ $0\leqq p\leqq 1$	$\mathrm{p}(x)=\binom{n}{x}p^x(1-\mathrm{p})^{n-x}$	np	$np(1-p)$
Poisson	$x=0,1,\ldots,\infty$	$\lambda>0$	$p(x)=\frac{\lambda^x e^{-\lambda}}{x!}$	λ	λ
Geometric	$x=1,2,\ldots,\infty$	$0\leqq p\leqq 1$	$p(x)=p(1-p)^{x-1}$	$\frac{1}{p}$	$\frac{1-p}{p^2}$
Negative binomial (Pascal)	$x=r,r+1,\ldots,\infty$	$r=1,2,\ldots$ $0\leqq p\leqq 1$	$p(x)=\binom{x-1}{r-1}p^r(1-p)^{x-r}$	$\frac{r}{p}$	$\frac{r(1-p)}{p^2}$

The range of a normally distributed random variable consists of all real numbers. The probability density function is defined by the equation

$$f(x) = \frac{1}{\sigma\sqrt{2\pi}} e^{-((x-\mu)^2/2\sigma^2)} \qquad -\infty \leq x \leq \infty$$

where the parameter μ is unrestricted and the parameter σ is positive.

The two parameters, μ and σ, used to specify the distribution happen to correspond to the mean and standard deviation, respectively, of the random variable. Any linear transformation of a normally distributed random variable is also normally distributed. That is, if X is normal with mean μ and variance σ^2, and if $Y = aX + b$, then Y is normally distributed with mean

$$\begin{aligned} E(Y) &= E(aX+b) \\ &= aE(X) + b \\ &= a\mu + b \end{aligned}$$

and with variance

$$\begin{aligned} V(Y) &= V(aX+b) \\ &= a^2 V(X) \\ &= a^2\sigma^2 \end{aligned}$$

The significance of these facts is that every normal distribution, whatever the values of the parameters, can be represented in terms of the *standard* normal distribution which has a mean of zero and variance of 1. The linear transformation required to convert a normally distributed random variable X with mean μ and variance σ^2 to the standard normal random variable Z is

$$Z = \frac{X-\mu}{\sigma}$$

The density function of the standard normal random variable is just

$$f(x) = \frac{1}{\sqrt{2\pi}} e^{-(x^2/2)} \qquad -\infty \leq x \leq \infty$$

Because it is so frequently used, the standard normal density function is granted the special notation $\phi(x)$. The cumulative distribution function also has its own notation, $\varphi(x)$. Unfortunately, integrals of the density function cannot be evaluated by ordinary methods of calculus, so there is no closed form expression for $\varphi(x)$, other than as an integral of $\phi(x)$. Extensive tables of $\varphi(x)$, obtained by means of numerical integration, are available. Once you have become familiar with the tables, virtually any desired probability can be evaluated with little trouble.

The normal distribution is reproductive; that is, the sum of two or more normally distributed random variables is itself normally distributed. The mean of the sum is, as always, the sum of the means. The variance of the sum is the sum of the variances, provided that the random variables are independent. Even if they are not, the variance of the sum can be expressed in terms of the variances and covariances of the constituents.

An even more remarkable result is established by the famous Central Limit Theorem, which states that (under certain broad conditions) the sum of a large number of independent *arbitrarily* distributed random variables will be (approximately) normally distributed. Since quite frequently a random variable of interest may be conceptualized as being composed of a large number of independent random effects, the Central Limit Theorem "explains" why the normal distribution appears so often in real-life applications. It also provides justification for *assuming* that certain random variables are normally distributed.

B.17 THE LOGNORMAL DISTRIBUTION

The lognormal distribution is the distribution of a random variable whose natural logarithm follows a normal distribution. The lognormal density function is given by

$$f(x) = \frac{1}{\sigma x\sqrt{2\pi}} e^{[(\ln x - \mu)^2/2\sigma^2]}$$

The range of the random variable is all $x > 0$. The parameters μ and σ may be interpreted from

$$\mu = E(\ln X)$$
$$\sigma^2 = V(\ln X)$$

but the mean and variance of X are, respectively,

$$E(X) = e^{\mu + (1/2\sigma^2)}$$
$$V(X) = e^{2\mu + \sigma^2}(e^{\sigma^2} - 1)$$

In practice, the best way to deal with a lognormally distributed random variable X is to transform it by taking its natural logarithm. That is, let $Y = \ln X$. Of course, Y would be normally distributed and is therefore easily handled. For example, suppose that you wish to evaluate

$$P(a \leqq X \leqq b)$$

where X is lognormally distributed. Using the fact that Y is normal,

$$P(a \leqq X \leqq b) = P(\ln a \leqq Y \leqq \ln b)$$
$$= \int_{\ln a}^{\ln b} \frac{1}{\sigma\sqrt{2\pi}} e^{-[(y-\mu)^2/2\sigma^2]} dy$$

which in turn would be evaluated by standardizing Y. Define $Z = (Y - \mu)/\sigma$, and

$$P(\ln a \leqq Y \leqq \ln b) = P\left(\frac{\ln a - \mu}{\sigma} \leqq Z \leqq \frac{\ln b - \mu}{\sigma}\right)$$
$$= \int_{(\ln a - \mu)/\sigma}^{(\ln b - \mu)/\sigma} \phi(z) dz$$

The lognormal distribution arises from the *product* of many independent nonnegative random variables, in contrast to the normal, which arises from the sum of independent random variables. It has been used to describe lifetimes of mechanical and electrical systems, incubation periods of infectious diseases, concentrations of chemical elements in geological materials, abundance of species of animals, and many other random phenomena occurring in both the social and natural sciences.

The demonstrated usefulness of the lognormal distribution immediately suggests the more general concept of fitting familiar distributions to *transformed* data rather than to the data itself. The normal distribution together with the logarithmic transformation is just one possible combination. Exponential, quadratic, square root, and trigonometric transformations could also be applied to the raw data in an attempt to find a simple distribution to describe them. This idea has become standard practice in modern data analysis.

B.18 THE NEGATIVE EXPONENTIAL DISTRIBUTION

Let X be a continuous random variable defined over the range $x \geqq 0$. If

$$f(x) = \lambda e^{-\lambda x} \quad x \geqq 0$$

where the parameter λ is positive, we say that X has the negative exponential distribution or, sometimes, just the exponential distribution. The cumulative distribution function has, in this case, a convenient expression

$$F(x) = 1 - e^{-\lambda x}$$

The complementary cumulative distribution function is even simpler

$$G(x) = e^{-\lambda x}$$

The expectation of a negative exponentially distributed random variable is the reciprocal of the parameter

$$E(X) = \frac{1}{\lambda}$$

and the variance is the square of the same value

$$V(X) = \frac{1}{\lambda^2}$$

The negative exponential distribution is used extensively to describe random variables corresponding to durations. In other words, it is a waiting time distribution. It has a number of useful properties, and they are explored fully in Chapter 6.

B.19 THE ERLANG DISTRIBUTION

A continuous random variable defined over the range $x \geqq 0$ is Erlang distributed if its density function is of the form

$$f(x) = \frac{\lambda^r x^{r-1}}{(r-1)!} e^{-\lambda x} \quad x \geqq 0$$

where the parameter λ is positive and r is an integer $\geqq 1$. When $r = 1$ the density function reduces to that of a negative exponential distribution, so the Erlang distribution can be thought of as a generalization of the negative exponential. In fact, if we had r independent negative exponential random variables, each with the parameter λ, then the sum of these random variables would be Erlang distributed with parameters λ and r. If each of the negative exponential random variables is a waiting time, the Erlang random variable can be thought of as the time until the rth event.

The expectation is most easily found as the sum of the expectations of the negative exponential random variables.

$$E(X) = \frac{r}{\lambda}$$

and the variance is found by a similar argument

$$V(X) = \frac{r}{\lambda^2}$$

In addition to its use as a waiting time for the rth event, the Erlang distribution is often considered as a candidate to fit empirical data in queuing, reliability, inventory, and replacement applications. In this case, r has no physical interpretation; it is just a parameter that may be adjusted to obtain a better fit.

B.20 THE GAMMA DISTRIBUTION

Let X be a continuous random variable defined over the range $x \geqq 0$. It is gamma distributed if the density function is of the form

$$f(x) = \frac{\lambda^r x^{r-1}}{\Gamma(r)} e^{-\lambda x} \quad x \geqq 0$$

where both λ and r are positive, and $\Gamma(r)$ is the gamma function, defined by

$$\Gamma(r) = \int_0^\infty x^{r-1} e^{-x} dx$$

The gamma function is tabulated, so $\Gamma(r)$ can be thought of as just a constant whose value can be easily found when r is given.

When r is an integer, $\Gamma(r) = (r-1)!$ In this case, the gamma distribution reduces to the Erlang. One way to think of the gamma is as a generalization of the Erlang in which r need not be an integer. The generalization permits somewhat greater flexibility in fitting empirical data.

When $\lambda = 1/2$ and $r = \nu/2$, where ν is an integer, the Γ density function becomes

$$f(x) = \frac{1}{2^{\nu/2}\Gamma(\nu/2)} x^{(\nu-2)/2} e^{-(1/2)x} \quad x \geqq 0$$

This special form is commonly called the *chi-square density function*, and the single parameter ν is referred to as the degrees of freedom associated with the chi-square random variable. A chi-square random variable with ν degrees of freedom results when a set of ν independent standard normal random variables are each squared and then summed. The distribution has many applications in statistical hypothesis testing, but is not often used as a descriptive distribution in modeling applications. For the latter purpose, the Erlang distribution is a much more useful and special case of the gamma function.

The expectation and variance of a gamma distributed random variable X are

$$E(X) = \frac{r}{\lambda}$$

$$V(X) = \frac{r}{\lambda^2}$$

B.21 THE WEIBULL DISTRIBUTION

A continuous random variable defined over the range $x \geqq 0$ has a Weibull distribution if its density function is given by

$$f(x) = \lambda\beta(\lambda x)^{\beta-1} e^{-[\lambda x]^\beta} \quad x \geqq 0$$

where λ and β are positive constants. Since, when $\beta = 1$, this density function reduces to that of the negative exponential distribution, the Weibull can be thought of as a generalization of the negative exponential. Since the Erlang family generalizes the negative exponential, and the gamma in turn generalizes the Erlang, one might momentarily suspect that the Weibull and gamma distributions are the same, or that one is a special case of the other. However, there are gamma distributions that are not Weibull, and there are Weibull distributions that are not gamma. Thus, these two families are definitely distinct despite the fact that they each generalize the negative exponential family.

The mean of a Weibull distributed random variable is

$$E(X) = \frac{1}{\lambda}\Gamma\left(1 + \frac{1}{\beta}\right)$$

The variance is

$$V(X) = \frac{1}{\lambda^2}\left\{\Gamma\left(1 + \frac{2}{\beta}\right) - \left[\Gamma\left(1 + \frac{1}{\beta}\right)\right]^2\right\}$$

The gamma function $\Gamma(\ )$ appearing in these expressions can be found tabulated in most handbooks of mathematical tables.

The Weibull distribution has been found to be useful for describing lifetimes and waiting times in reliability applications. If a system consists of a large number of parts, each of which has a lifetime distribution of its own (independent of the others), and if the system fails as soon as any one of the parts does, then the lifetime of the system is the minimum of the lifetimes of its parts. Under these circumstances, there is theoretical justification for expecting a Weibull distribution to provide a close approximation to the lifetime distribution of the system.

B.22 THE BETA DISTRIBUTION

A beta-distributed random variable is defined over the range $0 \leqq x \leqq 1$ and has the density function

$$f(x) = \frac{x^{r-1}(1-x)^{s-1}}{B(r, s)} \qquad 0 \leqq x \leqq 1$$

where $B(r, s)$ is the beta function, which is tabulated directly or may be found from tables of the gamma function from the relation

$$B(r, s) = \frac{\Gamma(r)\Gamma(s)}{\Gamma(r+s)}$$

The mean and variance of X are, respectively,

$$E(X) = \frac{r}{r+s}$$

$$V(X) = \frac{rs}{(r+s)^2(r+s+1)}$$

These relations can be inverted to obtain interpretations of the parameters r and s in terms of the mean and variance.

$$r = E(X)\left[\frac{E(X)[1-E(X)]}{V(X)} - 1\right]$$

$$s = [1-E(X)]\left[\frac{E(X)[1-E(X)]}{V(X)} - 1\right]$$

When it is desired to define a beta-distributed random variable, say Y, having a range $a \leqq y \leqq b$, the simplest approach would be to transform Y according to

$$X = \frac{Y-a}{b-a}$$

which yields a beta-distributed random variable defined on the range $0 \leqq x \leqq 1$, and therefore has the density function mentioned above.

The beta distribution is commonly used to describe random variables that are not uniformly distributed, but whose possible values lie in a restricted interval of numbers. Sometimes the range is restricted by the very nature of the random variable, such as when it represents a fraction or a percentage. Other times, the range is restricted by choice. In project-scheduling networks (PERT and CPM), the durations of activities are often assumed to follow a beta distribution, where the minimum and maximum possible times are supplied by people who are familiar with the activity.

Table B.2 summarizes the properties of common continuous probability distributions.

Table B.2
CONTINUOUS PROBABILITY DISTRIBUTIONS

Name	Range	Parameters	PDF	Expectation	Variance
Continuous uniform	$a \leqq x \leqq b$	a, b	$f(x) = \frac{1}{b-a}$	$\frac{a+b}{2}$	$\frac{(b-a)^2}{12}$
Normal	$-\infty < x < \infty$	μ			
		$\sigma^2 > 0$	$f(x) = \frac{1}{\sigma\sqrt{2\mu}} e^{-[(x-\mu)^2/2\sigma^2]}$	μ	σ^2
Lognormal	$0 \leqq x$	μ	$f(x) = \frac{1}{\sigma x\sqrt{2\pi}} e^{-[(\ln x-\mu)^2/2\sigma^2]}$	$e^{\mu+(1/2\sigma^2)}$	$e^{2\mu+\sigma^2}(e^{\sigma^2}-1)$
		$\sigma^2 > 0$			
Negative exponential	$0 \leqq x$	$\lambda > 0$	$f(x) = \lambda e^{-\lambda x}$	$\frac{1}{\lambda}$	$\frac{1}{\lambda^2}$
Gamma (Erlang when	$0 \leqq x$	$\lambda > 0$ $r > 0$	$f(x) = \frac{\lambda^r x^{r-1}}{\Gamma(r)} e^{-\lambda x}$	$\frac{r}{\lambda}$	$\frac{r}{\lambda^2}$
r is an integer; χ^2 when $\lambda = 1/2$, $r = \nu/2$)					
Weibull	$0 \leqq x$	$\lambda > 0$ $\beta > 0$	$f(x) = \lambda\beta(\lambda x)^{\beta-1} e^{-[\lambda x]\beta}$	$\frac{1}{\lambda}\Gamma\left(1+\frac{1}{\beta}\right)$	$\frac{1}{\lambda^2}\left[\Gamma\left(1+\frac{2}{\beta}\right) - \left\{\Gamma\left(1+\frac{1}{\beta}\right)\right\}^2\right]$
Beta	$0 \leqq x \leqq 1$	$r > 0$ $s > 0$	$f(x) = \frac{x^{r-1}(1-x)^{s-1}}{B(r, s)}$	$\frac{r}{r+s}$	$\frac{rs}{(r+s)^2(r+s+1)}$

IV. USING DISTRIBUTIONS

B.23 FITTING DISTRIBUTIONS TO DATA

Given the necessity of selecting a distribution to describe a particular random phenomenon, one would ordinarily attempt to acquire data representing a large number of independent samples of the random variable one has in mind. Sometimes, of course, the acquisition of adequate data may be economically infeasible or even physically impossible. In these cases, there may exist theoretical justification for believing that a certain distribution family is appropriate. For example, if the phenomenon can be thought of as the number of successes in a sequence of independent Bernoulli trials, a binomial distribution would be appropriate; or if it can be thought of as consisting of the sum of a large number of independent random variables, the Central Limit Theorem would suggest the normal distribution. On other occasions, the choice of distribution is influenced by a need for particular mathematical properties. The beta distribution in PERT and CPM applications and the negative exponential distribution in Markov process models are selected for reasons that have little to do with observed data.

Preferably, however, one would like to have real-world data to provide assurance that the distribution selected really does describe the real-world phenomenon. Because it is difficult to see any pattern in a raw list of values, one would ordinarily plot a histogram as a first step in identifying an appropriate distribution. The next step, that of selecting one or more candidate distribution types, requires a familiarity with the characteristics of various distribution families. In particular, one has to know what "shapes" a PDF is capable of assuming, to decide whether there is any hope of adjusting the parameters to get a PDF that looks like the histogram. The book by Derman et al. (3) provides especially good descriptions of all of the distribution types summarized only briefly here, as well as a number of others that have not even been mentioned. It also provides guidance on how to fit each distribution to particular data, and gives examples.

Once a distribution type is at least tentatively selected, the next problem is to set values for the parameters that fix the distribution within the family. Unless other, external factors intervene, one would usually use the data to estimate, in the formal statistical sense, values for the parameters. In a few cases, the statistics to use are obvious. For example, the parameter λ in a Poisson distribution is estimated by the sample mean, and μ and σ^2 in a normal distribution are estimated by the sample mean and sample variance, respectively. In other cases, however, the appropriate statistic is not so obvious. The parameters r in a gamma distribution, or β in a Weibull, require additional statistical work before proper estimation formulas become apparent. Again, the book by Derman et al. is useful in providing this kind of information.

After the parameters are adjusted so as to provide the best fit to the data that a selected distribution type can provide, one is still left with the question of whether the fit is good enough. In other words, you should validate your model by checking the goodness of fit. As a bare minimum, one could graph the precise PDF over the histogram (using vertical scales that permit comparison), and observe the discrepancies. A more formal procedure would be to perform any of several available statistical tests for goodness of fit. The chi-square and the Kolmogorov-Smirnov goodness-of-fit tests are probably the best known. Descriptions of these two tests can be found in almost all intermediate-level statistics textbooks.

One of the basic points to bear in mind about statistical goodness-of-fit testing is that the null hypothesis assumes that the candidate distribution is correct. Only if the discrepancies between the data and the candidate distribution are significantly large will the test lead you to reject the candidate. In other words, the test is, by its very nature, biased in favor of whatever distribution you have selected to test. The mere fact that the test does not reject the distribution should not be taken as strong evidence that the selected distribution is correct. Someone else might have selected a different distribution and come up with just as much confirmation that his choice was correct. This is particularly likely to occur when the data base is small.

The word to describe the capability of a statistical test to detect that a null hypothesis is false is *power*. Other factors being equal, a greater amount of data will make for a more powerful test. To obtain a very powerful test, however, may require truly enormous quantities of data—orders of magnitude greater than would be required for good hypothesis tests about parameters. It is easy to see why this is so if you think about how many total observations are required to provide enough information about the "tails" of a distribution to ensure that you have obtained a proper fit.

As a final philosophical point, it is well to keep in mind that no amount of data can confirm absolutely that you have selected the correct distribution. Ultimately, there is no escape from having to make assumptions. On the other hand, there is no need for a model to represent its real-world referent perfectly. It can be useful despite acknowledged imperfections if it represents the significant aspects adequately.

Recommended Readings

If any of the topics mentioned in this appendix seems hazy, or if you would just feel more confident about proceeding if you had worked some problems, you should by all means devote some time to an elementary textbook on probability. There are many fine ones available. Unfortunately for the purposes of this book, the orientation of many leans toward statistical, as opposed to modeling, applications. However, any of the books [Clarke (1), Cramer (2), Drake (4), Meyer (7), Neuts (8), Parzen (9), or Ross (10)] should serve the purpose adequately. If one does not suit your taste, feel free to select another. Feller's two volumes, (5) and (6), are classics familiar to everyone seriously interested in probability. Even beginners can find much of interest in them. The first volume deals with discrete, the second, with continuous distributions.

REFERENCES

1. Clarke, B., and R. Disney, *Probability and Random Processes: A First Course with Applications*, 2nd ed., Wiley, New York, 1985.
2. Cramer, H., *The Elements of Probability Theory and Some of Its Applications*, Wiley, New York, 1955.
3. Derman, C., L. J. Glaser, and I. Olkin, *A Guide to Probability Theory and Application*, Holt, Rinehart, Winston, New York, 1973.
4. Drake, A. W., *Fundamentals of Applied Probability Theory*, McGraw-Hill, New York, 1967.
5. Feller, W., *An Introduction to Probability Theory and Its Applications*, Vol. I, 2nd ed., Wiley, New York, 1957.
6. Feller, W., *An Introduction to Probability Theory and Its Applications*, Vol. II, Wiley, New York, 1966.
7. Meyer, P. L., *Introductory Probability and Statistical Applications*, Addison-Wesley, Reading, Mass., 1965.
8. Neuts, Marcel F., *Probability*, Allyn and Bacon, Boston, 1973.
9. Parzen, Emanuel, *Modern Probability Theory and Its Applications*, Wiley, New York, 1960.
10. Ross, Sheldon M., *A First Course in Probability*, Macmillan, New York, 1976.

APPENDIX C
RANDOM NUMBER GENERATION

C.1 RANDOM NUMBERS

The following sequence of random numbers will be used in this Appendix.

0.2537	0.6767	0.3273	0.2178	0.8016
0.5962	0.7084	0.1520	0.2100	0.3364
0.1900	0.3001	0.5401	0.5548	0.4654
0.9124	0.4687	0.5940	0.0994	0.4848
0.6162	0.5216	0.8403	0.4257	0.6029
0.1441	0.1607	0.2950	0.2061	0.7591
0.1523	0.8092	0.7789	0.9791	0.6894
0.1541	0.6542	0.8866	0.0280	0.7688
0.9264	0.9548	0.8912	0.3455	0.9875
0.4595	0.9688	0.6479	0.1016	0.0396
0.8296	0.3852	0.0360	0.4227	0.8848
0.5094	0.3384	0.5954	0.8163	0.0180
0.7222	0.9673	0.6753	0.7680	0.6065
0.7483	0.1145	0.2270	0.7541	0.4235
0.5356	0.7623	0.5197	0.7585	0.8820
0.2672	0.6601	0.1400	0.9107	0.3606
0.8566	0.1277	0.6068	0.8044	0.2180
0.3903	0.7651	0.0546	0.2404	0.2012
0.4146	0.4226	0.3791	0.4881	0.8793
0.7885	0.4261	0.1267	0.0337	0.8038

C.2 TESTING A UNIFORM RANDOM-NUMBER GENERATOR

Since a random deviate from a certain distribution is created by performing a transformation on a uniform variate, the main emphasis on statistical testing will concern the ability of a random-number generator to accurately generate sequences of numbers uniformly distributed on the (0, 1) interval. It is assumed that the transformations involved in transforming deviates of the uniform distribution to deviates of the particular distribution desired are mathematically correct. Hence, if anything is statistically wrong with the final distribution, it will be because of deficiencies in the original random-number generator. Therefore, the following statistical tests will be performed on the sequence of numbers generated by the uniform generator. Although there are many tests commonly used to determine the accuracy of a random-number generator [see, for example, Emshoff and Sisson

(1), Naylor et al. (3), and Schmidt and Taylor (5)] only four will be presented for illustrative purposes.

The Frequency Test

Probably the most important testing is concerned with the uniformity of the distribution and is generally carried out by using the chi-square test and the Kolmogorov-Smirnov test. Both tests are based on the grouping of sample data into classes over the interval (0, 1). The chi-square goodness of fit test allows us to determine whether the observed frequencies in each class are sufficiently close to the frequencies expected if the data did, in fact, come from the uniform distribution. The test statistic is given by

$$\bar{C} = \sum_{i=1}^{n} \frac{(O_i - E_i)^2}{E_i} \tag{C.1}$$

where

$$\begin{aligned} O_i &= \text{observed number in } i\text{th class} \\ E_i &= \text{expected number in } i\text{th class} \\ &= T/n \\ T &= \text{total number of observations} \\ n &= \text{number of classes} \end{aligned}$$

The value of $\bar{C}$ is to be compared with the value $X^2_\alpha(n-1)$, which comes from a chi-square distribution on $n-1$ degrees of freedom with a level of significance of α. If the test statistic given by the above summation is greater than $X^2_\alpha(n-1)$, then the uniform generator should be viewed with suspicion.

For example, consider the random-number table, Table C.1: there are 100 numbers generated, so $N = 100$. Now we shall divide the unit interval into n equal intervals; let $n = 10$. The expected number of random numbers in each subinterval (0.10 in length) is $N/n = 100/10 = 10$. By counting the number of generated numbers falling in each subinterval, we obtain the following results:

Actual	10	11	11	11	8	11	10	13	10	5
Expected	10	10	10	10	10	10	10	10	10	10

0 0.1 0.2 0.3 0.4 0.5 0.6 0.7 0.8 0.9 1.0

Using the chi-square test,

$$\bar{C} = \sum \frac{(\text{actual} - \text{expected})^2}{\text{expected}}$$

and summing over the ten subintervals, we have

$$\frac{(10-10)^2}{10} + \frac{(11-10)^2}{10} + \frac{(11-10)^2}{10} + \frac{(11-10)^2}{10} + \frac{(8-10)^2}{10} + \frac{(11-10)^2}{10} + \frac{(10-10)^2}{10} + \frac{(13-10)^2}{10} + \frac{(10-10)^2}{10} + \frac{(5-10)^2}{10} = 4.2$$

If we arbitrarily specify a value of $\alpha = 0.05$ (this signifies an error of Type I to be 5 percent), the computed value of $\bar{C} = 4.2$ can be compared to a critical value of $\chi^2_{0.05}(9) = 16.919$. Since $\chi^2_{0.05}(9) \gg \bar{C}$, the randomness of the numbers in Table C.1 might be accepted.

The other test under the general category of goodness of fit is the Kolmogorov-Smirnov test, which involves the use of a cumulative frequency distribution. Let $F(x_0) = x_0$ be the continuous cumulative distribution of a sample of T observations. For any given observation, x_0, $S_T(x_0) = m/T$ where m is the observed number of observations less than or equal to x_0. The Kolmogorov-Smirnov test statistic is given by D, which equals the largest deviation between $F(x_0)$ and $S_T(x_0)$ over the range (0, 1) at a specified number of equal in-

Table C.1
A RANDOM NUMBER TABLE

0.001213	0.898980	0.578800	0.676216	0.050106
0.499629	0.282693	0.730594	0.701195	0.182840
0.108501	0.386183	0.769105	0.683348	0.551702
0.557434	0.799824	0.456790	0.216310	0.876167
0.092645	0.589628	0.332164	0.031858	0.611683
0.762627	0.696237	0.170288	0.054759	0.915126
0.032722	0.299315	0.308614	0.833586	0.517813
0.352862	0.574100	0.265936	0.859031	0.433081
0.941875	0.240002	0.655595	0.385079	0,908297
0.199044	0.936553	0.888098	0.817720	0.369820
0.339548	0.543258	0.624006	0.091330	0.416789
0.155062	0.582447	0.858532	0.887525	0.337294
0.751033	0.239493	0.535597	0.333813	0.493837
0.634536	0.199621	0.650020	0.745795	0.791130
0.227241	0.191479	0.406443	0.081288	0.734352
0.721023	0.222878	0.072814	0.641837	0.442675
0.789616	0.052303	0.106994	0.558774	0.141519
0.760869	0.120791	0.277380	0.657266	0.792691
0.805480	0.826543	0.294530	0.208524	0.429894
0.585186	0.986111	0.344882	0.343580	0.115375

tervals. This value of D must be compared with the critical value of $D_{1-\alpha}$ from the Kolmogorov-Smirnov goodness-of-fit table for the sample size given. If D (data) is greater than $D_{1-\alpha}$ (table), then the hypothesis that the data came from a true uniform distribution might be rejected.

Other considerations regarding the applications of the above two tests can be found in Schmidt and Taylor (5), as well as Phillips (4).

Two rough tests for randomness of N uniform random deviates R_i, $i=1, 2, \ldots, N$ is to compute the mean and partial variance. The numbers are uniformly random if

$$\frac{1}{N}\sum_{i=1}^{N} R_i \approx \frac{1}{2}$$

$$\frac{1}{N}\sum_{i=1}^{N} R_i^2 \approx \frac{1}{3}$$

These are merely rough tests that may save time and further testing if the data does not approach these values.

The Gap Test

If the random numbers generated are considered as digits, then this test is performed by counting the number of digits that appear between successive occurrences of a particular specified digit. The number of gaps of each length k is recorded, then a chi-square test is utilized for cells defined at $k=0, 1, 2, \ldots$. For example, 93649 illustrates a gap of length 3 between the two 9's.

For any k the probability of a gap of length k is

$$P(k)=0.1(0.9)^k$$

This says that $P(0)=0.1$ or that one-tenth of the time pairs of digits will occur in a sequence of truly random digits, a somewhat surprising result.

This test can also be performed by treating the numbers generated as real numbers instead of treating them as digits. The logic behind such a test is as follows. A gap of length k

is said to occur if k successive digits appear that are not between α and β, and the $(k+1)$st digit is between α and β for some specified $0 \leqq \alpha < \beta \leqq 1$. Since the probability that a number drawn from a uniform (0, 1) population over the entire interval is one, the probability that a gap of length k occurs over a chosen interval is equal to the length of that interval. We will not illustrate this modification; the reader is referred to Knuth (2) for details. The standard gap test is given below.

EXAMPLE C-1

First, we multiply each of our 100 pseudorandom numbers by 10 to give integer values (we may multiply by a larger number than 10 if more significance is desired). Next, choose a suitable sequence for k. We shall let $k=0, 1, 2, \ldots, 10$. That is, we shall consider gaps of length k for each of the values 0 through 10. From the formula that gives the probability that a gap of length k will occur,

$$P(0)=0.1 \qquad P(6)=0.053$$
$$P(1)=0.09 \qquad P(7)=0.048$$
$$P(2)=0.081 \qquad P(8)=0.043$$
$$P(3)=0.072 \qquad P(9)=0.039$$
$$P(4)=0.066 \qquad P(10)=0.035$$
$$P(5)=0.059$$

To obtain the observed probabilities, we count the number of times a gap of length k, $(k=0, 1, \ldots, 10)$ occurs and divide by 100 (the number of digits sampled). The resulting table of actual and expected occurrences is shown below. Gap occurrences were calculated by rows in Table C.1.

Actual	0.11	0.07	0.02	0.1	0.09	0.1	0.05	0.06	0.08	0.13	0.12
Expected	0.1	0.09	0.08	0.07	0.07	0.06	0.05	0.05	0.04	0.04	0.04
Gap length	0	1	2	3	4	5	6	7	8	9	10

A chi-square test is now performed using 11 cells and Eq. C.1.

The Runs Test

A slightly modified special case of the second gap test with $\alpha=0$, $\beta=0.5$ is the test of runs above and below the mean. A binary sequence is created whose ith term is 0 if the ith random number generated is less than 0.5, and 1 if the ith random number generated is greater than 0.5. A sequence of 1's bracketed by zeros constitutes a run as does a sequence of zeros bracketed by 1's. The expected number of runs is $(N+1)/2$ and the expected number of runs of length k is $(N-k+3)*2^{-k-1}$ where N is the number of numbers generated.

Another version of the runs test is a test of runs up and down. A sequence of random numbers, $r_1, \ldots, r_N$ is generated. A binary sequence is created whose ith term is 0 if $r_i < r_{i+1}$, and 1 if $r_i > r_{i+1}$. In this case the expected number of runs is $(2N-1)/3$ and the expected number of runs of length k is

$$\text{for } k<N-1 \qquad E(k)=\frac{2[(k^2+3k+1)N-(k^3+3k^2-k-4)]}{(k+3)!}$$

$$\text{for } k=N-1 \qquad E(k)=2/N!$$

To illustrate this version of the runs test, consider the following binary sequence produced from the 100 random number samples in Table C.1.

0	1	0	1	0
1	0	1	1	1
0	0	1	1	1
0	1	1	0	1
0	1	1	0	0
1	1	1	0	1
0	0	0	1	1
0	1	0	1	0
1	0	1	0	1
0	1	1	1	1
0	0	1	0	1
0	0	0	1	0
1	0	1	0	0
1	0	0	0	1
0	0	1	0	1
1	1	0	1	0
0	0	0	1	0
1	0	0	0	0
0	1	1	0	0
0	1	1	1	

We shall give k an arbitrary value, say $k=1, 2, \ldots, 5$. Then

$$E(1)=\frac{2[(1)^2+3(1)+1]100-[1^3+3(1^2)-1-4]}{(1+3)!}$$

$$=41.7$$

$$=\text{approximately } 42$$

Similarly $E(2)=18$, $E(3)=5$, $E(4)=1$, $E(5)=0$. Using these values and counting the runs in the binary sequence, we obtain the following table:

Actual	41	11	8	2	1
Expected	42	18	5	1	0
k	1	2	3	4	5

For example, when $k=1$, there are 22 runs generated by a single 1 bracketed by zeros, and 19 runs generated by a single zero bracketed by 1's.

A chi-square goodness-of-fit test is again used. A common fault of random-number generators is an excess of long runs.

The Poker Test

The poker test examines individual digits of the pseudorandom numbers. Taking five digits at a time, the digits are classified as a particular poker hand. Probabilities can be calculated for the number of times each hand is expected to occur. This was done through the use of two correction factors and combinatorial formulas. The first factor represents the probability of a particular configuration of the type of hand. The second factor represents the number of ways the hand can be arranged. In the case of two pair the extra factor of 1/2 is needed because, for example, the arrangements of AABBC and BBAAC produce the same outcomes but are counted as two different occurrences by the first factor. The correction factors essentially account for permutations. The probabilities are given by the following calculations:

All different $\frac{10\cdot 9\cdot 8\cdot 7\cdot 6}{10^5}\cdot 1=0.30240$

One pair $\frac{10\cdot 1\cdot 9\cdot 8\cdot 7}{10^5}\binom{5}{2}=0.50400$

Two pair $\frac{1}{2}\frac{10\cdot 1\cdot 9\cdot 1\cdot 8}{10^5}\binom{5}{2}\binom{3}{2}=0.10800$

Three of a kind $\frac{10\cdot 1\cdot 1\cdot 9\cdot 8}{10^5}\binom{5}{3}=0.07200$

Four of a kind $\frac{10\cdot 1\cdot 1\cdot 1\cdot 9}{10^5}\binom{5}{4}=0.00450$

Five of a kind $\frac{10\cdot 1\cdot 1\cdot 1\cdot 1}{10^5}\binom{5}{5}=0.00010$

Full house $\frac{10\cdot 1\cdot 1\cdot 9\cdot 1}{10^5}\binom{5}{3}\binom{2}{2}=0.0090$

EXAMPLE C-2

For our sample of pseudorandom numbers, one first considers the first five digits in each number generated and classifies it as to the type of hand it represents. A count is kept of the total occurrences for each type of hand and then the total is divided by 100 (the number of digits considered). This gives the observed probabilities for the numbers generated. The expected values are obtained from the above formulas. A simple chi-square test is now performed with the chart below and the seven categories of possible hands (seven cells).

Actual	0.25000	0.5600	0.080	0.1000	0.0000	0.0000	0.0100
Expected	0.30204	0.5040	0.108	0.0720	0.0045	0.0001	0.0090
Categories	All different	One pair	Two pair	Three of a kind	Four of a kind	Five of a kind	Full house

In summary it should be noted that even if the numbers tested are truly uniform, the probability that they will fail at least one test may still be high. This is true since each test is actually a hypothesis test run at a level of significance (type I error) of α. By definition α is the probability of rejecting the null hypothesis even if it is true.

EXAMPLE C-3

Perform five tests, each at level of significance $\alpha=0.05$. The probability of failing at least one of the five tests is:

$$\begin{aligned} p(\text{fail at least one test}) &= 1-(1-\alpha)^n \\ &= 1-(0.95)^5 \\ &= 0.226 \end{aligned}$$

even though the numbers are truly acceptable!

There are other tests that can be run to test the "randomness" in a set of pseudorandom numbers. Additional tests are:

1. Serial tests—serial tests are used to check the randomness of successive numbers in a sequence.
2. Product tests—product tests measure the independence (correlation) between sequences of random numbers.

3. Tests for autocorrelation—autocorrelation tests examine the dependency of a particular number on another number in a sequence. These tests attempt to discern whether there are patterns or relationships among sets of random numbers.
4. The Maximum Test—examines sequences of numbers to detect abnormally high or low numbers.

REFERENCES

1. Emshoff, J. R., and R. L. Sisson, *Design and Use of Computer Simulation Models*, Macmillan, New York, 1970.
2. Knuth, D. E., *The Art of Computer Programming*, Vol. 2, Addison Wesley, Reading, Mass., 1968.
3. Naylor, T. H., J. L. Balintfy, D. S. Burdick and Kong Chu, *Computer Simulation Techniques*, Wiley, New York, 1966.
4. Phillips, D. T., "Applied Goodness of Fit Testing," AIIE Monograph Series, AIIE-OR-72-1, Atlanta, Ga., 1972.
5. Schmidt, J. W., and R. E. Taylor, *Simulation and Analysis of Industrial Systems*, Richard D. Irwin, Homewood, Ill., 1970.

INDEX